Electricity and Electronics

by
Howard H. Gerrish

William E. Dugger, Jr.
Director
Technology for All Americans Project

Richard M. Roberts
Assistant Principal for Technology
Tampa Bay Technical Center

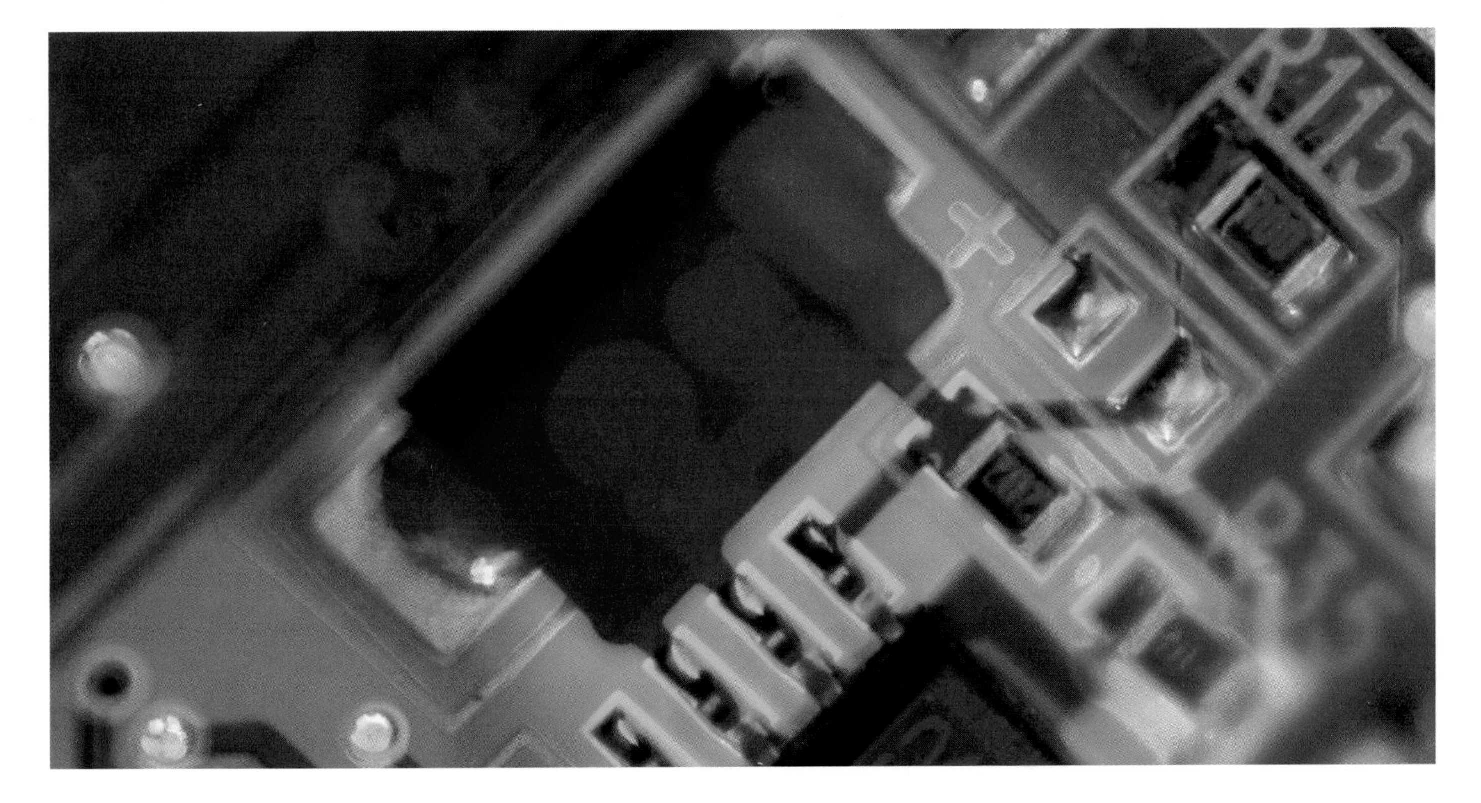

Publisher
The Goodheart-Willcox Company, Inc.
Tinley Park, Illinois

Manufactured in the United States of America.

Library of Congress Catalog Card Number 2002035438

International Standard Book Number 1-59070-207-7

2 3 4 5 6 7 8 9—04—08 07 06 05 04 03

Goodheart-Willcox Publisher Brand Disclaimer: Brand names, company names, and illustrations for products and services included in this text are provided for educational purposes only, and do not represent or imply endorsement or recommendation by the author or the publisher.

Library of Congress Cataloging-in-Publication Data

Gerrish, Howard H.

Electricity and electronics / by Howard H. Gerrish, William E. Dugger, Jr., Richard M. Roberts.

p.cm.

Includes index.

ISBN 1-59070-207-7

1. Electric engineering. 2. Electronics. I. Dugger, William. II. Roberts, Richard M. III. Title.

TK146.G44 2004

621.3—dc21 2002035438

Introduction

As a student interested in the rapidly changing and expanding field of electricity and electronics, you will find that *Electricity and Electronics* provides the fundamentals in easy-to-understand language.

Electricity and Electronics employs a three-pronged approach to learning:

1. Experimentation and demonstration to ensure thorough understanding of the principles. Experiments with step-by-step instructions are placed throughout the text and have been separated from the text flow by blue rules.
2. Illustration of the principles covered through the use of prevalent and familiar applications of those principles in the fields of electricity and electronics. These applications bring the concepts into an everyday focus for the reader. Many chapters contain *Applied E&E* topics, which relate the electrical principles of electricity and electronics to state-of-the-art technology. Familiar elements (copying machines, electric guitar) and less familiar elements (magnetic resonance imaging, night vision) are explored. *Applied E&E* topics have been separated from the text flow by green rules.
3. Practical application of developed principles and skills by constructing inexpensive, professional-appearing projects. The projects have proven appeal. Projects placed throughout this text have been separated from the text flow by orange rules.

Electricity and Electronics is divided into five sections. Each section begins with an overview of the chapters to be covered. Each chapter begins with a list of objectives and a list of the chapter's key words and terms. Each chapter concludes with a summary and two sets of questions over the material you have just learned. Solutions to all *Test Your Knowledge* questions can be found in the chapter or solved using the equations presented. Questions and problems presented in the *For Discussion* section involve independent thought or independent research, or both.

In the body of the chapters, important words and terms are placed in a bold/italic typeface. Formulas are highlighted when they are introduced, allowing you to locate them easily. In addition, each chapter is divided into subsections. At the end of each subsection, *Review Questions* are provided.

When you have successfully completed this textbook and your course, you should have a thorough understanding of the concepts of electricity and electronics. You will also have a basis for applying the principles learned to other, more advanced studies in these fields.

A bright outlook for the electronics industry in the future translates into opportunities for specialized education and a lifetime of challenging and rewarding employment. The world of electricity and electronics is a fascinating one, and we are pleased that you have chosen to explore it with us.

Howard H. Gerrish
William E. Dugger, Jr.
Richard M. Roberts

SAFETY PRECAUTIONS FOR THE ELECTRICITY AND ELECTRONICS SHOP

There is always an element of danger when working with electricity. Observe all safety rules that concern each project. Be particularly careful not to contact any live wire or terminal, regardless of whether it is connected to either a low voltage or a high voltage. Projects do not specify dangerous voltage levels. However, keep in mind at all times that it is possible to experience a surprising electric shock under certain circumstances. Even a healthy person can be injured or seriously hurt by the shock or what happens as a result of it. Do not fool around. Working with electricity can be fun, but it can also be dangerous!

Be concerned for everyone else's safety, too. Wherever you are working, check for electrical hazards. When you identify a hazard, report it to your teacher or supervisor. Also, watch how others work. If someone is working in an unsafe manner, stop them and explain what they are doing incorrectly. Generally, workers use unsafe techniques because they do not realize the danger. By making them aware of the risks, you may save a life. If you identify an unsafe process or procedure, alert someone of the problem. As you gain knowledge of electrical safety, you have a responsibility to use your knowledge to help ensure the safety of others.

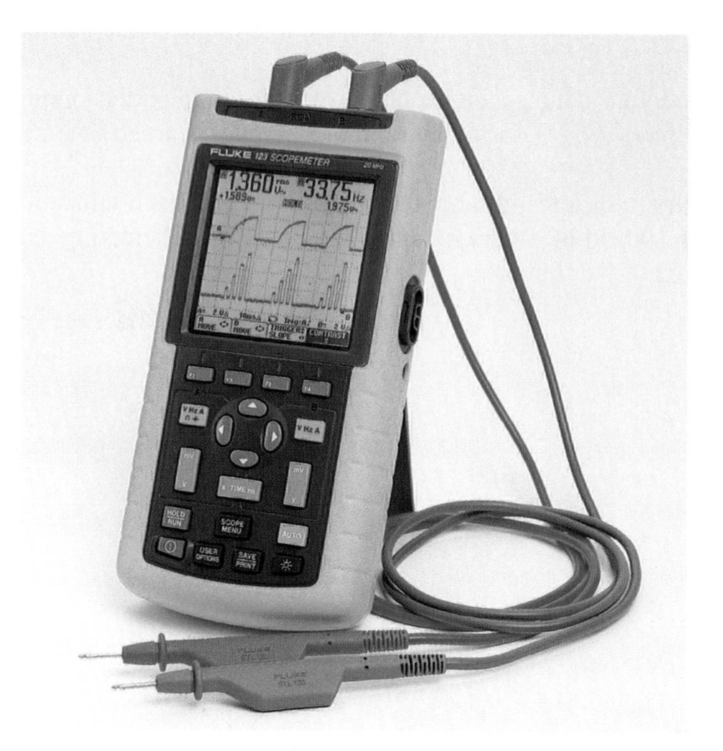

Table of Contents

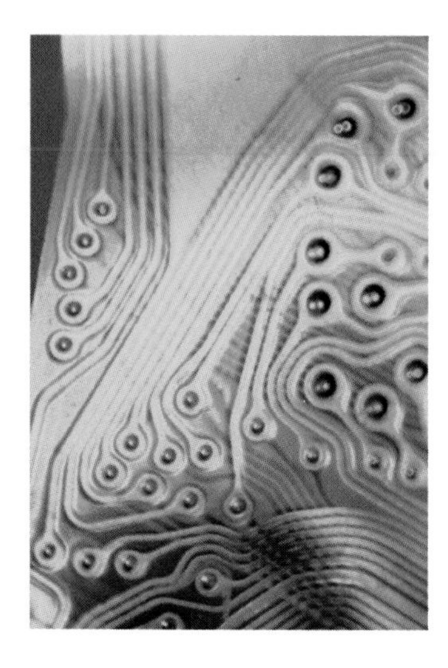

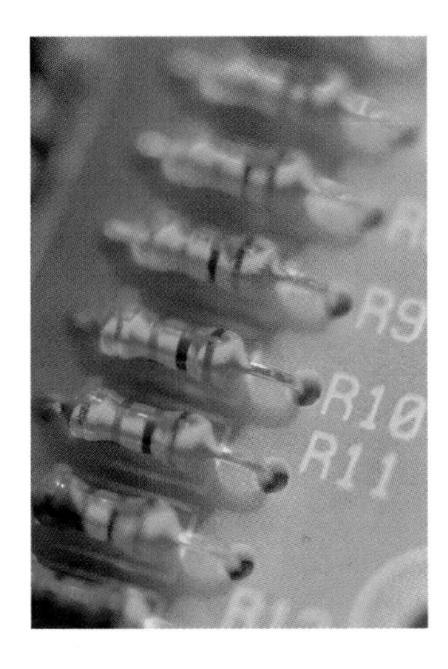
R9
R10
R11

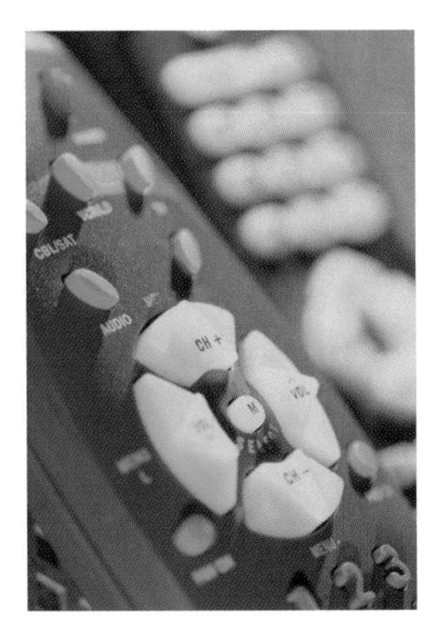

SECTION 1

Fundamentals of Electricity and Electronics

The first five chapters of this text provide the basic skills you will need to begin any electrical or electronics program. In these chapters, emphasis is placed on providing the skills needed to start working on electrical laboratory activities.

Chapter 1 is an introduction to electrical theory, voltage, current, resistance, and Ohm's law. Chapter 2 is an introduction to basic electrical meters such as the voltmeter, ammeter, and ohmmeter. Once you have learned how to use these electrical meters, laboratory activities can be started at once.

In Chapter 3, basic electrical materials are introduced. Again, the emphasis is to provide the necessary information to gain useful laboratory experiences. Chapter 4 introduces the principles of energy, both mechanical and electrical. Section I ends with Chapter 5 introducing the common sources of electricity.

The first five chapters, or Section I, cover all the necessities for conducting laboratory activities.

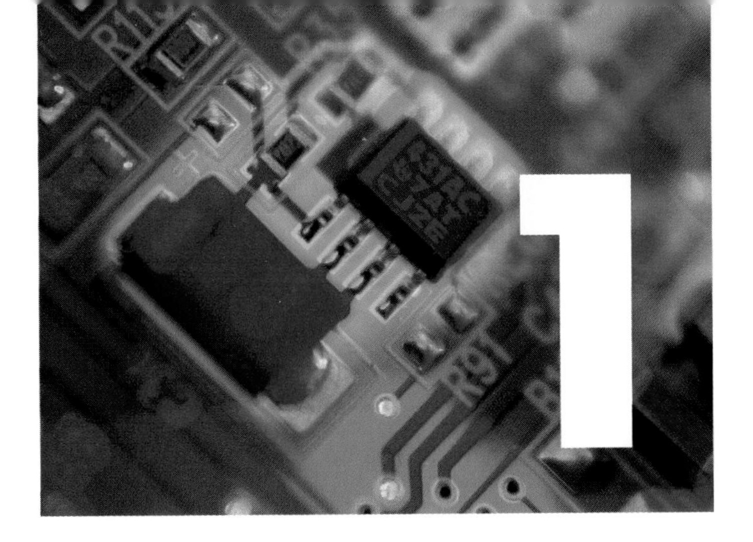

1 Science of Electricity and Electronics

Objectives

After studying this chapter, you will be able to:

- Identify the relationship between elements and compounds.
- Construct a model of an atom.
- Discuss the concepts of atomic weight and atomic number.
- State the law of charges and explain it using several examples.
- Explain what is meant by electric current, voltage, and resistance.
- Describe the two theories of current direction.
- Distinguish between conductors, insulators, and semiconductors.
- State and explain Ohm's law.

Key Words and Terms

The following words and terms will become important pieces of your electricity and electronics vocabulary. Look out for them as you read this chapter.

alternating current	insulator
ampere (A)	neutron
atom	Ohm's law
conductor	ohm
coulomb	potential difference
current (I)	proton
direct current	resistance
electromotive force (emf)	semiconductor
electron	volt (V)
element	voltage (*E* or *V*)

We are fortunate to live in an age in which the opportunity exists to study the electron. New discoveries, developments, and applications in electronics occur almost daily. These open a promising vista of unlimited opportunities for the creative scientist as well as for the skilled technician. We are living in a truly electronic age.

1.1 THE NATURE OF MATTER

Everything in the universe is made up of matter. ***Matter*** can be defined as anything that occupies space or has mass. Matter can be found in the form of solids, liquids, and gases. However, these states are subject to relative temperature. Water is usually found in liquid form. Yet water can be readily changed to a solid or a vapor form by changing its temperature. Matter can also be described by color, taste, and hardness, but these are only observable characteristics. They may not truly identify a substance. To truly identify a substance, the substance must be broken down into its smallest parts. The substance must be described in terms of its *atomic structure*. Only then can it truly be defined and its behavioral characteristics identified.

A substance has been broken down to its purest form when breaking it down further will change its atomic characteristics. This form is called an ***element.*** There are over 100 elements. Most of these elements occur naturally in our universe. Some of the elements do not occur naturally, but have been created in laboratories. Some common examples of naturally occurring elements are iron, copper, gold, aluminum, carbon, and oxygen. **Figure 1-1** is a periodic table. This table lists all of the known elements and describes them in the scientific terms that make them each unique. If two or more of these elements are mixed together, a ***compound*** is created. A compound can be reduced to its individual elements. An element can be reduced to its atomic structure. If the atomic structure is reduced, that element is changed to a different element.

Molecule and the Atom

If a crystal of table salt is cut in half, the result is two smaller crystals of common salt. The composition of the salt crystal does not change; it is simply smaller. Salt is a

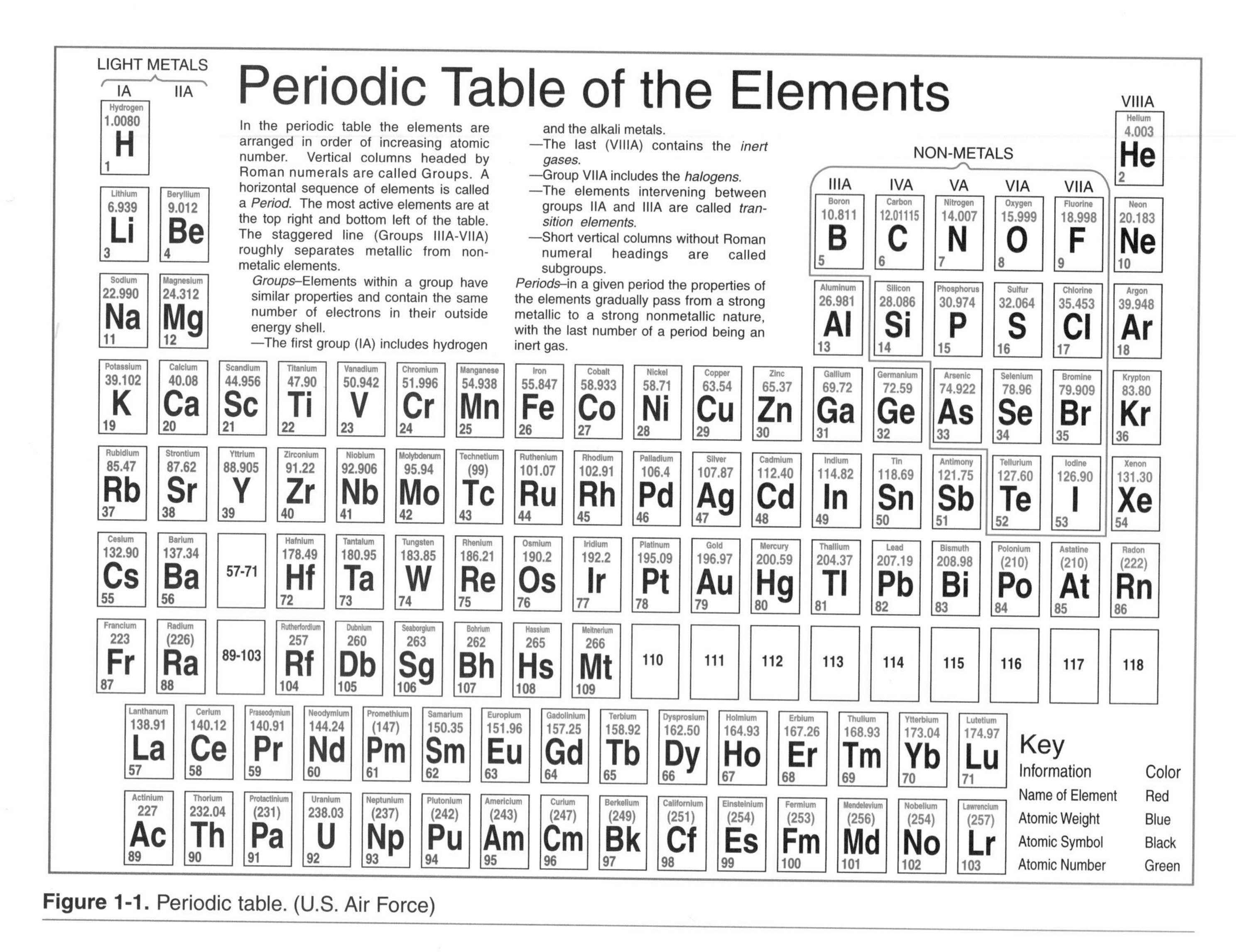

Figure 1-1. Periodic table. (U.S. Air Force)

chemical compound composed of two different elements. The elements are sodium (Na) and chlorine (Cl). Each of these elements, sodium and chlorine, are deadly poisons to the human body. But when combined, sodium and chlorine become the harmless compound known as common table salt, (NaCl). See **Figure 1-2.**

If it were possible for you to smash that crystal of salt into its smallest possible piece, you would have one molecule of salt. A ***molecule*** is the smallest part of a compound that still retains all the characteristics of that compound. If you reduce that single molecule of salt into its next smallest form, it is no longer salt. It is now broken into two different parts. These two parts are the basic elements sodium and chlorine. You have now created poisons out of the normally harmless salt. Do not worry, we have never known salt to be reduced at the dinner table.

The smallest form of an element is known as the ***atom***. The word atom is derived from the Greek word meaning indivisible. The atom is so small that it is difficult to visualize. If we attempted to fill a matchbox with atoms at a rate of ten million per second, it would take over a billion years to fill the box! The atom is the smallest form any material can assume without changing its characteristics. See **Figure 1-3** for a chart and illustrations showing the relationship of matter, compounds, elements, molecules, and atoms.

Electrons, Protons, and Neutrons

To understand the mystery of electricity, and especially the characteristics of solid state electronics, we must have a basic understanding of the structure and forces that make up the atom. Physicists have discovered that atoms are composed of many minute particles. We will be concerned with only the three basic parts of the atom.

The structure of the atom, **Figure 1-4,** is similar to our solar system. In our solar system, the planets (Earth, Venus, Mars, etc.) revolve around the Sun. The planets

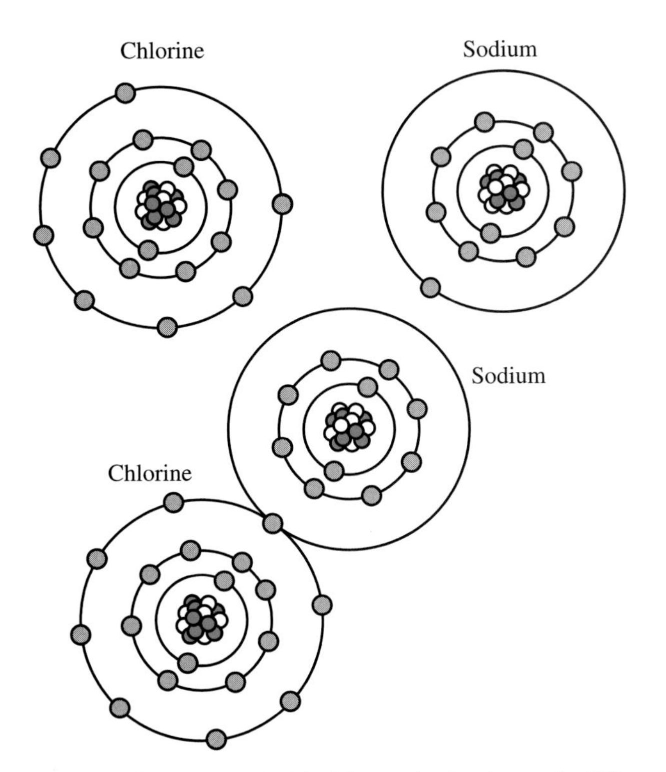

Figure 1-2. Sodium and chloride are both elements. When the atoms of these two elements combine they form a molecule of the compound commonly referred to as table salt.

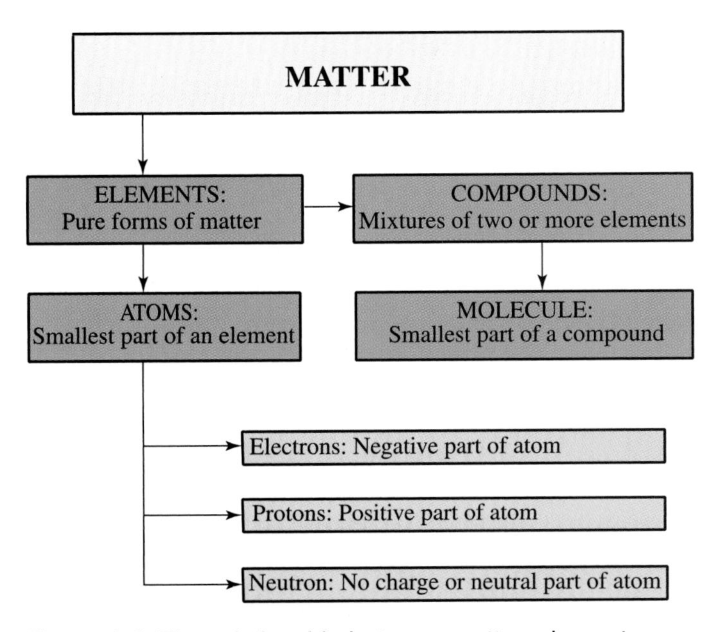

Figure 1-3. The relationship between matter, elements, compounds, atoms, and molecules.

whirl around the sun in their orbits suspended in space by the effects of centrifugal force (pushing them away from the sun) and gravitational attraction (pulling them toward the sun). In an atom, the sun's place is taken by the ***nucleus*** in the center. ***Electrons*** whirl around this nucleus.

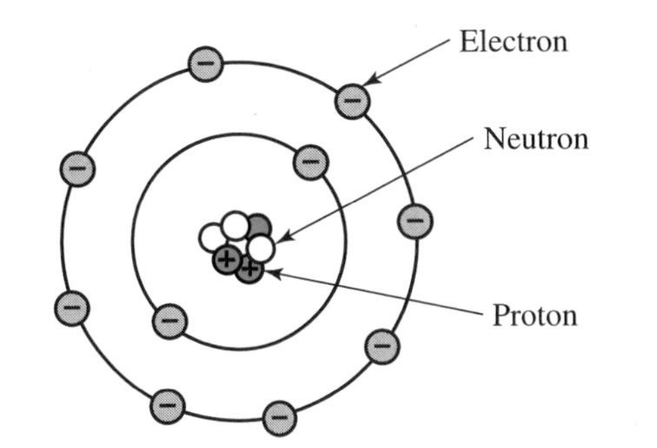

Figure 1-4. The atom consists of electrons, protons, and neutrons.

While planets are held in place with gravity, electrons are held in their orbit by their attraction to the nucleus, overcoming the centrifugal force. The electrons orbiting around the nucleus display a negative charge. The nucleus displays a positive charge because it is composed of positively charged ***protons*** and neutrally charged (neither negative or positive) ***neutrons***. The number of electrons and protons that make up a particular atom are usually equal in number. This equal number creates a canceling effect between the negative and positive charges. The atomic structure of each element can be described as having a fixed number of electrons in orbit. Examples of the atomic structure of two common elements are displayed in **Figure 1-5.**

All elements are arranged in the periodic Table of Elements according to their atomic number**.** The ***atomic number*** of an element refers to the number of protons or electrons that make up an atom of that element. The order of elements may also be arranged by ***atomic weight***. The atomic weight of an element refers to the approximate number of protons and neutrons in the nucleus. Referring to Figure 1-1, note that the atomic weight of hydrogen is one (scientifically 1.008), and its atomic number is one. The atomic weight of oxygen is sixteen; its atomic number is eight.

Ionization

Usually, an atom remains in its normal state unless energy is added by some exterior force such as heat, friction, or bombardment by other electrons. When energy is added to an atom, the atom becomes excited. If the exterior force is of sufficient strength, electrons in the atom's outer rings or orbits can leave their orbit. How tightly bound these outer electrons are to an atom depends on the element and the number of electrons in the outer orbit. If electrons leave the outer orbit, the atom becomes out of balance electrically. This concept is extremely important and will be repeated throughout this book and in your studies of electrical phenomena.

When the electron leaves the outer orbit, the atom becomes ***ionized,* Figure 1-6.** An ionized atom is electrically unbalanced. An atom that loses an electron from its outer orbit has more protons (positive particles) in the nucleus than electrons (negative particles) in orbit around the atom. The atom becomes a ***positive ion*** and displays positively charged characteristics. When an atom gains an extra electron, it becomes a ***negative ion.*** Negative ions display negatively charged characteristics. The electron that has broken out of its orbit is negatively charged. This concept of negative and positive ions is a key building block to understanding electronic theory.

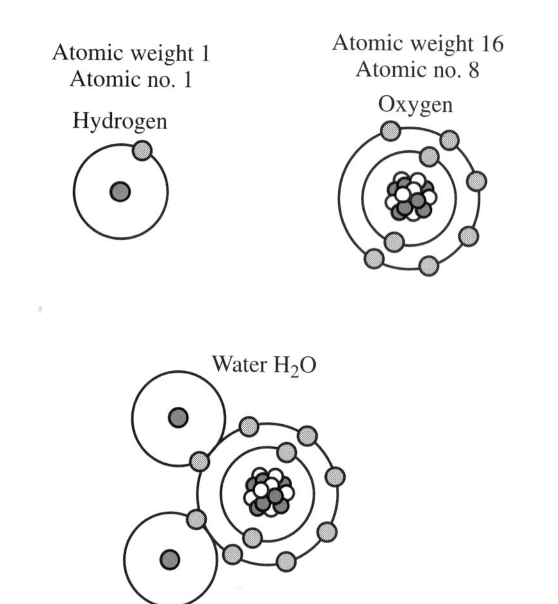

Figure 1-5. The atomic weight is the approximate number of protons and neutrons that compose the nucleus of the atom. The atomic number is the number of electrons or protons in an atom. When two hydrogen atoms combine with one oxygen atom, a single molecule of water is formed.

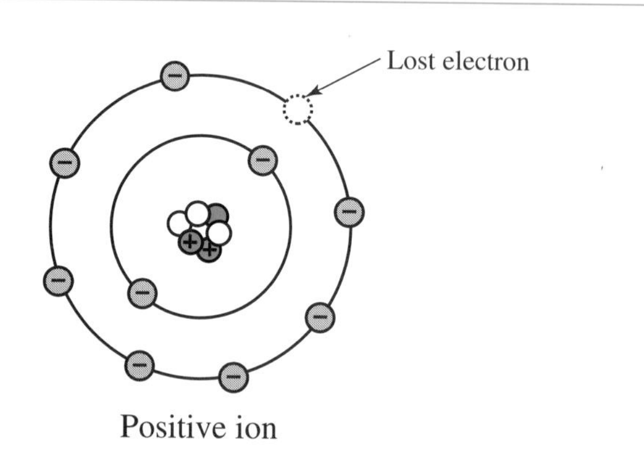

Figure 1-6. Ionization occurs when an atom gains or loses an electron. When there are more protons than electrons remaining, a positive ion is created.

Review Questions for Section 1.1

1. Define the following terms:
   a. Matter.
   b. Element.
   c. Compound.
   d. Atom.
   e. Molecule.
2. Name some common elements.
3. What is atomic weight?
4. Define atomic number.
5. What is the meaning of ionization?
6. Explain how a positive ion is formed.
7. Define a negative ion.

1.2 STATIC ELECTRICITY

The word *static* means at rest. Electricity can be at rest. The generation of static electricity can be demonstrated in many ways. When stroking the fur of a cat, you will notice that its fur is attracted to your hand as you bring your hand back over the cat. You will also hear a crackling sound. If this is done at night, you may see tiny sparks. The sound is caused by the discharge of static electricity. When we stroke the fur with our hand, the friction between the cats fur and our hand excites the atoms. Some atoms lose electrons while others gain electrons. The sparks are created as the atoms attempt to neutralize themselves by gaining back the lost electrons.

You can generate a static charge of electricity by walking across a wool or nylon rug with plastic-soled shoes. After walking across such a rug, you receive the surprising experience of discharging several thousand volts of static electricity to a metallic object such as a door handle. This condition is especially present on cold winter days when the humidity is quite low. You can also experience a similar discharge when sliding across the seat of a car covered with certain types of upholstery. The friction of your clothing against the seat leads to a discharge when you touch the earth and the metallic frame of the car at the same time.

Law of Charges

One of the fundamental laws in the study of electricity is the law of charges. The law of charges states that: *"Like charges repel each other and unlike charges attract each other."* The power of attraction can be seen when you run a comb through your hair several times. The comb will attract some of the hair towards itself because of the unbalanced electrical charge created by the friction between the hair and plastic comb.

EXPERIMENT 1-1: Demonstrating the Law of Charges

These tests demonstrate the law of charges.

Materials

2–stands with suspended pith balls
1–vulcanite rod
1–piece of fur
1–glass rod
1–piece of silk

1. Negatively charge the vulcanite rod with the piece of fur.
2. Bring the rod close to one hanging pith ball.
3. Observe that the ball is first attracted to the rod because of unlike charges. When the ball touches the rod, it is immediately repelled. Continue to attempt to touch the pith ball with the vulcanite rod. Examine **Exhibit 1-1A.** Why has the pith ball acted this way?

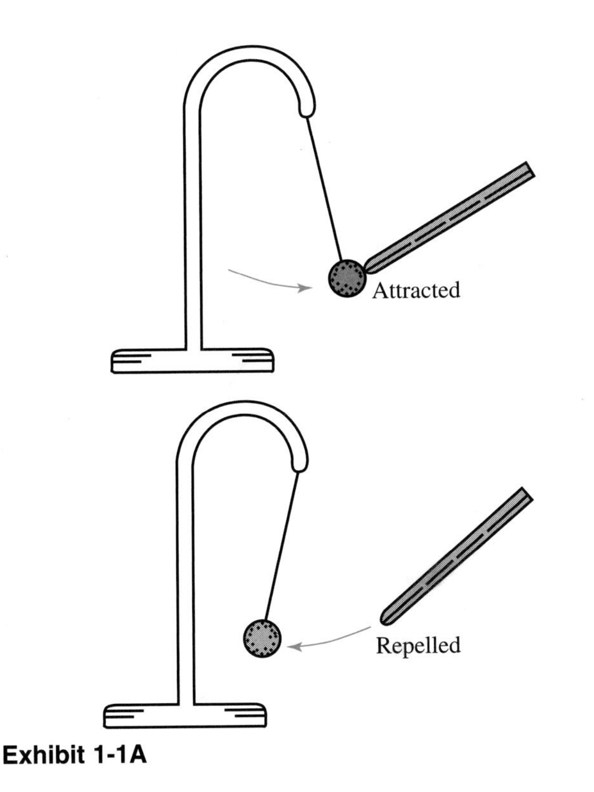

Exhibit 1-1A

4. Recharge the vulcanite rod and touch it to the second pith ball. You now have two negatively charged pith balls.
5. Bring the two pith balls near each other. Do your observations match **Exhibit 1-1B?**

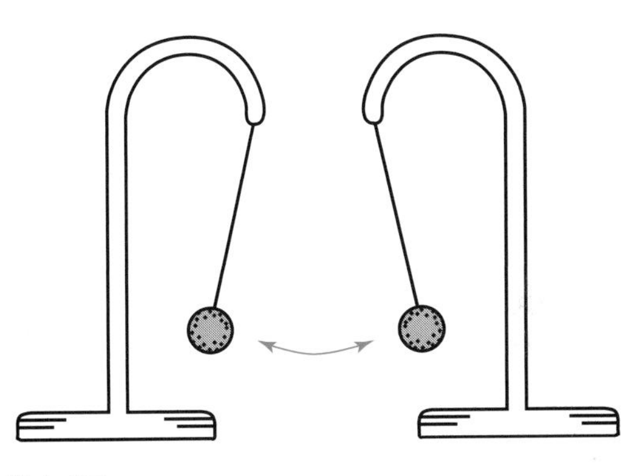

Exhibit 1-1B

6. Charge up the glass rod by rubbing it with a piece of silk. Touch the glass rod to one of the pith balls. Leave the other negatively charged.
7. Bring the two pith balls near each other. Do your observations match **Exhibit 1-1C?**
   Explain what you have just observed using the law of charges.

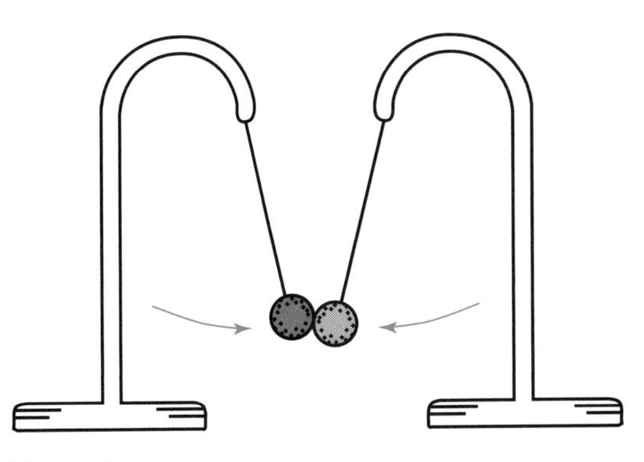

Exhibit 1-1C

Copy Machines

There are many different copying machines available today. The copying method explained here is based upon the xerography system of copying. The term xerography is derived from the Greek xeros and graphe, meaning dry writing. The xerography process uses powder toner, heat, light electrostatics, and photoelectric phenomena to produce a copy.

Recall that like charges repel and unlike charges attract. Now, let's introduce a new electrostatic principle. When a strong beam of light strikes a positively charged area or surface, the light photons will dissipate the positively charged surface areas. Photons exhibit a negatively charged characteristic. This is the underlying principle of xerography.

1. Original document

2. Static pattern on drum. Dark areas are negative or no charge. White areas are positive charge.

3. Toner pattern is controlled by the attractive force of the static electricity.

4. Copy

The original document is placed face down on the glass top of the copy machine. A positive charge is induced on the drum surface and a negative charge is placed on the blank paper. As the drum rotates, a strong light source is moved along the glass just below the original. The light source is directed toward the original and then reflected back from the white areas through a system of optical lenses.

The original document absorbs the light in the dark areas and reflects the light from the white or lighter areas. Shades of gray, or colors which are seen as gray, are partially reflected. The reflected light is directed toward the positively charged drum area through the optical lenses. The positive charge on the drum surface is neutralized by the light. Photon energy is negative.

The areas not affected by the light are left with a positive charge on the surface of the drum. The toner is dispersed on the drum and is attracted to the remaining positively charged areas. The negatively charged paper passes under the drum transferring the toner to the paper. Toner is fused to the paper as it passes by a heater. The paper is ejected to the outside collection tray.

A color copier uses the same principles as described above except it makes use of color filters and four shades of toner. The four shades of toner are magenta, cyan, yellow, and black. These four colors are mixed to reproduce the different colors on the original. The light and optic systems use color filters to transfer the image to the roller and add the colored toner in stages.

Other refinements of copier systems include the use of electronics and advanced optical lens systems. The size of the copy can be enlarged or reduced by adjusting the distance between the optical lens and the image. Lightening or darkening of the copy is achieved by changing the intensity of the light beam. Electronic counters are used to indicate the number of copies needed.

Diagnostic circuits can also be used to indicate conditions such as paper jams, low toner, or an empty paper tray. Many machines can switch between applications such as copying, faxing, scanning, and printing. Advances in design also allow the user to print on different types of material including mailing labels, transparencies, and even envelopes. Some machines print simultaneously on both sides of the paper, reducing printing time and paper jams.

The digital copier/printer illustrated above is compact. The digital design enables greater machine dependability, and its versatility makes it a very economical choice. (XEROX Corporation)

EXPERIMENT 1-2: Examining Electrical Induction and Conduction

These two short tests examine the law of charges using electrical induction and conduction.

Materials

1–electroscope
1–vulcanite rod
1–piece of fur

1. Quickly rub the vulcanite rod with the fur. You have just placed a negative charge on the rod. Electrons have been transferred to the rod from the fur by friction.
2. Bring the charge rod close to (but do not touch) the electroscope. The leaves in the electroscope expand.
3. Move the rod away from the electroscope. The leaves drop back down.
4. Examine **Exhibit 1-2A**. Can you explain what you have just observed?
5. Recharge the vulcanite rod with the fur.
6. Touch the ball on the electroscope with the rod. Observe that the leaves in the electroscope expand.
7. Move the rod away from the electroscope. Observe that the leaves do not drop back down.
8. Examine **Exhibit 1-2B**. Can you explain why the leaves behaved differently in this test?

Discuss the concepts of induction and conduction with your instructor.

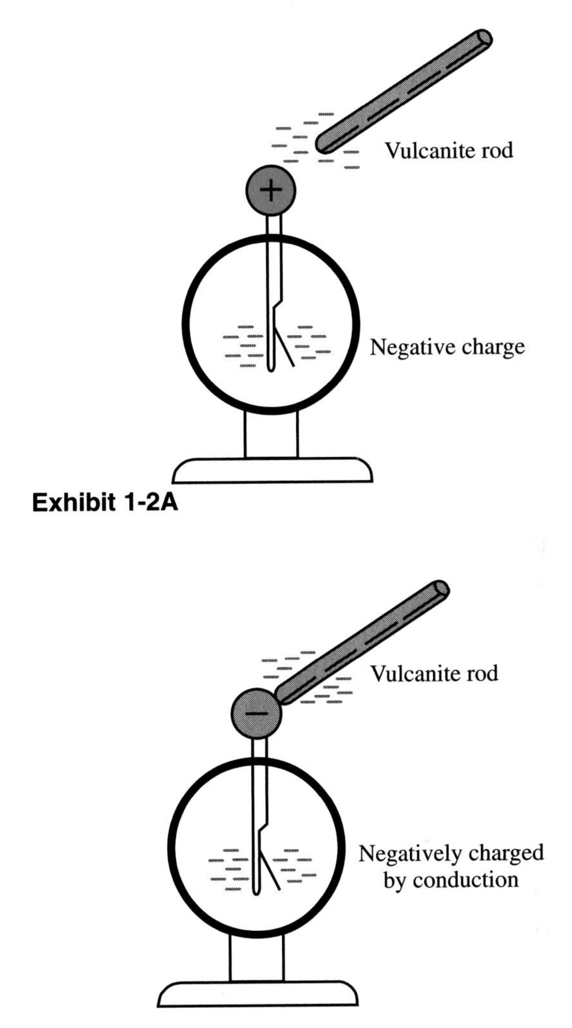

The Coulomb

The force of attraction and repulsion of charged particles was studied by the French scientist, Charles A. Coulomb. Because an atom, and electrons in particular, are so very small, a charge of just a few electrons, say a dozen, is almost impossible to measure. Consequently, Coulomb developed a practical unit for measurement of an amount of electricity. It is known as the ***coulomb.*** One coulomb represents approximately 6.24×10^{18} electrons (6,240,000,000,000,000,000). While the coulomb is used to describe the flow of electricity, it is not used to describe static charges. It is impractical to describe the very small difference in charges between two bodies using values so large.

Electrostatic Fields

The field of force surrounding a charged body is called the ***electrostatic field*** or ***dielectric field.*** The field can exhibit a positive or negative charge depending on a gain or loss of electrons. Two charged masses are shown in **Figure 1-7**. Lines represent the electrostatic fields of opposite polarity and the attractive force existing between the masses. In **Figure 1-8**, two charged masses are shown

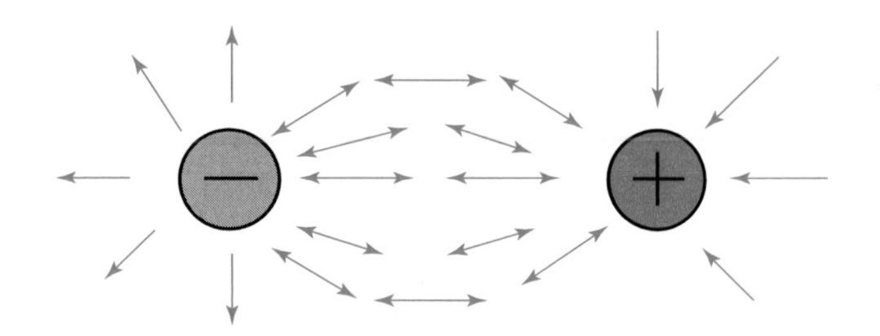

Figure 1-7. The electrostatic fields of unlike charged bodies show attractive forces.

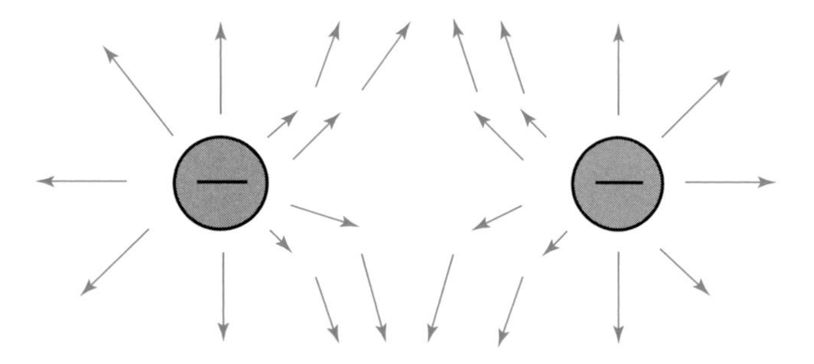

Figure 1-8. The electrostatic fields of like charged bodies repel each other.

with like polarities. A repulsive force exists between the charged masses due to the electrostatic fields. The field is strongest very close to the charged body. The field strength diminishes at a distance inversely proportional to the square of the distance. **Figure 1-9** illustrates the concept of strength being inversely proportional to the square of the distance.

When two electrostatic fields are joined together, the electrons flow from the mass with an excess of electrons to the mass that has a need of electrons. **Figure 1-10** illustrates this principle. The excess electrons flow from the body that is negatively charged to the positively charged body that has the electron deficiency. This transfer of electrons can be accomplished by touching the two bodies together or by connecting them with a material that supports the flow of electrons between the two bodies. This connecting material is known as a conductor because it "conducts" electricity.

Induction

Charges can be transferred in two ways. One way is by direct contact. When a charged body such as a glass rod touches another body such as the top of an electroscope, the electroscope takes on part of the charge of the rod. Another way of transferring a charge is ***induction.*** A charge is induced by bringing a charged object near another object. The glass rod need only be brought near the top of the electroscope to charge it. When an object is charged by induction, the object takes on the *opposite* charge as the rod. When the rod touches the object, the object takes the *same* charge as the rod. Refer back to Experiment 1-2.

Static Electricity Applications

The principles of static electricity are used in industry to reduce air pollution. One piece of equipment used

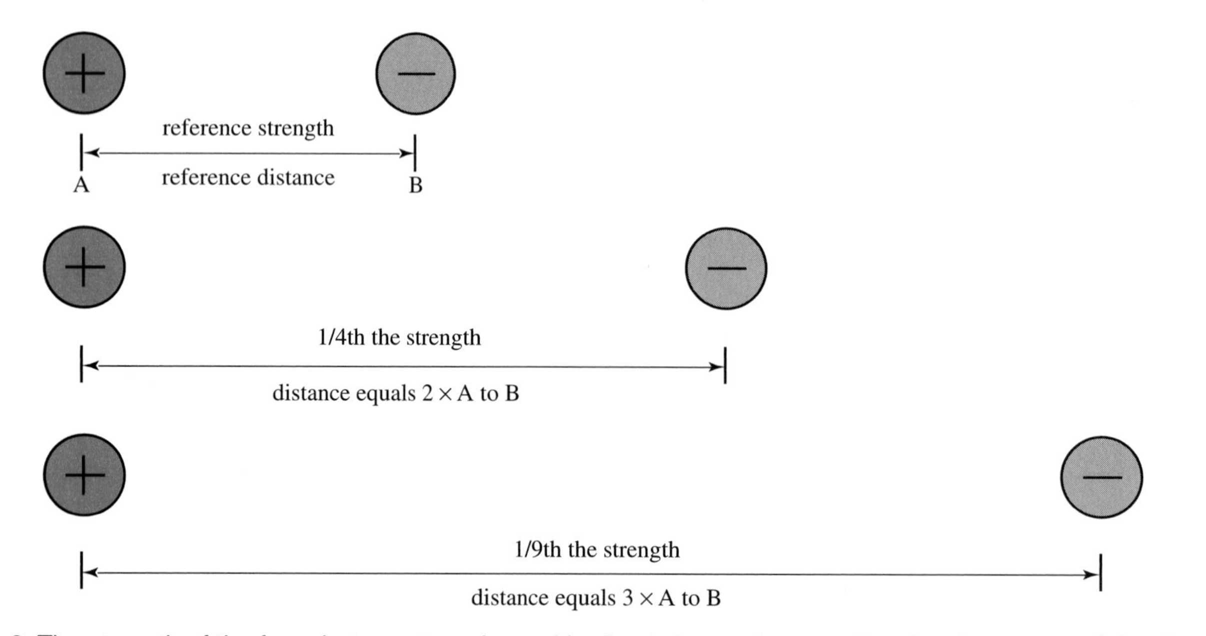

Figure 1-9. The strength of the force between two charged bodies is inversely proportional to the square of the distance.

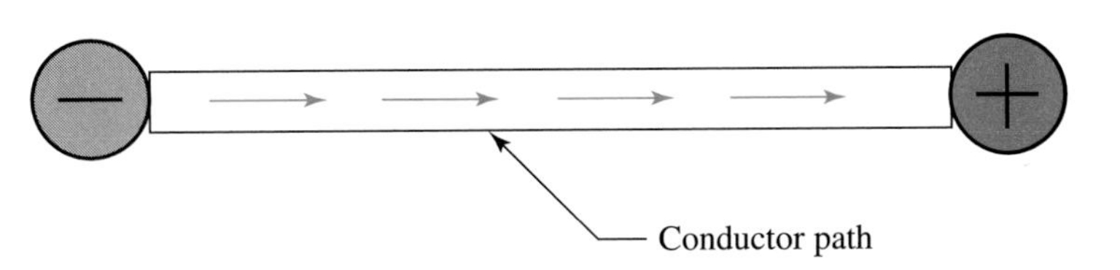

Figure 1-10. When two charged bodies are connected with a conductor, excess electrons will flow through the conductor from the mass having a surplus of electrons to the mass having a deficit of electrons.

in reducing pollution is called an ***electrostatic precipitator.*** Most precipitators are divided into two parts, a charging section and a collecting section. The charging section can be designed in many different ways. It can be an assembly of parallel rods or wires, a screen pattern, or a bank of piping. Regardless of the physical design, the electrical principle is the same in all cases.

Figure 1-11 illustrates the operation of an electrostatic precipitator. A grid of electrodes is installed into the stream of pollutants. The electrodes are charged between 45,000 and 75,000 volts forming an ionized field or a corona around each electrode. A corona is an area of ion or static charge surrounding a high voltage conductor. The pollution particles become negatively charged as they pass through the corona. The negatively charged particles are then attracted to the positively charged collection plates. After the collection plates become laden with pollution particles, they must be cleaned. This principle of removing particles (pollutants) from the air is used in industries such as concrete, paper, chemical, and coal-fired power generation.

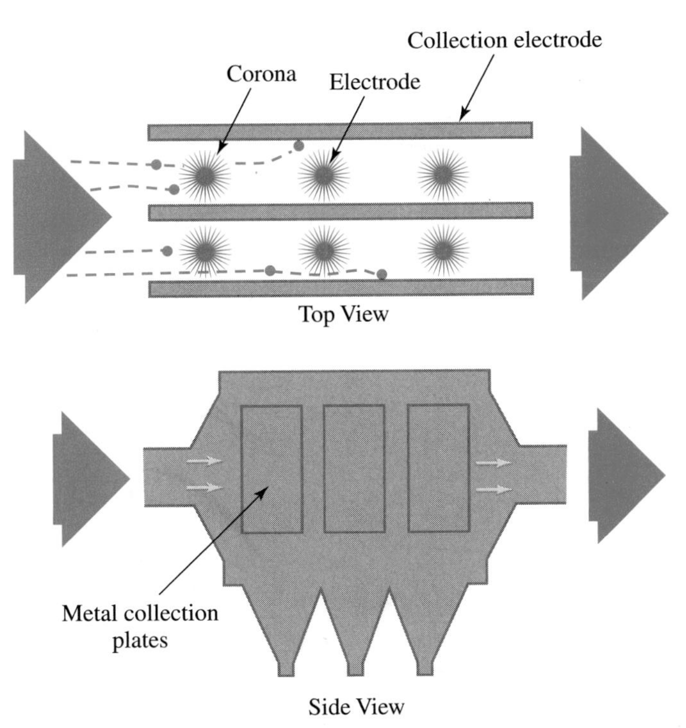

Figure 1-11 An electrostatic precipitator places an electrostatic charge on particles, and then collects the particles on an oppositely charged plate. (Powerspan Corp.)

The application of electrostatics is also gaining popularity in residential use. Electrostatic filters are a part of the latest air-conditioning systems.

The principle of attraction and repulsion is also applied in the painting industry and in the manufacture of sandpaper. The painting industry sprays a positively charged mist of paint onto a negatively charged surface such as the panels of an automobile. This procedure reduces the amount of overspray and saves paint. The sandpaper industry charges the backing paper with a positive static charge, and the silica crystals (sand or some other abrasive) with a negative charge. The result is an even spread of granules over the entire surface of the paper.

Review Questions for Section 1.2

1. Define static electricity.
2. What is a compound?
3. Explain what is meant by electrostatic field.
4. Like charges __________ each other.
5. Unlike charges __________ each other.
6. What is meant by a negatively charged body?
7. Define induction.

1.3 BASIC ELECTRICAL CIRCUIT

A basic electrical circuit consists of three main parts, a source of voltage, a load, and conductors. In **Figure 1-12,** a basic circuit is illustrated. This circuit consists of a battery as the source of electrical energy, a lamp as the electrical load, and two wires as the conductors connecting the battery to the lamp. In the source of this circuit, the battery, a chemical reaction takes place that results in ionization. This ionization produces an excess of electrons (negative charge) and a depletion of electrons (positive charge).

The battery has two terminals. These terminals are connection points for the two conductors. One terminal is

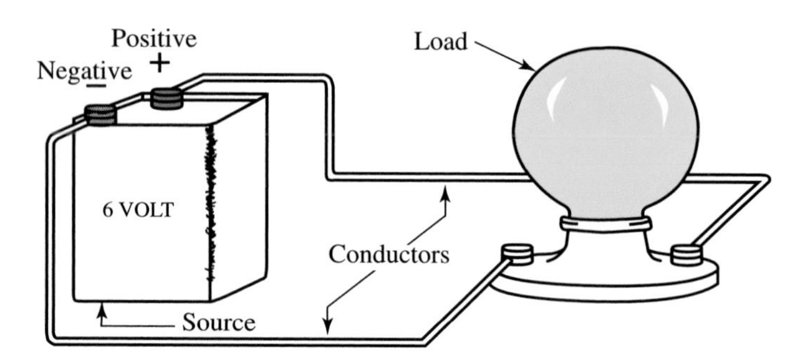

Figure 1-12. A basic electrical circuit consists of three main parts: the source, the load, and the conductors.

marked with a plus sign (+) and the other a negative sign (–). These two markings are referred to as *polarity markings*. Not all electrical devices have polarity markings. However, when polarity is a critical issue, it will be marked on the device. The proper polarity must be followed to avoid damage to equipment and/or personnel.

LESSON IN SAFETY:

There will be occasions when you become confused while working on electrical projects and with unfamiliar devices. Anytime you are uncertain about connecting any electrical device, check with your instructor. Damage from an improperly connected circuit is usually instantaneous and cannot be reversed.

A load is created when the electrical energy produced in a circuit is converted to some other form of energy such as heat, light, or magnetism. The load in the simple circuit of **Figure 1-12** is a lamp that produces light. The source and the load should match according to voltage rating. If the lamp is rated at 6 volts, then the battery should also be rated at 6 volts. If the battery is rated at a lower voltage rating, the lamp will appear dim or will not light. If the battery is rated at a much higher voltage, the lamp will be damaged by the excess electrical energy.

The conductors we are using are two copper wires covered with a plastic insulation coating. The copper wire provides a path through which the electrical energy can flow, while the plastic coating restricts the electrical energy to the copper wire. This makes the conductor pathway safe for personnel. This completes the description of the basic components of a circuit in which electrical energy is channeled by way of electrical conductors, through a device, where it is then converted to some useful form.

Voltage

Ionization can be caused by forces such as heat, light, magnetism, chemical action, or mechanical pressure. This results in the creation of an electrical voltage. What is voltage? ***Voltage*** is the force behind electron flow. In the simple circuit just described, the battery was the source of electrical energy. This battery has a rating of 6 volts. The ***volt (V)*** is the electrical unit used to express the amount of electrical pressure present, or the amount of electrical force produced by the chemical action inside the battery.

The term voltage is used to express the amount of electrical force in much the same way we use horsepower to express the amount of mechanical force for an automobile. Electrical pressure or voltage can also be expressed as ***potential, potential difference***, or as ***electromotive force (emf).*** For our purposes, these terms mean the same thing. Voltage is usually represented by the capital letter ***E*** or ***V***.

Current

Electrical ***current*** is the flow of electrons. The amount of electrons flowing past any given point in one second is rated in the electrical unit ***ampere (A)***. The ampere is expressed using the letter ***I.*** Remember that a coulomb is a quantity of electrons. The ampere describes the rate of flow of the electrons past any given point in a circuit. One ampere is equal to one coulomb of charge flowing past a point in one second.

Compare a balloon filled with air to an electrical battery. In **Figure 1-13**, the amount of air molecules in the balloon represents the amount of electrons or coulombs. The amount of air pressure inside the balloon is expressed as pounds per square inch (PSI) of air pressure. In the battery, the amount of electrical pressure inside the battery is expressed as the voltage rating of the battery. The rate of air flow out of the balloon is similar to electron flow, or current, from the battery. The current from the battery in the electrical circuit is the volume of electron flow past a given point, and is rated in amperes or amps. Just as the air will continue to escape from the balloon until the balloon is empty, the electron flow can continue as long as there is voltage or electrical pressure present in the battery.

Resistance

All electrical circuits have resistance. ***Resistance*** is the opposition to the flow of electrons. Resistance is

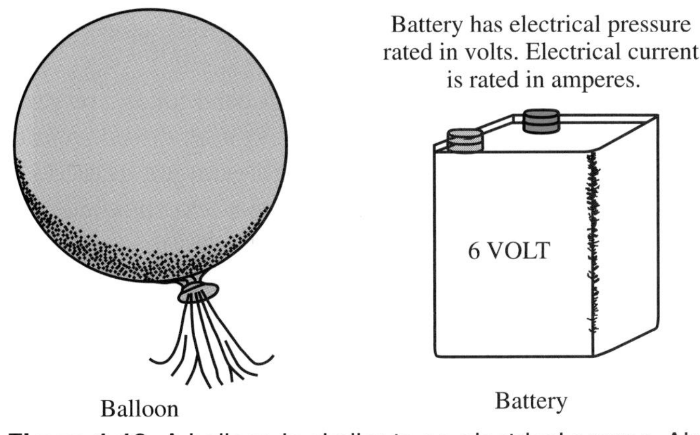

Balloon has air pressure rated in PSI. Air escaping from balloon is rated in cubic inches per minute.
Battery has electrical pressure rated in volts. Electrical current is rated in amperes.
6 VOLT
Balloon
Battery

Figure 1-13. A balloon is similar to an electrical source. Air escaping from the balloon is similar to electrons flowing from a source.

measured in ***ohms***, and the electrical symbol for ohm is Ω (the Greek letter omega). The resistance values of elements and compounds differ according to the atomic structure of the material. A good ***conductor*** of electricity is anything that permits the free flow of electrons. A poor conductor of electricity is a material that will not permit the free flow of electrons. Extremely poor conductors are referred to as ***insulators***. A ***semiconductor*** is a material that limits the flow of free electrons. A semiconductor is considered neither a good conductor nor poor conductor of electricity. Semiconductor materials are at the very heart of modern electronic applications and will be explored in depth in Chapter 17. Some examples of conductors and insulators are listed in **Figure 1-14.**

Note that the earth can be a good conductor of electricity. There are many factors that determine whether or not the earth will be a good conductor. The earth's conductivity is primarily dependent upon its organic composition and on the minerals found in the soil at any given place. The amount of moisture in the soil also determines the amount of resistance in the soil. Moisture can affect the electrical conducting ability of many materials. It can even cause an insulator to become a good conductor. Take wood as an example to illustrate this point. When wood is dry, it is classified as an insulator, but when wood becomes wet or moist, it behaves more like a semiconductor.

Conductors	Insulators
copper	Bakelite®
iron	glass
steel	mica
aluminum	porcelain
silver	air
tin	dry wood
damp earth	sand
salt water	distilled water
	some plastics
	paper
	rubber
	oil

Figure 1-14. Common conductors and insulators.

It is the outer ring of an atom that determines whether an element is a good or poor conductor. If the outer ring has only one electron, that electron can be freed from its orbit rather easily by an outside force. If there are many electrons in the outer orbit, the electrons are held tighter in orbit. They are harder to free from the atom. Elements that do not readily give up an electron are insulators.

Figure 1-15 is an illustration of the copper atom. Notice how this atom has only one electron in its outer orbit. This electron can be easily freed by an outside force. Copper is an excellent conductor of electricity.

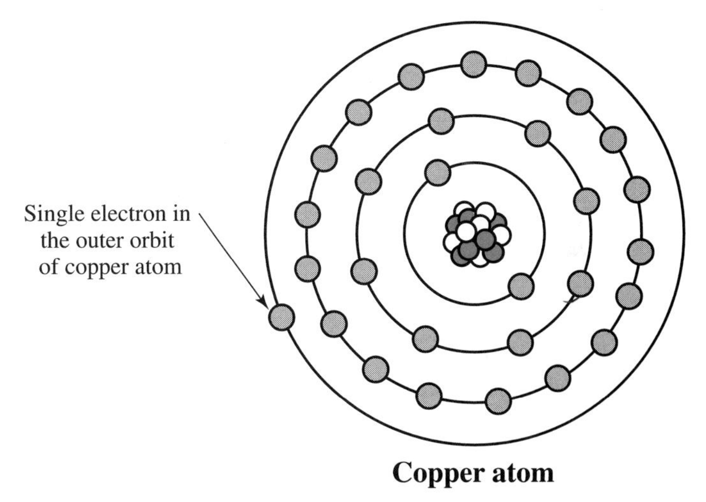

Figure 1-15. The element copper is an excellent conductor. It has only one electron in its outer orbit. This electron can be easily released from its orbit by an outside force.

Current, AC and DC

There are two types of electrical current, dc (direct current) and ac (alternating current). The difference between these currents is how they flow through an electrical circuit. ***Direct current*** flows in only one direction through an electrical circuit. An example of direct current is a standard battery. The battery has a set polarity (positive and negative terminals) and will produce an electric current in only one direction. On the other hand, ***alternating current,*** as its name implies, flows in both directions. First it flows in one direction, and then it reverses its flow to the opposite direction. See **Figure 1-16.**

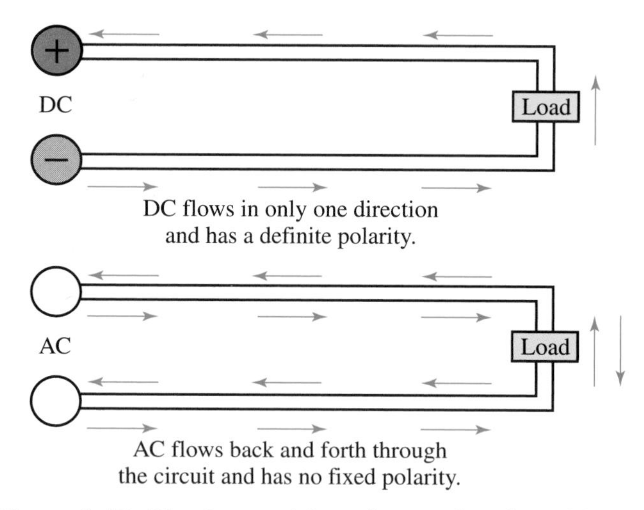

+
DC
Load
−
DC flows in only one direction
and has a definite polarity.
AC
Load
AC flows back and forth through
the circuit and has no fixed polarity.

Figure 1-16. Direct current flows in one direction while alternating current repeatedly alternates direction.

There are no positive or negative polarity markings in alternating current because the polarity changes so rapidly in the typical ac electrical circuit. The terms *cycle* and *hertz* are used to describe how fast the current is alternating or changing direction in the circuit. A 60 cycle ac circuit (operating at 60 hertz) changes direction 120 times per second. This is the standard for ac in the USA.

Conventional Current Flow vs. Electron Flow Theory

Approximately 200 years ago, scientists theorized that electricity had both positive and negative polarities. At that time they arbitrarily decided that electrical current flowed from positive to negative. While it was never actually proven as fact, this theory was accepted for quite some time. This theory is known as the ***conventional current flow theory.*** As our knowledge of science progressed, and with the discovery of the atom and semiconductor electronics, it became apparent that the conventional current flow theory was incorrect. It is widely accepted that it is the electrons that actually move, flowing from negative to positive, not from positive to negative. This newer theory is known as ***electron flow theory.*** The emergence of this new theory caused a controversy that is still in existence today. For over 150 years all circuit designs had been based upon the old, conventional current flow theory.

Many circuits and devices still used today are based on the conventional theory. This text uses the convention that will make the concepts in each example most easily understood. Most of the figures in this text show electron flow. Regardless of which theory is used to explain the phenomena of electronics, the most important point is *that the correct polarity must be maintained when building circuits with devices that require a definite polarity.* Examine each example for polarity markings. See **Figure 1-17.**

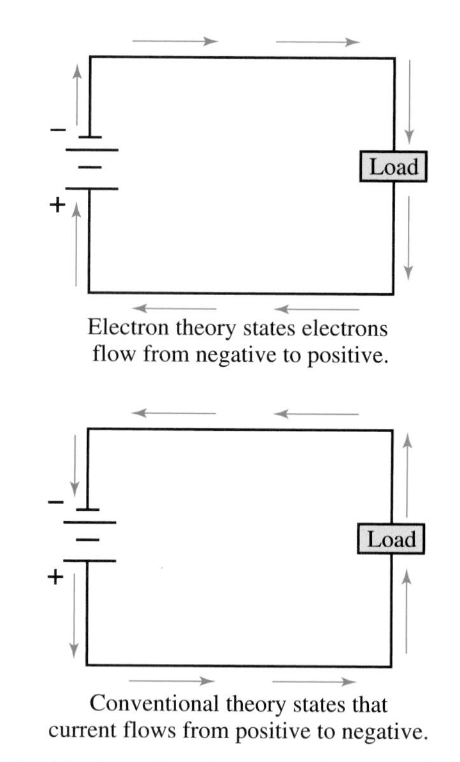

Figure 1-17. Electron flow theory and conventional current flow theory.

Series and Parallel

Series and parallel are two important concepts. They must be learned early to fully understand the next few chapters. There are two ways a component can be connected in a circuit, either series or parallel. **Figure 1-18** and **Figure 1-19** illustrate the two types of connections. The circuit in Figure 1-18 has three lamps connected to a

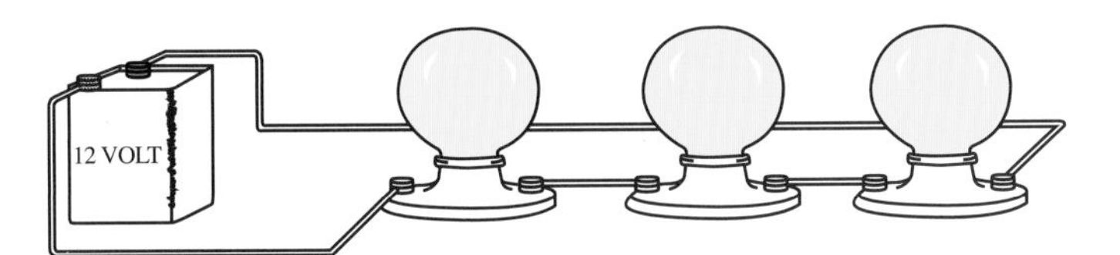

Figure 1-18. Three lamps connected in series.

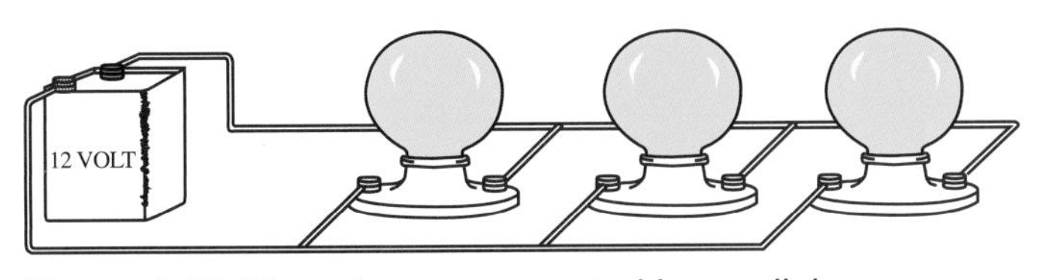

Figure 1-19. Three lamps connected in parallel.

battery. In this circuit, there is only one path over which the electrons can flow. When electrons only have one circuit path to follow, that circuit is called a *series* circuit. The lamps are said to be wired in series with respect to each other.

In Figure 1-19, there are three lamps connected in *parallel*. In this circuit, there are three different paths for the electrons to follow from battery terminal to battery terminal. Both the series and parallel circuits have advantages and disadvantages. These will be thoroughly covered in later chapters in this text. For now, be able to readily distinguish between the two types of circuits.

Review Questions for Section 1.3

1. What are the three main parts of an electrical circuit?
2. The __________ supplies the electrons that will flow through the circuit.
3. The __________ provide a path through which the electrical energy can flow.
4. The __________ is where the electrical energy is converted to another form of energy.
5. The source has __________ markings that are identified with a ___ or ___ symbol.
6. The movement of electrons is known as __________.
7. Opposition to current flow is called __________.
8. Opposition to current is measured in __________.
9. Electrical pressure is measured in __________.
10. __________ theory states that electrons flow from negative to positive.
11. __________ current flows only in one direction while __________ current constantly changes direction.
12. Connecting the correct __________ of an electrical device in an electrical circuit is more important than which theory of current flow is used.
13. A(n) __________ circuit provides only one path for electron flow.
14. A(n) __________ circuit provides more than one path for electron flow.

1.4 OHM'S LAW

Electrical circuits that are built correctly will be in perfect electrical balance. The current through the resistance is directly related to the amount of electrical pressure or voltage applied to the circuit. This balance of the three factors, voltage (E), resistance (R), and current (I), can be expressed by ***Ohm's law***. Ohm's law is named for the 19th century German scientist George Simon Ohm.

The relationship expressed by Ohm's law is the basic formula that is used more extensively than any other electrical formula you will encounter in your study of electricity and electronics. It is the basis for many other formulas and electrical relationships studied in this text. Ohm's law states that the current measured in amperes (I) in a circuit is equal to the applied voltage (E) divided by the resistance (R). Ohm's law is expressed in three formulas below.

$E = I \times R$ Applied voltage is equal to the current multiplied by the resistance.

$I = E/R$ Current is equal to applied voltage divided by the resistance.

$R = E/I$ Resistance is equal to the applied voltage divided by the current.

I = current (electron flow) in amperes
E = voltage (electrical pressure) in volts
R = resistance (opposition to current) in ohms

A memory device commonly used to assist you in learning Ohm's law is illustrated in **Figure 1-20.** To see how easy it is to use, simply cover the unknown quantity with your finger, and the remaining letters will show the solving equation. For example, when the voltage is the unknown quantity, cover E with your finger. The I and R remain. Thus $E = I \times R$. Now cover I with your finger. An E over R remains. Thus, $I = E/R$.

Let's look at an application of Ohm's law. In **Figure 1-21,** a lamp with a resistance of 4 ohms has been connected to a 12 volt source. The current is unknown. By applying Ohm's law you can determine the current to be equal to 3 amperes.

Figure 1-20. A memory device used to help solve Ohm's law problems. Cover the unknown quantity and the remaining letters show the correct equation. Example: *I* is equal to *E* divided by *R*.

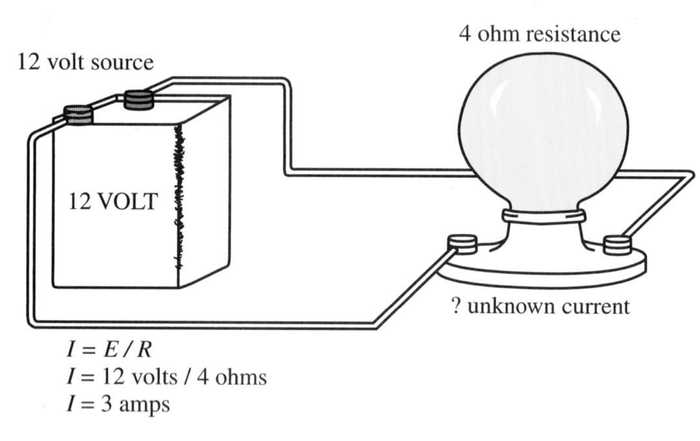

$I = E / R$
$I = 12 \text{ volts} / 4 \text{ ohms}$
$I = 3 \text{ amps}$

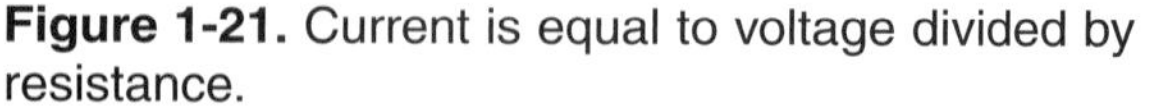

Figure 1-21. Current is equal to voltage divided by resistance.

In **Figure 1-22,** a 24 ohm resistance heater works most efficiently when using 5 amps of current. How much voltage is required for the heater to operate at 5 amps? By applying Ohm's law, the amount of electrical pressure needed to conduct 5 amps through a 24 ohm resistance is 120 volts.

In **Figure 1-23** a resistor is connected to a 24 volt source. A meter that measures current indicates there are 3 amperes present in the circuit. Again, by applying Ohm's law, the amount of resistance needed to limit the current value to 3 amperes is found to be 8 ohms. The

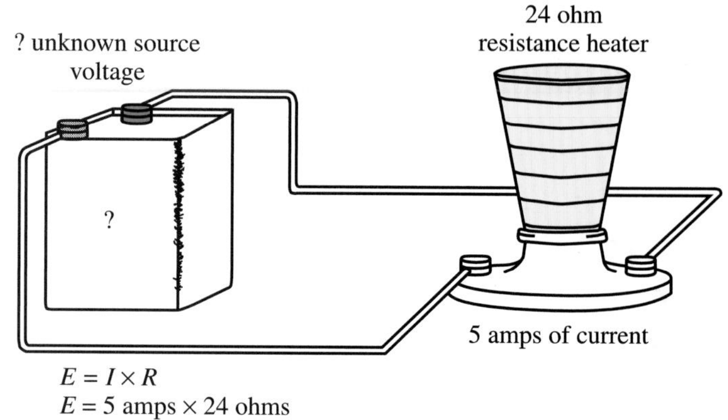

$E = I \times R$
$E = 5 \text{ amps} \times 24 \text{ ohms}$
$E = 120 \text{ volts}$

Figure 1-22. Voltage is equal to resistance multiplied by current.

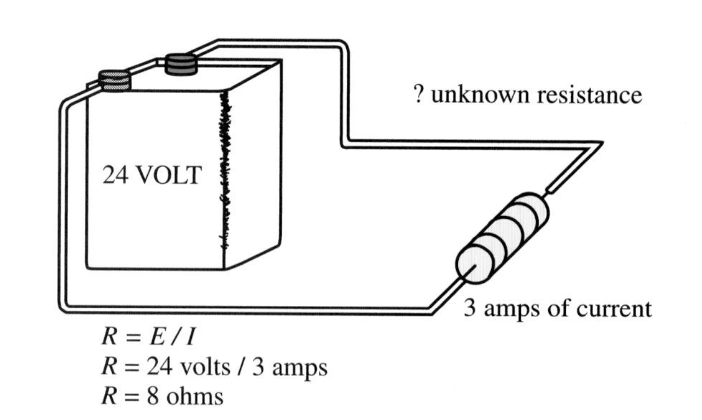

$R = E / I$
$R = 24 \text{ volts} / 3 \text{ amps}$
$R = 8 \text{ ohms}$

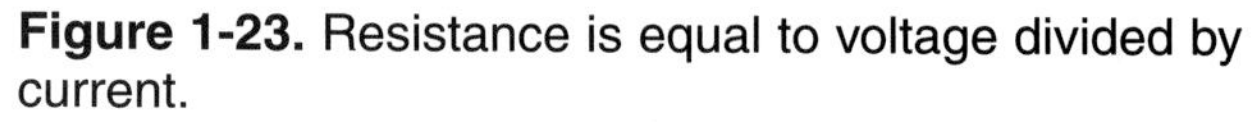

Figure 1-23. Resistance is equal to voltage divided by current.

application of Ohm's law may seem strange at first, but with a little practice, it will become second nature to you.

Electrical Prefixes

Measurements of electrical quantities vary from small amounts to large amounts. To make it easier to label electronic parts and equipment, and to make electronic calculations easier, a system of prefixes is used when expressing electrical quantities. Without this system we would need to use an excessive amount of zeros to the left and right of a decimal point. **Figure 1-24** is a listing of the most common prefixes used in the electronics industry today.

When an electrical quantity such as voltage is written, it is expressed in units such as kilovolt, megavolt, millivolt, and microvolt to avoid using an awkward numerical form. For example, the quantity 5,000,000 (five million) volts, would be written as 5 MV. If the quantity was 0.005 (five thousandths) volts, it would be

COMMON ELECTRICAL PREFIXES

Prefix	Symbol	Decimal Equivalent	Power of Ten
tera	T	1,000,000,000,000.	10^{12}
giga	G	1,000,000,000.	10^{9}
mega	M	1,000,000.	10^{6}
kilo	k	1,000.	10^{3}
basic unit		1.	
milli	m	.001	10^{-3}
micro	µ	.000 001	10^{-6}
nano	n	.000 000 001	10^{-9}
pico	p	.000 000 000 001	10^{-12}

Figure 1-24. Shown is a list of the electrical prefixes you will most often encounter.

written as 5 mV. It is important to note the use of the upper case letters and duplicate lower case letters in these units. When voltage is expressed with a capital letter M, it represents millions of volts, but when voltage is expressed using a lower case m, it represents the fractional unit of thousandths.

Review Questions for Section 1.4

1. A voltage of __________ is needed to produce a current of 2 amps through a resistance of 6 ohms.
2. When 12 volts are connected to 4 ohms, a current equal to __________ flows through the resistance.
3. A resistance of __________ produces a current equal to 8 amps when connected to 24 volts.
4. How many volts does 4 kV represent?
5. What is 8 MΩ equal to in kΩ?
6. Write 3 mV as a decimal unit of voltage.
7. One millivolt is how much larger than one microvolt?
8. How many milliamps are there in 20 amps?
9. Is 5×10^{-3} larger or smaller than 5×10^{-6}?
10. Express 6×10^{6} ohms with a prefix.

Summary

1. Matter is anything that occupies space or has mass.
2. Elements are basic or pure forms of matter.
3. Compounds are mixtures or combinations of two or more elements.
4. Atoms are the simplest forms of an element still having the unique characteristics of that element.
5. Molecules are the simplest form of a compound still having the unique characteristics of that compound.
6. The negatively charged particle of the atom is the electron and the positively charged particle is the proton.
7. Like charges repel each other while unlike charges attract each other.
8. Induction occurs when a charged body is brought close to another body.
9. The coulomb is a quantity of electrons (6,240,000,000,000,000,000, or 6.24×10^{18} electrons).
10. Current is the movement of electrons in a conductor.
11. Voltage is the force behind the electrons. It moves them along a conductor resulting in current.
12. Ohm's law can be stated three ways: $E = I \times R$, $I = E/R$, and $R = E/I$
13. There are two types of current, ac (alternating current) and dc (direct current).
14. Correct polarity must be observed when connecting electrical devices.

Test Your Knowledge

Please do not write in the text. Place your answers on a separate piece of paper.

1. An electron displays a(n) __________ charge, a proton displays a(n) __________ charge, and a neutron displays a(n) __________ charge.
2. What is the name for the electrical unit based on the number of electrons? How many electrons does it represent?
3. What are the two laws of electrostatic charges?
4. Electron movement is called __________.
   A. voltage
   B. current
   C. resistance
   D. ohm
5. The force behind electron movement is called __________.
   A. voltage
   B. current
   C. resistance
   D. ohm

6. When an atom loses an electron, it becomes __________.
   A. vaporized
   B. ionized
   C. resistance
   D. negative
7. The invisible line of force that surrounds a charged body is called the __________.
   A. electron force
   B. electrostatic force
   C. electric field
   D. None of the above.
8. What causes ionization to occur?
9. Matter is found in what three forms?
10. The smallest part of a compound is called a(n) __________.
11. The smallest part of an element is called a(n) __________.
12. What are the three basic parts of a circuit?
13. Which part of a circuit provides electrical energy?
14. Which part of a circuit converts the electrical energy to another form of energy?
15. Which part of the circuit provides a current path?
16. Write 16,000 volts with a prefix.
17. Write 0.005 volts with a prefix.
18. Write the three forms of Ohm's law.
19. How much voltage is required to force 3 amperes through a resistance of 36 ohms? (Include the formula with your answer.)
20. How much current will a 24-ohm resistor connected to a 6-volt source produce? (Include the formula with your answer.)
21. A meter indicates 5 amps in a simple circuit connected to a 12-volt source. How much resistance is present? (Include the formula with your answer.)
22. How much current will a 12 V car battery produce when connected to a lamp with a 600 ohm resistance value?
23. A 5 kΩ resistance connected to a 100-volt source will produce how much current?
24. A 6 V battery is connected to a 240Ω lamp. What is the expected amount of amperes produced by this circuit?
25. How much voltage is needed to produce a current of 50μ A through a resistance of 1 kΩ?

2 Basic Instruments and Measurements

Objectives

After studying this chapter, you will be able to:

- Explain the correct procedure for using an ammeter, a voltmeter, and an ohmmeter.
- Interpret a linear scale.
- Compute shunt resistor values.
- Compute multiplier resistor values.
- Interpret a nonlinear scale.
- Discuss the concept of meter sensitivity.
- Understand basic electrical diagrams.
- State and explain Ohm's law.

Key Words and Terms

The following words and terms will become important pieces of your electricity and electronics vocabulary. Look for them as you read this chapter.

ammeter
analog meter
common
D'Arsonval movement
digital meter
digital multimeter (DMM)
field-effect transistor-VOM (FET-VOM)
linear meter scale
multiplier resistor
nonlinear scale
ohmmeter
ohms-per-volt
resolution
root mean square (rms) value
schematic
shunt
volt-ohm-milliammeter (VOM)

Electricity and electronics technicians rely on instruments to judge the action and traits of a circuit precisely. The skillful use of instruments is the mark of a good technician and will enable one to quickly and efficiently troubleshoot a circuit. The student of electricity and electronics must know what he or she is trying to measure and how to measure it. In this chapter we will discuss the three most common types of meters used today. These meters are the ohmmeter, ammeter, and the voltmeter. Each of these meters will be covered thoroughly to give you the basic skills necessary to continue your studies. Think of an electrical meter as an electronic ruler that is used to measure electrical quantities such as voltage, current, and resistance.

The meters you will be using come in two formats. There are analog meters and digital meters. ***Analog meters,*** discussed first in this chapter, use a scale with continuous variable values. ***Digital meters*** give values in discrete amounts using the units 0 through 9. Digital meters are discussed later in this chapter.

2.1 BASIC ANALOG METER MOVEMENT

A common type of meter movement measures current and voltage. It is the ***D'Arsonval movement***, or stationary magnet, moving-coil galvanometer, **Figure 2-1**. The movement consists of a permanent-type magnet and a rotating coil in the magnetic field. An indicating needle is attached to the rotating coil, **Figure 2-2**.

When a current passes through the moving coil, a magnetic field is produced. This field reacts with the stationary field and causes rotation (deflection) of the needle. This deflection force is proportional to the strength of the current flowing in the moving coil. When the current ceases to flow, the moving coil is returned to its "at rest" position by hair springs. These springs are also connected to the meter coil. The deflecting force rotates the coil against the restraining force of these springs. See **Figure 2-3.**

CAUTION:
The coil that rotates in the magnetic field is mounted on precision-type jewel bearings much like a fine watch. The jewel type bearings and mount, known as a D'Arsonval movement, make the instrument very easy to damage if dropped or jarred. Extreme caution should be used when transporting or moving a meter with a D'Arsonval type of movement.

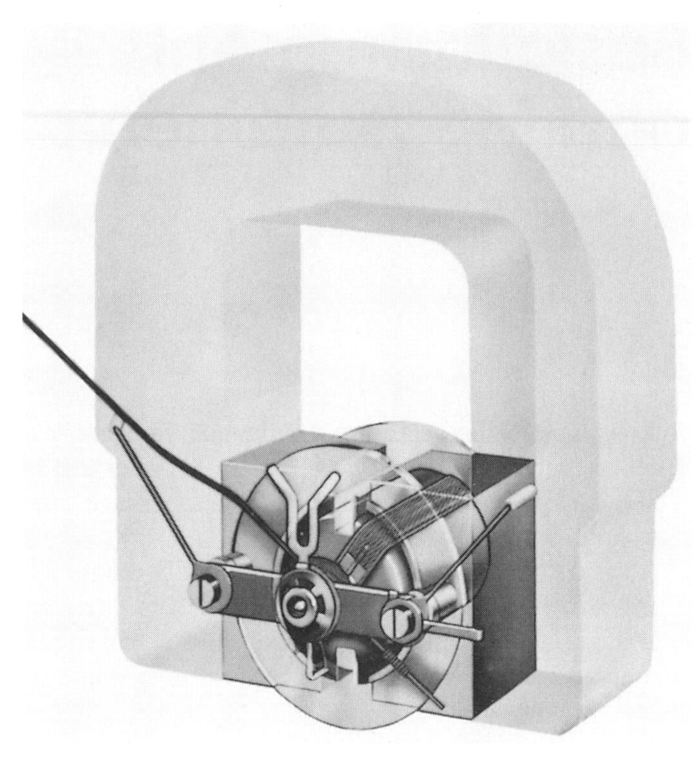

Figure 2-1. A phantom view of the D'Arsonval meter movement.

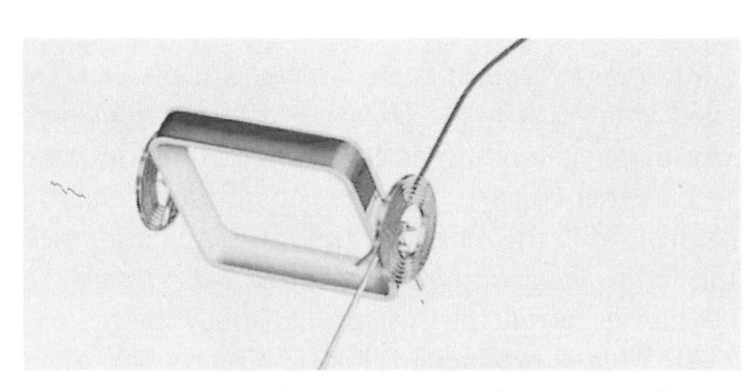

Figure 2-2. On the D'Arsonval, the indicating needle is attached to a rotating coil on the meter.

When connecting a meter to an electrical circuit, proper polarity must be maintained. The meter is equipped with polarity markings, usually a red plus sign (+) and a black negative sign (–). Some meters use the abbreviation COM, which stands for ***common***, for the negative polarity marking. The meter coil rotates inside the permanent magnet field. If proper polarity is not used, the coil will deflect in the direction opposite to that which it was designed. At the very least, the needle will not deflect, and there will appear to be no reading. At worst, this situation could possibly damage the meter. Some meters have circuit protection built into them. This protects the meter movement from damage that can be caused by improper connection.

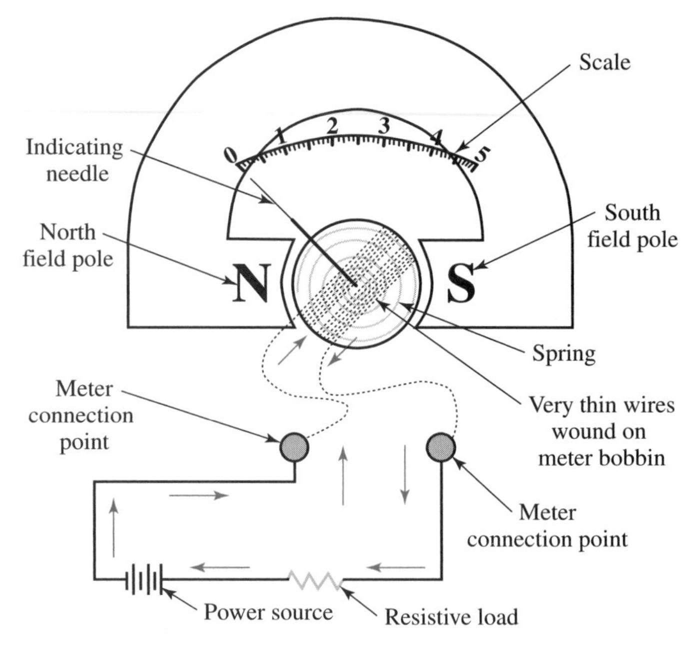

Figure 2-3. Current flowing through the ammeter must be limited by a resistance in the circuit being tested.

Iron Vane Meter Movement

The operation principle of the iron vane meter movement is shown in **Figure 2-4**. Two pieces of the iron are placed in the hollow core of a solenoid (a coil of wire). When the current passes through the solenoid, both pieces of metal become magnetized with the same polarity. Because like poles repel each other, the two pieces of iron are repelled from each other. One piece of metal is fixed in its position. The other piece of metal pivots. The pivoting piece can turn away from the fixed metal. An indicating needle is attached to the moving vane. The needle is equipped with hair springs so that the vane must move against the spring tension for accurate readings.

An applied voltage causes current to flow in the solenoid and creates the magnetic field. The moving vane is repelled against the spring according to the strength of the magnetic field. The needle may indicate either voltage or current. It is calibrated for the magnitude (average size) of the applied voltage or current.

When the iron vane movement is used for a voltmeter, the solenoid is commonly wound with many turns of fine wire. Proper multiplier resistance may be used to increase the range of the meter. A selector switch is used to select proper ranges. When used as an ammeter, the solenoid has a few turns of heavy wire. This is because the coil must be connected in series with the circuit and carry the circuit current.

Regardless of the polarity of the applied voltage or current, the iron vane meter movement always deflects in

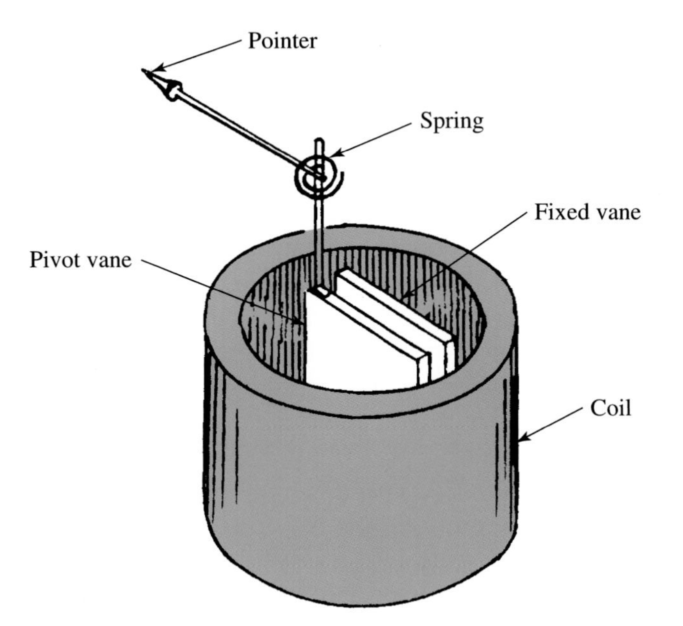

Figure 2-4. Operating principle of the iron vane meter movement.

the same direction. Either ac or dc may be measured with this instrument. Generally, this type of meter is best suited for high power circuit measurements.

Meter Scales

The meter scale used to interpret ampere and voltage values is the linear type. A ***linear meter scale*** has evenly spaced marks used to indicate the amount of current flowing, or voltage present, in the meter movement. **Figure 2-5** shows a typical linear scale for an ammeter.

The scale illustrated in Figure 2-5 is marked from 0 to 5 with ten smaller marks between each major numbered mark. To determine the value of each mark between the major divisions (the scale factor), divide the value of the first major division by the number of spaces in that division. The dial to the right of each scale in Figure 2-5 is the *range selector*. The range selector must be correlated to the scale to determine full scale deflect.

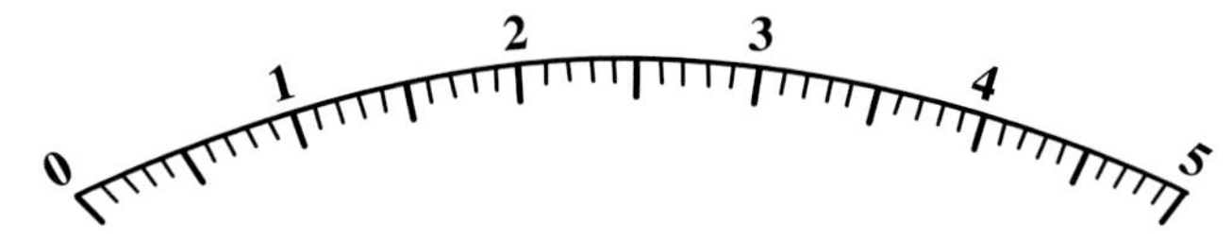

Full scale deflection equals 5 amps.
Major numbered divisions equal 1 amp each.
The small divisions equal 0.1 each.

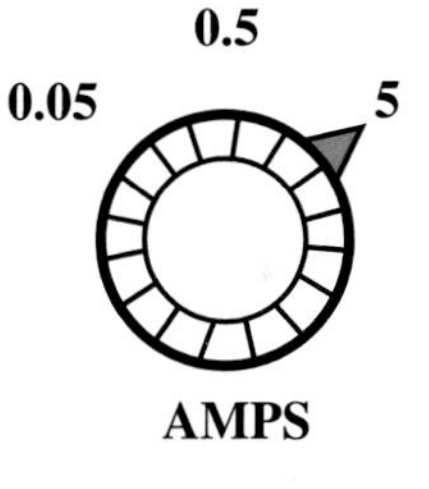

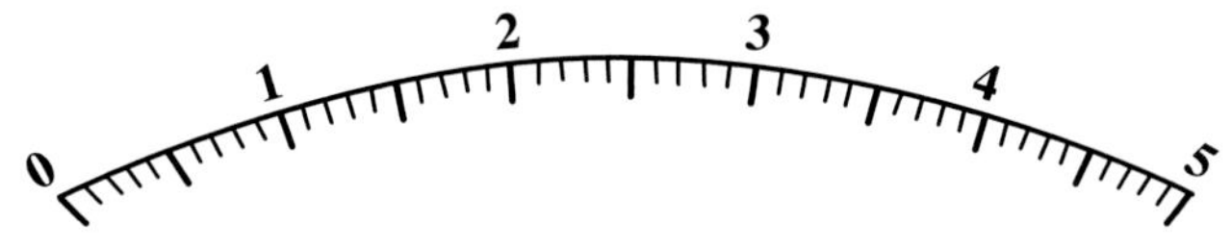

Full scale deflection equals 0.5 amps.
Major numbered divisions equal 0.1 amp each.
The small divisions equal 0.01 each.

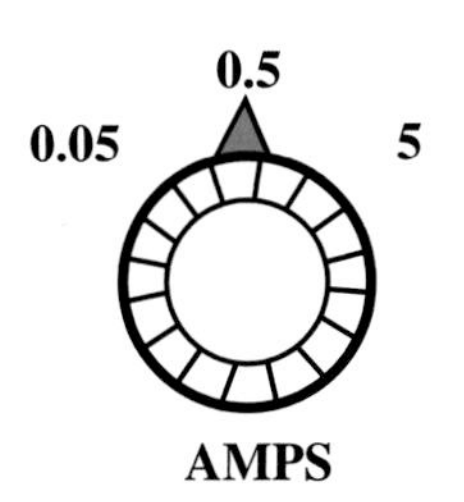

Full scale deflection equals 0.05 amps.
Major numbered divisions equal 0.01 amp each.
The small divisions equal 0.001 each.

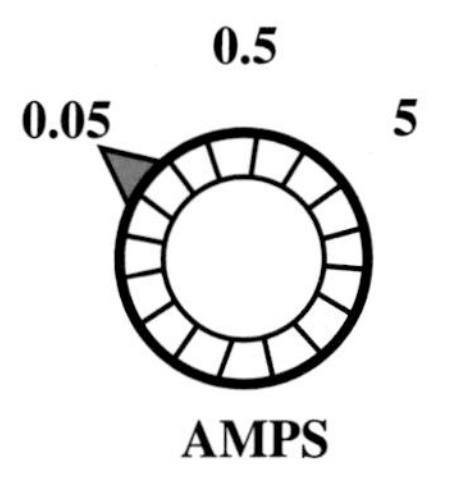

Figure 2-5. Typical linear meter scale.

The formula for scale factor is as follows.

$$\text{Scale factor} = \frac{\text{Value of major division}}{\text{Number of spaces}}$$

Study Figure 2-5. Note that the value of each division changes as the range selector changes. On the scales, the first major division is marked with a one (1). In the top example, the range selector is set on 5 amps. This means that the full scale deflection is 5 amps. On this scale the one (1) represents 1 amp. There are ten spaces between the one and the zero. By dividing one by ten, we can conclude that each space is equal to one tenth (0.1) of the first major mark or 0.1 amp.

The second example has a full scale deflection equal to 0.5 amperes. Therefore, the major scale markings are equal to 0.1 ampere each. Since there are ten spaces between each major division, each small mark is equal to 0.01 amperes each (10 milliamps).

In the third example the range selector switch is set at 0.05 amps. This makes the full scale deflection equal to 0.05 amperes. Each major numbered division is equal to 0.01 ampere. Since there are ten equal spaces between each major division on the scale, each small mark is equal to 0.001 amperes or 1 milliamp.

Review Questions for Section 2.1

1. Another name for a stationary-magnet, moving-coil meter is the __________ movement.
2. Explain how a moving-coil meter operates.
3. Why must you be careful when handling a moving-coil meter?
4. A linear meter scale has __________ spaced marks used to indicate circuit values.

2.2 AMMETER

An ***ammeter*** measures electrical current in a circuit. An ammeter will usually measure in amperes, milliamperes, or microamperes, depending on the scale or design of the instrument. The coil in the meter movement of an ammeter is wound with many turns of fine wire. If a large current is allowed to flow through this coil, the ammeter will quickly burn out. In order to measure larger currents, a ***shunt***, or alternate path, is provided for current. Most of the current flows through the shunt, leaving only enough current to safely work the meter movement coil. The shunt is a precision resistor connected in parallel with the meter coil. The use of shunts is illustrated in **Figure 2-6.**

In **Figure 2-7** you see the proper way to connect an ammeter to an electrical circuit. When an ammeter is connected into the circuit, it becomes part of the circuit in order to allow the current to flow through the meter coil. To connect an ammeter in a circuit, one usually has to make an open by disconnecting some device in the circuit. This allows you to insert the meter into the circuit. Notice that you are connecting the meter in *series* with the circuit or device you are trying to measure.

Example: The specification of a certain meter movement requires 0.001 A, or one milliampere of current, for full scale deflection of the needle. The ohmic resistance of the meter movement coil is 100 ohms. Compute the shunt resistor values for a meter that will measure four different ampere ranges. The ranges are as follows: 0–1 mA, 0–10 mA, 0–50 mA, and 0–100 mA.

Step 1. First calculate the voltage required for full scale deflection on the lowest setting which is 0–1 mA.

$$E = I \text{ (full scale current)} \times R \text{ (resistance of coil)}$$
$$E = 0.001\ \text{A} \times 100$$
$$E = 0.1\ \text{V}$$

The meter will read from 0–1 mA without a shunt. For full scale deflection 0.1 volts is required.

Step 2. To convert this same meter to read from 0–10 mA, a shunt must be connected that will carry 9/10 of the current. Thus, 9 mA of current will travel through the shunt, leaving one milliampere to operate the meter. The first step in the calculation determined that 0.1 V is required for full scale deflection. The shunt is connected in parallel with the coil, so it will also have 0.1 V applied to it. Since 0.1 V must be applied across the shunt, and the shunt must also account for 9/10 of the current, you can apply Ohm's law to calculate the shunt's resistance.

$$R_S = \frac{E}{I}$$

$$R_S = \frac{0.1\ \text{V}}{0.009\ \text{A}}$$

$$R_S = 11.1\ \Omega$$

The meter will require a shunt with a resistance value of 11.1 ohms for the 0–10 mA scale.

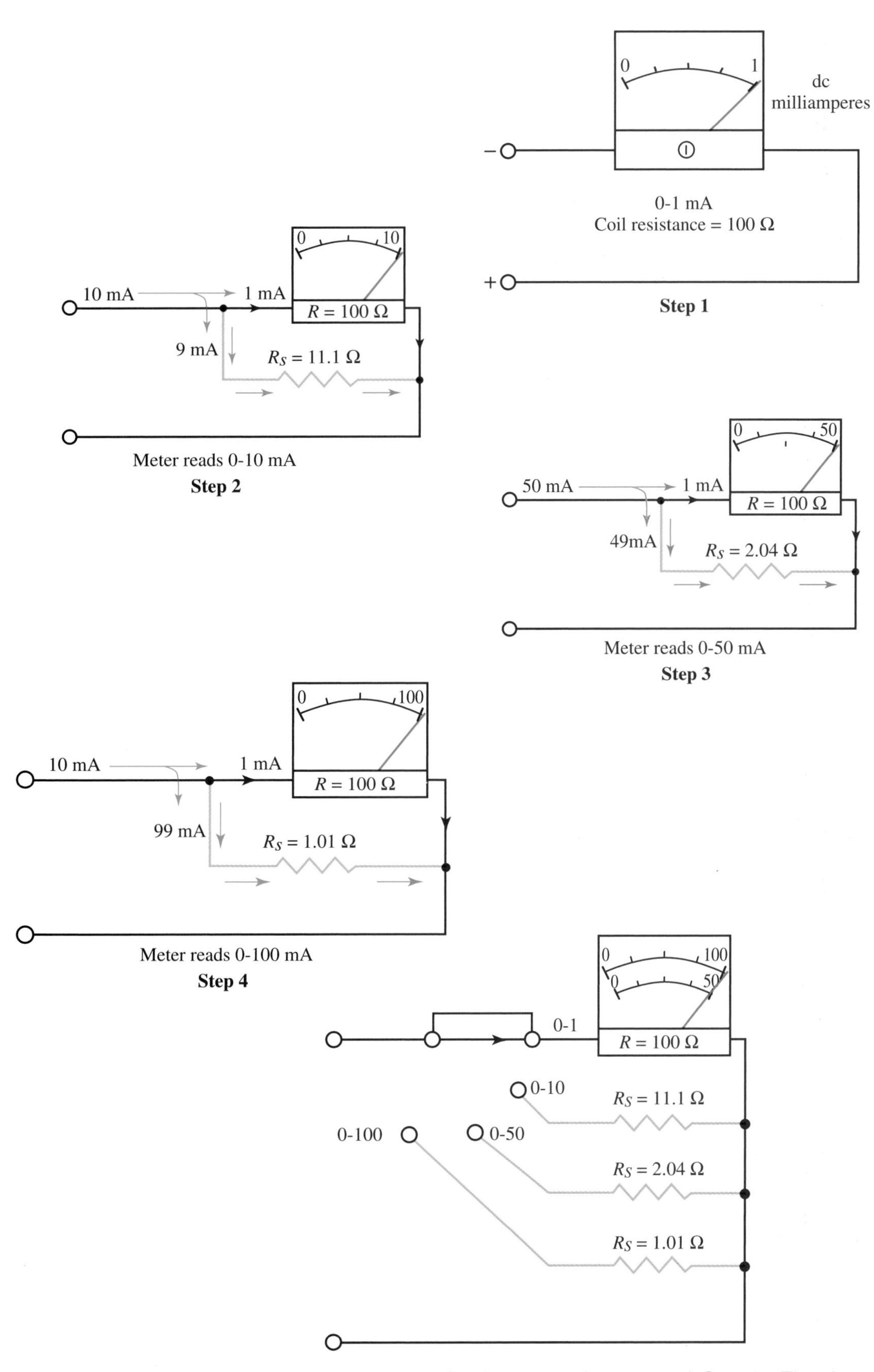

Figure 2-6. Step 1—The voltage that causes the full scale deflection current is computed. Step 2—The shunt carries 9/10th of the current. Step 3—The shunt carries 49/50th of the current. Step 4—The shunt carries 99/100th of the current. Bottom—Basic setup of an ammeter with three shunt resistors. A switch selects the range.

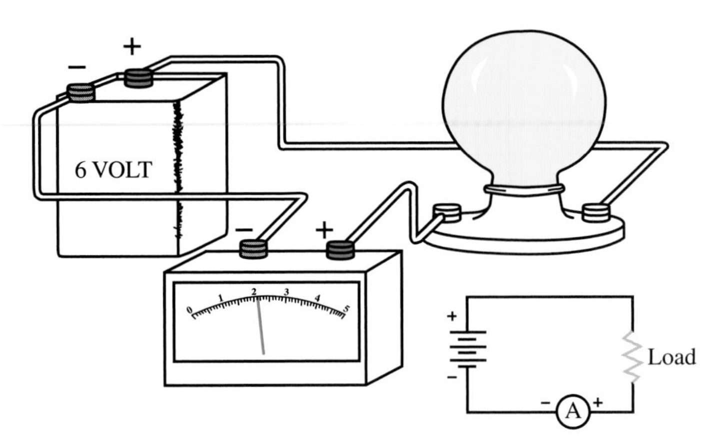

Figure 2-7. An ammeter is always connected in series with the circuit device being measured. The meter must be connected with proper polarity.

Step 3. To convert this meter for the 0–50 mA scale, a shunt must be used that will carry 49/50 of the current, or 49 mA. The computation is the same as in Step 2.

$$R_S = \frac{0.1\ \text{V}}{0.049\ \text{A}} = 2.04\ \Omega$$

Step 4. To convert the meter for the 0–100 mA scale, a shunt must be used that will carry 99/100 of the current, or 99 mA.

$$R_S = \frac{0.1\ \text{V}}{0.099\ \text{A}} = 1.01\ \Omega$$

A shunt with an ohmic value of 1.01 is required for the meter to safely use a 0–100 mA range. Look again at **Figure 2-6**. Notice the switching device used to change the ranges of the meter. The correct scale on the range dial must be used to correspond to the selected range.

CAUTION:

There are two important things to remember for the safety of your ammeter. First, an ammeter must always be connected in series with a circuit device or the power supply. Never connect an ammeter in parallel with the power supply or circuit devices, Figure 2-8. As you can see through the meter shunt calculations, the applied voltage to the meter movement coil only required 0.1 V for full scale deflection. If a voltage greater than 0.1 is used, it will cause excessive current to flow through the coil. This will result in damage to the coil. To make a series connection usually requires breaking the circuit open or disconnecting a device in order to insert the meter. This allows the current to flow through the meter. The second thing to remember, is when the current value you are testing is unknown, start at the highest meter range. This way you will not exceed the highest value on the meter scale during the reading of a circuit.

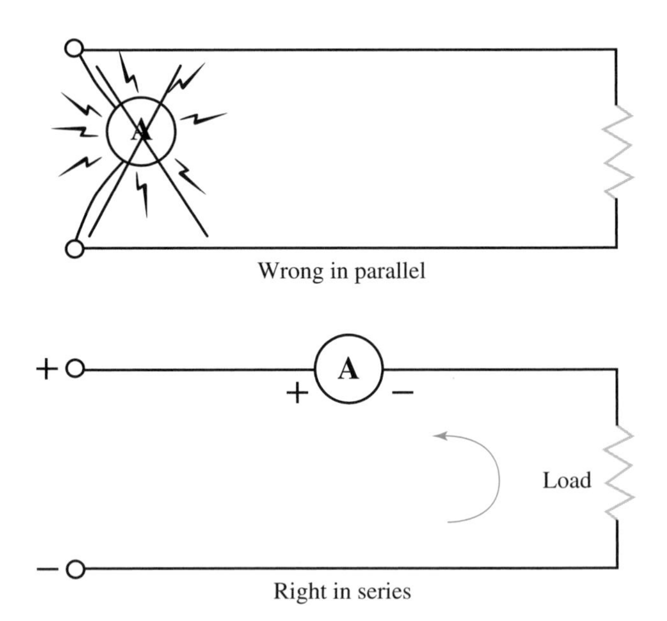

Figure 2-8. Connecting an ammeter. Top—An incorrect way. Bottom—The correct way.

Review Questions for Section 2.2

1. An ammeter is used to measure __________.
   a. voltage
   b. current
   c. resistance
   d. electrical potential
2. In order to measure larger currents, a(n) __________, or alternate path, is provided for current.
3. An ammeter should always be connected __________.
   a. in parallel with a load
   b. in series with the load
   c. either series or parallel with the load
   d. directly across the source

2.3 VOLTMETER

The same basic meter movement that is used in an ammeter is also used to measure voltage. This is providing that the impressed voltage across the coil never exceeds 0.1 volt, as computed, for full scale deflection. To arrange the meter to measure higher voltages,

multiplier resistors are placed in series with the meter movement coil using a switch. A meter similar to the meter that measured current is used. Refer to **Figure 2-9**. Voltmeters are always connected in parallel with the device being measured.

Example: Follow the steps as the multipliers are computed so that the meter can measure voltages from 0–1 V, 0–10 V, 0–100 V, and 0–500 V.

Step 1. Remember that no more than 0.1 volt is allowed across the meter coil at any time. Therefore, a resistor that will cause a voltage drop of 0.9 V must be placed in series with the meter if the meter is used to measure one volt. Also, the meter only allows 0.001 A for full scale deflection. This is the highest current allowed in the coil circuit. The multiplier resistor must produce a 0.9 V drop when 0.001 A flows through it.

$$R_M = \frac{E}{I}$$

$$R_M = \frac{0.9\ \text{V}}{0.001\ \text{A}}$$

$$R_M = 900\ \Omega$$

Step 2. To convert the 0–10 volt range, a resistor must be selected to produce a 9.9 volt drop.

$$R_M = \frac{9.9\ \text{V}}{0.001\ \text{A}} = 9900\ \Omega$$

Step 3. To convert the 0–100 volt range, a resistor must be selected to produce a 99.9 volt drop.

$$R_M = \frac{99.9\ \text{V}}{0.001\ \text{A}} = 99{,}000\ \Omega$$

Step 4. Finally, to use the 0–500 volt range, the resistor must cause a 499.9 volt drop.

$$R_M = \frac{499.9\ \text{V}}{0.001\ \text{A}} = 499{,}900\ \Omega$$

Again, a switching device is used to select the correct multiple resistor for the range in use. Read the scale on the dial that corresponds to the range selected. The dial on a meter is generally referred to as the range selector switch.

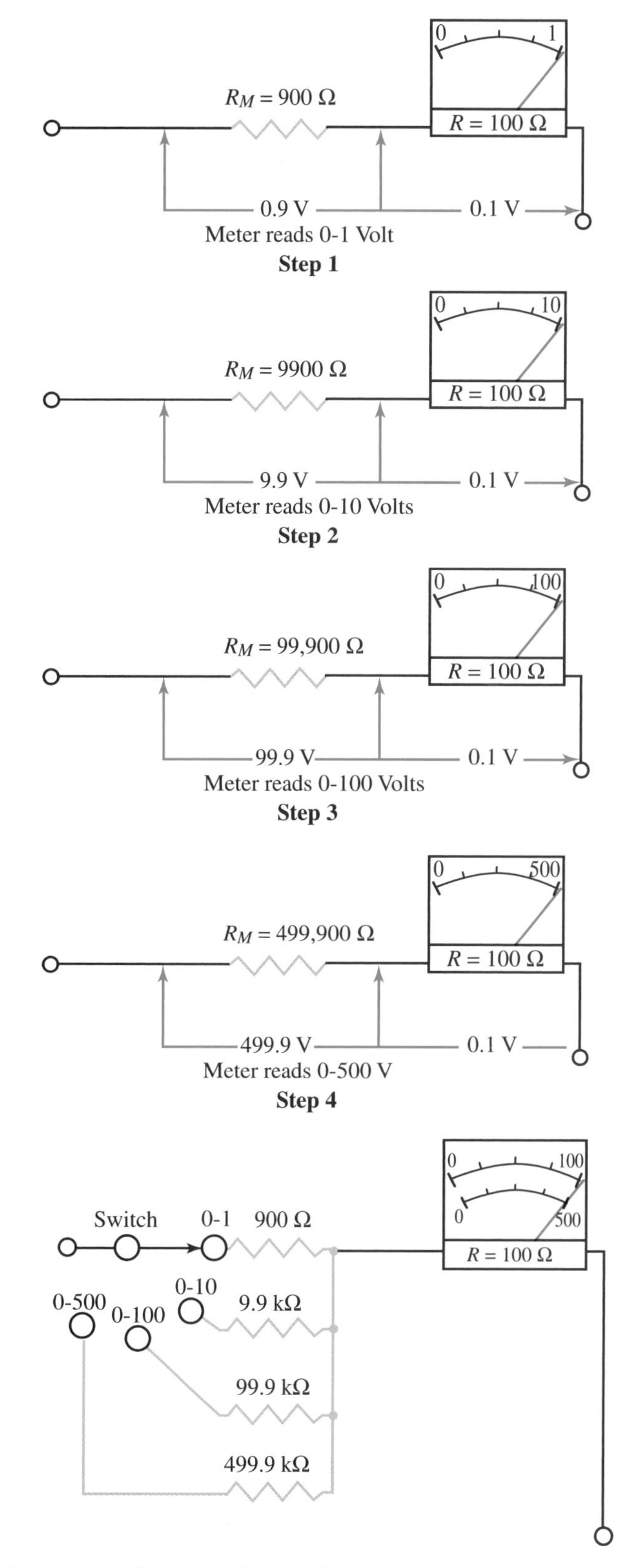

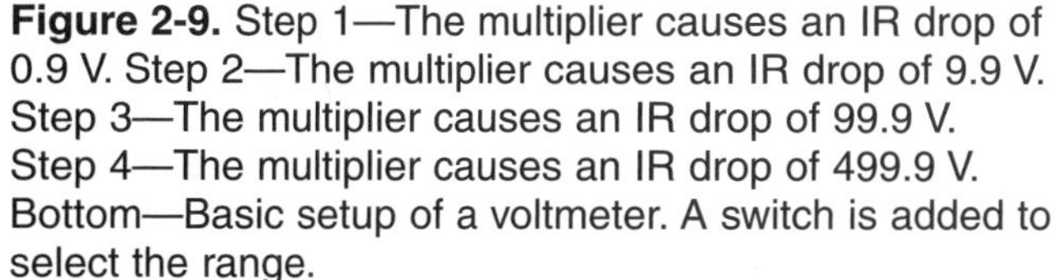
Figure 2-9. Step 1—The multiplier causes an IR drop of 0.9 V. Step 2—The multiplier causes an IR drop of 9.9 V. Step 3—The multiplier causes an IR drop of 99.9 V. Step 4—The multiplier causes an IR drop of 499.9 V. Bottom—Basic setup of a voltmeter. A switch is added to select the range.

CAUTION:

A voltmeter is always connected in parallel or across the circuit. To measure a voltage, the circuit does not have to be broken. See Figure 2-10. As with the ammeter, when measuring an unknown voltage, always start measuring with the meter set on its highest range. Adjust downward to the proper range to avoid damaging the meter. In addition, be sure that the leads are connected with the correct polarity. The black lead is negative and the red lead is positive.

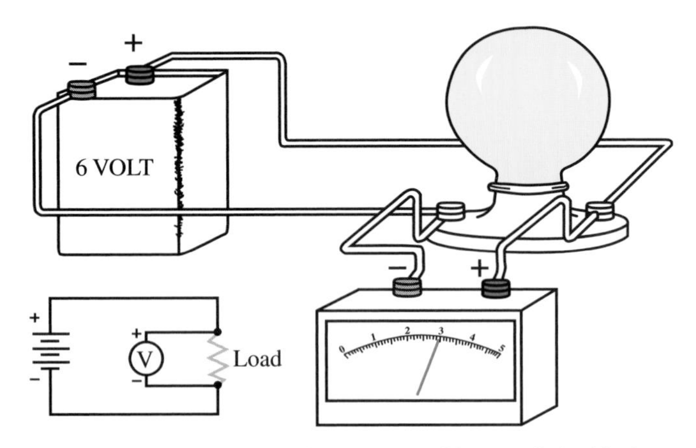

Figure 2-10. A voltmeter is connected in parallel with the device when taking a voltage reading.

Voltmeter Sensitivity

The sensitivity of a meter is a sign of quality. ***Ohms-per-volt*** is the unit for measuring sensitivity. In Step 4 of the previous example, the total resistance of the meter and its multiplier resistance is:

$$\begin{array}{r} 499{,}900\ \Omega\ (\text{in } R_M) \\ +\ 100\ \Omega\ (\text{meter resistance}) \\ \hline 500{,}000\ \Omega\ (\text{total resistance}) \end{array}$$

The total amount of resistance in the 500 V range is equivalent to the following:

$$\frac{500{,}000\ \Omega}{500\ \text{V}} = 1000 \text{ ohms per volt}$$

Using Ohm's law, $I = E/R$. The reciprocal of I is R/E. This is the same as the meter sensitivity. Therefore, the sensitivity is equal to the reciprocal of the current required for full scale deflection. For the meter used in the above example:

$$\text{Sensitivity} = \frac{1\ \text{V}}{0.001\ \Omega\ (\text{the coil resistance})}$$

$$= 1000 \text{ ohms/volt}$$

The sensitivity of a meter can be used to gauge meter quality. A quality meter has a sensitivity of at least 20,000 ohms/volt. Precision laboratory meters measure as high as 200,000 ohms/volt. Accuracy of the meter is commonly expressed as a percentage, such as 1 percent. This means that the true value will be within one percent of the scale reading. Another system of rating meters is the accuracy expressed as a percentage of full scale reading. A meter may have a rating of ±0.05 percent or less. In general, the smaller the percentage, the higher the quality of the meter.

Loading a Circuit

When a voltmeter is connected across a circuit to measure a potential difference, it is in parallel with the load in the circuit. This situation can introduce errors in voltage measurement. In meters with low sensitivity, this is very common. It is very important to keep this in mind.

In **Figure 2-11**, two 10,000 ohm resistors form a voltage divider circuit across a ten volt source. The voltage drops across both R_1 and R_2 are 5 volts each. If a meter with a sensitivity of 1000 ohms/volt on the ten volt range is used to measure the voltage across R_1, the meter resistance will be in parallel with R_1. Solving parallel circuits is explained in depth in Chapter 7. For now, it is enough to know that the addition of this meter cuts the effective resistance of R_1 in half. The combined resistance of the meter and R_1 is equal to:

$$\frac{R_1 + R_M}{2} = R_{effective}$$

$$\frac{10{,}000\ \Omega}{2} = 5000\ \Omega$$

With the meter connected, the total circuit resistance becomes:

$$R_{effective} + R_2 = R_{total}$$

$$5000\ \Omega + 10{,}000\ \Omega = 15{,}000\ \Omega$$

Using Ohm's law, the current can be calculated at approximately 0.00067 amps. Using Ohm's law again, $E_{R_1} = 3.35$ V and $E_{R_2} = 6.7$ V. The meter has caused an error of more than one volt due to its shunting effect. To avoid an excess of errors resulting from this effect, a more

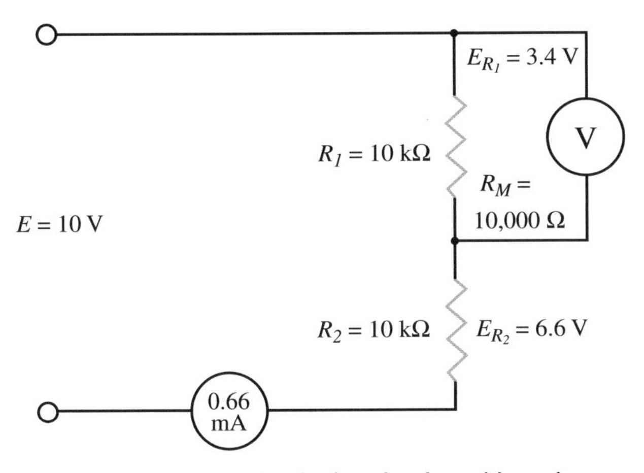

Figure 2-11. The meter loads the circuit and introduces an error in the voltage reading.

sensitive meter should be used. In **Figure 2-12** a 5000 ohms/volt meter is used.

In this case, the combined resistance of the meter and R_1 equals 8333 ohms. The total circuit resistance is 18,300 ohms. Using Ohm's law, $I = 0.00055$ amps, $E_{R_1} =$ 4.6, and $E_{R_2} = 5.5$ volts. An error of 0.4 volts still exists, but the increased sensitivity of the meter has reduced the error. Even more costly meters, with a sensitivity of 20,000 ohms/volt, can reduce the error to an amount that would be barely noticed.

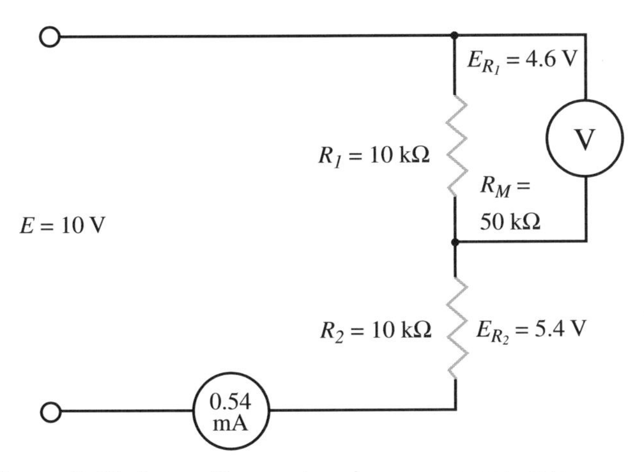

Figure 2-12. A sensitive meter gives more accurate readings.

Review Questions for Section 2.3

1. A voltmeter is used to measure __________.
   a. voltage
   b. current
   c. resistance
   d. wattage
2. To measure higher voltages, __________ __________ are placed in series with the meter movement coil.
3. A voltmeter is connected in __________ with a device.
4. The black lead of a meter is connected to the __________ polarity of the device being read while the red lead is attached to the __________ polarity.
5. A quality meter has a sensitivity of at least __________ ohms/volt.
6. Explain what is meant by loading a circuit as it relates to an electrical meter.

2.4 OHMMETERS

A meter used to measure the value of an unknown resistance is called an ***ohmmeter***. The same meter movement that was used in the volt and ammeter can be used for the ohmmeter. A voltage source and a variable resistor are added to the ohmmeter's circuit. A series type ohmmeter is shown in **Figure 2-13**.

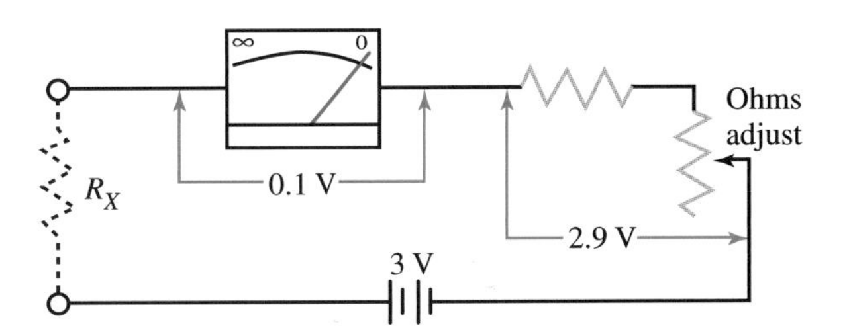

Figure 2-13. Schematic diagram of a series ohmmeter.

A three volt battery is used as the source for the ohmmeter. The battery is built into the meter case. The meter movement permits only 0.1 volt for a current of 0.001 amps for full scale deflection. Therefore, a multiplier resistor is placed in series with the meter coil to reduce the voltage applied to the meter coil.

$$R_M = \frac{E}{I}$$

$$R_M = \frac{2.9 \text{ V}}{0.001 \text{ A}}$$

$$R_M = 2900\ \Omega$$

The 2900 ohm multiplier resistor, plus the meter coil resistance, is equal to 3000 ohms. Part of this resistance is made up of a variable resistor to allow the total resistance to vary. Because temperature changes or weak batteries can affect the total resistance of the circuit, the ohmmeter must be calibrated (adjusted for zero resistance) in order to ensure the most accurate reading possible.

The knob used for adjusting the pointing needle position to zero is usually marked "Zero Adjust," or with an omega (Ω) symbol near it. To use the ohmmeter, first short the test leads together. This applies 0 ohms across the meter. Adjust the ohms adjustment knob until the needle points at zero. The needle should deflect from its position at rest on the left to the zero resistance indication on the right side of the scale. If the needle does not deflect, it is possible that the battery is dead or extremely weak. After the ohmmeter has been calibrated to read zero ohms when the leads are shorted, you can make a reading of an unknown resistance by placing the unknown resistance between the test leads.

CAUTION:

Before connecting an ohmmeter to any electrical circuit to read an unknown value, be sure that the circuit is not energized. An energized circuit will damage the meter and can be harmful to you. The electrical energy in a circuit is not needed to operate the meter movement coil as it is when using a voltmeter or an ammeter. The batteries inside the case provide the source of power for the ohmmeter. Connecting the ohmmeter to an energized circuit will apply the circuit voltage directly to the coil and battery, which can result in damage to the meter and possible harm to you.

A shunt ohmmeter is connected as shown in **Figure 2-14**. In this circuit, the unknown resistance R_X is shunted (connected in parallel) across the meter. Low values of R_X cause lower currents through the meter. High values of R_X cause high meter currents. When the ohmmeter is connected in the shunt position, the indicating needle deflects from right to left in the manner of the ammeter and voltmeter. Zero resistance is on the left. The scale increases from left to right.

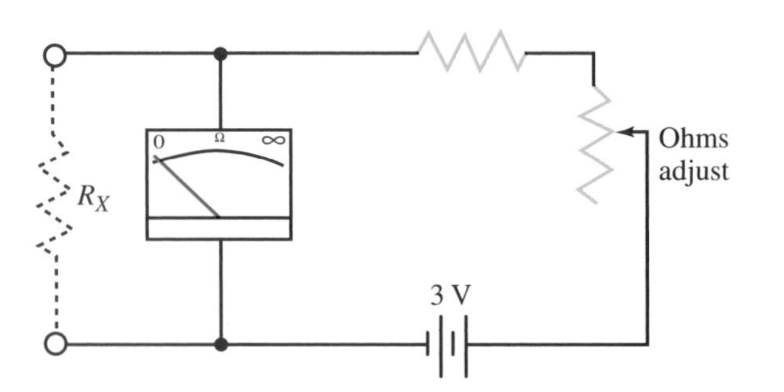

Figure 2-14. Schematic diagram of a shunt ohmmeter. Note that the meter reads in the opposite direction of other ohmmeters.

Ohmmeter Scales

The resistance value is indicated on the ohms scale, which is a nonlinear scale. A ***nonlinear scale*** has markings that are not evenly spaced. The nonlinear scale factor increases as the needle travels from zero resistance to infinite resistance. In **Figure 2-15**, a typical ohmmeter scale is represented. On the right side of the scale is zero. On the left side is infinity (∞). An infinity reading means that the resistance value is so high that it exceeds the capabilities of the ohmmeter to read it.

Notice how the scale factor changes along the ohmmeter scale. On the right side, the small marks between the numbers 0 and 2 represent 0.2 ohms each. On the left side, between the 50 and 70 ohm marks, the small marks represent 5 ohms each. To take accurate readings of unknown resistance values, it is recommended that the range selector switch be changed until the reading falls in the mid-third of the scale. A ohmmeter comes with a selection of ranges that can be changed by rotating the selector switch. Typical range values are R × 10, R × 100, R × 1k, R × 10k. These markings mean that the reading indicated on the ohm scale should be multiplied by 10, 100, 1000, and 10,000 respectively.

Review Questions for Section 2.4

1. An ohmmeter is used to measure __________.
   a. voltage
   b. current
   c. resistance
   d. wattage
2. An ohmmeter uses a(n) __________ scale, while a volt or ammeter uses a(n) __________ scale.
3. Why must you calibrate an ohmmeter before use?
4. Find the value of the multiplier resistor (R_X) in the following circuit. It should read 50 volts across points A to B.

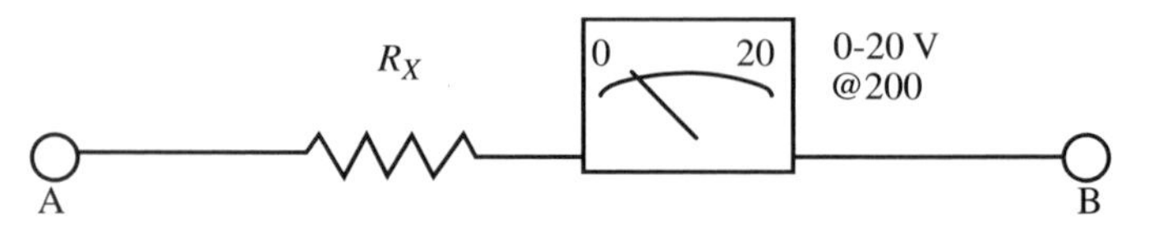

2.5 THE VOLT-OHM-MILLIAMMETER (VOM)

A common and simple multimeter used in electronic circuits is the ***volt-ohm-milliammeter*** or ***VOM***. A VOM is a voltmeter, ammeter, and ohmmeter, all in one. The

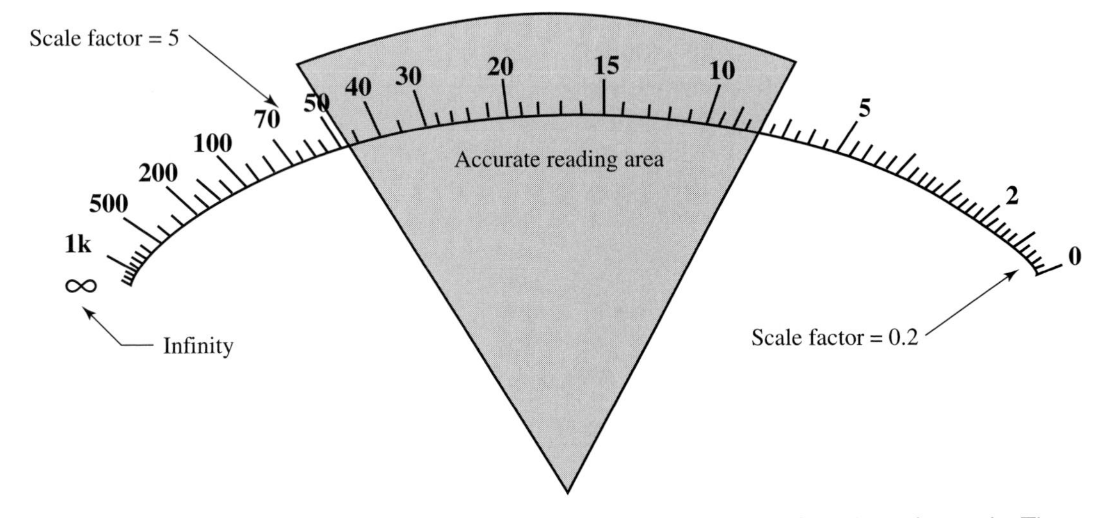

Figure 2-15. The ohmmeter scale is a nonlinear scale. The scale factor varies in value along the scale. The most accurate readings are taken along the shaded area (the middle 1/3) of the scale.

VOM has the advantages of being inexpensive and portable. It does, however, usually have a low input resistance (in ohms per volt) on the lowest voltage range. This factor can cause accuracy problems.

When an electronic device called a field-effect transistor was developed, a VOM was designed to overcome the low input impedance problem. The ***field-effect transistor-VOM (FET-VOM)*** measures ac and dc voltage, ac and dc current, resistance, and decibel ratings.

Some multimeters are also equipped with accessories such as temperature probes. The leads of the temperature probe are inserted into the meter while the probe itself can be placed in front of an air-conditioning duct, near a furnace heater, or submerged in a hot liquid. The scale on the meter reflects the temperature in Celsius and/or Fahrenheit. Other optional accessories for VOMs include adapters for reading higher than normal voltages and larger than normal meter current values.

Digital Multimeters

Digital multimeters (DMM) are the most commonly used meters in the electronics field today. They are rapidly replacing the analog meter, which operates on the principle of magnetism and rotating coil discussed in the previous examples. The DMM uses modern electronic circuitry to take electrical measurements and display values, usually on a liquid crystal display screen. This circuitry is beyond the scope of this chapter. See **Figure 2-16**.

Digital meters are more rugged and smaller in size than analog meters. They are also very accurate and very portable. However, some technicians still prefer the analog meters for taking certain types of readings involving solid state circuits. This will be explained in more depth when solid state electronic devices are explored.

Figure 2-16. DMM using a liquid crystal display. (Knight Electronics)

The liquid crystal display shows the meter reading in digits rather than on a scale. Some DMMs simultaneously display digits as well as a bar graph that simulates a linear scale reading, **Figure 2-17**. Digital meters not only measure volts, ohms, and current, but can also test electronic components such as transistors and diodes, **Figure 2-18**. The digital multimeter can come with a rotary dial (like the analog meter) to select functions or a key pad that is pressed with the fingertips. Most digital meters use an international standard of labels to indicate various meter functions such as AC, DC, and combinations of symbols, **Figure 2-19**. The graphic symbols for AC and DC are often combined with metric prefixes to identify the function or range of the meter setting. Many

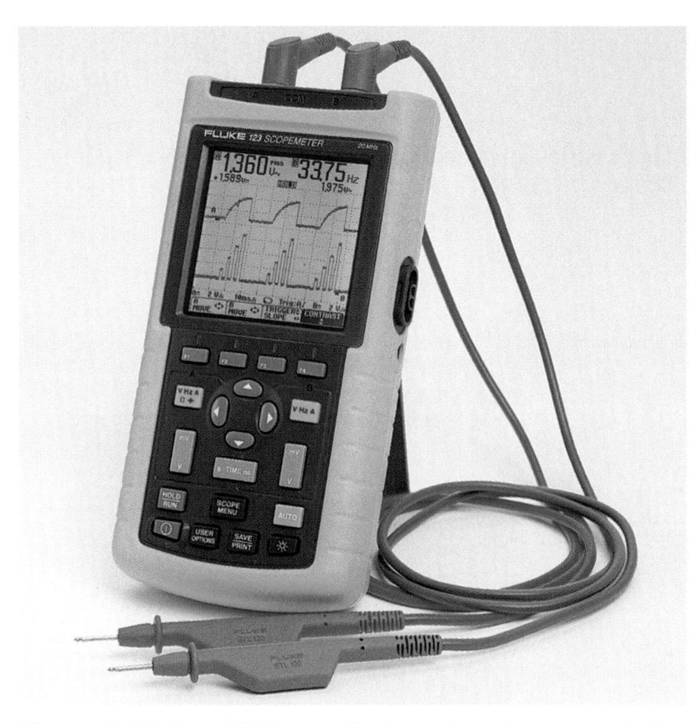

Figure 2-17. This DMM can display values in a numerical form or in a grafical form. (Fluke Corp.)

digital multimeters are equipped with protective circuitry to prevent accidental damage when the wrong function is used taking readings. Not all digital meters have this capability, but it is available.

Polarity is usually not an issue when using a digital meter. The meter will automatically adjust for an incorrect polarity, or it will flash a message or symbol on the liquid crystal display, warning the user of the wrong polarity. Some digital meters have an autorange feature. This means that it is not necessary to determine which range to select when using the meter. On these meters, this function is done automatically through the internal electronic circuitry, **Figure 2-20**. Resistance reading using a DMM still requires you to disconnect the circuit from the power source to prevent damaging the meter.

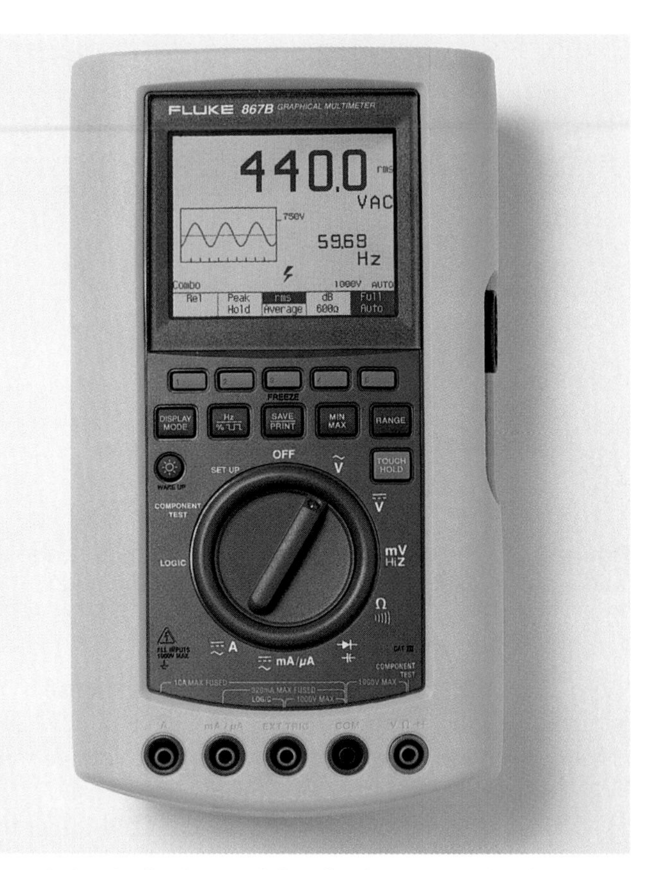

Figure 2-18. At the lower right of selector switch of this DMM is a setting to test diodes and capacitors. (Fluke Corp.)

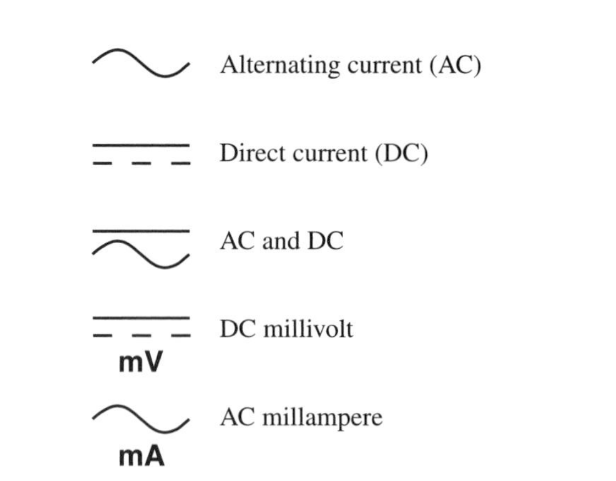

Figure 2-19. The international graphic symbols for AC and DC combine with other electrical prefix symbols to indicate the meter setting.

Computer Display Meters

Some manufacturers offer interface cards that can be installed in a personnel computer and have test leads similar to meter leads. After the interface board is installed in an expansion slot in the computer, software is loaded. Now the computer will display simulated meters on the monitor. The computer can then be used to take

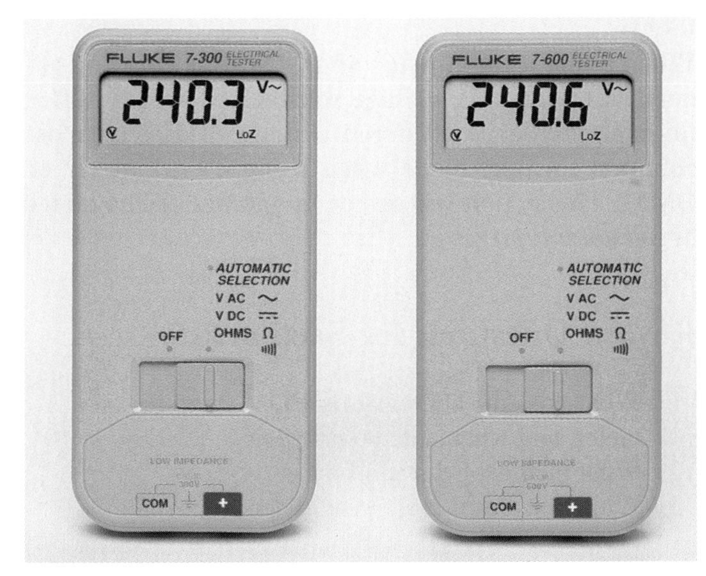

Figure 2-20. These meters select range automatically. (Fluke Corp.)

voltage, current, and resistance readings. Using the computer is very similar to using a digital meter, but with the computer you can use the memory and hard drive system to store measurements and retrieve them later. This type of metering equipment is commonly installed in new industry applications to monitor high speed assembly equipment or to test electrical products.

AC Meter Readings

When ac is applied to the meter movement, the needle does not deflect. Remember that ac rapidly changes direction. The meter coil current changes direction in pace with the applied ac voltage. The result is that the magnetic field rises and collapses, and then reverses so rapidly that the coil cannot deflect the needle. The coil simply vibrates under the influence of applied ac voltage. To remedy this, a meter changes the ac voltage into dc by means of a *rectifier*. The rectifier is covered in detail in Chapter 17. For now it is sufficient to say that a rectifier converts the applied ac current to a dc current of equal value.

When the applied ac current is rectified to an equal dc value, the value is referred to as the ***rms value***. The abbreviation, rms, stands for ***root mean squared***. This is a formula used to equate ac voltage to dc voltage. The rms value is the equivalent dc value of an ac waveform. Many meter scales, and some DMMs, use this abbreviation. Some meters have a special location to plug in the meter lead when reading ac voltage or current. Many meters have a scale marking printed as rms. This simply means that the readings taken on that scale are equal for dc or ac voltages.

Resolution is a term that describes the degree of change that must take place before the meter will display the value. For example, if a meter has a resolution of 1/1000th, it can measure a voltage down to 1 millivolt. In general, the better the meter resolution, the more expensive the meter. In digital meters, resolution is also determined by how many digits the meter can display. A digital meter with a five-digit readout has a better resolution than a digital meter with a four-digit readout.

Important Meter Information

Multimeters are very useful tools. There are a number of important details to remember when using these meters.

- A meter is a delicate instrument. Handle it with care and respect. Jarring, dropping, and other rough treatment can damage a meter, especially coil meter movement.
- When measuring voltage, the meter must be connected in *parallel* to the device being read. Start on the *highest* range when measuring an unknown voltage and move slowly to a lower range for increased accuracy as needed. *Remember to observe correct polarity!* The red, or positive, lead goes to the positive side of the circuit. The black, or negative, lead goes to the negative side of the circuit.
- When measuring current, an ammeter must be connected in *series* with the circuit. A wire must be disconnected to insert the meter. It is wise to make a rough current calculation using Ohm's law to determine the proper current range on the meter. When troubleshooting a circuit, it is best to start on the *highest* possible setting. A faulty component can cause higher currents than would be normally expected. Again, observe correct polarity: red to the positive side of the circuit and black to the negative side of the circuit.
- When measuring resistance, be certain that no power is applied to the circuit. It is best to disconnect the voltage source before taking resistance measurements. In general, it is not necessary to observe polarity when taking resistance measurements. However, as you advance through your studies, polarity must be observed when checking certain solid state devices. Always adjust and zero the meter on proper range before measurements are made. *The ohmmeter should be readjusted after changing ranges or after prolonged use.* An open circuit will have an infinite reading.

- On all meter measurements, make a flash check before permanently connecting the meter to the circuit. What does "flash check" mean? First, a decision should be made on how to connect the meter to the circuit. Then, only the negative lead should be connected with the positive meter lead left disconnected. While observing the meter, quickly touch and then remove the positive lead to the circuit. Did the needle move in the wrong direction? If so, the polarity must be changed. Did the needle move too violently? If so, the meter range selector should be changed to a higher range. Remember, you should start on the highest range when you are not positive of the voltage or current values. The flash check will save you many dollars in meter replacement and repairs as well as wasted time.
- A meter has its greatest accuracy at about two thirds deflection on the meter scale. Use the range that reads as close to this deflection as possible.
- Often electronic circuits are quite compact. Be sure that the test leads do not cross over two or more connection points, as this could result in a short circuit.
- Be sure that the test leads are in good working condition. There should be no frayed or bare wiring. Make it a habit to keep your fingers from touching any exposed metal part of the test lead tips and/or the circuit being tested. Most of the time you will be working with voltages that are harmless, but sooner or later you will work on circuits that can produce severe or fatal shock. The habits you develop while testing low voltage circuits will be carried with you when you are working with higher voltages. *Make a habit of testing circuits safely!*
- Never work alone when dangerous voltages are present.
- Do not make the common mistake of connecting the meter to a voltage source without first changing the selector mode switch after checking resistance or amperage. In addition, some meters have a separate input for the test leads when in the current mode, especially when high currents are to be measured. You may have selected the correct mode of operation using the selector switch, but left the test lead plugged into the wrong meter jack.
- It is very important not to attempt reading from the back of a television picture tube or a computer monitor. There are extremely high voltages present in these locations, and the meter usually requires special high voltage test leads specifically designed for this purpose.

CAUTION:
When using a multimeter or DMM, it is easy to connect the meter to a voltage source immediately after taking a resistance or current reading. This is the most common mistake made when using a multimeter or DMM. This action will result in damage to the meter or *personal injury*.

Review Questions for Section 2.5

1. What does the abbreviation COM stand for on a meter, and what is its polarity?
2. What does DMM stand for?
3. What does resolution mean when referring to a DMM?
4. How do you connect a ammeter to a circuit?
5. How do you connect a voltmeter to a circuit?
6. Which of the three meters should not be used while the circuit is energized?
   a. Voltmeter.
   b. Ammeter.
   c. Ohmmeter.

2.6 ELECTRICAL DIAGRAMS

Electrical diagrams convey specific information to the technician. They illustrate such items as the size, type, component part number, and component location in relationship to the other circuit components. Diagrams can be used for installation, fabrication, troubleshooting, or to explain the circuit's operation or purpose. Symbols are used to represent circuit components. Wires or conductors are usually shown as lines. Their connections can be shown a number of ways. See **Figure 2-21**.

One primary type of electrical drawing you will encounter is the ***schematic*** diagram. See **Figure 2-22**. This is a typical schematic diagram. It shows what parts are needed and how they connect to one another. The distance between the components do not represent the actual

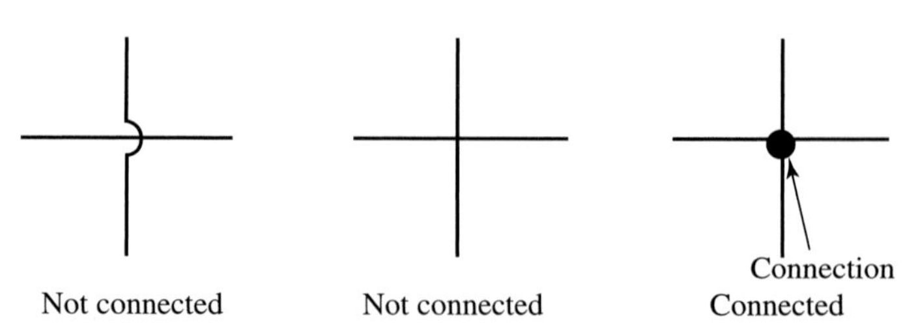

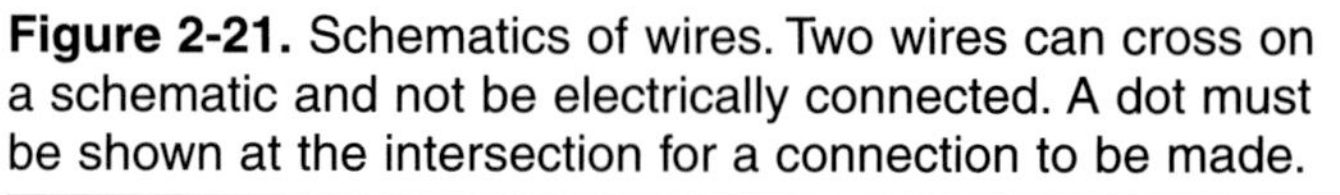

Figure 2-21. Schematics of wires. Two wires can cross on a schematic and not be electrically connected. A dot must be shown at the intersection for a connection to be made.

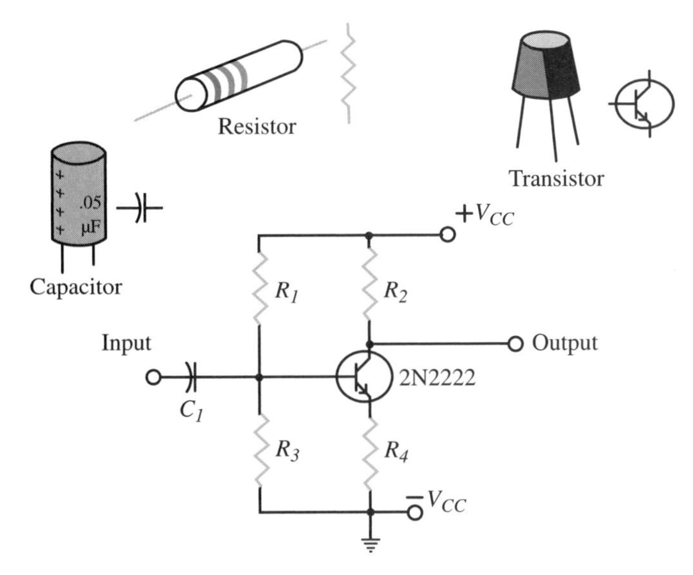

Figure 2-22. A typical schematic illustrates the location of the components and how they relate to one another.

distances. The main purpose of the schematic is to show how the components relate to each other. The diagrams show which components are in series or parallel with each other. Schematics are an extremely valuable troubleshooting tool.

The combination of meters, wiring diagrams, schematics, and electronic theory allow a technician to find circuit problems. Many circuits are impossible to troubleshoot without the aid of schematics and the application of electronic theory.

Figure 2-23 is a comparison of an elementary line diagram and a wiring diagram. This illustration shows the operation of a typical stop-start motor control system. The elementary line diagram on the left is similar to a schematic. It is used primarily in industrial processes to illustrate how a system's electrical controls relate to each other. On the right is the actual wiring diagram. This would be used to connect the control system. The elementary diagram clearly illustrates how the circuit operates, while the wiring diagram illustrates the relative positions of the connection points and components as they would actually be found in the equipment. Each diagram has its own purpose.

Sometimes a block diagram is used to show how an overall system works. Look at **Figure 2-24** to see a block diagram of a typical am radio. The components, such as the amplifier, are grouped together in stages.

Figure 2-25 is a typical plan of the electrical circuits to be installed in one room of a residence. The drawing indicates the general location of switches, outlets, and lighting. Descriptions of wire sizes, switch amperages,

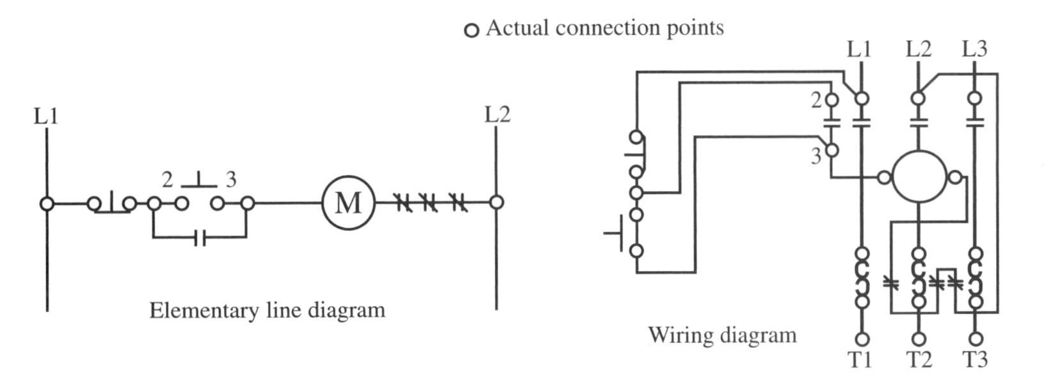

Figure 2-23. Both the elementary line diagram and the wiring diagram shown here are of the same electrical system, but they are presented two different ways. The elementary line diagram is used to clearly express how the circuit works. The wiring diagram is used to install the system.

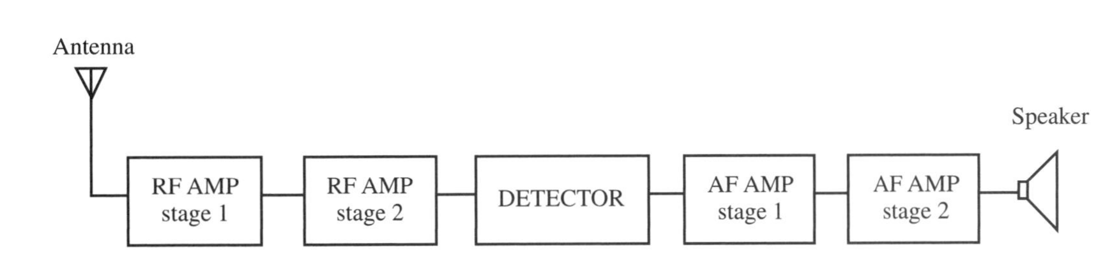

Figure 2-24. A block diagram is used to illustrate how major electrical systems relate to each other.

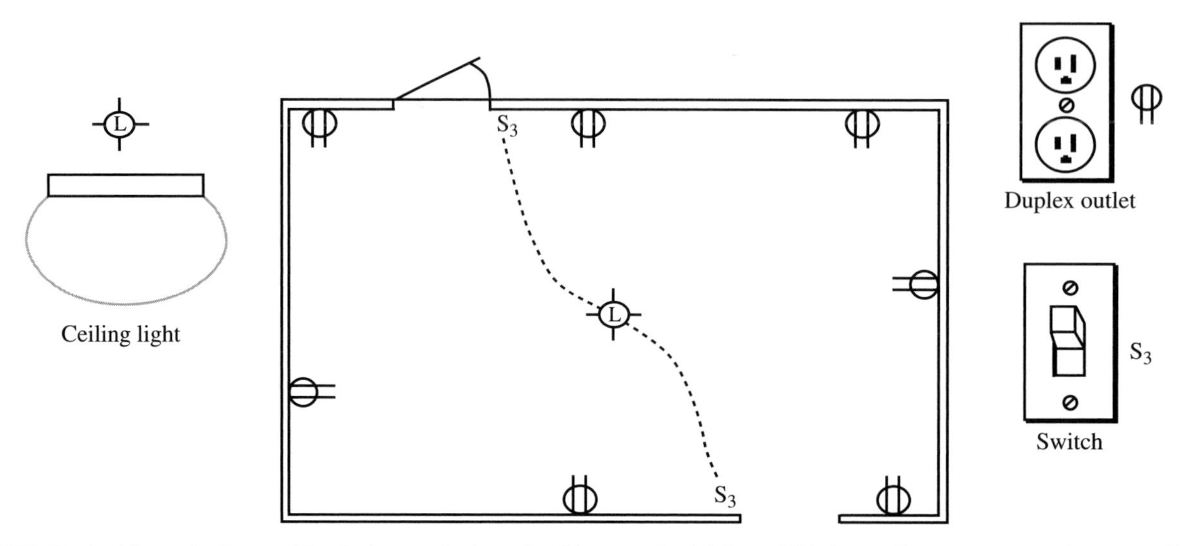

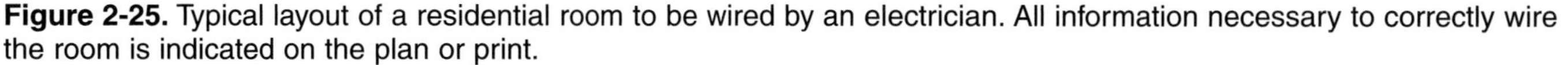

Figure 2-25. Typical layout of a residential room to be wired by an electrician. All information necessary to correctly wire the room is indicated on the plan or print.

and breaker sizes are not shown on this type of plan because the electrician is trained to be familiar with electrical codes dealing with these factors.

As you progress through the text, you will gain a large symbol vocabulary. This will help you interpret many different types of electrical drawings. This vocabulary, combined with a complete understanding of schematics and meters, will help you troubleshoot, repair, or construct any electrical system rapidly.

When constructing an electrical system, you may find using a circuit design software program beneficial. Circuit designers rely heavily on computers and software for modern electronic circuit design. See **Figure 2-26**. In these software programs, components can be selected from menus and placed on the drawing area. Electronic characteristics for each component, such as resistance values, current ratings, and voltage limits, can also be added.

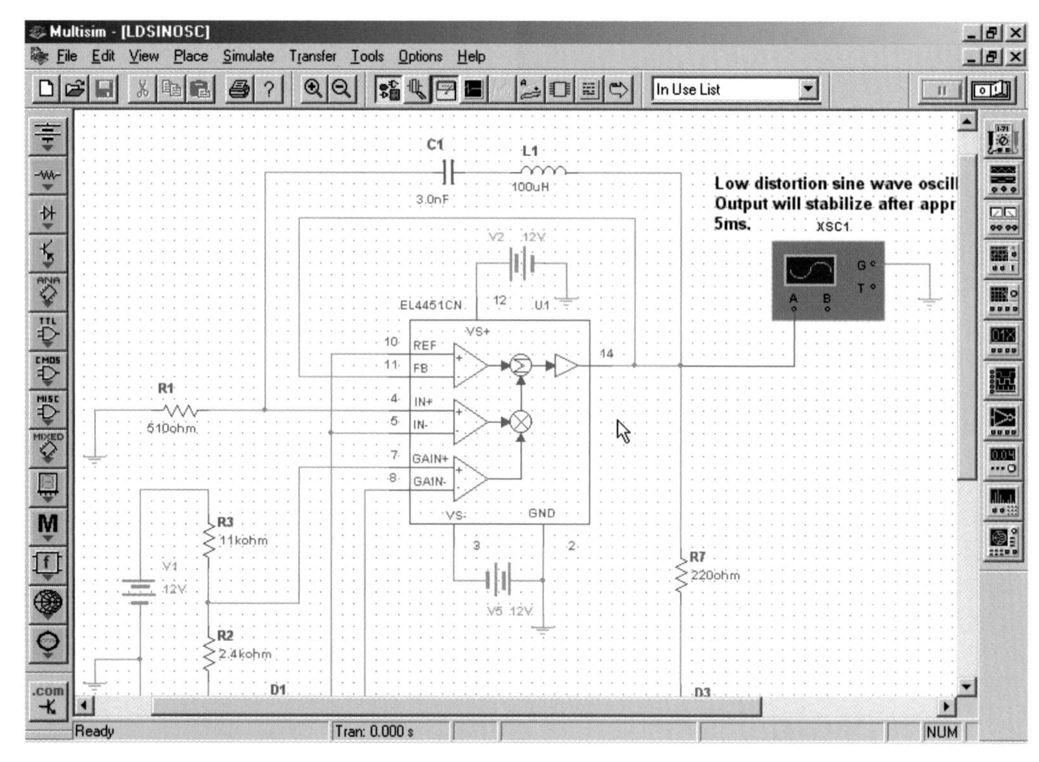

Figure 2-26. Screen capture of Multisim Electronics Workbench.

Software systems not only can be used to draw out electronic circuitry, they can actually be used to simulate the circuit as though it was constructed with electronic components. Virtual meters can be connected to different points in the circuit for experimentation and testing. A complete list of materials can be generated from the circuit design. The pattern required for a printed circuit board can be printed. This makes the design and testing process quicker and easier than if the circuit was built using actual components. Once the circuit design is tested to satisfaction, the circuit can be built using actual components.

Review Questions for Section 2.6

1. What is a schematic?
2. What are symbols used for?
3. Name three typical electrical drawings and the use of each.

Summary

1. Analog meters use a scale with continuous variable values. Digital meters give values in discrete amounts in units 0 through 9.
2. The basic meter movement used for many analog instruments is the moving-coil galvanometer, or the D'Arsonval movement.
3. It is vital to observe correct polarity in the use of analog meters.
4. Linear meter scales have evenly spaced marks. Ammeters and voltmeters use linear scales.
5. Nonlinear scales do not have evenly spaced marks. Ohmmeters use a nonlinear scale.
6. Ammeters measure current and are connected in series in a circuit. Shunts are resistors connected in parallel with ammeters to increase the range of the meter.
7. Voltmeters measure voltage and are connected in parallel with the component to be read. Multiplier resistors are connected in series with voltmeters to increase the range of the meter.
8. The sensitivity of a meter is an indication of its quality. Sensitivity is measured in an ohms-per-volt rating. Another system of rating accuracy is based on percentage of full scale reading.
9. When a voltmeter is connected across a circuit to measure a potential difference, it is in parallel with the load in the circuit. This situation can introduce errors in voltage measurement.
10. A meter used to measure the value of an unknown resistance is called an ohmmeter. Ohmmeters are connected across the resistance being measured.
11. A multimeter is one instrument that will measure a number of different types of values, such as current, voltage, and resistance.
12. Digital meters are rugged, small in size, accurate, and portable.
13. There are a number of different types of electrical diagrams. They include: schematics, wiring diagrams, line diagrams, and block diagrams.

Test Your Knowledge

Please do not write in the text. Place your answers on a separate sheet of paper.

1. The D'Arsonval movement measures __________ and __________.
2. A(n) __________ is an alternate path, supplied for measuring large currents.
3. A meter movement has a moving coil resistance of 50 ohms and requires 0.001 amperes for full scale deflection. Compute the shunt values for the meter to read in the following ranges:
   a. 0–1 mA.
   b. 0–10 mA.
   c. 0–50 mA.
4. Compute the multipliers for the meter in question 3 to use on the following ranges:
   a. 0–1 volt.
   b. 0–10 volts.
   c. 0–100 volts.
5. A meter used to measure the value of an unknown resistance is called a(n):
   a. ammeter.
   b. ohmmeter.
   c. voltmeter.
   d. None of the above.
6. The ∞ symbol on an ohmmeter scale represents what?
   a. Low resistance.
   b. High resistance.
   c. High voltage.
   d. High current value.

7. When reading an unknown voltage value, you should set the range selector position to __________.
   a. the lowest possible voltage range
   b. the lowest possible resistance range
   c. the highest possible ampere range
   d. the highest possible voltage range
8. What should you do when using an ammeter?
   a. Hold the test probes tightly to the circuit component using your fingertips.
   b. Connect the meter in parallel with the power source to be sure it is working correctly.
   c. Connect the meter in series with the component to be tested.
   d. All of the above.
9. Which meter must be zeroed before use to assure accurate readings?
   a. The ohmmeter.
   b. The voltmeter.
   c. The ammeter.
   d. The DMM.
10. What is meant by "loading the circuit" with a meter?

For Discussion

1. Discuss the safety rules and precautions to follow when working with meters.
2. Why is a rectifier necessary in measuring an ac current or voltage?
3. Would it be practical to use pictorial diagrams for wiring circuits rather than a schematic?

Meter Maintenance

- Store test leads properly to avoid damaging them. Using damaged test leads on an energized circuit or device could be hazardous to you and to the meter.
- Make sure the meter's battery is fully charged. A meter with a low battery can give false readings. A false reading may result in harm to you or the meter.
- Do not use water to clean the meter. Instead, wipe the meter with a damp cloth and detergent. Water on the inside and on the outside of the meter or on the probes can result in electrical shock.

Introduction to Basic Electrical Circuit Materials

Objectives

After studying this chapter, you will be able to:

- Identify different conductor types.
- Determine the cross-sectional area of a conductor.
- List the factors affecting resistance.
- Identify various insulation materials.
- List three special conductor pathways.
- Explain the manufacture of printed circuit boards.
- Identify various switching devices.
- Identify various lighting devices.
- Identify different types of resistors.
- Determine the value of color coded resistors.

Key Words and Terms

The following words and terms will become important pieces of your electricity and electronics vocabulary. Look for them as you read this chapter.

actuator
ambient temperature
branch circuit
candela
chassis
circuit breaker
circular mil (cmil)
fuse
lumen
lumens per watt (lpw)
mil
National Electrical Code (NEC)
potentiometer
resistor
siemens
single-pole
single-throw
superconductor
thermistor
tolerance

All of electricity and electronics is based on the fundamental concept of moving electrons through some type of material and then applying the electrons at a load. At the load, the electrical energy is transformed into another form of energy such as heat, light, or magnetism. In this chapter, we shall explore methods to control the flow of electrons using common devices and materials. In addition, we will explore the conversion of electrical energy for some common uses, such as lighting.

3.1 CONDUCTORS

Conductors are the pathways that allow electrons to flow through an electrical circuit. Conductors can be made in various shapes and sizes. They can be made from a number of different materials. The most common conductor you will encounter is copper wiring. Copper is an excellent conductor and can be formed into many different shapes. Copper comes in round, square, stranded, solid, flat ribbon, and bar shapes. Each shape has its own unique qualities that make it best for use in some specific application.

Conductors will vary depending on the type of material from which they are made. Copper is an excellent conductor. Silver is an even better conductor. However, silver is too expensive to be used in common circuits. Many high-voltage lines are made from aluminum because of its light weight, and they have a center core made of stranded steel for added strength. Brass is used for electrical-mechanical equipment such as switch parts because of its strength and conductive properties.

The actual conduction of electricity is done by transferring electrons from one atom to the next atom in the conductor. Assume that a piece of copper wire is neutral. If an electron is forced into one end of the wire, then an electron will also be forced out from the other end, **Figure 3-1**.

The original electron did not flow through the conductor. Yet the energy was transferred by the interaction between the electrons in the conductor. The actual

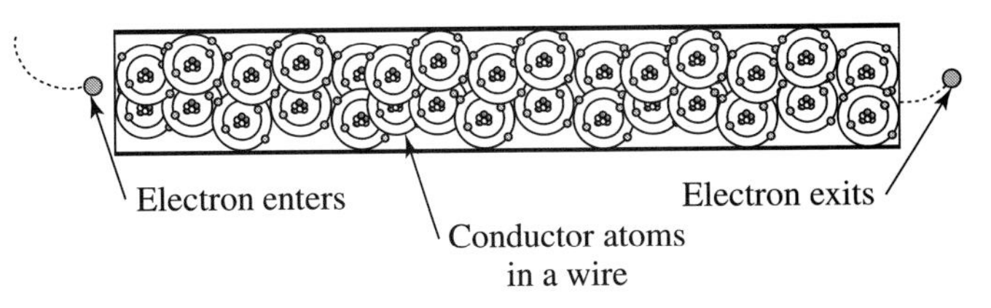
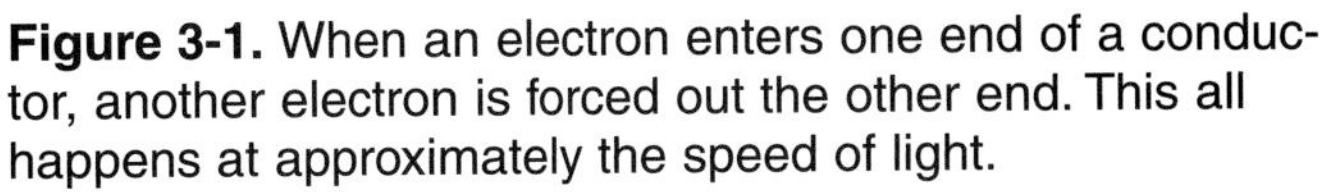

Figure 3-1. When an electron enters one end of a conductor, another electron is forced out the other end. This all happens at approximately the speed of light.

transfer of electrical energy occurs at an amazing speed. The speed has been accurately measured, and it approaches the speed of light: 186,000 miles per second! In metric terms, this speed is 300,000 kilometers (or 300,000,000 meters) per second.

The unit for measuring conductance is the ***siemens***. This unit was named after the German inventor Ernst von Siemens, who did a great deal of work in the development of telegraphy use. The abbreviation for the siemens is the capital letter S. The unit formula for computing electrical conductance is as follows:

$$G = \frac{1}{R}$$

where G is conductance (in siemens) and R is resistance (in ohms).

The conductance of a material is the reciprocal of the resistance of the material.

LESSON IN SAFETY:

Your body is a good conductor of electricity. Never touch a circuit conductor or component unless you are sure that it is not energized. An electric current flowing in one hand, across your chest and heart, and out your other hand, is dangerous and can be fatal. The smart technician uses only one hand when working on high voltage circuits. The other hand is kept in a pocket. Technicians who work on high voltage lines work in insulated bucket trucks or on insulated platforms. Standing on a rubber mat while working on electrical systems is a standard safety (and a wise) practice. While the technician is standing on an insulated material, there is no direct path to ground in case of accidental contact with conductors.

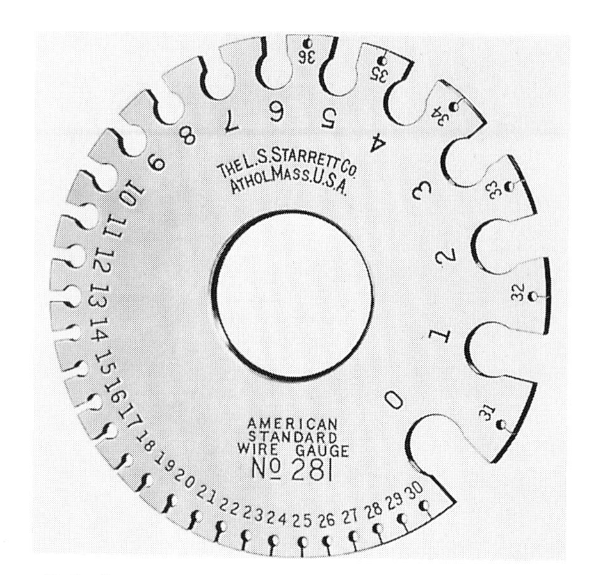

Figure 3-2. A gauge is used to find wire size.

Conductor Sizes

Conductors are sized by their cross-sectional area. The amount of cross-sectional area a conductor has determines how much current it can handle without overheating. Circular conductors are arranged according to size by the American Wire Gauge System. The larger the gauge number, the smaller the cross-sectional area a wire will have. For example, a No. 14 wire is larger in diameter than a No. 20 wire. See **Figure 3-2.**

This system of wire sizes was developed over a hundred years ago. At that time, no one could foresee that the demand for electrical products would make this system obsolete. Some wire sizes are now larger than the No. 1 conductor. A temporary solution to this problem created the next set of larger sizes. The next four sizes that were developed are 0 (pronounced "ought"), 00, 000, and 0000–or one through four ought (1/0 through 4/0).

When it became apparent that even larger sizes would be needed, the final solution to the sizing problem, a third set of sizes, was developed. The final system of sizing is the circular mil system. The circular mil system is based on the diameter of the conductor measured in mils (1/1000 of an inch). Typical conductors larger than four ought are from 250 kcm to 2000 kcm. The letters kcm represent 1000 circular mils, so a 250 kcm represents 250,000 circular mils. **Figure 3-3** contains all the wire sizes you would normally encounter.

Circular Mils

Circular mil area is the common way to express the cross-sectional area of a conductor. As previously noted, one ***mil*** is equal to 1/1000th of an inch (0.001 inches). The number of ***circular mils (cmil)*** in a conductor is equal to the diameter of a round conductor squared (D^2), where D is in mils. Therefore, a wire with the diameter of 1 mil has an area of 1 circular mil (1 cmil). A wire with a diameter of 2 mils has an area of 4 cmil ($2 \times 2 = 4$). A wire with a diameter of 15 mils gives us an area of 225 cmil ($15 \times 15 = 225$). See **Figure 3-4.**

The circular mil is a more convenient method of expressing the size of a conductor than πr^2, which is used to compute the area of a circle. If we must find the area of a square or rectangular conductor, their equivalent in cmils is not difficult to find. To find the circular mils

Gauge no.	Diameter mils	Circular mil area	Resistance ohms per 1000 ft. at 25 C (68°F) copper wire	aluminum wire
4/0	460.0	211,600	0.04901	0.07930
3/0	409.6	167,800	0.06182	0.1000
2/0	364.8	133,100	0.07793	0.1261
1/0	324.9	105,600	0.09825	0.1590
1	289.3	83,690	0.1264	0.2005
2	257.6	66,370	0.1593	0.2529
3	229.4	52,640	0.2009	0.3189
4	204.3	41,740	0.2533	0.4021
5	181.9	33,100	0.3195	0.5072
6	162.0	26,250	0.4028	0.6395
7	144.3	20,820	0.5080	0.8060
8	128.5	16,510	0.6405	1.016
9	114.4	13,090	0.8077	1.282
10	101.9	10,380	1.018	1.616
11	90.74	8234	1.284	2.04
12	80.81	6530	1.619	2.57
13	71.96	5178	2.042	3.24
14	64.08	4107	2.575	4.08
15	57.07	3257	3.247	5.15
16	50.82	2583	4.094	6.50
17	45.26	2048	5.163	8.18
18	40.30	1624	6.510	10.3
19	35.89	1288	8.210	13.0
20	31.96	1022	10.35	16.4
21	28.46	810.1	13.05	20.7
22	25.35	642.4	16.46	26.2
23	22.57	509.5	20.76	32.9
24	20.10	404.0	26.17	41.5
25	17.90	320.4	33.00	52.4
26	15.94	254.1	41.62	66.4
27	14.20	201.1	52.48	83.2
28	12.64	159.8	66.17	106.0
29	11.26	126.7	83.44	131.0
30	10.03	100.5	105.2	168.0
31	8.928	79.70	132.7	212.0
32	7.950	63.21	167.3	262.0
33	7.080	50.13	211.0	333.0
34	6.305	39.75	266.0	422.0
35	5.6	31.52	335.0	536.0
36	5.0	25.00	423.0	671.0
37	4.453	19.83	533.4	____
38	3.965	15.72	672.6	____
39	3.531	12.47	848.1	____
40	3.145	9.88	1069	____

Figure 3-3. Wire chart.

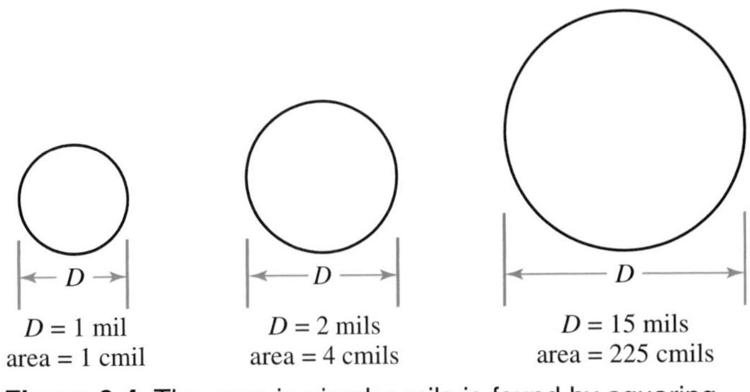

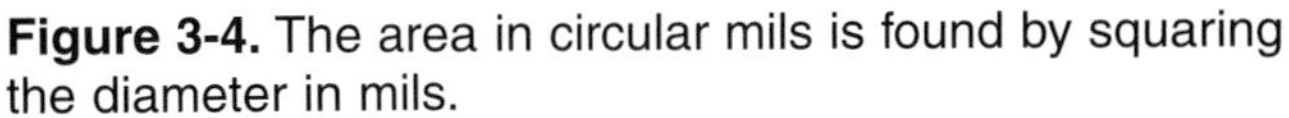

Figure 3-4. The area in circular mils is found by squaring the diameter in mils.

equivalent, compute the square/rectangular conductor's area in square mils and then divide that area by 0.7854.

For example, if a 1" by 1/4" bar of copper is to be used as a conductor, first convert the inch measurements to mils (1" = 1000 mils and 1/4" = 250 mils). Then multiply the dimensions in mils to get the area in square mils (1000 × 250 = 250,000 sq. mils). Finally, divide the square mil area by 0.7854 (250,000/0.7854 = 318,309). This results in the original 1" by 1/4" copper bar having a surface area equivalent to a round conductor of 318,309 cmils.

Wire charts commonly express the electrical values for conductors solely based on their size in cmils. To determine the current carrying capacity, or resistance value, of a rectangular conductor, the rectangular conductor must be converted to circular mils.

Conductor Insulation

Materials with only a few free electrons do not conduct electrons well. These materials are called ***insulators***. To keep the electron flow contained to the conductor path, and to prevent contact with other conductors and people, conductors are coated. This protective coating is called ***insulation***. As the name implies, insulation is made of insulating material such as rubber, plastic, or other synthetic materials. There are many types of insulation commonly used today. These materials include thermoplastic, neoprene, Teflon™, nylon, and polyethylene.

The type of conductor insulation is determined by the application of the electrical system. In determining insulation type, many factors must be taken into consideration. Some questions one might ask would include:

- Will the insulation be exposed to extreme heat generated by an industrial furnace or the extreme cold of a freezer in a food processing plant?
- Will the insulation need to withstand exposure to the acidic vapors in a chemical plant or to wet, acidic soil?
- Will the cable be exposed to oil from manufacturing equipment?
- Will the insulation give off toxic fumes if burned?

- Will the insulation in a building affect the breathable air of its occupants?

Insulation codings

Insulations are marked with code letters indicating their approved uses. Examples of these markings are HH for high heat resistant, M for oil resistant, UF for underground installation, etc. When selecting conductor insulation, all conditions must be considered. A brief list follows:

R Rubber
H Heat
HH High heat
A Asbestos
T Thermoplastic
M Oil resistant
UF Underground feeder
C Corrosion resistant

Conductor insulation may be color coded to assist a technician in tracing it throughout a building or throughout some other electrical application such as an automobile. At times, the colors represent certain voltages, polarities, or grounding conductors. The color used is generally governed by building codes or manufacturing associations. Uniform standards in the color coding of insulators assist the technician in troubleshooting electrical and electronic systems.

Types of Conductors

Conductor wiring comes in many shapes and assemblies. See **Figure 3-5**. The conductor can be solid or stranded. There can be one or more conductors inside one cable assembly. The insulation jacket sometimes contains a metallic sheath to provide added protection against physical damage or to prevent interference from other electrical systems. The conductor assembly can be flat to accommodate special installations, or it can be designed to be installed in electrical pipe, referred to as *conduit*.

Conductors have many applications that require a unique design to match the type of special electronic application. Some of the applications that require unique designs are telephone communication cables, cable television, computer lines, high voltage transmission cable, fiber optics, motor and relay windings, and marine applications. There are hundreds of types of conductors available today.

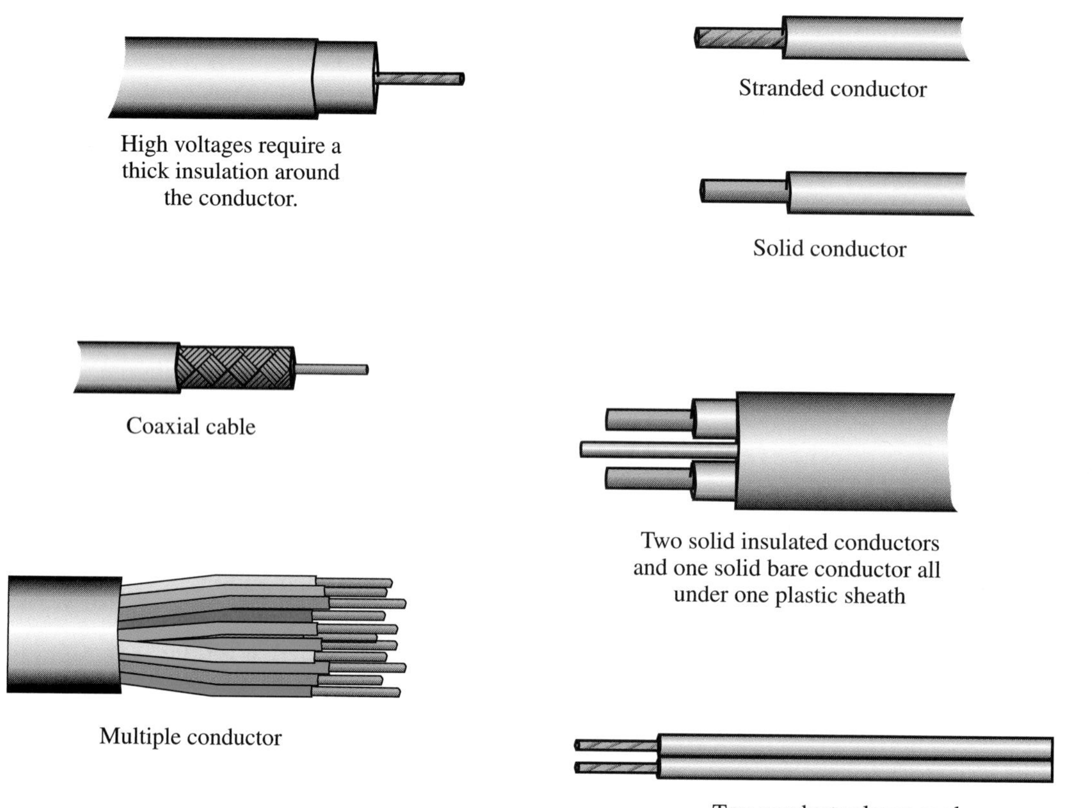

Figure 3-5. Typical conductors found in electrical and electronics work.

Conductor Resistance

Even though conductors provide a low resistance path for electron flow, they still have some resistance. This resistance must be considered when long distances are involved. There are four factors directly relating to the resistance of conductors.

- *Cross-sectional area of the conductor.* The larger the surface area or diameter of a conductor, the lower the resistance.
- *Type of conductor material.* Different materials have different resistance values.
- *Length of conductor.* The longer the conductor, the greater the resistance.
- *Temperature of material.* Resistance of a material rises with rising temperature.

Cross-sectional area of a conductor

Increasing the cross-sectional area of a conductor increases the amount of current that can flow. To help visualize this, use the flow of water through a pipe as an example, **Figure 3-6**. A large diameter pipe can carry more gallons of water per minute than a small diameter pipe. An electric conductor operates in the same fashion. The large diameter conductor carries more electrons per minute than a small diameter conductor.

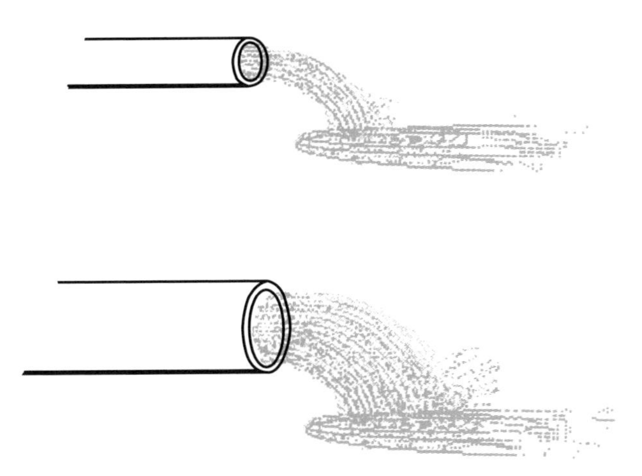

Figure 3-6. The larger the diameter of a pipe, the more gallons flow per minute. The larger the diameter of a conductor, the greater the current and the lower the resistance.

Type of material

As discussed in Chapter 1, some materials are better conductors than others. The type of material affects conductance and resistance. **Figure 3-7** shows one example. A No. 12 copper wire has 1.619 ohms resistance per 1000 feet. An aluminum conductor of the same diameter and length offers 2.57 ohms of resistance.

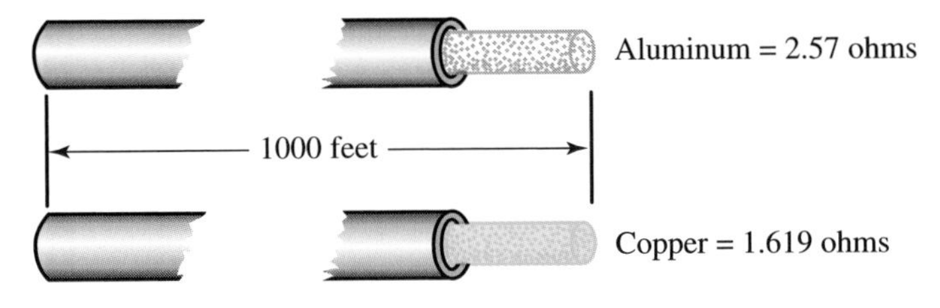

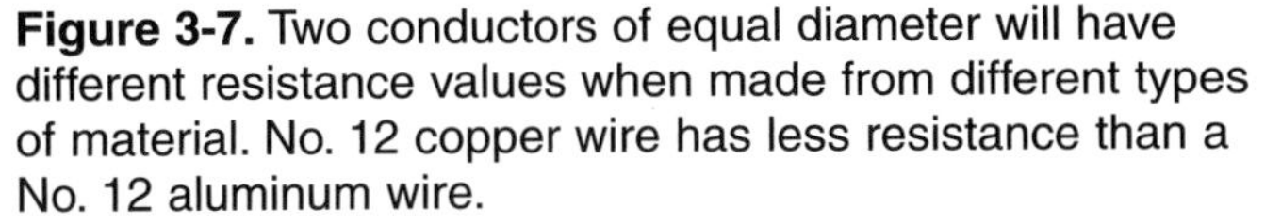

Figure 3-7. Two conductors of equal diameter will have different resistance values when made from different types of material. No. 12 copper wire has less resistance than a No. 12 aluminum wire.

Length of conductor

The length of a conductor greatly affects its total resistance. If one foot of wire has a certain resistance, then ten feet of the same wire will have ten times more resistance. Fifty feet of the wire will have fifty times more resistance, and so on. As a conductor becomes longer it creates a voltage drop in a circuit. A circuit using short lengths of wire, such as an electrical lab project using three to six inches of No. 22 wire, does not create a major problem. But long runs of wiring can create electrical problems.

In **Figure 3-8**, a 10 amp load is connected to a circuit using No. 22 copper wire. The load is at a distance of 100 feet from the source. When the switch to the motor is closed to connect the motor, the motor will heat up internally because an insufficient voltage is being applied. The low voltage is a result of the voltage drop along the length of the conductor. The loss of voltage can be computed using Ohm's law.

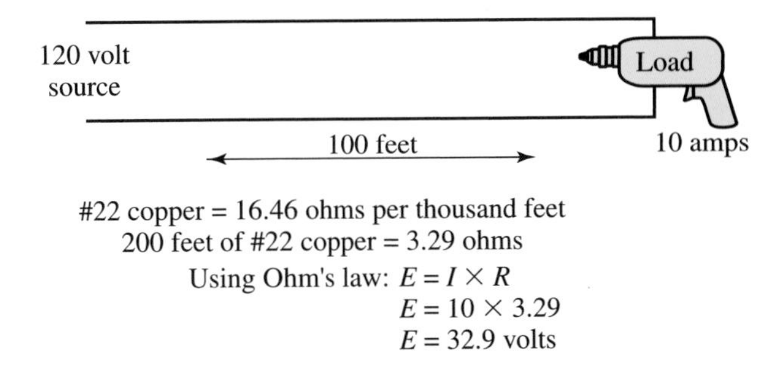

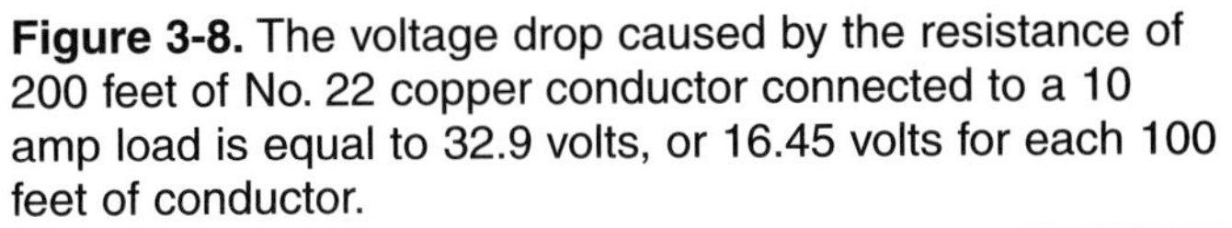

Figure 3-8. The voltage drop caused by the resistance of 200 feet of No. 22 copper conductor connected to a 10 amp load is equal to 32.9 volts, or 16.45 volts for each 100 feet of conductor.

The electrons must travel a total distance of 200 feet to make a complete circuit. The total resistance of the wiring is equal to 3.29 Ω. The current through the load is 10 amps. By applying Ohm's law to these conditions, a voltage drop equal to 32.9 volts has been created ($E = 10 \text{ A} \times 3.29 \ \Omega$). Subtracting the 32.9 volt drop from the source voltage of 120 leaves only 87.1 volts for the load. This amount of voltage is undesirable, and will no doubt cause equipment failure, especially for an electrical motor load, such as a drill.

LESSON IN SAFETY:

When using long extension cords to operate lights and tools, be sure that they have sufficient conductor size. The larger the wire size used, the smaller the voltage drop created. Small diameter extension cord conductors may heat up and burn. Motor driven tools at the end of a long extension cord can heat up and operate inefficiently.

Temperature of material

A fourth consideration in conductivity is temperature. Most metals used in conductors, such as copper and aluminum, increase in resistance as the ambient temperature rises. ***Ambient temperature*** is the temperature of the material surrounding the conductor, such as air, water, or soil. In many electronic products, careful design is necessary to ensure proper ventilation and radiation of heat from the current carrying conductors. Many electrical products such as computers have a fan installed inside the appliance to circulate heat away from electronic components. These components can be damaged by heat easily.

When many conductive materials are exposed to extremely low temperatures, their resistance value approaches zero. These materials become ***superconductors***. **Figure 3-9** is a picture of a superconducting cable. Magnets can be produced from these superconductors. Early experiments with supermagnets, using superconductors, produced magnetic fields so powerful that the magnet's housing was crushed by the tremendous force. Since electrical motors operate on the principles of magnetism, the more powerful the magnet, the more powerful the electric motor will be. Superconductive material is also used in some magnetic imaging medical equipment. The superconductive materials have made it possible to peer deep into the human body. Magnetism and electromagnetism will be explained in detail in Chapter 9.

To make a superconductor, the temperature of the conductive material must be lowered to nearly absolute zero, –273°C or –460°F. Liquid nitrogen is used to cool conductive materials to such a low temperature. Today, some materials display superconductive properties at much higher temperatures. In 1987, a team of scientists produced a superconductor at approximately –270°F. New applications for superconductors are appearing everywhere.

Figure 3-9. Superconducting cable with insulating wrapping exposed. (Fermi National Accelerator Laboratory)

Safety Standards

The ***National Electrical Code (NEC)*** is a collection of electrical standards that must be followed to ensure safety to personnel and prevent the possibility of an electrical fire. There are many regulations contained within the code book pertaining to the design of electrical systems installed in homes, industry, and commercial establishments. The guidelines take into consideration all aspects of wiring. These aspects include ambient temperature, the type of material from which the conductor is made, the type of insulation and length of conductor, as well as the location of installation. The minimum code standards should always be followed when installing conductors in electrical systems.

The National Electrical Code does not deal with the installation of conductors on the inside of electrical equipment such as in stereos, televisions, or other appliances. A not-for-profit organization, Underwriters Laboratories, tests these products for safety standards. Always use equipment and appliances that have the UL label to ensure that the equipment meets or exceeds national safety standards.

Conductor voltage drop

The National Electrical Code has set a maximum voltage drop standard for branch circuits. A ***branch circuit*** is the wiring from an electrical circuit panel to the last device connected in that circuit. A maximum of a 3%

voltage drop is permissible by NEC standards. A formula for calculating the size conductor needed to stay within the 3% limit is below.

$$CMA = \frac{K \times I \times L}{VD}$$

CMA is the area in circular mils, *K* is the material constant, I is the amperage, *L* is the length of wire, and *VD* is the voltage drop.

Example: First, decide the type of material to be used in wiring the circuit, copper or aluminum. The material constant (*K*) for copper is 12. The material constant for aluminum is 18. This is an NEC standard based on the resistance of the material. To find the size of a copper conductor needed for a 10 amp load located 150 feet from the circuit breaker panel, we multiply the material constant times the amperage times the length ($K \times I \times L$). Remember, the length is twice the distance from the panel. Next, divide by the permissible 3% voltage drop. The voltage in this case is 120 volts. Three percent of 120 volts is 3.6 volts.

$$CMA = \frac{12 \times 10 \times 300}{3.6}$$

$$CMA = \frac{36000}{3.6}$$

CMA = 10,000 (or a No. 10 copper conductor is needed)

This formula is very handy when trying to determine a conductor size that will not cause a significant voltage drop. Excessive voltage drops prevent electrical equipment from operating properly.

Review Questions for Section 3.1

1. A half inch is equal to __________ mils.
2. The cross-sectional area of a conductor is measured in __________.
3. The unit measure for conductance is __________.
4. What is the circular mil area of No. 22 copper wire?
5. List four factors that affect conductor resistance.
6. What is the resistance value of 500 feet of No. 4 copper conductor?

3.2 SPECIAL CONDUCTOR PATHWAYS

There are several special conductor pathways used in electrical and electronic work. They are the breadboard, printed circuit board, and metal chassis. These pathways serve special needs. Yet, their main purpose is the same as any other conductor, to provide a good conductive path for electron flow.

Breadboards

A breadboard is a useful device for learning about circuits. A ***breadboard*** consists of a series of holes aligned in rows across the entire surface of an insulation material, such as plastic. Copper strips are run in parallel under the rows of holes and are used as conductor pathways. Electronic devices, such as resistors and transistors, are inserted into the holes. Jumper wires are then used to make additional connections between the devices. The breadboard provides an easy system for constructing circuits quickly. The boards are commonly used in experiments or to make a prototype of a circuit before the circuit is soldered or constructed in mass assembly systems. Some boards are called *proto boards.* See **Figure 3-10**.

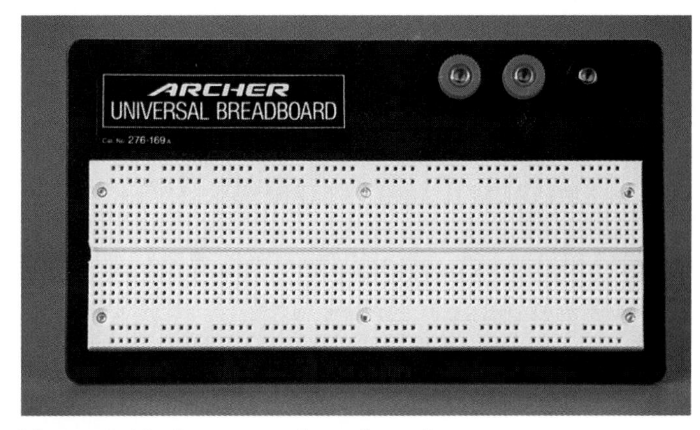

Figure 3-10. A common breadboard.

Printed Circuit Boards

A ***printed circuit board (PCB)*** is made from a thin layer of conductor material, usually copper foil, cut in strips and attached to an insulated board. The strips act as the circuit paths. Components are inserted through holes in the insulated board and then soldered to the conductive strips at circular spots called *connection pads.* The connection pads provide the extra area required for making good solder connections. **Figure 3-11** shows a typical circuit board layout. The *heat sink area* is a wide copper surface that is used for a common connection point for many devices, as well as to dissipate the heat generated by the electronic devices. *Edge connectors* provide the electrical point of attachment to other electrical equipment.

The conductor paths can be less than 0.001 inch in thickness. These strips can be on one side or both sides, depending on the design of the board. The currents used on most circuit boards are very low and usually do not require heavy wiring.

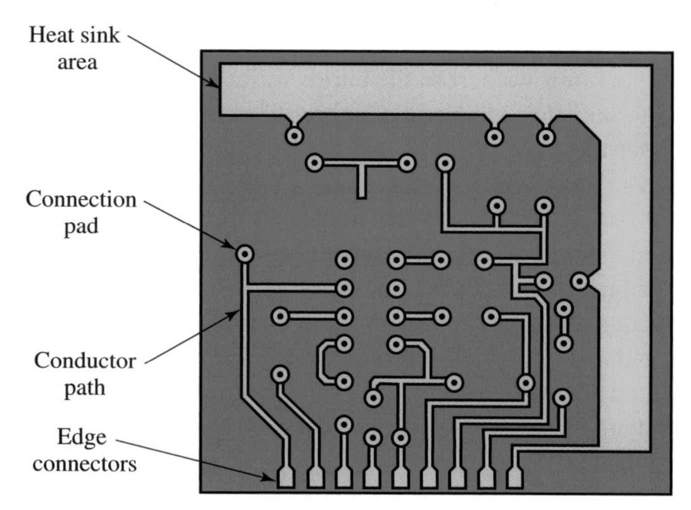

Figure 3-11. A typical printed circuit board layout consists of thin foil copper ribbons on the surface of an insulated board.

Construction of printed circuit boards

A printed circuit board starts out with a thin sheet of copper foil covering the entire surface of an insulating material. If the PCB is to be double sided, both sides will be covered with a thin sheet of foil. There are a couple of common methods used to produce the circuit design on the foil. The layout of the circuit can be put on the copper using a *resist* material. The board is then dipped in an *etchant*. The etchant removes all of the copper surface not protected by the resist material. The board is then cleaned to remove any excess resist or etchant. Holes are then drilled at the connection pads.

Printed circuit boards can also be made using a photographic process. Light-sensitive material covers the surface of the board. A negative of the circuit board layout is placed over the board. The board is then exposed to light and developed in a manner similar to that used in photography. After developing, the unexposed areas are washed away with a solvent. The exposed area, the pattern of the circuit, remains. The board is then etched as was described in the first method.

Circuit board components can be installed by hand or by automated systems. The choice is usually based on economics. If a low number of boards are to be produced, hand assembly is less costly. If thousands of boards are to be produced and assembled, automated systems are used.

Chassis

In the early years of electronics, components were mounted on metallic surfaces called ***chassis***. The chassis itself served as part of the circuitry. Use of a chassis as a conductor is not as common today, but it is still in great use. One of the most visible uses of the chassis as a conductor is in the automobile. The metal frame and body parts serve as a chassis providing a conductor path for the negative side (ground) of the circuit. See **Figure 3-12**.

The positive line leaves the positive battery terminal and connects to a device such as the car's headlamps. A fuse or circuit breaker is in series with the lamps for protection. A switch is also in series for turning the lamps on and off. The return to the negative side of the battery is through the frame or car chassis itself. The negative conductor for the lamp simply terminates at the lamp location, connecting to the metallic shell of the car. The electrons return to the battery through the metallic frame of the car.

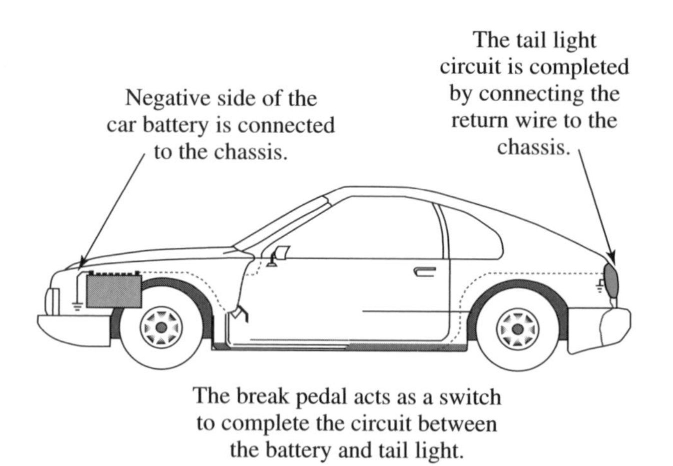

Figure 3-12. The frame of the car is used as the return path for the tail light circuit. This type of wiring is referred to as chassis wiring.

Review Questions for Section 3.2

1. A temporary circuit is usually wired using a __________.
2. The most economical way to mass produce an electronic circuit is using __________.
3. The automobile is an example of __________ wiring.
4. Connection of devices to a PCB is usually made by __________.
5. What other types of transportation could use the frame for part of the electrical circuit?

3.3 COMMON CIRCUIT DEVICES

There are several common circuit devices that are present in most electrical and electronic circuits. They provide a means of controlling electron flow through the conductor paths and provide for safe operation of circuits. The three important items discussed here are switches, connectors, and circuit protection devices.

Switches

Switches are installed in circuits to control the flow of electrons through the circuit. They can be categorized by their actuator and electrical switching path. The ***actuator*** is the mechanical device that causes the circuit to open and close. Various types of switches are illustrated in **Figure 3-13**. The schematic symbol associated with each type of switch is also included.

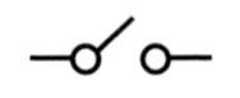

Toggle switch

Slide switch

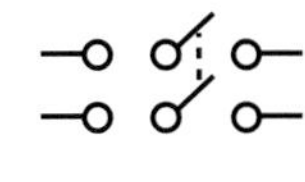

Double pole double throw switch

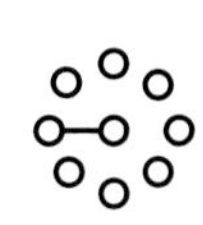

Rotary switch

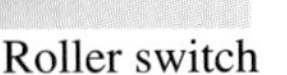

Roller switch

Figure 3-13. Typical switch types. Notice how many have the same electrical symbol but different actuators.

Some of the most common actuators are the slide, toggle, rotary, and push button. As you can see, the name of the switch indicates the type of actuator used to turn the circuit on and off. The electrical circuit inside the switch is described in terms of poles and throws. The simplest type of switch is the single-pole single-throw switch, which is abbreviated SPST. The term ***single-pole*** means that the switch provides one path for the electron flow and that it can be turned on or off. The term ***single-throw*** means that the switch controls only one circuit.

A single-pole double-throw switch (SPDT) has one common connection point and can complete a circuit path to two different circuits. However, only one circuit can be completed at a time. There are many possibilities and combinations for a switch of this type.

A useful application of the single-pole double-throw switch is its ability to control a load, such as a lamp, from two different locations. In residential wiring systems, the single-pole double-throw switch is referred to as a three-way switch. In **Figure 3-14**, two single-pole double-throw switches are connected to a lamp. Either switch is capable of turning the lamp on or off.

A double-pole double-throw switch (DPDT) has two common connection points and can provide two circuit paths simultaneously. The DPDT switch is like having two SPST switches connected in parallel.

Switch ratings

Switches are rated for ampacity and voltage. The ***ampacity rating*** of a switch is an indication of how much

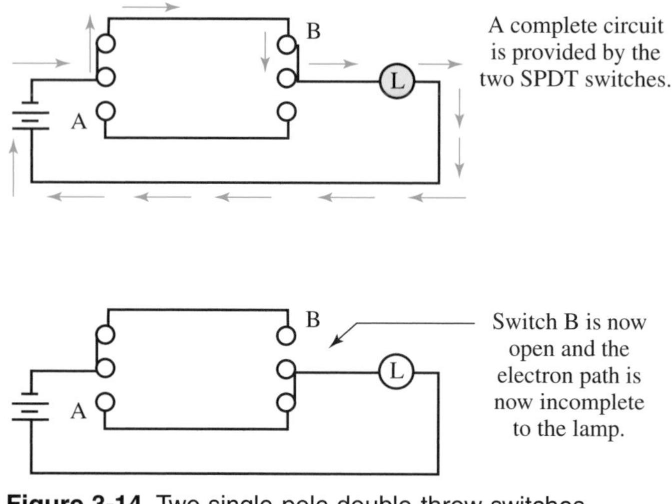

B
A
L
A complete circuit is provided by the two SPDT switches.
B
A
L
Switch B is now open and the electron path is now incomplete to the lamp.

Figure 3-14. Two single-pole double-throw switches (SPDT) can be used to control a lamp from two different locations.

current it can safely handle. The ***voltage rating*** is the maximum voltage for which a switch is designed. Exceeding the maximum voltage rating will cause the electrical-mechanical circuitry inside the switch to fail. For example, if a toggle switch is rated as one amp and 24 volts, a current in excess of one amp will burn out the switching circuitry inside the switch. If the 24-volt switch is connected to a 240-volt circuit, it may fail to open the circuit sufficiently to stop the flow of electrons. This action will result in a dangerous situation that can melt the switch's insulation and short circuit the switch.

Connectors

There are many types of connectors used with electrical conductors. The type of connection used varies according to the type and size of the conductor, the

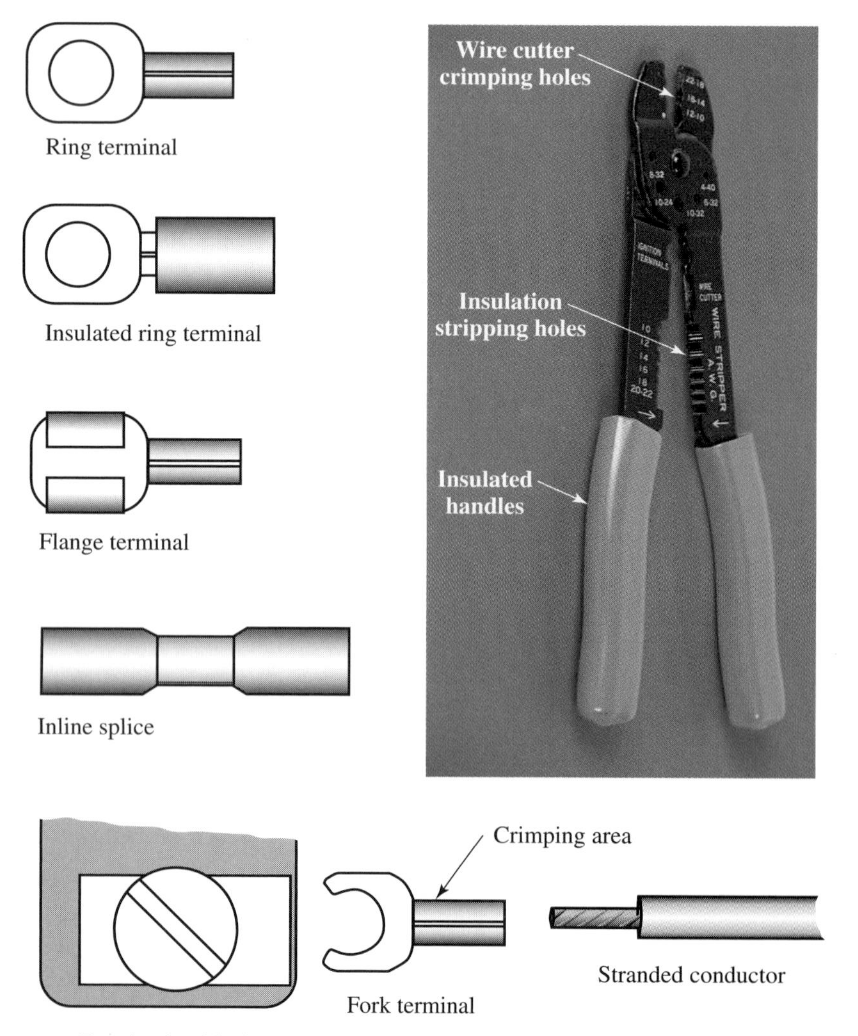

Figure 3-15. Some wire connectors are made to be crimped on the end of stranded conductors. After the connector is crimped on the wire, the wire can be easily secured under a termination block screw.

purpose served by the connection, and the type of device to be connected. Look at **Figure 3-15**. You will see many common types of connectors. One general classification is solderless connectors. A solderless connector does not require the use of solder to make the connection. These connectors generally require a crimping tool. The crimping tool squeezes the connector to the conductor. **Figure 3-15** shows common wire crimps on terminals and splices.

Some types of connectors use screws and bolts to form the mechanical connection to conductors. These connectors are used primarily for larger conductors. See **Figure 3-16**.

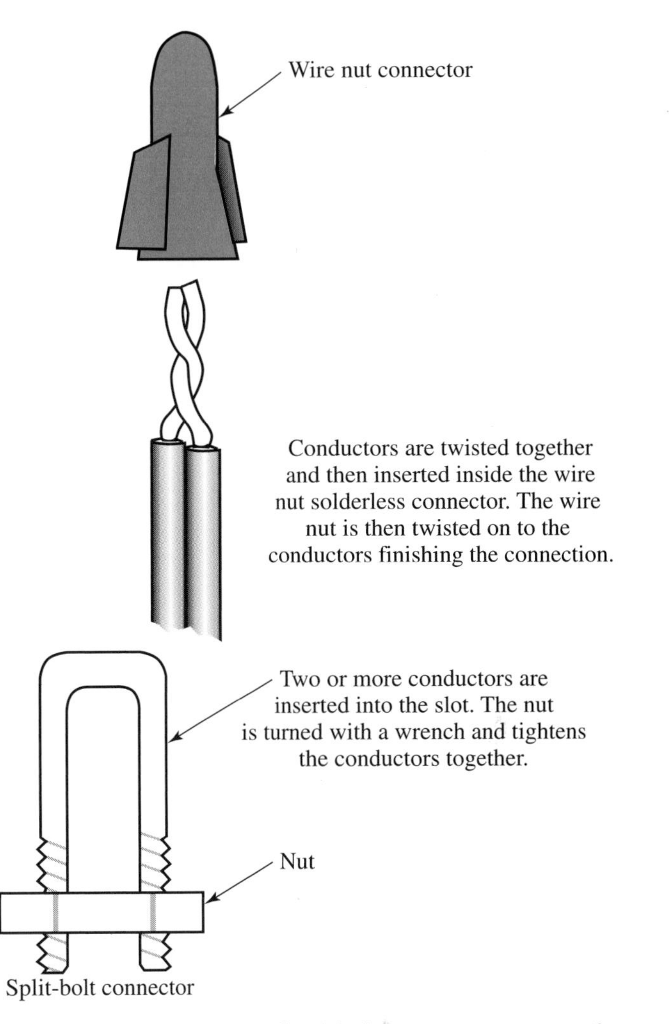

Wire nut connector
Conductors are twisted together and then inserted inside the wire nut solderless connector. The wire nut is then twisted on to the conductors finishing the connection.
Two or more conductors are inserted into the slot. The nut is turned with a wrench and tightens the conductors together.
Nut
Split-bolt connector

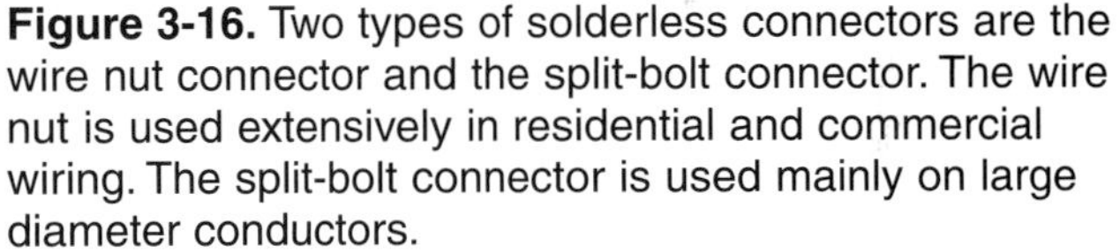

Figure 3-16. Two types of solderless connectors are the wire nut connector and the split-bolt connector. The wire nut is used extensively in residential and commercial wiring. The split-bolt connector is used mainly on large diameter conductors.

Circuit Protection Devices

Common circuit protection devices are fuses and circuit breakers. ***Fuses,*** such as those shown in **Figure 3-17,** are constructed from small, fine wire. This wire is engineered to burn if certain amperages are exceeded. Fuses are sized by their voltage and current capacity, primarily current.

For example, a three amp fuse is designed to burn and open the circuit when the current exceeds three amps. A load that draws three amps or greater will generate sufficient heat in the fuse to melt the fuse link inside the glass tube. *The time required to melt the fuse link is inversely proportional to the amount of overload.* This means that the higher the overload current, the faster the melting action occurs. When a fuse melts, it must be replaced.

A ***circuit breaker,*** sometimes called a ***reset,*** is another device used to protect a circuit from overload and short circuit conditions. See **Figure 3-18**. The main advantage of a circuit breaker over the fuse link is that the circuit breaker need not be replaced after tripping. It can be reset by moving the handle to the *off* position and then returning it to the *on* position. Some circuit breakers have an actuator similar to a push button switch. These breakers are pushed *in* to reset after tripping. Circuit breakers and resets require a waiting period to allow the internal trip mechanism to cool down. Most homes today use circuit breakers as the safety device to prevent overloads. Overloads could result in house fires.

Circuit breakers are produced with two different tripping methods. One method uses bimetallic strips. A bimetallic strip is a metal strip made of two different types of metal. Different metals expand at different rates. Heat generated from the overload condition causes the bimetal trip to expand. The different metals expand at different rates. This causes the breaker's trip mechanism to bend and break contact. Some trip mechanisms are adjustable to allow for a more precise trip current.

A second tripping mechanism uses magnetism to operate. The circuit current runs through a coil. As the current increases through the coil, the amount of magnetism in the coil increases. When a predetermined point is reached, the tripping mechanism operates and opens the circuit. The magnetic circuit breaker is much faster and

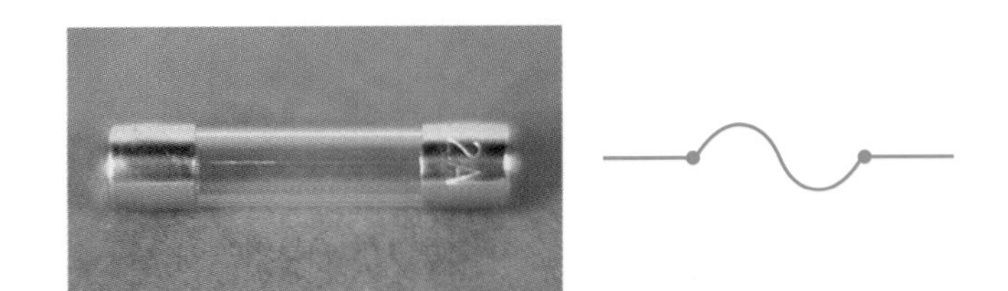

Figure 3-17. A typical fuse and the schematic symbol that represents the fuse.

Figure 3-18. Two typical circuit breakers.

more accurate than the bimetallic circuit breaker. The magnetic circuit breaker, however, is more expensive than the bimetallic type.

Review Questions for Section 3.3

1. What type of switch is best for controlling lighting from two different locations?
2. What does the abbreviation DPDT mean?
3. What advantage does a circuit breaker have over a fuse?
4. How are fuses sized?

3.4 LIGHTING

There are many different methods in which light is produced. Light is generated when burning a variety of gasses (methane, propane), liquids (gasoline, fuel oil), and solids (wood, paper). Certain chemicals will luminesce when mixed together. Necklaces and bracelets filled with these chemcials are often sold at nighttime celebrations. Various minerals and glow-in-the-dark toys supply light through *phosporescence*. They absorb energy (from sunlight or a lamp) and release this energy in the form of light when the energy source is removed. However, for most of our lighting systems, the energy source is electricity. There are two very different types of electric lighting systems. One is the incandescent lamp and the other is the discharge lamp.

Incandescent Lamp Principles

In 1879, Thomas Edison developed the first successful incandescent lamp. In an ***incandescent lamp***, light and heat are created from current flowing through a filament. The first lamp was of simple construction with a carbon filament inside a glass envelope. The air was removed from the glass envelope to prevent the carbon from igniting. Soon afterward the tungsten filament replaced the carbon filament. The tungsten filament proved to be a much more efficient method of producing light than the carbon filament. In **Figure 3-19**, the basic construction of the incandescent lamp is illustrated.

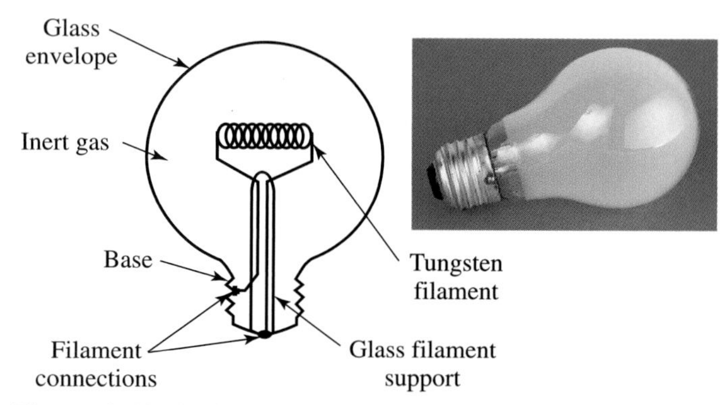

Figure 3-19. An incandescent lamp consists of a tungsten filament inside a glass envelope filled with an inert gas.

Halogen lamp

In a halogen lamp, a tungsten filament is inserted through a glass tube, and the tube is filled with halogen gas. This type of lamp will economically produce a great deal of light. It is commonly used in automobile headlights. The halogen gas makes the tungsten filament last longer. The tungsten filament in most lamps evaporates over a long period of time due to the tremendous heat generated by tungsten filament lamps. The halogen returns the boiled off tungsten filament particles back to the filament, thus causing the filaments to last longer, **Figure 3-20**.

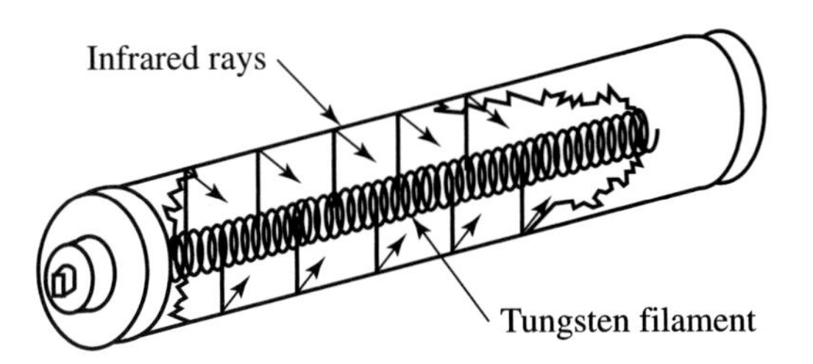

Figure 3-20. Halogen—IR uses a coating inside the tube that causes infrared rays to be reflected back towards the filament.

Discharge Lamp Principles

A discharge lamp operates differently than an incandescent lamp. A ***discharge lamp*** produces light by energizing a gas such as argon, neon, helium, or a vapor of mercury or sodium. The gas or vapor is ionized by the electrical pressure and will glow, thus emitting light. A discharge lamp can easily produce 20 times more light than a conventional filament bulb, while using the same amount of electrical energy. Descriptions of several types of discharge lamps follow.

Fluorescent lamp

The fluorescent lamp consists of a long glass tube coated on the inside with a phosphor. See **Figure 3-21**. A filament or electrode is inserted at each end of the tube. The air is removed from the tube, and then it is filled with an inert gas and a small amount of mercury. When the tube is energized, the filaments at the end will glow. However, the filaments do not provide very much light. The filaments are used to produce heat, which vaporizes the mercury inside the tube. Once the mercury has been vaporized, electrons flow along the mercury vapor.

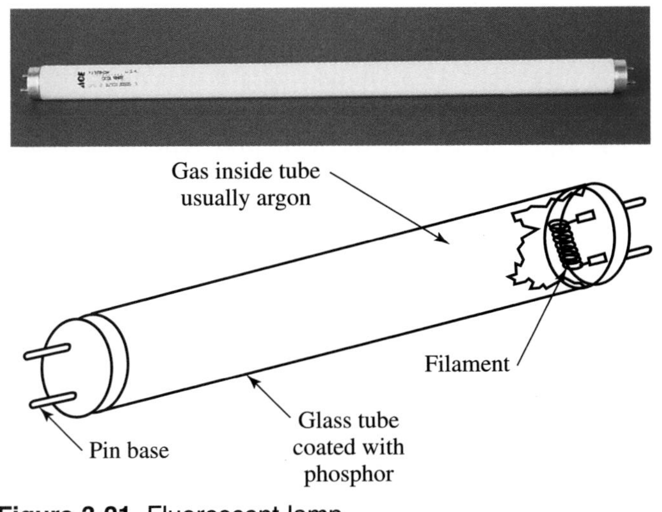

Figure 3-21. Fluorescent lamp.

Ultraviolet light is produced inside the tube. The ultraviolet light strikes the phosphor coating and causes it to glow. This produces the familiar phosphorous light. A ***ballast*** is used to limit current inside the tube. The ballast consists of many turns of fine wire. The ballast is also used to produce a higher than usual voltage, which is applied to the filaments in the ends of the tube.

An older style of fluorescent lighting used starters in the circuitry between the ballast and lamp. The purpose of the starter was to allow the filament inside the lamp to conduct and produce the necessary heat to vaporize the mercury. Once this was accomplished, the starter would open and cease the heating action of the filament. Newer fluorescent lamps incorporate a device in the end of the tube that replaces the need for a starter.

Mercury vapor lamp

Figure 3-22 shows the construction of a typical mercury vapor lamp. It consists of two electrodes connected at opposite ends of the arc tube. The arc tube contains a small amount of mercury and argon gas. When the lamp is energized, the starting electrode generates heat to vaporize the mercury and ionize the argon. After considerable time, the starting electrode opens allowing current to flow through the arc tube. The disadvantage of this type of lamp is the long delay required when starting or when the power has been temporarily interrupted.

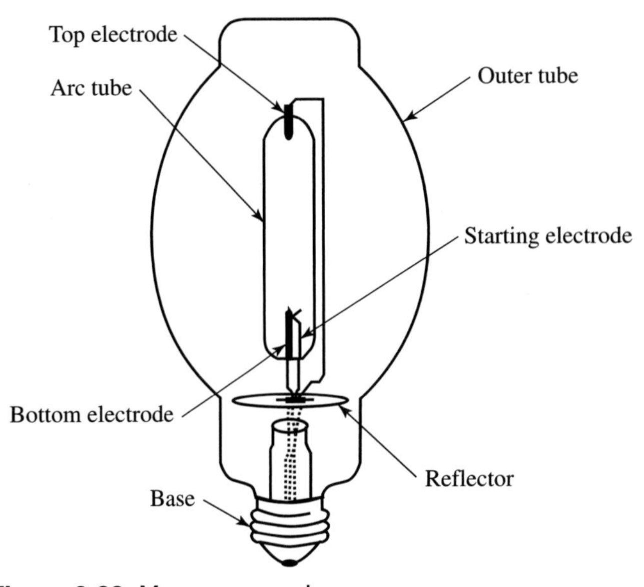

Figure 3-22. Mercury vapor lamp.

Neon lamp

A neon lamp consists of two electrodes inserted in the ends of a long glass tube. The tube is often heated and shaped into words or pictures for commercial

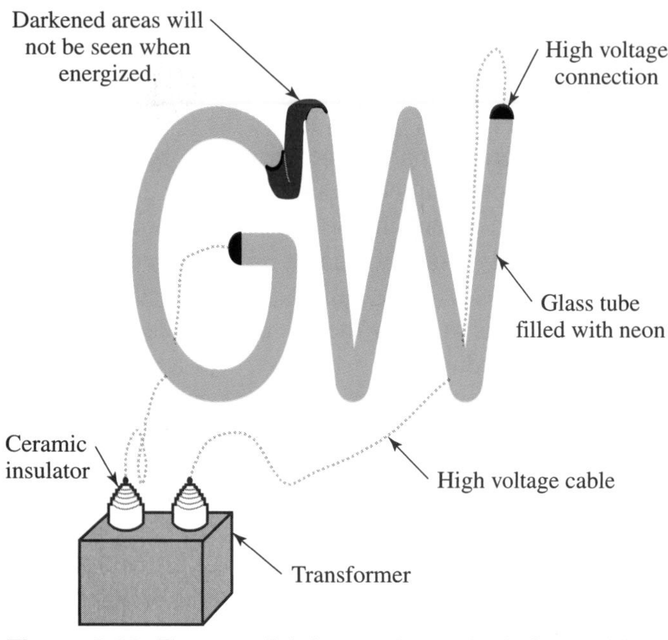

Figure 3-23. For neon lighting, a glass tube is filled with neon gas. Then an electrical charge, in excess of 10,000 volts, is passed through the tube.

applications, **Figure 3-23**. After being shaped to achieve a desired effect, the tube is filled with neon gas.

A neon light transformer is used to produce a high voltage (10,000 volts or more), which is needed to create a current through the neon gas. Because of the high voltage present, special high voltage insulation is used to insulate the conductors. After the light is energized, the neon tube will glow. Parts of the tube may be darkened out by black paint to achieve varied designs. To create a variety of colors, other gases such as argon and helium can be used. Although these lamps are filled with gases other that neon, they are often refered to as "neon lamps" out of custom.

Glow lamp

The glow lamp is very similar in construction to a neon lamp. It consists of two electrodes inside a short glass tube filled with neon or argon. The electrodes in a glow lamp are quite close in comparison to the neon light. Consequently, a glow lamp does not require the same high voltages required in typical neon signs.

Strobe lamp

A strobe lamp may vary from a very short to a very long piece of glass tubing. The long strobe tubes are usually manufactured in a spiral shape, such as that shown in **Figure 3-24**. The strobe lamp operates by discharging a high dc voltage directly through the tube. What is observed is the electrical arc flashing through the tube.

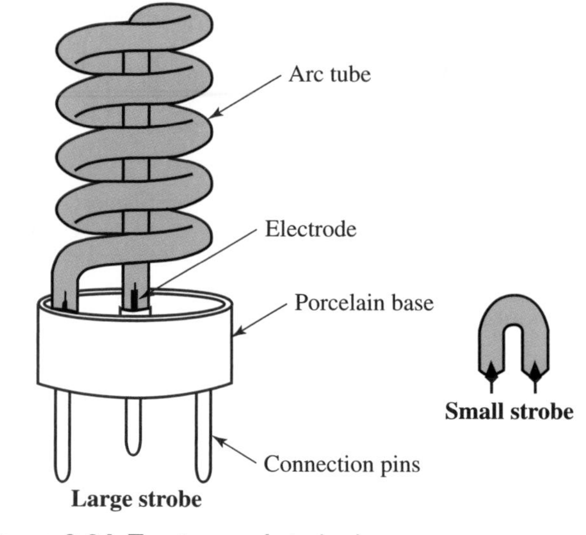

Figure 3-24. Two types of strobe lamps.

Sometimes there are exciters placed along the length of the tube to help attract the electrons through the long tube. ***Exciters*** are charged lengths of wire that assist the arc along the length of the glass tube. Strobes are capable of producing flashes of such great intensity that they can be used for airfield approach systems. The flash of a strobe can generate an excess of 50,000 watts and can be observed at distances greater than ten miles during the light of day.

Lumen

Most people rate lighting brightness in wattage because they are familiar with the incandescent light. This is misleading. The amount of light produced by a lamp is rated in candelas or lumens. The ***candela*** is based upon the amount of light generated by one candle. The ***lumen*** is the term used to measure the amount of light generated by lighting systems. Look at **Figure 3-25** to see an illustration comparing the candela and lumen. A candela will produce 12.57 lumens at a distance of one foot from its center.

A 100 watt incandescent light (a typical value you will find in your home) can produce as much as 4000 lumens of light. The chart that follows shows a comparison of light intensities. ***Lumens per watt (lpw)*** gives the amount of light produced for each watt of energy used. *The higher the lpw, the more light you receive for each watt of electricity.*

Edison's first lamp	1.4 lpw
Indescent lamps	10–40 lpw
Fluorescent lamps	35–100 lpw
Halogen lamps	20–45 lpw
Mercury vapor lamps	50–60 lpw

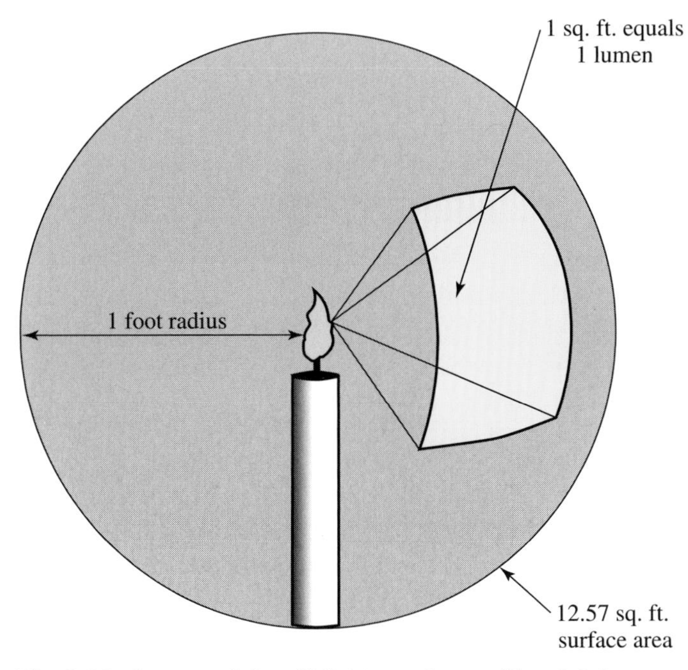

Fig 3-25. One candela will light a sphere with a total surface area of 12.57 square feet.

Metal halide lamps	80–125 lpw
High pressure sodium lamps	100–140 lpw

Review Questions for Section 3.4

1. What type of lamp was invented first?
2. Give several examples of electrical discharge lamps.
3. What lamp provides the most economical lighting based on the amount of electricity used to create the light?
4. Explain how a gas like neon becomes a conductor?
5. How is light intensity measured?

3.5 RESISTORS

One of the most common components encountered in the study of electronics is the resistor. The ***resistor*** is used to create desirable voltage drops and limit current values in electronic circuitry. **Figure 3-26** shows several molded composition, fixed resistors. They are manufactured in many sizes and shapes. The schematic symbol for a fixed resistor is also shown.

The chemical makeup that causes resistance is accurately controlled in the resistor manufacturing process. Resistor values can be purchased in a range of values from less than 1 ohm to over 22 MΩ. The physical size and material used for resistance is rated in watts. A resistor's ***wattage rating*** refers to the resistor's ability to safely dissipate heat. Heat is generated by electrons flowing through the resistor. Common wattage sizes range from 1/4 watt to 25 watts. Resistors are grouped by ohms and watt sizes. When purchasing a resistor, the desired resistance and wattage rating must be specified. For example: 1000 ohms and the watt size, 1/4 watt, 1/2 watt, or 2 watts, etc. In each watt size, the resistance value would be the same. Electrical power and wattage will be explained in detail in Chapter 4. See **Figure 3-27** to examine the construction of a molded composition resistor.

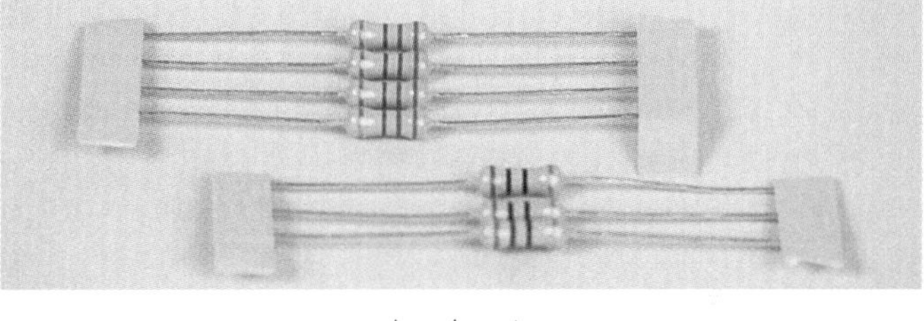

Figure 3-26. Group of carbon composition resistors and the fixed resistor symbol.

SOLDER-COATED LEADS
Suitable for soldering and welding even after long periods in stock.
SOLIDLY EMBEDDED LEADS
Load wires are formed to provide large contact area and high pull strength.
PERMANENT COLOR CODING
Bright, baked on colors are highly resistant to solvents, abrasion and chipping. Colors remain clearly readable after long service.
RUGGED CONSTRUCTION
Resistors are hot-molded. Resistance material, insulation material and lead wires are molded at one time into a solid integral structure.
SOLID RESISTANCE ELEMENT
Resistance material has large cross section resulting in low current density and high overload capacity. Uniformity of material eliminates "hot spots."

Figure 3-27. Cutaway of a carbon composition resistor. (Allen-Bradley)

Another type of small wattage, fixed value resistor is the thin film resistor. The thin film resistor is similar to the molded composition resistor in appearance and function. However, the thin film resistor is made by depositing a resistance material on a glass or ceramic tube. A photographic process is used to deposit this film. Leads with caps are fitted over each end of the tube to make the body of the resistor. Thin film resistors are usually color coded.

Film resistors are another type of resistor and are produced in two styles. One style is made with a very thin coat of carbon over the entire surface area of an insulator. The other style coats the outside of an insulator material with narrow spiral bands of metal film. The ends of the resistor have metallic caps that connect the metal film to small terminal leads. The advantage of this type of construction is the creation of very precise resistance values.

Chip resistors are made using a similar process as computer chips. They have an extremely thin carbon coating over an insulator material, usually in thin line patterns. These resistors are formed into rectangular shapes with leads made to fit into a printed circuit board, **Figure 3-28**.

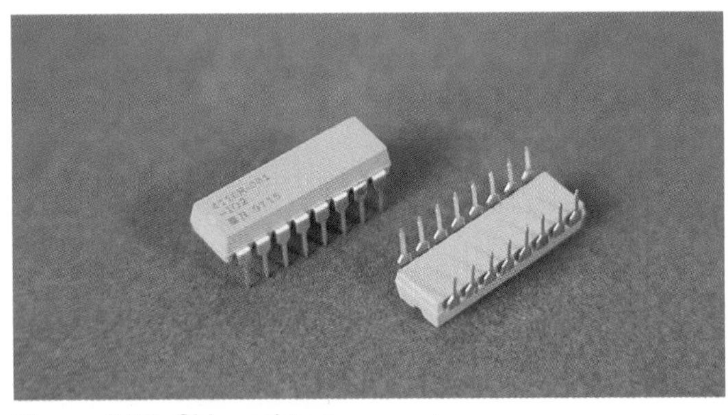

Figure 3-28. Chip resistors.

For higher current uses, resistors are wire wound. A thin wire is wound on a ceramic core. The wire has a specific fixed-value resistance. The entire component is insulated by a coat of vitreous (opaque) enamel. One of these resistors is shown in **Figure 3-29**. Wire wound resistors are commonly manufactured in sizes from 5 to 200 watts. The wattage chosen depends on the heat dissipation required during operation. Metal oxide resistors are also used for high voltage and wattage requirements.

Another type of wire wound resistor is the adjustable resistor. Unlike the standard wire wound resistor, the ***adjustable resistor*** is not entirely covered by enamel material. Instead, a portion of one side of the wire is exposed. An adjustable ***sliding tap*** is attached to move across the exposed surface. This allows the resistance value to be varied. Adjustable resistors may have two or more taps for providing various resistance values in the same circuit. An adjustable resistor and symbol are shown in **Figure 3-30**.

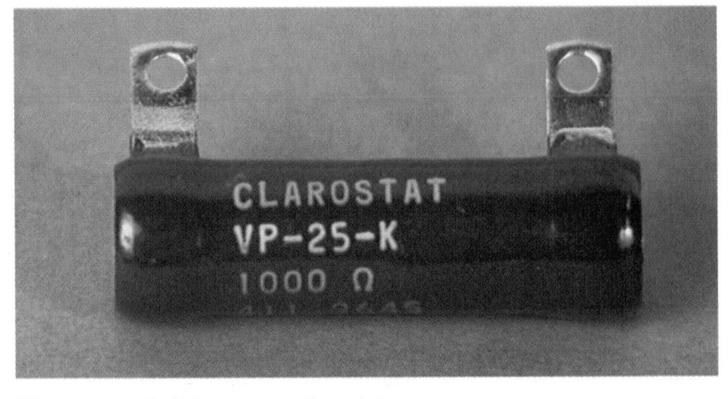

Figure 3-29. Wire wound resistor.

Potentiometers

Most electronic equipment requires the use of variable resistance parts. A ***potentiometer*** is a very common type of variable resistor found in electronic projects. The potentiometer has a rotary knob that varies the resistance value as it is turned. The variation in resistance is provided by a contact that is attached to a ring of resistive material inside the device. This device is similar to the wire wound resistor. Many potentiometers are constructed with thin wire inside as the source of resistance. Various styles of potentiometers are illustrated in **Figure 3-31**. Also shown is the potentiometer's schematic symbol.

Thermistors

A special type of resistor is called a thermistor. In comparison to other types of resistors, the ***thermistor*** is unusual due to its ability to change resistance value rapidly as its temperature changes. It is commonly used to prevent high inrush currents in electrical circuits.

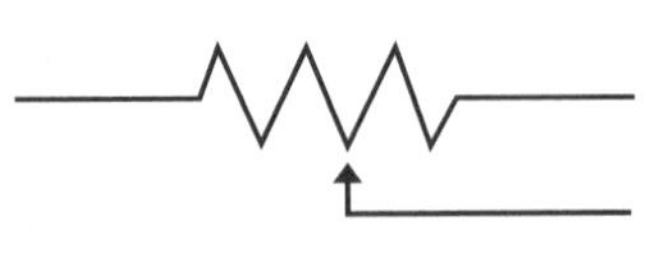

Figure 3-30. This adjustable resistor provides a sliding tap for voltage divider uses. On the right is the schematic symbol for an adjustable resistor.

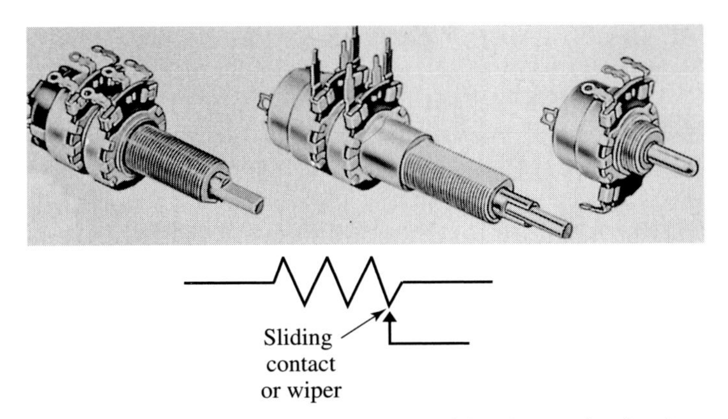

Figure 3-31. Potentiometers are used in electronic circuitry for fixed and variable resistance. Notice that the schematic symbol for the potentiometer is the same as for the adjustable resistor. (Centralab)

An example of a thermistor use can be seen in a blow dryer. A common blow dryer has heating elements composed of tungsten wire. The tungsten wire has a very low resistance value when cold, and a high resistance value when red hot. The thermistor is placed in series with the heating elements to prevent a high current value when the dryer is first turned on. As the blow dryer heats up, the resistance value goes down. The result is a fairly consistent current value as the dryer's heating element changes from low resistance (cold) to high resistance (hot). Early blow dryer models caused a dimming and flickering of lights and other electronic equipment in the home because inconsistent current draw. The thermistor eliminates this problem.

Resistor Color Code

Larger resistors are usually marked with their numerical resistance value printed directly on the side of the resistor. However, this type of labeling is not always practical, especially on small resistors. The resistor color code system was developed for this purpose. The color code marking system has been adopted by the Electronics Industries Association (EIA) and the United States Armed Forces. This system of color coding is recognized throughout the world. Refer to **Figure 3-32**. Note how the color codes are printed, or banded, around the entire body of the resistor. This method of coding permits the value of the resistor to be read regardless of the mounting position. To see how to read the color coded bands refer to **Figure 3-33**.

Resistors commonly have three or four (and sometimes five) bands. Each band has a unique meaning. The first band represents the value of the first digit of the resistance value. The second band represents the second digit of the resistance value. The third band is called the *multiplier*. The multiplier gives the factor of ten that the first two digits should be multiplied by. The fourth band

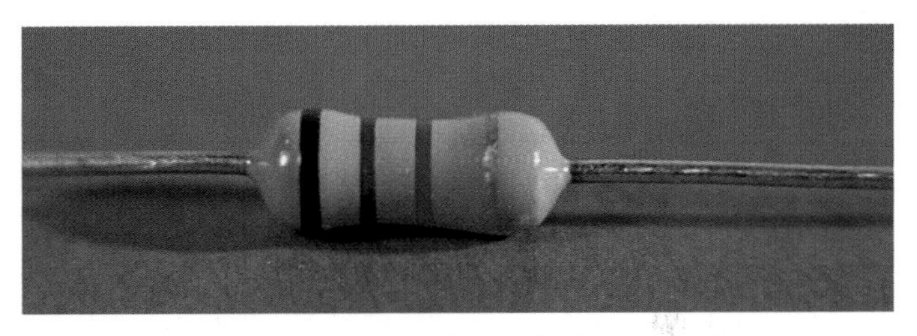

Figure 3-32. Color code bands encircle the resistor.

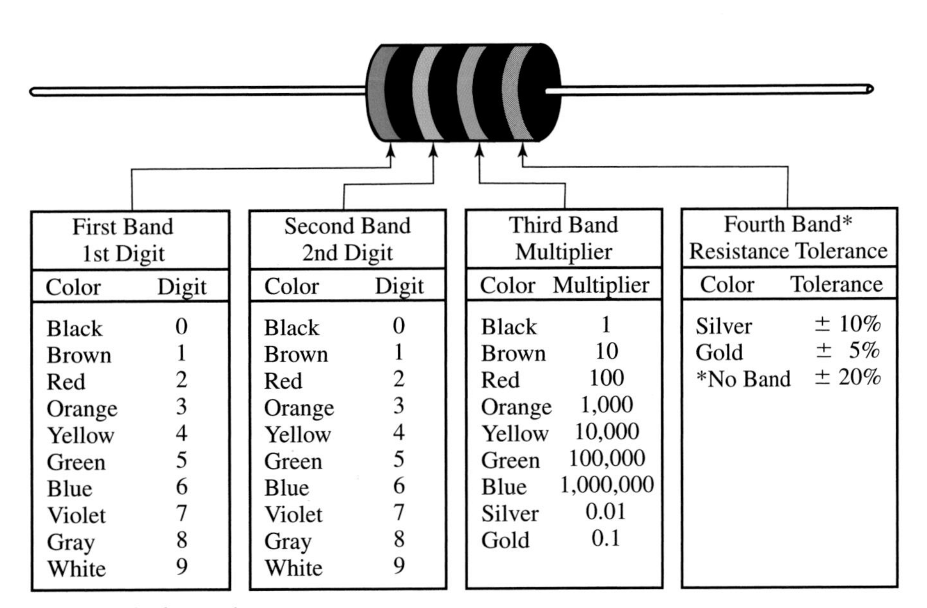

First Band 1st Digit	
Color	Digit
Black	0
Brown	1
Red	2
Orange	3
Yellow	4
Green	5
Blue	6
Violet	7
Gray	8
White	9

Second Band 2nd Digit	
Color	Digit
Black	0
Brown	1
Red	2
Orange	3
Yellow	4
Green	5
Blue	6
Violet	7
Gray	8
White	9

Third Band Multiplier	
Color	Multiplier
Black	1
Brown	10
Red	100
Orange	1,000
Yellow	10,000
Green	100,000
Blue	1,000,000
Silver	0.01
Gold	0.1

Fourth Band* Resistance Tolerance	
Color	Tolerance
Silver	± 10%
Gold	± 5%
*No Band	± 20%

Figure 3-33. Standard color code for resistors.

represents the tolerance of the resistor. Resistor ***tolerance*** is a reflection of the precision of resistor's value. If a 20 ohm resistor has a 10% tolerance, the resistor's value can vary by ± 2 ohms. In this case the resistor can have a true resistance value of 18 to 22 ohms. A fifth band is sometimes used to indicate resistor reliability or expected failure rate.

See **Figure 3-34**. In the first sketch, the first bar is red and the second bar is violet. Checking the color code chart shows the first two digits in the resistor's value to be 2 and 7 (or 27). This number is then multiplied by the third band. The third band is brown (× 10). This indicates that the value 27 is multiplied by 10, for a final value of 270 ohms. This is what the resistor value would be if the resistor was perfect. However, the fourth band is silver. This indicates that this resistor has a tolerance of 10 percent. In this example, the tolerance is a ±27 ohms (270 × 0.1 = 27). Thus the value of this resistor is somewhere between 243 ohms and 297 ohms.

Work out the stated value, maximum value, and minimum value for the other two sketches in Figure 3-34 on your own.

Review Questions for Section 3.5

1. Write the color codes for the following resistor values (first color, second color, multiplier, tolerance).
   a. 2400 Ω, 10%
   b. 680 Ω, 5%
   c. 91,000 Ω, 20%
   d. 27 Ω, 10%
   e. 3600 Ω, 5%
   f. 100 Ω, 10%
   g. 5.1 Ω, 5%
   h. 9.1 MΩ, 10%
2. Draw the symbols for the following.
   a. Fixed resistor.
   b. Tapped resistor.
   c. Potentiometer.

Summary

1. Conductors provide a low resistance path while insulators provide a high resistance path.
2. Four factors that affect the resistance of a conductor are length, cross-sectional area, temperature, and type of material.
3. A mil equals 1/1000 of an inch.
4. A circular mil equals the diameter (expressed in mils) squared, or D^2, for a round conductor.
5. Conductor insulation varies according to application.
6. Switches are used to control the flow of electrons through a circuit.
7. Switches are classified according to their actuator, internal electrical circuitry (poles and throws), ampacity, and voltage.
8. Electric lamps may use filaments or ionized gas as a conductor.
9. Resistors are manufactured in many different ways, and they come in fixed and adjustable packages.
10. Color codes are used to identify resistor resistance values.

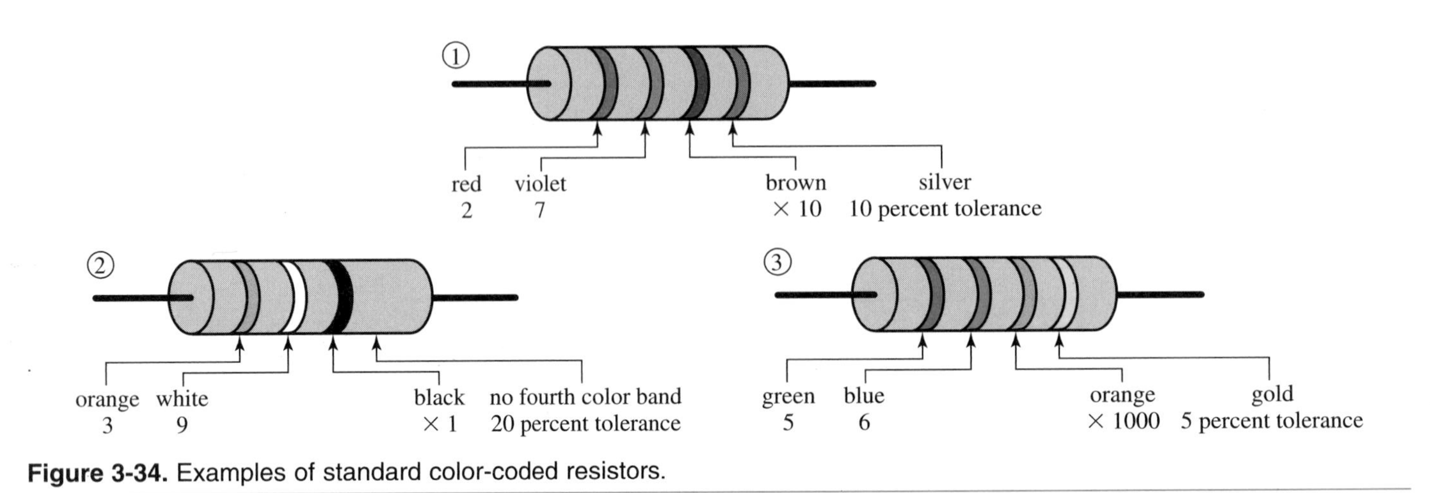

Figure 3-34. Examples of standard color-coded resistors.

Test Your Knowledge

Do not write in this text. Place your answers on a separate sheet of paper.

1. What is the standard unit for measuring conductor cross-sectional area?
2. A mil equals __________ inches.
3. A round conductor that is 1/8th inch in diameter has a cmil area of __________.
4. Which conductor has the larger cross-sectional area, a No. 12 or a No. 10 conductor?
5. Which conductor has more resistance per foot, a No. 22 or a No. 6 conductor?
6. A 20-amp lighting load is connected to a 120-volt source 300 feet away. How much voltage drop or loss is caused by using a No. 12 aluminum wire? (Remember that the circuit would be 600 feet long! Also, be sure to use the aluminum wire resistance column and not the copper column.)
7. You would most likely use a(n) __________ for assembling a circuit temporarily.
8. To mass produce printed circuits, a(n) __________ would be the most economical system.
9. To control a lamp from two different locations, you would use __________ switches.
10. To open two different circuits simultaneously, you would use __________ switches.
11. A fuse or circuit breaker provides protection from __________.
12. Fuses and circuit breakers are rated in __________ and __________.
13. The first electric lamp had a(n) __________ filament.
14. Name three lamps that conduct through gas rather than using a tungsten filament.
15. What keeps the current limited in a fluorescent lamp?
16. Name four different types of resistors.
17. What color code would be on a 1200 ohm, 2 watt resistor?
18. What color code would be used on a 1.2 ohm resistor?
19. What color code would be present on a 1.2 MΩ resistor?
20. Record the resistance values for the following resistors.

a. _______

b. _______

c. _______

d. _______

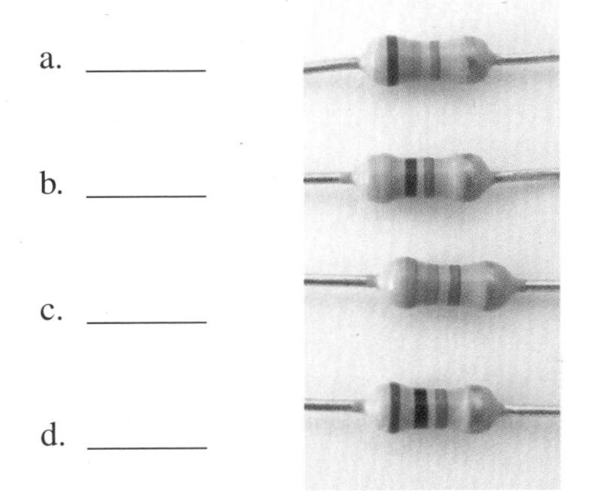

For Discussion

1. Other than an automobile, what other items could use the chassis as a conductor?
2. A good understanding of the relationship between conductors, insulators, and semiconductors is important in the study of electricity and electronics. Discuss the reasons why this is so.
3. What factors would influence the selection of a specific type of lighting for a particular job?

Impressive energy distribution systems stretch out across North America.

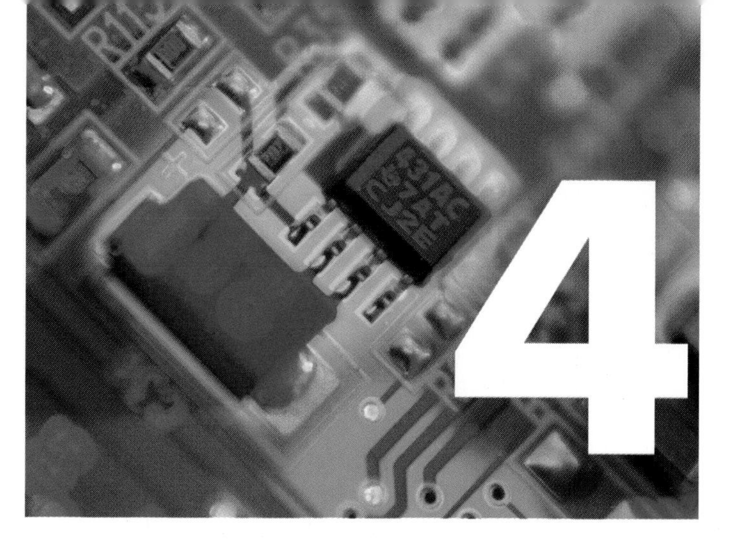

Energy

Objectives

After studying this chapter, you will be able to:

- Define work, power, and horsepower.
- Calculate electrical power in watts.
- Convert horsepower to watts.
- Combine Ohm's law with Watt's law to find unknown currents, voltages, resistances, and powers.
- Read a wattmeter.
- Determine efficiency.
- Determine gear and pulley ratios and power.
- State and explain Ohm's law.

Key Words and Terms

The following words and terms will become important pieces of your electricity and electronics vocabulary. Look for them as you read this chapter.

conservation of energy	power (*P*)
efficiency	Watt's law
electrodynamometer movement	watt (W)
energy efficiency rating (EER)	watt-hour meter
horsepower (hp)	wattmeter
kilowatt-hour (kWh)	work

An understanding of electrical power will enable you to select the proper size components needed to build or maintain a circuit. All electrical circuits convert electron flow into some form of energy such as heat, light, or mechanical energy (magnetism). Many industrial systems are custom designed. They are one-of-a-kind or prototypes. These systems must be laid out according to predetermined loads. Once the amount of power required is determined, we can easily calculate such factors as conductor and fuse size.

4.1 WORK, POWER, AND HORSEPOWER

Work is force applied over a distance. Work is expressed in units such as foot-pounds (ft.-lb.) or inch-pounds (in.-lb.). For example, if a 10 pound weight is lifted one foot, the work accomplished equals 10 foot-pounds. See **Figure 4-1.**

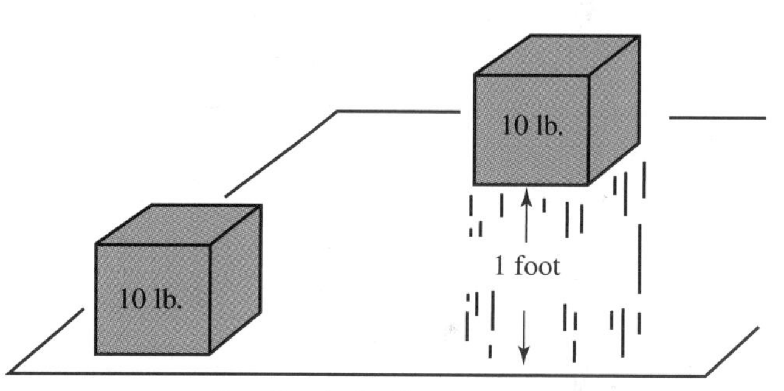

Figure 4-1. When a 10 pound weight is lifted one foot, 10 foot-pounds of work is done.

$$F \times D = \text{Work}$$

If F is the force in pounds and D is the distance in feet, the work is calculated in foot-pounds.

10 lb. × 1 ft. = 10 foot-pounds (ft.-lb.) of work

It is important to note that no reference is made to time when calculating work. It may take five seconds or 10 minutes to move the 10 pounds the one foot in the example.

Power (P) is the amount of work done based on a time period such as seconds or minutes. Power is the time rate of doing work. In the last example, 10 pounds was moved one foot with no time restrictions. If the same 10 pounds were moved one foot in 1/2 (0.5) second, the power expended would equal 20 foot-pounds per second. The formula for power is:

$$\text{Power} = \frac{\text{Work}}{\text{Time}} = \frac{\text{Force} \times \text{Distance}}{\text{Time}}$$

Horsepower is used for stating mechanical power. Electrical machines such as motors are rated in horsepower. One ***horsepower (hp)*** is defined as a work rate of 550 ft.-lb./second. Also, 33,000 ft.-lb./min. equals one horsepower.

In electricity, the unit of power is the ***watt (W).*** It was named in honor of James Watt, who is credited with the invention of the steam engine. When *one volt* of electrical pressure moves *one coulomb* of electricity in *one second,* the work accomplished is equal to *one watt* of power.

Recall the definition of the ampere: *the movement of one coulomb of electrons past a given point in a circuit in one second.* Thus, power in an electrical circuit can be written in terms of voltage and current. The formula for electrical power is:

$$P\text{ (watts)} = E\text{ (volts)} \times I\text{ (amperes)}$$

To convert electrical power, in watts, into mechanical power, in horsepower, you can use the following conversion factor.

746 watts = 1 horsepower

The power formula, sometimes called ***Watt's law,*** can be arranged algebraically. If two quantities are known, the third unknown can be found. Use the device in **Figure 4-2** to help memorize the following formulas.

$$P = I \times E \text{ or } I = \frac{P}{E} \text{ or } E = \frac{P}{I}$$

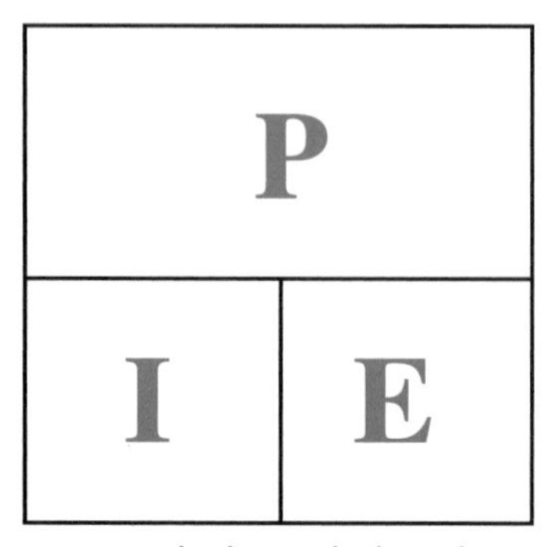

Figure 4-2. A memory device to help solve power problems. Cover the unknown quantity and the remaining letters give the correct equation. For example, if *E* is unknown, cover *E* and *P*/*I* remains. Thus, *E* = *P* divided by *I*.

Example: A circuit with an unknown load has an applied voltage of 100 volts. The measured current is 2 amperes. How much power is consumed?

$$P = I \times E$$
$$P = 2\text{ A} \times 100\text{ V}$$
$$P = 200\text{ W}$$

Example: An electric toaster, rated at 550 watts, is connected to a 117 volt source. How much current will this appliance use?

$$I = \frac{P}{E}$$
$$I = \frac{550\text{ W}}{117\text{ V}}$$
$$I = 4.7\text{ A}$$

LESSON IN SAFETY:

An electric current produces heat when it passes through a resistance. Heaters and resistors will remain hot for some time after the power is removed. Handle them carefully. Burns should be treated immediately. See your instructor.

Review Questions for Section 4.1

1. Define work.
2. Define power.
3. What is the symbol for power?
4. What is the unit of electrical power?
5. Define horsepower.
6. State Watt's law. Use it in a formula with a given voltage and current.
7. Compute the voltage and the power for the following figure.

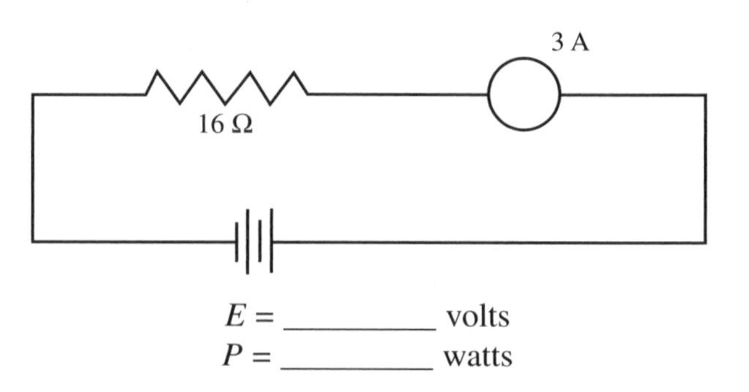

E = __________ volts
P = __________ watts

8. An electromechanical load is rated at 2238 watts. What is the equivalent rating in horsepower?

4.2 OHM'S LAW AND WATT'S LAW

It is possible to combine Ohm's law and Watt's law to produce simple formulas that permit you to solve for current, voltage, resistance, or power if any two of those quantities are known. In **Figure 4-3,** these equations are listed with an explanation of how they were created from the two laws.

These formulas can be arranged in a wheel-shaped memory device for ready reference. Refer to **Figure 4-4.** To use this device, find your *unknown* quantity in the smaller center circle. This is the left half of your equation. Next, choose one of the three variable combinations in outer circle that falls in the same quarter of the chart as

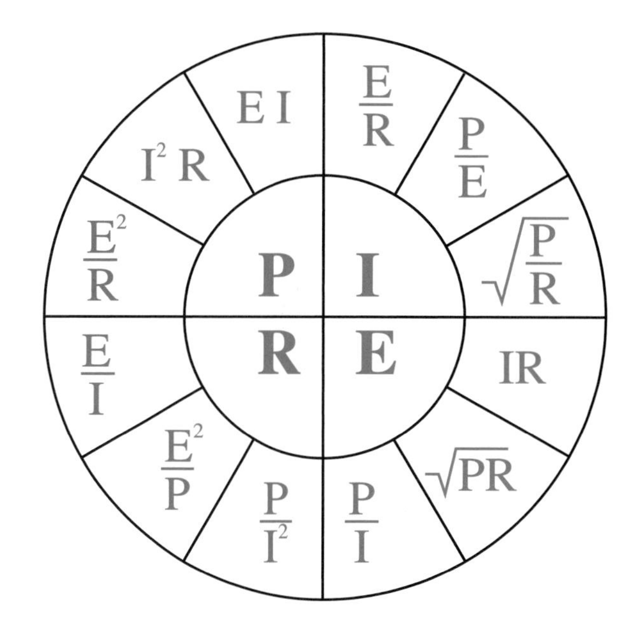

Figure 4-4. A memory device combining Ohm's law and the power formulas. It can be a great help when solving problems.

1. $E = I \times R$	Ohm's law
2. $E = \frac{P}{I}$	Watt's law
3. $E = \sqrt{PR}$	By transposing equation 12 and taking the square root.
4. $I = \frac{E}{R}$	Ohm's law
5. $I = \frac{P}{E}$	Watt's law
6. $I = \sqrt{\frac{P}{R}}$	By transposing equation 9 and taking the square root.
7. $R = \frac{E}{I}$	Ohm's law
8. $R = \frac{E^2}{P}$	By transposing equation 12.
9. $R = \frac{P}{I^2}$	By transposing equation 11.
10. $P = I \times E$	Watt's law
11. $P = I^2 \times R$	By substituting $I \times R$ from equation 1, for E.
12. $P = \frac{E^2}{R}$	By substituting $\frac{E}{R}$ from equation 4, for I.

Figure 4-3. This table states a variety of basic formulas that are needed to solve problems.

your unknown quantity. Choose the variable combination that uses two *known* quantities in your circuit. This becomes the right half of your equation.

For example, you are given a situation where current and resistance are known and power is unknown. Find the P in the upper left corner of the center circle. Then, find the variable combination that involves current and resistance (the middle combination of the three choices, I^2R). Thus your final equation is $P = I^2R$. Plug in your known values, and solve for the power.

Review Questions for Section 4.2

1. Give the correct missing letter to finish the formulas that follow:

$$E = \frac{P}{?} \quad I = \sqrt{\frac{P}{?}} \quad P = \frac{?}{R} \quad R = \frac{P}{?}$$

2. Solve for the unknown values.

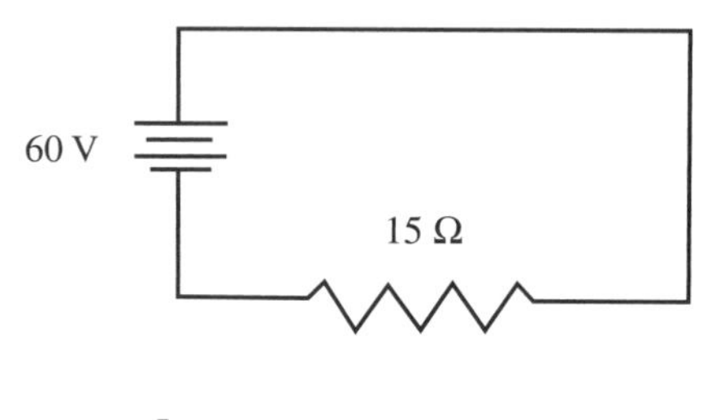

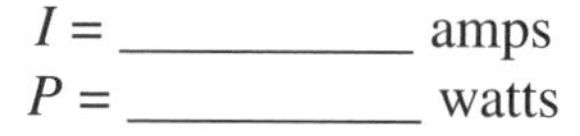

$I =$ __________ amps
$P =$ __________ watts

3. Solve for the unknown values.

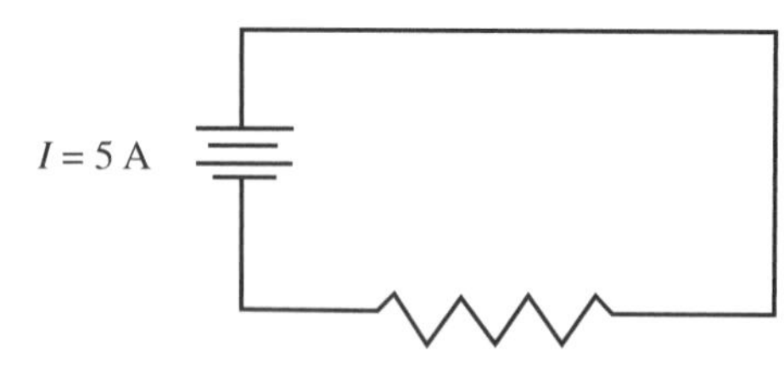

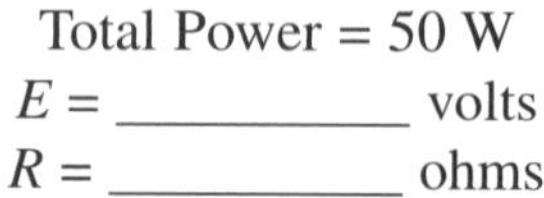

Total Power = 50 W
$E =$ __________ volts
$R =$ __________ ohms

4. A 1/4 watt resistor is connected to a six volt source. How much current is required to reach the maximum 1/4 watt rating of the resistor?
5. A 1/2 watt 1200 ohm resistor is connected to a power source. What is the maximum voltage that can be applied to match its power rating?
6. A circuit design specification states that all carbon composite resistors used shall not exceed 50 percent of their rated wattage. What is the maximum current a 1.2 kΩ, 1/4 watt carbon composite resistor can safely carry and meet the specifications?

4.3 WATTMETER AND WATT-HOURS

Power in an electric circuit is equal to the product of the voltage and the current. To devise a meter that measures watts, a movement similar to the D'Arsonval movement can be used. The permanent magnetic field found in the D'Arsonval movement, however, is replaced with coils from an electromagnet. This type of meter is referred to as an ***electrodynamometer movement.***

A circuit diagram of a simple electrodynamometer movement is shown in **Figure 4-5.** This arrangement creates a ***wattmeter,*** a meter that measures the instantaneous power. A moving coil, with the proper multiplier resistance, is connected across the voltage in the circuit. The coils of the electromagnet are connected in series with the circuit under measurement. The action between the two magnetic fields is proportional to the product of the voltage and the current. Deflection of the indicating needle is read on a calibrated scale in watts.

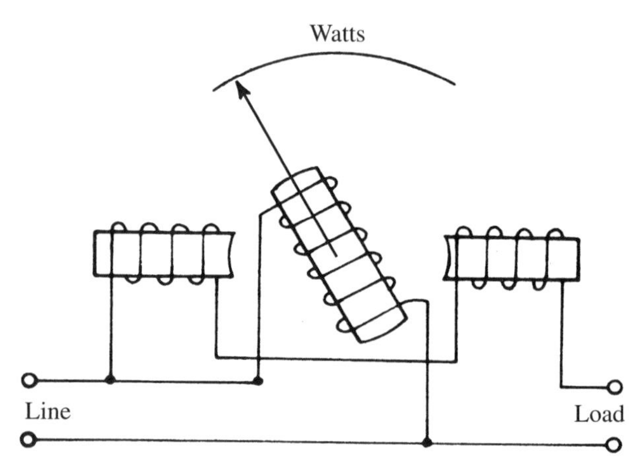

Figure 4-5. Diagram of a wattmeter.

While wattmeter measures the instantaneous power used in a circuit, a ***watt-hour meter*** measures the amount of power used in a given time. It is installed by a power company on the outside of a home or business. Since a watt-hour is a very small unit, standard utility meters read in ***kilowatt-hours (kWh),*** or 1000 watt-hours.

$$\text{number of kW} = \frac{E \times I}{1000}$$

Electric power consumed is purchased at current rates per kWh.

The watt-hour meter is a complicated type of induction motor. It uses field coils in series with the line current and also field coils connected across the line voltage. An aluminum disk rotates within these fields at a rate proportional to the power consumed. The disk is geared to an indicating dial. The dial shows the amount of power used.

To read the watt-hour meter, see **Figure 4-6.** Dial A, on the right, reads from 0–10 units of kilowatt hours. For each revolution of dial A, dial B advances one. For each revolution of dial B, dial C advances one. For each revolution of dial C, dial D advances one. Mathematically, dial A represents units of one; dial B represents units of ten; dial C represents units of one hundred; dial D represents units of one thousand.

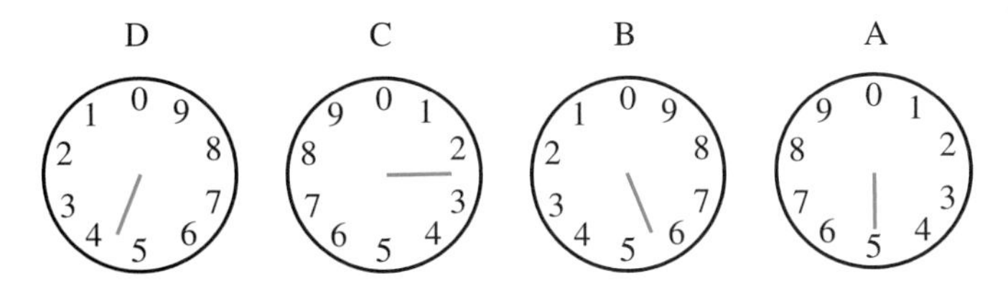

Figure 4-6. The kilowatt hours indicated equal 4255. When the dial indicator rests between numbers, the lower number is selected.

Be careful when reading the meter. Note that two of the dials read clockwise (A and C), and two of the dials read counterclockwise (B and D). This arrangement is because the gears for the dials are connected directly to each other.

Read the meter illustrated in Figure 4-6:

Dial A points to 5.
Dial B points between 5 and 6.
Dial C points between 2 and 3.
Dial D points between 4 and 5.

Therefore, the correct reading would be 4255 kWh. Notice that when the indicating arrow is between numbers, the *lower* number should be used. At times it is difficult to determine if the arrow has reached the number to which it is pointing or if the arrow has just *approached* the number. To make your determination, look to the dial immediately to the right of the dial in question. If the arrow on the dial to the right is between 9 and 0, then the number is still the lower value. If the arrow is past the 0, then the number is the larger value.

Use **Figure 4-7** as an example. The arrow in dial D appears to be pointing to the number 4. To determine if it a 4 value or still a 3, look at the dial immediately to its right, dial C. Dial C has not quite reached 0. It is still between the 9 and the 0. Thus, dial D is actually still a 3.

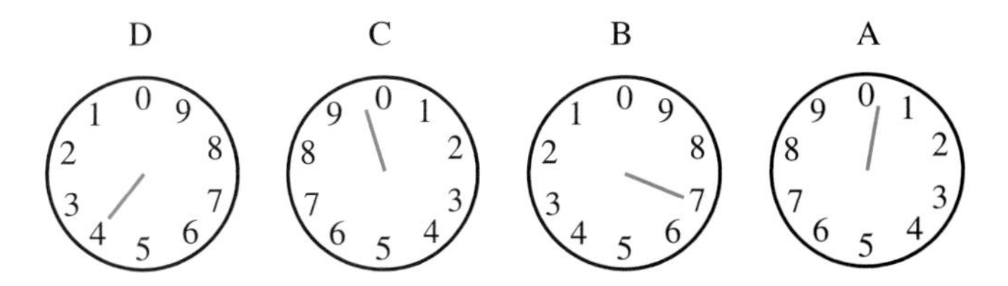

Figure 4-7. This watt hour meter reads 3970 kWh. To determine if dial D is really a four, you must look at dial C. Since dial C is between the nine and the zero, dial D is still three. It will become a four when the dial C indicator passes zero and rests between the zero and one. Dial B indicates seven. Dial A verifies that dial B is a seven and not still a six because dial A indicator is at rest between the zero and one.

To practice reading meters, why not read the meter at your home each day for several days and compute the power used. The power consumed is found by subtracting the previous reading from the present reading. The difference between the two readings is the number of kilowatt-hours of power used over the length of time between your two readings.

Review Questions for Section 4.3

1. A car radio draws 1.5 amps when connected to a 12 volt source. How much power is consumed by the radio?
2. What is the difference between a wattmeter and a watt-hour meter?
3. When the dial pointer rests directly over a number on a watt-hour meter, how do you determine if the dial has truly reached that number or if the value is still the lower number?
4. A 240 volt, 15 amp heater uses how many kilowatts of power?

4.4 EFFICIENCY

Efficiency can best be defined as a ratio of electrical power. This ratio is a comparison of the input power to the output power of the device. Efficiency can be expressed as a percentage.

$$\%\ \text{efficiency} = \frac{\text{Power out}}{\text{Power in}} \times 100$$

In the previous chapter, the efficiencies of electric lamps were determined by comparing their input powers (watts) to their output light (lumens). The more output in lumens per watt—the more efficient the lamp. Most electrical equipment compares the input (volts and amps) to the output power, which can be volts and amps, watts, or horsepower.

To determine the efficiency of a dc motor, the input power is compared to the output power. Look at **Figure 4-8.** A 12 volt, one horsepower dc motor is connected to a one horsepower mechanical load. Compare the input power to the output power. Multiplying the input voltage (12 volts) times the input amperage (90 amps) gives the input power rating of 1080 watts. The output power of the motor under a mechanical load equals one horsepower (746 watts). Dividing the output power, 746 watts, by the input power, 1080 watts, gives 0.69. Multiply by 100, and the efficiency is 69 percent.

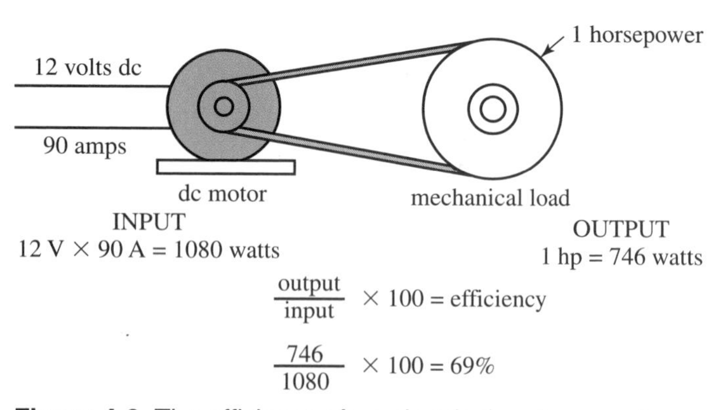

Figure 4-8. The efficiency of an electrical system can be determined by comparing the output power to the input power.

Energy Efficiency Rating

The federal government tests products to ensure that the ***energy efficiency rating,*** or ***EER,*** is correct. Electrical appliances are required to display their EER on the product label. EERs are expressed as a percentage. The higher the percentage, the more efficient the unit. The more efficient the unit, the cheaper it is to operate. Manufacturers of equipment such as heating, cooling, cooking, and other appliances have designed electrical systems for operation at peak efficiency.

Many times it is the technician's responsibility to determine the best equipment to purchase. When selecting equipment, it is important to consider cost efficiency of operation as well as the actual purchase price of the product. When choosing an electromechanical device such as an electric motor, lighting system, air-conditioning unit, or welder, consideration of the operational cost is critical. Determining the efficiency of the equipment can save the owner thousands of dollars over several years of use.

Gears, Pulleys, and Power

By using gears and pulleys in an electromechanical system, we can manipulate such factors as torque, work, and rpm while maintaining the same horsepower. **Figure 4-9** shows a motor with pulley A, which is connected to a pulley B. The ratio of the circumferences of these two pulleys is three to one. This means that pulley A must make three complete revolutions for pulley B to complete one revolution. This also means that the two pulleys have a three to one speed ratio. Once this ratio is known, it is easy to find the speed and work ratio between the two pulleys.

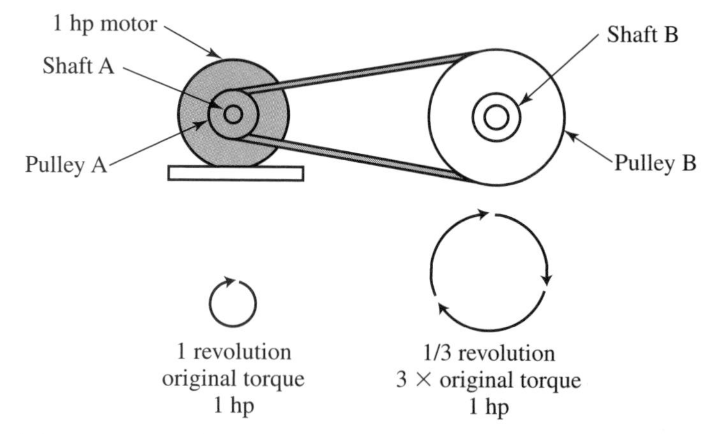

Figure 4-9. Pulley A and pulley B are at a 3 to 1 ratio. Thus, pulley B is rotating at 1/3 the speed of pulley A, and pulley B has three times the torque of pulley A. The horsepower is equal.

The study of physics includes the law of conservation of energy. The law of ***conservation of energy*** states that *energy can neither be created nor destroyed.* Energy can only be converted from one form to another. In the pulley example, the energy (power) of pulley A is equal to that of pulley B (with the exception of a small energy loss due to friction).

Look carefully at the ratio of the pulleys in the example. While the speed of pulley B is decreased, its torque (twisting force) is increased. Using this pulley setup, it is possible to move great weights long distances with a relatively small motor. However, the time period it takes to move the weight is increased.

Look at **Figure 4-10,** the speed of pulley C is 1/3 the speed of pulley B and 1/9th the speed of pulley A. The speed in this example has been reduced nine times from pulley A to pulley C, but the torque has been increased nine times. This nine fold increase in torque is how it is possible to move extremely heavy loads with a relatively

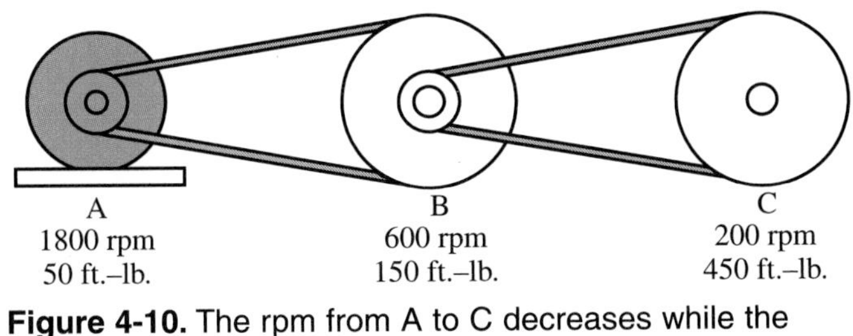

Figure 4-10. The rpm from A to C decreases while the torque increases. The horsepower remains the same.

small motor. Note the fact that the horsepower will be the same throughout the mechanical system even though the amount of work has changed.

To avoid the safety problems associated with pulleys and chains, many applications use gear reducers to obtain great twisting forces. The gear reducers lower the rpm of the motor which in turn increases the twisting force of the gear reducer shaft. See **Figure 4-11.** Gear reducers with a ratio of 50 to 1 or higher are common.

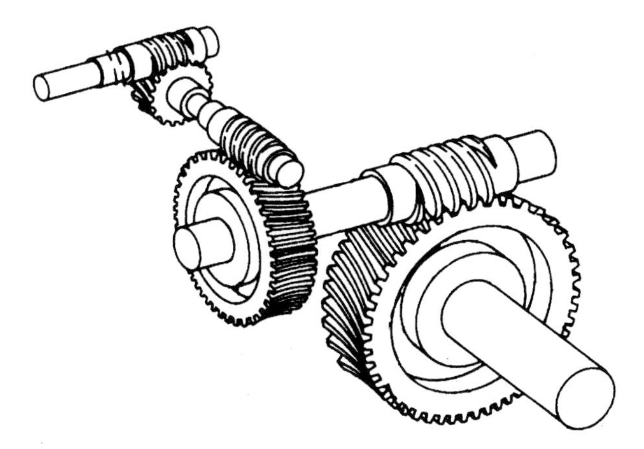

Figure 4-11. An example of an internal arrangement of a double-reduction, helical-worm gear reducer unit. There are many different configurations of gear reducers.

Review Questions for Section 4.4

1. Efficiency is a comparison of what two factors?
2. What is the highest possible efficiency rating of any appliance?
3. A fully loaded 230 volt, 7 horsepower motor draws a current equal to 24 amps. What is the efficiency of the motor?
4. What is an EER?
5. Describe the relationship of speed and torque in a pulley or gear system.

Summary

1. Work is equal to force times distance.
2. Power is the time rate of doing work.
3. Mechanical power is measured in horsepower.
4. Electrical power is measured in watts.
5. One horsepower is equal to 746 watts.
6. Watt's law is a mathematical formula stating the relationship of the power, voltage, and current in a circuit. It is $P = I \times E$.
7. There are definite relationships between Ohm's law and Watt's law, as shown in Figures 4-3 and 4-4.
8. A wattmeter measures the instantaneous power used in a circuit, and a watt-hour meter measures the amount of power used in a circuit over a period of time.
9. Efficiency is expressed as a percentage and indicates how well one form of energy is converted to another form of energy. Efficiency is equal to the power out divided by the power in.
10. The amount of work done can be manipulated by gears and pulley systems while maintaining the same horsepower.
11. The law of conservation states that energy can neither be created nor destroyed. It can only be converted from one form to another.

Test Your Knowledge

Please do not write in the text. Place your answers on a separate sheet of paper.

1. If you move 10 pounds a distance of 20 feet, you will accomplish __________ of work.
2. What is the difference between power and work?
3. One horsepower is equal to __________ ft.-lb./min. or __________ watts.
4. A certain two horsepower electric motor is 50% efficient. What wattage is needed to operate the motor?
5. The input power and the output power of a pulley system is approximately __________.
6. A machine that works at 550 ft.-lb./second is equivalent to __________ horsepower.
7. Which of the following is Watt's law?
   a. $P = I \times E$
   b. $I = P/E$
   c. $E = P/I$
   d. All of the above.

8. Use Figures 4-3 and 4-4 to solve the following problems.
   a. $P = 10$ W, $I = 2$ A,
      $E =$ ______________________________.
   b. $E = 100$ V, $I = 0.5$ A,
      $P =$ ______________________________.
   c. $P = 500$ W, $E = 250$ V,
      $I =$ ______________________________.
   d. $P = 100$ W, $I = 2$ A,
      $R =$ ______________________________.
   e. $E = 10$ V, $P = 10$ W,
      $R =$ ______________________________.
   f. $E = 100$ V, $R = 1000\ \Omega$,
      $P =$ ______________________________.
   g. $I = 0.5$ A, $R = 50\ \Omega$,
      $P =$ ______________________________.
   h. $I = 4$ A, $R = 10\ \Omega$,
      $P =$ ______________________________.
   i. $I = 10$ mA, $E = 50$ V,
      $P =$ ______________________________.
   j. $P = 10$ W, $I = 1$ A,
      $R =$ ______________________________.
   k. $I = 100$ mA, $R = 100\ \Omega$,
      $P =$ ______________________________.
   l. $P = 500$ W, $E = 100$ V,
      $I =$ ______________________________.
   m. $E = 100$ V, $R = 100\ \Omega$
      $P =$ ______________________________.
   n. $P = 10$ W, $R = 10\ \text{k}\Omega$,
      $E =$ ______________________________.
   o. $P = 50$ W, $R = 2\ \Omega$,
      $I =$ ______________________________.

For Discussion

1. State Ohm's law and Watt's law in your own words. What is the relationship between these two laws?
2. What are some of the ways you can conserve energy (and save money) in your own home?
3. Which would cost more and why? Moving 100 pounds 1000 feet in one minute or moving 100 pounds 1000 feet in 10 minutes?

5 Sources of Electricity

Objectives

After studying this chapter, you will be able to:

- List the six basic sources of electricity.
- Explain the chemical action that creates electricity in various types of cells.
- Define polarization.
- Explain the differences between primary cells and secondary cells.
- Distinguish between series and parallel connections in batteries.
- Calculate the outputs of batteries connected in series and parallel.
- Demonstrate proper use of a hydrometer and explain its use.
- Calculate the theoretical capacity of a battery.

Key Words and Terms

The following words and terms will become important pieces of your electricity and electronics vocabulary. Look for them as you read this chapter.

ampere-hours (Ah)	local action
anode	magnetohydrodynamic (MHD)
battery	photoresistive cell
capacity	photovoltaic cell
cathode	piezoelectric effect
dry cell	primary cell
electrolyte	secondary cell
fuel cell	specific gravity
galvanometer	thermocouple
generator	voltaic cell

Every student knows the story of Benjamin Franklin and his kite. For centuries before Franklin, scientists and philosophers had observed lightning. It was through the experimentation and research of Franklin that the relationship between lightning and static electricity was confirmed. What is electricity and where does it come from? Years before the discovery of the electron theory by J. J. Thomson, it was suggested by Franklin that electricity consisted of many tiny particles, or electric charges. He further theorized that electrical charges were created by the distribution of electrical particles in nature.

We have learned that a potential difference or electromotive force is created when electrons are redistributed. A body might assume a charge; its polarity is determined by the deficiency or excess of electrons. People have turned their scientific interests and research to the development of machines and processes that cause an electrical imbalance and an electrical pressure.

There are six basic sources of electricity or electromotive force. They are friction, chemical action, light, heat, pressure, and magnetism.

In this chapter, we will discuss in detail producing electricity from chemical action or batteries. You will also learn how electricity is produced using light, solar batteries, pressure, and heat.

5.1 CHEMICAL ACTION

One of the more familiar sources of an electrical potential or voltage is the battery. In 1790, the Italian scientist, Luigi Galvani, observed a strange phenomena during the dissection of a frog supported on copper wires. Each time he touched the frog with his steel scalpel, its leg would twitch. Galvani reasoned that the frog's leg contained electricity.

Alessandro Volta, another Italian scientist, invented the electric cell, named in his honor, called the ***voltaic cell.*** The unit of electrical pressure, the volt, is also named in his honor. Volta discovered that when two dissimilar elements were placed in a chemical that acted upon them, an electrical potential was built up between them. Thus, electricity can be produced by chemical action.

The student can construct several voltaic cells to demonstrate this action. Cut a one inch square of blotting paper and soak it in a strong salt solution. Place the wet paper between a penny and a nickel as shown in **Figure 5-1.** If a sensitive meter is connected to the coins, it will indicate that a small voltage is present.

In **Figure 5-2** electricity is created using a grapefruit. Make small cuts in the skin of a grapefruit. In one cut, place a penny. In the other cut, place a nickel. Once again, a meter will indicate that a small voltage is present. A better cell can be made by placing a carbon rod (these may be removed from an old dry cell) and a strip of zinc in a glass jar containing an acid and water solution, **Figure 5-3.** *Follow all safety precautions when performing experiments.*

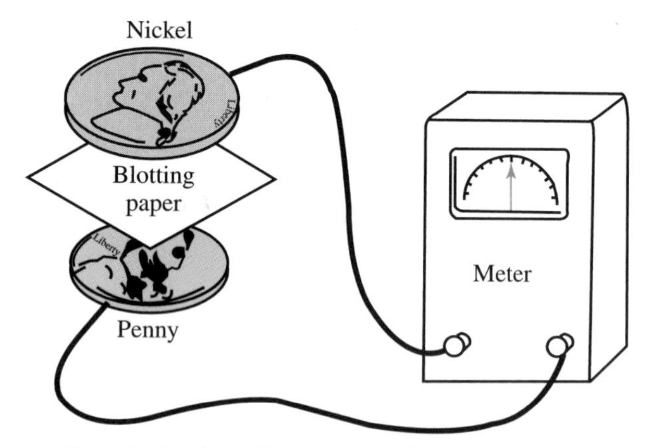

Figure 5-1. A simple cell is produced using a nickel, a penny, and a salt solution.

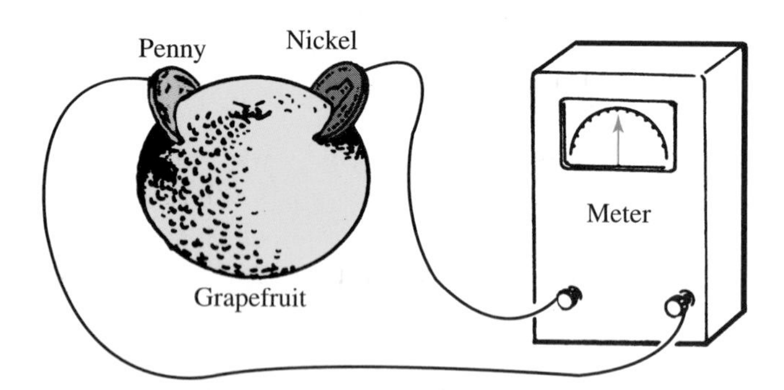

Figure 5-2. A grapefruit can be used to produce enough electricity to operate a small radio.

LESSON IN SAFETY:

When mixing acid and water, always pour acid into water. Never pour water into acid. Acid will burn your hands and your clothing. Wash your hands at once with clean water if you spill acid on them. Acid may be neutralized with baking soda. See your instructor for first aid!

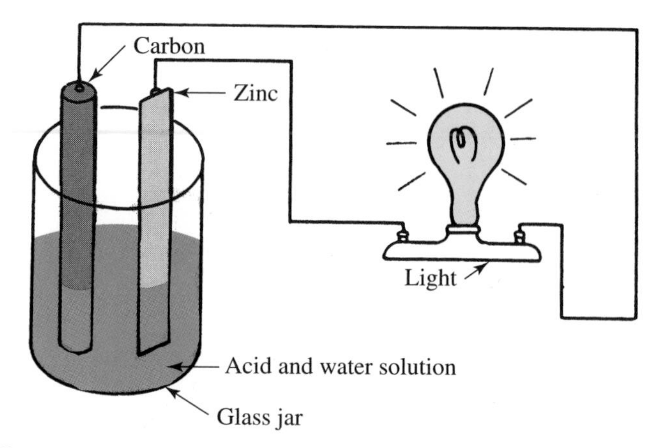

Figure 5-3. This experimental cell, made with zinc, carbon, and acid, produces enough electricity to power the light.

When the polarity of the carbon rod is tested, it will be positive. The zinc strip is negative. If a wire is connected between these elements, or electrodes, a current will flow. A *voltaic cell* can be described as a way of converting chemical energy into electrical energy.

In the zinc-carbon example of a voltaic cell, the sulfuric acid (H_2SO_4) and water (H_2O) solution is also known as an ***electrolyte.*** When the electrodes are placed in this acid electrolyte, a chemical action takes place. The sulfuric acid breaks down into positive ions (2H+) and negative ions (SO_4^{2-}). The negative ions move toward the zinc electrode, and combine with it by making zinc sulfate ($ZnSO_4$). The positive ions move toward the carbon electrode. This action creates a potential difference between the electrodes. The zinc will be negative. The carbon will be positive. This cell will develop about 1.5 volts.

If a load, such as a light, is connected to the cell, a current flows and the light glows, as seen in Figure 5-3. As the cell is used, the chemical action continues until the zinc electrode is consumed. The chemical equation for this action is:

$$Zn + H_2SO_4 + H_2O \rightarrow ZnSO_4 + H_2O + H_2 \uparrow$$

Zinc plus sulfuric acid plus water chemically reacts to form zinc sulfate and water and free hydrogen gas. This cell cannot be recharged because the zinc has been consumed.

Primary Cells

The zinc-carbon cell just described is what is known as a primary cell. A ***primary cell*** is a cell in which the chemical action cannot be reversed. A primary cell cannot be recharged.

Defects in the primary cell

One might think that the chemical action of the zinc-carbon primary cell would continue to produce a voltage as long as the active ingredients of the cell were present. In studying the equation for the discharge of the cell, you will observe the formation of free hydrogen gas. Since the carbon electrode does not enter into chemical action, the hydrogen forms gas bubbles. These collect around the carbon electrode. As the cell continues to discharge, an insulating blanket of bubbles forms around the carbon. This reduces the output and terminal voltage of the cell. The cell is said to be ***polarized.*** The action is called ***polarization.***

To overcome this defect in the simple voltaic cell, a ***depolarizing agent*** can be added. Compounds that are rich in oxygen, such as manganese dioxide (MnO_2), are used for this purpose. The oxygen in the depolarizer combines with the hydrogen bubbles and forms water. This chemical action appears as:

$$2MnO_2 + H_2 \rightarrow Mn_2O_3 + H_2O$$

The free hydrogen has been removed, so the cell will continue to produce a voltage.

One might assume that when current is not being used from the cell, the chemical action would also stop. However, this is not true. During the smelting of zinc ore, not all impurities are removed. Small particles of carbon, iron, and other elements remain. These impurities act as the positive electrode for many small cells within the one large cell. This chemical action adds nothing to the electrical energy produced at the cell terminals. This action is called ***local action.*** It can be reduced by using pure zinc for the negative electrode, or by a process called amalgamation. With ***amalgamation*** a small quantity of mercury is added to the zinc during manufacturing. As mercury is a heavy liquid, any impurities in the zinc will float on the surface of the mercury, causing them to leave the zinc surface. This process increases the life of a primary cell.

Types of Primary Cells

There are many different primary cells. What follows are details on some of most common primary cells you might encounter.

Zinc-carbon cell

Although the primary cell has been described as a liquid cell, the liquid type is not in common use. Rather, the primary cell is often a dry cell. In a ***dry cell,*** the electrolyte is in a paste form as opposed to a liquid form. A dry cell averts the danger of spilling liquid acids.

Flashlight batteries (cells) are examples of dry cells. The dry cell consists of a zinc container that acts as the negative electrode. A carbon rod in the center is the positive electrode. Surrounding the rod is a paste made of ground carbon, manganese dioxide, and sal ammoniac (ammonium chloride), mixed with water. The depolarizer is the MnO_2. The ground carbon increases the effectiveness of the cell by reducing its internal resistance. During discharge of the cell, water is formed.

You may recall having difficulty removing dead cells from a flashlight. This is because the water produced caused the cells to expand. Although this problem has been solved by improved manufacturing techniques, it is still not advisable to leave cells in your flashlight for long periods of time. You should keep fresh cells in your flashlight, so it will be ready for emergency use.

LESSON IN SAFETY:

Improper battery use can cause leakage and explosion. Therefore, obey the following precautions.

1. **Install the batteries with the positive (+) and negative (–) polarities in the proper direction.**
2. **Do not use new and old batteries together.**
3. **Never attempt to short circuit, disassemble, or heat batteries. Do not throw batteries into a fire.**
4. **Batteries contain dangerous materials that should be recycled or disposed of properly. Contact your local recycling facility or fire department for more information.**

Alkaline cell

The alkaline battery uses manganese dioxide for the positive activating substance. Zinc powder is used as the negative activating substance. A caustic alkali is used for the electrolyte. Recent progress in electronic product design has demanded more compact supply sources. The number of products needing a large current and a long battery life have increased. This required the development of more advanced batteries. Cylindrical alkaline batteries are now widely used to supply power for electronic products. They can be used with common manganese dioxide batteries, **Figures 5-4** and **5-5.**

Mercury cell

A relatively new type of dry cell is shown in **Figure 5-6.** It is called a mercury cell. It creates a voltage of 1.34 volts from the chemical action between zinc (–) and mercuric oxide (+). It is costly to make. However, the mercury cell is better in that it creates about five times more current than the conventional dry cell. It also maintains its terminal voltage under load for longer periods of operation. The mercury cell has found wide use in

Figure 5-4. AA size alkaline cells.

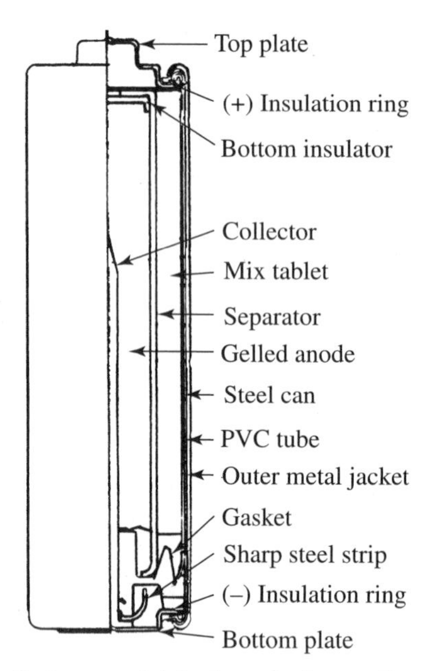

Figure 5-5. Cutaway of AA size alkaline cell. (Panasonic Battery Sales Division)

powering field instruments and portable communications systems.

Lithium cell

Lithium has the highest negative potential of all metals. It is, therefore, the best substance for an anode. Many battery makeups are possible by mixing lithium with various cathode substances. Energy densities of these batteries can be computed by *respective reaction equations*. **Figure 5-7** shows the energy densities of lithium batteries compared with those of conventional batteries. Lithium is the most suitable anode for production of high voltage and lightweight batteries. Refer to **Figure 5-8.**

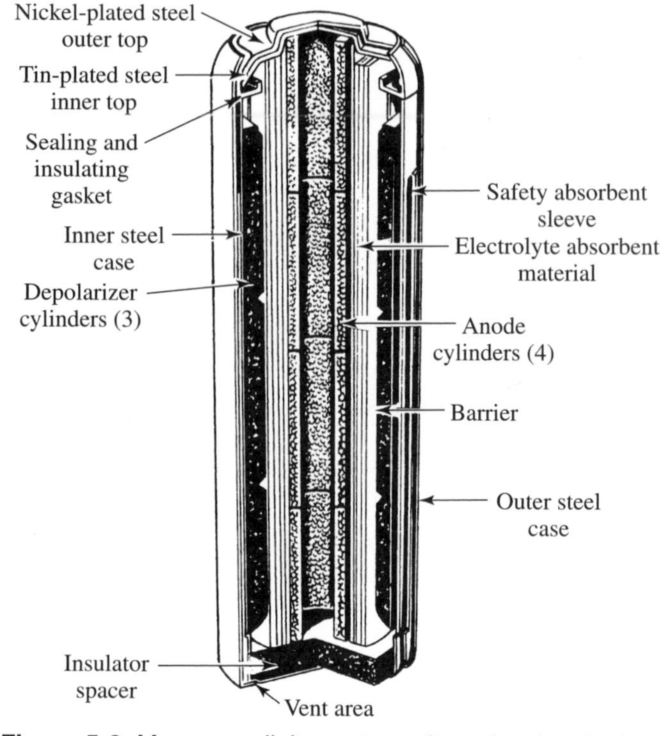

Figure 5-6. Mercury cell. It creates voltage by chemical action between zinc and mercuric oxide.

Features of lithium batteries, such as voltage and discharge capacity, are determined by the type of cathode substance used. Fluorocarbon is an ***intercalation*** (inserted between or among existing elements) compound. It is produced through reaction of carbon powder and fluorine gas. It is expressed in (CF)n.

Silver oxide cell

Silver oxide cells have several advantages over other types of cells. These advantages include:

- Very stable discharge voltage.
- Excellent high discharge characteristics.
- High energy density per unit volume.
- Wide range of operating temperatures.
- Compact, thin size.

Compact silver oxide batteries have the highest electrical volume and leakage resistance of any battery of that size. They are commonly used in watches. Two types of silver oxide batteries are made for use in watches. One type uses caustic potash for electrolyte. The other uses caustic soda. The caustic potash battery has the symbol W on the bottom of the battery. It is for high drain use, where more power is needed. It is used in wristwatches with liquid crystal displays and multifunction analog watches. The caustic soda battery has the symbol SW on the bottom of the battery. It is for low drain use. It is used mostly in single function analog watches. **Figure 5-9** shows a cutaway of a silver oxide cell.

Reaction	E	(Wh/kg)	Battery type
$nLi + (CF)n \longrightarrow nLiF + nC$	3.2	2,260	Polycarbonmonofluoride lithium battery
$8Li + 3SOCl_2 \longrightarrow 6LiCl + Li_2SO_3 + 2S$	3.61	1,877	Thionyl chloride lithium battery
$2Li + CuF_2 \longrightarrow 2LiF + Cu$	3.54	1,646	———
$2Li + NiF_2 \longrightarrow 2LiF + Ni$	2.83	1,370	———
$2Li + 2SO_2 \longrightarrow Li_2S_2O_4$	2.95	1,114	Sulfur dioxide lithium battery
$2Li + 2MnO_2 \longrightarrow Li_2O + Mn_2O_3$	2.69	768	Manganese dioxide lithium battery
$2Li + Ag_2CrO_4 \longrightarrow Li_2CrO_4 + 2Ag$	3.35	520	Silver chromate lithium battery
Conventional batteries			
$Zn + 2MnO_2 \longrightarrow Zn^{2-} + 2MnOOh$	1.7	234	Carbon zinc battery
$Zn + HgO \longrightarrow Zn(OH)_4^{2-} + Hg$	1.4	266	Mercury battery
$Zn + Ag_2O \longrightarrow ZnO + 2Ag$	1.6	287	Silver oxide battery

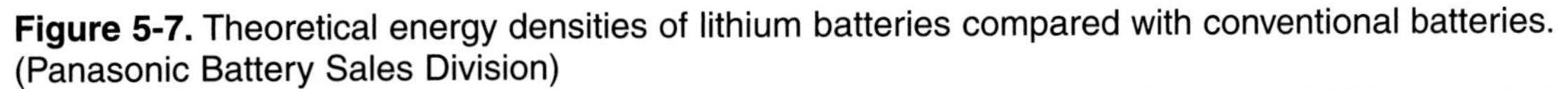
Figure 5-7. Theoretical energy densities of lithium batteries compared with conventional batteries. (Panasonic Battery Sales Division)

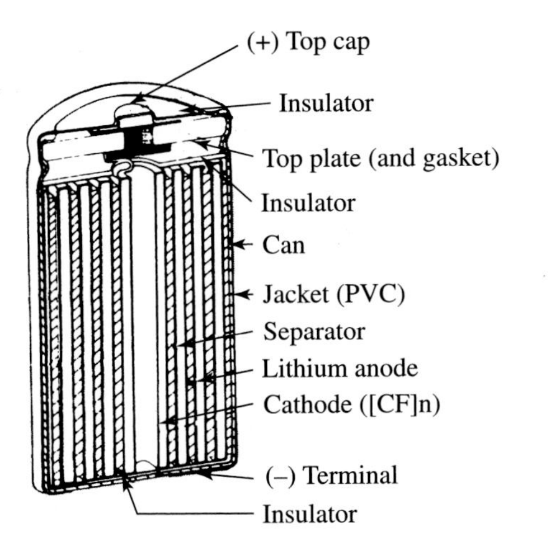

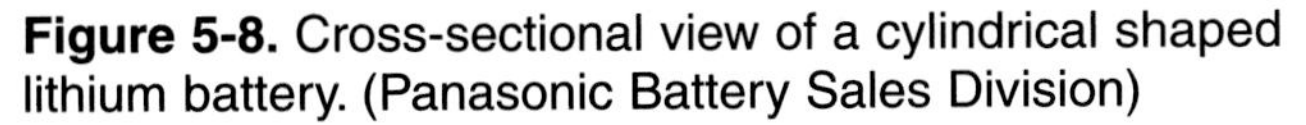
Figure 5-8. Cross-sectional view of a cylindrical shaped lithium battery. (Panasonic Battery Sales Division)

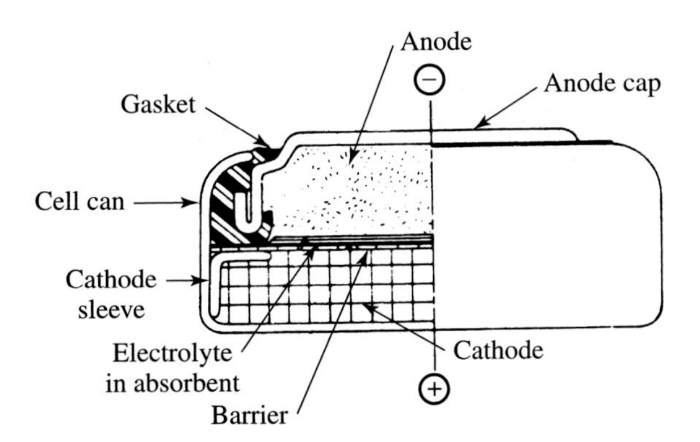

Figure 5-9. Cutaway view of a silver oxide cell. (Panasonic Battery Sales Division)

Secondary Cells

A ***secondary cell*** can be recharged or restored. The chemical reaction that occurs on discharge may be reversed by forcing a current through the battery in the opposite direction. This charging current must be supplied from another source, which can be a generator or a power supply. **Figure 5-10** shows one type of battery charger used for recharging automobile and motorcycle batteries. An alternating current, which will be studied in a later chapter, must be *rectified* to a direct current for charging the battery.

Figure 5-10. This type charger is called a trickle charger. It slowly brings a battery back to full charge.

Lead acid cell

A common type of lead acid cell is the car storage battery. A storage battery does not store electricity. Rather, it stores chemical energy, which in turn produces electrical energy. The active ingredients in a fully charged battery are lead peroxide (PbO_2), which acts as the positive plate, and pure spongy lead (Pb) for the negative plate. The liquid electrolyte is sulfuric acid (H_2SO_4) and water (H_2O). The positive plates are a reddish-brown color. Negative plates are gray.

The chemical reaction is rather involved. However, study the information given in **Figure 5-11.** Notice that

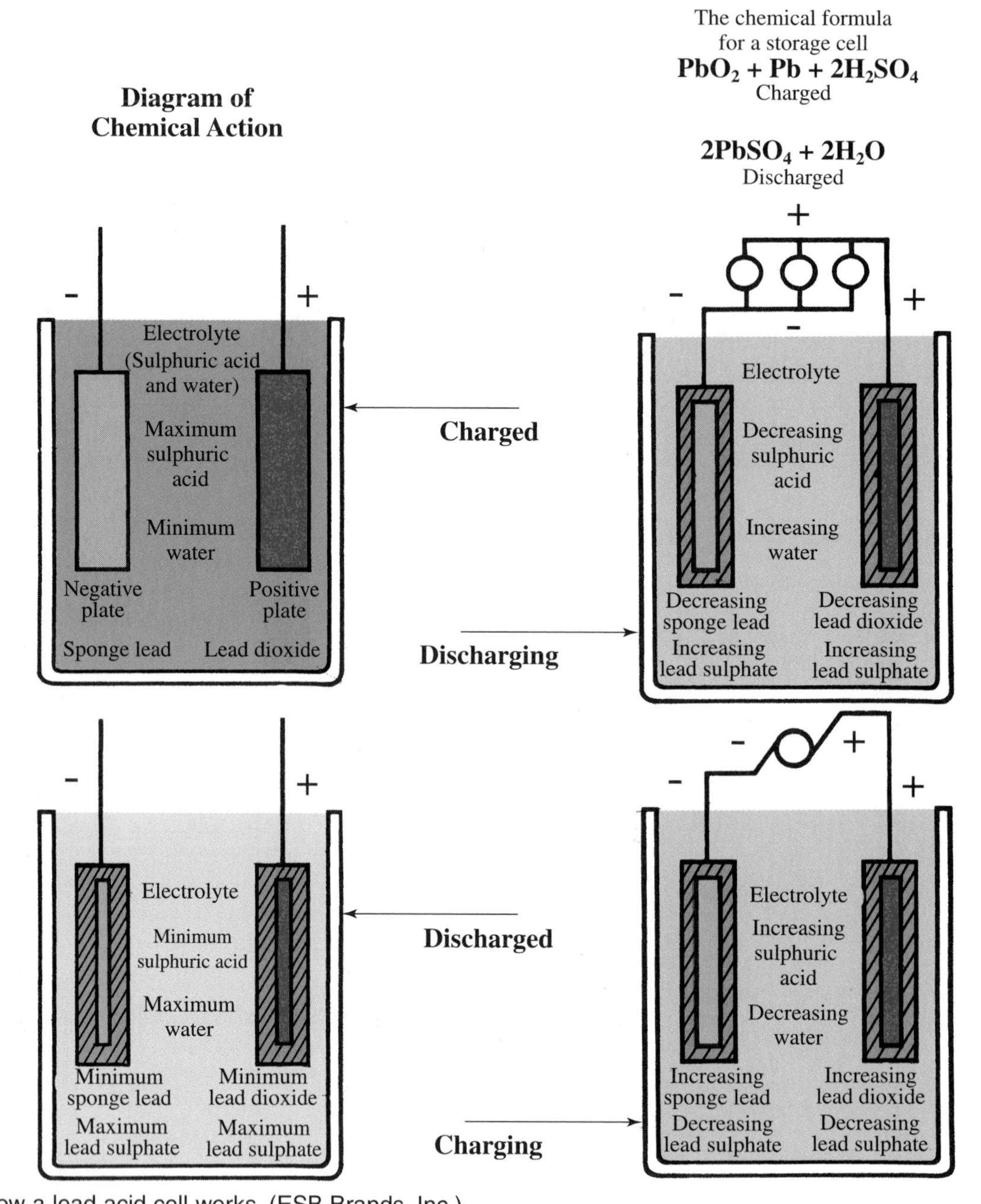

Figure 5-11. How a lead acid cell works. (ESB Brands, Inc.)

during discharge, both the spongy lead and the lead peroxide (also called lead dioxide) plates are being changed to lead sulfate, and the electrolyte is being changed to water. When the cell is recharged, the reverse action occurs. The lead sulfate changes back to spongy lead and lead peroxide; the electrolyte to sulfuric acid.

The electrolyte of a fully charged battery is a solution of sulfuric acid and water. The weight of pure sulfuric acid is 1.835 times heavier than water. This is called its specific gravity.

LESSON IN SAFETY:

During the charging process of a storage battery, highly explosive hydrogen gas may be present. Do not smoke or light matches near charging batteries. Charge only in a well ventilated room. Batteries should be first connected to the charger before the power is applied. Otherwise, the sparks made during connection might ignite the hydrogen gas and cause an explosion.

Specific gravity is the weight of a liquid as it compares to water. The specific gravity of water is 1.000. The acid and water mixture in a fully charged battery has a specific gravity of approximately 1.300 or less. As the electrolyte changes to water when the cell discharges, the specific gravity becomes approximately 1.100 to 1.150. Therefore, the specific gravity of the electrolyte can be used to determine the state of charge of a cell.

The instrument used to measure the specific gravity is a ***hydrometer.*** The principle of the hydrometer is based on Archimedes principle in physics. This principle states that *a floating body will displace an amount of liquid equal to its own weight.* If the cell is in a fully charged state, the electrolyte liquid is heavier, so the float in the hydrometer will not sink as far. The distance that the float does sink, is calibrated in specific gravity on the scale. This can be read as the state of charge of the cell.

CAUTION:

Expensive storage batteries may be destroyed by excessive vibration and rough handling. Chemicals may break off from the plates and cause internal short circuits and dead cells. Handle a battery gently and be sure it is securely clamped and bolted in your car.

In the 12-volt automotive battery, six lead acid cells are placed in a molded hard rubber case. Each cell has its own compartment. At the bottom of each compartment a space, or sediment chamber, is provided. This is where particles of chemicals broken from the plates due to chemical action or vibration can collect. Otherwise, these particles would short out the plates and make a dead cell. The individual cells are connected in series by lead alloy connectors.

CAUTION:

Automotive batteries contain large amounts of lead. Consequently, they should never be disposed of in landfills. Stores that sell automotive batteries are required by law to accept old batteries for recycling.

Nickel-cadmium cell

The nickel-cadmium cell is a rechargeable dry cell. Basically, these are nickel-cadmium alkaline batteries with paste rather than liquid for the electrolyte. The ability to be recharged is just one of their advantages. Other advantages include long life, high efficiency, compactness, and lightweight. The nickel-cadmium cell produces a high discharge current due to its low internal resistance. Other uses include the powering of small radios, burglar alarm systems, camera flashes, and aircraft instruments.

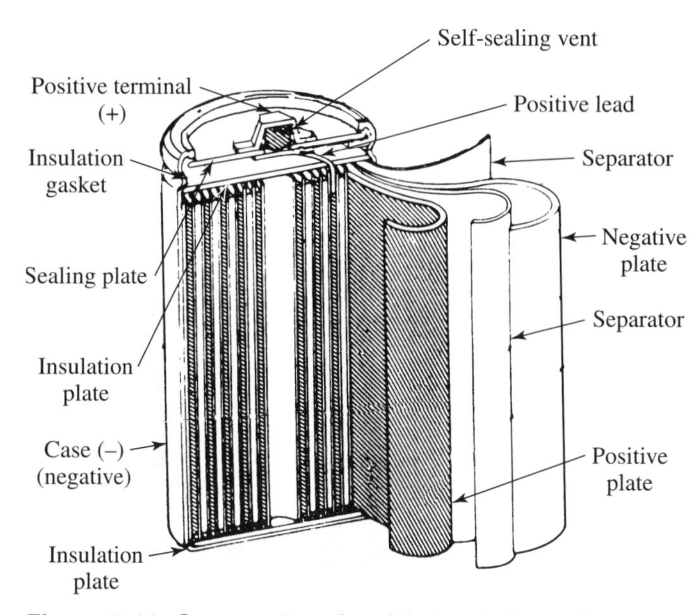

Figure 5-12. Construction of a nickel-cadmium cell. (Panasonic Battery Sales Division)

One type of nickel-cadmium cell uses positive and negative plates, a separator, alkaline electrolyte, a metal case, and a sealing plate with self-resealing safety vent. It is shown in **Figure 5-12.** The positive plate of this battery is a porous, powdered nickel base plate. It is filled with nickel hydroxide. The negative plate is a punched plate of thin steel, coated with cadmium active material. The separator is made of a polyamide fiber. For high temperature uses, it is made of a nonwoven polypropylene fiber. The positive plate, separator, and negative plate are pressed together, wound into a coil, and inserted in the metal case.

The electrolyte is an alkaline aqueous solution. It is totally absorbed into the plate and separator. The metal case is constructed of nickel-plated steel. It is welded on the inside to the negative plate. It becomes the negative pole. The sealing plate uses a special liquid sealing agent to form a perfect seal. The positive plate is welded on the inside to the sealing plate. It becomes the positive pole. The self-resealing safety vent permits the discharge of gas in the event of an abnormal increase of internal pressure. This prevents against rupture or other damage. The vent is made of a special alkaline and oxidation resistant rubber. This ensures that operating pressure and safety features will be retained over a long period of time.

The electromechanical processes of a nickel-cadmium alkaline cell are outlined below.

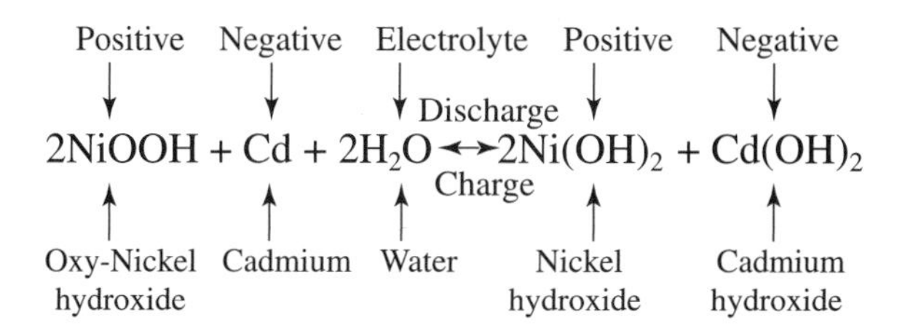

In this process, charging and discharging are reversed in a very efficient manner. The electrical energy used during discharge is regained during recharge. During the final charging stage, an oxygen gas is created with the reaction occurring at the positive.

$$\text{Positive} \quad 4OH^- \rightarrow O_2\uparrow + 2H_2O + 4e^- \quad ...(1)$$

(Hydroxide ions → $4OH^-$; Oxygen → O_2; Water → $2H_2O$; Electrons → $4e^-$)

This oxygen passes through the separator to the negative. After this, an absorption reaction takes place at the negative and absorption occurs.

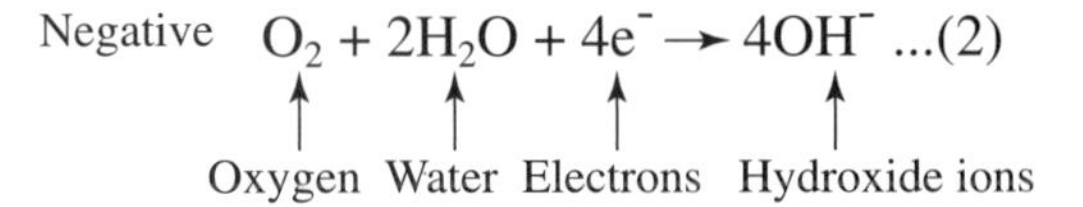

Batteries in Series and Parallel

Often, a single cell is called a battery. By strict definition, however, a ***battery*** consists of two or more cells connected together. These cells are enclosed in one case.

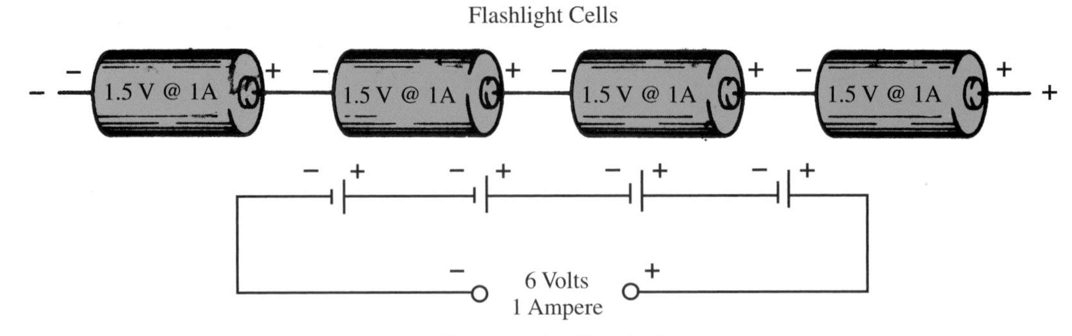

Figure 5-13. Pictorial and schematic diagram of four cells connected in series. Voltages add when connected in series.

EXPERIMENT 5-1: Voltage in Series

In this experiment you will demonstrate and observe how voltages add in series.

Materials

4–flashlight cells (D-type)
4–cell holders
1–multimeter

1. Connect the four cells in series. See the schematic in **Exhibit 5-1A.**

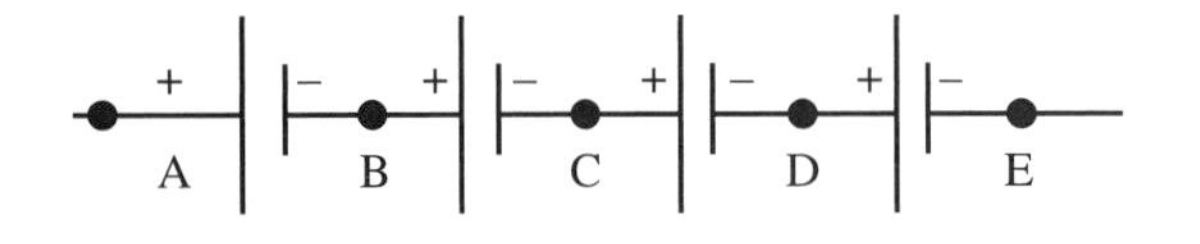

Exhibit 5-1A.

2. Set your multimeter to read voltages in the 0 V to 10 V range.
3. Measure and write down the voltage between points A and B.
4. Measure and write down the voltage between points A and C.
5. Measure and write down the voltage between points A and D.
6. Measure and write down the voltage between points A and E.

Do your results follow the formula for voltage sources in series? What conclusion can you draw from this experiment?

Series connection

In the study of electricity, it is important to understand the purpose and results of connecting cells in groups. First, consider the *series connection.* In this method, the positive terminal of one cell is connected to the negative terminal of the second cell. In **Figure 5-13,** four cells are connected in series. The output voltage will equal:

$$E_{out} = E_{one\ cell} \times N$$

where N equals the number of the cells. So in the case of the batteries shown,

$$E_{out} = 1.5\ \text{V} \times 4 = 6\ \text{V}$$

Notice that the voltage has increased four times. However, the capacity of the battery to supply a current is the same as one cell. Cells are connected in this manner to supply higher voltages for many uses. A flashlight may use two or more cells in series. Batteries for portable equipment have many cells in series to produce 6, 9, or 24 volts. The amperage does not increase by connecting cells in series.

Parallel connection

In **Figure 5-14,** the positive terminals have been connected together, and the negative terminals have been connected together. These cells are connected in *parallel.* The total voltage across the terminals of the battery is the same as one cell only. Although the voltage has not increased, the life of the battery has been increased because the current is drawn from all cells instead of just one. The amperage is added by connecting cells in parallel.

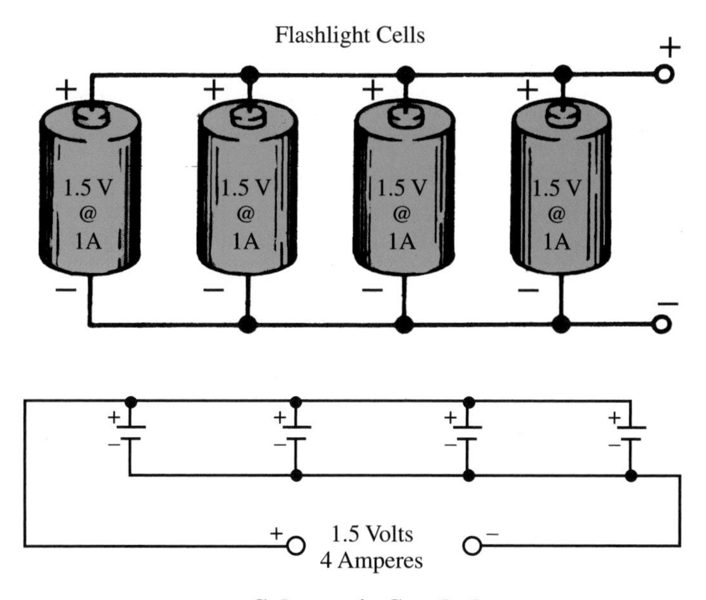

Figure 5-14. Pictorial and schematic diagram of four cells connected in parallel. This connection only provides 1.5 volts, but with 4 amps of current.

EXPERIMENT 5-2: Voltages in Parallel

In this experiment you will demonstrate and observe how voltages add in parallel.

Materials

4–flashlight cells (D-type)
4–cell holders
1–multimeter

1. Connect the four cells in parallel. See the schematic in **Exhibit 5-2A.**

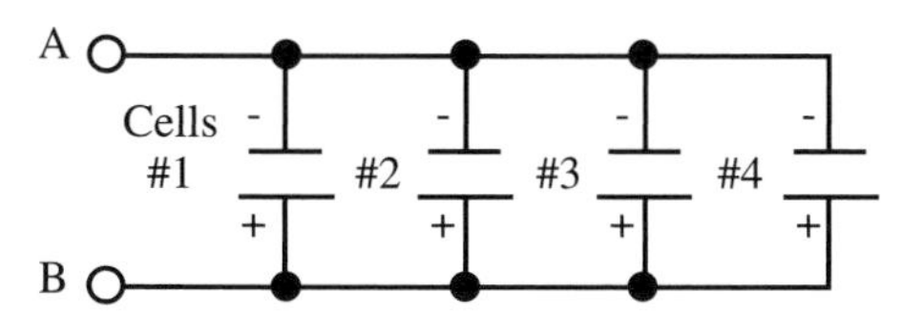

Exhibit 5-2A.

2. Set your multimeter to read voltages in the 0 V to 10 V range.
3. Measure and write down the voltage between points A and B.
4. Remove cell #4. Measure and write down the voltage between points A and B.
5. Remove cell #3 (do not replace cell #4). Measure and write down the voltage between points A and B.
6. Remove cell #2 (do not replace cells #3 or #4). Measure and write down the voltage between points A and B.

Do your results follow the formula for voltage sources in parallel? What conclusion can you draw from this experiment?

Sound Navigation Ranging (Sonar)

Sonar is an underwater detection system based on horizontally directed sound waves and their echo reception. Sonar operates on a principle similar to that of ***radar,*** but, instead of electromagnetic waves, it uses sound waves.

There are two main types of sonar, *passive* and *active.* Passive sonar listens for sound given off by possible targets and can determine the direction in which an object is located, but not its distance. Submarines generally use passive sonar systems because they do not give off any sound that another sonar might detect. Active sonar sends sound waves through the water and receives echoes from any object struck by the outgoing waves.

Most surface ships must use active sonar because noise from the ship itself interferes with passive sonar. Recall the familiar *ping* of the sonar in movies featuring submarines. This ping is an electrically-produced sound wave transmitted by a sonar ***transducer***. The transducer uses a ***crystal*** and pulses of high-frequency voltage to produce the vibration that is transmitted through the water. When the crystal is subjected to pulses of high frequency voltage, the crystal will distort at the same rate as the frequency. It is the crystal's distortion that produces the vibration.

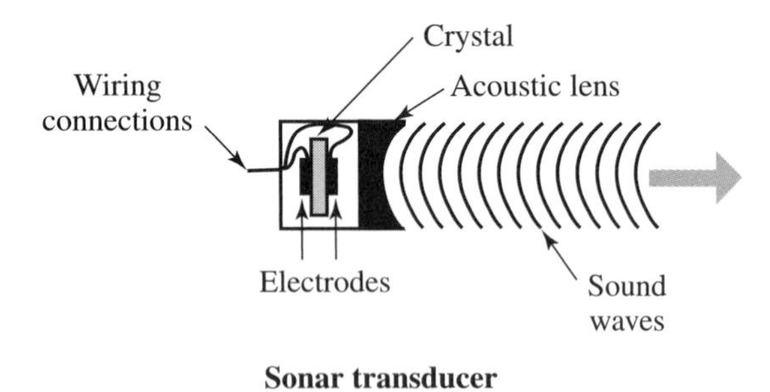

Sonar transducer

The transducer is lowered through the bottom of the ship's hull and sound waves are sent out through the water at regular intervals in bursts. If the outgoing waves strike a target, an echo is returned to the sonar receiver. The transducer rotates in complete circles, allowing the sonar operator to determine the direction from which an echo comes by noting the transducer's position when the echo is received.

Since the speed of sound through water is known, the range or distance from the sonar-equipped vessel to the target can be determined by measuring the time interval between the ping, or outgoing sound wave, and the return of its echo.

Advances in underwater sound transmission and increased understanding of natural sea noises have helped to make the use of sonar more effective in everything from submarine warfare to mapping the ocean floors. New transducers and electronic scanning and searchlight techniques have made possible an extensive underwater communication system to assist in fleet operations and in the guidance of submarines. Sonar has proved to be more effective than radar in warning against the presence of icebergs and smaller fragments of ice dangerous to navigation. Fishermen are able to detect schools of fish and observe the shape and pattern of the nets and the movements of fish entering or avoiding the net. Sonar can also be used to find sunken vessels and aircraft.

In the medical field, sonar technology is used for various ultrasonic tests. The images produced by transmitting and receiving sound echoes in the human body are ***sonograms***. A sonogram does not provide the high quality picture of the body as some other imaging techniques, but it is very safe for the patient. Ultrasonic imaging is cost effective, and the images produced can be enhanced through the use of a computer.

Series-parallel connection

Cells can also be connected in a mixed grouping. In **Figure 5-15,** two groups of batteries, each with a six-volt terminal voltage, are connected in parallel. The total voltage is still six volts. But the capacity has been increased by this *series-parallel* method of connecting cells. The total current is added, giving two amperes.

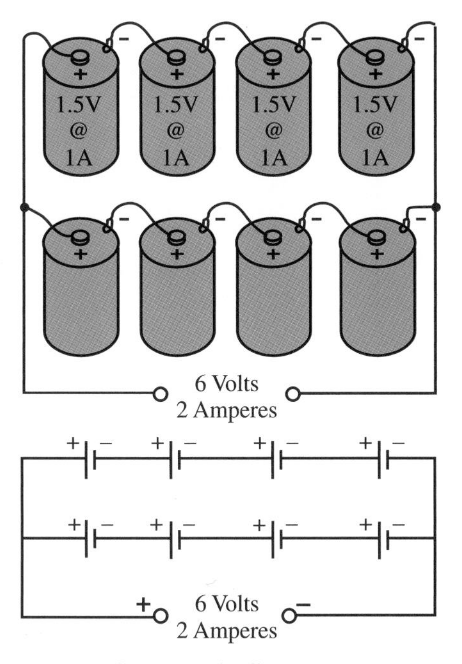

Figure 5-15. Pictorial and schematic diagram of two groups of cells connected in series and those two groups connected in parallel.

Battery Capacity

It is important that you understand the term capacity as it relates to batteries. The ***capacity*** of a battery is its ability to produce a current over a certain period of time. It is equal to the product of the amperes supplied by the battery and the time. Capacity is measured in ***ampere-hours (Ah).*** The description of an automotive battery might indicate a capacity of 100 ampere-hours. This would mean that the battery could supply:

100 amps for	1 hour	$100 \times 1 = 100$ Ah
50 amps for	2 hours	$50 \times 2 = 100$ Ah
10 amps for	10 hours	$10 \times 10 = 100$ Ah
1 amp for	100 hours	$1 \times 100 = 100$ Ah

A battery will not perform exactly by this schedule, as the battery's *rate of discharge* must always be considered. A rapidly discharged battery will not give its maximum ampere-hour rating. A slowly discharged battery may exceed its rated capacity. The Society of Automotive Engineers (SAE) has set standards for the rating of automotive batteries. A manufacturer must meet these standards in order to advertise a battery as a specific ampere-hour capacity.

Several factors determine the capacity of a storage battery:

- The number of plates in each cell. An increased number of plates provides more square inches of surface area for chemical action. Automotive batteries are commonly made with 13, 15, and 17 plates per cell. The number of plates is a determination of the life and quality of a battery.
- The kind of separators used has an effect on the capacity and life of a battery.
- The general condition of the battery with respect to its state of charge, age, and care will influence the capacity rating of any given battery.

To compare primary and secondary cells and batteries, refer to **Figure 5-16.** Note that their most common uses are given.

Review Questions for Section 5.1

1. Explain how a primary cell operates.
2. Name four types of primary cells.
3. Name two types of rechargeable secondary cells.
4. What are the advantages of a silver oxide cell?
5. Connecting cells in series increases their __________ rating.
6. Connecting cells in parallel increases their __________ rating.
7. Give some typical uses for zinc carbon and alkaline batteries.

5.2 OTHER SOURCES OF ELECTRICAL ENERGY

Batteries are a very common source of electrical energy, however, there are many other sources of electrical power. Devices that convert energy in the forms of light, heat, and mechanical pressure are found everywhere. If you have used a solar powered calculator or a crystal microphone, you have seen these conversions in action.

Electrical Energy from Light

Thanks to the United States' space program, we can convert the sun's light directly into electricity. The

Category	Type	Configuration			(V)					Storage life	Number of times it can be recharged	Typical applications:
		Positive-activating substance	Electrolyte	Negative-activating substance	Nominal voltage	Strong current	Voltage stability	Low temper-ature	High temper-ature			
Primary batteries	Carbon Zinc batteries	MnO_2	$ZnCl_2$ NH_4Cl	Zn	1.5	C B during intermi-tence	C	B	B	A	Non-rechargeable	Lightning equipment, radios, tape recorders, electronic calculators, hearing aids, toys, TVs
	Alkaline-manganese dioxide batteries — Cylinder	MnO_2	KOH(ZnO)	Zn	1.5	B	B	A	A	A	Non-rechargeable	Electric shavers, camera strobes, tape recorders, electronic calculators
	Alkaline-manganese dioxide batteries — Button									AA	Non-rechargeable	Clocks, watches, toys, electronic calculators, cameras, lighters
	Mercury batteries	HgO	KOH(ZnO)	Zn	1.35	B	AA	B	A	A	Non-rechargeable	Hearing aids, electronic calculators, exposure meters, EE cameras, wireless microphones, measuring instruments
		HgO	NaOH(ZnO)	Zn	1.35	C	AA	C	A	AA	Non-rechargeable	Watches
		HgO+ MnO_2	KOH(ZnO)	Zn	1.4	B	A	A	A	A	Non-rechargeable	Medical equipment, radios, cameras, electronic shutters, hearing aids, pagers
	Silver oxide batteries	Ag_2O	KOH(ZnO)	Zn	1.55	B	AA	B	A	A	Non-rechargeable	Cameras, watches, electronic calculators, lighters, hearing aids, radios, thermometers, remote control units
		Ag_2O	NaOH(ZnO)	Zn	1.55	C	AA	C	A	AA	Non-rechargeable	Watches
	Lithium batteries	(CF)n	Organic tetrachloride solvent	Li	3.0	C~B	A	AAA	AA	AAA	Non-rechargeable	Memory backup, power for cameras, illuminated buoys, watches, com-munication equipment, water and gas meters, computers
		MnO_2	Organic tetrachloride solvent	Li	3.0	C~B	B	A	A	AA	Non-rechargeable	Electronic calculators, memory backup, power sources, cameras, watches, toys
Secondary batteries	Sealed nickel-cadmium (Ni-Cd) batteries	NiOOH	KOH	Cd	1.2	AA	A	AA	A	B	300~1000	Lighting equipment, electric shavers, electric tools, electronic calcula-tors, ECRs, emergency lights, guide lights
	Sealed lead-acid batteries	PbO_2	H_2SO_4	Pb	2.0	A	A	A	A	A	100~300	VCRs, TVs, measuring instruments, ECRs, emergency equipment

Figure 5-16. Comparison of primary and secondary batteries and their uses.

invention of the *photovoltaic cell,* also called solar cell and photocell, made this conversion possible. The first photovoltaic cell was made from selenium, but today crystalline silicon is used. Photovoltaic cells made from crystalline silicon have a much higher efficiency than the original selenium cells.

A ***photovoltaic cell,*** **Figure 5-17,** is constructed of two thin layers of crystalline silicon each injected with impurities to form a negative and a positive semiconductor material. When secured together and then exposed to light, an electrical potential is developed. The layers are connected to thin wires that allow the photovoltaic cell to be connected to an electrical circuit. See **Figure 5-18.** While a typical photovoltaic cell produces approximately 1 watt and 0.5 volts, the cells can be connected into arrays. Arrays consist of many cells connected in series and parallel to increase the voltage and current capabilities to a level sufficient to power lights and equipment.

One application of power supplied by photovoltaic cells is power for a residential home. As a demonstration

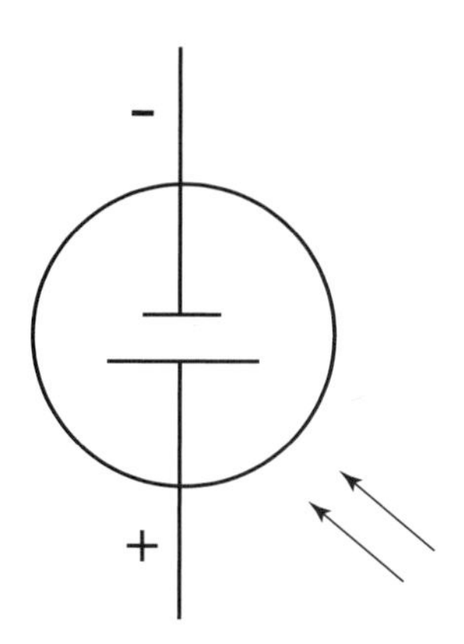

Figure 5-17. Schematic symbol for a photovoltaic cell.

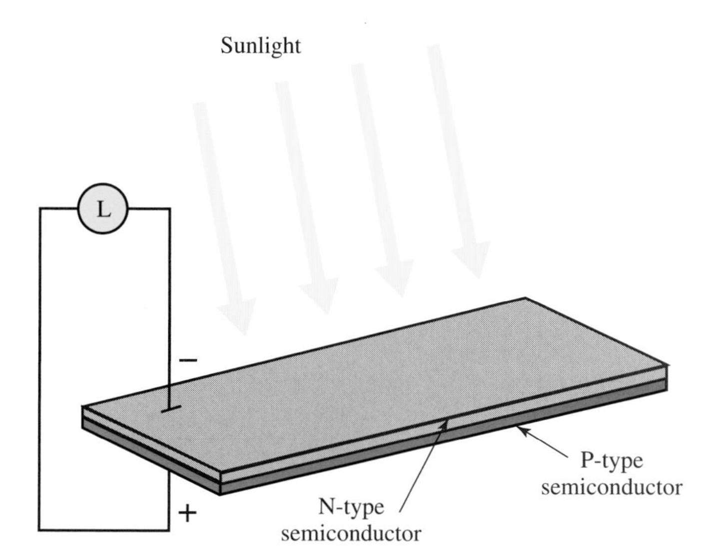

Figure 5-18. Two semiconductor materials are sandwiched together to form a photovoltaic cell. Electrical energy is produced when light shines on the cell.

of this type of power, the Boston Edison Power Company built a model home powered by solar cells. The home is powered by two arrays of solar cells mounted on the roof. Each array contains 12 modules with 432 individual cells. The solar cells provide approximately 45 percent of the family's electricity. Today, solar cells are commonly used to partially power homes and heat swimming pools. The cells can also supply the complete power needs for items such as calculators, watches, and satellites. In addition, solar cells supply power for remote locations where there are no accessible power lines. See **Figure 5-19.** **Figure 5-20** shows a common solar array for providing power.

Photoelectric control

The same principle of generating electricity from light is used to produce a device called a photoresistive

Figure 5-19. Solar powered radio transmitter used for highway construction.

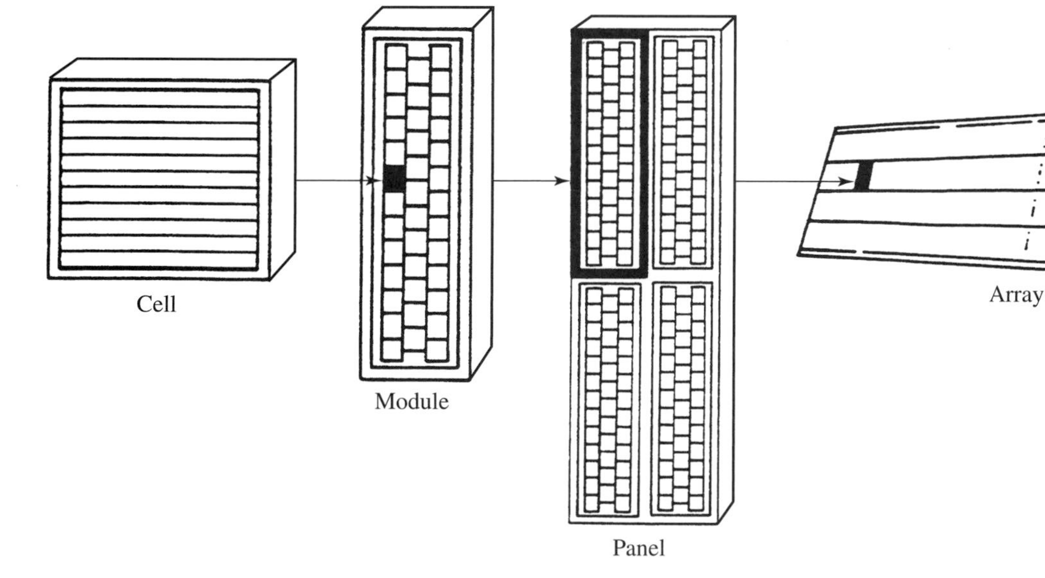

Figure 5-20. Solar array.

PROJECT 5-1: Photoelectric Controller

Try building your own photoelectric control. A photoelectric control device, using a photocell, is diagrammed in the project schematic.

When light shines on the cell, the resistance of the circuit changes. A cadmium-sulfide photocell is a light variable resistor. It is most sensitive in the green to yellow portion of the light spectrum. With it you can use light to control many electronic devices. Photocells are used in counting operations, burglar alarms, door opening mechanisms, and in many other devices.

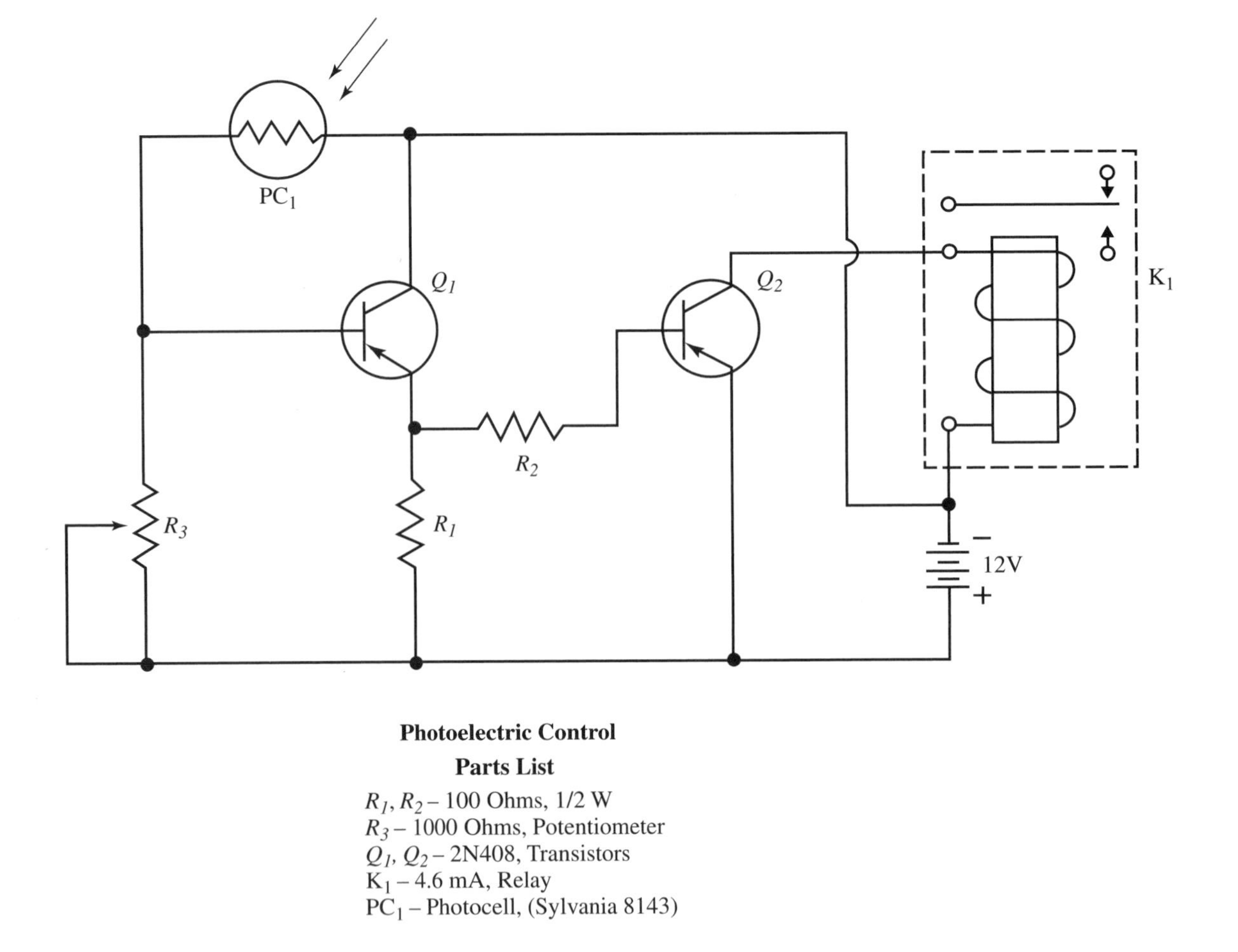

cell. ***Photoresistive cells*** are light sensitive resistors. Instead of providing a direct supply of electricity, this device is used to vary the amount of current that can pass through it, much like a variable resistor would do. The photoresistive cell increases circuit resistance when there is no light. It decreases resistance when there is light. The symbol for a photoresistive cell is shown in **Figure 5-21.**

Electrical Energy from Heat

A device used to indicate and control the heat of electric ovens and furnaces is shown in **Figure 5-22.** This device is called a ***thermocouple.*** When two dissimilar metals in contact with each other are heated, a potential difference develops between the metals, **Figure 5-23.**

EXPERIMENT 5-3: Building a Thermocouple

In this experiment you will demonstrate and construct a simple thermocouple.

Materials

1–piece of copper wire
1–piece of iron wire
1–pack of matches
1–multimeter

1. Twist one of the copper wires together with one end of the iron wire.
2. Connect the leads of the multimeter to the opposite two ends of the copper and iron wire. Set the meter to read voltage. (Be sure to observe the proper measuring technique.)
3. Light a match and heat the junction of the two wires.
4. Observe the results on the meter. Construct a thermocouple by twisting the ends of a copper and an iron wire together. Connect the wires to a voltmeter. Heat the junction with a match. Observe the voltage developed on the multimeter.

Explain what you have just observed. Could you use materials other than copper and iron?

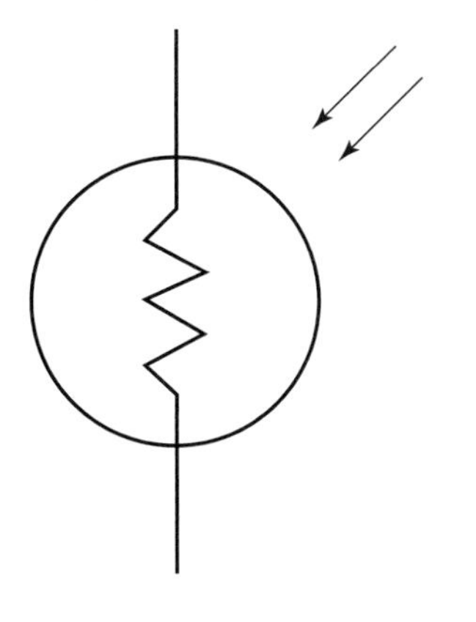

Figure 5-21. Schematic symbol for a photoresistive cell.

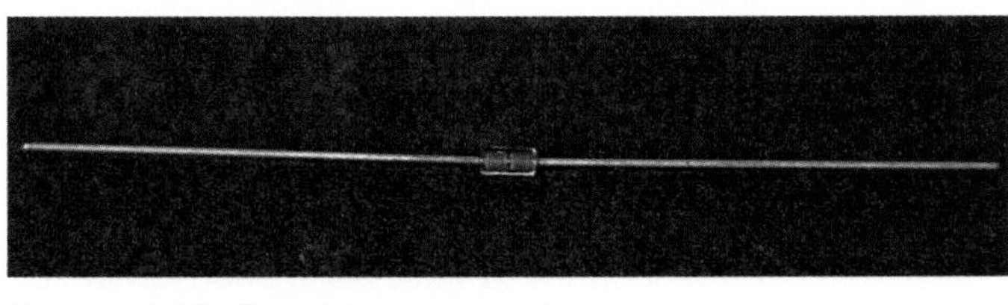

Figure 5-22. Small thermocouple.

In the demonstration of Figure 5-23, an iron wire and a copper wire are twisted tightly together. Their ends are connected to a sensitive meter, such as a galvanometer. A ***galvanometer*** is a device that is capable of measuring very small currents. When the flame of a lit match heats the twisted joint, a reading on the meter can be observed. This indicates that an electromotive force is present. The output voltage of a thermocouple can be strengthened and used to work large motors, valves, controls, and recording devices.

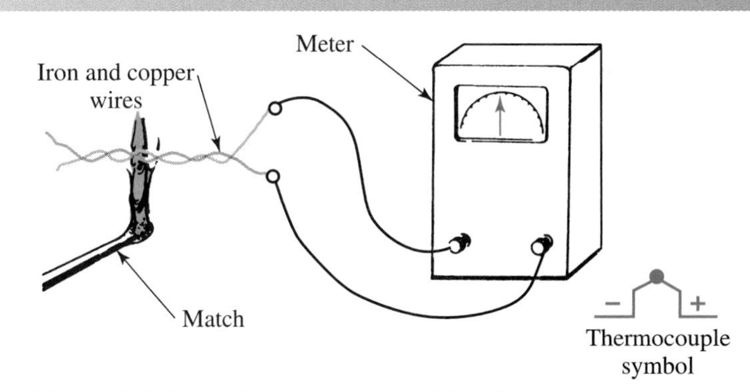

Figure 5-23. The basic principle of the thermocouple can be demonstrated by heating two dissimilar wires that have been twisted together.

Commercial types of thermocouples employ various kinds of dissimilar metals and alloys such as nickel-platinum, chromel-alumel, and iron-constantan. These unfamiliar names apply to alloys specially developed for thermocouples.

The combination indicating device, including a meter and a thermocouple, is called a ***pyrometer.*** For an instrument that must be sensitive to temperature change, a large number of thermocouples can be joined in series. Such a group is known as a ***thermopile.***

Electrical Energy from Mechanical Pressure

Many crystalline substances such as quartz, tourmaline, and Rochelle salts have a peculiar characteristic. When a voltage is applied to the surfaces of the crystal, the crystal becomes distorted. The opposite is also true. If a mechanical pressure or force is applied to the crystal surface, a voltage is developed. The crystal microphone,

Figure 5-24, is a familiar example of this process. Sound waves striking a diaphragm, which is mechanically linked to the crystal surfaces, cause distortion in the crystal. This develops a voltage across its surfaces. Thus, sound waves are converted to the electrical energy.

Creating electricity by the mechanical distortion of a crystal is known as the ***piezoelectric effect.*** Crystals can be cut for particular operating characteristics. In a later chapter, the use of crystals as frequency controls for radio transmitters will be discussed.

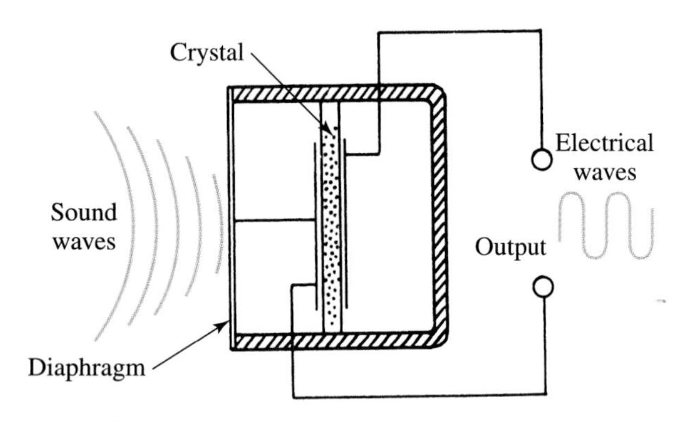

Figure 5-24. A crystal microphone converts sound waves to electrical energy.

Fuel Cells

A ***fuel cell*** is constructed much like a battery cell. Two metallic electrodes are designed to allow hydrogen and oxygen gases to combine with the electrolyte of potassium hydroxide KOH. See **Figure 5-25.**

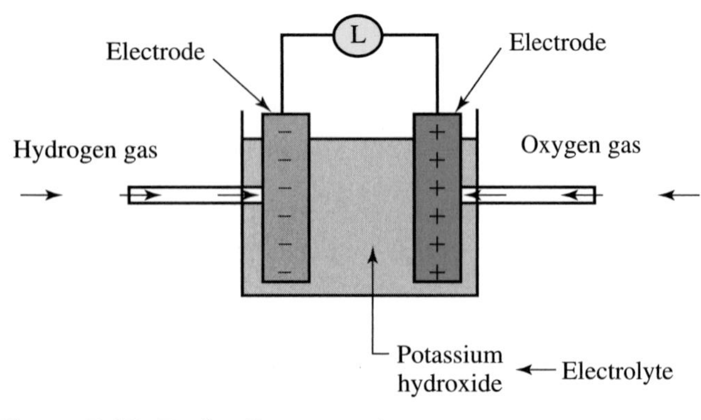

Figure 5-25. Fuel cell construction.

The two metallic electrodes are not part of the chemical reaction, but rather a means to allow the gases to combine with the electrolyte. Once the gases combine with the electrolyte, ionization occurs. The electrode attached to the oxygen line develops a positive potential, while the electrode connected to the hydrogen line develops a negative potential. The chemical reaction in the fuel cell takes place when the gases combine with the potassium hydroxide. The ionization will continue as long as the two gases are supplied to the electrolyte. Typically, the cell develops only 1.23 volts. However, when used in the space program, these cells have been designed to develop over 2 kW of energy. The theoretical efficiency of the fuel cell is 100%. There is no heat loss due to chemical reactions. The only by-product of the fuel cell is water, resulting in virtually no pollution.

Magnetohydrodynamic Power Generation

Electricity is generated when an ionized gas is passed through a magnetic field. This method of producing electricity is called ***magnetohydrodynamic (MHD).*** **Figure 5-26** is an illustration of the MHD converter.

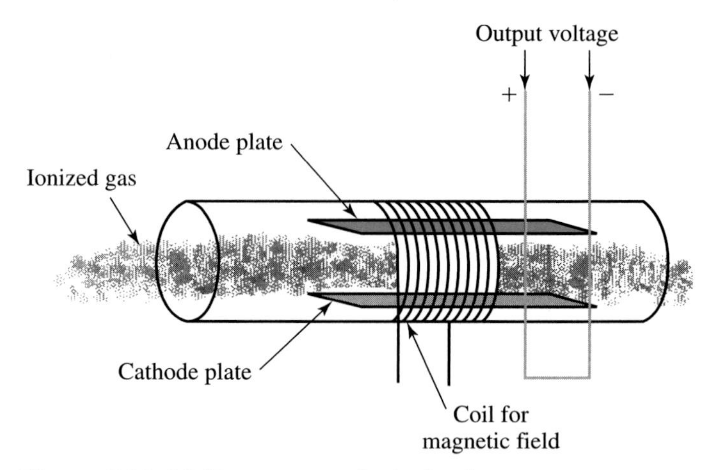

Figure 5-26. MHD converter. An ionized gas vapor passes through a magnetic field. The two plates (anode and cathode) collect the ions and pass them to the outside to provide electric power.

A gas, like argon or helium, is heated by solar energy and reaches a temperature in excess of 2000°F. At this temperature the gas ionizes. The ionized gas is forced through tubing and passed through a magnetic field. The magnetic field is produced by winding conductors into a coil around a pipe and then energizing the coil with a small amount of direct current. As the gas passes through the magnetic field, the ions are collected on two conductive plates. The negative plate is referred to as the ***cathode,*** and the positive plate is referred to as the ***anode.*** The terms anode and cathode are used throughout the study of electronics and are directly associated with the polarity described above. This type of power is in limited production in Middle Eastern countries. Using the sun to produce this type of energy is relatively cheap, but unfortunately not constantly available.

Electricity from Magnetism

A common source of electrical energy is the dynamo or generator. Generators prove that magnetism can produce electricity. A ***generator*** is a rotating machine that converts mechanical energy into electrical energy. This source of electricity requires detailed study. Chapter 10 is devoted to generators, generator types, and controls.

Review Questions for Section 5.2

1. The __________ produces electricity from light.
2. A light variable resistor is called a(n) __________.
3. When two different metals are joined together and heated, they produce electricity. This is an example of a(n) __________ device.
4. A(n) __________ will produce electricity when pressure is applied to its surface.
5. A(n) __________ converts sunlight directly into electricity.
6. An ionized gas passes through a magnetic field and produces electricity. This form of electrical production is made by a(n) __________ device.
7. Explain how a fuel cell works.
8. What is the polarity of an anode?
9. What is the polarity of a cathode?

Summary

1. Electricity can be produced by chemical action.
2. Cells are the basic unit for producing electricity by chemical action. Batteries are two or more cells connected together.
3. Primary cells cannot be recharged, while secondary cells can be recharged.
4. Some popular primary cells include the zinc-carbon cell, the alkaline cell, the mercury cell, the lithium cell, and the silver oxide cell.
5. Nickel-cadmium cells and lead-acid cells are secondary cells.
6. Connecting cells in series (– to +) increases their voltage rating, while connecting them in parallel (– to – and + to +) increases their current rating.
7. Battery capacity is a current producing rating measured in ampere-hours (Ah).
8. A photovoltaic cell produces electricity from light.
9. Photoresistive cells are light sensitive resistors.
10. The device used to produce electricity by heat is the thermocouple.
11. Electricity can also be produced by applying pressure to certain objects such as quartz, tourmaline, and Rochelle salts.

Test Your Knowledge

Please do not write in the text. Place your answers on a separate sheet of paper.

1. Name six basic sources of electricity or electromotive force.
2. What safety precautions should be taken when mixing acid and water?
3. A(n) __________ __________ is a way of converting chemical energy into electrical energy.
4. A cell that cannot be recharged is a:
   a. one-way cell.
   b. secondary cell.
   c. primary cell.
   d. None of the above.
5. What is polarization?
6. List five advantages silver oxide batteries have over other types of batteries.
7. What is the major advantage of the nickel-cadmium cell?
8. In a(n) __________ connection, the positive terminal of one cell is connected to the negative terminal of the second cell.
9. In a(n) __________ connection, all positive terminals are connected and all negative terminals are connected.
10. What is the composition of the electrolyte in a lead acid storage battery?
11. Specific gravity:
    a. can be measured with a hydrometer.
    b. can be used to determine the state of charge.
    c. is the weight of a liquid as it compares to water.
    d. All of the above.
12. What is the measurement of battery capacity?
13. A(n) __________ is a device used to indicate and control the heat of electric ovens and furnaces.
14. What is the piezoelectric effect?

For Discussion

1. How does chemical action operate our nervous system? Research this question. Write a brief report on your findings.

2. Do research on one of the following topics. Give a report in class.
   a. The electrical system of a lightning bug.
   b. The electric eel.
3. How can electricity from light be used to solve some of our energy problems? What are the restrictions?
4. Why do materials such as quartz have a piezoelectric effect?

Large array of photovoltaic cells.

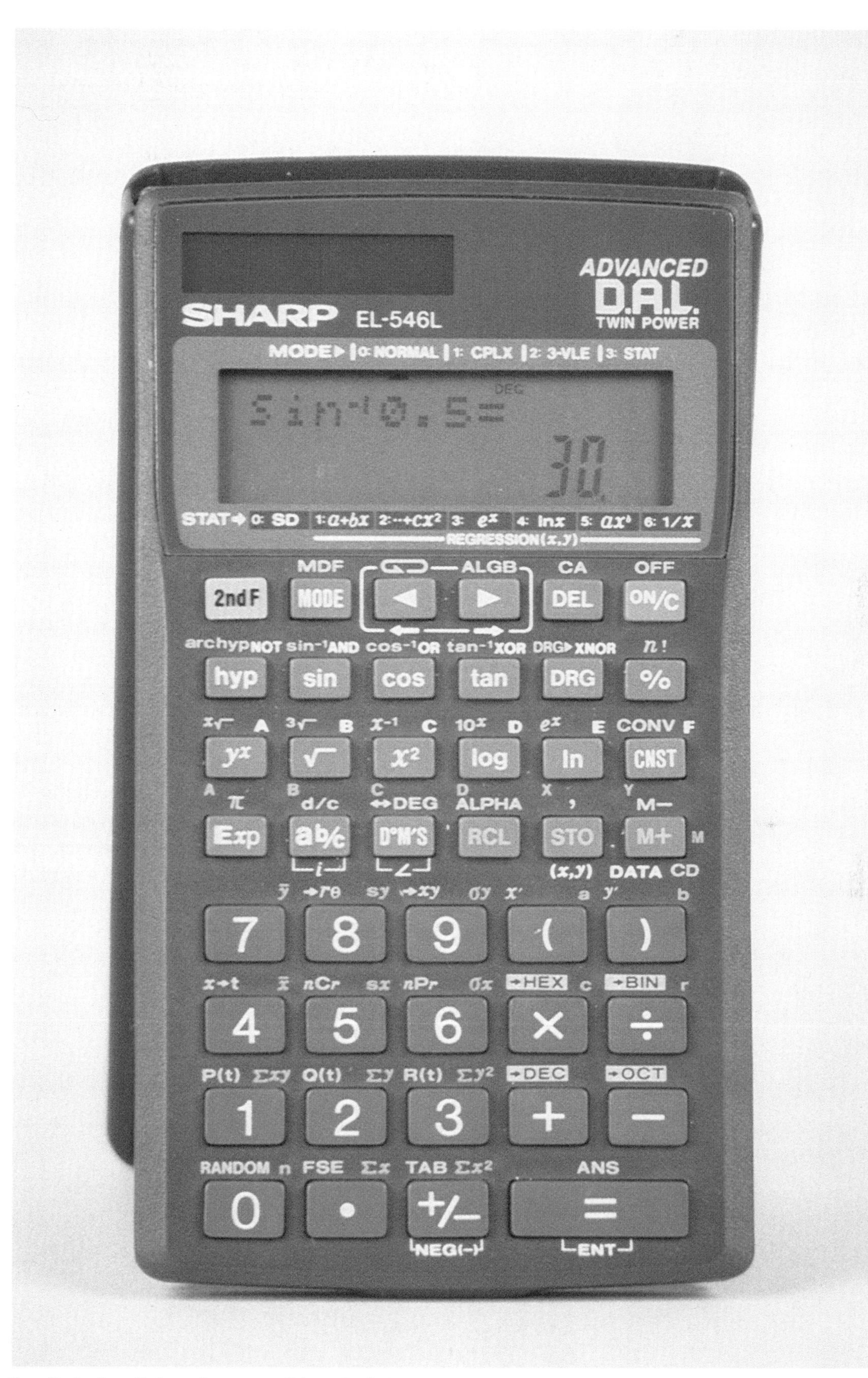

Small photovoltaic cells power this calculator.

Basic Circuits

Chapter 6
Series Circuits

Chapter 7
Parallel Circuits

Chapter 8
Combination Circuits (Series-Parallel)

The three chapters that make up Section II provide a thorough understanding of electrical circuitry. Chapter 6 explains the principles of series circuits. You will take complex series circuits and reduce them into simple, solvable circuits. Chapter 7 explains several different methods for reducing and solving parallel circuits.

The final chapter in this section, Combination Circuits, takes you step-by-step through a number of examples. When you have finished this chapter, you will be able to reduce any series-parallel circuit into a simple series or parallel circuit.

In all three chapters, typical circuits are presented with illustrations of proper troubleshooting techniques. These illustrations show how to use meters to locate open and short circuits. These sections should prove to be valuable for laboratory work, and they will assist you in comprehension.

Series Circuits

Objectives

After studying this chapter, you will be able to:

- Determine the total resistance of a series circuit.
- Determine the voltage drops in a series circuit.
- Determine the current values of a series circuit.
- Determine the wattage values of a series circuit.
- Apply Ohm's law to solve for unknown voltage, current, and resistance in a series circuit.
- Apply series circuit theory to assist in troubleshooting a series circuit.

Key Words and Terms

The following words and terms will become important pieces of your electricity and electronics vocabulary. Look for them as you read this chapter.

Kirchhoff's voltage law
series circuit

Chapters 6, 7, and 8, cover series, parallel, and combination circuits. Series, parallel, and combination are the only three ways an electrical circuit can be constructed. The combination circuit is simply a combination of a series circuit and a parallel circuit. These three chapters, when mastered, will provide you with the basic skills to construct, analyze, and troubleshoot many electrical systems.

6.1 SERIES CIRCUIT PRINCIPLES

A ***series circuit*** has only one path for electron flow through the devices wired in the circuit. There are three formulas used to explain the laws of series circuits.

$$E_T = E_1 + E_2 + E_3 ... + E_N$$
$$R_T = R_1 + R_2 + R_3 ... + R_N$$
$$I_T = I_1 = I_2 = I_3 ... = I_N$$

Where N is the total number of voltage sources, resistances, or current sources in a circuit.

Voltage in a Series Circuit

Gustav R. Kirchhoff was a mid-nineteenth century scientist who discovered that the total voltage applied to a series circuit is equal to the total number of individual voltage drops in a series circuit.

$$E_T = E_1 + E_2 + E_3 ... + E_N$$

This is known as ***Kirchhoff's voltage law*** as applied to series circuits. The source voltage of a series circuit is equal to the total value of each individual voltage drop. **Figure 6-1** has three resistors wired in series with a voltage drop indicated at each resistor. When the three voltage drops are added together, they equal the source voltage.

$E_T = E_1 + E_2 + E_3$
$E_T = 3\text{ V} + 5\text{ V} + 4\text{ V}$
$E_T = 12\text{ V}$

$E_1 = 3\text{ V}$ $E_2 = 5\text{ V}$ $E_3 = 4\text{ V}$

Figure 6-1. The total voltage is equal to the sum of the individual voltage drops.

Current in a Series Circuit

The current in a series circuit is equal throughout the circuit. In **Figure 6-2,** three resistors are connected in series to the source. Since there is only one path for the electron flow, the current *must* be the same value at any point in the circuit.

$$I_T = I_1 = I_2 = I_3 ... = I_N$$

The formula for series circuit current value is a mathematical way of expressing that a current value measured at any point in a series circuit is equal to the current value at any other point in that same circuit. If there are 5 mA flowing through R_1, then there must be 5 mA flowing through any resistor or conductor in that circuit.

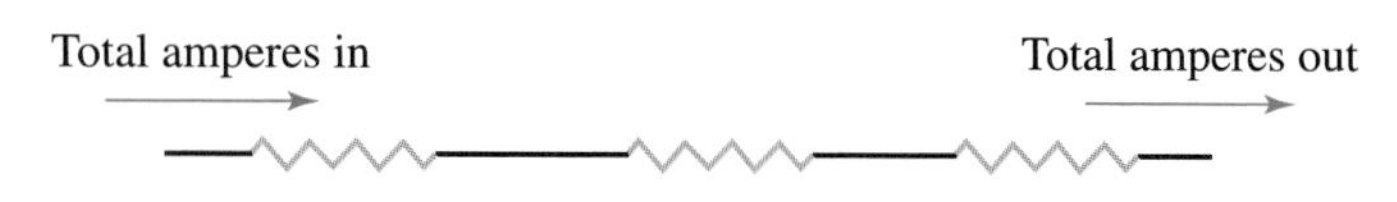

Figure 6-2. In a series circuit, there is only one path for the current.

Resistance in a Series Circuit

Total resistance in a series circuit is equal to the total circuit resistance combined. By adding the value of the individual resistors together, we find the total resistance of the circuit.

$$R_T = R_1 + R_2 + R_3 ...+ R_N$$

Figure 6-3 consists of 20 ohm, 40 ohm, and 60 ohm resistors connected in series. The total resistance value for this circuit is 120 ohms.

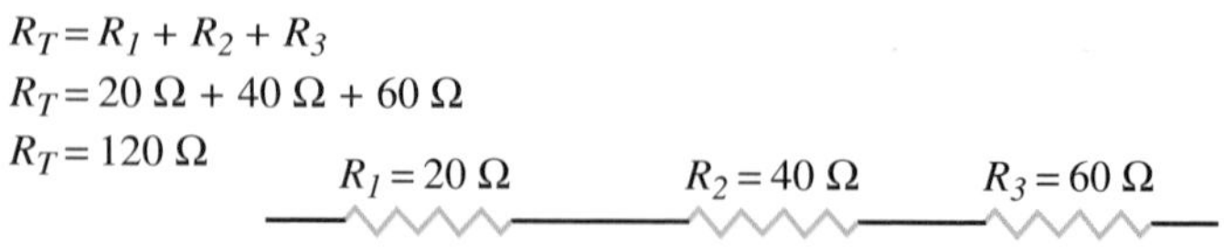

$R_T = R_1 + R_2 + R_3$
$R_T = 20\ \Omega + 40\ \Omega + 60\ \Omega$
$R_T = 120\ \Omega$

$R_1 = 20\ \Omega$ $R_2 = 40\ \Omega$ $R_3 = 60\ \Omega$

Figure 6-3. The total resistance in a series circuit is equal to the sum of the individual resistances.

Determining an Unknown Voltage

To find a single unknown voltage value, subtract the known resistance voltage drop values from the source. In **Figure 6-4,** two resistors are wired in series. The value of R_1 is 8 volts and the source is equal to 12 volts. To apply

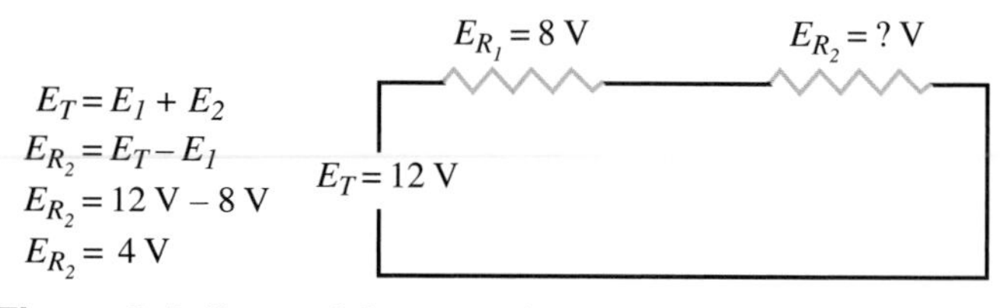

Figure 6-4. Determining an unknown voltage drop.

the law of voltages in a series circuit, subtract the known circuit voltage drops from the source.

Power in a Series Circuit

The total amount of power consumed in a series circuit is equal to the source voltage multiplied by the circuit current.

$$P_T = E_T \times I_T$$

Each device in the circuit that consumes power is part of the total load. An example of consumed power is the amount of wattage dissipated as heat by each resistor. See **Figure 6-5.** Devices that are not normally considered to consume power in a lab circuit are switches, fuses, and conductors.

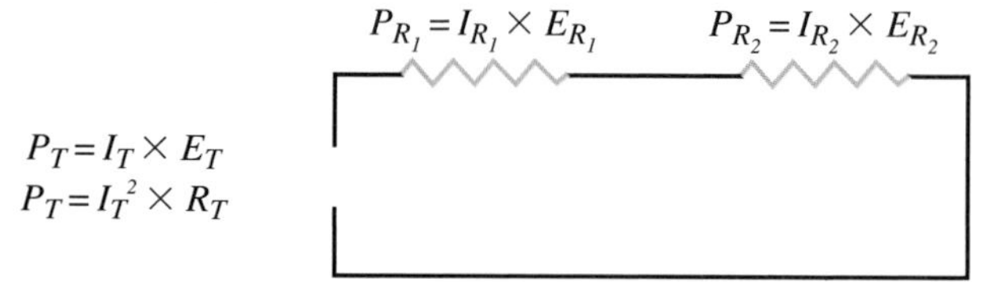

Figure 6-5. Calculating power in a series circuit.

There are several methods of determining total power rated in watts consumed by a series circuit. In the first method we multiply the source voltage times the current value of the circuit ($E_T \times I_T$). Another easy method is to multiply the current squared by the total circuit resistance ($I_T^2 \times R_T$).

It is very important to remember that the wattage rating of a resistor is *not* the amount of energy it uses. The wattage rating of a resistor is the amount of energy, in the form of heat, the resistor can safely dissipate without being damaged. In other words, a 2.2 kΩ resistor rated at 1/4 watt does not consume 0.250 watts. The wattage (1/4 W) is the maximum amount of heat energy it can safely dissipate without damage.

Review Questions for Section 6.1

1. Find the value of the voltage source.

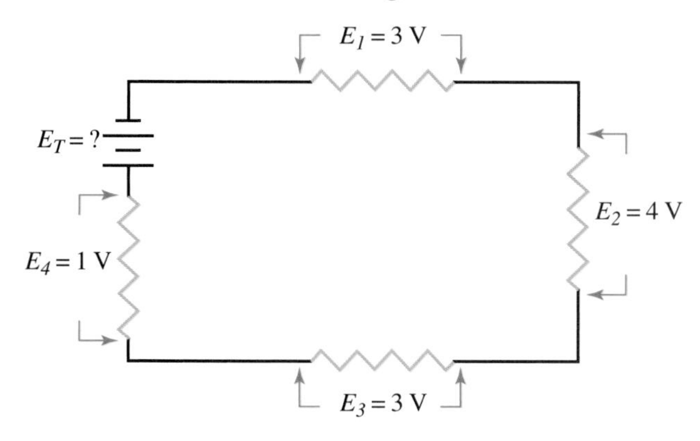

2. Find the total current.

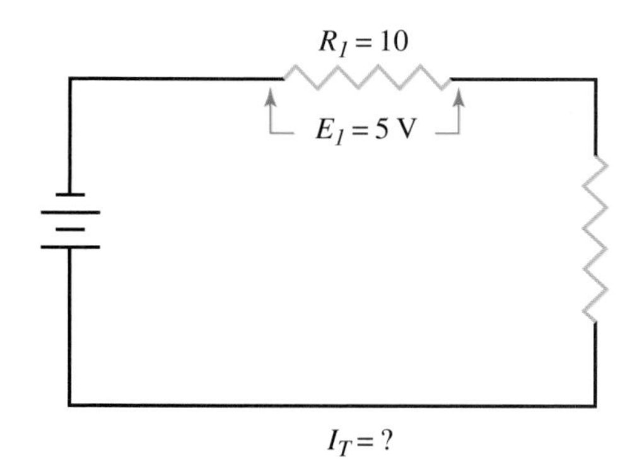

3. Find the missing resistance.

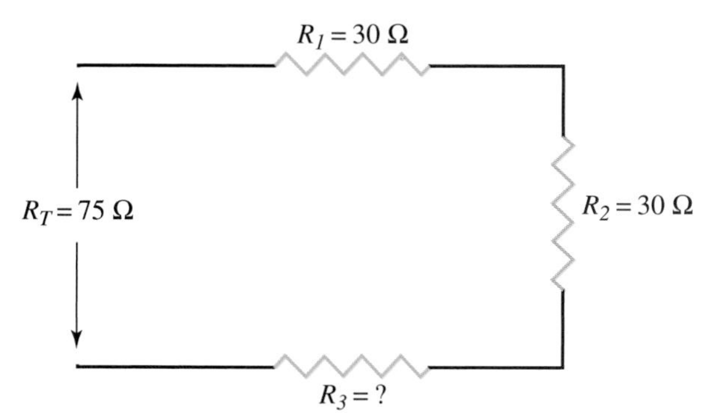

4. Find the power in the circuit shown.

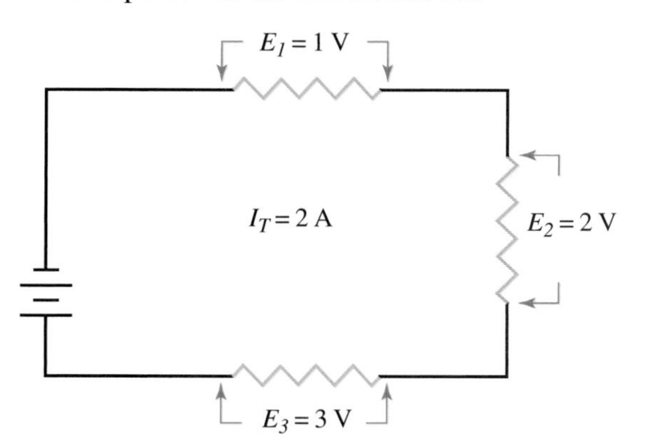

6.2 APPLICATIONS AND TROUBLESHOOTING SERIES CIRCUITS

There are numerous applications for the principles of series circuits. This section demonstrates how to apply those principles and shows you the specific example of an airfield lighting system. In addition, troubleshooting series circuits is discussed.

Applying Ohm's Law to a Series Circuit

Ohm's law can be applied to any individual component of a series circuit. Look at **Figure 6-6.** The Ohm's law formulas are noted at each location. If you know any two values at an individual location, you can apply Ohm's law to find the third value.

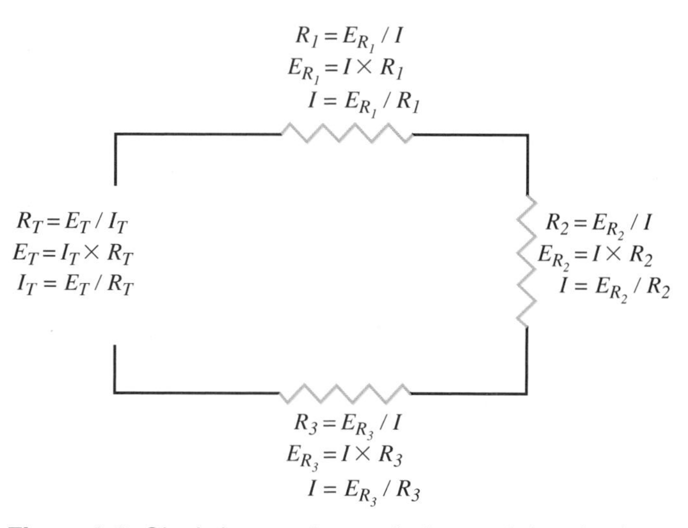

Figure 6-6. Ohm's law can be applied at each location in the circuit. Note that when the current value is found, it can be applied anywhere in the circuit.

Example: If R_1 has a resistance value of 10 ohms and a voltage drop of five volts, then applying Ohm's law to the R_1 location will give you a current value of 0.5 amperes.

$$I_1 = \frac{V_1}{R_1}$$

$$I_1 = \frac{5\text{ V}}{10\ \Omega} = 0.5\text{ A}$$

Could you go on from this point and calculate the amount of power used by R_1?

Airfield Lighting System

An airfield lighting system has many miles of circuit cable. The system has miles of cable originating from the source of power to the runway, plus the distance around the runway itself. The only practical way to light a runway and taxiway system is to use the series circuit principles. **Figure 6-7** illustrates how large the system may be. The voltage losses caused by the resistance of the copper lines do not permit a practical application of a parallel circuit. The lighting consists of a series circuit with a transformer located at each individual light location. The circuit is connected to a voltage regulator that maintains a constant amperage applied to the circuit. Since the circuit has a constant amperage, and each transformer and lighting unit has equal resistance, each lamp will have the same voltage drop value.

At times, the number of lamps in the circuit will change due to lamp failure. When the number of lamps change, the amount of applied voltage from the regulator will change in order to maintain the constant 6.6 amperes applied to the circuit. By maintaining a constant 6.6 amperes to the circuit, each lamp will burn at the same brightness, regardless of how many lamps are lit at once, and regardless of the voltage drop along the length of the conductor. See **Figure 6-8.**

Troubleshooting a Series Circuit Using a Voltmeter

A series circuit is really easy to troubleshoot. In **Figure 6-9,** three resistors are connected in series to the power supply. A fuse is used for circuit protection and a SPST switch is used to control the current to the resistors. One of the most common circuit faults in a series circuit is an open. A voltmeter, combined with the knowledge of the laws of voltages, is a quick and easy method of analysis. There are several ways to locate a circuit fault. The following step-by-step presentation is not the only way to locate the fault.

Figure 6-7. An airport lighting system can be over 50 miles in length for all circuits.

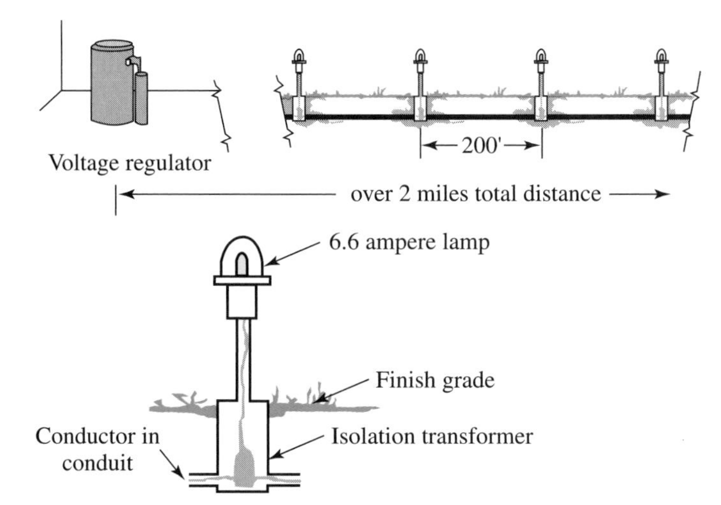

Figure 6-8. A typical airfield lighting system consists of a series of lamps connected to a voltage regulator. Each lamp will glow at equal brightness.

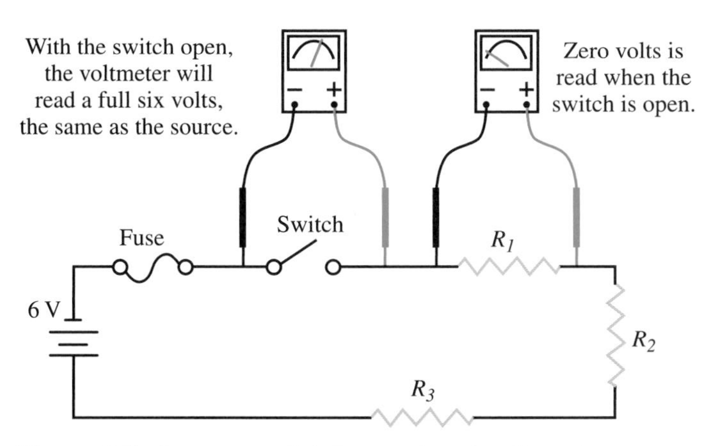

With the switch open, the voltmeter will read a full six volts, the same as the source.
Zero volts is read when the switch is open.
Fuse
Switch
6 V
R1
R2
R3

Figure 6-9. An open switch in a series circuit produces a reading on a voltmeter equal to the source voltage.

LESSON IN SAFETY:

Remember good safety is a habit that needs to be developed over a period of time. When troubleshooting even low voltage circuits, practice all safety techniques. Safety must become second nature to a technician when working on electrical circuits.

To begin the troubleshooting process, measure to determine if there is voltage at the supply. *Remember, a bad battery can still have a high voltage reading when it is not connected to a load.* When checking the battery for proper voltage, make sure the switch (S_1) is closed. If the voltage at the battery is good, move to the fuse. Take a voltage drop reading across the fuse. A good fuse will not produce a voltage drop. A blown fuse will produce a voltage drop that is equal to the source voltage. Also check

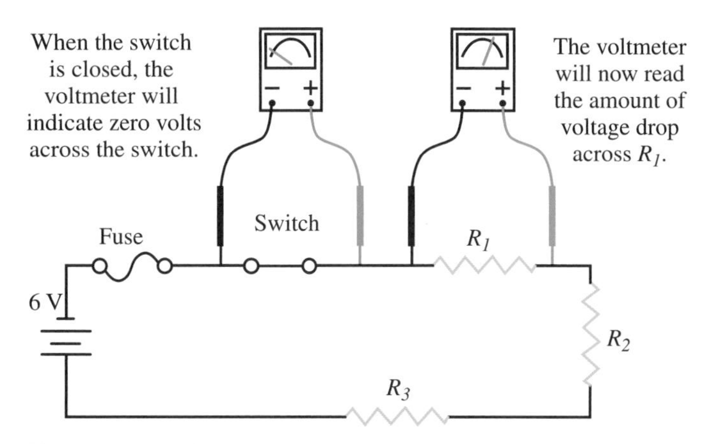

Figure 6-10. When the series circuit is complete, there will be voltage drops across each load component in the circuit.

the voltage drop across the switch. The closed switch should not produce a voltage drop, **Figure 6-10.** A single-pole single-throw (SPST) switch that produces a voltage drop when closed is defective. Next, check to see if there is a voltage drop across the individual resistors. If a resistor connected in a series circuit is open, it will have a voltage drop equal to the source voltage. The other resistors will have no voltage drop at all.

A short circuit is another possible problem. If one of the resistors is shorted, that resistor will show a voltage drop of zero. The other resistors in the circuit will drop the entire source voltage.

Review Questions for Section 6.2

1. Find the missing values of the three resistors.

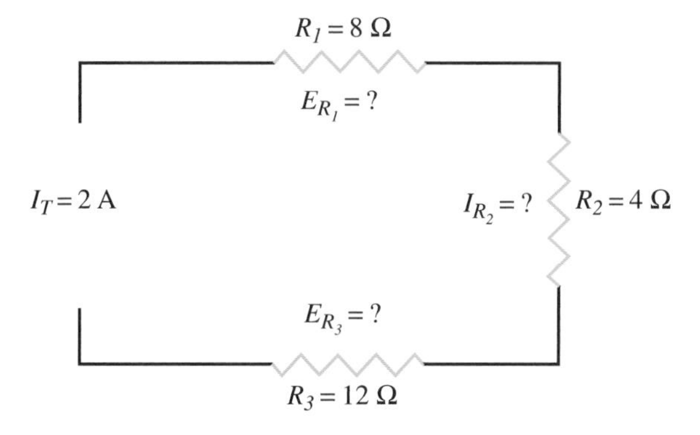

2. What does a voltage regulator do for the lights in an airfield lighting system?
3. How much voltage will be found across a good fuse?
4. How much voltage will be found across a blown fuse?

Summary

1. A series circuit provides only one path for current flow.
2. In a series circuit, current is equal throughout the circuit ($I_T = I_1 = I_2 = I_3 ... = I_N$).
3. The total resistance in a series circuit is equal to the sum of all the resistance values in the circuit ($R_T = R_1 + R_2 + R_3 ... + R_N$).
4. The source voltage is equal to the sum of the voltage drops in a series circuit ($E_T = E_1 + E_2 + E_3 ... + E_N$).
5. Total power consumed in the series circuit is equal to the sum of the individual power consummation ($P_T = P_1 + P_2 + P_3 ... + P_N$).

Test Your Knowledge

Please do not write in the text. Place your answers on a separate sheet of paper.

1. In a series circuit, the current values are __________ everywhere in that circuit.
2. In a series circuit, the source voltage is equal to the __________ of the individual voltage drops.
3. __________ is the total circuit resistance for two 25 ohm resistors connected in series.
4. When resistors of equal value are connected in series, they will develop voltage drops of __________ value.
5. Two 25 ohm resistors connected in series will require a __________ voltage source to produce a 6 volt voltage drop across one 25 ohm resistor.
6. Using the circuit below, solve for the unknown values.

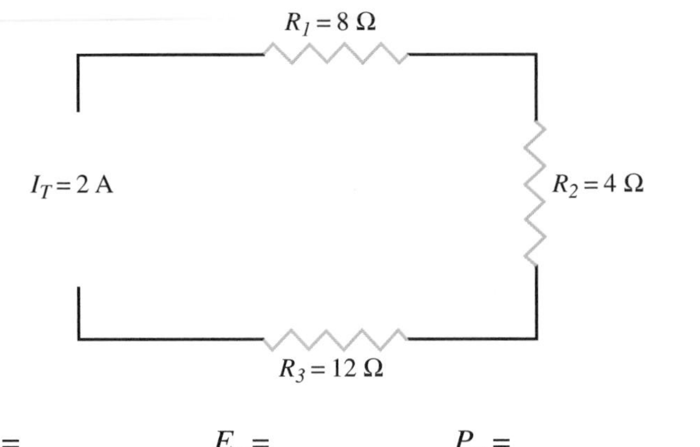

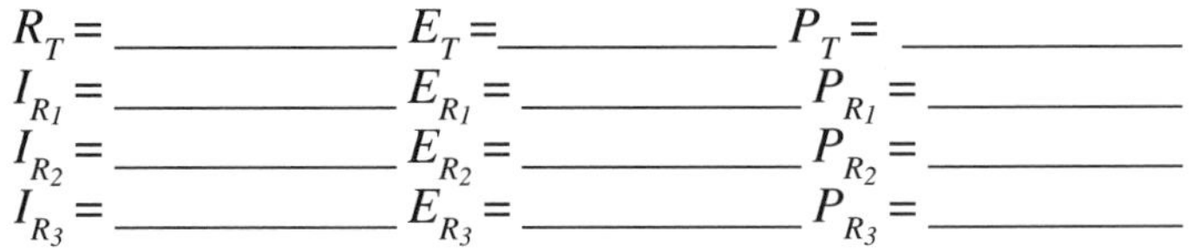

R_T = __________	E_T = __________	P_T = __________
I_{R_1} = __________	E_{R_1} = __________	P_{R_1} = __________
I_{R_2} = __________	E_{R_2} = __________	P_{R_2} = __________
I_{R_3} = __________	E_{R_3} = __________	P_{R_3} = __________

7. Using the circuit below, solve for the unknown values.

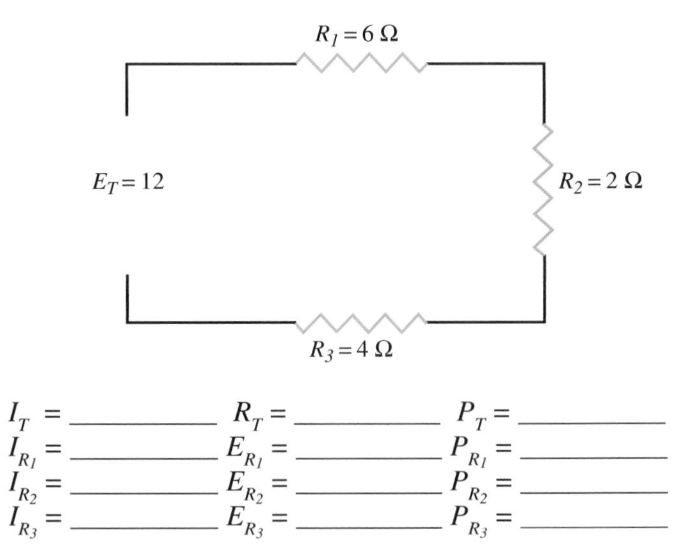

Parallel Circuits

Objectives

After studying this chapter, you will be able to:

- Determine the total resistance of a parallel circuit.
- Determine the voltage drops in a parallel circuit.
- Determine the current values of a parallel circuit.
- Determine the wattage values of a parallel circuit.
- Apply Ohm's law to solve for unknown voltage, current, and resistance in a parallel circuit.
- Apply parallel circuit theory to assist in troubleshooting a series circuit.

Key Words and Terms

The following words and terms will become important pieces of your electricity and electronics vocabulary. Look for them as you read this chapter.

branch currents	Kirchhoff's current law
mainline current	parallel circuit
reciprocal	

7.1 PARALLEL CIRCUIT PRINCIPLES

In simple terms, a ***parallel circuit*** provides two or more paths for electron flow through a circuit. Look at **Figure 7-1.** Take note of how each of the three resistors are connected to the source with each electron flow path independent of the other.

When three resistors are connected in series and one resistor is removed, the entire circuit appears to be dead and the electrons cease flowing. There is no complete path for the current. If the same three resistors are connected in parallel, one of the resistors can be removed from the circuit without stopping the flow of electrons in the circuit. This unique condition of the parallel circuit is the basis of the first parallel circuit formula we will study.

Parallel Circuit Voltage

The voltage in a parallel circuit is equal to the source voltage. If two or more components are connected to the

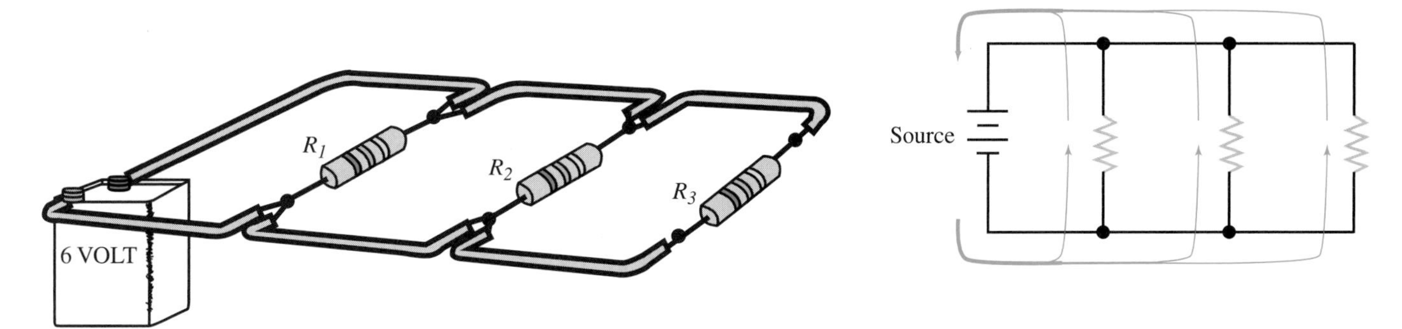

Figure 7-1. Three resistors connected in parallel.

source in parallel, the voltage drop across each component is equal to the source voltage. The formula representing the voltage condition of a parallel circuit in words is: The total voltage is equal to any individual voltage drop in the parallel circuit. The formula can be written:

$$E_T = E_1 = E_2 = E_3 ... = E_N$$

where N is the total number of voltages.

In **Figure 7-2,** there are three resistors connected in parallel. A six volt source is connected to the circuit. Each resistor has a full voltage potential applied across it. Each resistor voltage is equal to the source voltage.

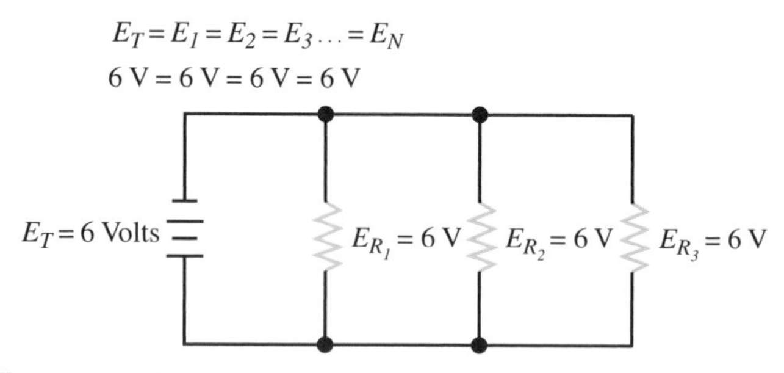

Figure 7-2. The voltage applied to each component is equal to the source voltage in a parallel circuit.

Parallel Circuit Current

Current in a parallel circuit follows some simple principles. These principles are summed up in Kirchhoff's current law. ***Kirchhoff's current law*** states that *the algebraic sum of all currents entering any point will equal the sum of all currents leaving that point.*

Simply stated, the current flowing into a junction of parallel resistance is equal to the current flowing out of that same junction. This leads to the conclusion that the total current in a parallel circuit is equal to the sum of the individual currents caused by each individual resistance. The individual currents are called ***branch currents.*** The total current is called the ***mainline current,*** see **Figure 7-3.**

Since the total current in a parallel circuit is equal to the sum of the branch currents caused by each individual resistance, a formula can be written:

$$I_T = I_1 + I_2 + I_3 ... + I_N$$

where N is the total number of currents.

Work through the circuit shown in Figure 7-3. You can see that the sum total of the three individual resistor currents equals the total source current of nine amperes.

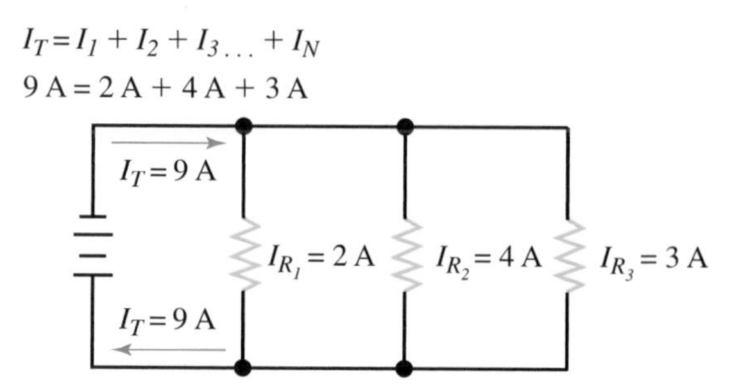

Figure 7-3. The total circuit amperes of a parallel circuit is equal to the sum total of the individual branch currents.

Parallel Circuit Resistance

When resistors are connected in a parallel circuit, the *total resistance* of the parallel circuit *is always less than the smallest resistance* in the parallel group. In other words, if a six ohm and a four ohm resistor are connected in parallel, the total resistance will be less than four ohms. This principle is extremely important and should be memorized. There are several methods to determine the total resistance of a parallel circuit. These methods include:

- The product over the sum method.
- The reciprocal method.
- The equal resistances method.
- The graph method.

Product over the sum method

The product over the sum method is used as a quick way to solve for the total resistance when there are two resistances of unequal value connected in parallel. The product over the sum method derives its name from the mathematical terms used in the formula. The product is the term used for the answer to a multiplication problem, and the term sum is used for the answer to an addition question. The formula for this method is written:

$$R_T = \frac{R_1 \times R_2}{R_1 + R_2}$$

Follow the equations in **Figure 7-4.** A three ohm and a six ohm resistor are connected in parallel. The total resistance for the circuit is two ohms.

Reciprocal method

The reciprocal method can be used to find the total circuit resistance when there are three or more resistances connected in parallel. The reciprocal method gets its name because the reciprocal of the resistance values is used in finding the total resistance. A ***reciprocal*** of a

$$R_T = \frac{R_1 \times R_2}{R_1 + R_2}$$

$$R_T = \frac{3\ \Omega \times 6\ \Omega}{3\ \Omega + 6\ \Omega}$$

$$R_T = \frac{18}{9}\ \Omega = 2\ \Omega$$

$R_T = 2\ \Omega$ = $R_1 = 3\ \Omega$ $R_2 = 6\ \Omega$

Figure 7-4. The product over the sum method is used primarily for two unequal resistors connected in parallel.

number is equal to 1 divided by that number. For example, the reciprocal of 2 is 1 ÷ 2, or 1/2. The reciprocal of 3/4 is 1 ÷ 3/4, or 4/3. The formula used for the reciprocal method is:

$$\frac{1}{R_T} = \frac{1}{R_1} + \frac{1}{R_2} + \frac{1}{R_3} \cdots + \frac{1}{R_N}$$

where N is the total number of resistances.

Notice that when using the reciprocal method, each resistance value is expressed as a fraction (the resistance value's reciprocal). The number 1 is the numerator for the fraction and the value of individual resistance is the denominator.

Figure 7-5 has a four, a six, and a twelve ohm resistor connected in parallel. Since the values are expressed as fractions, a common denominator must be found. Then the values can be added together. In this problem the common denominator is twelve. The total resistance is equal to six over twelve. The six over twelve is the reciprocal of the total resistance. The fraction is inverted and twelve is then divided by six to find the total resistance value of two ohms (6/12: Reciprocal = 12/6 = 2 = R_T).

Unfortunately, most circuits do not have resistance values that lend themselves as neatly as our example to this method of solution. However, the reciprocal method can be used in these messy situations by converting the individual values to decimals. In **Figure 7-6,** resistors of 35, 18, and 7 ohms are connected in parallel. The problem is set up exactly as before, but we simply use a calculator to convert each fraction into a decimal number so that they can be added together easily. The three resistor values total to 0.2271. The combined total of the three individual resistances is the reciprocal of the total resistance value of the circuit. The total resistance is equal to one divided by 0.2271 or 4.4.

Equal resistances method

When a parallel circuit consists of two or more resistors of equal value, the total resistance is equal to the

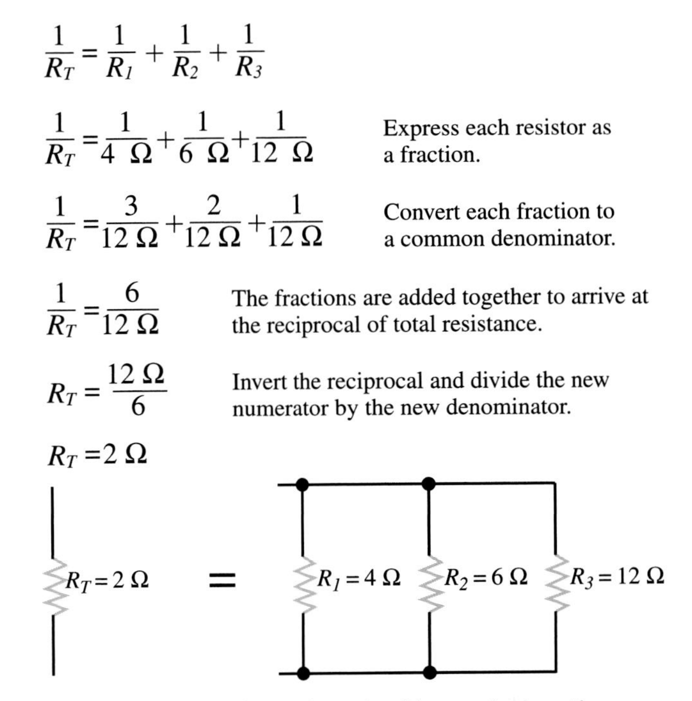

$\frac{1}{R_T} = \frac{1}{R_1} + \frac{1}{R_2} + \frac{1}{R_3}$
$\frac{1}{R_T} = \frac{1}{4\ \Omega} + \frac{1}{6\ \Omega} + \frac{1}{12\ \Omega}$
Express each resistor as a fraction.
$\frac{1}{R_T} = \frac{3}{12\ \Omega} + \frac{2}{12\ \Omega} + \frac{1}{12\ \Omega}$
Convert each fraction to a common denominator.
$\frac{1}{R_T} = \frac{6}{12\ \Omega}$
The fractions are added together to arrive at the reciprocal of total resistance.
$R_T = \frac{12\ \Omega}{6}$
Invert the reciprocal and divide the new numerator by the new denominator.
$R_T = 2\ \Omega$
$R_T = 2\ \Omega$
=
$R_1 = 4\ \Omega$
$R_2 = 6\ \Omega$
$R_3 = 12\ \Omega$

Figure 7-5. The reciprocal method is used when there are three or more unequal resistors connected in parallel.

$$\frac{1}{R_T} = \frac{1}{R_1} + \frac{1}{R_2} + \frac{1}{R_3}$$

$$\frac{1}{R_T} = \frac{1}{35\ \Omega} + \frac{1}{18\ \Omega} + \frac{1}{7\ \Omega}$$ Express each resistor as a fraction.

$$\frac{1}{R_T} = 0.0286\ \text{S} + 0.0556\ \text{S} + 0.1429\ \text{S}$$ Convert each fraction to a decimal number.

$$\frac{1}{R_T} = 0.2271\ \text{S}$$ The decimal numbers are totaled.

$$R_T = \frac{1}{0.2271\ \text{S}}$$ The one over R_T is cleared as a fraction by placing the one over the decimal equivalent of the resistance values.

$$R_T = 4.4\ \Omega$$

$R_T = 4.4\ \Omega$ = $R_1 = 35\ \Omega$ $R_2 = 18\ \Omega$ $R_3 = 7\ \Omega$

Figure 7-6. The reciprocal method can also be used with decimals rather than fractions. Notice during some intermediate steps, while working with the reciprocal of the total resistance, the values are in siemens.

value of any one resistor divided by the total number of equal resistors in the parallel circuit. The formula is written:

$$R_T = \frac{R}{N}$$

where N is the total number of resistances.

In **Figure 7-7,** three twelve ohm resistors are connected in parallel. The total resistance for the circuit in Figure 7-7 is four ohms (12 Ω ÷ 3 = 4 Ω).

$$R_T = \frac{R \text{ (the value of any one resistor in the parallel circuit)}}{N \text{ (the number of resistors in the parallel circuit)}}$$

$$R_T = \frac{12}{3} = 4\ \Omega$$

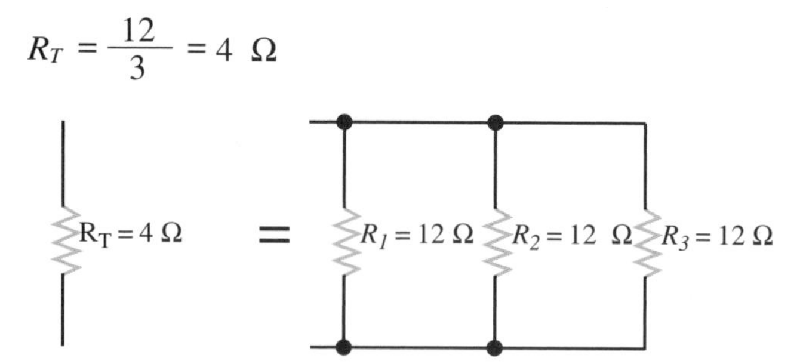

Figure 7-7. When all resistances are the same value, simply divide the value of one resistance by the number of resistances.

Note that the equal resistances method can be used in combination with other methods to simplify calculations. If a number of resistors are of an equal value, but not all of them, the equal resistances method can be used to produce an equivalent resistance from the equal value resistors.

> **Example:** You have a circuit with 10 resistors in parallel. Eight of the resistors have a value of 80 Ω. The other resistors have values of 20 Ω and 40 Ω. How do you find the total resistance? Rather than having a total resistance equation with 10 reciprocals, the eight like-valued resistors can be turned into one equivalent resistance. Eighty ohms divided by eight produces an equivalent resistance of 10 Ω. Now add the two other resistors to your equivalent resistance using the reciprocal formula.

Graph method

An interesting method of estimating the value of total resistance is the graph method. The values of two resistances are expressed on a bar graph, like the one shown in **Figure 7-8.** A line is then drawn from the base of the first resistance to the top of the second resistance. A second line is then drawn from the base of the second resistance to the top of the first resistance (making a crisscross). The total resistance value of these resistances may be read where the two lines intersect. A third resistance may be added and the process repeated. This method is not considered to be very accurate, but it is a quick and easy method to check your work.

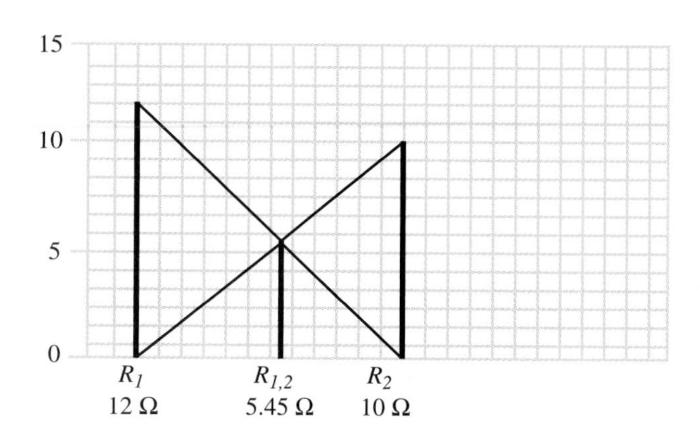

The total resistance of R_1 and R_2 is equal to 5.45 ohms.

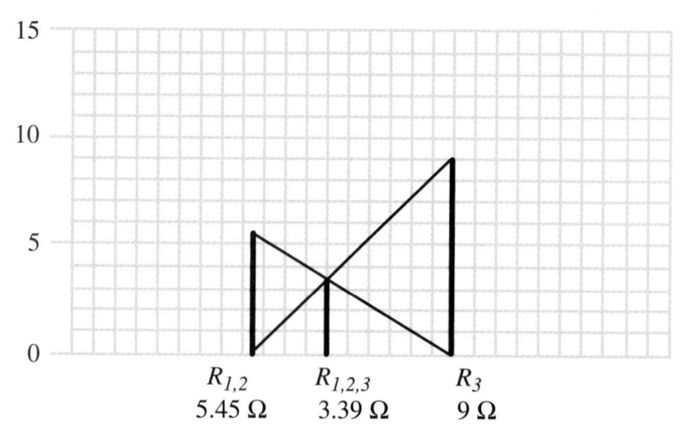

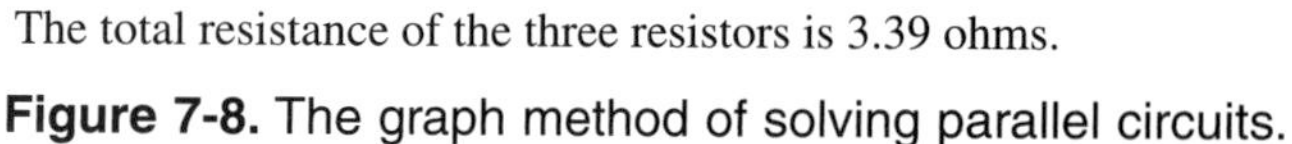

The total resistance of the three resistors is 3.39 ohms.

Figure 7-8. The graph method of solving parallel circuits.

Power in a Parallel Circuit

The total power consumed in a parallel circuit is equal to the sum of all the individual powers in the circuit. The formula is written:

$$P_T = P_1 + P_2 + P_3 ... + P_N$$

where N is the total number of powers.

In **Figure 7-9,** you can see that the total power consumed (72 W) is equal to the sum total of the individual electrical power units (R_1 = 12 W, R_2 = 36 W, R_3 = 24 W) consumed at each resistance.

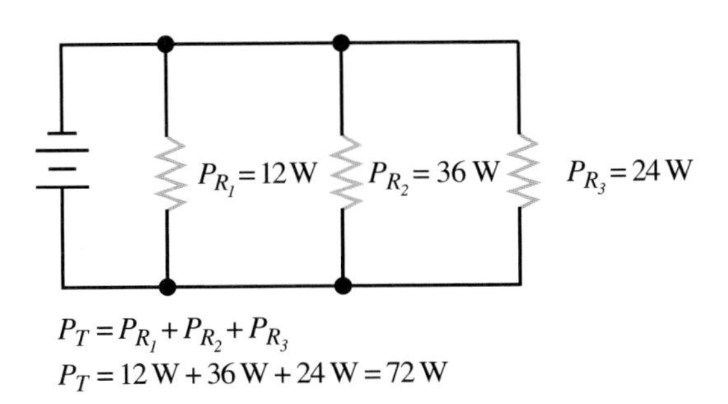

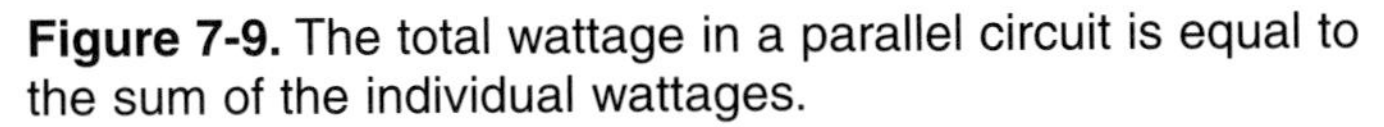

Figure 7-9. The total wattage in a parallel circuit is equal to the sum of the individual wattages.

Review Questions for Section 7.1

1. If the source voltage in a parallel circuit is equal to 10 volts, what is the voltage drop across resistor R_3? Assume R_3 is equal to 50 Ω.
2. A circuit containing three resistors in parallel has branch currents of 3 A, 5 A, and 10 A. What is the mainline current in the circuit?
3. List three formulas that can be used to find the total resistance in a parallel circuit.
4. What formula for finding the total resistance in a parallel circuit will work for any number of resistors of any value?
5. A parallel circuit has two resistors. Each consumes 6 W. Calculate the total power for the circuit.

7.2 APPLICATIONS AND TROUBLESHOOTING PARALLEL CIRCUITS

Parallel circuits are found in the home and industry alike. They allow the remaining bulbs in a string of lights to stay lit when one bulb burns out, and they allow computers to work on many parts of a problem at one time.

Applying Ohm's Law to Parallel Circuits

Ohm's law can be applied to the individual components of a parallel circuit to find unknown values such as current. The total circuit resistance in a parallel circuit is always equal to the source voltage divided by the sum total of the individual resistor currents.

An example is shown in **Figure 7-10.** You can see that the total of the individual circuit current values ($I_1 + I_2 + I_3$) equals six amperes. The source voltage of twelve volts is divided by six amperes. The total resistance is found to be two ohms.

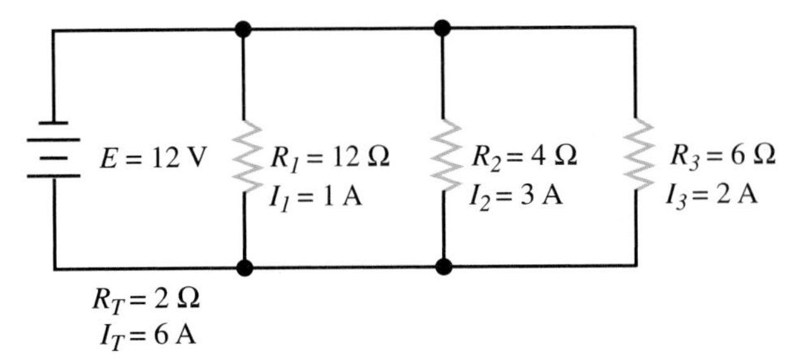

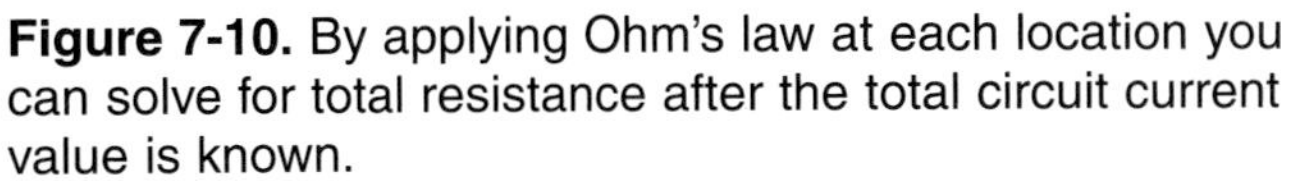

Figure 7-10. By applying Ohm's law at each location you can solve for total resistance after the total circuit current value is known.

Troubleshooting a Parallel Circuit

In a parallel circuit, the voltage is the same everywhere in the circuit. Checking voltage drops is not a very practical method for finding a bad component. One method of finding a problem is by comparing a total resistance reading, taken with an ohmmeter, to the calculated total resistance value. Looking at **Figure 7-11,** we can see that the total resistance value of the ohmmeter is 20 ohms. The calculated total resistance value is 10 ohms (20 Ω ÷ 2). Using a little deduction, you can determine that one of the resistors is open. If one of the resistors were shorted, it would cause the fuse to blow.

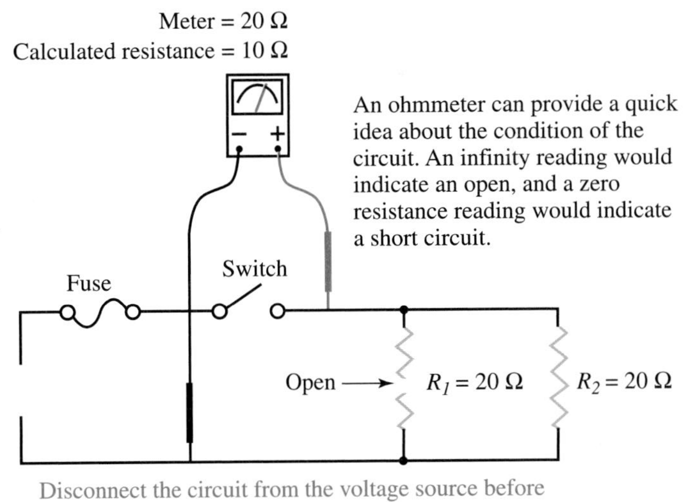

Meter = 20 Ω
Calculated resistance = 10 Ω
An ohmmeter can provide a quick idea about the condition of the circuit. An infinity reading would indicate an open, and a zero resistance reading would indicate a short circuit.
Fuse
Switch
Open
$R_1 = 20\,\Omega$
$R_2 = 20\,\Omega$

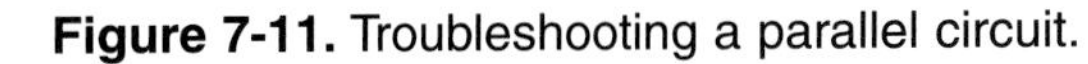

Disconnect the circuit from the voltage source before measuring the resistance with an ohmmeter!

Figure 7-11. Troubleshooting a parallel circuit.

Another method of troubleshooting uses current readings taken at each resistor location. First, the total current for the circuit is calculated. Then, this value is verified with an ammeter. See **Figure 7-12.** Using the ammeter, a current reading can be taken across an open switch. This reading is then compared to the calculated total current. An open resistor would draw no current while a good resistor would. See **Figure 7-13.** Troubleshooting a parallel circuit will become easier as we learn more about the current, voltage, and resistance characteristics.

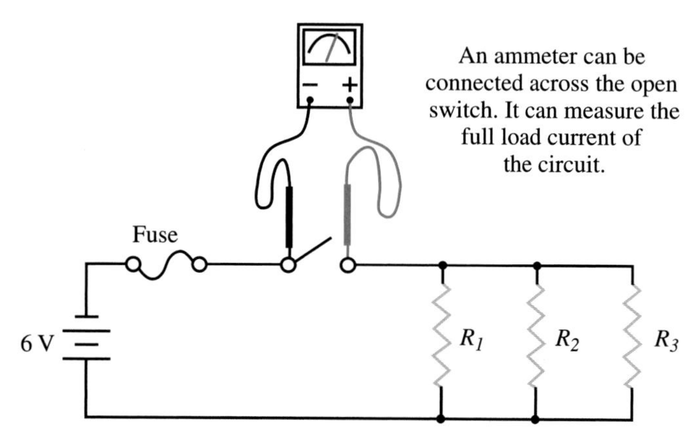

Figure 7-12. The switch location of a circuit is an ideal location to connect an ammeter to the circuit.

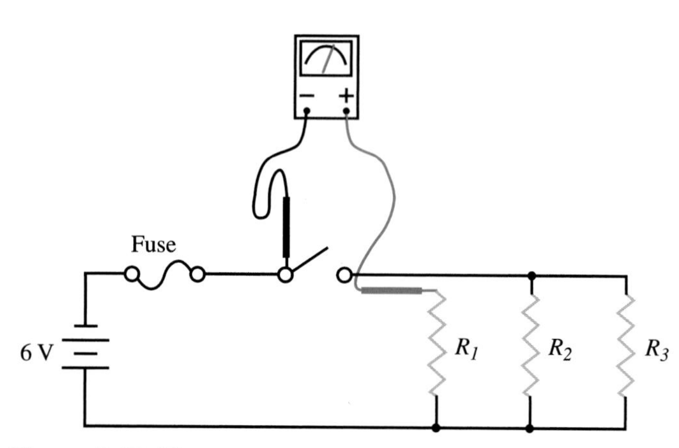

Figure 7-13. The ammeter must be placed in series with an individual component to indicate the current of only the individual component.

One factor that assists us in determining the best method to troubleshoot a circuit is determining how the circuit has been assembled. When a breadboard is used to construct circuits, troubleshooting is much simpler than when the components are soldered in place and must first be desoldered to take readings. In addition, desoldering components can damage them giving the technician an even more difficult troubleshooting problem. The technician may believe the problem has been located, and corrected, only to find another defective component. Unfortunately, the additional defective component may have been damaged while desoldering the component in order to obtain a meter reading.

Review Questions for Section 7.2

1. A parallel circuit consists of three resistors. All three resistors are 20 Ω. If the current through the third resistor is 7 amps, what is the source voltage?
2. Checking voltage drops is a __________ (practical/impractical) method of troubleshooting a parallel circuit.

Summary

1. A parallel circuit provides more than one path for current flow.
2. In a parallel circuit, the voltage is equal throughout the circuit ($E_T = E_1 = E_2 = E_3 ... = E_N$).
3. Kirchhoff's current law states that the total current entering a junction or parallel circuit is equal to the current leaving that junction or parallel circuit.
4. The total circuit current value in a parallel circuit is equal to the sum of the individual current values ($I_T = I_1 + I_2 + I_3 ... + I_N$).
5. The total circuit resistance is always less than the smallest resistance value in the parallel circuit.
6. Total power consumed in the circuit is equal to the sum of the individual power consumptions ($P_T = P_1 + P_2 + P_3 ... + P_N$).

Test Your Knowledge

Please do not write in the text. Place your answers on a separate sheet of paper.

1. What is the total circuit resistance of two 10 ohm resistors connected in parallel?

2. What is the total resistance of a 12 ohm and an 8 ohm resistor connected in parallel?
3. Solve for the unknown values in the parallel circuit below.

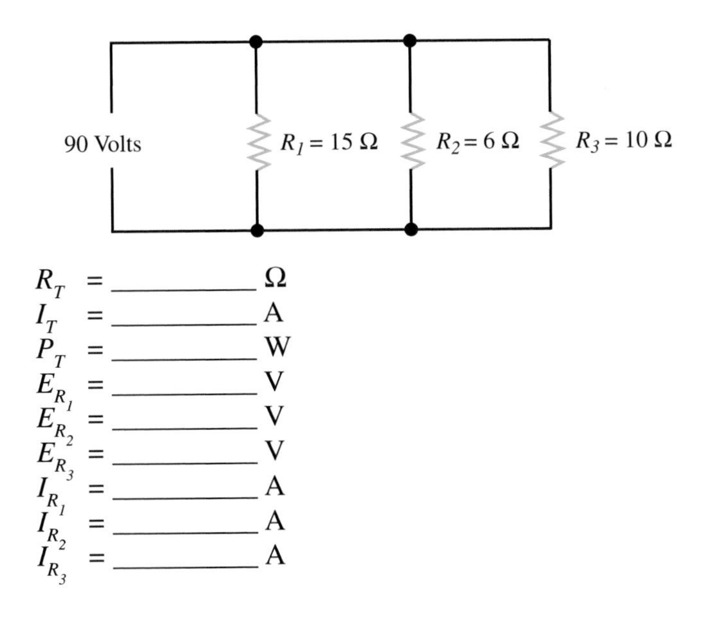

R_T = ________ Ω
I_T = ________ A
P_T = ________ W
E_{R_1} = ________ V
E_{R_2} = ________ V
E_{R_3} = ________ V
I_{R_1} = ________ A
I_{R_2} = ________ A
I_{R_3} = ________ A

4. Solve for the unknown values below.

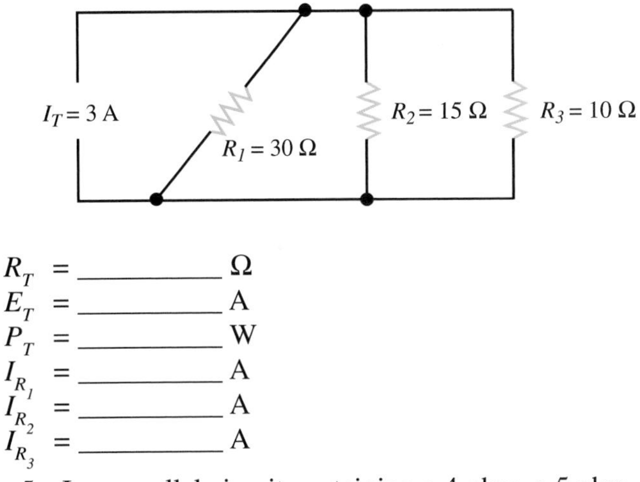

R_T = ________ Ω
E_T = ________ A
P_T = ________ W
I_{R_1} = ________ A
I_{R_2} = ________ A
I_{R_3} = ________ A

5. In a parallel circuit containing a 4 ohm, a 5 ohm, and a 6 ohm resistor, the current flow is:
   a. highest through the 4 ohm resistor.
   b. lowest through the 4 ohm resistor.
   c. highest through the 6 ohm resistor.
   d. the same for all three resistors.

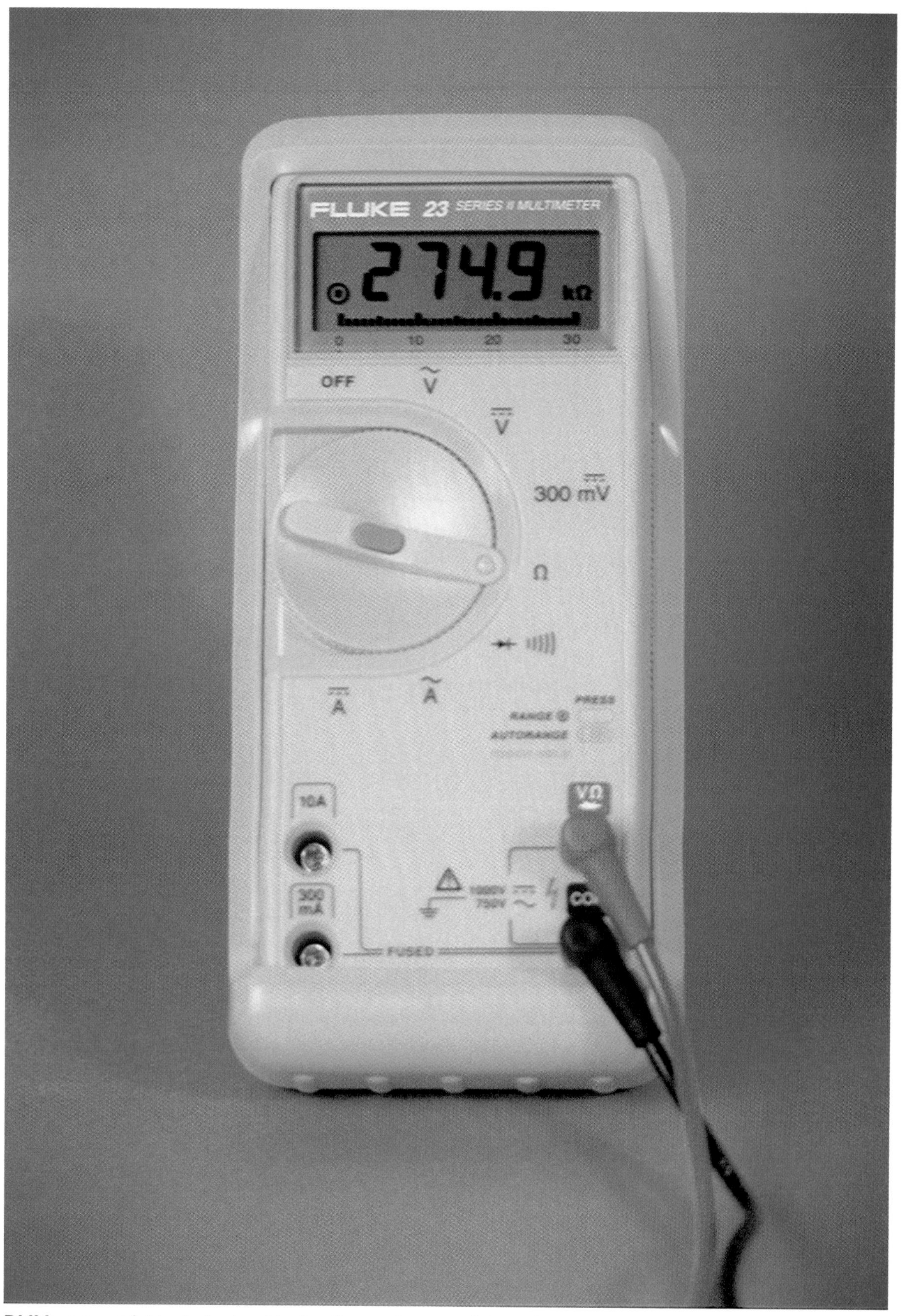

DMMs are useful tools for troubleshooting series, parallel, and combination circuits.

8 Combination Circuits (Series-Parallel)

Objectives

After studying this chapter, you will be able to:

- Determine the equivalent circuit resistance for a given combination circuit.
- Determine the voltage drops in a combination circuit.
- Determine the current values in a combination circuit.
- Determine the wattage values in a combination circuit.
- Apply combination circuit theory to troubleshoot a combination circuit.

Key Words and Terms

The following words and terms will become important pieces of your electricity and electronics vocabulary. Look for them as you read this chapter.

combination circuit
equivalent resistance
series-parallel circuit

The technician and engineer are often required to compute the total resistance of series-parallel circuits. This total resistance is referred to as the ***equivalent resistance*** of the circuit. It can be solved by applying the electrical characteristics of series and parallel circuits. No matter how complex the combination circuit may appear, remember that it is simply a composite of simple series and parallel circuits and can be reduced to one equivalent resistance.

8.1 REDUCING A COMPLEX CIRCUIT

When solving a ***series-parallel circuit,*** often called a ***combination circuit,*** you will use all of the formulas and techniques that you learned in Chapters 6 and 7. When one branch of a parallel circuit has resistors in series, you will use the series circuit formulas to turn those resistors into one equivalent resistor. When you come across parallel branches in series with other resistors, you will use your parallel circuit formulas to turn those resistors into an equivalent series resistor. The simplest way to understand this process is to work through some examples. Two examples follow. Be sure to work through all the steps.

Reducing to a Simple Series Circuit

Example one: Look at the top of **Figure 8-1.** There are two 8-ohm resistors in parallel. In series with the two 8-ohm resistors is a 5-ohm resistance. To find total resistance for this circuit, we reduce the combination circuit to a simple series circuit. This is done by applying Kirchoff's laws of series and parallel circuits.

Using the equal resistances method learned in Chapter 7, you know that two 8-ohm resistors connected in parallel have a total resistance value equal to 4 ohms. This step is shown in the middle of Figure 8-1. The combination circuit can be redrawn as a simple series circuit with a new resistor identified as $R_{2,3}$. The new resistor $R_{2,3}$ is the equivalent value of R_2 and R_3 in parallel. The equivalent value of R_2 and R_3 in parallel is 4 ohms.

Once the new resistor ($R_{2,3}$) is drawn in series with the 5-ohm resistor R_1, you can see that what is left is a simple series circuit. The resistances in the new series circuit are then added, and the total resistance of 9 ohms is calculated.

In the second example, the circuit is just a little more complicated. One more step must be taken when reducing the circuit. In example two, there are two resistors in series in one of the parallel branches.

Example two: Look at the top of **Figure 8-2.** This combination circuit has two resistors in series in one

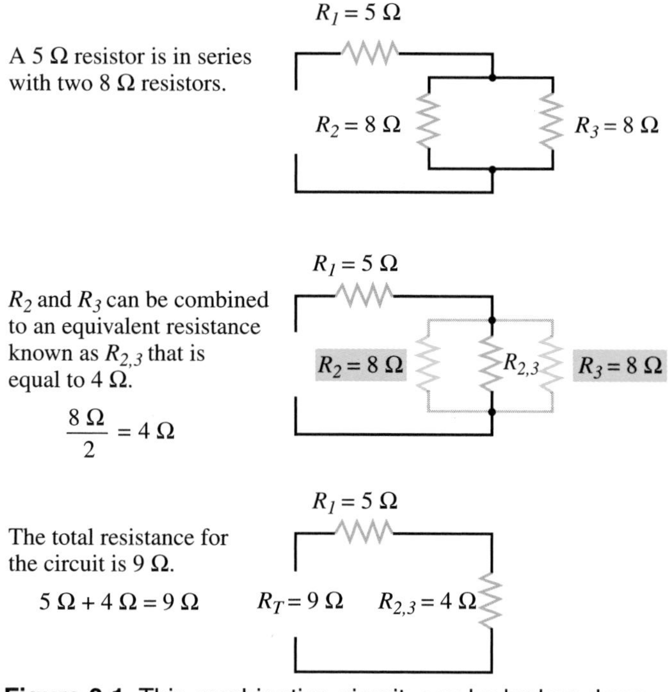

Figure 8-1. This combination circuit, can be broken down to a simple series circuit.

branch of the parallel portion of the combination circuit. These series resistors must be added together before the parallel circuit can be reduced into an equivalent series resistance.

R_2 and R_3 can be combined using the series circuit resistor formula to create an equivalent 12-ohm resistor known as $R_{2,3}$, upper right in Figure 8-2. Resistor R_4 (12 Ω) is in parallel with $R_{2,3}$ (12 Ω). When the resistors are combined using the equal resistances method for parallel circuits, you create an equivalent resistance of six ohms. This equivalent resistance is noted as $R_{2,3,4}$, bottom of Figure 8-2. All that is left then is a simple series circuit with a total resistance value of 10 ohms.

Reducing to a Simple Parallel Circuit

A complex circuit can also be reduced to a simple parallel circuit. Look at the top of **Figure 8-3.** The fastest method of solving this problem is to first combine resistor R_1 and R_2. This leaves you with a simple parallel circuit that can be solved easily using the product over the sum method or the equal resistances formula. Work through this example on your own.

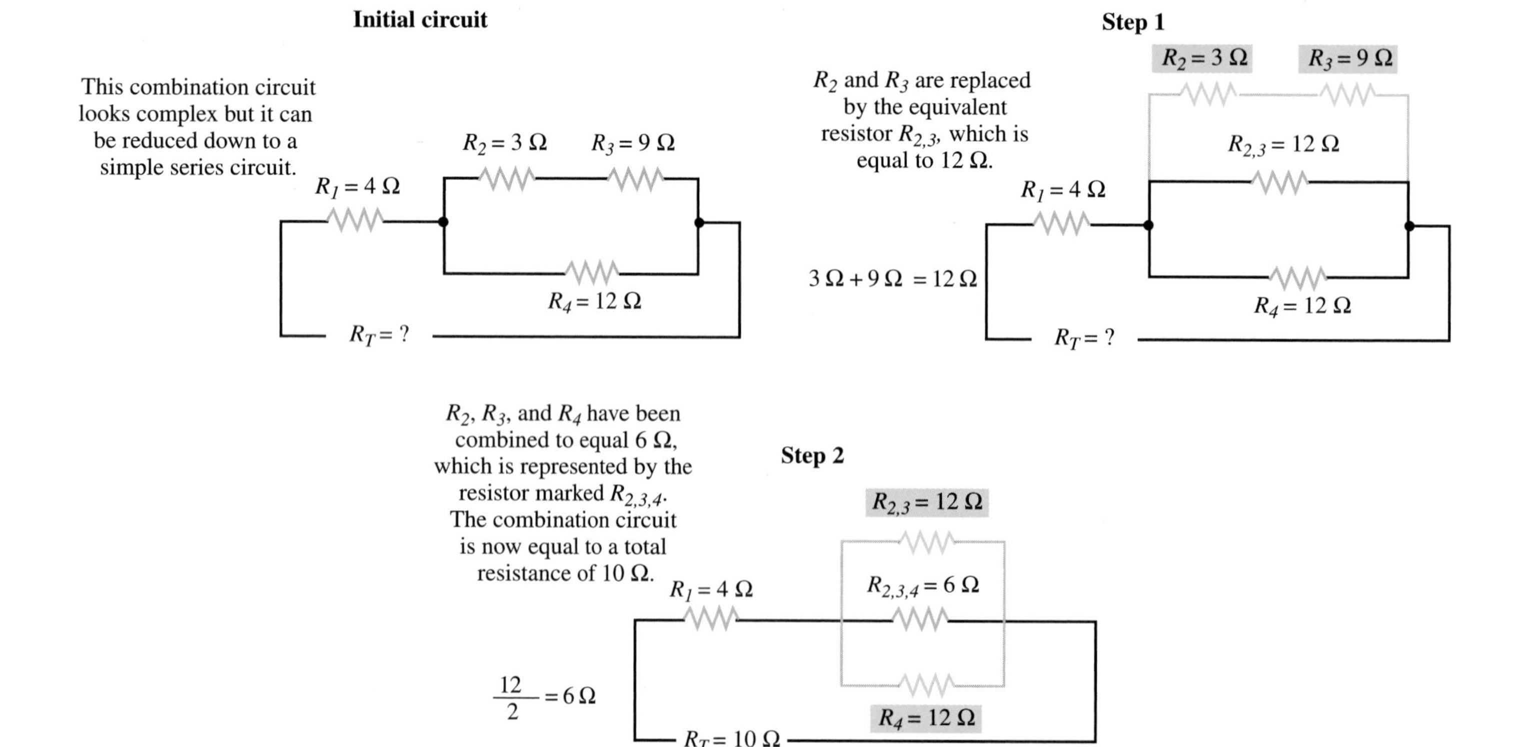

Figure 8-2. This combination circuit can be reduced to a simple series circuit with a total resistance value of 10 ohms.

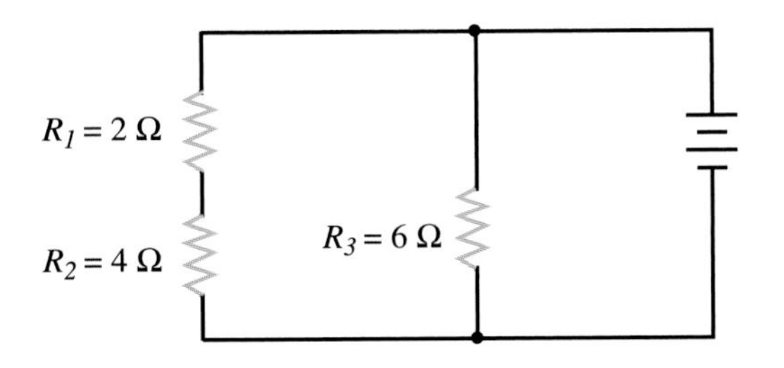

Figure 8-3. Change this series-parallel circuit into a simple parallel circuit, and then solve for the total resistance.

Review Questions for Section 8.1

1. Reduce the following circuit to its simplest form.

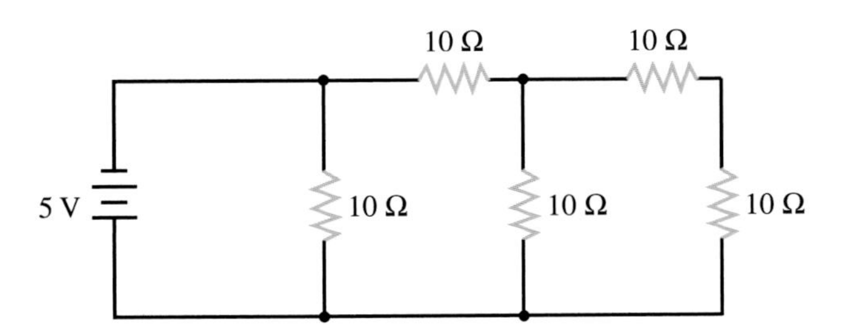

2. Reduce the following circuit to its simplest form.

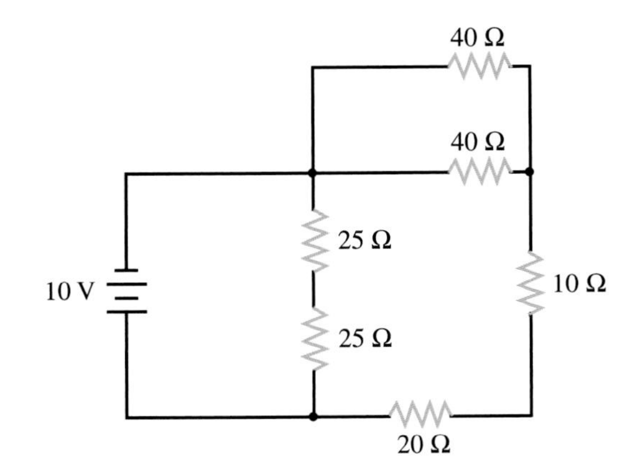

8.2 SOLVING FOR VOLTAGE AND CURRENT VALUES

Finding unknown voltages and currents for individual resistors is a little more complicated in combination circuits than it was for simple series or parallel circuits. However, you have just learned the first step in finding those values, reducing the circuit. Once the combination circuit has been reduced to a simple circuit, the source voltage can be applied and total circuit current can be determined using Ohm's law. If the total current is given, the source voltage can be calculated using Ohm's law. Once this information is calculated, the steps used to reduce the combination circuit can be reversed. This reversal allows you to now determine the individual voltages and the currents through all resistor paths.

Once again, the simplest way to learn this process is through examples. Work through the following two examples. Be sure to work through all the steps.

Example three: Examine **Figure 8-4.** It shows the same circuit as Figure 8-1, but a source voltage of 18 volts has been added to the circuit. Using Ohm's law and the total resistance value previously calculated (9 Ω):

$$E_T = I_T \times R_T$$

$$18 = I_T \times 9\ \Omega$$

$$I_T = 2\ \text{A}$$

The total current value of the circuit is two amperes.

Once the total circuit current value is determined, it can be applied to the simple series circuit to determine the voltage drop values. Remember that the current anywhere in a series circuit is the same. Thus, two amperes of current flow through R_1 and $R_{2,3}$. The voltage drop across R_1 is 10 V (2 A × 5 Ω). The voltage drop across $R_{2,3}$ equals 8 V (2 A × 4 Ω). See the upper-right circuit of Figure 8-4.

Reversing the simplification process, $R_{2,3}$ is restored to the two separate resistors, R_2 and R_3. The eight-volt drop is applied to each resistor as is done in parallel circuit applications. Since the voltage drop of each resistor is known, Ohm's law can be applied to find the current values of each remaining resistor.

$$I_2 = \frac{E_2}{R_2} \quad I_3 = \frac{E_3}{R_3}$$

$$I_2 = \frac{8\ \text{V}}{8} \quad I_3 = \frac{8\ \text{V}}{8}$$

$$I_2 = I_3 = 1\ \text{A}$$

R_2 and R_3 both have current values of one ampere.

Example four: The second example, **Figure 8-5,** is based on Figure 8-2. The total resistance value of Figure 8-2 was calculated in example two to be 10 ohms.

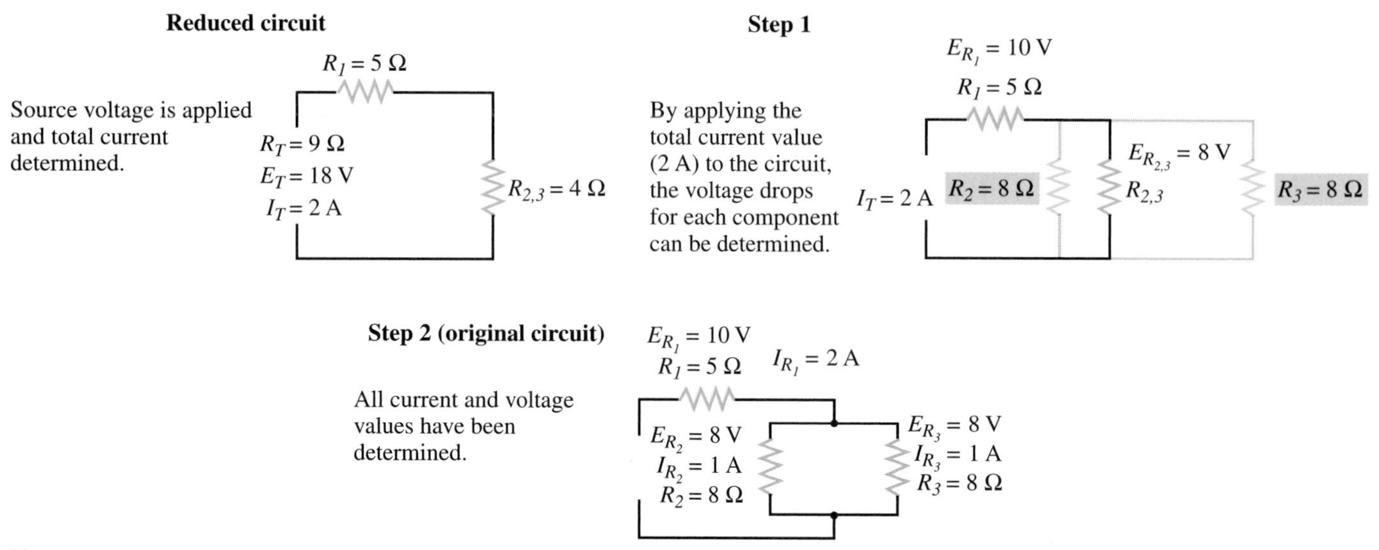

Figure 8-4. The original circuit shown in Figure 8-1 is used to apply voltage and current values.

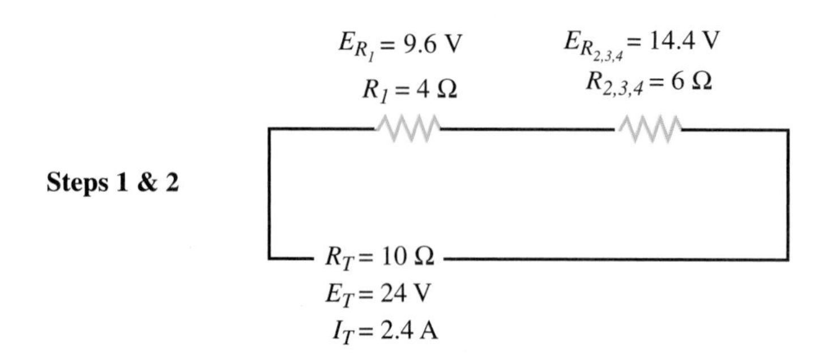

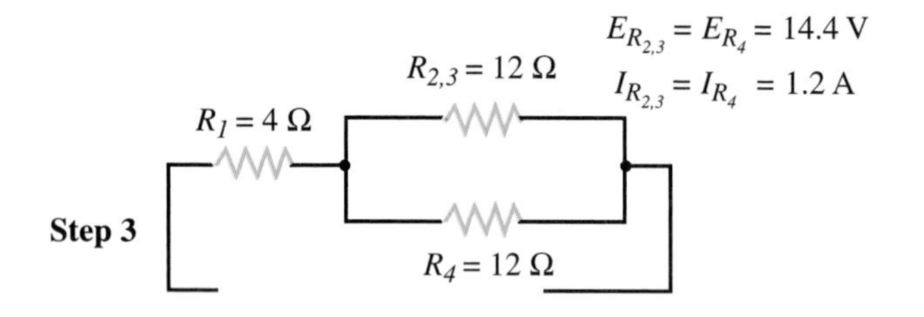

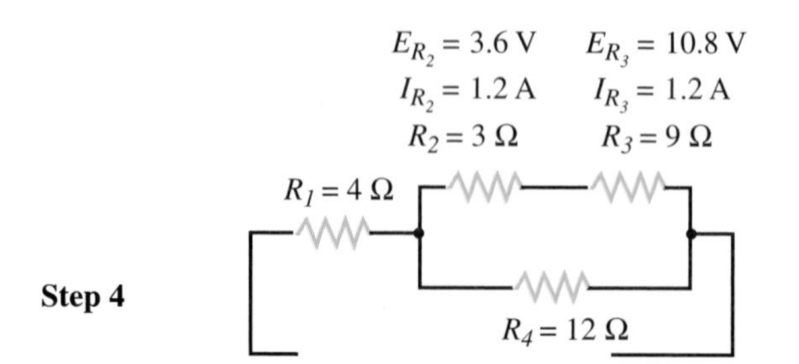

Figure 8-5. The illustration above is based on Figure 8-2, but now a 24-volt power supply has now been applied. Solutions for voltage drops and current values can now be found by reversing the solution process.

The first step is to apply the source voltage of 24 volts to the circuit. Using Ohm's law:

$$E_T = I_T \times R_T$$

$$24 = I_T \times 10\ \Omega$$

$$I_T = 2.4\ \text{A}$$

The total current is 2.4 amperes.

The second step applies the total current of 2.4 A to the simple series circuit. By applying series circuit principles:

$$E_{R_1} = I_T \times R_1 \quad \text{and} \quad E_{R_{2,3,4}} = I_T \times R_{2,3,4}$$

$$E_{R_1} = 2.4\ \text{A} \times 4\ \Omega \quad \text{and} \quad E_{R_{2,3,4}} = 2.4\ \text{A} \times 6\ \Omega$$

$$E_{R_1} = 9.6\ \text{V} \quad \text{and} \quad E_{R_{2,3,4}} = 14.4\ \text{V}$$

The voltage across R_1 has been found to be 9.6 V. To solve for the voltage drops across R_2, R_3, and R_4 you must further reverse the reducing process. The resistor $R_{2,3,4}$ is expanded into resistors $R_{2,3}$ and R_4 in step three. According to parallel circuit theory, the voltage across $R_{2,3}$ and R_4 are both equal to $E_{R_{2,3,4}}$ (14.4 V). The currents through $R_{2,3}$ and R_4 can be found using Ohm's law once again.

$$I_{R_{2,3}} = \frac{E_{R_{2,3}}}{R_{2,3}} \quad I_{R_4} = \frac{E_{R_4}}{R_4}$$

$$I_{R_{2,3}} = \frac{14.4\ \text{V}}{12\ \Omega} \quad I_{R_4} = \frac{14.4\ \text{V}}{12\ \Omega}$$

$$I_{R_{2,3}} = I_{R_4} = 1.2\ \text{A}$$

The final step is to find the voltage drops across R_2 and R_3. Using the current of 1.2 amperes and Ohm's law:

$E_{R_2} = I_{R_{2,3}} \times R_2$ and $E_{R_3} = I_{R_{2,3}} \times R_3$

$E_{R_2} = 1.2\text{ A} \times 3\ \Omega$ and $E_{R_3} = 1.2\text{ A} \times 9\ \Omega$

$E_{R_2} = 3.6\text{ V}$ and $E_{R_3} = 10.8\text{ V}$

Now every voltage drop and current for each resistor is known.

Practice is the only way to ensure understanding of combination circuits. Always look for two or more resistors connected in series or parallel as a starting point for reducing the circuit to an equivalent resistance. Once the total circuit has been reduced to an equivalent series or parallel circuit, the voltage source can be applied and the total current can be found using Ohm's law. To change each of the equivalent resistances back to their original state, the process is reversed. As this is being accomplished, the principles of series and parallel circuits can be applied to determine the voltage drops and currents. With a little practice, you will master this simple process and gain an in-depth understanding of circuit characteristics. This will assist you in dealing with complicated circuits.

Power in a Combination Circuit

Power in a combination circuit can be calculated by using either the formula $P = I^2R$ or the formula $P = E \times I$. These formulas can be used for each individual resistance or for the circuit as a whole. Total power in a combination circuit is equal to the sum of the individual wattages *regardless of whether they are connected with each other in series or parallel.* **Figure 8-6** illustrates how power is determined in a combination circuit.

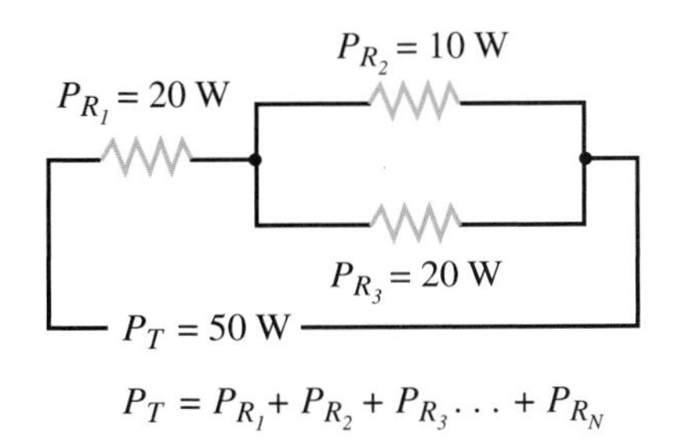

Figure 8-6. Total power is equal to the sum of the individual powers.

Review Questions for Section 8.2

1. Calculate the total current in the following circuit.

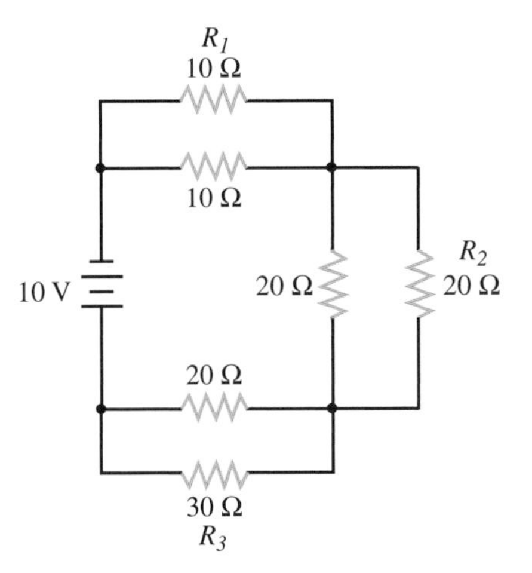

2. What is the value of the voltage source in the following circuit.

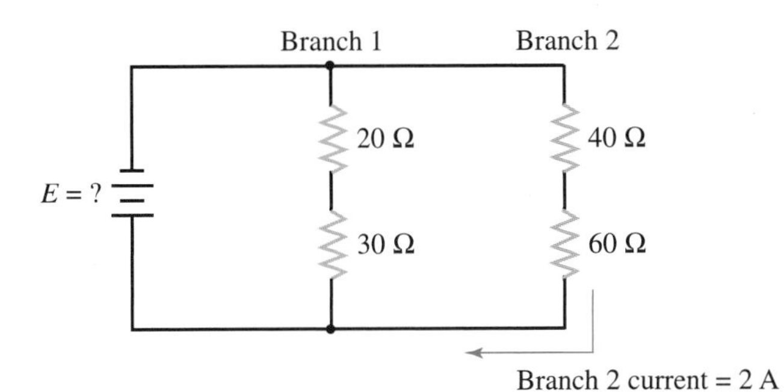

8.3 TROUBLESHOOTING A COMBINATION CIRCUIT

Once the electrical concepts of combination circuits have been mastered, troubleshooting such a circuit is a relatively easy task. In most cases, taking voltage readings of individual components is the most convenient method to use. Look for an open or short circuit condition, as indicated by a voltmeter.

Troubleshooting a combination circuit uses the techniques developed for troubleshooting both series and parallel circuits. The methods used depend on what portion of the circuit is being tested. Look at **Figure 8-7.** In part A, the voltmeter indicates an open circuit at series resistor R_1 (a voltage reading equal to the source). However, if R_2 or R_3 is open, you will not see a full voltage reading across them, see part B. If R_2 or R_3 is open there will still be a complete path for current through the other parallel resistor. To determine if the resistors are open or shorted,

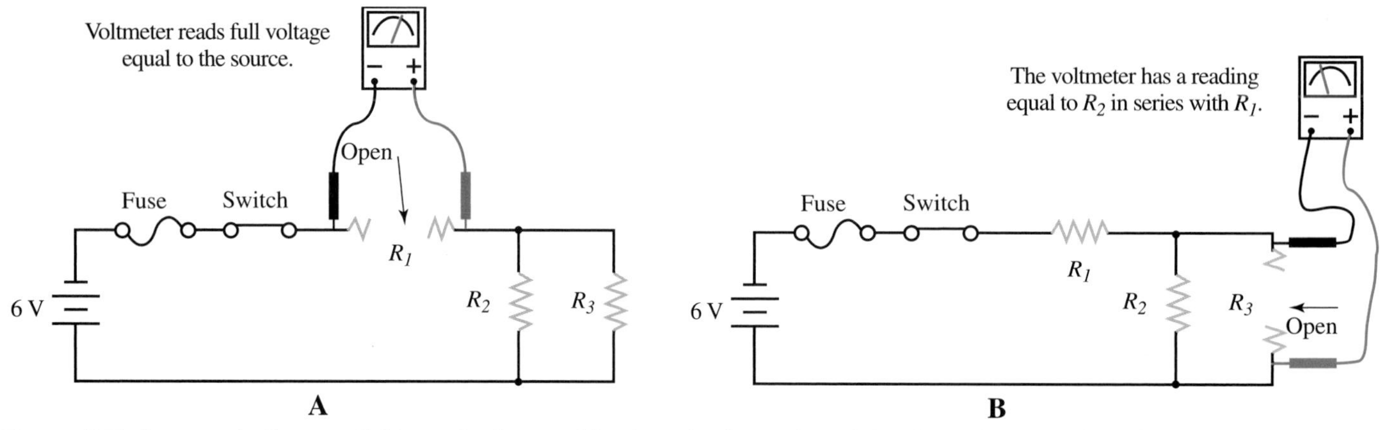

Figure 8-7. An open in the parallel branch of a combination circuit can be misleading when checking for continuity. Always know the expected values to be measured.

the expected voltage drop value must be calculated and compared to the actual voltage drop reading.

Look at **Figure 8-8.** Current readings are an excellent way to determine if the circuit components are operating correctly. But this method should be used as a last resort if it involves desoldering the circuit to take readings on individual components.

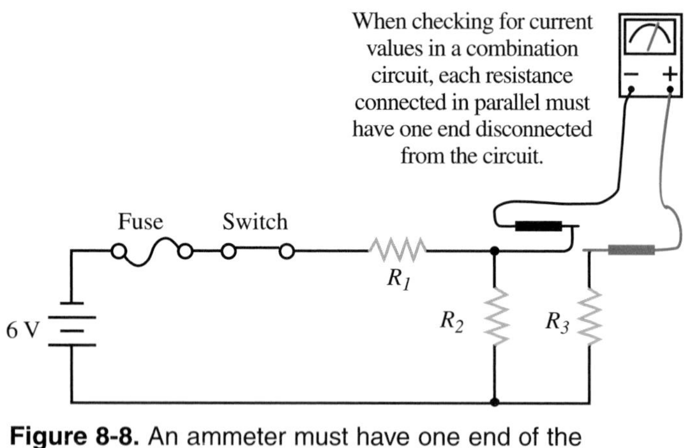

Figure 8-8. An ammeter must have one end of the component disconnected before a reading can be taken.

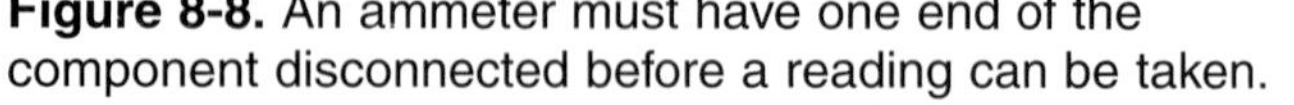

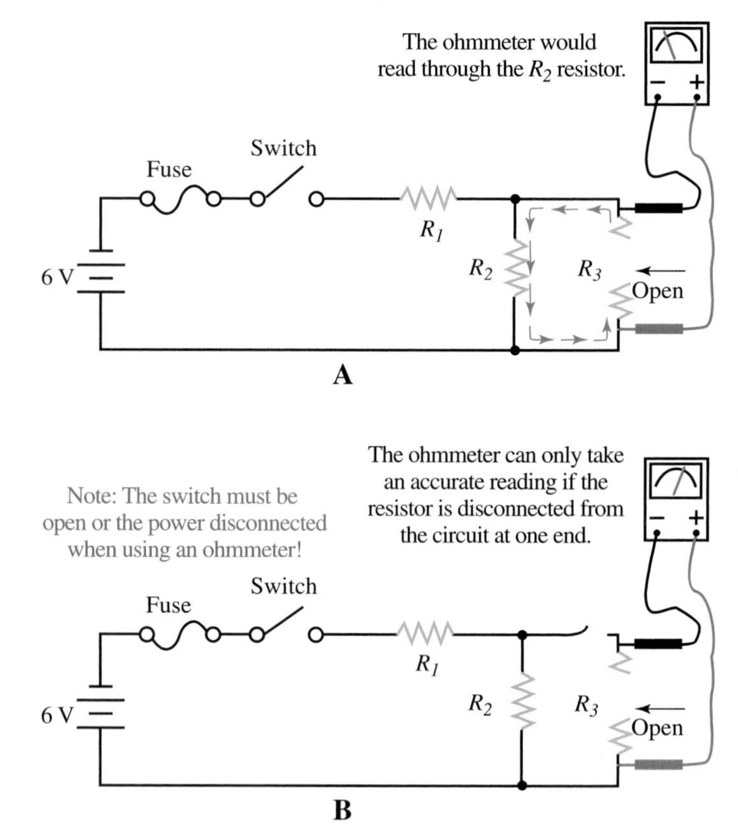

Figure 8-9. As when using the ammeter, one end of the component being tested must be disconnected when using the ohmmeter.

Total resistance can help determine if a circuit is operating correctly, but care must be taken to avoid a false reading caused by reading a component in parallel with another component. In part A of **Figure 8-9,** the ohmmeter is connected across R_3. The resistance on the meter value should be equal to only R_3. But notice that R_2 is connected in parallel with R_3, hence the resistance value on the meter will be equal to R_2. To take an accurate reading with an ohmmeter, one end of the component to be read should be disconnected from the circuit, part B, to prevent reading other components in the circuit. Compare parts A and B of Figure 8-9.

A component connected in parallel with another component or device in a complex circuit design can be very difficult to detect on a printed circuit board. A false conclusion about the component being tested can be reached easily. A schematic of a complex combination circuit is essential in order to properly identify which components are connected in series or parallel with other components. Most electronic systems will have a schematic available. Good technicians use them as an integral part of the troubleshooting process.

Review Questions for Section 8.3

1. Current readings should be used as a last resort in troubleshooting a combination circuit if what is involved?
2. What might cause a false reading when checking a resistance in a combination circuit?

Summary

1. Any complex combination circuit can be reduced to an equivalent resistance or simple circuit.
2. After the total resistance of a combination circuit is reduced to a simple series or parallel circuit, the source voltage can be applied and the total current value determined.
3. After total current is found in a combination circuit, the original circuit can be reconstructed and individual voltages and currents found by applying the characteristics of series and parallel circuits.
4. Total power consumed by a combination circuit is equal to the sum of all individual component power consumption wattages.
5. Troubleshooting a combination circuit uses the techniques developed for troubleshooting both series and parallel circuits.

Test Your Knowledge

Please do not write in the text. Place your answers on a separate sheet of paper.

1. A combination circuit can be reduced to a simple __________ or __________ circuit.
2. Two or more resistances in a combination circuit can be reduced to a(n) __________ resistance.
3. Using the circuit that follows, solve for the unknown values.

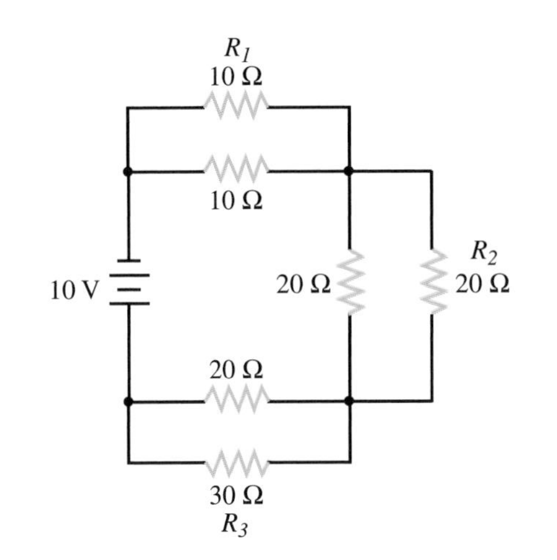

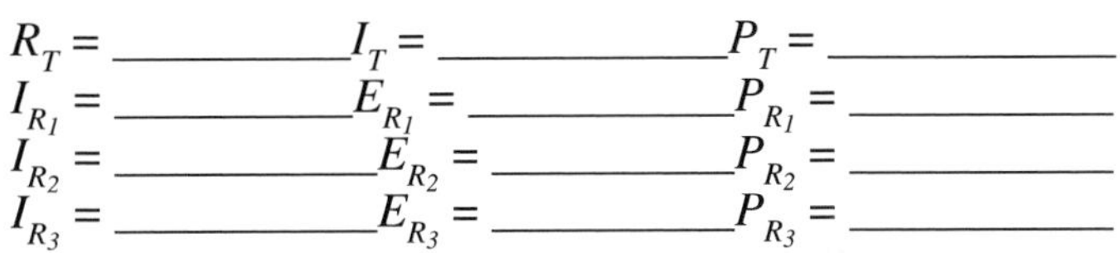

R_T = __________ I_T = __________ P_T = __________
I_{R_1} = __________ E_{R_1} = __________ P_{R_1} = __________
I_{R_2} = __________ E_{R_2} = __________ P_{R_2} = __________
I_{R_3} = __________ E_{R_3} = __________ P_{R_3} = __________

4. Using the circuit below, solve for the unknown values.

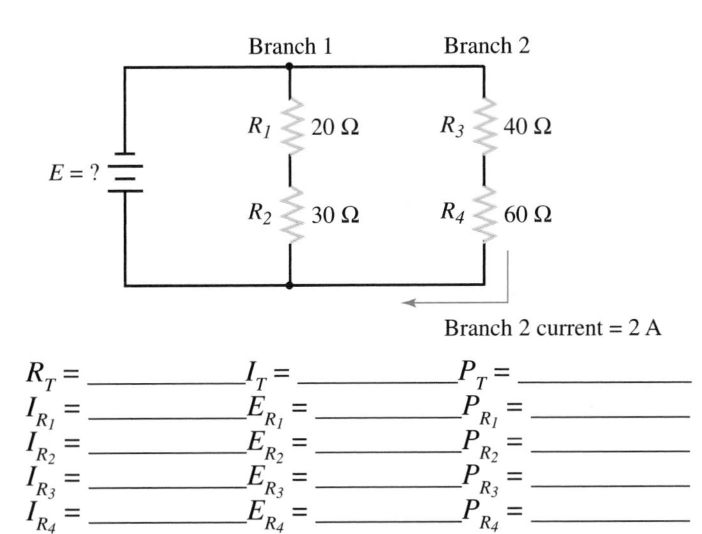

For Discussion

1. What are the implications of Kirchhoff's two laws for electrical circuits?
2. Why does the resistance of a conductor decrease as its cross-sectional area increases?
3. A 6-volt car radio uses 1.5 amps. You want to install it in a 12-volt car. Draw the necessary circuit, and give the values of the components needed to accomplish the installation.

Motors, Generators, and Power Distribution

In Section III of this text, emphasis is placed on providing the skills needed to start working with motors and generators. The five chapters comprising Section III contain most of the electromagnetic devices found today.

Chapter 9 introduces you to the principles and to the practical application of magnetism. Chapters 10 and 11 introduce generators and motors. You will learn about a variety of types of these devices as well as their extensive similarities. Chapter 12, Transformers, introduces the principles of induction and mutual inductance. The concepts of induction and mutual inductance directly relate to ac motor theory of operation, which follows in Chapter 13.

Magnetism

Objectives

After studying this chapter, you will be able to:

- Explain the basic magnetic principles.
- State the three laws of magnetism.
- Describe the link between electric current and magnetism.
- Explain Roland's law.
- Discuss various types of relays and the manner in which they work.
- Describe the use of magnetic shields.

Key Words and Terms

The following words and terms will become important pieces of your electricity and electronics vocabulary. Look for them as you read this chapter.

ampere-turns (At or NI)	magnetic flux
electromagnet	magnetic shielding
field intensity	normally closed (NC)
flux density	normally open (NO)
gauss	permeability
gilbert	relay
left hand rule for a coil	reluctance
left hand rule for a conductor	residual magnetism
lodestone	Rowland's law
magnet	solenoid

9.1 BASIC MAGNETIC PRINCIPLES

For centuries, magnetism has been of interest to people of all sorts. Shepherds, tending their flocks in ancient days, were mystified by small pieces of stone that were attracted to the iron tip on the shepherd's staff. The ancient Chinese navigators discovered that a small piece of this odd stone, attached to a string, would always turn in a northerly direction. These small stones were iron ore. They were called magnetite by the Greeks because they were found near Magnesia in Asia Minor. Since mariners used these stones in the navigation of their ships, the stones became known as "leading stones" or ***lodestones.*** These were the first forms of natural magnets. Today, a ***magnet*** can be defined as a material or substance that has the power to attract iron, steel, and other magnetic materials.

Through laboratory experiments, it was discovered that the greatest attractive force appeared at the ends of a magnet. These concentrations of magnetic force are called ***magnetic poles.*** Each magnet has a ***north pole*** and a ***south pole.*** It was also discovered that many invisible lines of magnetic force existed between poles. Each line of force was an *independent* line. None of the lines cross or touch a bordering line.

Figure 9-1 shows the magnetic lines of force. This field picture was created by placing a bar magnet beneath a sheet of paper, and sprinkling it with fine iron filings. Notice the pattern of lines existing between the poles. These lines of filings reflect the lines of force. Note the concentration of lines at each end of the magnet, or its poles. The lines of force are more concentrated at the poles.

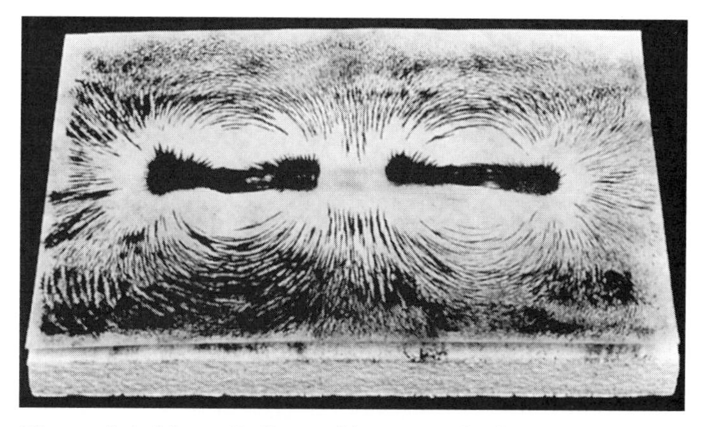

Figure 9-1. Magnetic lines of force can be "seen" by sprinkling iron filings across the lines of force.

Each magnetic line of force travels from the north pole to the south pole through space. The line returns to the north pole through the magnet itself. These closed loops of the magnetic field can be described as *magnetic circuits.* Compare the magnetic circuit to the electrical circuit. The magnetizing force can be compared to voltage, and the magnetic lines of force can be compared to current.

Further scientific investigation showed that the earth acts as one enormous magnet. The earth's magnetic poles are close to the north and south geographic poles. Refer to **Figure 9-2.** You can observe that magnetic north and the north geographic pole do not coincide. A compass would not necessarily point toward true north. This angle between true north and magnetic north is called the ***angle of declination*** or the ***angle of variation.***

There is, however, a line around the earth where the angle of declination is zero. When standing on this line, your compass would point to true north as well as magnetic north. At all other locations on the surface of the earth, the compass reading must be corrected to find true north.

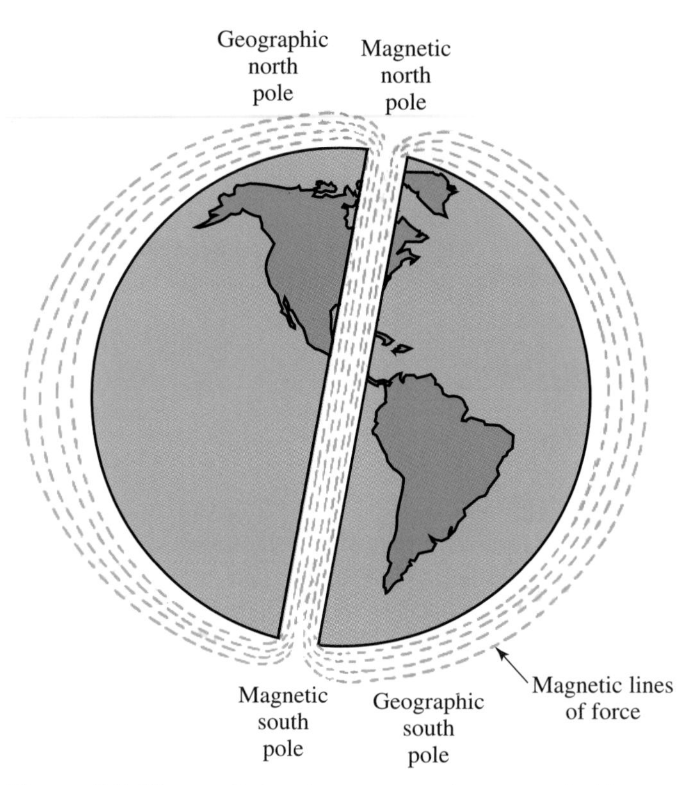

Figure 9-2. The earth is a large magnet, surrounded by a magnetic field.

Laws of Magnetism

The power of a magnet to attract iron has already been discussed. In **Figure 9-3,** two bar magnets have been hung in wire saddles and are free to turn. Notice that when the north (N) pole of one magnet is close to the south (S) pole of the other, an attractive force brings the two magnets together. If the magnets are turned so that two N poles or two S poles are close to each other, there is a repulsive force between the two magnets. This demonstrates two of the laws of magnetism. These laws are stated:

Like poles repel each other.
Unlike poles attract each other.

Place the two magnets under a sheet of paper. Using the iron filings to detect the invisible fields, observe that in the attractive position (N and S together) the invisible field is very strong between the two poles, **Figure 9-4** (top). However, when the magnets are placed with like

EXPERIMENT 9-1: Observing Magnetic Lines of Force

In this experiment, you will observe the placement of the lines of force emanating from permanent magnets.

Materials

2–bar-shaped permanent magnets
1–piece of cardboard
–small quantity of iron filings

1. Place magnets on a table facing so that the poles attract. Keep the magnet ends from touching. (You may need to place a nonmagnetic object between the magnets to keep them separated).
2. Place the cardboard on top of the magnets. Sprinkle iron filings on cardboard and observe pattern. Copy the pattern onto a sheet of paper. Remove the cardboard and the filings.
3. Place magnets on the table so that the poles repel.
4. Place the cardboard on the magnets and sprinkle the filings on the cardboard. Observe the pattern. Copy the pattern, and then remove the cardboard and the filings.
5. Place magnets on the table so that two attracting ends touch.
6. Place the cardboard on the magnets and sprinkle the filings. Observe and copy the pattern.

What conclusions can you draw from these observations?

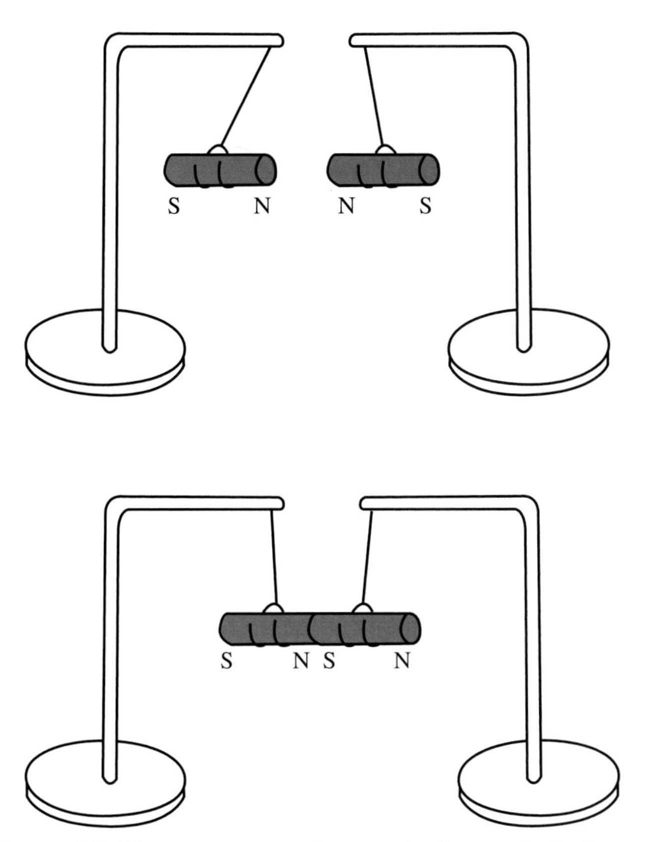

Figure 9-3. These permanent magnets demonstrate the laws of magnetism.

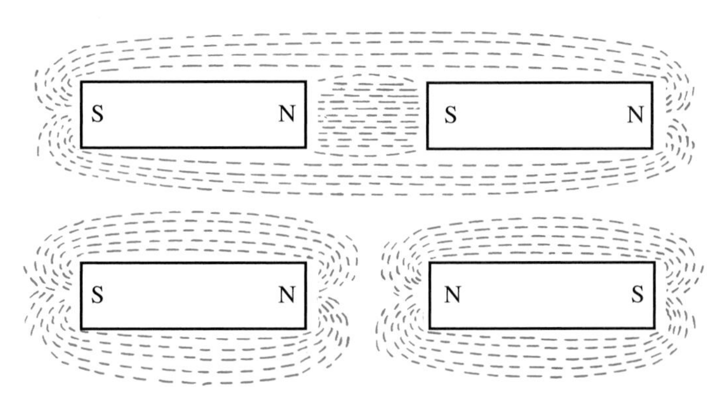

Figure 9-4. These sketches show the magnetic fields of attracting and repelling magnets.

poles together, the repulsive force is shown by an absence of lines between the poles, Figure 9-4 (bottom).

CAUTION:

Certain precautions should be observed during the handling and storage of permanent magnets. Magnetism is a result of molecular alignment. Any rough handling, such as the dropping or pounding of the magnets, can upset the molecular alignment and weaken the magnet.

What causes a substance to become magnetized? The molecules in an iron bar act as tiny magnets. If these tiny magnets are in a random order, **Figure 9-5** (top), the bar does not act as a magnet. However, when these tiny magnets are arranged so that their north and south poles are in line, Figure 9-5 (bottom), the iron is magnetized. This can be demonstrated by breaking a piece of magnetized iron into several pieces. *Each of the broken pieces acts as a separate magnet.* **Figure 9-6** shows a broken magnet. When the iron is demagnetized, these molecules are placed back in random positions.

This molecular action is further demonstrated by the way a magnet is made. For example, take an unmagnetized iron bar. Rub it a few times in the same direction with a permanent magnet. A test (bring the bar near some iron filings) will show that the bar is now magnetized. Rubbing the iron bar with the magnet lines up the molecules and causes the iron to become magnetized. Permanent magnets are made by placing the material to be magnetized in a very strong magnetic field.

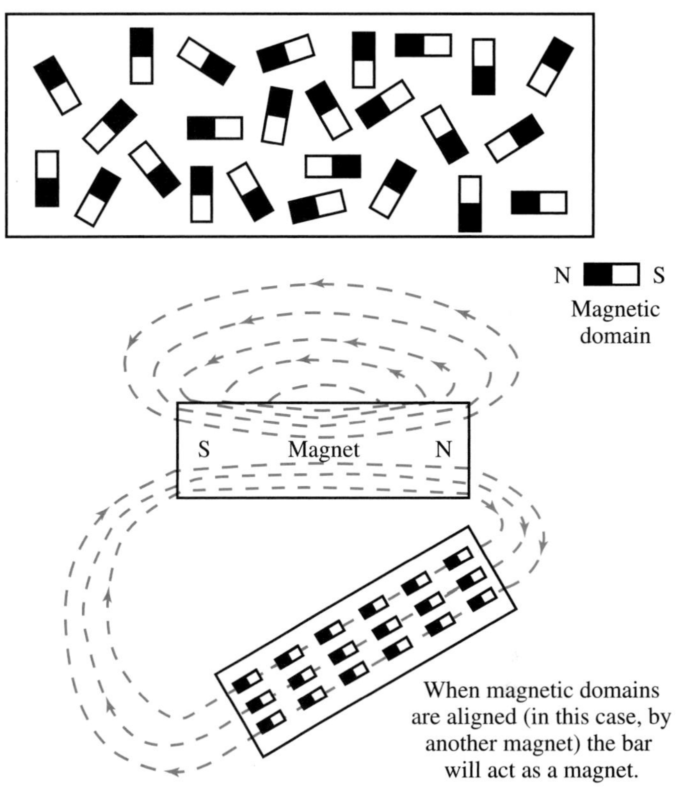

N
S
Magnetic domain
S
Magnet
N
When magnetic domains are aligned (in this case, by another magnet) the bar will act as a magnet.

Figure 9-5. Top—Molecules are not aligned, no magnetic force. Bottom—The molecules have been aligned by a nearby magnet. A magnet is created.

Figure 9-6. A long magnet may be broken into several smaller magnets.

CAUTION:

Bar magnets should be stored in pairs with north and south poles together to ensure their long life. A single horseshoe magnet may be preserved by placing a small piece of soft iron, called a keeper, across its poles.

Permanent magnets are made in a number of shapes and sizes, **Figure 9-7.** Flexible magnets may be used as a visual teaching aid. They are held in place on sheet iron by magnetism. Industry uses these boards for traffic control, production flow, temporary notices, and other announcements. You will find similar (but smaller) flexible magnets on refrigerators in almost any home, **Figure 9-8.** These magnets are created by mixing steel filings in with the rubber. These flexible rubber/steel combinations are then placed in a magnetic field.

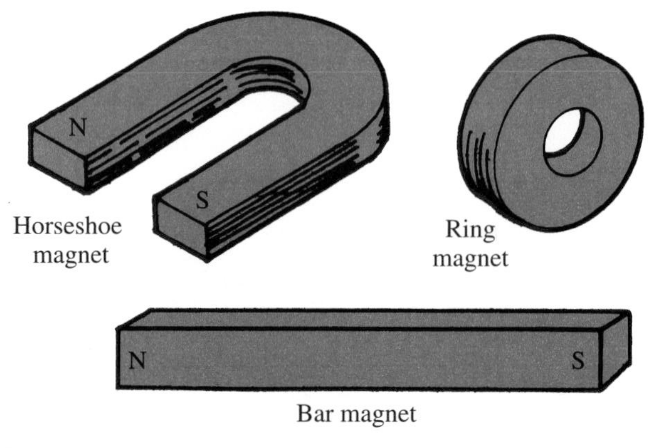

Figure 9-7. Magnets are made in a number of styles, shapes, and sizes. Note that N and S poles of the ring magnet cannot be identified.

CAUTION:

Heat will destroy a magnet. Heat energy causes an increase in molecular activity and expansion. This permits the molecules to return to their random positions on the unmagnetized piece of iron.

Recording tape used in cassettes, videotapes, and computers, stores information through the use of magnetic principles. Magnetized iron oxide particles are adhered to a plastic tape or data disk using a binder. By magnetizing these particles with a recording head, audio or video information, photographs, or data can be stored. Likewise, the information can be "read" from a tape or data disk.

Figure 9-8. Flexible magnets are made of a mix of steel and rubber.

Magnetic Flux

The many invisible lines of magnetic force surrounding a magnet are called the ***magnetic flux.*** If a magnet is strong, these lines of flux will be more dense. So, the strength of a magnetic field can be determined by the field's ***flux density,*** or the number of lines per square inch or per square centimeter. Flux density is expressed by the equation:

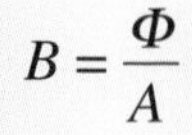

$$B = \frac{\Phi}{A}$$

where B equals flux density, Φ (the Greek letter phi) equals the number of lines, and A equals the cross-sectional area. The cross-sectional area can be measured in square inches or square centimeters. If the cross-sectional area is measured in square centimeters, then the flux density is given in the unit gauss. A ***gauss*** is the number of lines per square centimeter. The flux, B, is usually given in webers per square meter.

Third Law of Magnetism

A simple experiment helps to demonstrate a third law of magnetism. Place one bar magnet on a table. Slowly slide one pole of a second bar magnet toward the opposite pole of the first bar magnet sitting on the table. When the two magnets get close enough together, the magnet on the table will be drawn to the second magnet. This demonstrates that:

The attractive force increases as the distance between the magnets decreases.

When the attractive force gains enough strength to overcome the force of friction holding the first magnet to the table, the first magnet slides toward the second. This magnetic force, either attractive or repelling, varies inversely with the *square of the distance* between the poles.

For example, if the distance between two magnets with like poles is increased to *twice* the distance, the repulsive force reduces to *one-quarter* of its former value. This valuable information explains the reason for accurately setting the gap in magnetic relays and switches. The term ***gap*** is the distance between the core of the electromagnet and its armature.

Review Questions for Section 9.1

1. The two poles of a permanent magnet are the __________ pole and the __________ pole.
2. Unlike poles __________ each other.
3. List three common shapes for magnets.
4. Name six uses of magnets that you come in contact with each day.
5. The invisible lines of magnetic force are called __________ __________.
6. Flux density is measured in __________.

9.2 ELECTRIC CURRENT AND MAGNETISM

During the eighteenth and nineteenth centuries, a great deal of research was directed toward discovering the link between electricity and magnetism. A Danish physicist, Hans Christian Oersted, discovered that a magnetic field existed around a conductor carrying an electric current.

You can perform an experiment that shows the magnetic field around the current carrying conductor. Pass a current carrying conductor through a sheet of cardboard. Place small compasses close to the conductor. The compasses will point in the direction of the magnetic lines of force, **Figure 9-9.** Reversing the current will also reverse

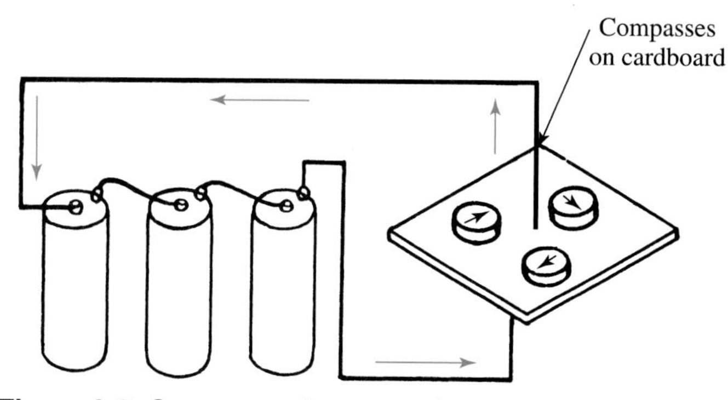

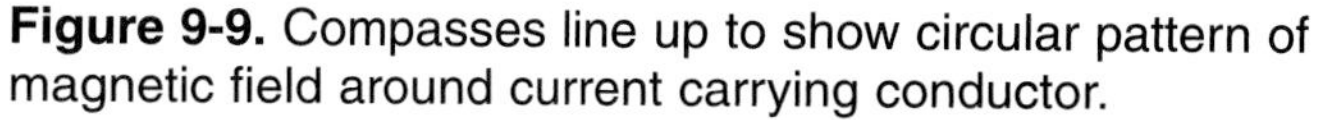

Figure 9-9. Compasses line up to show circular pattern of magnetic field around current carrying conductor.

the direction of the compasses by 180 degrees. This shows that the direction of the magnetic field depends upon the direction of the current.

The ***left hand rule for a conductor*** can be used to determine the direction of the magnetic field around a current carrying conductor. Grasp the conductor with your left hand, extending your thumb in the direction of the current. Your fingers curling around the conductor indicate the circular direction of magnetic field, **Figure 9-10.**

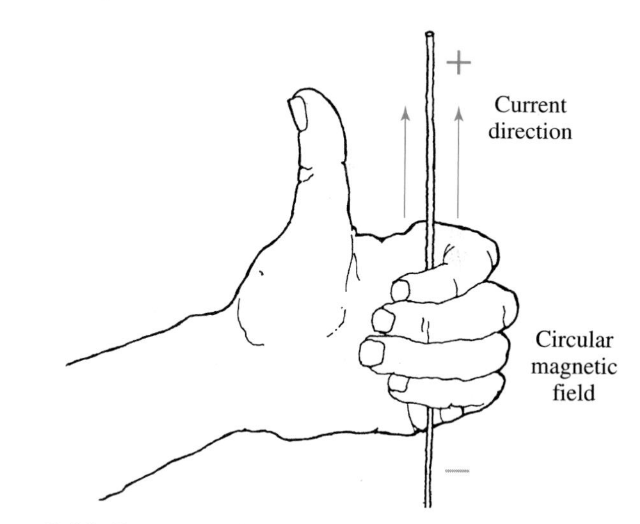

Figure 9-10. Demonstration of the left hand rule for conductors.

In **Figure 9-11,** picture the dot in the center of the conductor on the left as the point of an arrow. This shows that current is flowing toward you. Circular arrows show the direction of the magnetic field. This principle is very important when electrical wires carry alternating currents. This is because the placement of wires, or *lead dress*, has an influence on the workings of a circuit. Conductors are grouped in pairs whenever possible to

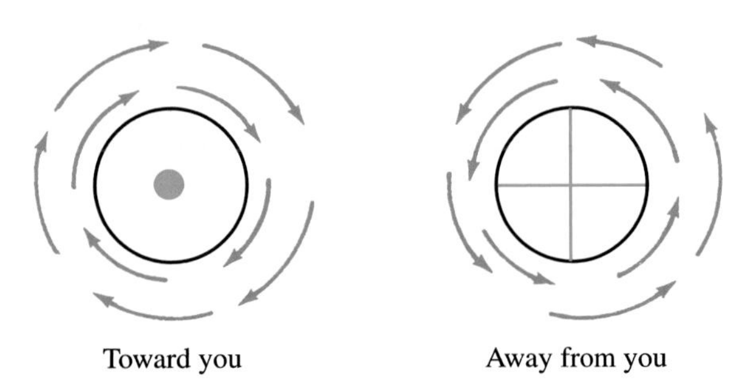

Figure 9-11. These conventions are used to show the link between current flow and the magnetic field. The dot represents a current arrow heading toward you. The cross on the right represents the tail end of the current arrow heading away from you.

eliminate heating effects and radio interference caused by the magnetic field created by the current flow. The National Electric Code requires that wires be run in pairs to help eliminate this heating effect.

Magnetic Circuits

A detailed study of magnetic circuits is beyond the scope of this text. However, you should know the terms used to describe quantities and characteristics of magnetic circuits.

Reluctance

It has already been mentioned that a likeness exists between magnetic circuits and electrical circuits. In addition, a law for magnetic circuits exists that is comparable to Ohm's law for electrical circuits. This law is known as Rowland's law. ***Rowland's law*** states that the number of lines of magnetic flux (comparable to current in electrical circuits) is in direct proportion to the force (comparable to voltage in electrical circuits) producing them. This force (F) is in units of magnetomotive force called ***gilberts.*** Also, there is resistance to the magnetic flux in a magnetic circuit. The resistance to magnetic flux is called ***reluctance.*** Reluctance is measured in gilberts per maxwell or in ampere-turns per weber. One ampere-turn per weber is sometimes called a rel. In equation form:

$$\Phi = \frac{F}{\mathcal{R}}$$

where Φ = total number of lines of magnetic force, F = force producing the field, $\mathcal{R}$ = resistance to the magnetic flux (reluctance).

The relationship between F in gilberts and ampere-turns is a factor of 1.257. This relationship is stated:

$$F = 1.257\, I \times N$$

where F is the force in gilberts, I is the current, and N is the number of turns.

The term reluctance has been defined as the unit of measurement of the resistance to the passage of magnetic lines of force. Experimentation proves that, as the flux path increases in length, the reluctance increases. Likewise, as the cross-sectional area of the flux path decreases, the reluctance increases.

Permeability and field intensity

Reluctance also depends upon the material used as the path. The ability of a material or substance to conduct magnetic lines of force is called ***permeability.*** The permeability of air is considered to be one. Materials with permeabilities higher than one conduct the lines of force better than air. Materials with permeabilities lower than one offer more resistance to magnetic lines of force. Permeability is expressed by the Greek letter μ, (pronounced mew). The factors determining the resistance in a magnetic circuit can be compared to the factors affecting the resistance of electron flow in a conductor (review Chapter 4). For the magnetic circuit:

$$\mathcal{R} = \frac{l}{\mu \times A}$$

where $\mathcal{R}$ is the reluctance, l is length of path, μ is the permeability of the material in the path, and A is the cross-sectional area of the path.

There is a distinct link between the density of a magnetic field, the force producing the field, and the permeability of the substance in which the field is produced. Mathematically:

$$\mu = \frac{B}{H} = \frac{\text{Flux lines per cm}^2 \text{ or gauss}}{\text{Gilberts per cm}}$$

The symbol H is the ***field intensity*** of a magnetizing force through a unit of length. It is commonly expressed in gilberts per centimeter. Mathematically:

$$H = \frac{F}{l} = \frac{1.257 \times I \times N}{l}$$

The term l is, again, the length of path expressed in centimeters. When making calculations the B, H, and μ of common magnetic materials can be found in an electricity handbook.

The Solenoid

When a current carrying conductor is wound into a coil, or ***solenoid,*** the magnetic fields circling the conductors seem to merge or join together. A solenoid will appear as a magnetic field with a N pole at one end, and a S pole at the opposite end. This solenoid is shown in **Figure 9-12.**

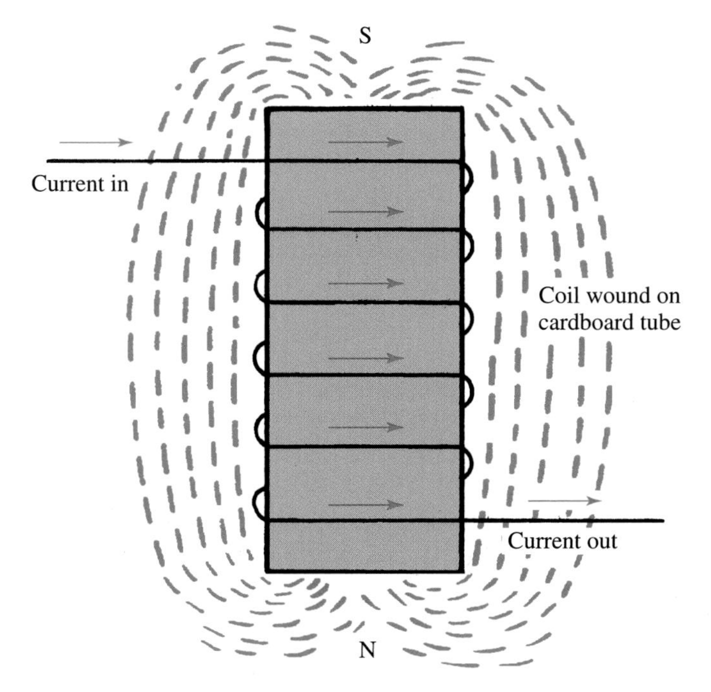

Figure 9-12. A wire wound into a coil is a solenoid and has a polarity set by the direction of current flow.

The ***left hand rule for a coil*** can be used to determine the polarity of the coil. Grasp the coil with your left hand in such a manner that your fingers circle the coil in the direction of current flow. Your extended thumb will point to the N pole of the coil, **Figure 9-13.**

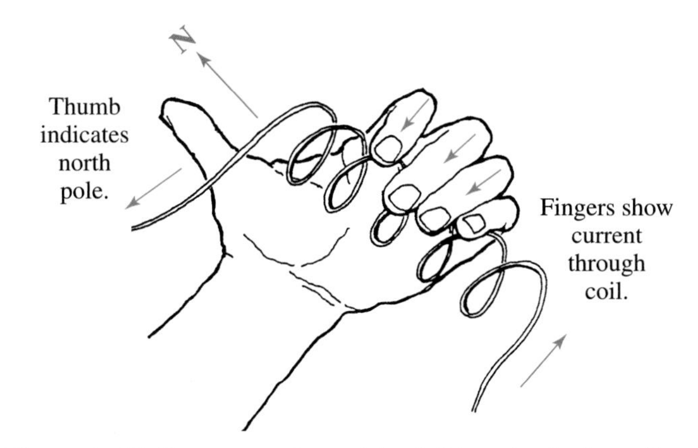

Figure 9-13. Demonstration of the left hand rule for coils.

The strength of the magnetic field of a solenoid depends upon the number of turns of wire in the coil and the size of the current in amperes flowing through the coil. The product of the amperes and turns is called the ***ampere-turns*** (***At*** or ***NI***) of a coil. This is the unit of measurement of field strength. If, for example, a coil of 500 ampere-turns will produce the field strength required for some situation, any combination of turns and amperes totaling 500 will work. Examples:

50 turns × 10 amps = 500 At
100 turns × 5 amps = 500 At
500 turns × 1 amp = 500 At

Electromagnets

In the solenoid just discussed, air only is the conductor of the magnetic field. Other substances conduct magnetic lines of force better than air. These materials would be described as having greater permeability.

To demonstrate this, a soft iron core can be inserted in the solenoid coil, **Figure 9-14.** The strength of the magnetic field is greatly increased. There are two reasons for this increase. First, the magnetic lines have been concentrated into the smaller cross-sectional area of the core. Secondly, the iron provides a far better path (greater permeability) for the magnetic lines. This device (solenoid with an iron core) is known as an ***electromagnet.*** The rules used to learn the polarity of an electromagnet are the same as those for the solenoid. Use the left hand rule.

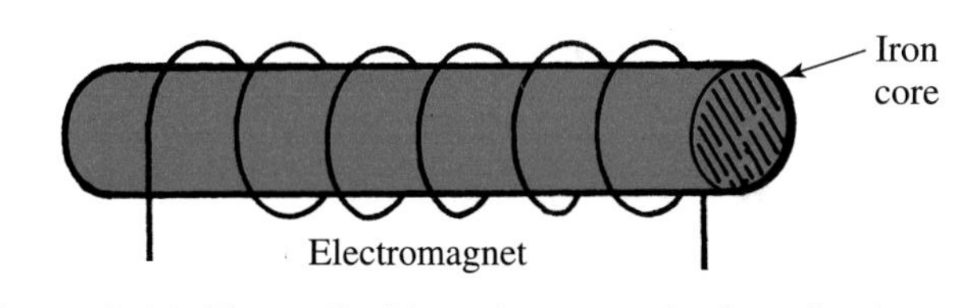

Figure 9-14. The coil with an iron core is described as an electromagnet.

When an electromagnet is energized it is a powerful magnet. When the electrical energy is disconnected, the electromagnet loses most of its magnetism, but not all of it. If the depowered magnet is brought near some iron filings, the filings will be attracted to the core because the iron core has retained a small amount of its magnetism. This magnetism is called ***residual magnetism.***

If very little magnetism remains, the core would be considered as having *low retentivity.* ***Retentivity*** is the ability of a material to retain magnetism after the magnetizing field has been removed. If a core retains a good deal of magnetism, it is said to have *high retentivity.* A soft iron core shows low retentivity. A steel core has high retentivity.

EXPERIMENT 9-2: Testing an Electromagnet

Use an electromagnet to examine the effect of current direction on a magnetic field.

(Note: You may use a prewound solenoid or build one yourself. Instructions for building a solenoid and other parts for electromagnet experiments can be found in Appendix 6.)

Materials

1–Battery, 9-volt
1–Prewound solenoid coil with iron core (or see Appendix 6)
1–Compass
–Pieces of insulated connection wire

1. Set up the circuit shown in the schematic in **Exhibit 9-2A.** The electromagnet is connected to the 9-volt battery.

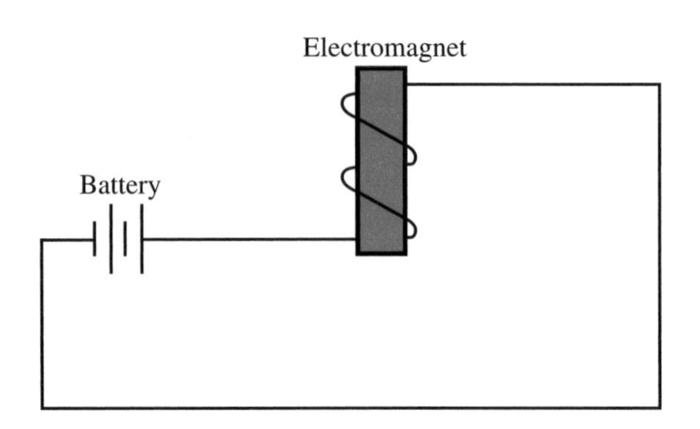

Exhibit 9-2A.

2. Bring the tip of the iron core of the electromagnet near the compass. What do you observe? See **Exhibit 9-2B.** Notice the polarity of the coil.

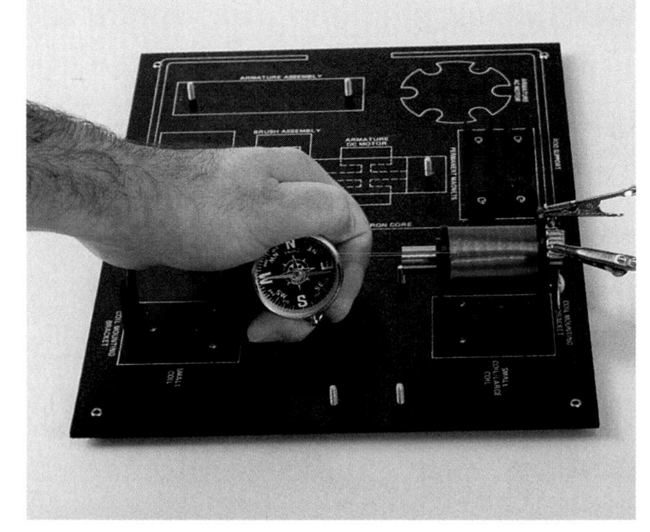

Exhibit 9-2B.

3. Reverse the connections to the battery. Observe the compass action.

Remember the direction of electron flow in a circuit. Does this experiment prove the left hand rule?

Review Questions for Section 9.2

1. What is an electromagnet?
2. Explain the left hand rule for a coil.
3. A coiled conductor carrying current is known as a(n) __________.
4. What is permeability?
5. The amount of magnetism left in a magnetic core is called __________ __________.
6. The unit of measure for magnetomotive force is the:
   a. flux.
   b. gilbert.
   c. gauss.
   d. None of the above.

9.3 THE RELAY

The ***relay*** is a device used to control a large flow of current by means of a low voltage, low current circuit. A relay is a magnetic switch. When a relay's coil is magnetized, its attractive force pulls the lever arm, called an ***armature,*** toward the coil. The contact points on the armature will open or close depending on their normally at rest position. The phrase *normally at rest position* refers to the position of the contacts before the solenoid is energized. If the normally at rest position of the contacts is touching, the large current will flow until the relay is activated. Activating the relay opens the circuit. If the normally at rest position of the contacts is open, then activating the relay will close the circuit. Current will then flow until the relay turns off.

Examine the schematic in **Figure 9-15.** In this circuit a light bulb is connected to a 115 volt power source. This setup is an example of controlling a 115 volt ac circuit with a 1.5 volt dc power source. The two different voltages are said to be mechanically connected by the relay. They are *not* considered electrically connected.

The advantages of this device are clear.

- From a safety point of view, the operator touches

EXPERIMENT 9-3: Residual Magnetism

Using the electromagnet from Experiment 9-2, examine magnetism and residual magnetism.

Materials

1–Battery, 9-volt
1–Prewound solenoid coil with iron core (or see Appendix 6)
1–Switch
–Small steel nails
–Iron filings
–Pieces of insulated connection wire

1. Set up the circuit shown in the schematic in **Exhibit 9-3A.** This schematic has the electromagnet connected in series with the 9-volt battery and the switch.
2. Close the switch and bring the tip of the iron core of the electromagnet near the small nails. What do you observe? See **Exhibit 9-3B.**
3. As the nails are held by magnetic force, open the switch. Do the nails stay attached or fall?
4. Leaving the switch open, bring the tip of the iron core near the iron filings. What do you observe?

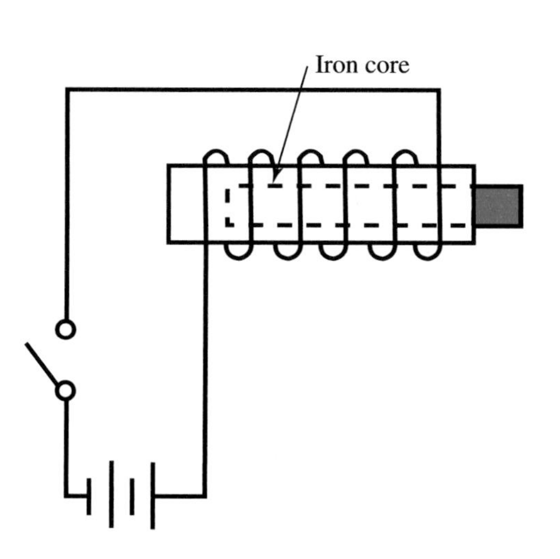

Exhibit 9-3A.

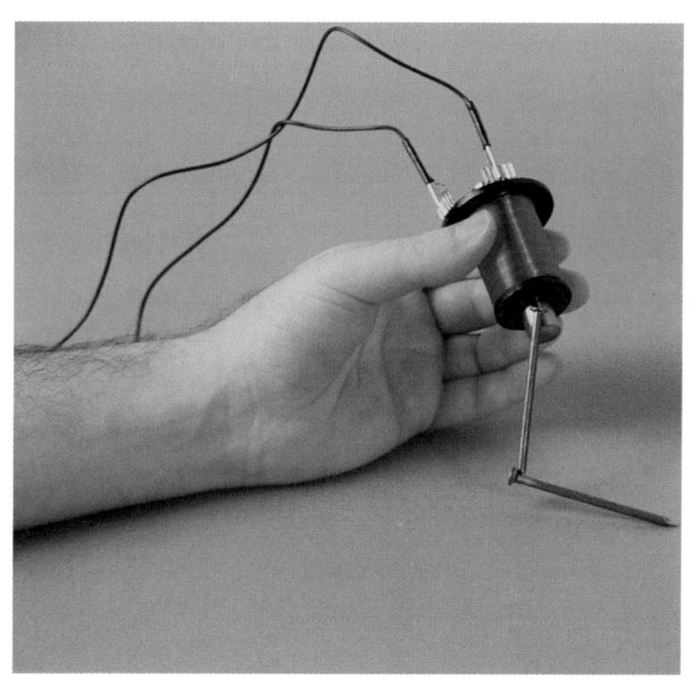

Exhibit 9-3B.

only a harmless, low-voltage circuit, yet controls perhaps several hundred volts by means of the relay.

- Heavy current machines can be controlled from a remote location without any need to run heavy wires to the controlling switch.
- Switching action by means of relays can be very rapid.

Some common relay applications include: large lighting loads, heavy currents in automobiles (such as headlights), and the control of electric motors. Relays are commonly found around the home as part of the air-conditioning and heating systems. They are used to control the starting and stopping of electrical motors used in manufacturing systems. Relay logic is used in industry to control the proper sequence and control of electrical automation systems. Relays are also a vital part of electrical appliances, office machines (such as copiers), and

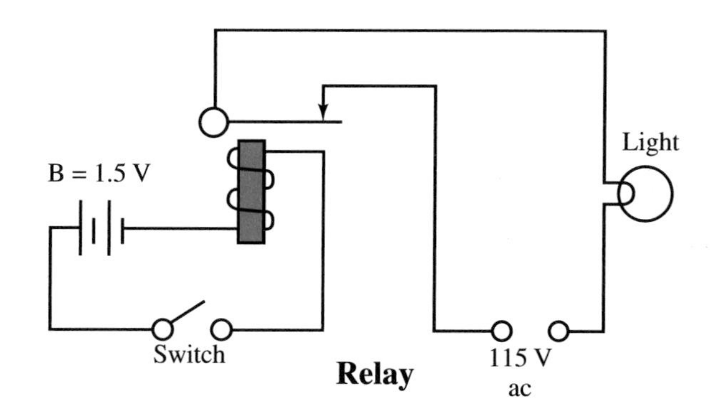

Figure 9-15. Setup of a simple relay you can construct.

EXPERIMENT 9-4: Testing a Solenoid

Experiment 9-4: Testing a Solenoid

In this experiment, you will demonstrate the magnetic attraction developed by a solenoid.

Materials

1–Battery, 9-volt
1–Prewound solenoid coil with iron or steel core (core needs small hole drilled through one end)
1–Switch
1–Metal plate with stand
1–Small rubber band
–Pieces of insulated connection wire

1. Set up the circuit shown in the schematic in **Exhibit 9-4A.** Leave the core out of the electromagnet.

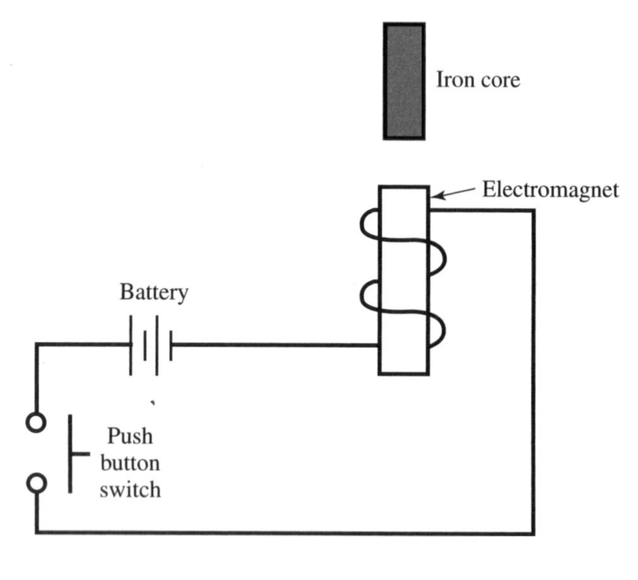

Exhibit 9-4A.

2. Energize the solenoid coil by closing the switch.
3. Bring the iron core close to one end of the coil. You should feel the pulling magnetic force.
4. Release the core from your fingers. The core will be drawn into, and come to rest at, the center of the coil.
5. Refer to **Exhibit 9-4B.** It shows a simple door chime. It is built by arranging the coil and parts as shown.

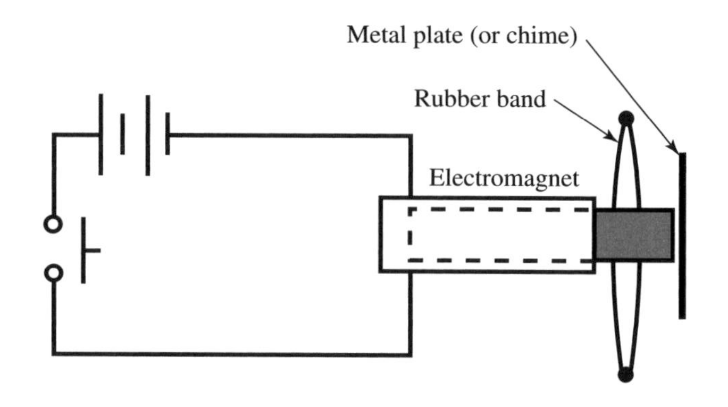

Exhibit 9-4B.

6. Set up this door chime circuit. **Exhibit 9-4C** shows a completed setup.
7. Energize the coil. What happens to the core.
8. Open the switch. Did your chime work?

These experiments show converting magnetism into mechanical motion. Solenoid magnetic attraction principle has many uses in industry. How many can you come up with?

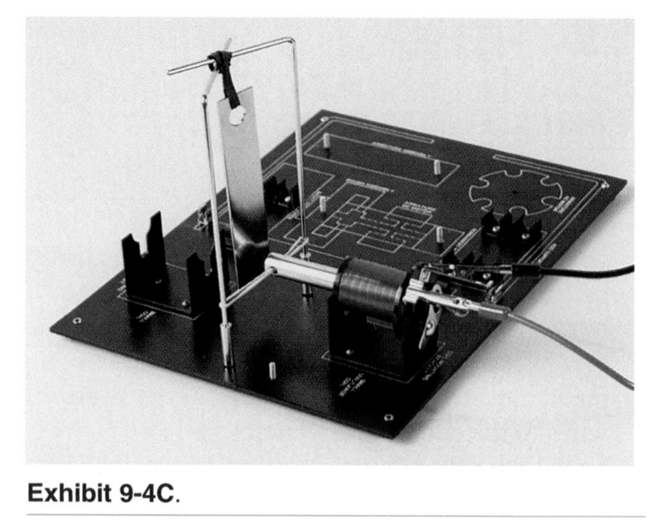

Exhibit 9-4C.

high voltage switching systems used by the power company. There are thousands of electrical applications involving relays. However, relay logic is rapidly being replaced by solid state electronic systems and computers.

Looking through a parts catalog, you will notice various relay designs, **Figure 9-16.** When choosing a relay for a special purpose, there are several things to be considered. Three important considerations are the num-

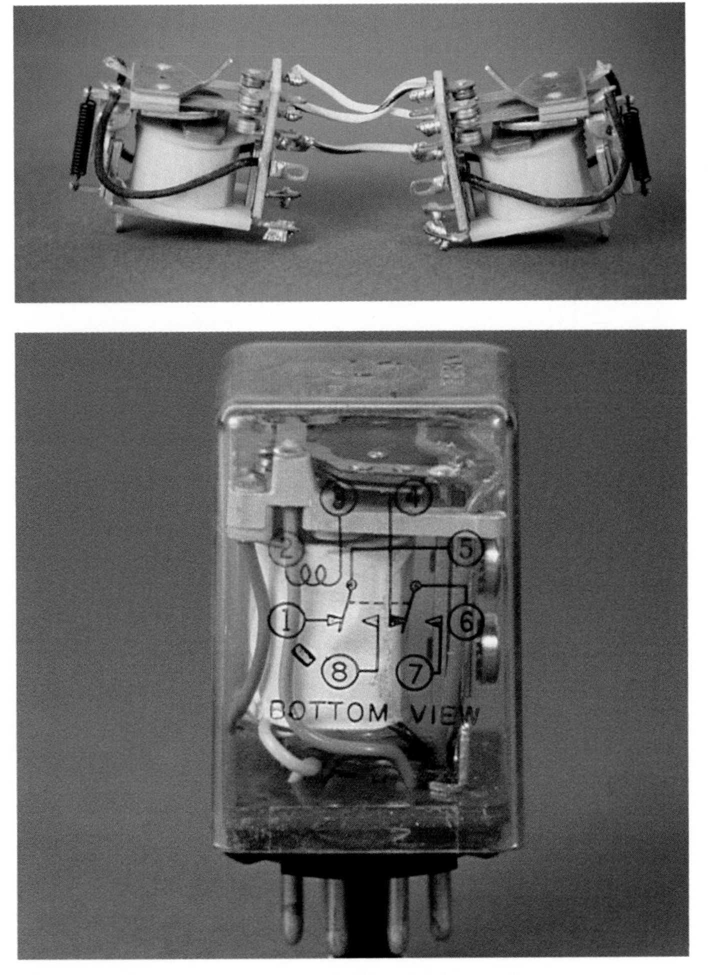

Figure 9-16. Relays found in electronic equipment.

ber of contacts, the amount of current the relay must carry, and their de-energized position. Well-designed relays have points made of silver, silver alloys, tungsten, or other alloys. A relay can be chosen to open or to close a circuit. The relay even can have both ***normally closed*** (closed in the de-energized position) and ***normally open*** (open in the de-energized position) contacts. The terms normally open and normally closed are usually abbreviated as ***NO*** and ***NC.*** The number of contacts and their de-energized position should be specified.

The coil is the most vital specification. The coil chosen needs to produce a large enough magnetic field at its rated voltage to ensure positive contact of switch points at all times. However, some relays are so sensitive they require only a milliampere or less to energize, so care must be taken.

In addition, relays are designed with coils that work on a direct current or an alternating current. You cannot energize a dc coil with ac voltage or an ac designed coil with a dc voltage. Doing this may result in an electrical fire. A detailed explanation of these principles is found in Chapter 14, dealing with inductive reactance.

Time Relays

Certain relays are designed to have an adjustable time delay incorporated into the coil action of the relay. The time-delay can be either a *time-delay on* or *time-delay off.* A ***time-delay on*** means that after the relay is energized, there is a pause before the contacts close. The relay can also be designed as a time-delay off. ***Time-delay off*** means that after the coil is de-energized, there is a time delay before the relay contacts change their position. Today, many time-delay relays are designed to function as both time-delay on or off.

The Reed Relay

The reed relay is another use of magnetism much like the ones we have already discussed. Refer to **Figure 9-17.** Two magnetically sensitive switch contacts are enclosed in a glass tube. If a permanent magnet is brought close to the glass tube, it causes the switch contacts to close. The reed relay can also be operated by an electromagnet. The operating coil is placed around the reed relay.

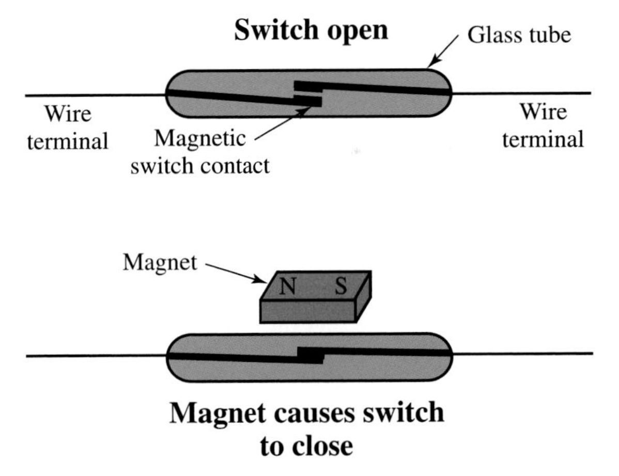

Figure 9-17. A sketch of a reed relay. The magnet causes the reed contacts to close.

Magnetic Circuit Breaker

When referring to electromagnets, the strength of a magnetic field depends upon the *ampere-turns.* If a coil is made with a fixed number of turns, then its field strength may be varied by controlling the current in the coil. The magnetic trip breaker is devised upon this principle. The magnetic trip breaker is usually referred to as an *instantaneous* breaker because there is very little time delay in its tripping action during an overload condition. The instantaneous breaker will trip in approximately 1/20th of one second if the current exceeds the rating of the breaker. Magnetic circuit breakers are usually adjustable. This means that a current range from 10 to 20 percent of the breakers rated current value can be set.

Look at the electromechanical coil and armature parts in **Figure 9-18.** Compare this device to the relay. Notice that, in the circuit breaker, the coil and armature points are in series, and the entire device is connected in series in the line carrying the load. If the line current exceeds a preset value, the coil will build up enough magnetic force to overcome the spring tension of the armature. The armature releases the points and will open the circuit. The contact points must be reset or closed by hand.

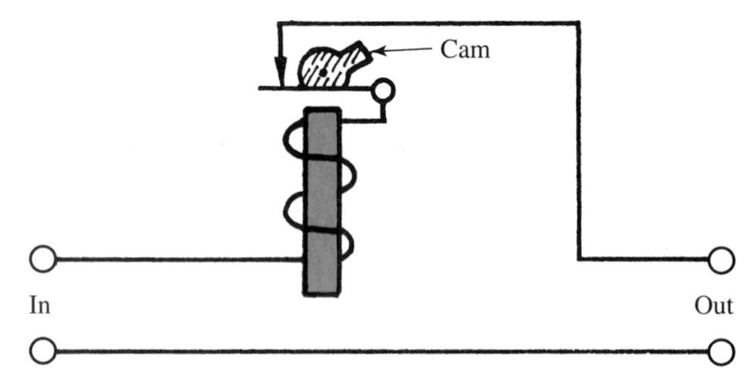

Figure 9-18. Circuit breaker circuit.

This device prevents a circuit from being overloaded beyond the safe capacity of the wires. An electrical code usually requires this type of device when there is a higher-than-normal danger of fire caused by overload. The use of circuit breakers in a petroleum refinery is an example of their importance. In a location such as a refinery, an electrical fire could result in a dangerous explosion.

LESSON IN SAFETY:

The circuit breaker is installed for safety. It prevents overloaded circuits and the danger of fire. The breaker should never be disabled or bypassed. If the breaker opens a circuit repeatedly, locate the condition that is causing the excessive current, and take immediate corrective action.

Buzzer and Doorbell

A common device that uses electromagnets is the buzzer. Arrange the coils and armature as in **Figure 9-19.** Notice that the coil and the contact points are again connected in series. When the coil is energized, the armature is attracted toward the coil. The contact points open and break the circuit. The attraction of the coil then falls to

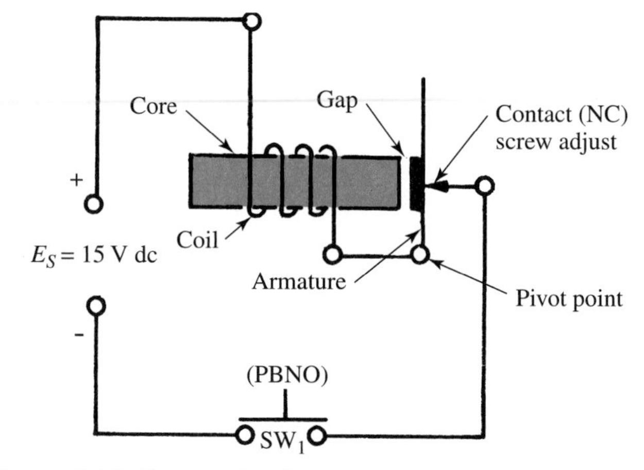

Figure 9-19. Buzzer circuit.

zero. The armature spring pulls the contact points closed. This, once again, energizes the coil and opens the contacts. This action continues as long as a voltage is applied. The device "buzzes" due to vibration of the armature. An extension may be placed on the vibrating armature to which a striker is attached. This striker hits a bell. This is the principle of the doorbell.

More examples of electromagnetic uses are discussed later in this text. A thorough understanding of the action of magnetic fields is vital. The principles of magnetism are the basis for many electrical and electronic applications such as radio, television, motors, generators, transformers, speakers, and many others. Learn the basic principles well and the other applications will be very easy to grasp.

Magnetic Shields

A magnetic field has no bounds. To the first-time student, it is hard to understand that the force of the magnet will pass through any type of substance. It does not matter whether the substance is a concrete wall, glass, or wood. However, an instrument or a circuit can be shielded from magnetic lines of force. This shielding is done by using the *permeability* of some other substance. *The magnetic lines of forces will flow through the path of least resistance.*

If a piece of iron is placed in a magnetic field, the lines of force will follow through the iron rather than through the air. This is because iron has a much greater permeability. The iron acts as a low resistance path for the magnetic lines. In **Figure 9-20** this action can be observed. By this method, magnetic lines may be conducted around the object that needs to be shielded. This is called ***magnetic shielding.*** Many shields are used in electronic equipment to prevent magnetic fields from interfering with the working of circuits.

Electric Guitar

The electric guitar uses a transducer called a pickup. There are several variations of pickups on the market but they all have similar operating principles. The pickup on a guitar is usually constructed of a winding and a magnet or a piezocell crystal.

The steel strings of the guitar vibrate across the pickup. This causes a slight voltage to be produced in the pickup coil. The ac wave of the voltage in the pickup coil is a frequency equal to a pitch of musical sound. If properly tuned, a guitar string that vibrates 440 times a second will produce a 440 Hz electrical frequency. When a 440 Hz electrical frequency is amplified and wired to a speaker, a perfectly pitched musical note is produced from the speaker.

Musicians run their fingers up and down the fret board of their guitars. This causes the length of the string to vary. This action varies the pitch or frequency of the note. Some frequencies sound great and are in perfect pitch. Other frequencies sound flat or out-of-tune. In these cases, the musician may blame the guitar. However, a tuned electric guitar *will* recreate the exact vibration produced by the musician.

Gibson Guitar Corp.

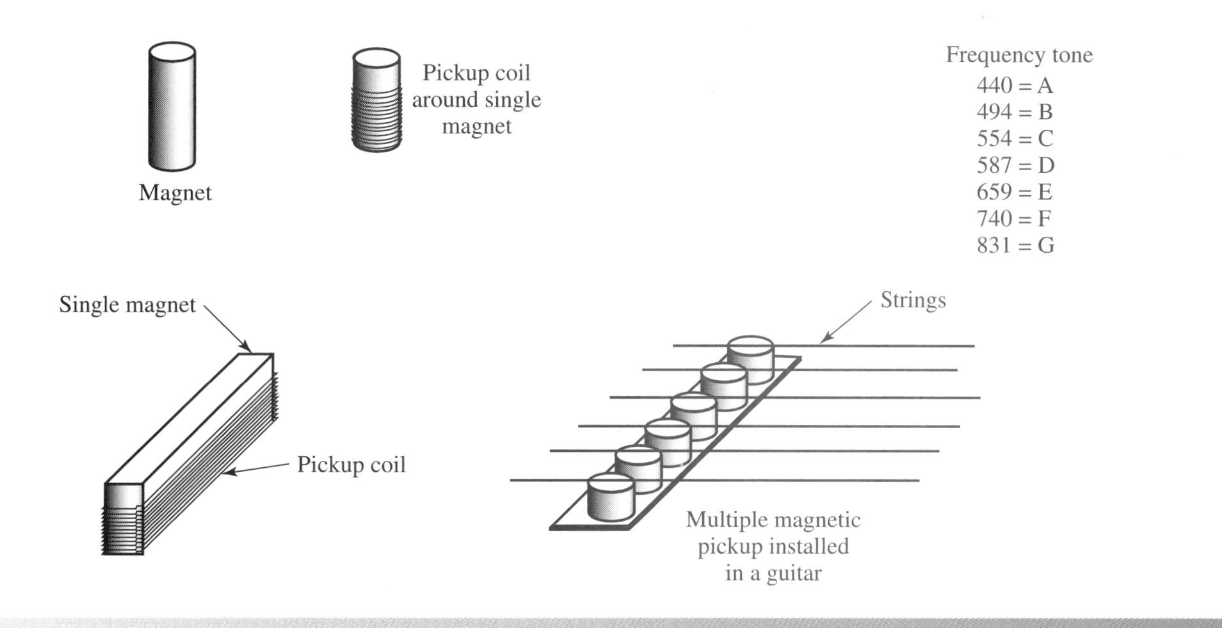

Magnetic Levitation Transportation

One of the fastest growing areas of magnetic application is the use of magnetism to elevate a train car and then propel it on the rails using magnetism. The rails are made up of a long series of solenoids mounted along the track. When the solenoids are energized, the train is elevated above the track. This construction eliminates the need for wheels, axles, or electric motors, which are the main causes of speed reducing friction within the system.

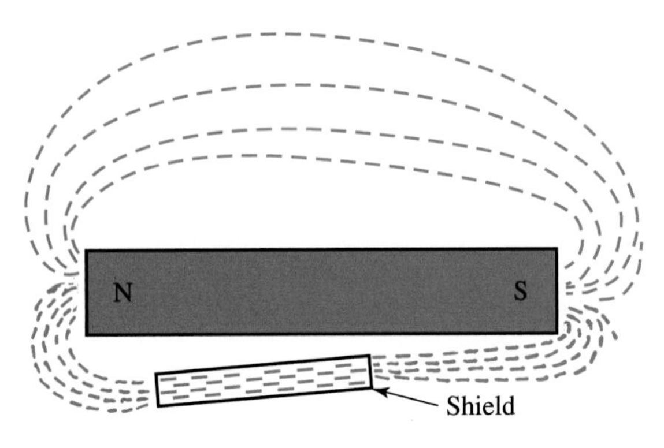

Figure 9-20. Magnetic lines of force may be conducted around a device by using a high permeability material as a shield.

Once elevated above the track, the solenoids receive signals that are sequenced by a computer. The computer energizes the solenoids, one by one, in such a way that the car can gradually increase its speed to over 200 miles per hour as it is propelled along the rail. This propulsion is based on the magnetic attraction and repulsion principles learned in this chapter.

Review Questions for Section 9.3

1. What is a relay?
2. The strength of a magnetic field in an electromagnet depends on its __________-__________.
3. A circuit breaker:
   a. is installed for safety.
   b. opens overloaded circuits.
   c. is often used in place of a fuse.
   d. All of the above.
4. Explain how a door buzzer operates.
5. What is magnetic shielding?

Summary

1. Magnets are materials or substances which have the power to attract iron, steel, and other magnetic material.
2. Magnets have poles. Every magnet has a north pole and a south pole.
3. The earth is a large magnet.
4. Like poles repel each other; unlike poles attract each other.
5. Temporary magnets lose their properties quickly. Permanent magnets keep their magnetic properties for a long period of time.
6. The invisible lines around a magnet are called magnetic flux lines.
7. Electromagnets are created when current flows through a coil of wire.
8. Magnetic resistance is reluctance.
9. Permeability is the ability of a material to conduct magnetic lines of force.
10. Residual magnetism is left in a core after an electromagnetic field is turned off.
11. The relay is an electromagnetic switch.
12. An electromagnetic circuit breaker can be used to prevent an overload of current.
13. Magnetic shielding uses materials with high permeabilities to draw magnetic flux away from sensitive devices.

Test Your Knowledge

Please do not write in the text. Place your answers on a separate sheet of paper.

1. Concentrations of magnetic force are called __________ __________.
2. What is the name of the angle between true north and magnetic north?
3. State the first two laws of magnetism.
4. What is the equation for flux density? What do each of the letter symbols in the equation stand for?
5. State the third law of magnetism.
6. Compare a magnetic circuit to an electrical circuit. Use Rowland's law.
7. List three advantages of using relays.
8. In your own words, state the meaning of the following letter symbols.
   a. H.
   b. μ.
   c. NI.

For Discussion

1. What are some uses of magnets in our everyday living at home?
2. Explain the differences and similarities between magnetism and gravitation.
3. How does a magnetic compass work in the northern hemisphere? How does it work in the southern hemisphere?
4. How can relays be used in elementary computers?

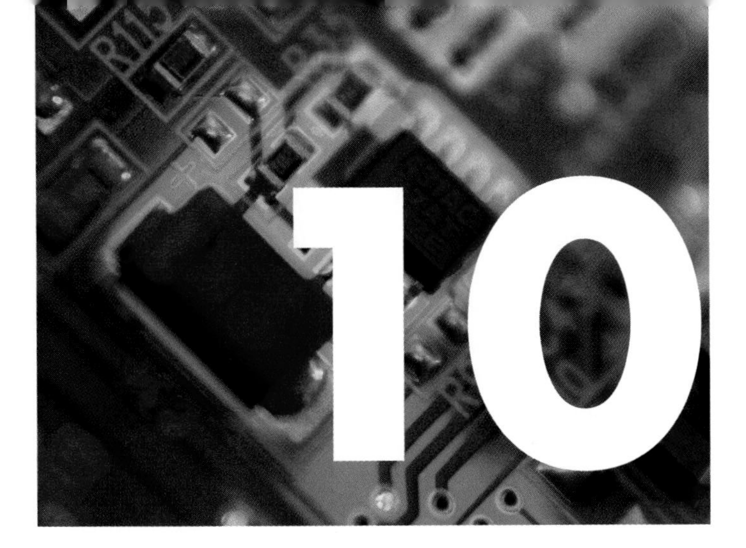

Generators

Objectives

After studying this chapter, you will be able to:
- Explain how mechanical energy can produce electrical energy.
- State Lenz's law.
- Discuss the difference between direct current and alternating current.
- Explain the construction and function of a generator.
- List three types of generator losses.
- Identify the types of generators.
- Discuss voltage and current regulation.
- Calculate average and effective values of ac signals.
- Explain the construction of an oscilloscope.
- Determine voltage using an oscilloscope display.
- Determine frequency using an oscilloscope display.

Key Words and Terms

The following words and terms will become important pieces of your electricity and electronics vocabulary. Look for them as you read this chapter.

amplitude	Lenz's law
average value	magnetic induction
cathode-ray tube (CRT)	oscilloscope
cycle	peak
eddy current	period
effective value	phase displacement
frequency	regulation
generator	rms value
hertz (Hz)	sine wave
instantaneous polarity	vector

10.1 ELECTRICAL ENERGY FROM MECHANICAL ENERGY

In Chapter 9, the discoveries of Hans Christian Oersted were cited as proof that electric current produces a magnetic field. In 1831, scientist Michael Faraday wondered: If electricity produces magnetism, can magnetism produce electricity?

Magnetic Induction

Based on his research, Faraday learned that magnetism could, in fact, produce electricity. He found that moving a conductor through a magnetic field *induced* (brought about) a voltage in the conductor. This is known as ***magnetic induction.*** Three things are required to induce a voltage. There must be:
- A magnetic field.
- A conductor.
- Relative motion between the field and conductor. See **Figure 10-1.**

In the following experiment, you can watch the principles of magnetic induction at work. Refer to **Figure 10-2.** A coil is attached to a galvanometer. A ***galvanometer*** is a meter that measures the size and direction of small electric currents. A permanent bar magnet is placed in the hollow coil. As the magnet moves into the coil, the needle on the meter shows current flowing in one direction. As the magnet is taken out of the coil, the needle shows current flowing in the other direction. When the magnet is still, no current is produced. The same meter movements are produced when the magnet is held still and the coil is moved over the magnet.

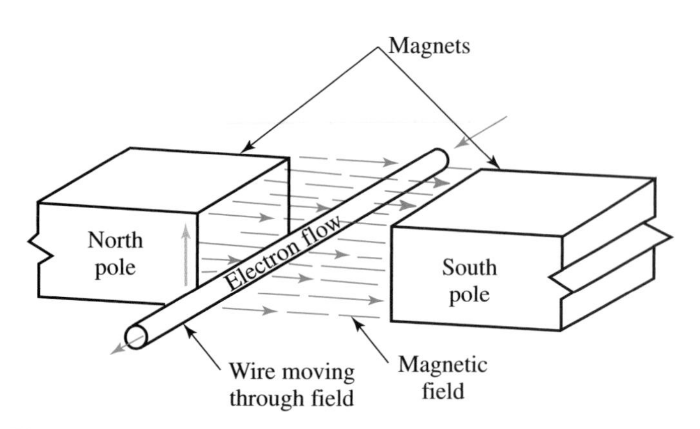

Figure 10-1. Magnetic induction. Passing a conductor through a magnetic field displaces electrons in the conductor. The electrons move through the conductor.

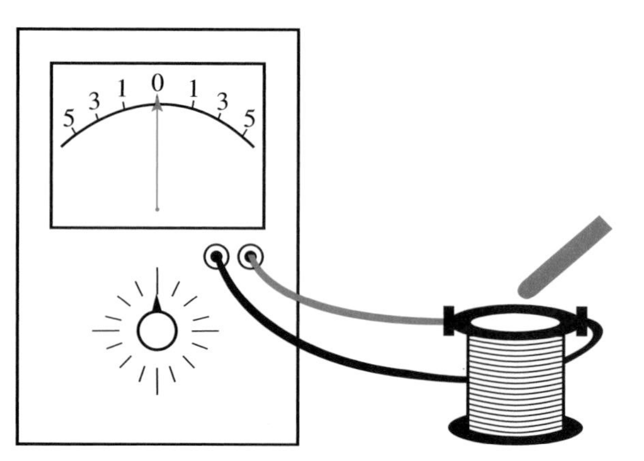

Figure 10-2. Use a galvanometer, a coil, and a magnet to observe the principles of magnetic induction.

Figure 10-3 shows the action of a coil turning in a magnetic field. In position A, the coil top moves parallel to field of magnetism. No voltage is produced. In position B, both sides of the coil are cutting the field at right angles. The highest voltage is produced at this right angle. Position C is like position A, the voltage drops to zero. In position D, the coil is again cutting the field at right angles, where the highest voltage is induced. However, in position D the voltage is in the opposite direction of that produced at position B. The curve in Figure 10-3 shows the voltage induced in the one turn of the coil.

How is this induced voltage created? **Figure 10-4** shows single conductors passing through a magnetic field. At the top, the conductor is pushed downward through the magnetic field. On the bottom, the conductor is pushed back upward. The current generated is shown by the arrows. In each case, the induced current in the conductor forms a magnetic field around the conductor.

The field around the conductor opposes, and is repelled by, the fixed field. This phenomenon is stated in ***Lenz's law:*** *The polarity of an induced electromagnetic force is such that it produces a current. The magnetic field of this current always opposes the change in the existing magnetic field.* More simply, Lenz's law says that: the field induced around the conductor is opposed by the existing field. Therefore, in order to produce electricity, some form of mechanical force must be applied to overcome this opposition and turn the coils. For example, water and steam supply the mechanical force to turn turbines in large power plants.

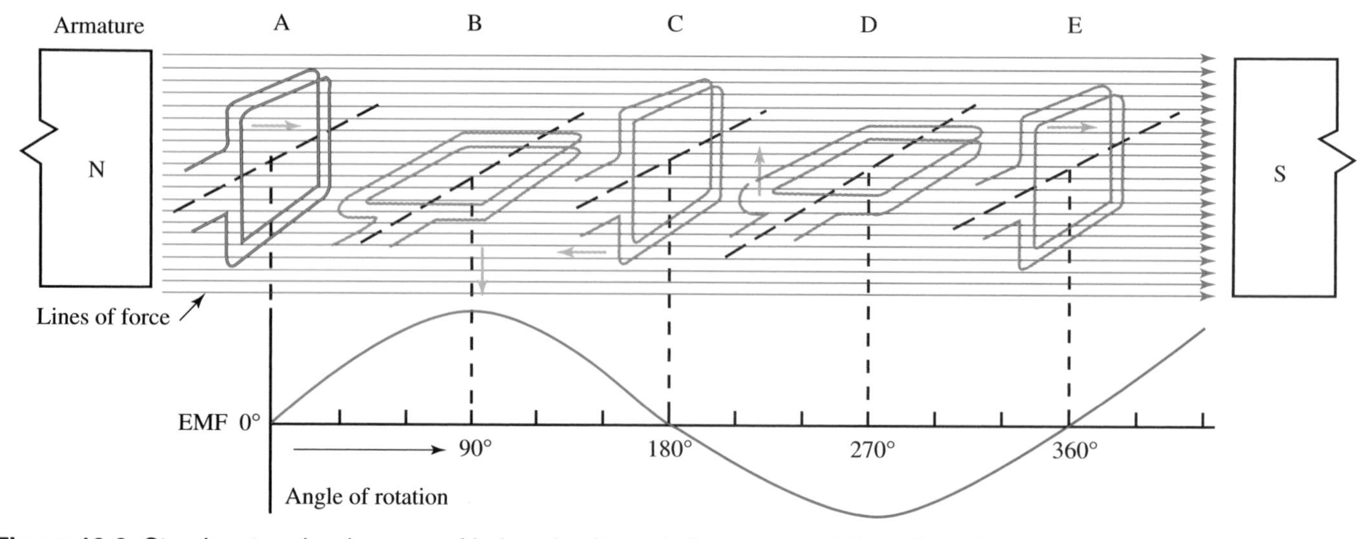

Figure 10-3. Step-by-step development of induced voltage during one revolution of a coil.

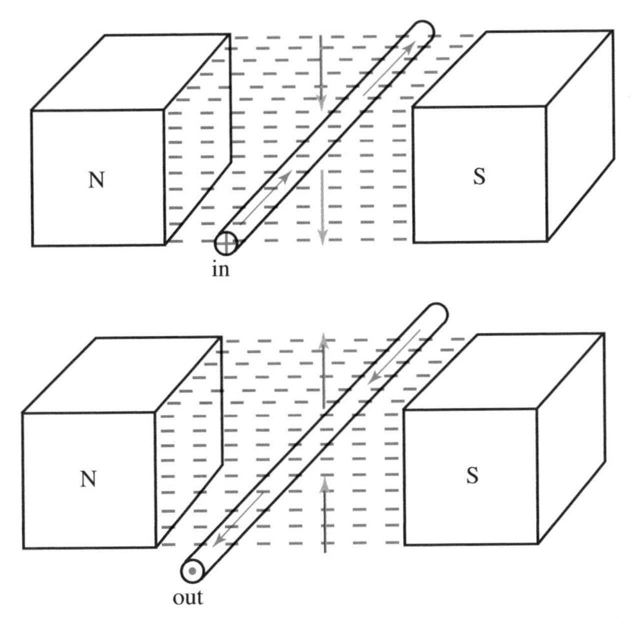

Figure 10-4. The direction of the current through a conductor is determined by the direction the conductor cuts across a magnetic field.

The strength of the induced voltage in a rotating coil depends on:

- The number of magnetic lines of force cut by the coil.
- The speed at which the conductor moves through the field.

When a single conductor cuts across 100,000,000 (10^8) magnetic lines in one second, one volt of electrical pressure is produced. This voltage can be increased by winding the armature with many turns of wire, by increasing its speed of rotation, or both. Reviewing what was learned in Chapter 6, this link can be expressed by the mathematical equation:

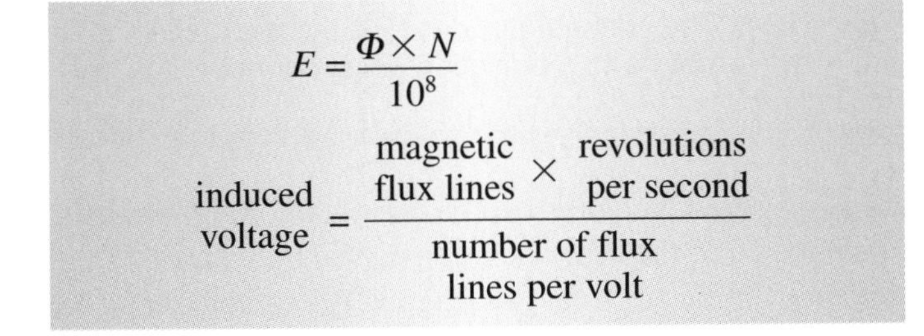

$$E = \frac{\Phi \times N}{10^8}$$

$$\text{induced voltage} = \frac{\text{magnetic flux lines} \times \text{revolutions per second}}{\text{number of flux lines per volt}}$$

where E equals the induced voltage, Φ equals the lines of magnetic flux, and N equals revolutions per second.

For example, if a fixed magnetic field consists of 10^6 lines of magnetic flux and a single conductor cuts across the field 50 times per second, the induced voltage would equal:

$$E = \frac{10^6 \times 50}{10^8} = 50 \times 10^{-2} = 0.5 \text{ V}$$

One half volt is produced by this generator.

DC versus AC

Direct current (dc) flows in only one direction through a conductor. If the flow of electrons is constant it is called ***pure direct current.*** If the flow is intermittent it is known as ***pulsating direct current.*** See **Figure 10-5.**

Alternating current (ac) flows in alternating directions (back and forth) through a conductor within a given time period. Alternating current is shown at the bottom of Figure 10-3. This curve is called a ***sine wave.***

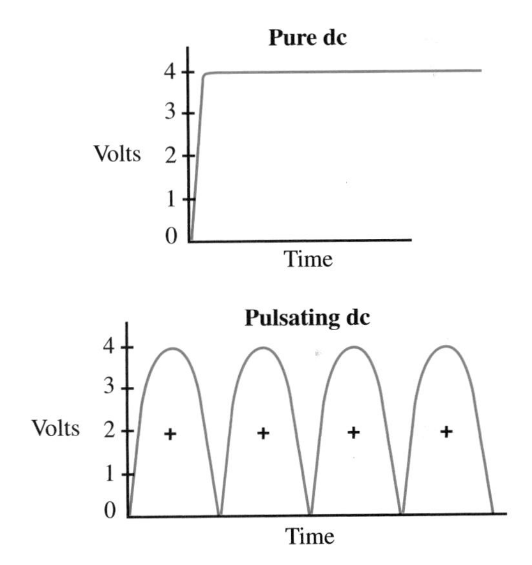

Figure 10-5. Graphs showing pure dc and pulsating dc.

Construction of a Generator

A ***generator*** is a device that changes mechanical energy into electrical energy. You saw this change in Figure 10-3. The revolution of the coil (mechanical energy) was changed to induced current (electrical energy). This action is an example of a very simple generator. The electricity produced in this simple generator is not very useful because it is not powerful enough to do work. The system requires some improvements.

A stronger magnetic field can be created for this improved generator by replacing the permanent magnets with electromagnets. Field coils can be placed over pole pieces or shoes that are fastened to the steel frame or generator case. The revolving coil, or ***armature,*** is suspended in the case resting on the proper bearings. The single coil is replaced by wire coils of many turns on the armature. The rotating armature is connected to the outside circuit through commutators or slip rings. The commutators and slip rings touch brushes in the outside circuit. Generator ***brushes*** are constructed mainly of simple, soft carbon.

The major difference between an ac generator and dc generator is the use of slip rings in the ac generator and the use of a commutator (split ring) in the dc generator. *Both slip rings and split rings provide the electrical current connections from the armature to the generator load circuit.*

Two ***slip rings*** are used on ac generators. The slip rings provide a mechanical means of maintaining the connection between the armature circuit and the outside circuit. See **Figure 10-6.** In the ac generator, the slip rings are in constant contact with the brushes. Since an alternating current is produced in the armature circuit, the outside circuit is also ac. Both the dc generator and ac generator produce an ac current in the armature windings.

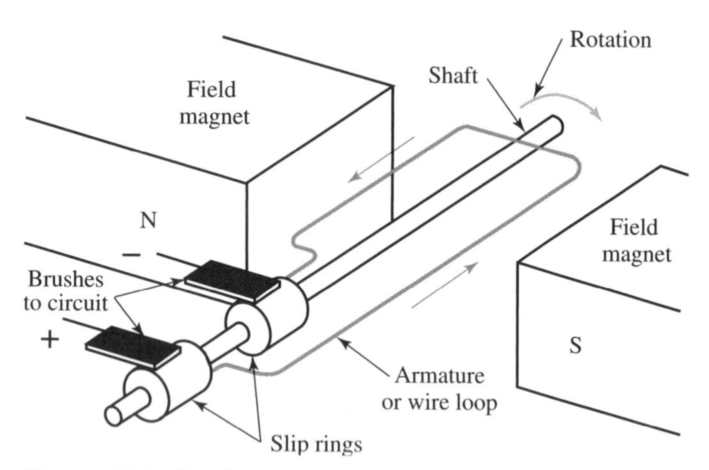

Figure 10-6. Simple ac generator. A wire loop carries the induced current. Electrons flow out one brush, through the circuit, and back in through the other brush.

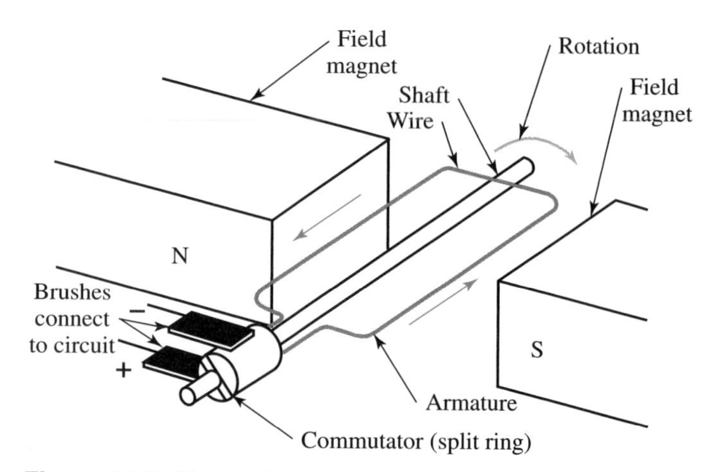

Figure 10-7. Simple dc generator.

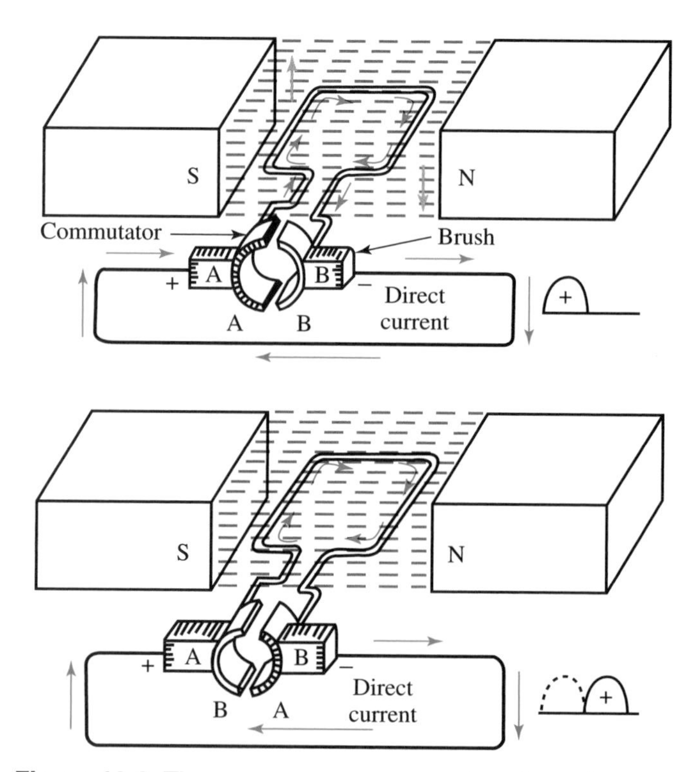

Figure 10-8. The commutator changes the alternating current in the armature to a pulsating direct current in the outside circuit.

When dc is desired in the outside circuit, a set of commutator segments and a set of brushes are used. A ***commutator,*** or ***split ring,*** is a device that reverses electrical connections and is used on dc generators. See **Figure 10-7.** The mechanical connection between the outside circuit and the armature constantly changes because of the brush and commutator connections. The action of the commutator and brushes maintains a constant flow in one direction toward the outside circuit. Study **Figure 10-8** closely to see how the direction of the current is maintained in the outside or load circuit. The drawings show the action of the commutator and brushes. The polarity of the brushes is constant. The polarity of the slip rings in the ac generator, Figure 10-6, changed as the direction of the current changed with each half revolution. In the dc generator, the alternating current in the armature is changed to a pulsating direct current.

Note that the current in the outside circuit of the dc generator always flows in one direction. The output of the generator is shown in **Figure 10-9.** The voltage rises and falls from zero to a maximum to zero, but always in the same direction.

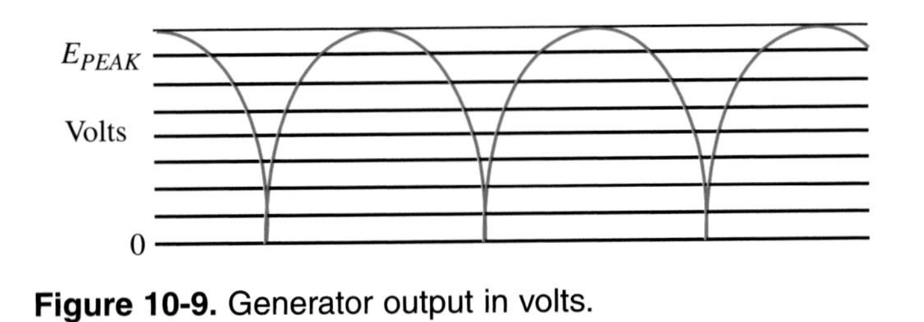

Figure 10-9. Generator output in volts.

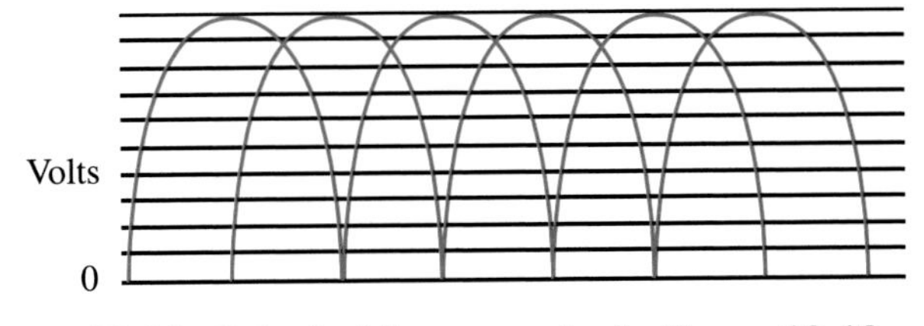

Figure 10-11. Output of the generator in Figure 10-10.

Follow the action in Figure 10-8. Brush A is in contact with commutator section A, and brush B is in contact with commutator section B. The first induced wave of current flows through the armature out of brush B, around the external circuit and into brush A, completing the circuit. When the armature revolves one-half turn, the induced current will reverse its direction. However, the commutator sections have also turned with the armature. The induced current flowing out of commutator section A is now in contact with brush B. This current flows through the external circuit to brush A into commutator section B, completing the circuit. *The current flows through the external circuit in the same direction both times.* The commutator has acted as a switch. It reversed the connections to the rotating coil when the direction of the induced current was reversed.

The current in the outside circuit is pulsating direct current. The output of this generator is not a smooth direct current. The weakness of pulsating dc can be improved two ways. The number of rotating coils on the armature can be increased and commutator sections can be supplied for each set of coils. To help you understand how the coils are added to the armature, examine **Figure 10-10.** Each coil has its own induced current. As the current starts to fall off in one coil, it is replaced by an induced current in the next coil.

The current is created as the coils cut across the magnetic field. A graph of the output from the generator of Figure 10-10 is shown in **Figure 10-11.** It is still a pulsating current. However, the pulses come twice as often and are not as large. The output of the two coil generator is much smoother. By increasing the number of coils, the output will closely duplicate a pure direct current with only a slight ripple variation.

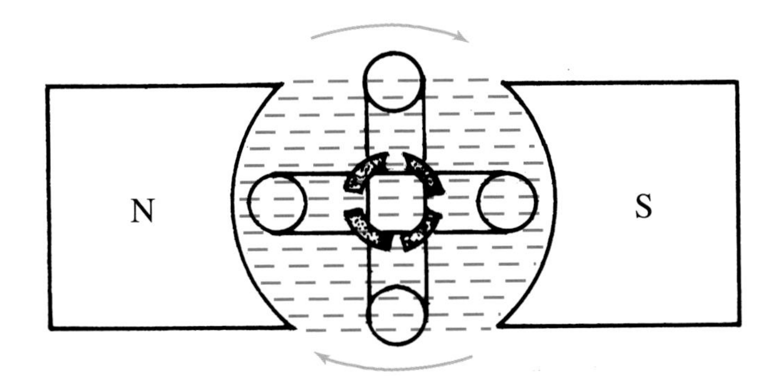

Figure 10-10. A simple generator with two coils at right angles to each other. They are rotated in magnetic field.

Generator Losses

All of the mechanical power used to turn the generator is not converted into useful electrical power. There are some losses.

I^2R losses

You will recall that all wires have some resistance. The size of this resistance depends on the wire size, material, and length. Resistance uses power. In the generator coils there are many feet of copper wire. The resistance of this wire must be overcome by the induced voltage. Voltage used in this manner supplies nothing to the external circuit. It only creates heat in the generator windings. Power loss in any resistance is equal to:

$$P = I^2R$$

Of special concern is the fact that power loss increases by the *square* of the current. To be specific, if the current in the generator doubles, the power loss is four times (2^2) more. The limiting factor in generator output is usually the wire size and current carrying capacity of the armature windings. Losses resulting from resistance in the windings are called ***copper losses*** or ***I^2R losses.***

Eddy currents

The armature windings in the generator are wound on an iron core, which is slotted to hold the coils. Like a conductor, as the solid iron core rotates, it induces voltages causing an alternating current to flow in the core. This alternating current is known as an ***eddy current.*** It produces heat, which is a loss of energy. To reduce this loss, cores are made of built up thin sections or laminations, **Figure 10-12.** Each section is insulated from the next section by lacquer or, at times, only oxide and rust. These laminations reduce the voltage in the core and increase resistance to the power draining eddy currents.

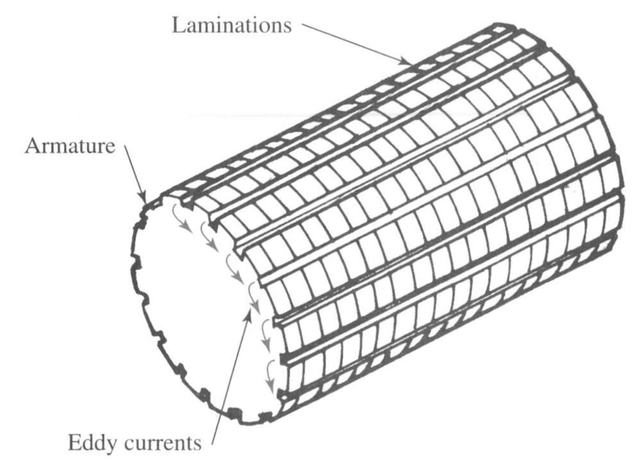

Figure 10-12. Lamination of the metal core of an armature reduces the flow of eddy currents.

Eddy currents increase in the core as the speed of rotation increases or the field density increases. When iron cores are used in rotating machines and transformers, they are laminated to reduce eddy current losses.

Hysteresis

The third loss occurring in a generator is called ***hysteresis loss.*** Hysteresis is also called molecular friction. As the armature rotates in the fixed magnetic field, many of the magnetic particles in the armature core remain lined up with the fixed field. These particles then rotate against those not lined up with the field. This rotation causes internal friction between the magnetic particles and creates a heat loss. Generator manufacturers now use a silicon steel which has a low hysteresis loss. Annealing the core (a heat process), further reduces this loss.

Review Questions for Section 10.1

1. Voltage is highest when the armature coil moves __________ (parallel, at a right angle) to the magnetic field pole.
2. In your own words, explain Lenz's law.
3. What two major factors determine the strength of induced voltage?
4. Define direct current and alternating current. Draw a wave pattern for each and label positive, zero, and negative voltage peaks.
5. A(n) __________ is used to maintain a constant direction of current from the armature.
6. Describe the difference between slip rings and split rings.
7. Sketch a dc generator. Use arrows to show current.
8. Name three types of generator losses.

10.2 TYPES OF GENERATORS

There are a number of different types of generators. Several of these generator types will be discussed briefly. Study their similarities as well as their differences. Generators can be distinguished by their method of excitation. Self excited generators can be separated further into the categories of shunt, series, and compound.

One feature that separates generators is the excitation method, the method that is used to start the generator running. Some generators require a separate power source during the starting of the generator. These are called ***separately excited field generators.*** Other generators use the generator's own leftover magnetism in place of that power source. These are ***self excited generators.***

Separately Excited Field Generator

Generator output is determined by the strength of the magnetic field and the speed of rotation. Field strength is measured in ampere-turns. So, an increase in current in the field windings will increase the field strength. Output voltage is in direct proportion to field strength times the speed of rotation. Therefore, most output regulating devices depend on varying the current in the field.

The field windings can be connected to a separate, or independent, source of dc voltage, **Figure 10-13.** This is the separately excited field generator. With the speed constant, the output may be varied by controlling the exciting voltage of the dc source. This is done by inserting *resistance in series* with the source and field windings.

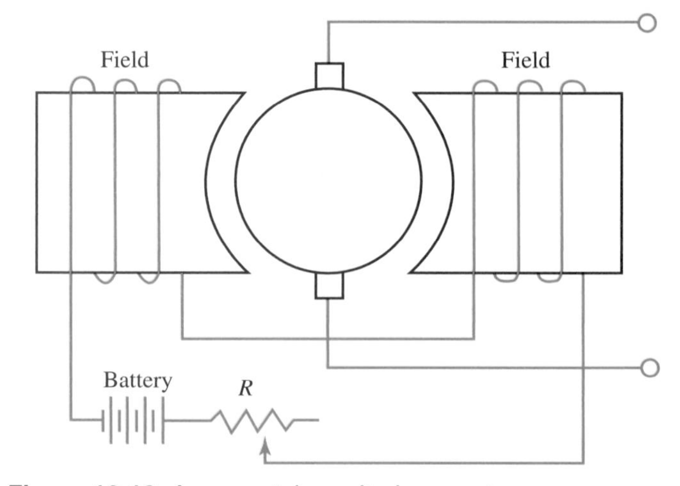

Figure 10-13. A separately excited generator.

Motion Simulators

The basic operation of pneumatic (air) or hydraulic (liquid) cylinders is quite simple. A pump is used to pump air or some hydraulic fluid into a tank to produce a constant supply of pressure. The release of air or hydraulic fluid pressure through the system can be controlled manually or through an electrical system. As a typical example, we shall examine an electrically-controlled pneumatic system.

When the circuit to solenoid A is completed, air is released into the cylinder. As the air pressure flows into the cylinder, the piston slides out. This action moves an object attached to the piston rod. When the solenoid is no longer in the open position, the cylinder stops. The piston rod and attached object will remain in position until the air pressure is released from the cylinder. Activating solenoid B releases the air pressure from inside the cylinder. The operational principle is very simple. Let's look at how these principles might be used in a flight simulator or an amusement park ride.

An aircraft simulator usually consists of a cockpit including all controls found in the actual aircraft model. The craft is mounted on a system of cylinders such as those described above. The combined action of these cylinders can cause the aircraft to simulate left and right banks, climbs, dives, and vibrations.

The purpose of the simulator is to simulate flight conditions without requiring the pilot to leave the ground. This method is used to train pilots on both the basics of flight and on emergency conditions. The pilot can practice maneuvers over and over again without endangering himself or the general public. This type of training is also very cost efficient.

Simulators are used extensively by airline companies and the space industry. New, virtual-motion simulators are now being used in amusement rides and other simulations of motion such as downhill skiing. The combination of cylinders with computer programs has caused the use of motion simulators to expand rapidly.

Courtesy of Mechtronix Systems Inc.

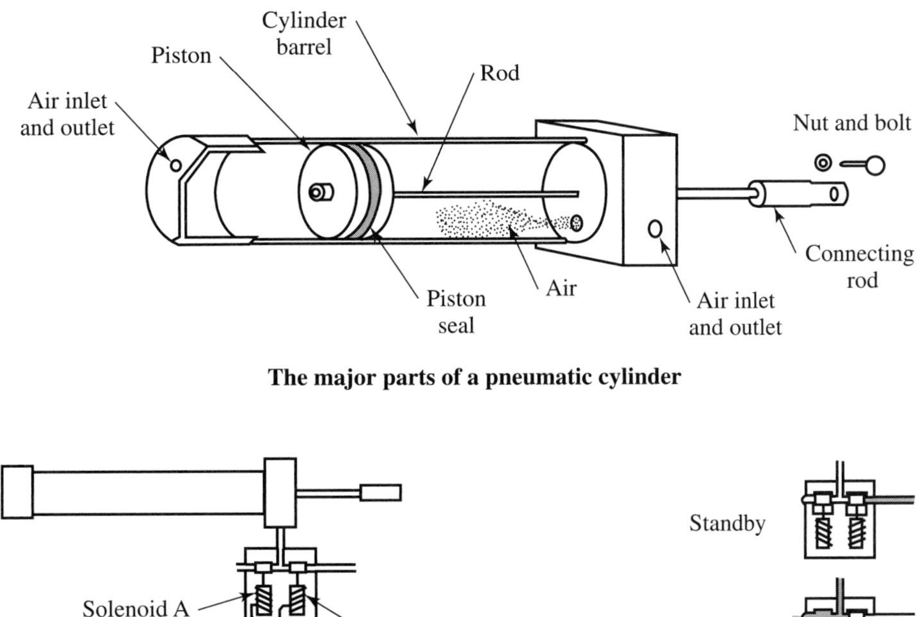

The major parts of a pneumatic cylinder

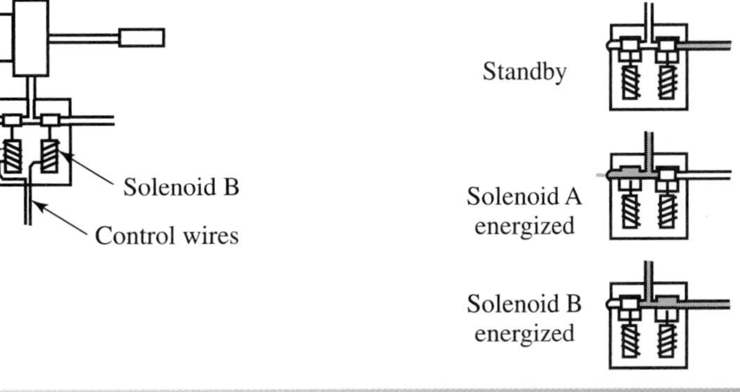

Self Excited Generator

A self excited generator uses no separate source of voltage to excite the generator field winding. The self excited generator produces a small voltage when the armature windings cut across a weak magnetic field. This weak magnetic field is caused by magnetism left over in the pole shoes or field coil cores after the voltage and current have ceased to flow. The magnetism left in a magnet after the magnetizing force has been removed is called *residual magnetism.* Residual magnetism was first discussed in Chapter 9.

Look ahead to the diagram of the shunt generator shown in **Figure 10-14.** A residual magnetic field will cause a small voltage to be produced as the armature conductors rotate past the field poles. The small voltage produced will, in turn, cause the current to increase through the field poles. An increase in field pole magnetism will cause a further increase in output voltage. The relationship of the current produced by the armature directly increasing the amount of magnetism in the field poles is how the self excited generator works. The magnetism produced by the armature voltage will increase until the field poles reach saturation, the point where the poles cannot contain any more magnetic lines of force.

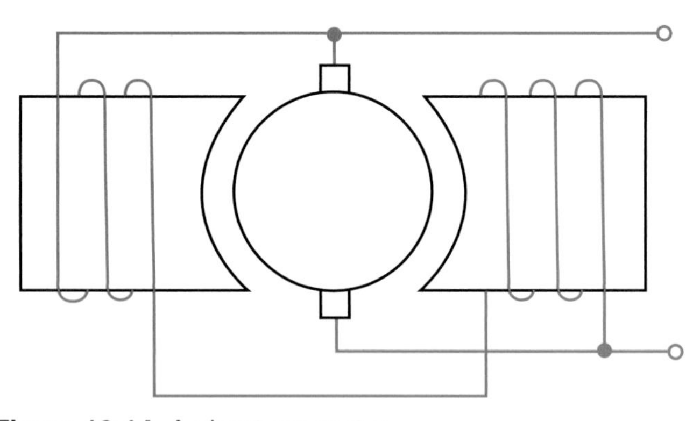

Figure 10-14. A shunt generator.

Shunt generator

The ***shunt generator*** derives its name from the way the field pole coils are connected in parallel to the armature, Figure 10-14. Another way of saying parallel is the term shunt. The field windings consist of many turns of small wire. They use only a small part of the generated current to produce the magnetic field in the pole's windings. The total current generated must, of course, be the sum of the field excitation current and the current delivered to the load. Thus, the output current can be thought of as varying according to the applied load. The field flux does not vary to a great extent. Therefore, the terminal voltage remains constant under varying load conditions. This type of generator is considered a *constant voltage machine.*

All machines are designed to do a certain amount of work. If overloaded, their lives are shortened. As with any machine, the life of a generator can be shortened by an overload condition. When overloaded, the shunt generator terminal voltage drops rapidly. Excessive current causes the armature windings to heat up. The heat can cause the generator to fail by destroying the thin coat of insulation covering the armature wires.

Series generator

The ***series generator*** is so named because its field windings are wired in series with the armature and the load. Such a generator is sketched in **Figure 10-15.** A series winding by itself will provide a *fluctuating voltage* to the generator load. As the current increases or decreases through the load, the voltage at the generator output terminals will greatly increase or decrease. Because of the wide difference in output voltage, it is not a very practical generator to use if the load varies.

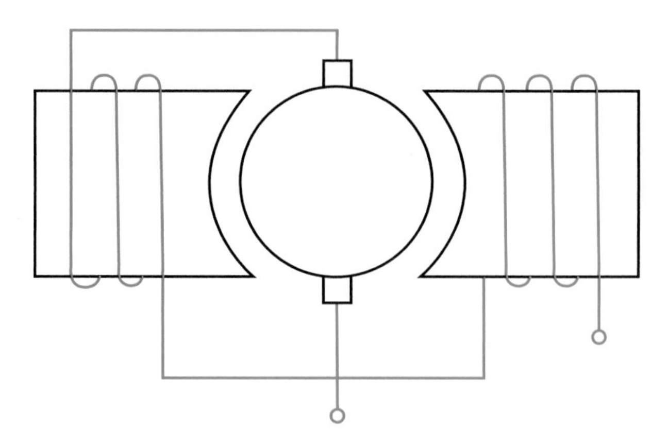

Figure 10-15. A series wound generator.

Compound generator

The ***compound generator*** uses both series and shunt windings in the field. The series windings are often a few turns of large wire. The wire size of the series winding is usually the same size as the armature conductors. These windings must carry the same amount of current as the armature since they are in series with each other. The series windings are mounted on the same poles with the shunt windings. Both windings add to the field strength of the generator field poles. If both act in the same direction or polarity, an increase in load causes an increase of current in the series coils. This increase in current would increase the magnetic field and the terminal voltage of the output. The fields are said to be *additive.* The resulting field would be the sum of both coils. However, the current through the series winding can produce magnetic saturation of the core. This saturation results in a decrease of voltage as the load increases.

The way terminal voltage behaves depends on the degree of ***compounding.*** A compound generator which maintains the same voltage either at no-load or full-load conditions, is said to be a ***flat-compounded*** generator. An ***overcompounded*** generator, then, will increase the output voltage at full-load. An ***undercompounded*** generator will have a decreased voltage at full-load current.

A variable load may be placed in parallel with the series winding to adjust the degree of compounding. **Figure 10-16** shows schematic diagrams of the shunt, the series, and the compound generator.

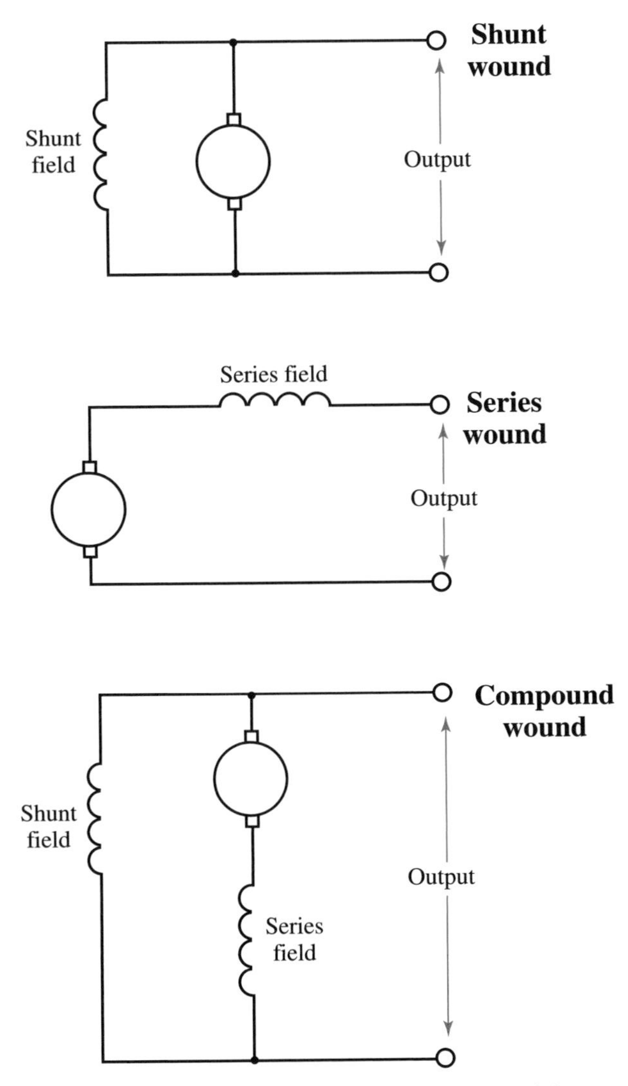

Figure 10-16. Compare these wiring diagrams of the shunt, series, and compound generator.

Voltage and Current Regulation

The ***regulation*** of a power source, whether a generator or a power supply, can be defined as the percentage of voltage drop between no-load and full-load. Mathematically, it can be expressed:

$$\frac{E_{(no\text{-}load)} - E_{(full\text{-}load)}}{E_{(full\text{-}load)}} \times 100 = \%\ \text{regulation}$$

To explain this formula, assume that the voltage of a generator with no-load applied is 100 volts. Under full-load the voltage drops to 97 volts. Filling in the equation:

$$\frac{100\ \text{V} - 97\ \text{V}}{97\ \text{V}} = \frac{3\ \text{V}}{97\ \text{V}} \times 100 = 3.1\%\ \text{(approximately)}$$

In most uses, the output of the generator should be maintained at a fixed voltage value under varying load conditions. The output voltage of the generator depends upon the field strength. The field strength depends upon the field current. Current, according to Ohm's law, varies inversely with resistance. Therefore, a device that would vary the resistance in the field circuit would also vary the voltage output of the generator. This regulator is shown in **Figure 10-17.** It was often used in automobiles.

The generator output at terminal G is joined to the battery and the winding of a magnetic relay. The voltage produced by the generator causes a current to flow in the

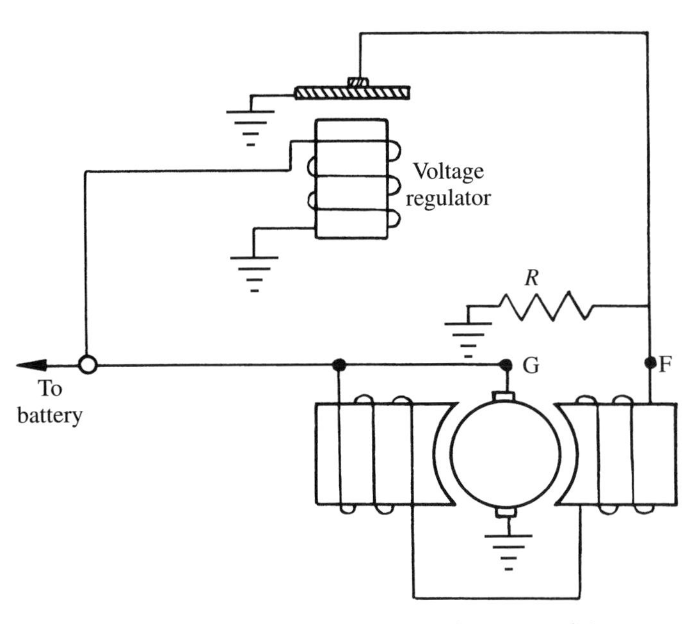

Figure 10-17. Circuit for a generator voltage regulator.

relay coil. If the voltage exceeds a preset value, the increased current provides enough magnetism to open the relay contacts.

Notice that the generator field is grounded through these contacts. When they open, the field current must pass through resistance *R* to ground. This resistance reduces the current, which reduces the field strength and reduces the terminal voltage. When the terminal voltage is reduced, the relay contact closes permitting maximum field current. The terminal voltage rises.

In operation, these contact points vibrate. They alternately cut resistance in and out of the field circuit and maintain a constant voltage output of the generator.

Mechanical-magnetic relays have served this purpose for many years. Now, however, electronic devices are being used on cars. An electronic regulator using an integrated circuit for the switching functions is shown in **Figure 10-18.** A study of integrated circuits is found in Chapter 19 of this text.

Review Questions for Section 10.2

1. Generator output is determined by two items. Name these items.
2. This generator requires an external power source to provide voltage to the field windings.
   a. Series generator.
   b. Compound generator.
   c. Shunt generator.
   d. Separately excited field generator.
3. This generator is considered a constant voltage device.
   a. Series generator.
   b. Compound generator.
   c. Shunt generator.
   d. Separately excited field generator.
4. This generator has both series and shunt windings in the field.
   a. Series generator.
   b. Compound generator.
   c. Shunt generator.
   d. Separately excited field generator.
5. State the mathematical equation for regulation of a power source.

10.3 ALTERNATING CURRENT

Direct current flows in only one direction. Alternating current changes its direction of flow at times in the circuit. In dc, the source voltage does not change its polarity. In ac, the source voltage changes its polarity between positive and negative.

Figure 10-19 shows the magnitude and polarity of an ac voltage. Starting at zero, the voltage rises to maximum in the positive direction. It then falls back to zero. Then it rises to maximum with the opposite polarity and returns to zero.

The current wave is also plotted on the graph. It shows the flow of current and the direction of the flow.

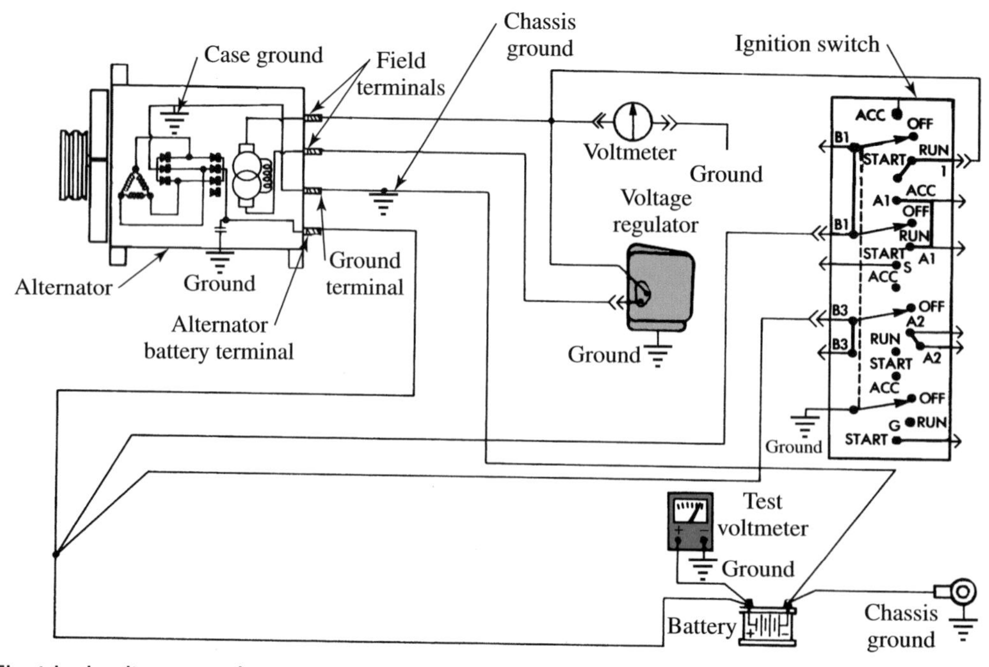

Figure 10-18. Electrical voltage regulators are used on all new cars. (Chrysler)

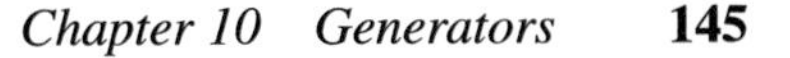

Figure 10-19. Current and voltage of alternating current.

Above the zero line, current is flowing in one direction. Below the zero line, the current is flowing in the opposite direction.

The graph in Figure 10-19 represents instantaneous current and voltage at any point in the cycle. But what is a cycle? A cycle is a sequence or chain of events occurring in a period of time. An ac ***cycle*** can be described as a complete set of positive and negative values for ac. The alternating current in your home changes direction 120 times per second. It has a frequency of 60 cycles per second (60 cps). ***Frequency,*** measured in cycles per second, or ***hertz (Hz),*** is the number of complete cycles occurring per second. If 60 cycles occur in one second, then the time period for one cycle is 1/60 of a second, or 0.0166 seconds. This is the ***period*** of the cycle. Refer again to Figure 10-19. The maximum rise of the waveform shows the ***amplitude*** of the wave, including the ***peak*** (highest point) voltage and current.

We learned that induced current in a rotating wire in a magnetic field flowed first in one direction and then in the other direction. This was defined as an alternating current. Two points to remember are:

- The frequency of this cycle of events increases as rotation speed increases.
- The amplitude of the induced voltage depends on the strength of the magnetic field.

Vectors

When solving problems involving alternating currents, vectors are used to depict the magnitude and direction of a force. A ***vector*** is a straight line drawn to a scale that represents units of force. An arrowhead on the line shows the *direction* of the force. The length of the vector shows the *magnitude.*

The development of an ac wave is shown in **Figure 10-20.** This wave is from a single coil armature, represented by the rotating vector, making one revolution through a magnetic field. Assume that the peak induced voltage is 10 volts. Using a scale in which one inch equals five volts, the vector is two inches, or 10 volts, long. Vectors of this nature are assumed to rotate in a counterclockwise direction.

The time base in Figure 10-20 is a line using any convenient scale. It shows the period of one cycle or revolution of the vector. The time base is grouped into segments that represent the time for certain degrees of rotation during the cycle. For example, at 90 degrees

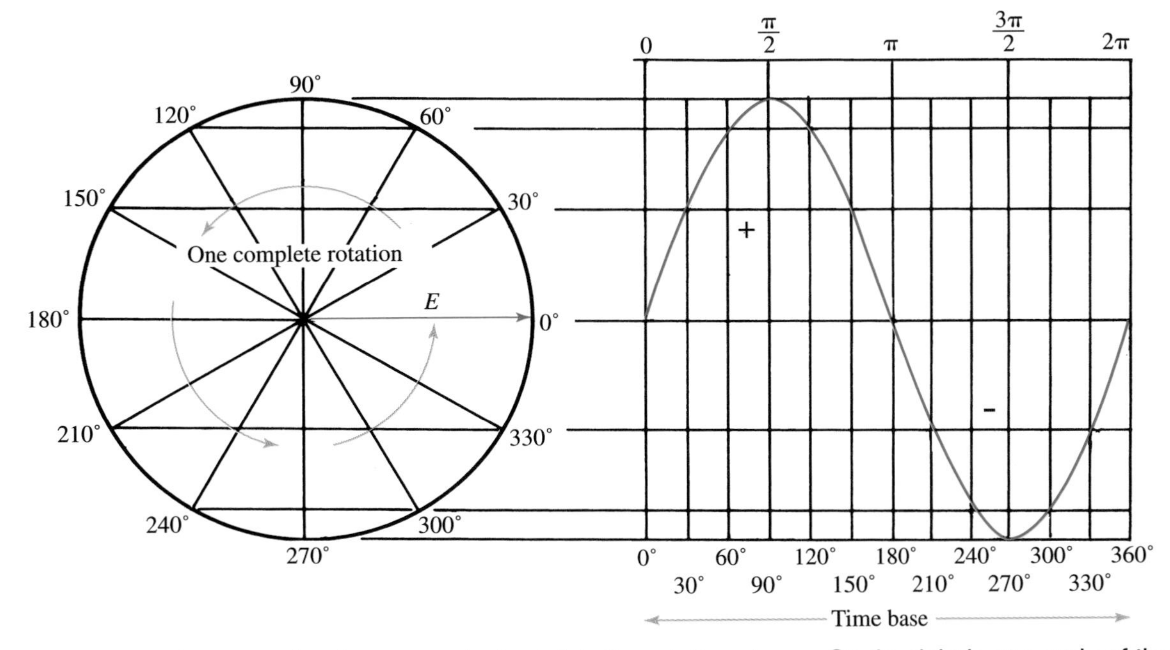

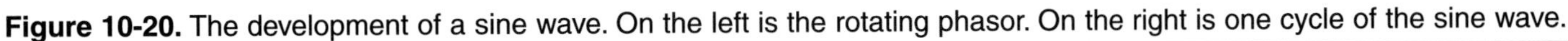
Figure 10-20. The development of a sine wave. On the left is the rotating phasor. On the right is one cycle of the sine wave.

rotation, one quarter of the time period is used. At 270 degrees rotation, three quarters of the time period is used.

The wave is developed by plotting voltage amplitude at *any instant* of revolution against the time segment. The developed wave is called a sine wave. The instantaneous induced voltages are proportioned to the sine of the angle θ (theta) that the vector makes with the horizontal. (Refer to the appendix for explanation of trigonometric functions and tables.) The instantaneous voltage may then be found at any point of the cycle by making use of the following equation:

$$e = E_{max} \times \sin \theta$$

(Notice that the lowercase e was used to represent the instantaneous voltage instead of the usual upper case. By convention, instantaneous values are represented with lower case variables.)

To apply this equation, assume that an ac generator is producing a peak voltage of 100 volts. What is the instantaneous voltage at 45 degrees of rotation?

$$e = 100 \text{ V} \times \sin 45°$$
$$e = 100 \text{ V} \times 0.707 = 70.7 \text{ V}$$

Average and Effective Values

A study of the differences between an ac wave and a direct current raises a key question. What is the actual value of the ac wave? The voltage and current vary constantly and reach peak value only twice during a cycle.

Often, the average value of the wave is needed. The ***average value*** is the mathematical average of all the instantaneous values during one half-cycle of the alternating current. The formulas for computing the average value from the peak value (max) of any ac waves are:

$$E_{avg} = 0.637 \times E_{max}$$

or

$$I_{avg} = 0.637 \times I_{max}$$

If E_{avg} or I_{avg} is known, the conversion to find E_{max} or I_{max} can be made using the following equations.

$$E_{max} = 1.57 \times E_{avg}$$

or

$$I_{max} = 1.57 \times I_{avg}$$

A more useful alternating current value is the effective value. The term effective value comes from scientists finding the ac heating effect equivalent of a direct current. A specified volume of water was heated using a specified voltage level of dc. Then, the same quantity of water was heated using ac. The ac voltage that produced equivalent heating to the dc voltage was the ***effective value.*** The formulas to find the effective value of any ac voltage or current are:

$$E_{eff} = 0.707 \times E_{max}$$

or

$$I_{eff} = 0.707 \times I_{max}$$

where E_{max} and I_{max} are the peak values of the ac signal. If E_{eff} or I_{eff} are known, the conversion to find the peak values can be made by using the following equations.

$$E_{max} = 1.414 \times E_{eff}$$

or

$$I_{max} = 1.414 \times I_{eff}$$

The effective value is also called the ***rms value*** (root mean square). It gets this name because the value represents the square root of the average of all currents squared between zero and maximum of the wave. The currents are squared, so the power produced can be compared to direct current. Watt's law states: $P = I^2R$. Using the 0.707 factor, the value of a direct current can be found that will equal the alternating current. For example, a peak ac current of 5 amperes produces the same heating effect in a resistance as a dc current of 3.53 amperes. Plugging the values into the equation:

$$I_{eff} = 0.707 \times 5 \text{ amperes of ac} = 3.53 \text{ amperes dc}$$

Note that average and effective values can be applied to either voltage or current waves.

Phase Displacement

A few waveforms can be drawn on the same time base to show the phase relationship between them. In **Figure 10-21,** waveforms *E* and *I* show the voltage and current in a given circuit. The current and voltage rise and fall at the same time. They cross the zero line at the same point. The current and voltage are ***in phase.*** The *in phase* condition only exists in the purely resistive circuit. More will be discussed on this topic later.

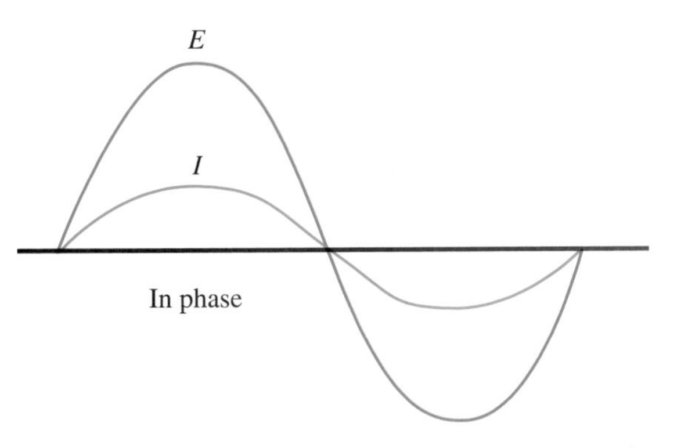

Figure 10-21. These current and voltage waves are in phase.

Many times the current will lead or lag the voltage, **Figure 10-22.** When the current wave leads or lags the voltage wave, the two waves are said to be ***out of phase.*** This creates a phase displacement between the two waves. Displacement is measured in degrees. The ***phase displacement*** is equal to the angle θ between the two polar vectors.

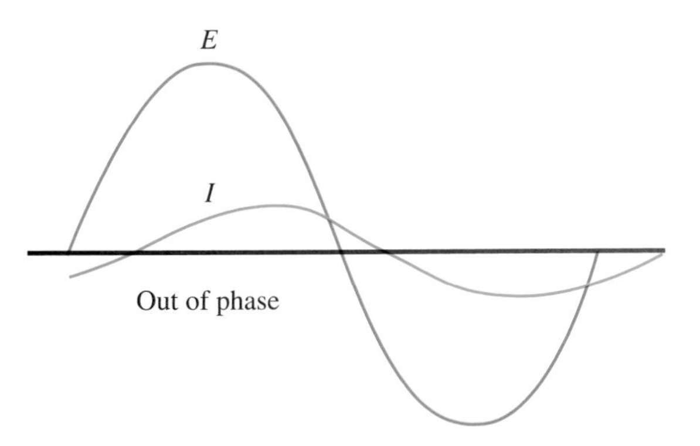

Figure 10-22. These current and voltage waves are out of phase.

Alternating Current Generator

The ac generator is like the dc generator in many respects with one key exception. The commutator is omitted. The ends of the armature coils are extended out to *slip rings*. Brushes sliding on the slip rings provide connection to the coils at all times. The current in the externally connected circuit is an alternating current.

In large commercial generators, the magnetic field is rotated and the armature windings are placed in slots in the stationary frame, or stator, of the generator. This method allows for the generation of large currents in the armature while avoiding sending these currents through moving or siding rings and brushes.

The rotating field is excited through slip rings and brushes by a small dc generator mounted on the same shaft as the rotating magnetic field. This small dc generator is called the ***exciter.*** The dc voltage is needed for the magnetic field. Commercial power generators convert many different items (such as moving water, coal, oil, wind, nuclear energy) into electricity. The force mechanism that is used to turn the generator is called the ***prime mover,*** **Figure 10-23.**

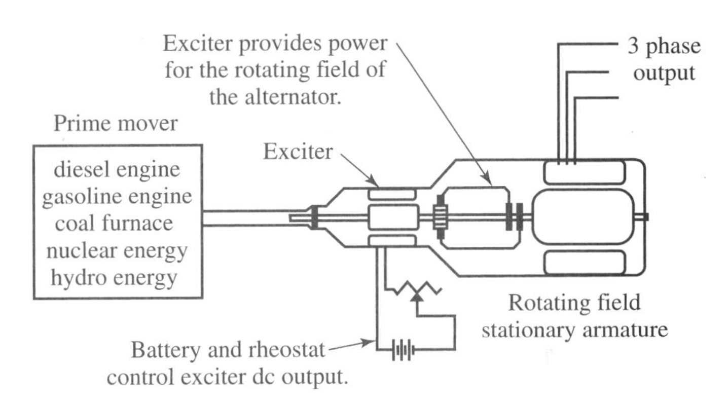

Figure 10-23. The prime mover, exciter, and three-phase alternator share a common rotating shaft. The exciter provides electrical energy for the alternator.

The alternator

The ac generator (also called an alternator) is used in the charging system of all U.S. automobiles. **Figure 10-24** shows the inside of the unit, including a built-in voltage regulator to control output. The output is rectified from alternating current to direct current for charging the battery and other electrical devices in the car. Manufacturers say the alternator has some advantages over the dc generator. These advantages include higher output at lower speeds, as well as trouble-free service.

Review Questions for Section 10.3

1. Define alternating current cycle.
2. __________ is the number of cycles per second.
3. What is a vector?
4. If the effective is 1.75 V, the maximum voltage is __________.
5. Another name for effective voltage is __________ __________.
6. Explain the terms, "in phase" and "out of phase."

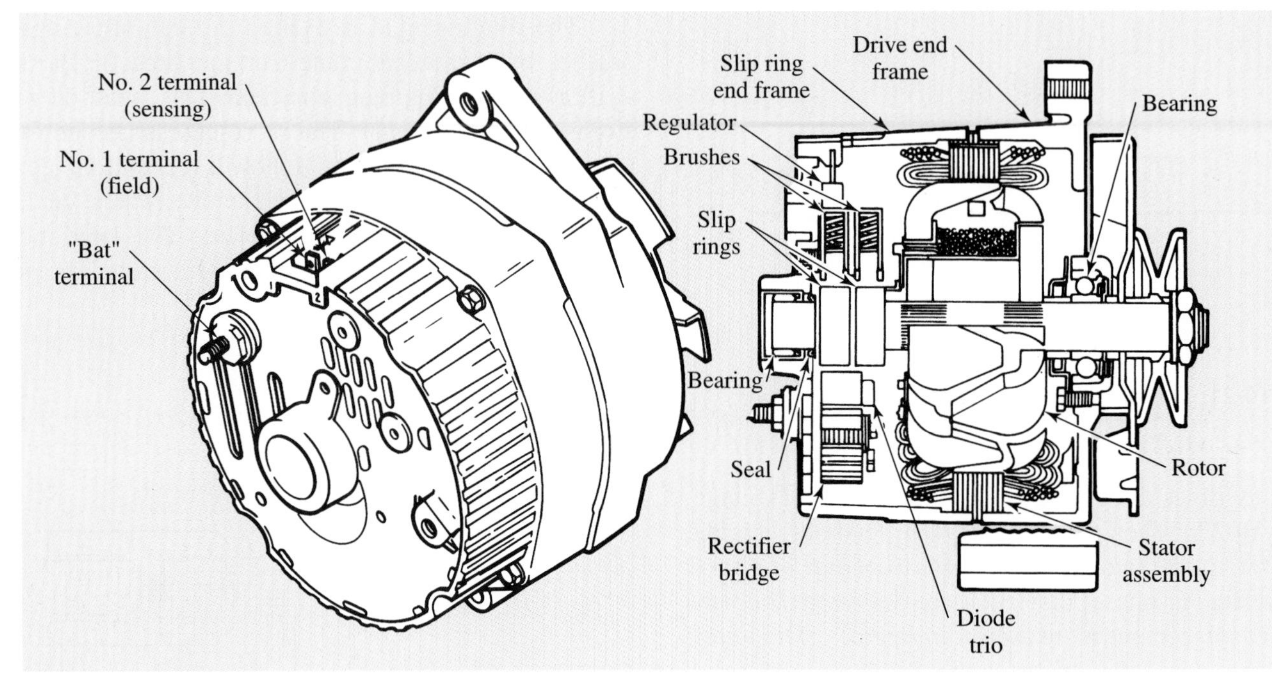

Figure 10-24. A typical ac generator (alternator) is shown in external and cutaway views. (Delco-Remy Div., General Motors Corp.)

EXPERIMENT 10-1: Building a Generator

In this experiment, you will construct and test a simple generator.

Materials

1–small motor (possibly from a toy automobile)
1–variable resistor, 0–500 Ω
1–DPDT switch
1–multimeter
1–galvanometer
1–roll of friction tape

1. Wrap the shaft of the small motor with friction tape.
2. Assemble the circuit shown in **Exhibit 10-1A.** Use your meter to set the variable resistor to 100 Ω. When hooking the meter into the circuit, the multimeter should be set to read milliamps.
3. Rotate the shaft of the small motor. (Note the switch in the circuit allows you to take measurements of rotation in either the clockwise or counterclockwise direction. Make sure the switch is in a position to give you a positive reading.)

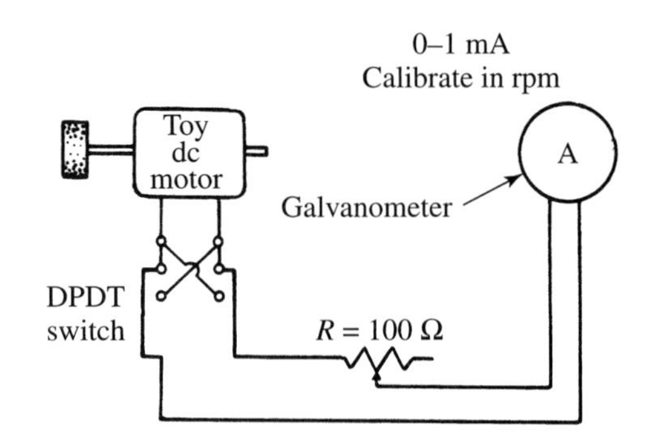

Exhibit 10-1A.

4. Take readings with the shaft rotating in both directions. If possible, take readings with the shaft rotating at different speeds.
Discuss your results with your class.

10.4 THREE-PHASE GENERATORS

The most common generator used in production of electrical power is the ***three-phase generator.*** It consists of a rotating magnetic field inside three sets of windings. See **Figure 10-25,** part A. The three sets of windings are referred to as the armature of the generator. As discussed earlier, the armature provides power for the load circuit. In this case the generator has a stationary armature and a rotating magnetic field. See Figure 10-25, part B.

Understanding how three-phase power is developed will be of great help when developing an understanding of ac motors in Chapter 13. Look again at Figure 10-25, part A. The sine wave that develops from three-phase electrical systems is unique. The three windings of the armature are connected as pairs. They are called phase A, phase B, and phase C. As the magnetic field rotates inside the generator armature windings, three different sine waves develop. The three waves are 120 electrical degrees apart. See Figure 10-25, part C. It is important to note that each of the three phases develops the same frequency and same peak voltage. The difference between the three phases is that each phase develops positive and negative voltage peaks at *different times,* making each phase unique.

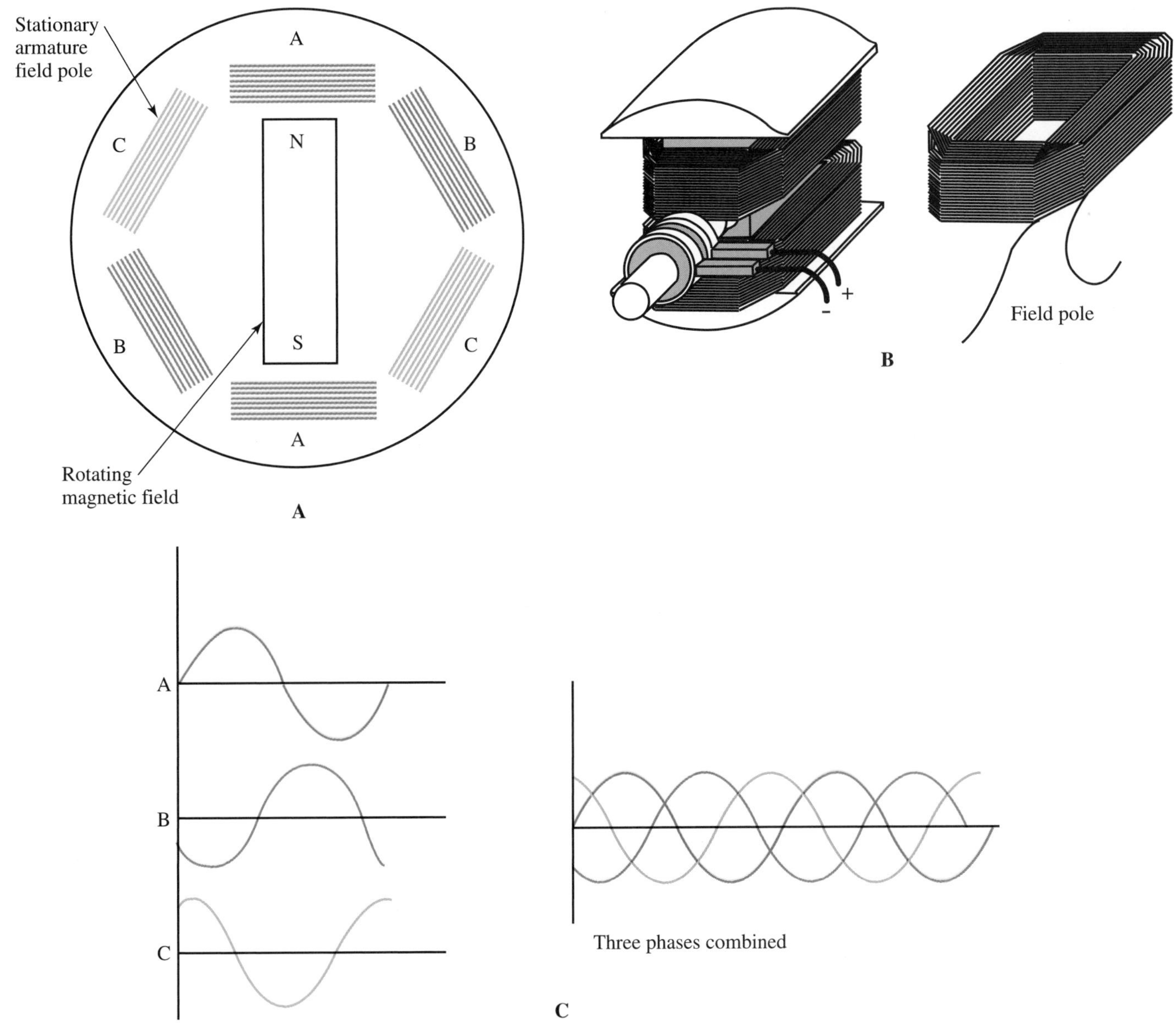

Figure 10-25. A—As the magnetic field rotates inside the generator housing, a three-phase electrical output develops. B—The rotating field inside the stationary armature ac generator consists of coils formed and connected to slip rings. The fields are powerful electromagnets. C—Each phase has the same voltage peak and frequency. The peak positive polarity occurs 120 electrical degrees apart.

The coils of the three-phase generator can be connected in series or parallel. When connected in series it is called a ***delta*** connection. When connected in parallel, it is called a ***wye*** or ***star*** connection. The reasons for their names can be seen in **Figure 10-26.**

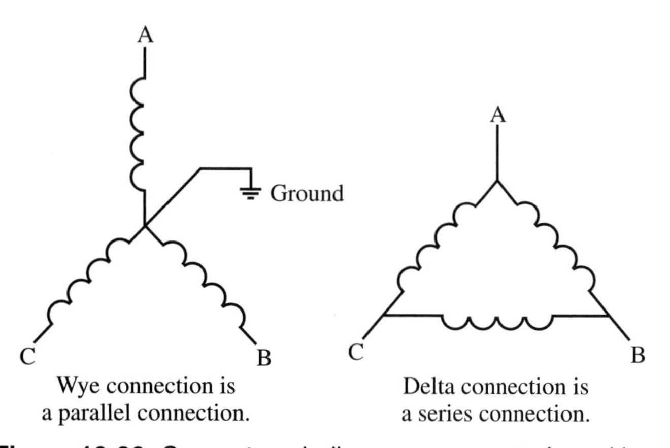

Figure 10-26. Generator windings are connected as either delta or wye.

You may have heard the terms 120/240 volt and 120/208 volt and wondered which is correct. Both terms are correct. The voltage present is determined by the type of generator connection used. A *wye connection* provides 120/208 volts, and *delta connection* provides 240 volts with no common, or neutral, connection. Delta is used mainly for heavy industrial loads, while the wye connection is used to support both power and lighting circuits. The delta and wye connections are covered in detail in Chapter 12. While some equipment will operate on either voltage, it is best to connect a piece of machinery to the voltage level specified in its operator's manual.

Paralleling Generators

Large generator plant systems that must provide millions of watts of power consist of several generators connected in parallel. When generators are connected in parallel, three electrical characteristics of each generator must match. These characteristics are the generator's frequency, voltage, and instantaneous polarity. If all three items do not match, and the switch is closed to connect the two generators, an electrical fire and explosion are likely to occur.

If the voltages of two parallel generators do not match, they can be made to match by adjusting one of the field pole rheostats. A slightly trickier part in paralleling generators is matching the correct frequencies of the two generators. The frequency or hertz of each generator can be compared using a frequency meter. If the frequencies do not match, we need to speed up or slow down one of the generators until they match.

Figure 10-27 shows sine waves of matching and non-matching instantaneous polarity. Each generator is producing a 120 volt output at 60 Hz. However, in the waves shown on the left, the third requirement, instantaneous polarity, does not match. The two sine waves in the figure are slightly off. Notice how their peaks do not align on the time axis. The ***instantaneous polarity*** of the ac generators refers to that part of the sine wave that is represented at the output terminals of each generator at that instant of time. These outputs could both be positive, or one positive and one negative. Ideally, if both sine wave patterns match, the generators have identical instantaneous polarity. These generators can be connected together electrically.

A simple way to tell if two three-phase generator outputs match is called the "two dark–one light method."

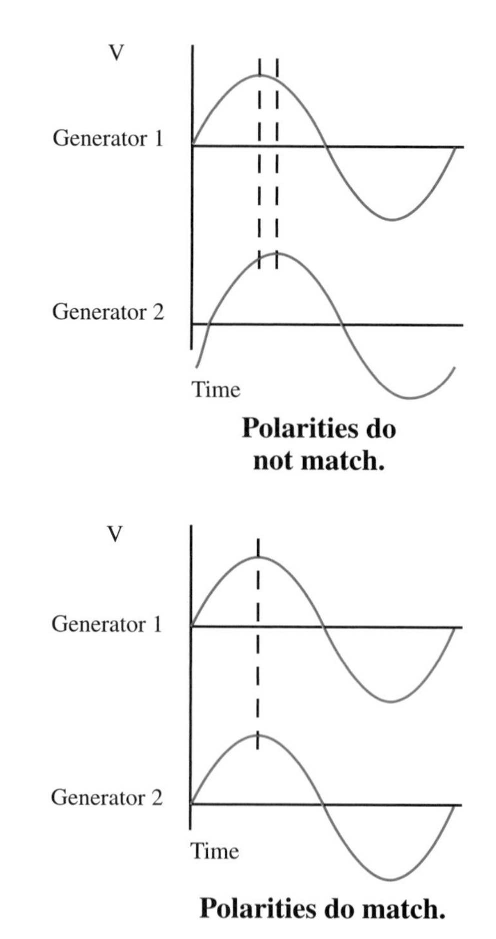

Figure 10-27. The two generators outputs shown on the top have the same voltage peaks and the same frequency, but their instantaneous polarities do not match. Notice how their peaks come at different times. The two outputs on the bottom match.

Two lights are connected across the phase A and B outputs of both generators in such a way that they match in polarity, voltage, and frequency. Both lamps should not light since there would be no difference in potential across the lamps. Another lamp is connected across two *out of phase* phases of the generator. This lamp should light indicating that the two phases are out of phase with each other. See **Figure 10-28.**

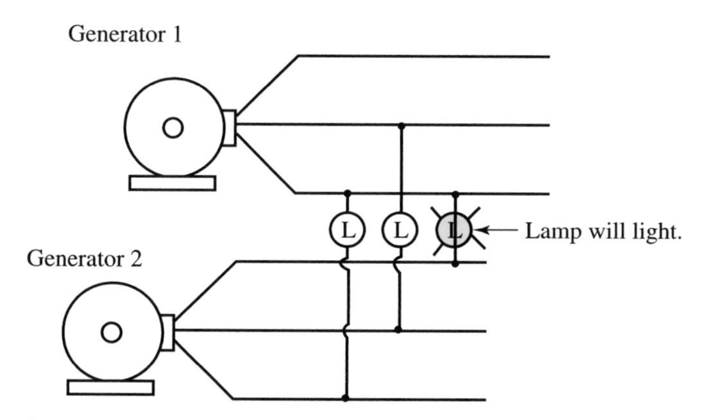

Figure 10-28. The two dark–one light method of paralleling generators. The two lamps connected across in phase lines from the two generators remain dark. The third lamp is connected across two out of phase lines. This lamp will glow. These two generators can be connected in parallel safely.

Troubleshooting Generators

Generators generally fail due to one of three things:

- Excessive brush wear.
- Excessive bearing wear.
- Electrical overload.

Excessive brush wear

As the brushes wear down, more arcing is produced between the brushes and commutator segments or slip rings. Excessive wear will damage the surface of either the commutator segments or the slip rings. A bluish burn pattern will be evident. Excessive brush wear can be determined by physical inspection. The measurement for minimum brush length for a generator can be found in the owner's manual. The brushes should be replaced if excessive wear is found. Brushes should be dressed when being replaced.

Dressing brushes refers to shaping the end of the brush to match the surface of the commutator or slip ring. A fine grade of sandpaper or emery cloth can be used for this purpose. The sandpaper is placed between the brush and commutator with the rough side facing the brush. Rotating the commutator causes the end of the brush to wear down to match the shape of the surface it is riding against. See **Figure 10-29.**

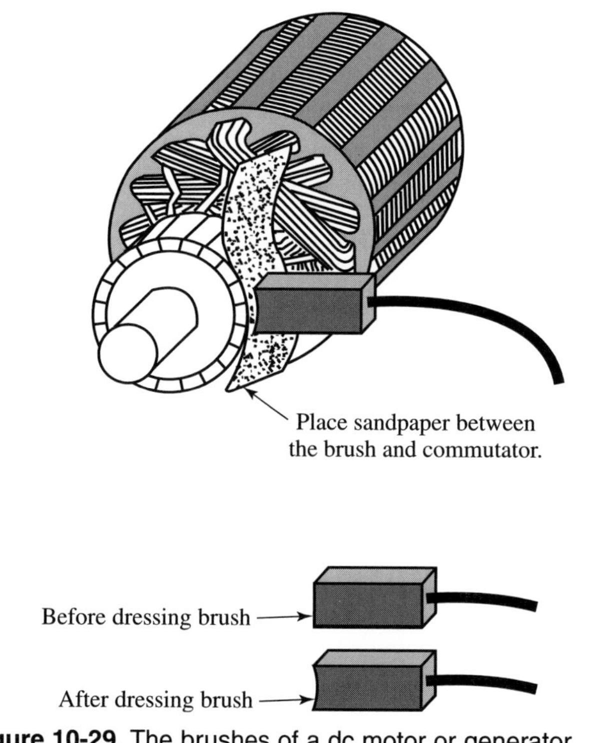

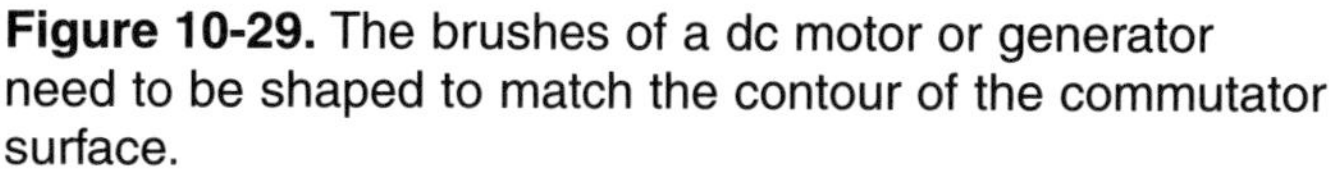

Figure 10-29. The brushes of a dc motor or generator need to be shaped to match the contour of the commutator surface.

Excessive bearing wear

Worn bearings will cause a roaring sound. Excessive vibration can also be present. Another sign of excessive bearing wear is a shiny spot appearing on the armature. This spot is caused by the armature rubbing against a pole piece as the bearing wears out. Bearings should be lubricated on a regular schedule as indicated in the owner's manual. Some bearings are permanently sealed and require no lubrication. If brushes and bearings are properly maintained, an electrical generator can easily provide more than 20 years of service.

Electrical overload

Another cause of failure is electrical overload. Electrical overload causes the electrical insulation to breakdown resulting in a short circuit or ground condition. Most times the damage can be detected by the naked eye. A dark discoloration is a sign of excessive overload on a conductor. Test instruments are needed to determine if the armature or field windings are good or bad. Usually a voltage equal to the generator voltage is used to check the windings for grounds. Most multimeters only use a 9-volt battery and cannot detect a ground created when 120 or 240 volts are present during generator operation.

Heat from an excessive overload can also cause the generator pole pieces to lose their residual magnetism. This is very damaging to the self excited generator. If the residual magnetism is lost, the generator will fail to build up sufficient voltage. Residual magnetism can be restored by connecting a separate dc power supply, such as a battery, to the field windings.

LESSON IN SAFETY:

Before connecting the separate dc power supply to the field windings, the armature leads must be disconnected to prevent the generator from rotating. When voltage is applied to a generator, it will revolve as a motor.

Review Questions for Section 10.4

1. Describe in your own words what is meant by three-phase power.
2. A delta connected generator has its winding connected in series or parallel?
3. If a three-phase power generator supplied 120/208 volts, would it most likely be from a delta or wye connection generator?
4. What three characteristics must match before you can connect two generators in parallel?
5. What are the three major causes of generator failure?
6. How may a self excited generator lose its residual magnetism?

10.5 THE OSCILLOSCOPE

The ***oscilloscope*** is a test instrument that permits observation of waveform patterns representing voltages in an electronic circuit, **Figure 10-30.** It is a priceless service instrument. At first, the oscilloscope can appear to be a very complex instrument, but after a little familiarization, you will see that it is really very easy to use. All oscilloscopes have similar features, but there is some degree of variation among different models. The basic principles and operating controls covered in this unit will be the same for most oscilloscopes.

Oscilloscopes display waveforms on a ***cathode-ray tube (CRT).*** An electron beam is emitted from the cathode gun in the CRT tube. The beam travels between two sets of beam deflector plates. One deflector plate is for horizontal deflection of the electron beam, and the other plate is for vertical deflection. The electron beam strikes the inside surface of the CRT display. This surface is coated with a luminous material such as phosphorus. When the luminous display screen is struck by the electron beam, it produces light. The light leaves a pattern on the display screen that represents the amplitude (height) and time period (width) of the electrical wave being observed and studied. See **Figure 10-31.**

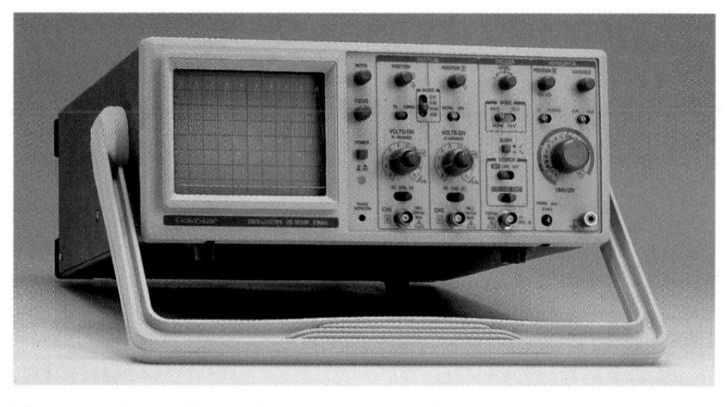

Figure 10-30. Typical oscilloscope. (Knight Electronics)

If an ac voltage is applied to the scope, the vertical deflection circuit controls the vertical deflection plate magnetic intensity. The vertical plate causes the beam of electrons to produce a wave of light on the display screen that is similar to the amplitude of the input voltage. This wave represents the instantaneous voltages during the cycles of the ac input.

The horizontal sweep oscillator can be adjusted through a wide range of frequencies until it matches the frequency of the input voltage. Once the horizontal sweep is matched appropriately to the input voltage frequency, a series of light waves, with a width that graphically represents the frequency of the input voltage, is displayed on the CRT screen. Assume that the horizontal sweep is adjusted at 60 Hz. It is then producing 60 straight lines per second across the face of the CRT. The time for one sweep would be 1/60th of a second.

Now, a 60 Hz ac voltage is applied to the vertical input terminals. This wave starts at zero, rises to maximum positive, and returns to zero. It then decreases to maximum negative and returns to zero. *All of this occurs during one cycle.* The period of this cycle is 1/60th of a second. Therefore, the horizontal sweep and the input signal voltage are synchronized. One wave appears on the screen. If the horizontal sweep were set at 120 Hz, which is twice as fast as the 60 Hz input signal, then two complete waves would appear on the screen. The horizontal frequency can be adjusted so that the waveforms appear *stationary* (not moving).

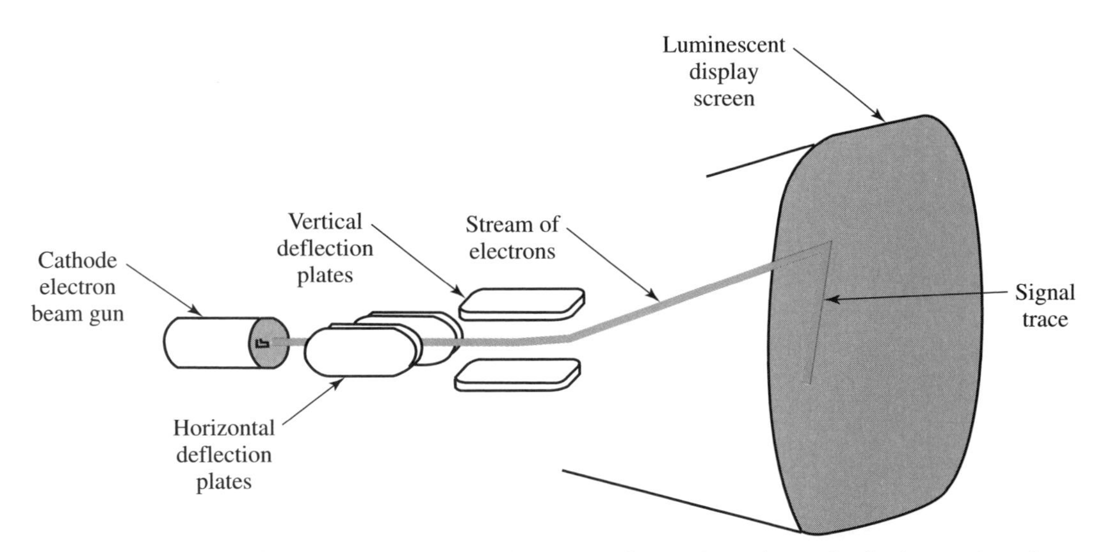

Figure 10-31. A cathode-ray tube (CRT) is used in an oscilloscope. It consists of a cathode that emits a beam of electrons that strike a luminescent display screen. The vertical deflection plates and horizontal deflection plates control where the beam hits the display screen.

The display screen of a typical oscilloscope is a grid pattern. The grid pattern is used to aid the observer in measuring or comparing the wave pattern. Look at **Figure 10-32.** A sine wave pattern is displayed on the grid. Using the grid and the setting of the vertical range selector switch, a voltage measurement can be made. The selector switch is used in conjunction with the grid to determine the voltage amplitude of the waveform. The grid is also used in conjunction with the time sweep setting of the scope.

Voltage Measurement

To measure the voltage of the wave displayed on the screen simply use the grid as you would use a common ruler. If the voltage selector switch (usually marked volts/div. or volts/cm.) is set on 0.2, then the distance across each grid square from top to bottom is equal to 0.2 volts. If the volts/div. is set on 0.005, then the same grid space is equal to five millivolts. The small lines running up and down the center of the display are used to assist in making accurate measurements. The small lines divide each grid square into five equal parts. Look again at Figure 10-32. The sine wave pattern covers a distance of four grid spaces from top to bottom. If the volts/div. selector switch is set on 0.2, then the voltage amplitude of the waveform is equal to 0.8 volts peak to peak.

A probe is used to connect the oscilloscope to the test circuit. A typical oscilloscope probe is shown in **Figure 10-33.** It is constructed of a shielded cable with a connector on one end and the probe tip on the other. Shielded cable is used to prevent interference from stray electromagnetic fields. Most probes are also equipped with a switch for amplifying the reading being taken. The two switch settings are usually marked X1 and X10, which means "times" one and "times" ten. When the slide switch on the probe is set at X10, the voltage reading on the grid should be multiplied by ten. When the probe is equipped with a multiplier such as times ten, then the probe is referred to as an ***attenuator*** probe. The times ten reading is due to an internal resistor that is connected in series with the probe lead.

Time Period and Frequency

The time period for the wave can be determined in a similar fashion to that used for finding the voltage. The time period of the sweep is set using the time/div. selector switch. The time period selector switch is marked in whole seconds, milliseconds, and microseconds. If the time/div. selector switch is set at 0.1 ms, then each grid space is equal to 0.1 millisecond. If the time/div. selector switch is set to 2 s, then each grid spacing is equal to two whole seconds. Look again at Figure 10-32. The selector switch time/div. is set at 0.5 milliseconds. The sine wave pattern covers four grid spaces to complete one full cycle. The time period is equal to 0.002 seconds. To find the frequency of the wave, we divide one by the time period. In the example in Figure 10-32, the frequency is equal to 500 Hz ($1 \div 0.002 = 500$).

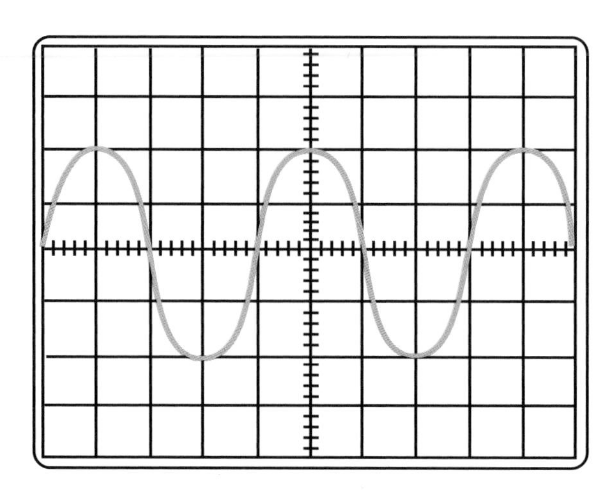

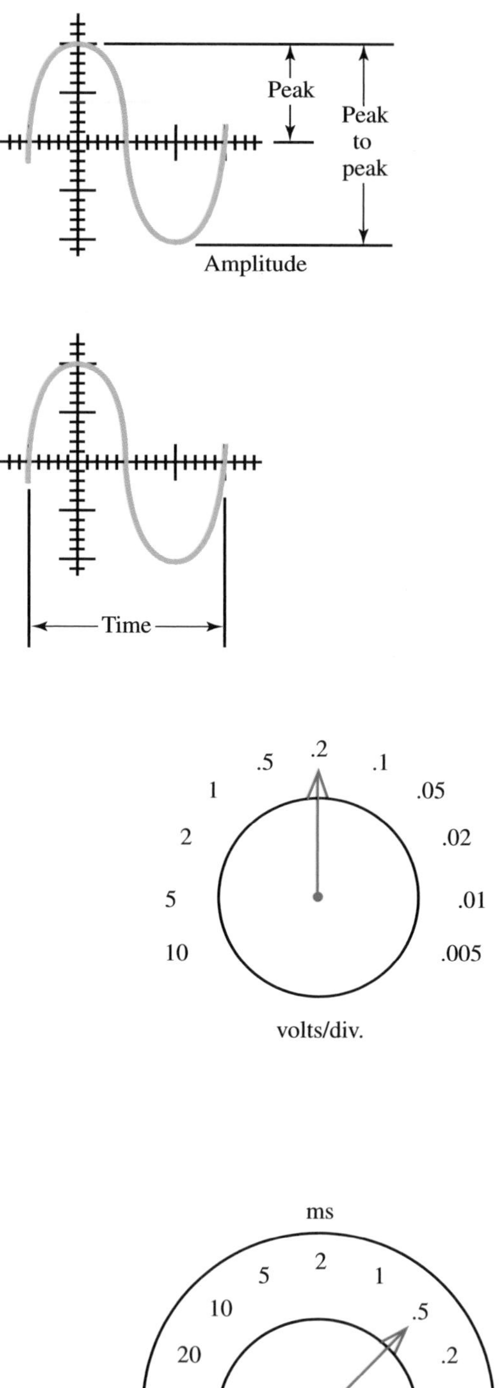

To determine peak to peak voltage

Grid distance × Volts/div. setting = Amplitude

$4 \times 0.2 = 0.8$ volts peak to peak

To determine time period

Horizontal grid distance × Time/div. = Time period

4×0.5 milliseconds $= 0.002$

To convert to frequency

Frequency = 1 / Time period

$1 / 0.002 = 500$ Hz

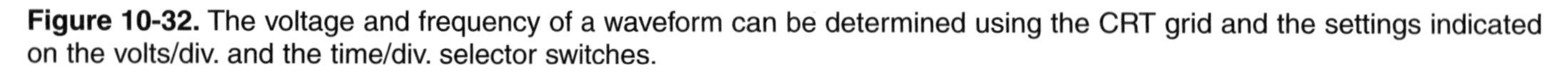

Figure 10-32. The voltage and frequency of a waveform can be determined using the CRT grid and the settings indicated on the volts/div. and the time/div. selector switches.

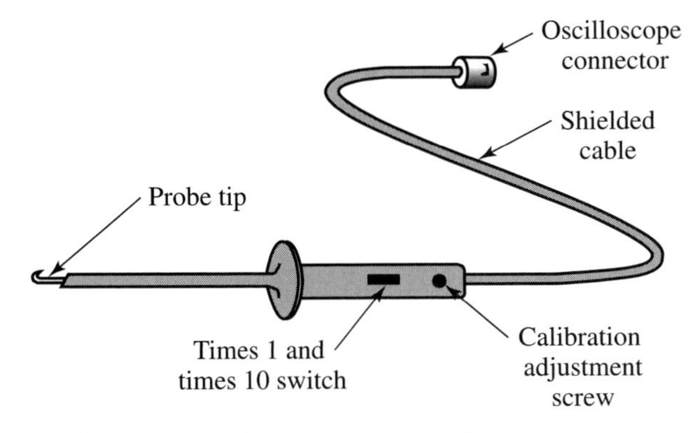

Figure 10-33. Typical oscilloscope probe.

Calibration

To ensure accurate readings are taken from the oscilloscope, it should be calibrated before use. The exact method of calibration will vary from scope to scope. The calibration technique described here is generic in nature and may not exactly match the scope you will be using. Always read the owner's manual to learn how to properly calibrate your oscilloscope.

A square wave pattern is generally placed on screen to calibrate the amplitude and time period. The square wave can also be used to correct any distortion caused by the oscilloscope's probes. First locate a point on the face of the scope usually marked CAL., which is the abbreviation for calibration. This point is where you will connect the tip of the probe. The voltage, time period, and wave shape is provided by the scope as the perfect reference for the calibration operation. A square wave will appear on the screen. From this point the appropriate adjustments can be made to make the wave on screen match the expected reference wave. Most scopes use 0.5 volts at 1000 Hz square wave as the reference.

See **Figure 10-34.** The three examples of the square wave are shown. The rounded off or sharp peaked corners of the square wave pattern indicate that the probe needs adjustment. There is usually a small screw head in the side of the probe or at the connector. Adjust the screw until the square wave has a nice squared off corner appearance as indicated in frame on the left side of Figure 10-34.

Intensity and Focus

The focus and intensity control the appearance of the wave on the CRT screen. ***Focus*** is used to make the wave appear sharper. It eliminates any fuzzy appearance the wave may have.

The ***intensity*** controls the brightness of the light beam striking the front of the screen. The intensity should never be set to a level higher than is needed to comfortably observe the wave pattern. If the setting of the intensity is too high and the scope is left unattended, the display screen can be permanently damaged. Reduce the intensity when you are not using the scope. Before the scope is turned on, and while it is warming up, the intensity should be set to its lowest setting. After the warm-up period (check owners manual for recommended warm-up time), it is safe to adjust the intensity level.

Additional Features

Today, many oscilloscopes in use are completely digital and incorporate a computer as part of the scope system. The wave pattern can be observed, as well as saved to a battery-powered computer memory. Many scopes are also equipped to print a paper copy of the wave pattern. Some scopes used today are a combination VOM and oscilloscope with a display screen. The display screen not only displays the wave pattern but also digitally displays the frequency and voltage values.

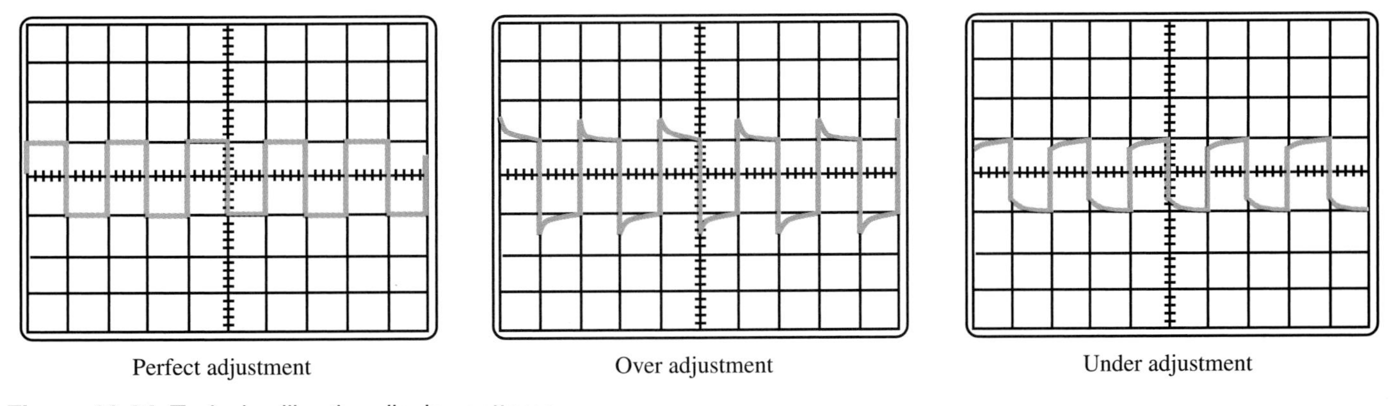

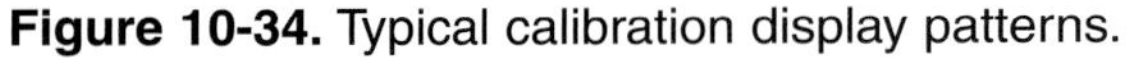
Figure 10-34. Typical calibration display patterns.

Oscilloscopes can also be connected directly to a personal computer. This is referred to as ***data acquisition and transfer.*** This application usually requires an expansion card for the computer, software to run the program, and an interface cable to connect the oscilloscope to the computer. These peripherals are often provided with the scope. A personal computer used in this way will display the wave patterns, voltage, and frequency of the waveform. The information can be downloaded directly into documents for use in reports or training manuals.

Review Questions for Section 10.5

1. The oscilloscope is used to display __________ and __________ in the circuit being tested.
2. The X10 on a oscilloscope probe means that the reading on the screen is __________.
3. If the oscilloscope volts/div. is set at 0.2 and the wave shape covers 4 grid squares, what is the voltage?
4. Before using an oscilloscope, you should __________ it.
5. The height of the waveform is controlled by the __________ selector switch.
6. The number of cycles displayed on the CRT screen is controlled by the __________ selector switch.
7. If the __________ is set too high, the display screen could be damaged.

Summary

1. To produce an induced current with magnetism, three factors must exist:
   a. There must be a magnetic field.
   b. There must be a conductor (or coil) in a closed circuit.
   c. There must be a relative movement between the field and the conductor.
2. A generator is a device that converts mechanical energy into electrical energy.
3. Lenz's law states that the polarity of an induced emf is such that it sets up a current, the magnetic field of which always opposes the change in the existing field.
4. Direct current (dc) is current flowing in a single direction through a conductor. Alternating current (ac) is current flowing in more than one direction through a conductor.
5. Direct current generators have commutators while ac alternators have slip rings.
6. Generator losses include copper losses, eddy currents, and hysteresis losses. These losses reduce their efficiency.
7. Generator types include the shunt, series, and compound. They can be independently excited or self excited.
8. Regulation of a power source, whether it is a generator or a power supply, can be defined as the percentage of voltage drop between no-load and full-load.
9. A cycle is a complete set of positive and negative values for alternating current.
10. Frequency is the number of cycles that occur each second. It is measured in hertz.
11. The value of ac that does actual work is referred to as the effective or root-mean-square value.
12. Formulas for converting the effective value to the peak value (and vice-versa) are:

$$E_{eff} = 0.707\ E_{max}$$
$$E_{max} = 1.414\ E_{eff}$$

13. Two sine waves are in phase when they are of the same frequency and they go through the zero points at the same time.
14. The three-phase generator consists of a rotating magnetic field inside three sets of windings.
15. To connect generators in parallel each generator's frequency, voltage, and instantaneous polarity must match.
16. Generators generally fail because of excessive brush wear, excessive bearing wear, or electrical overload.
17. The oscilloscope is a test instrument that permits observation of waveforms in an electronic circuit.
18. To ensure accurate readings are taken from the oscilloscope, it should be calibrated before use.

Test Your Knowledge

Please do not write in the text. Place your answers on a separate sheet of paper.

1. Highest induced voltages are produced when the coils cut the magnetic field at __________ angles.
2. Commutators are used on __________ current generators. Slip rings are used on __________ current generators.
3. In what way can a pulsating current be improved?

4. What does the strength of induced voltage depend on? (Choose all that apply.)
   a. The number of magnetic lines of force cut by the coil.
   b. The size of the armature.
   c. The speed at which the conductor moves through the field.
   d. None of the above.
5. State the mathematical equation used to find the strength of induced voltage in a rotating coil.
6. Losses resulting from resistance in the windings are classified as __________ losses.
7. An eddy current:
   a. produces heat.
   b. flows in the core of an armature winding.
   c. causes a loss of energy.
   d. All of the above.
8. Explain the cause of hysteresis loss.
9. Identify each of the following generator diagrams.
   a.

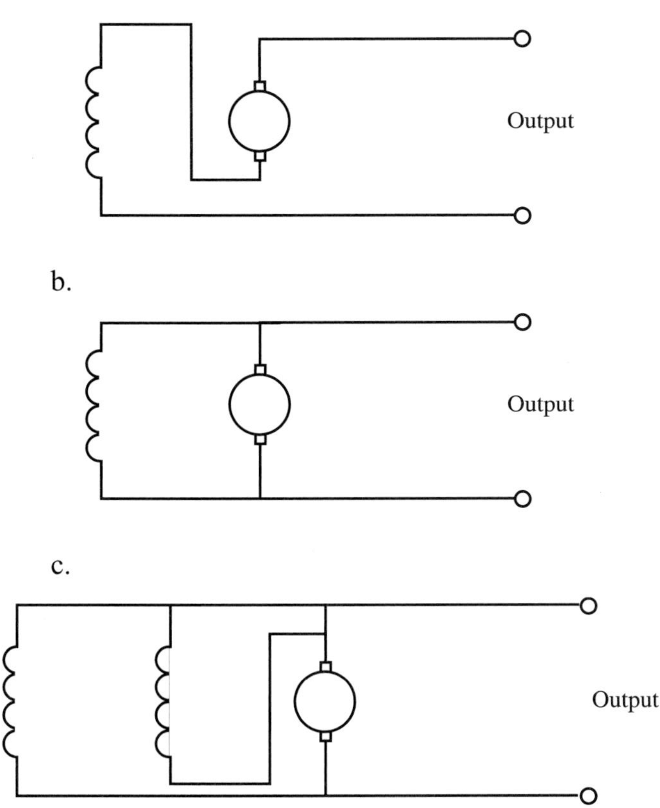

10. A generator has a no-load voltage of 25 volts. When load is applied, terminal voltage drops to 24 volts. What is the percent of regulation?
11. A generated voltage has a peak value of 240 volts. What is the instantaneous voltage at 60 degrees?
12. What is effective value of the generated voltage in Problem 11?
13. Assume that a wall outlet in your home has an effective value of 117 V. What is the peak value of the alternating current coming from that outlet?

For Discussion

1. Explain how sources of energy can be converted into electricity.
2. Why do we need both ac and dc electricity in electrical and electronic systems?
3. How does a generator differ from an alternator?
4. Why is ac not as efficient as dc?

Oscilloscope Maintenance

- Use a dry, soft cloth or one dampened with isopropyl or denatured alcohol to remove dust from the exterior of the oscilloscope. Do not use an abrasive cleaner or one with chemical agents as these may damage the plastic on the equipment.
- To prevent the oscilloscope from overheating, keep the oscilloscope in a well-ventilated area. Do not place other test equipment or any other items on top of or near its cooling vents.
- When the oscilloscope is not in use, cover the oscilloscope with a protective covering to prevent dust from getting inside the unit. A sufficient layer of dust inside the unit can cause overheating and component breakdown.

Motors are an integral part of industrial electronics. (Fluke Corp.)

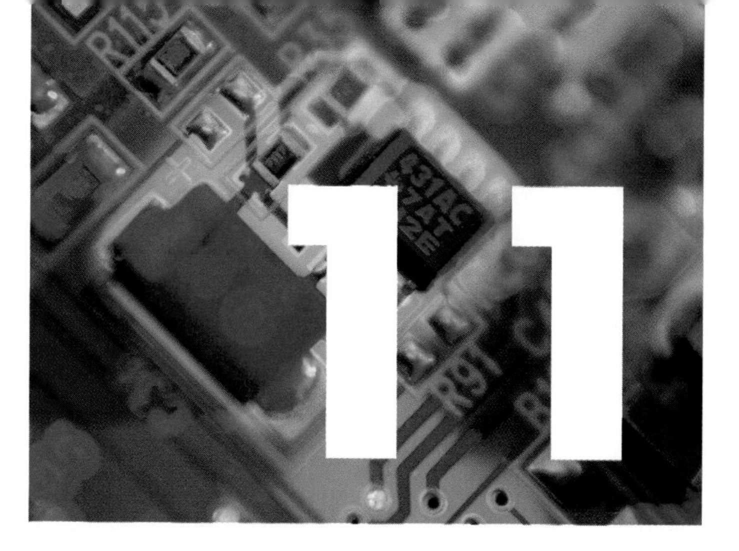

DC Motors

Objectives

After studying this chapter, you will be able to:
- Explain the operating principles of dc motors.
- Explain counterelectromotive force.
- Identify various dc motors.
- Discuss the purpose for, and operation of, motor starting circuits.
- Identify and explain the operation of various dc motors.

Key Words and Terms

The following words and terms will become important pieces of your electricity and electronics vocabulary. Look for them as you read this chapter.

commutating pole
constant speed motor
counterelectromotive force (cemf)
cumulative compound motor
dc machine
differential compound motor
digital encoder
electric motor
field windings
interpole
locked rotor
percent of speed regulation
push-button starter
servo motor
thermo-overload
torque
universal motor

One of the most important developments in the field of electricity is the electric motor. The ***electric motor*** converts electrical power into rotating mechanical power. Motors are used for such items as refrigeration and air conditioning, food mixers, vacuum cleaners, grinders, pumps, power bench saws, lathes, various wood and metal machines, as well as hundreds of other useful machines.

Throughout the world today, there are high speed manufacturing and assembly lines that would not be possible without the application of rotating devices such as the dc motor. No matter how sophisticated the automated assembly system, the heart of the system is the control of the rotation of the electric motor. The first mass assembly plants used a common shaft or pulley system to drive all of the equipment in a plant. Today, individual motors are placed throughout the system. These motors, combined with electrical control systems, allow the manufacture and assembly of products limited only by the imagination.

11.1 MOTOR OPERATION PRINCIPLES

The dc motor is simply an application of magnetic principles. Motor rotation depends on the interaction of magnetic fields. You will recall that the laws of magnetism state that:

Like poles repel each other.
Unlike poles attract each other.
or
A north pole repels a north pole.
A south pole repels a south pole.
but
North and south poles attract each other.

The theory of the simple dc motor is detailed in **Figures 11-1** through **11-9.** Figures 11-1 and 11-2 diagram the basic parts, the fields and the armature. Figures 11-3 and 11-4 put the motor parts together. Figures 11-5 through 11-9 take you through the motor action. Examine these figures.

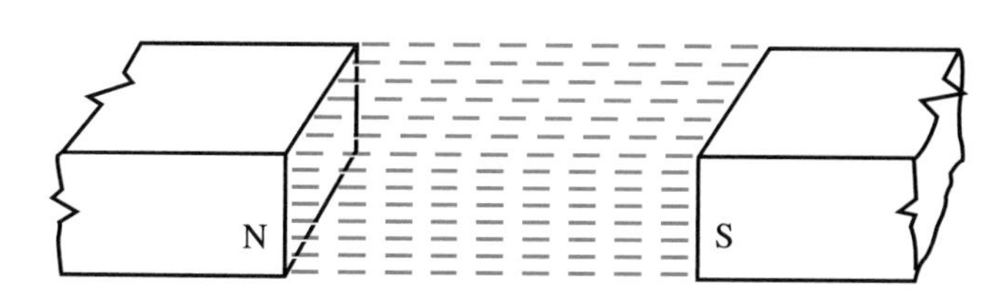

Figure 11-1. A magnetic field exists between the north and south poles of a permanent magnet.

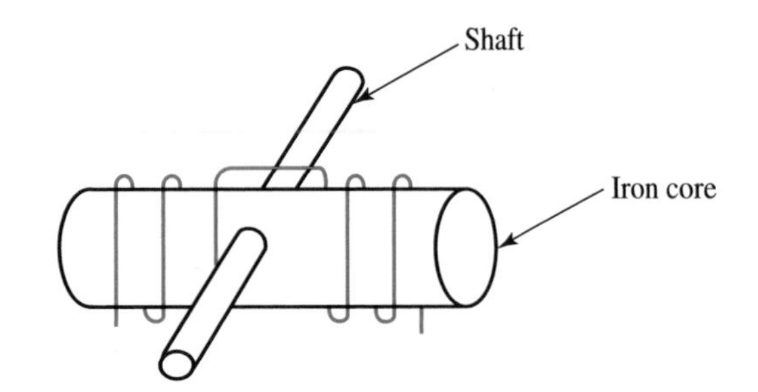

Figure 11-2. An electromagnet is wound on an iron core and the core is placed on a shaft so it can rotate. This assembly is called the armature.

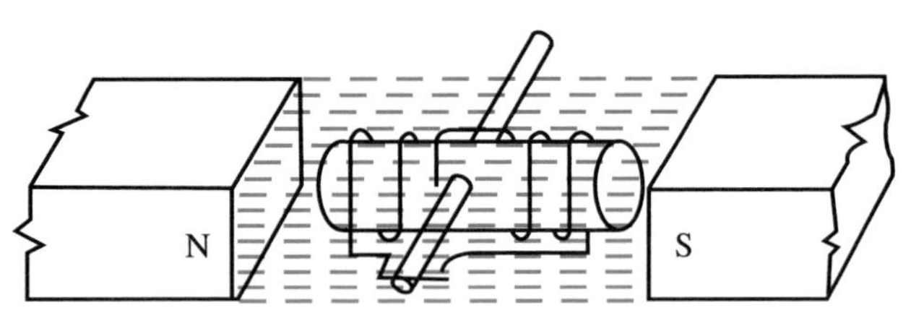

Figure 11-3. The armature is placed in the permanent magnetic field.

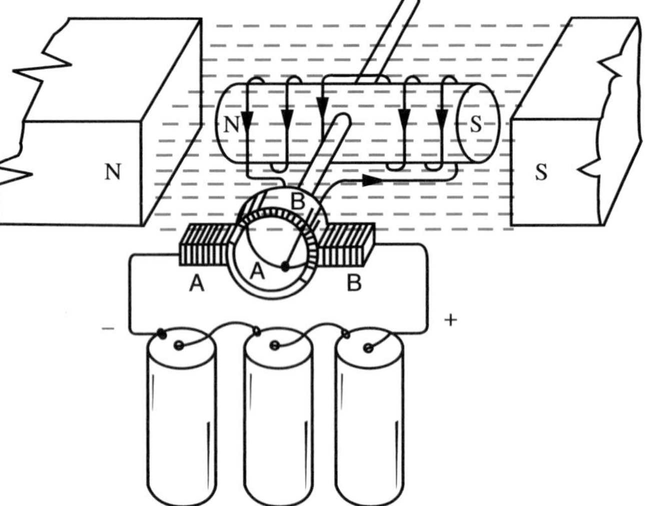

Figure 11-4. The ends of the armature coil are connected to semicircular sections of metal called commutators. Brushes contact the rotating commutator sections and energize the armature coil from an external power source. (Recall that the polarity of the armature electromagnets depends on the direction of the current flowing through the coil.) A battery is connected to the brushes. Current flows into brush A to commutator section A, through the coil to section B, and back to the battery through brush B, completing the circuit. The armature coil is magnetized as indicated in the sketch.

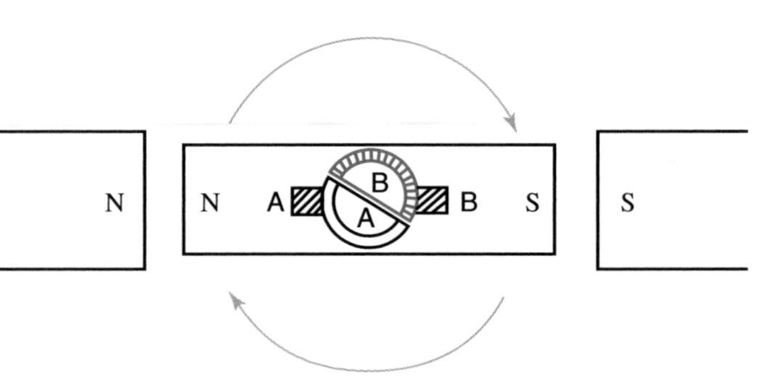

Figure 11-5. The north pole of the armature is repelled by the north pole of the field magnet. The south pole of the armature is repelled by the south pole of the field magnet. The armature turns one quarter revolution, or 90 degrees.

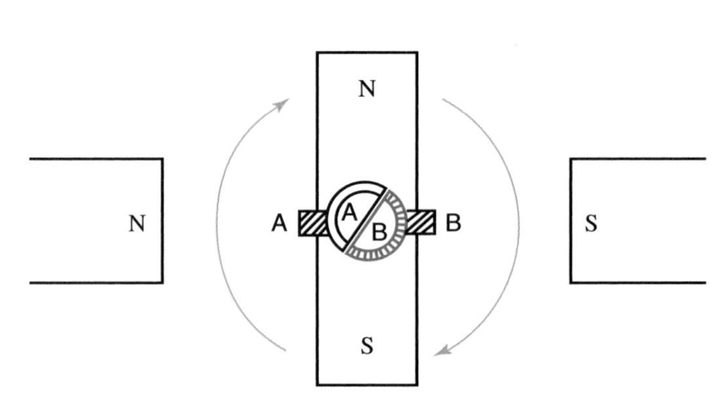

Figure 11-6. The north pole of the armature is attracted by the south pole of the field magnet. The south pole of the armature is attracted by the north pole of the field magnet. The armature turns another quarter turn. It has now turned one-half revolution.

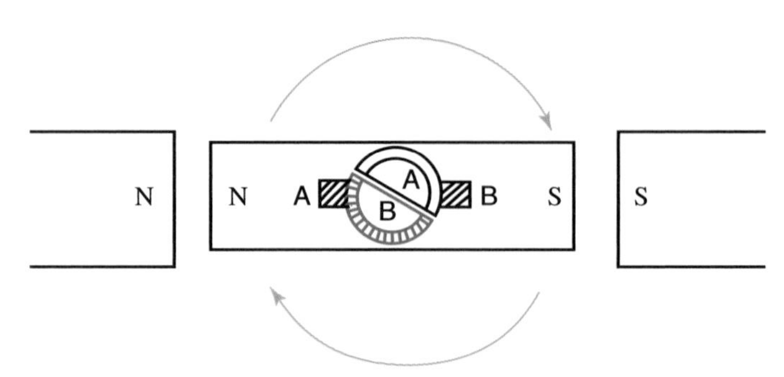

Figure 11-7. As the commutator sections turn with the armature, section B contacts brush A and section A contacts brush B. The current now flows into section B and out section A. The current has been reversed in the armature due to commutator switching action. The current reversal changes the polarity of the armature, so that unlike poles are next to each other.

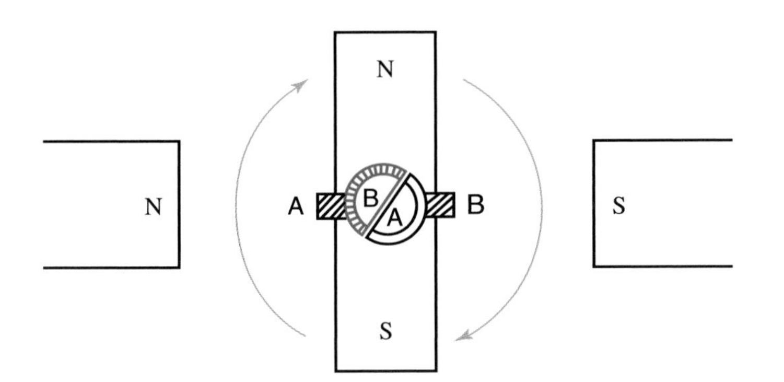

Figure 11-8. Like poles repel each other and unlike poles attract each other. The armature turns another quarter turn.

Figure 11-9. Unlike poles attract each other and the armature turns the last quarter turn, completing one revolution. The commutator and brushes are now lined up in their original positions, which causes the current to reverse in the armature again. The armature continues to rotate by repulsion and attraction. The current is reversed at each one-half revolution by the commutator.

The construction of a simple dc motor is very similar to a dc generator. *In fact, a dc generator and motor are often interchangeable in use.* In these cases, they are referred to as ***dc machines.***

As with the generators discussed in Chapter 10, to make the motor more powerful, permanent field magnets can be replaced by electromagnets called ***field windings.*** The field winding is placed over a soft iron pole piece. It consists of many turns of enamel covered copper wire. Like the generator, the field windings can have an independent source of voltage connected to them. Or, the field windings can be connected in series or parallel with the armature windings to a single voltage source, examine **Figure 11-10.**

Construct a trial motor, **Figure 11-11.** Connect the motor first in series, then in parallel as a shunt motor. Compare the speed and power of the two motors.

A Practical Motor

In industry, motors are made in a slightly different manner than discussed. Rotational force comes from the interaction between the magnetic field found around a current carrying conductor and a fixed magnetic field. A conductor carrying a current has a magnetic field around it. The direction of the field depends on the direction of the current. When this conductor is placed in a fixed magnetic field, the interaction between the two fields causes motion. Study **Figures 11-12** through **11-16.**

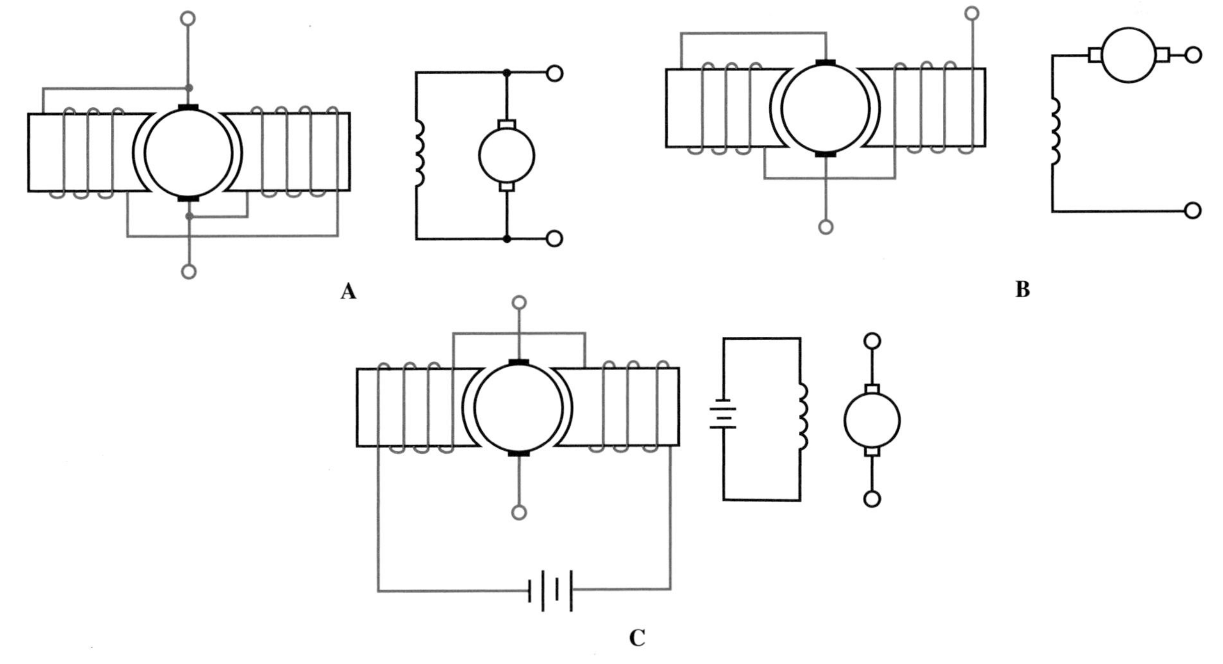

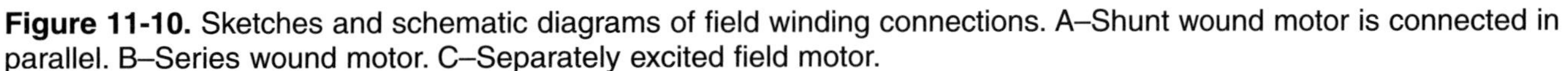

Figure 11-10. Sketches and schematic diagrams of field winding connections. A–Shunt wound motor is connected in parallel. B–Series wound motor. C–Separately excited field motor.

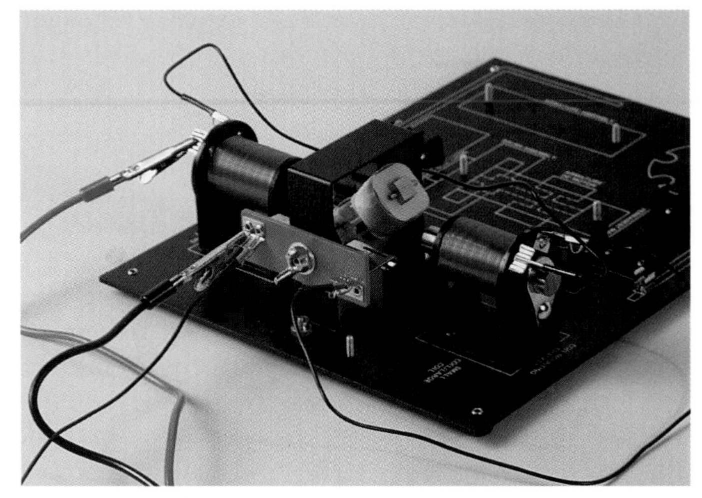

Figure 11-11. Examine the trial motor with series and parallel connections.

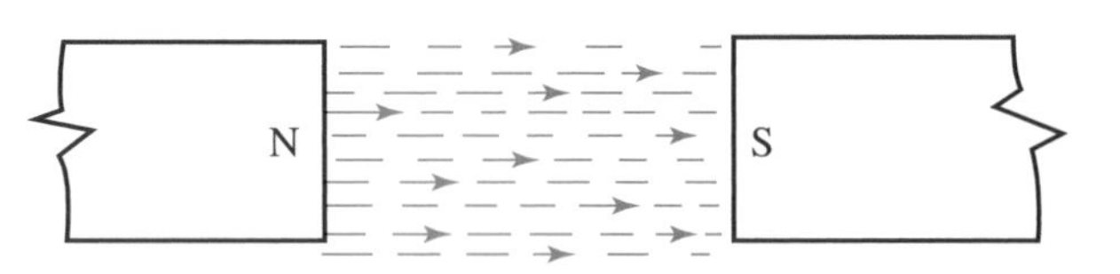

Figure 11-12. A magnetic field exists between the poles of a permanent magnet. The arrows indicate the direction of the field.

Figure 11-13. A current carrying conductor has a magnetic field; its direction depends on the direction of the current. Use the left hand rule to determine direction.

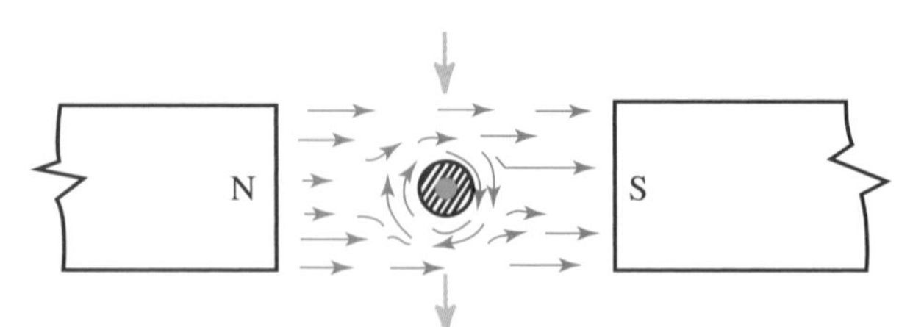

Figure 11-14. The field around the conductor flows with the permanent field above the conductor but opposes the permanent field below the conductor. The conductor will move toward the weakened field.

Armature coils on industry motors are connected to commutator sections, as in the trial motor. The theory of operation is similar. A practical motor has several armature coils wound in separate slots around the core. Each coil has a commutator section. Increasing the number of field poles gives the motor greater power.

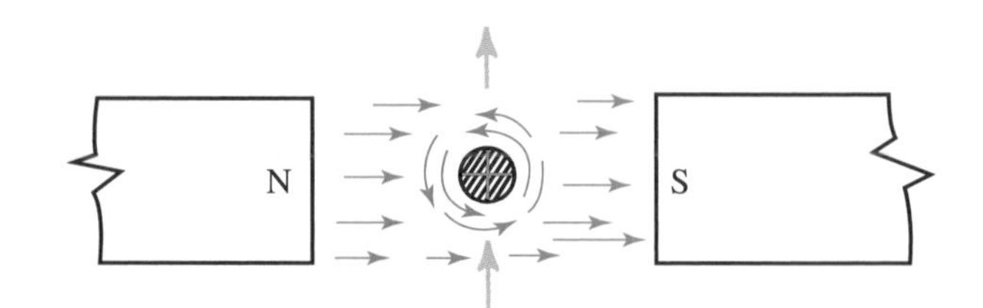

Figure 11-15. The current has been reversed in the conductor, causing the conductor field to reverse. Now the field is reinforced below the conductor and weakened above the conductor. The conductor will move up.

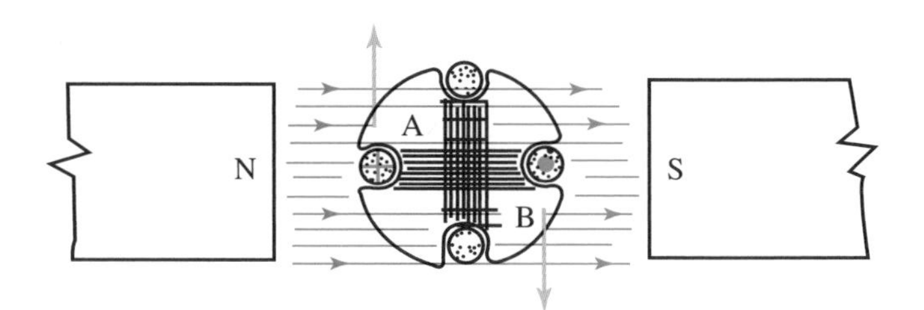

Figure 11-16. The single conductor is replaced by a coil of conductors wound in the slots of an armature core. Notice how the interaction of the two fields will produce rotation. Coil side A moves up and coil side B moves down. The rotation is clockwise.

A four-pole motor is sketched in **Figure 11-17.** The current divides into four parts. The current flowing in windings under each field pole produces rotation. This then increases the turning power, or ***torque,*** of the motor.

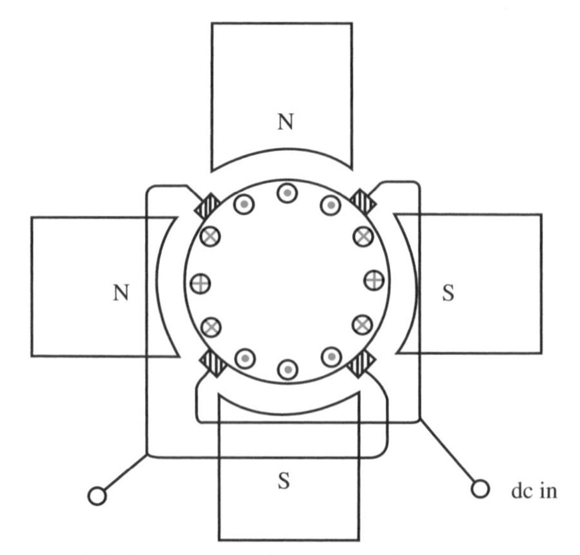

Figure 11-17. The torque of the motor is increased by adding armature coils and field coils.

Counterelectromotive Force

In Chapter 10, you learned that when a conductor cuts through a magnetic field, voltage is induced in the moving conductor. And while a motor is meant to convert electrical energy into mechanical energy, when the armature begins to rotate, the motor also becomes a generator.

The generated electrical force that opposes the applied emf is called ***counterelectromotive force.*** Counterelectromotive force is often written as ***counter emf*** or ***cemf.*** It is a result of the generator action of the motor. If the motor were connected to a prime mover and rotated in the same direction as the dc motor, it would produce a voltage with the opposite polarity. See **Figure 11-18.**

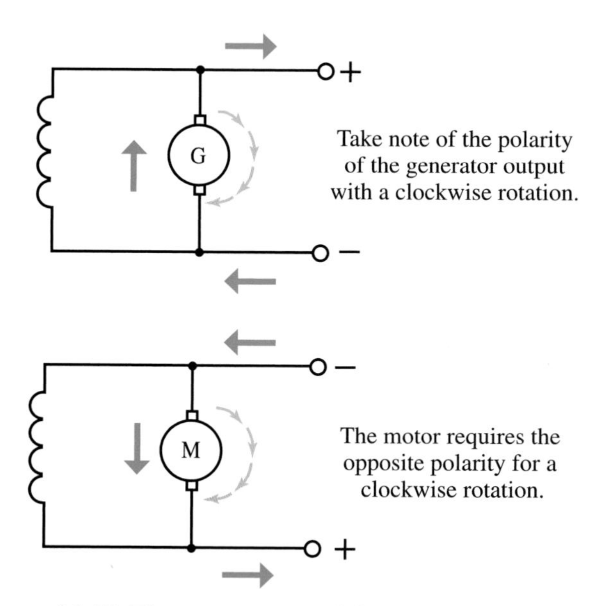

+
G
Take note of the polarity of the generator output with a clockwise rotation.
–
–
M
The motor requires the opposite polarity for a clockwise rotation.
+

Figure 11-18. The generator and the motor are rotating clockwise. The dc generator develops a polarity opposite of the motor polarity for the same clockwise rotation. This is the basis of counter emf.

The counter emf magnitude increases as the rotational speed and field strength increase. Therefore:

$$\text{Counter emf} = \text{Speed} \times \text{Field strength} \times \text{K}$$

where K equals some constant. This constant will vary in different motors. It is affected by things such as the number of windings. The actual effective voltage when applied to the windings in the armature must equal:

$$E_{source} - E_{counter} = E_{armature}$$

The current flowing in the armature windings at any given instant can be found using Ohm's law when the ohmic resistance of the windings is known:

$$I_{armature} = \frac{E_{armature}}{R_{armature}}$$

It is important to note that, as rotation of the motor armature slows down, less counter emf is generated. As a result of less counter emf being produced, there will be an increase in the current through the armature circuit. The current will continue to increase until the motor stops rotating as it does when physically overloaded. When the motor stalls, maximum current through the armature circuit is limited only by the resistance of the armature. This condition results in extremely high current values. A dc motor must be properly protected against overload conditions.

Overload Protection

Overload protection can be provided through one of several methods. The method used depends upon the size, type, and application of the motor. The circuit feeding power to the motor is usually protected by a fuse or circuit breaker. A fuse or circuit breaker provides the best method of protection against damage from short circuit or locked rotor conditions. ***Locked rotor*** is a term that means the rotor is not turning because of a physical impedance while the power is applied to the motor.

Actual overload protection is usually provided by a thermo-overload. A ***thermo-overload*** device is a simple ratch wheel device held in place by a metal alloy such as solder. When the overload condition generates sufficient current flow to melt the solder, the wheel is free to rotate, causing the circuit to open. This allows the motor to safely shut down before any damage occurs to the equipment or personnel. See **Figure 11-19.**

According to the National Electric Code, dc motors over one horsepower should be protected by a fuse or a breaker that is no more than 150 percent of the full load current. The size of the conductor feeding the motor should be able to carry at least 125 percent of the full load current. The thermo-overload device is sized closely to the maximum full load current rating of the motor. It is usually sized at 115 to 125 percent of the full load current depending on the exact type of motor and its application.

Commutation and Interpoles

As the motor armature rotates, the current in the armature windings routinely reverses. This is caused by commutator action. Due to the self-inductance of the windings, however, the current does not instantly reverse. This results in sparking at the commutator brushes.

Submarine

DC motors are the normal propulsion unit for submarines today. The following illustration shows one possible system of power conversion from nuclear energy to the dc motor system that drives the submarine.

The nuclear energy creates heat that will boil water and turn it into steam. The heating of the water takes place in the heat exchange unit. The steam passes through the turbine and forces the generator to rotate, producing electrical power. The generator supplies electrical energy that charges storage batteries and powers the two dc motors that drive the submarine's propellers.

Notice the battery storage system that is charged by the dc generators. The battery bank can be used to power the ship when the primary power source is not available such as under maintenance conditions. The ship's electrical control panel houses the electrical instrumentation necessary to monitor the production and distribution of power.

Even conventional diesel engine submarines and many surface ships use electric motors to power the ship's propellers. Electric motors provide a wide range of rpm that would not be readily available to diesel craft without the addition of a gear-based transmission. Even with the addition of a gear box, most engines could not produce the same high rpm that is available from a dc electric motor.

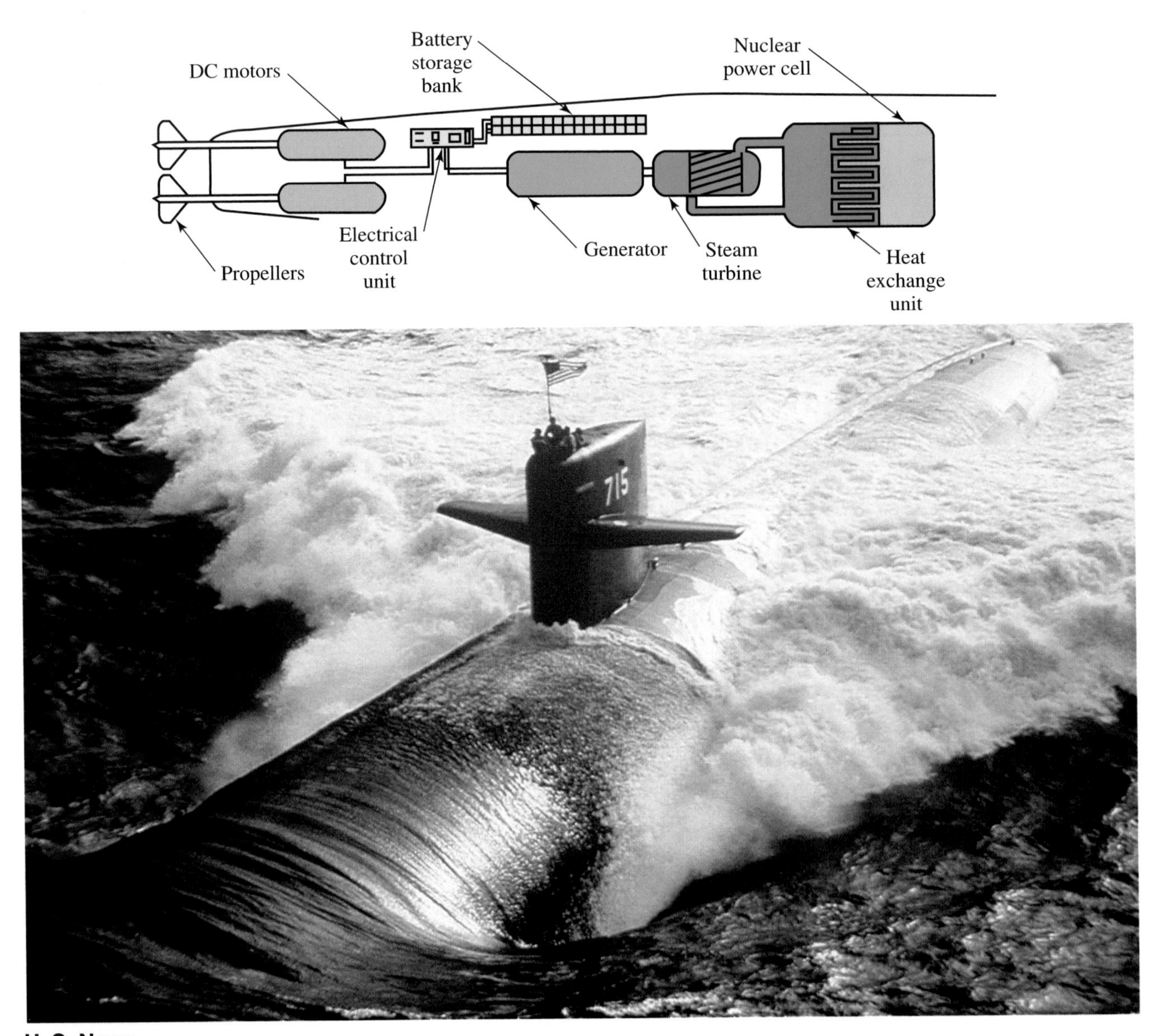

U. S. Navy

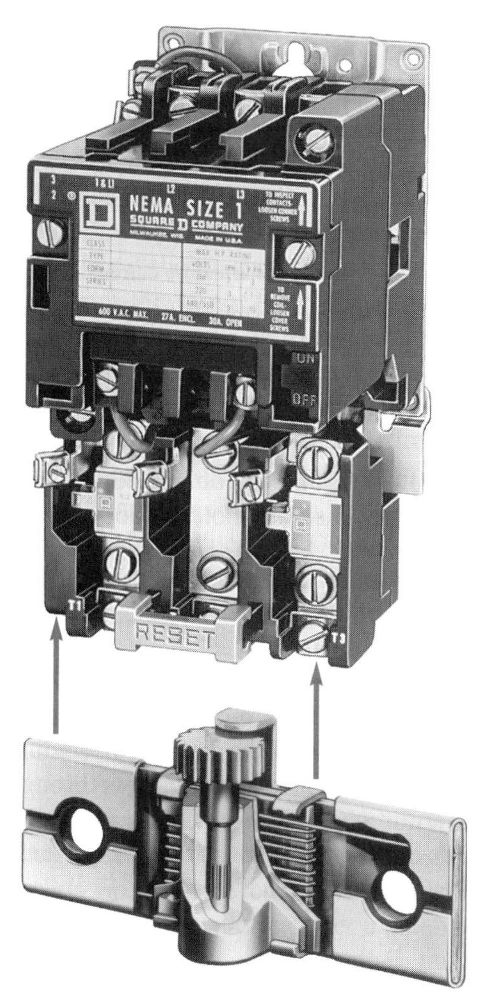

Figure 11-19. The thermo-overload mounts to the bottom of a motor starter and provides protection for motor overloads.

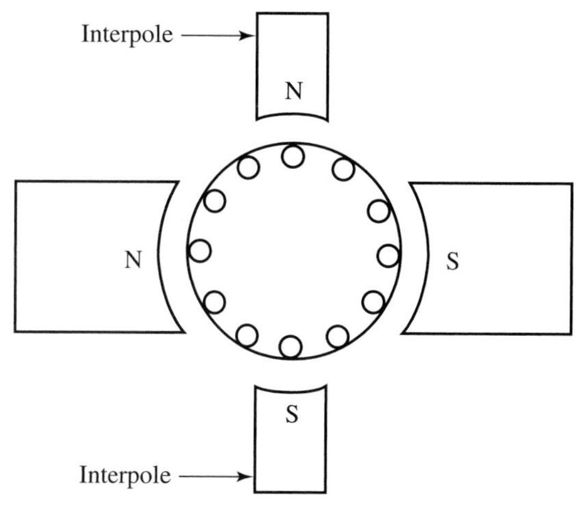

Figure 11-20. Interpoles reduce sparking at the commutator.

There are a number of methods for preventing these sparks. Changing the position of the brushes is one method. With this method, the brushes are moved slightly against the direction of rotation, and the counter emf is used to induce the previous pole. The counter emf opposes the self-induction caused by the decreasing current in the coil. Sparking is eliminated.

This method, however, is not a practical method for preventing sparks on large motors used in a varying load condition. As the load varies on the motor, the brush position must be changed by hand. Instead, larger motors use interpoles to reduce the sparking. An ***interpole*** is a smaller field pole placed midway between main field poles, **Figure 11-20.** The interpole has the same polarity as the main field poles and follows the main pole in direction of rotation. Interpoles are also called ***commutating poles.***

A counter emf is developed as the armature passes the interpole. This counter emf overcomes the emf caused by self-induction in the armature windings. The windings of the interpole are connected in series with the armature and carry the armature current. Thus, interpole field strength varies as the load varies, and it provides automatic control of commutator sparking.

Speed Regulation

Many motors are designed for special purposes. Some develop full power under load, while others must be brought up to speed before the load is applied.

When the speed for a motor is determined on the job, the motor should maintain that speed under varying load conditions. A ratio of the speed under no-load conditions to the speed under full-load can be expressed as a percentage of the full-load speed. This is called the ***percent of speed regulation.*** The equation is written:

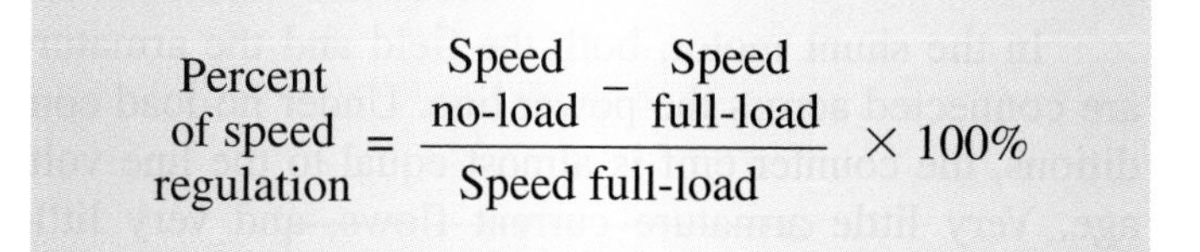

$$\text{Percent of speed regulation} = \frac{\text{Speed no-load} - \text{Speed full-load}}{\text{Speed full-load}} \times 100\%$$

A low speed regulation percentage means that the motor operates at a somewhat constant speed, regardless of load applied.

Review Questions for Section 11.1

1. The motor is used to convert __________ energy into __________ energy.
2. Outline how a simple dc motor operates.
3. The turning power of a motor is called the __________.

Example: Using Ohm's law, assume the armature resistance is 0.1 ohm, and the applied voltage is 100 volts. The initial armature current would be:

$$I_{armature} = \frac{100\ \text{V}}{0.1\ \Omega} = 1000\ \text{amps}$$

This excessively high current would burn the insulation off the wires, and the motor would be destroyed.
When the motor is up to speed and cemf has developed, then the voltage across the armature is equal to:

$$E_{line} - \text{cemf} = E_{armature}$$

If the motor develops a cemf of 95 volts as it approaches full speed, the armature current would be:

$$I_{armature} = \frac{100\ \text{V} - 95\ \text{V}}{0.1\ \Omega} = 50\ \text{amps}$$

This is a safe current for this motor armature to carry. Therefore, we should start the motor with the full series resistance in the armature circuit. Then, the resistance should be gradually decreased until the motor can limit the current by its own cemf.

Manual starters

Most manual starters have been replaced by automatic starter systems. However, some manual starters can still be found in industry. With a manual starter, the starting resistance is adjusted by hand using a lever. The lever decreases the resistance step-by-step, until the motor reaches full speed, **Figure 11-26.**

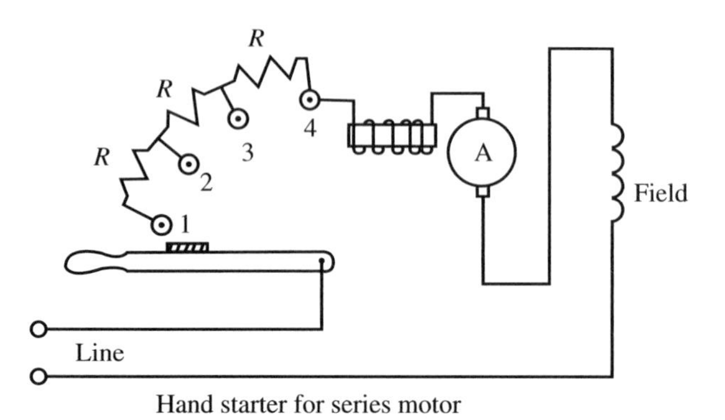

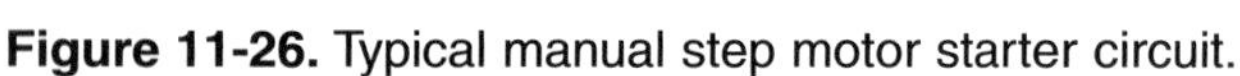
Hand starter for series motor

Figure 11-26. Typical manual step motor starter circuit.

In step 1, the maximum resistance is in series. With each following step, the resistance decreases. At step 4, the lever arm is held magnetically by the holding coil, which is also in series with the circuit. If the line voltage should fail, the lever arm will snap back to the off position, and the motor will have to be restarted. This protects the motor from an applied full voltage when the power comes back on.

Operators must use good judgment when starting a motor. They must not move the lever to the next lower resistance position until the motor has gained sufficient speed.

Automatic starters

Automatic starters remove the possibility of human error and also permit remote starting of the motor. Several devices are used. **Figure 11-27** shows one type of automatic starting switch. When line switch S is closed, voltage is applied to the field across the line. Voltage is also applied to the armature through R_1, R_2, and R_3 in series with the armature contactors C_1, C_2, and C_3. The contactors will not close until the current has decreased to a preset value.

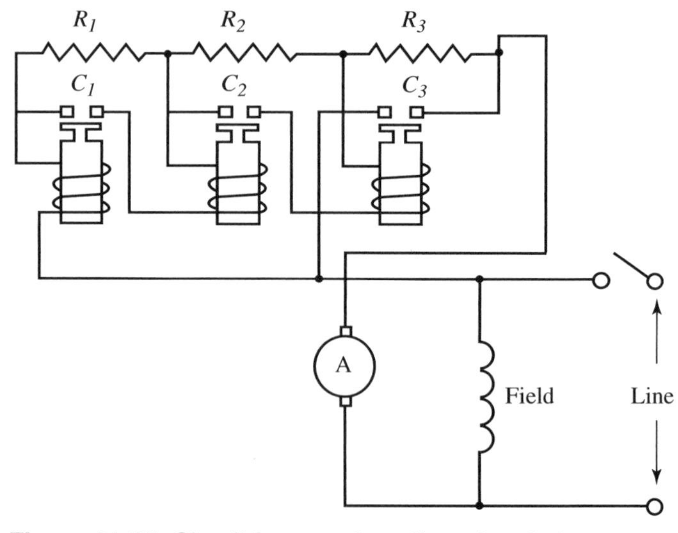

Figure 11-27. Circuit for an automatic motor starter.

The first surge of current in the armature circuit flows through the series coil of C_1, which holds it open. As the motor increases speed and the armature current decreases, C_1 contactor closes and cuts R_1 out of the circuit. Similar action occurs in C_2 and C_3. When the motor reaches full speed, all contactors will be closed. The protective resistance is no longer needed.

Another type of motor starter is the ***push-button starter,*** **Figure 11-28.** Motors with push-button starters are equipped with starting boxes containing push buttons marked stop and start. The circuit diagram is shown in **Figure 11-29.**

This starter is a heavy-duty relay operated by momentary contact push-button switches. Look at Figure 11-29 as the operation of the starter circuit is explained.

Figure 11-28. This type of push-button starter pulls to start and pushes to stop. There is no separate button.

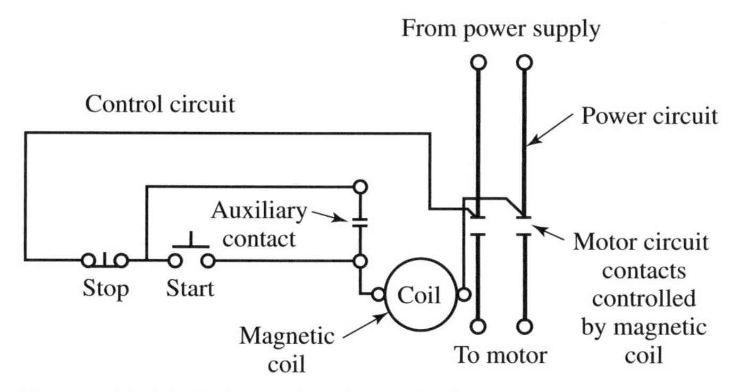

Figure 11-29. Schematic of a typical dc motor starter with a push-button control circuit.

1. To start the motor, the push button marked "start" is pushed closed.
   a. The action of closing the start button will energize the magnetic coil through the normally closed stop button.
   b. When the magnetic coil is energized, the two motor contacts and the auxiliary contact close.
   c. The motor circuit contacts now supply power to the motor.
   d. The auxiliary contact keeps the circuit completed to the coil so that the coil will remain energized when the start button is released. The circuit remains energized until the stop button is pushed.
2. When the stop button is pushed, the circuit to the magnetic coil is opened.
   a. When the coil is no longer energized, all three contacts open.
   b. Power is disconnected from the motor.

This is one of the most common motor control devices in use today.

In many locations the magnetic starter has been replaced by the solid state motor starter. The solid state motor starter will be covered in detail later in the textbook.

Thyristor Motor Controls

The thyristor is a family of semiconductor devices that includes the silicon controlled rectifier (SCR) and the triac. These devices are used in many circuits, including motor controls, to control the speed of a motor. Examine **Figure 11-30.**

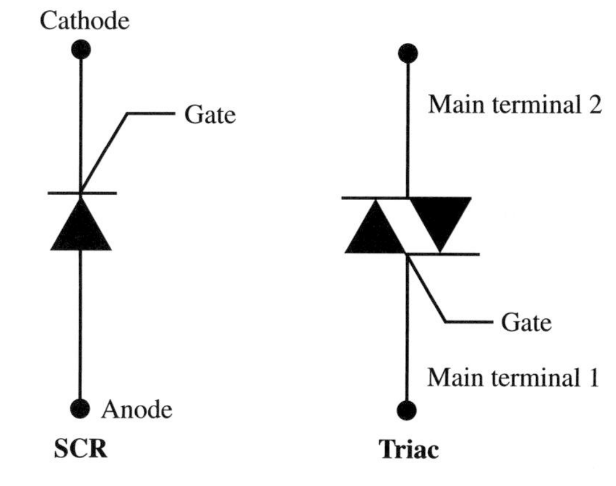

Figure 11-30. Schematic symbols for two types of thyristors, the SCR and the triac.

A half-wave SCR motor control circuit is shown in **Figure 11-31.** This circuit controls up to seven amps in a universal motor. These circuits are often found in hand drills and hand saber saws. **Figure 11-32** shows a typical SCR package.

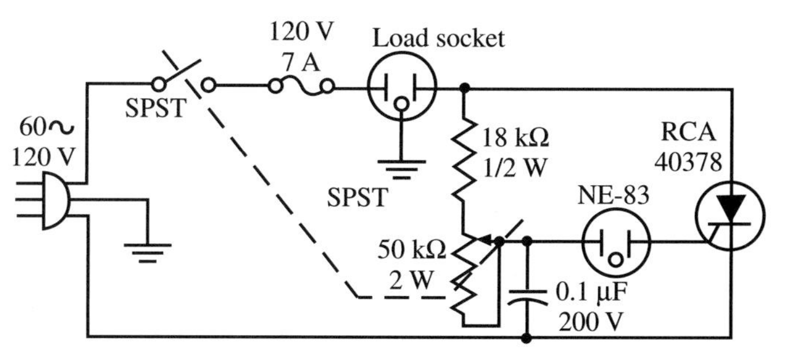

Figure 11-31. Schematic for a half-wave SCR motor control (RCA circuit).

Figure 11-32. SCR package.

The triac controls the full ac wave, while the SCR controls only half the wave. A full-wave triac circuit is shown in **Figure 11-33.** Larger triacs can control up to 2000 amps.

SCRs and triacs are discussed in more detail in Chapter 18.

The Universal Motor

The laws of magnetism were used to explain the operation of the dc motor. But, will a dc motor operate on alternating current? The answer is *yes,* to a limited extent. Motors that operate on either ac or dc power are called ***universal motors.***

With an alternating current, the poles of both the field and armature windings will periodically reverse. However, since two north poles repel each other, as do two south poles, motor action continues in the dc motor when ac is applied. For best results, a series motor should be used. When the shunt motor is connected to ac, the inductance of the field windings causes a phase displacement. This impairs motor action.

When a universal type motor is used in industry, the series wound type is preferred. These motors are not used for heavy-duty purposes because of the large amount of sparking at the brushes. Commercial motors of these types are used for small fans, drills, and grinders.

Permanent Magnet Motor

Permanent magnet, dc motors are used widely in industry today. They range in output from 1/50th horsepower to 5 horsepower. These motors are of simple design and require voltage to the armature circuit only. The permanent magnet, dc motor uses permanent magnets for field poles in place of electrical coils. Look at **Figure 11-34** to become familiarized with the parts of a dc motor.

DC Servo Motor

A ***servo motor*** is any motor that is modified to give feedback information concerning the motor's speed, direction of rotation, and sometimes the number of revolutions it has made. This feedback is used to determine position of the mechanism it is driving, such as a robotic arm. Look at **Figure 11-35.** Note the location of the tach and tach magnets. The tach generates voltage as the motor rotates. The tach voltage signal to the speed control unit gives feedback about the motor's revolutions per minute (rpm). Look at the closed loop feedback speed control diagram in **Figure 11-36.**

As an illustration of the servo motor, let us assume that a motor is required to operate at a constant 1000 rpm. After the motor has been running for a period of time, an unusually heavy mechanical load is exerted on the motor.

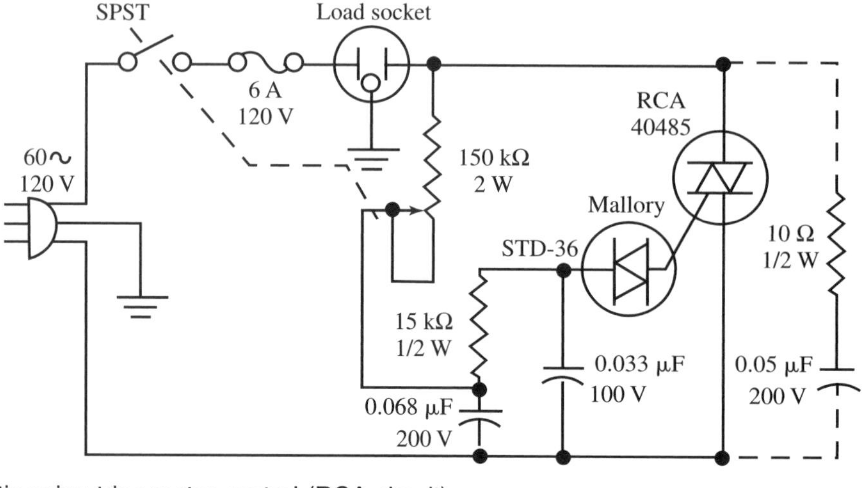

Figure 11-33. Schematic using triac motor control (RCA circuit).

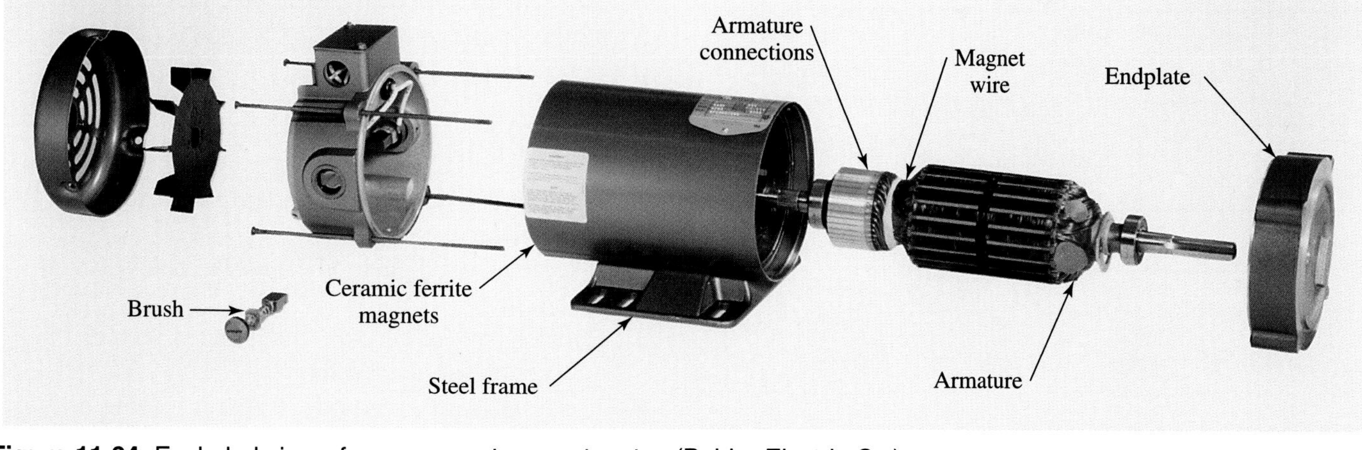

Figure 11-34. Exploded view of a permanent magnet motor. (Baldor Electric Co.)

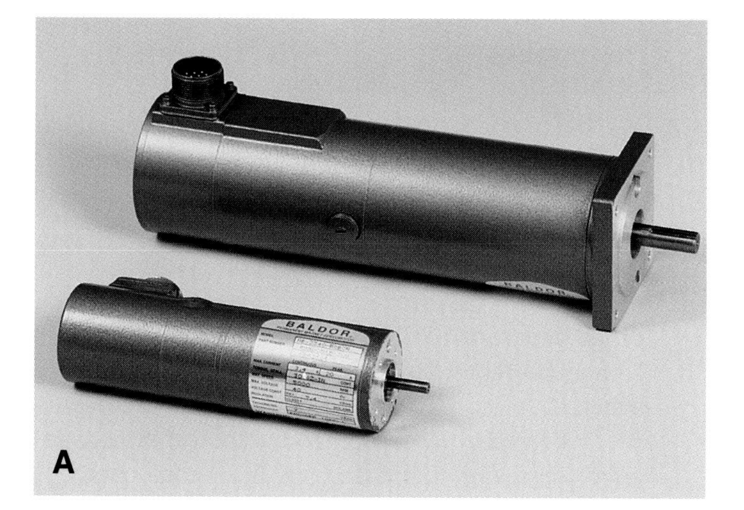

As a result, the motor's rpm drops to 900 rpm. As the motor slows, the tach also slows down and sends less voltage back to the electronic speed control. The electronic circuitry now increases the applied voltage to the dc motor until 1000 rpm is regained. When the mechanical load is released from the motor, the motor rpm will increase above the required 1000 rpm. The tach now sends a higher voltage to the speed control unit. This causes the control unit to apply lower voltage to the dc motor until it slows back to 1000 rpm.

Another type of feedback system is the digital encoder. The ***digital encoder*** sends pulses back to the control unit rather than varying the voltage. An electronic counter counts the pulses generated by the encoder. By

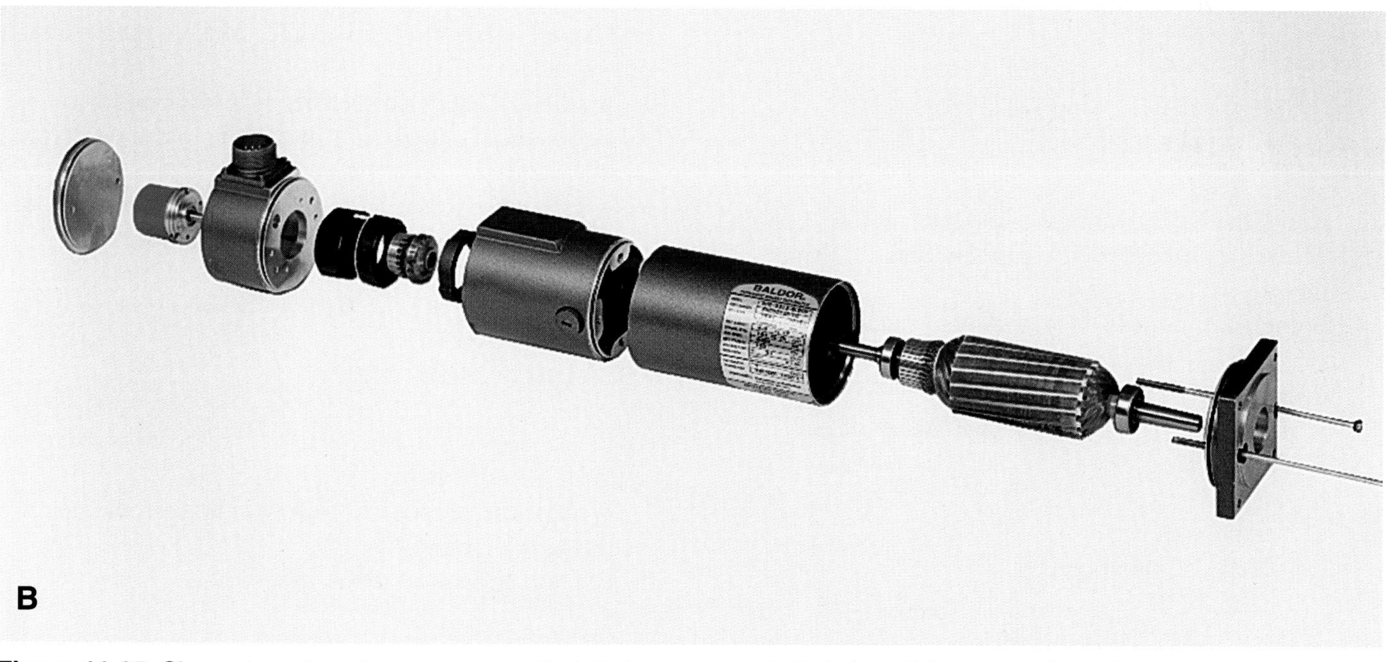

Figure 11-35. Shown here is a dc servo motor. Part B shows an exploded view of the motor shown in part A. (Baldor Electric Co.)

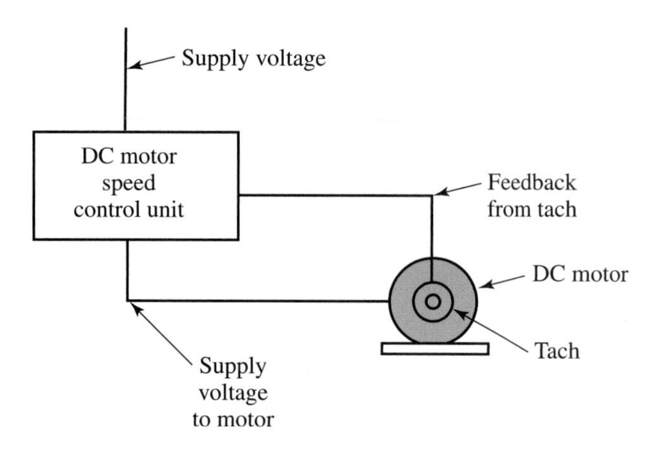

Figure 11-36. A simple block diagram of a servo motor system.

counting the pulses, the system can accurately determine the number of revolutions completed by the motor and the position of the device the motor is driving. The device could be an assembly line conveyer belt or even a dc motor driven, robotic arm.

Review Questions for Section 11.2

1. Name the type of motor shown in the following schematic.

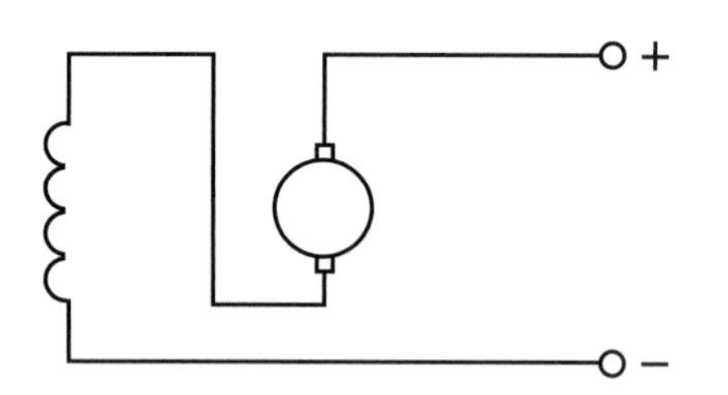

2. Name the key advantage of a series motor.
3. There are two types of compound motors. Name them.
4. Some automatic starting switches contain __________ that will not close until the current has decreased to a preset value.
5. A(n) __________ controls a half-wave of ac.
   a. SCR
   b. triac
   c. universal motor
   d. None of the above.
6. How does a tach differ from an encoder in a servo motor control system?

Summary

1. A motor is a device for changing electrical energy to mechanical energy.
2. Like poles repel each other; unlike poles attract each other.
3. Counter emf, or cemf, is the induced voltage that opposes the applied voltage.
4. The rotation of the motor produces turning or twisting power called torque.
5. Percent of speed regulation can be calculated by:

$$\text{Percent of speed regulation} = \frac{\text{Speed no-load} - \text{Speed full-load}}{\text{Speed full-load}} \times 100\%$$

6. Shunt motors have their field windings connected in parallel with the armature. They are stable motors under varying loads.
7. The series motor has the field windings connected in series with the armature. They develop a high torque under load.
8. Compound motors have their field windings connected in series and parallel with the armature. There are two types of compound motor: the cumulative compound motor and the differential compound motor.
9. There are many types of motor starting circuits. Those discussed in the chapter include both manual and automatic types.
10. A silicon controlled rectifier is a semiconductor device that controls half of the sine wave while a triac controls the full wave.
11. A servo motor is a motor that provides feedback of data about the motor's rpm or position for the motor control system.

Test Your Knowledge

Please do not write in the text. Place your answers on a separate sheet of paper.

1. What is the purpose of field windings in the operation of a dc motor?
2. Explain generator action in a motor.

3. Interpoles are used to prevent __________ at the commutator brushes.
4. Draw diagrams of a shunt motor and a series motor.
5. Motor starting circuits are intended to protect motors until they build up __________ and __________.
6. What are two advantages of automatic starting circuits?
7. The __________ is a family of semiconductor devices.

For Discussion

1. Why does a dc motor increase in speed when its field strength is decreased?
2. Why does a motor get hot when overloaded?
3. If a load is removed from a series motor, it will destroy itself by centrifugal force. Explain this occurrence.
4. Why are motor starters necessary on heavy-duty motors?
5. Will your experimental dc motor run on 6 volts ac? Explain your answer.

Construction of an electrical transmission tower. Transformers step the voltage up prior to transmission, and down for use in homes and businesses.

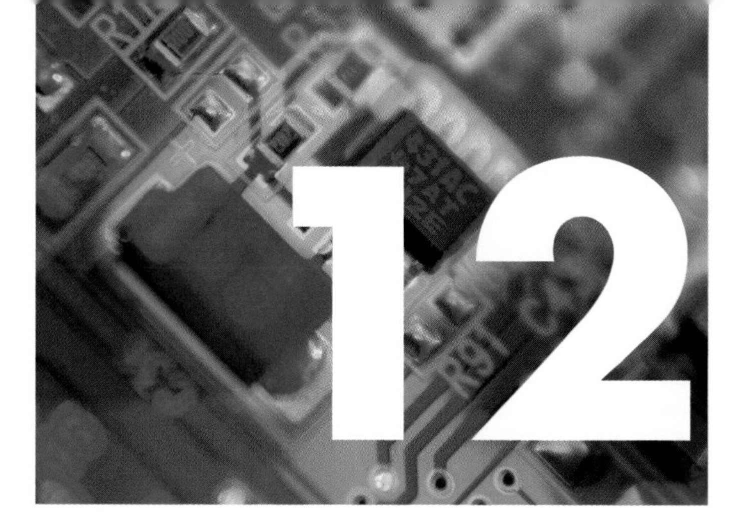

Transformers

Objectives

After studying this chapter, you will be able to:

- Explain the operation of a transformer.
- Discuss the relationship between mutual inductance and transformers.
- Describe the effect of self inductance.
- Calculate the various values of currents and voltages in transformer circuits.
- List three types of transformer losses.
- Identify delta and wye transformer connections.
- Discuss grounding an electrical circuit.
- Explain troubleshooting procedures for transformers.
- Describe several special transformer applications.

Key Words and Terms

The following words and terms will become important pieces of your electricity and electronics vocabulary. Look for them as you read this chapter.

copper losses
current ratio
eddy current losses
high leg
hysteresis loss
impedance
induction
isolation
megohmmeter
mutual induction
primary
secondary
self induction
step-down transformers
step-up transformers
tap
transformer
turns ratio
voltage ratio

The invention of the transformer made possible the distribution of electrical energy to homes and factories all over America. In this chapter, you will study the basic principles of transformer operation. You will learn how power companies use transformers to distribute electricity to your home. You will also learn how transformers are used to convert 120 volt household voltage to a lower voltage for use in many electronic devices such as stereos, televisions, computers, and video games.

12.1 TRANSFORMER THEORY

A ***transformer*** is a device used to transfer energy from one circuit to another using electromagnetic induction. A transformer consists of two or more coils of wire wound around a common laminated iron core. Transformers have no moving parts and require very little care. They are simple, rugged, and efficient devices, **Figure 12-1.**

Figure 12-1. A typical transformer. It is used in the power supply for electronic circuits.

The construction of a simple transformer is shown in **Figure 12-2,** along with its schematic symbol. The first winding, which is the input winding, is called the ***primary.*** This winding receives the energy from the source. The second winding, which is the output winding, is called the ***secondary.*** The output load is attached to the secondary.

The energy in the secondary is the result of the changing magnetic field generated by the primary windings. In a transformer, the varying magnetic field of the

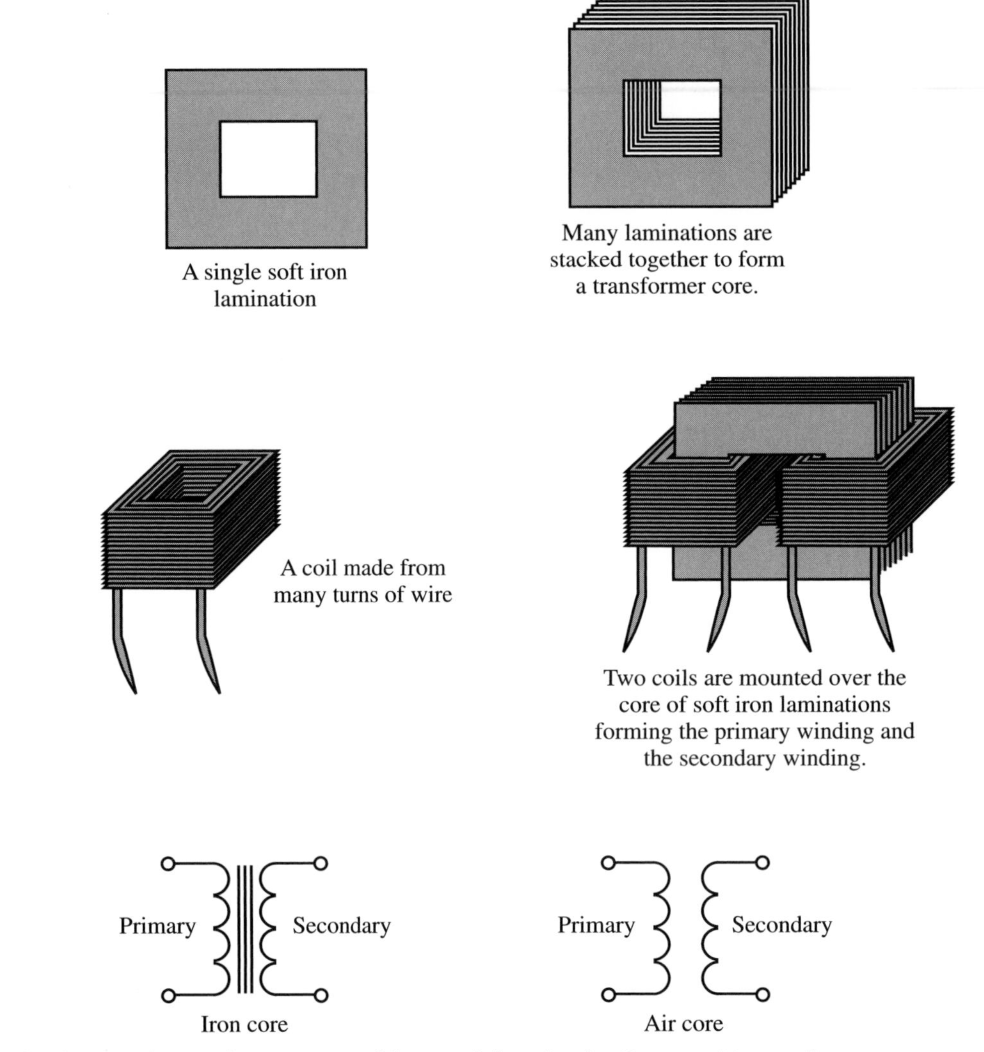
A single soft iron lamination
Many laminations are stacked together to form a transformer core.
A coil made from many turns of wire
Two coils are mounted over the core of soft iron laminations forming the primary winding and the secondary winding.
Primary
Secondary
Iron core
Primary
Secondary
Air core

Figure 12-2. A simple transformer is constructed from soft iron laminations and two coils.

primary cuts across the windings of the secondary. As you will recall from your study of Chapters 10 and 11, a changing magnetic field cutting across a conductor induces a voltage. Consequently, the changing primary voltage induces a voltage in the secondary. Therefore, the transformer is a device that must work on an alternating current or a pulsating direct current. The primary field must be a moving magnetic field in order for the transfer of energy to take place.

In **Figure 12-3,** two types of construction are shown: core construction and shell construction. In the core type, the coils surround the core. In the shell type, the core surrounds the coil.

In either type of construction, a strong coupling must exist between the primary and secondary windings to ensure that very little loss of power will occur. In core construction, each winding is wound on a separate leg of the core. In shell construction, the windings are in alternate layers, with one winding on top of the other.

Induction

The ability to produce electrical energy in a conductor without making physical contact with the conductor is referred to as ***induction.*** In your study of static electricity you found that a static charged body could produce a static charge on another body simply by being placed near it. The first body induced a static charge on the second body. A voltage can be induced on a conductor by using a magnetic field and motion. In Chapter 10, Generators, you learned that when a coil is moved past a magnetic field, a voltage is induced on the coil. You also learned that a magnetic field moved past a coil produces

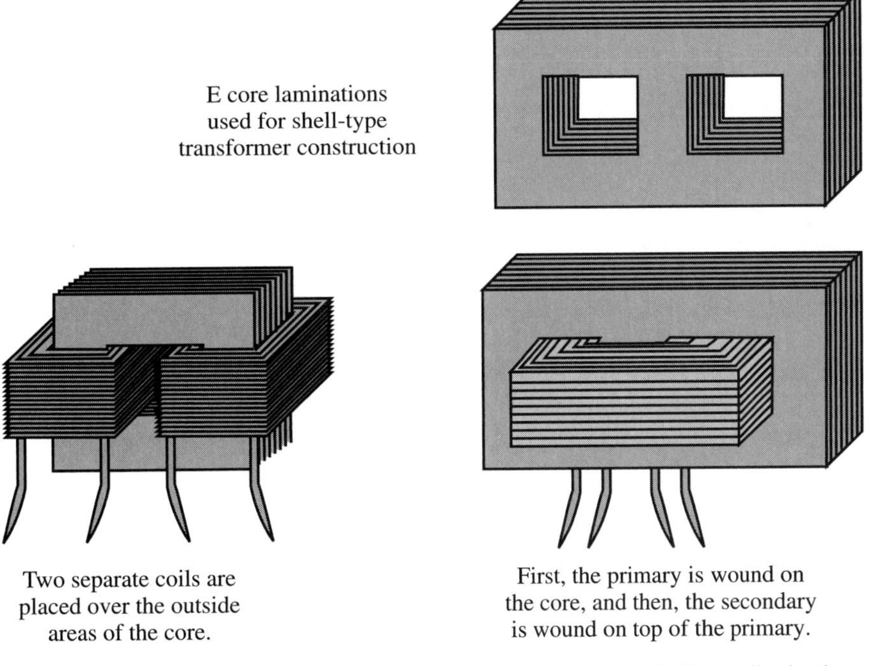

E core laminations
used for shell-type
transformer construction
Two separate coils are
placed over the outside
areas of the core.
First, the primary is wound on
the core, and then, the secondary
is wound on top of the primary.

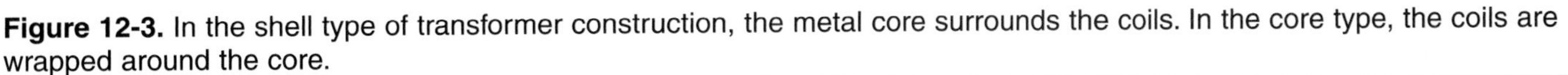

Figure 12-3. In the shell type of transformer construction, the metal core surrounds the coils. In the core type, the coils are wrapped around the core.

the same result. Basically, if there is a magnetic field, a coil, and motion, an emf is produced.

Examine **Figure 12-4.** When coil A is energized by an ac voltage source, a current is established in the coil. The alternating current flowing through coil A produces a rising and collapsing magnetic field. The magnetic field rises and collapses at the same rate as the frequency of the ac voltage source. If the ac voltage source frequency is equal to 60 Hz, then the magnetic field rises and collapses 120 times a second.

A transformer applies the same principle as generating emf. The difference is that there is no physical motion in the transformer. Instead, the magnetic field rising and collapsing provides the needed motion. Coil A produces a rising and collapsing magnetic field that cuts across the conductors in coil B. Coil B has a voltage induced by the electrical action of coil A. This is the principle behind all transformers. This principle is called ***mutual induction.***

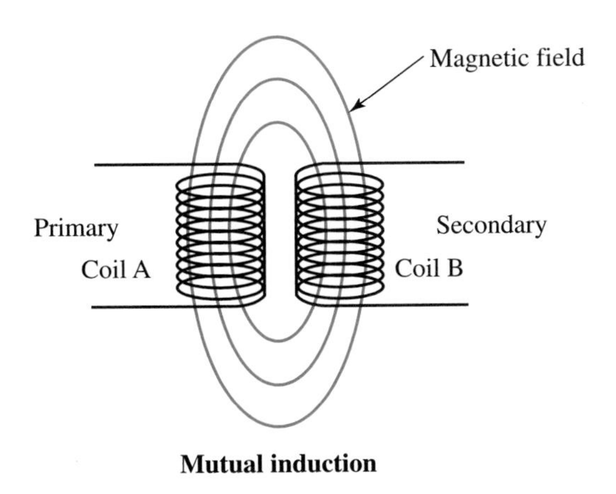

Figure 12-4. Two coils coupled together by a rising and collapsing magnetic field is an example of mutual induction.

Another electrical phenomena occurs at the same time. This phenomena is called self induction. ***Self induction*** occurs when the magnetic field surrounding one conductor cuts across the conductors left and right of it. In **Figure 12-5,** you can see that when a coil is energized, it produces a magnetic field. Each winding produces a separate magnetic field. These separate fields combine to form an overall magnetic field. This action induces an emf in the direction *opposite* to the force generating the magnetic field. Self induction will occur any time a conductor is wound into a loop and passes an ac current.

The theory of induction and its application to other electronic circuits is covered in greater detail in Chapter 14, Inductance and RL Circuitry. For now, only the basic concepts of induction, mutual induction, and self induction that apply to the transformer will be covered.

Turns Ratio Principle

One key use of transformers is to increase or decrease a voltage. This is done by increasing or decreasing the number of turns on the secondary winding. The

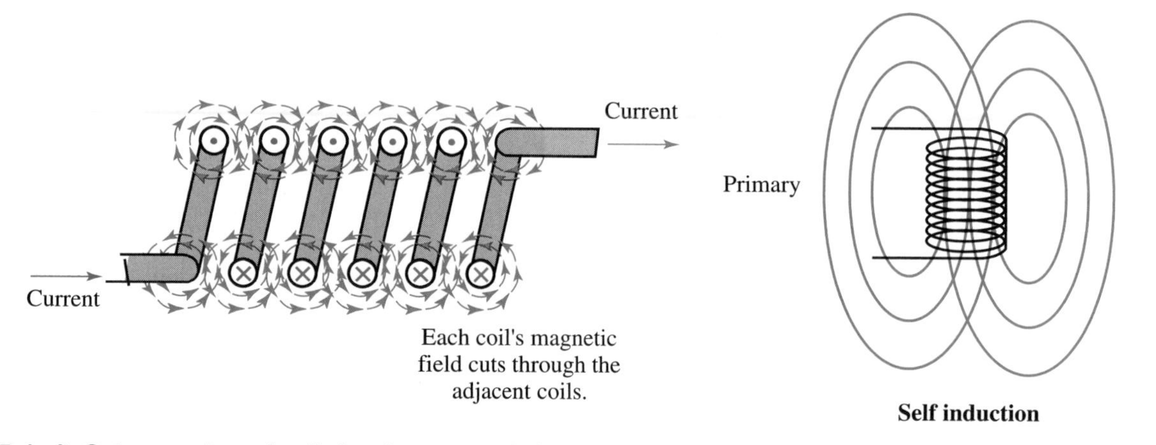

Figure 12-5. Left. Cut-away view of coil showing magnetic field around each coil. This is an example of self induction. Right. The magnetic field rising and collapsing around coil A is an example of self induction.

amount the voltage is raised or lowered by is based upon the *ratio of the number of turns of wire used in the transformer's primary versus its secondary windings.* As described earlier, the primary winding is the winding closest to the source. The primary winding induces an emf into the secondary winding. The secondary winding is connected to the load. See **Figure 12-6.**

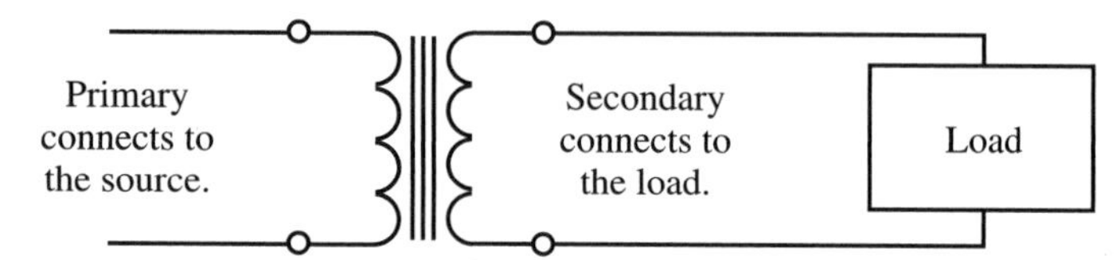

Figure 12-6. It is important to remember that the primary of a transformer is the side that connects to the source and that the secondary side of the transformer is the side that connects to the load.

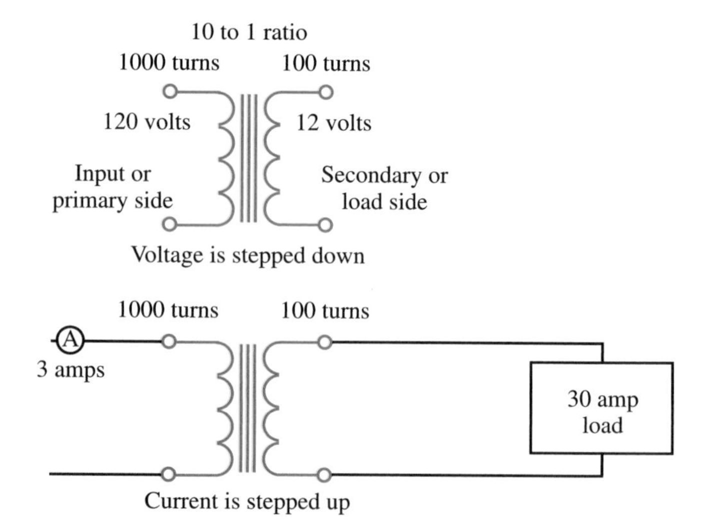

Figure 12-7. Turns ratio explains the relationship between the primary and secondary voltages and current values.

Let's look at the formulas for the transformer turns ratio and expected voltage, current, and power. The ratio between the number of turns in the primary and secondary is called the ***turns ratio.*** The turns ratio is simply the number of turns in the primary divided by the number of turns in the secondary. Written as a formula:

$$\text{Turns ratio} = \frac{N_P}{N_S}$$

where N equals the number of turns in the primary (P) or the secondary (S).

Using the transformer in **Figure 12-7,** the turns ratio can be calculated as:

$$\text{Turns ratio} = \frac{N_P}{N_S} = \frac{10}{1} \text{ or 10 to 1}$$

There are 1000 turns of conductor in the primary of the transformer, and the secondary has 100 turns of wire. The ratio here is 10 to 1.

The ***voltage ratio*** is the ratio between the voltages of the primary and secondary. It is in the same proportion as the turns ratio:

$$\text{Voltage ratio} = \frac{E_P}{E_S} = \frac{N_P}{N_S}$$

The ***current ratio*** is the ratio between the currents in the primary and secondary. It is in *inverse* proportion to the turns ratio:

$$\text{Current ratio} = \frac{I_S}{I_P} = \frac{N_P}{N_S}$$

Combining the three ratios:

$$\frac{E_P}{E_S} = \frac{N_P}{N_S} = \frac{I_S}{I_P}$$

In Figure 12-7, there are ten turns in the primary winding for every one turn in the secondary winding. In a 10 to 1 ratio, the voltage on the secondary side will be ten times lower than the primary voltage. If the primary side of the transformer is connected to a 120 volt supply, there will be 12 volts present on the secondary.

Examining the ratios, you can see that the current ratio is the *opposite* of voltage ratio. If the voltage ratio is 10 to 1, the current ratio will be 1 to 10. The current value of the primary is based on the current in the secondary. The current in the secondary is determined by the load connected to the secondary. The current value for the primary in Figure 12-7 will be 1/10th that of the current flowing in the secondary.

Transformers that are used to raise or lower a voltage are known as step-up transformers and step-down transformers. See **Figure 12-8.** ***Step-up transformers*** have more turns in the secondary than the primary. The voltage is increased, it steps up. ***Step-down transformers*** have fewer turns in the secondary than the primary. The voltage is decreased, it steps down.

Example: Refer to **Figure 12-9.** The transformer has 200 turns in the primary and 1000 turns in the secondary. If the applied voltage is 117 volts ac, what is the secondary voltage?

$$\frac{E_P}{E_S} = \frac{N_P}{N_S}$$

$$\frac{117\text{ V}}{E_S} = \frac{200}{1000}$$

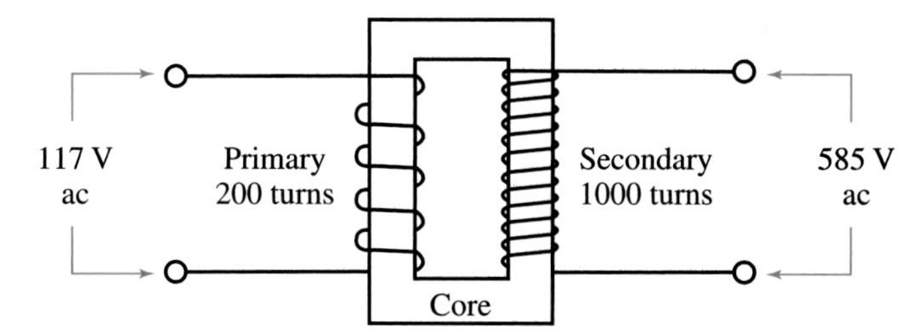

Figure 12-9. Turns ratio helps determine output voltage.

Transposing the equation:

$$E_S = \frac{117\text{ V} \times 1000}{200} = \frac{170{,}000\text{ V}}{200} = 585\text{ V}$$

This is an example of a step-up transformer.

Example: What if this transformer were made with a 10 turn secondary? What would the secondary voltage be?

$$\frac{117\text{ V}}{E_S} = \frac{200}{10}$$

$$E_S = \frac{117\text{ V} \times 10}{200} = 5.85\text{ V}$$

This is a step-down transformer.

Transformer Power

Calculating the power in the secondary of a transformer should take into account a transformers efficiency. However, for now we will consider the transformer an ideal piece of equipment working at a 100 percent efficiency. *If we make this assumption, then the power on the*

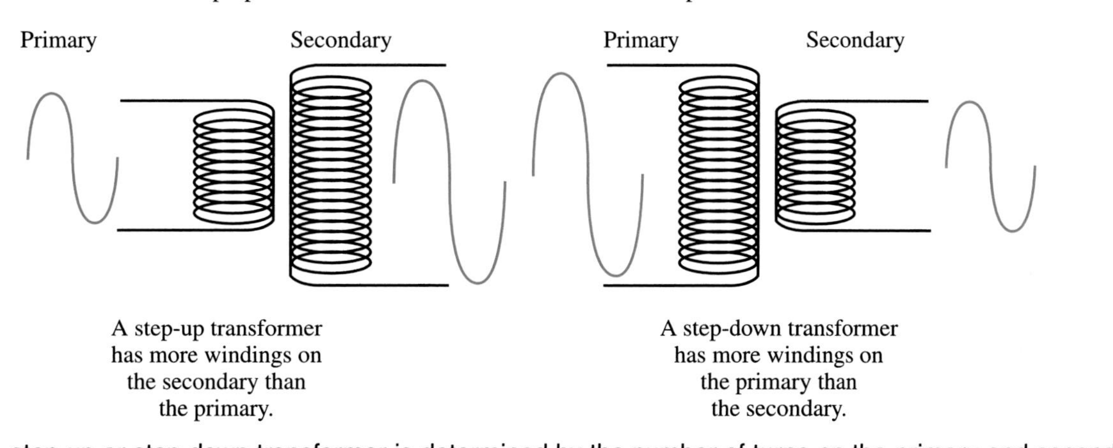

Figure 12-8. A step-up or step-down transformer is determined by the number of turns on the primary and secondary.

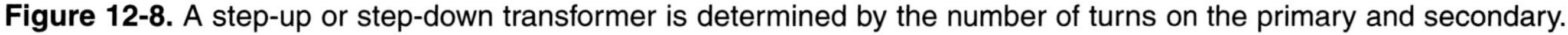

primary side should be equal to the power on the secondary side.

In the form of a formula:

$$P_P = P_S$$

The power rating for a transformer is usually expressed as VA (volt-amps) or kVA (kilovolt-amps) rather than watts. The reason for this terminology will become evident in Chapter 14.

The power used in the secondary circuit must be supplied by the primary. Assuming that the transformer is 100 percent efficient, the power in the secondary, $I_S \times E_S$, must equal the power in the primary, $I_P \times E_P$.

Example: A step-up transformer produces 300 volts in the secondary when 100 volts ac is applied to the primary. A 100-ohm load is applied to the secondary. What is the power in the primary? See **Figure 12-10.**

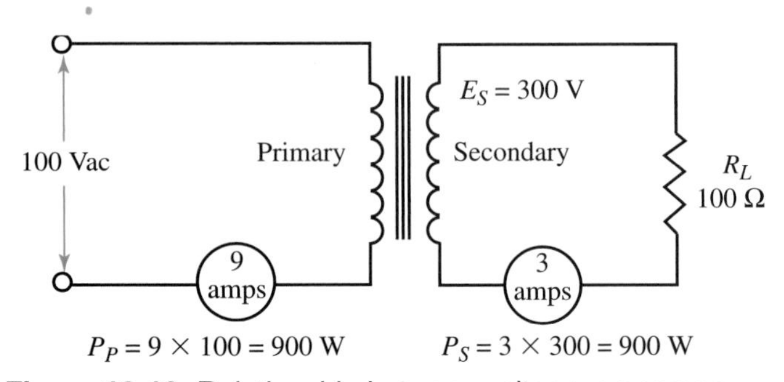

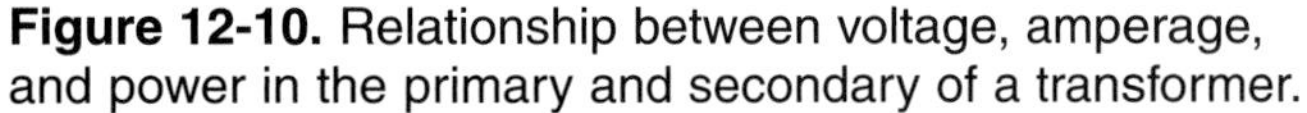

Figure 12-10. Relationship between voltage, amperage, and power in the primary and secondary of a transformer.

Using Ohm's law:

$$I = \frac{E_S}{R} = \frac{300\ \text{V}}{100\ \Omega} = 3\ \text{A}$$

A current of three amperes flows. The power used in the secondary can be found using:

$$P = I_S \times E_S$$
$$P = 3\ \text{A} \times 300\ \text{V} = 900\ \text{VA}$$

Since the primary must supply this power:

$$I_S \times E_S = I_P \times E_P = 900\ \text{VA}$$

and

$$I_P = \frac{P_P}{E_P} = \frac{900\ \text{VA}}{100\ \text{V}} = 9\ \text{A}$$

The key principle of transformer action is that as *voltage increases* in the secondary, *current decreases* in the secondary. Power companies use this key principle as the basis for distribution of electricity. In the study of conductors, you learned that all wires have some resistance. You will also recall that the power loss in a circuit or conductor is $P = I^2R$. *Power loss varies as the square of the current.*

A length of wire having a resistance of two ohms and carrying a 10-ampere current will have a power loss of $P = I^2R = (10\ \text{A})^2 \times (2\ \Omega) = 200\ \text{W}$. If the current is doubled from 10 to 20 amperes, then the power loss would be $(20\ \text{A})^2 \times (2\ \Omega) = 800\ \text{W}$. *This is four times the loss of power as there was at 10 amperes.*

It is more economical to raise the voltage and decrease the current in electrical distribution systems to reduce power loss in the transmission lines. Transformers are used for this purpose. Examine **Figure 12-11.** An electric appliance in your home uses 10.25 amperes of current. How much power is consumed when the appliance is turned on?

$$P = I \times E = 10.25\ \text{A} \times 117\ \text{V} = 1200\ \text{W}$$

In the 12,000 volt city lines, a current of 0.1 amp would flow.

$$0.1\ \text{A} \times 12{,}000\ \text{V} = 1200\ \text{W}$$

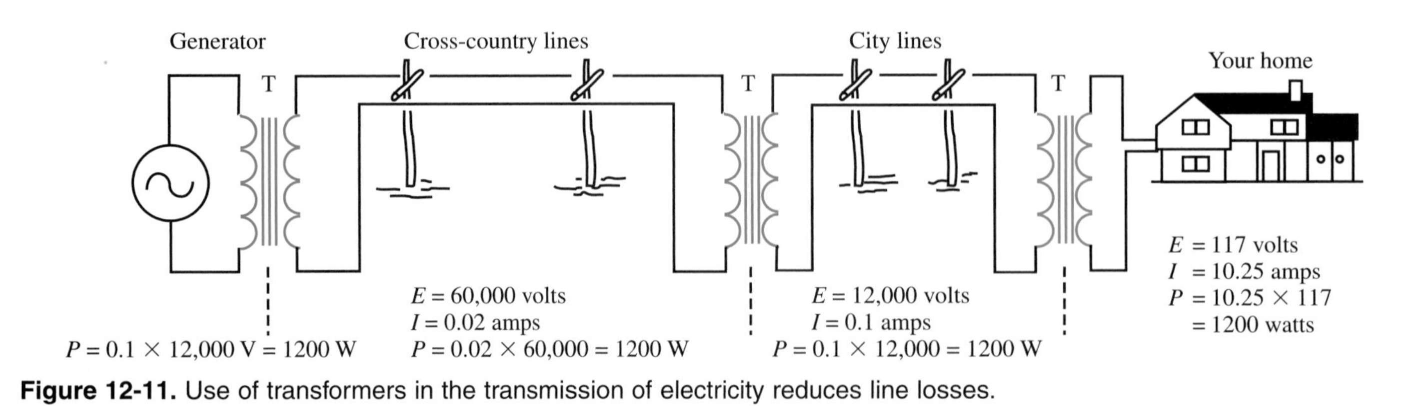

Figure 12-11. Use of transformers in the transmission of electricity reduces line losses.

In the cross-country power lines, a current of 0.02 amp would flow.

$$0.02\ \text{A} \times 60{,}000\ \text{V} = 1200\ \text{W}$$

By raising the voltage, power companies are able to supply large cities and industries with electricity using small wires for transmission lines. And while smaller wires do have greater resistance per foot, the loss from this resistance is only in direct proportion to the current used. Power companies select a wire that has the least resistance, yet is large enough to carry the expected current. These wires must also be strong enough to withstand high winds, ice, and snow.

Transformer Losses

Transformer losses are very small when compared to other types of electrical equipment. Most theory calculations involving transformers are based upon 100 percent efficiency, but there are, in fact, some losses with which you need to be familiar. Three types of losses associated with transformers are copper losses, eddy current losses, and hysteresis losses. These losses all result in unwanted heating of the transformer and prevent it from being 100 percent efficient. These types of losses were first introduced to you in Chapter 10, when discussing losses in generators. The causes of these losses in transformers are the same as with the generators, and the methods of loss prevention are very similar.

Copper losses are the result of resistance of wire used in the transformer windings. These are also called I^2R losses. As current increases in the windings so does the heat loss according to Ohm's law and the power law. Larger diameter wire has lower losses.

Eddy current losses are caused by small whirlpools of current induced in the core material. These losses are reduced by using laminated core construction. Each lamination is insulated from its bordering layer by varnish. This insulation cuts the number of paths on which currents can flow and concentrates the magnetic field more efficiently.

Hysteresis loss, or molecular friction, is the result of magnetic particles changing polarity in step with induced voltage. Special alloys and heat-treating processes can be used to make core materials that reduce hysteresis loss.

Impedance

The total opposition to current in an ac circuit is referred to as ***impedance.*** The impedance of a transformer is a combination of the conductor resistance and the self induction. An ohmmeter reading of the primary winding of a transformer would reveal only the copper resistance value. The self inductance value would have to be calculated, but for now, think of the impedance of the transformer as the total circuit resistance value.

Phase Relationship in Transformers

The output voltage of a transformer can be in phase or 180 degrees out of phase with the primary voltage. Whether the output is in phase or out of phase depends on the direction of the windings and the method of connection.

Restating Lenz's law, the polarity of the induced voltage will be opposite to the voltage producing it. A diagram of a transformer and the waveforms of the primary and secondary voltage are shown in **Figure 12-12.** The direction of the magnetic flux is indicated by arrows within the core. This type of transformer inverts the alternating voltage wave applied to the primary.

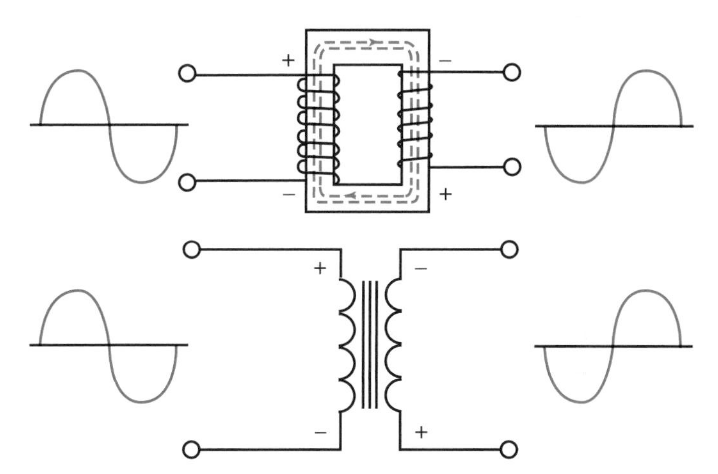

Figure 12-12. Phase relationship between the primary and secondary of the transformer. The schematic symbol is shown.

Often in schematics the phase alignment is indicated by phasing dots, **Figure 12-13.** When the dots are on the same side, the waves in the primary and secondary are aligned. When the dots are on opposite sides, the wave is inverted as it is transferred to the secondary.

Taps

Power transformers often have a number of taps in their secondary winding. A ***tap*** is a fixed electrical connection made to a winding at a point other than its terminals. Taps allow one transformer to provide a number of different voltages for use.

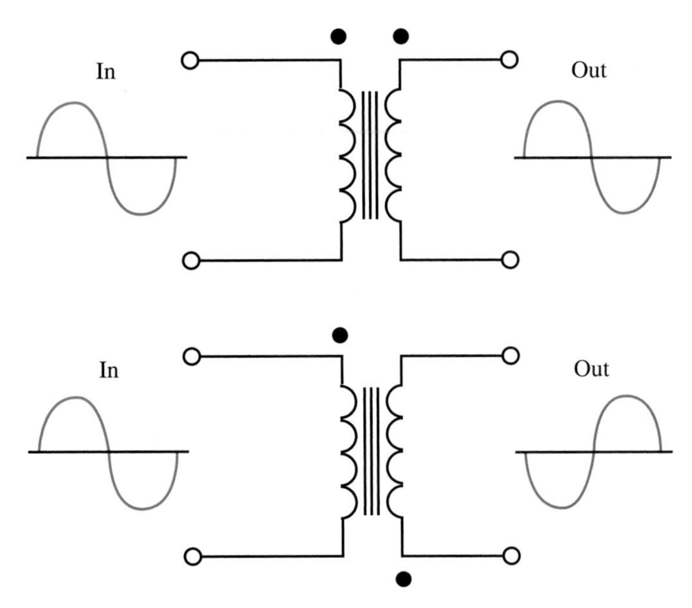

Figure 12-13. Phasing dots show how the primary and secondary are wired with respect to each other.

If a tap is made on the secondary winding as shown in **Figure 12-14,** two voltages can be received from the output. These voltages are 180 degrees out of phase with each other. Voltage phase is discussed in detail in Chapter 15.

The primary of a power transformer can also be tapped. This action allows the use of varying input voltages. This feature is useful when electrical or electronic equipment is moved from one location to another location that has a different ac voltage.

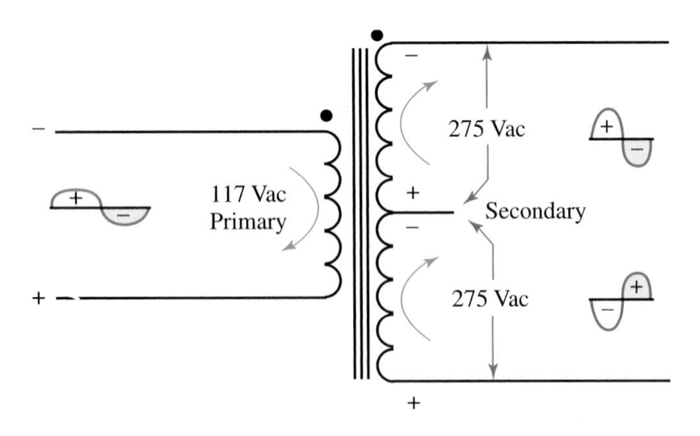

Figure 12-14. Tap on secondary winding.

Review Questions for Section 12.1

1. Define mutual inductance.
2. Define self inductance.
3. What is the difference between inductance and resistance?
4. The total opposition to current in an ac circuit is called __________.

12.2 TYPICAL DISTRIBUTION SYSTEM

Electrical power is generated at the power plant. Typical voltages are 2400 to 4160 volts, but for the distribution of electrical energy to be economical, transformers are utilized. After the generator produces the initial 2400 volts, the energy enters its first set of transformers. These are step-up transformers and they significantly step up the voltage. The voltage might be stepped up to 130,000 volts or higher.

Next, the electrical energy is routed through the area to the next set of transformers called a substation. Here, the voltage is stepped-down to a lower voltage such as 13,800 volts. Again, it is typically routed throughout the city and country to other transformers. The next transformers usually step the voltage down to a more desirable level such as 120 and 240 volts. This is the voltage used in most homes and small business locations. Some industries use a higher voltage such as 277 and 480 volts.

The pattern of voltage distribution is engineered not only with efficiency, but also safety in mind. Higher voltage distribution systems typically run on higher poles and towers. Residential areas usually have lower voltage distribution systems such as 2400 and 4160 volts. These voltages are still very deadly, but it is much safer than utilizing systems of over 50,000 volts in a residential neighborhood.

The power company wants to use the highest voltage possible for distribution because it results in lower current. If the system uses lower current values, much smaller wire sizes can be used and the smaller wire sizes result in a tremendous reduction in installation cost. Using less current also results in less power loss in the

Electric Resistance Welding

Electric resistance welding is based on the principle that when an electrical current is passed through metal, friction caused by *resistance* to this current heats the metal. When two pieces of metal are placed in contact and a strong electrical current is applied, the majority of the heat develops where the two pieces of metal meet. With enough current, extremely high temperatures can be reached. Welding occurs when these temperatures reach the *fusion* (melting) temperature.

Electric resistance welding is well-suited to automatic production because it is fast, causes little warpage of the metal, and can be accurately controlled. Common resistance welding processes include *spot welding, projection welding, seam welding,* and *upset welding.* Resistance spot welding (RSW) is the most common.

RSW consists of lapping two pieces of metal and clamping them between two electrodes. A current is passed between the two electrodes and a small molten spot is formed. After the current stops, the electrodes continue to hold the metal, allowing the spot to solidify. The two pieces of metal are now fused together by a *spot weld* or *weld nugget.*

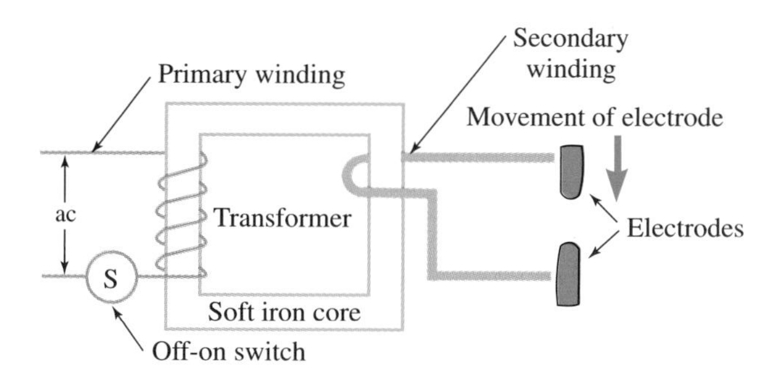

The welding current used can be either ac or dc, but most RSW machines develop an alternating current. An ac RSW machine is basically an electric transformer requiring high current (amperage) at low voltage. A *step-down-transformer* is used to decrease the voltage and increase the current. To accomplish this, the transformer's primary winding has many turns and the secondary has only one to three. Insulated conductors run from the secondary winding to the two electrodes.

The time required to make a spot weld is divided into three separate periods: *squeeze time, weld time,* and *hold time.* These three time periods make up one *weld sequence.* Electronically programmed machines complete the weld sequence automatically while a manual machine requires the operator to control the times of the weld sequence. In a large product such as an automobile, the sequence is repeated many times.

Robotic spot welding is commonly used in industry for fabrication of sheet metal parts. (FANUC Robotics North America, Inc.)

lines. Based on the formula I^2R, the resistance of the wire will generate heat causing a power loss. The less current applied to the system, the lower the losses.

Delta and Wye, Three-Phase Power Systems

Figure 12-15 is an illustration of a typical single-phase transformer. It has a primary winding that is connected to high voltages such as 2400 V, 4160 V, or 13,200 V depending on the rating and turns ratio of the transformer. The secondary of the transformer provides 120/240 volts. Notice that the neutral is indicated by the N. It is derived from the tap in the middle of the secondary and is common to both of the 120 volt lines. The neutral is also grounded at the transformer. This makes the potential of the neutral connector the same as that of the earth. Also note that the secondary windings can be connected in series or parallel.

In Chapter 10, Generators, you learned how three-phase power was generated. Power companies distribute electrical energy by using a three-phase distribution system. This system is made of either a three-phase transformer with three sets of windings inside or three individual single-phase transformers that are connected together. When three separate transformers are used, they can be connected in different configurations to obtain different voltages.

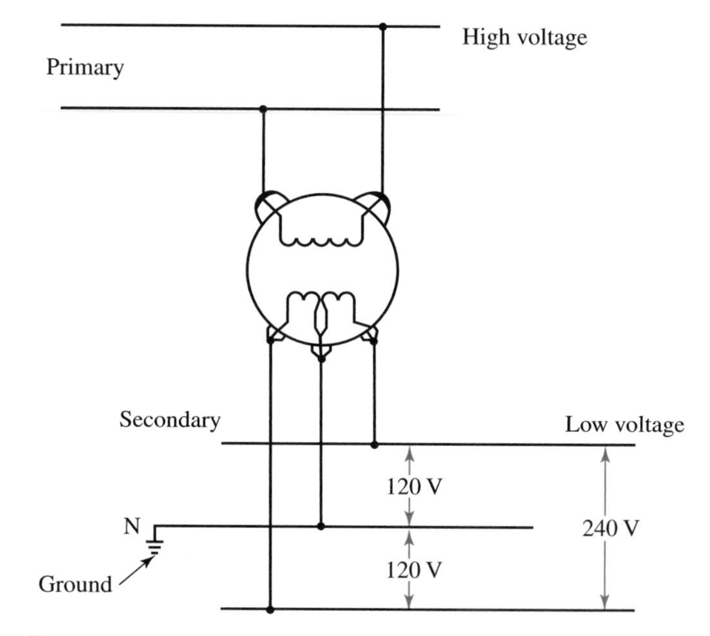

Figure 12-15. A typical single phase transformer used to supply 120/240 volts for residential use.

Like generators, the windings of the transformers can be connected in either series or parallel. As with generator connections, when the transformer windings are connected in series they are referred to as *delta connected.* When windings are connected in parallel they are called *wye connected.* In **Figure 12-16** you will see that the primary is delta connected and the secondary is wye connected.

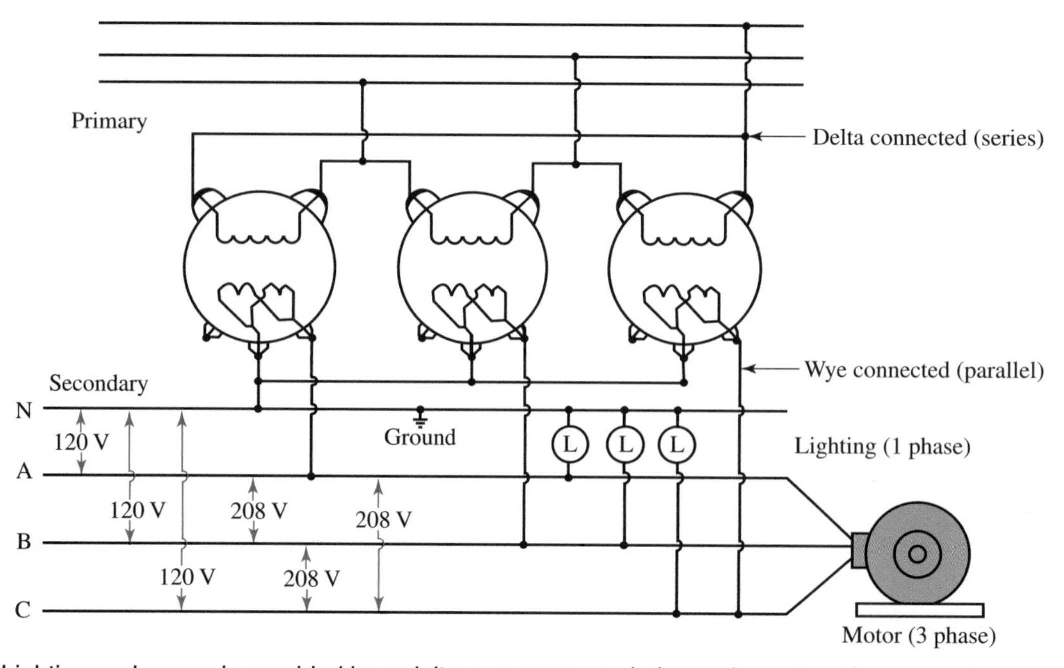

Figure 12-16. Lighting and power is provided by a delta-wye connected, three-phase transformer system.

Generator windings are connected in a similar manner. For our purpose we will not be too concerned as to which way the power company connects the primary windings of the distribution transformers they use. More importantly here, we must know how the secondary is connected so we can safely predict the voltages serving the building and equipment inside.

When a secondary side of a three-phase system is connected as a wye system, we can expect the phase to phase voltage to be equal to 1.73 times the phase to neutral or ground. In Figure 12-16, a phase to phase voltage would be taken between lines A, B, and C (A to B, A to C, or B to C). Because the phase relationship of the two voltages are 120 degrees apart, the resulting voltage will be 1.73 times the voltage rating of one phase. Notice that the voltage between line A and ground is 120 V. The voltage between lines A and B is 1.73 × 120 V = 208 V.

When the secondary side of the transformer is delta connected, the transformer is used mainly for three-phase power systems such as electric motors. See **Figure 12-17.** A delta secondary does not always have a neutral conductor. When a neutral conductor is provided, then the voltage is equal to 120 volts to neutral from phase A or B. Phase C is over 200 volts to ground. The current flowing through the delta-connected windings is equal to 1.73 times the current in one winding, not double the current. Again, this condition is a result of the phase relationship of the series connected delta system.

Now, let's note one of the most important facts concerning delta connected transformers with a neutral: *Only two lines of the delta connection are equal to 120 volts to ground or neutral, the other is equal to approximately 200 volts.* Look at Figure 12-17. See how the neutral is common to both the A and B windings in the center transformer. The line marked C is referred to as the high leg. The ***high leg*** has 1.5 full windings separating it from the neutral conductor. This condition, and the phase relationship, causes a much higher potential to exist between phase C and the neutral.

When properly installed, all lighting and receptacles are connected using only the voltage provided by A and B. The line marked C is used only as part of a three-phase equipment system. A technician must be careful when connecting circuits to a three-phase panel not to connect 120-volt equipment to the high leg. The high leg is usually distinguished by an orange marking.

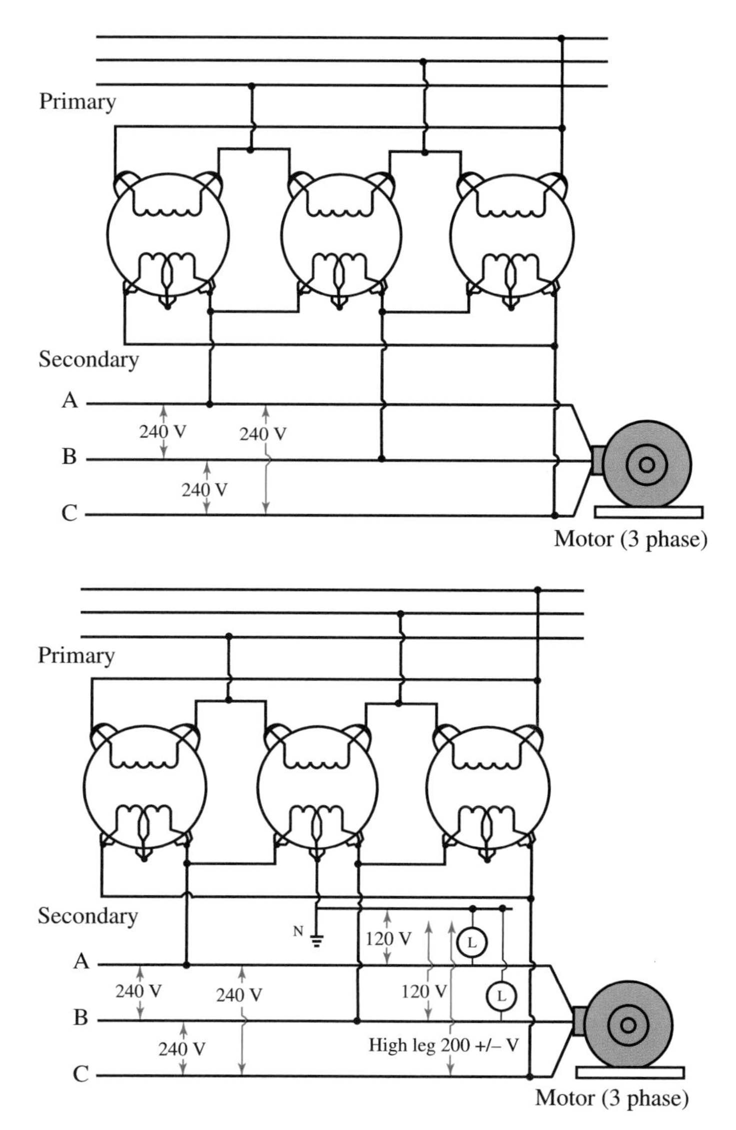

Figure 12-17. The top drawing is a delta connected secondary used only for power. The bottom delta connection has a neutral to provide for lighting.

Grounding

Grounding an electrical system provides a safe path for fault current. The fault current is the amount of amperage flowing to earth from the electrical fault. The amount of fault current is limited by only the conductor resistance and the transformer impedance. A minimum ground fault current can be in excess of 10,000 amps.

Look at **Figure 12-18.** When a transformer converts electrical energy by magnetic means, the secondary side of the transformer becomes electrically isolated from the rest of the system as far as the system ground is concerned. Until one of the conductors on the secondary side is grounded, no conductor on the secondary side will conduct electrons to earth. If one of the conductors is not connected to earth, a conductor to ground voltage reading will behave like an open circuit.

A ground is established by driving a rod into the ground and connecting it to one of the conductors. Once this is accomplished, any conductor on the secondary side

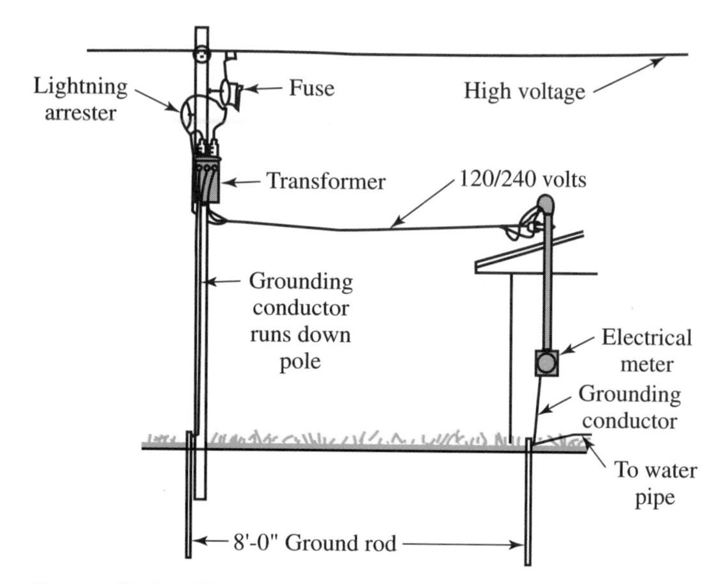

Figure 12-18. The ground for the electrical system is derived by two ground rods, one at the pole and the other at the residence.

of the transformer is capable of completing a circuit through to the earth. In addition to driving a rod into the ground at the transformer, the electrical system is also grounded at the electrical meter or service panel. Residential and commercial electrical systems are grounded by a long rod driven into the earth and by a connection to a water pipe. This dual connection helps to ensure a good ground.

Troubleshooting Transformers

When troubleshooting transformers it is important to remember that ohmmeter readings of resistance *do not* give the technician a good diagnosis of the condition of the primary and secondary windings. In general, the larger the transformer, the lower the resistance reading. However, this resistance reading is a purely dc reading and does not take into consideration the inductive reactance. Most step-down transformers have a very low dc resistance reading on the primary and an even lower reading on the secondary.

A better way to tell if a transformer is good is by using a megohmmeter. A ***megohmmeter*** is an ohmmeter that uses higher than normal voltages and accurately reads resistance into the megohm values. Most transformer failures are caused by an overload condition and will have a discolored or burnt appearance and a burnt odor.

Another method of testing a transformer is by disconnecting the load, insulating the secondary leads with tape or wire nut connectors, and then fusing the primary with a small value fuse. The size of fuse used depends largely on the size of the transformer. A one amp fuse will usually suffice. After making sure that all persons are clear, the transformer is then energized. If the fuse blows, it is safe to assume that the transformer is bad. There has to be a short somewhere in the transformer for the fuse to blow.

Review Questions for Section 12.2

1. What are the advantages of using high voltages for power distribution?
2. Transformer windings connected in series are referred to as ________ ________.
3. What is a high leg?
4. A(n) ________ is a meter that can be used to troubleshoot a transformer.

12.3 SPECIAL TRANSFORMER APPLICATIONS

There are a quite a number of special transformer applications. The following text details only a few of these applications.

Autotransformers

The common transformer consists of two windings: primary and secondary. While the two windings are not physically joined, a magnetic coupling exists between them. The electrical separation between the windings is called ***isolation.*** Isolation reduces the chance for shock.

An autotransformer has only one winding. The primary and secondary windings are physically joined at some point. There is no isolation. This condition increases the risk of electrical shock. However, the autotransformer is most often used in relatively low-voltage applications.

Autotransformers can be of either the step-up or step-down type, **Figure 12-19.**

Induction Circuit Breaker

One common use of the induction coil is in circuit breakers. Circuit breakers, discussed in Chapter 6, use an inductor to trigger a switch to shut off the current in a circuit. Current traveling through the coil of a circuit breaker

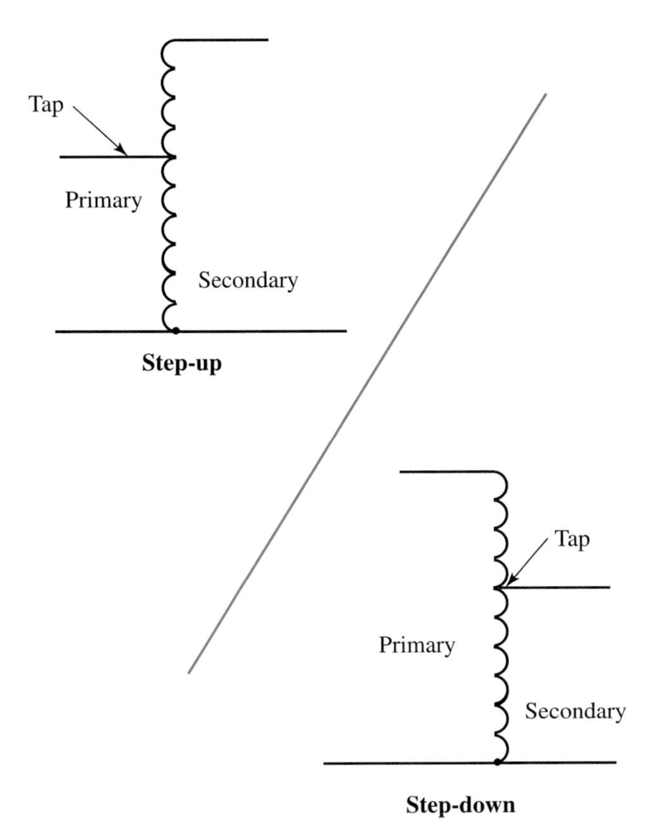

Figure 12-19. Autotransformers are less costly than regular transformers. Taps may be made variable so that the secondary voltage can be changed.

produces a magnetic field. When a certain level of current is reached, the strength of the magnetic field breaks the circuit. The circuit must be reset manually.

Many circuit breakers are designed to protect equipment or prevent fires. The circuit breaker in **Figure 12-20** is designed to prevent fires caused by a short in an electrical device. The button on the side allows the breaker to be reset.

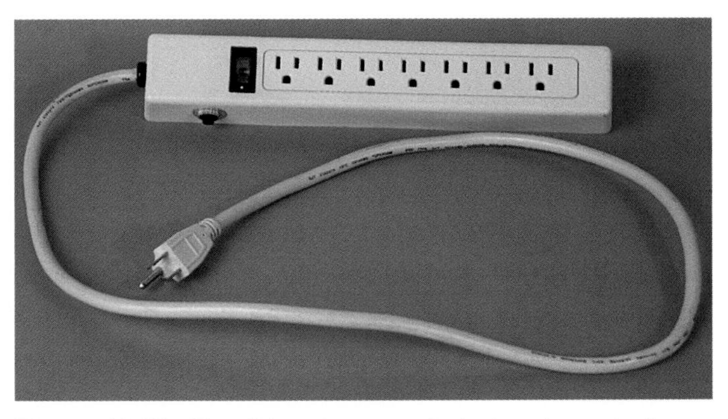

Figure 12-20. Circuit breakers use inductors to protect equipment, property, and people.

Lighting Ballast

Electrical discharge lighting systems such as fluorescent lighting and mercury vapor use a special classification of transformer known as a ballast. A ballast serves two purposes in electric discharge lighting. The ballast is used to start the lamp by producing the necessary voltage levels to create an arc through the lamp. It also limits current through the lamp.

When the electric discharge lamp is in the conductive mode, the lamp offers almost no resistance to the current. The windings in the lamp ballast consist of many turns of small wire that create a high resistance path for current when energized. Typically, a dead short could be created on the secondary side of the ballast circuit without overloading the ballast, although it is not recommended. The ballast used in lighting circuits are typically very inductive. Most ballast systems use either an internal or external capacitor to help reduce the amount of induction caused by the ballast. The capacitor helps to reduce the amount of power loss caused by the induction of the windings. Many lighting circuits use an autotransformer style of ballast. See **Figure 12-21.**

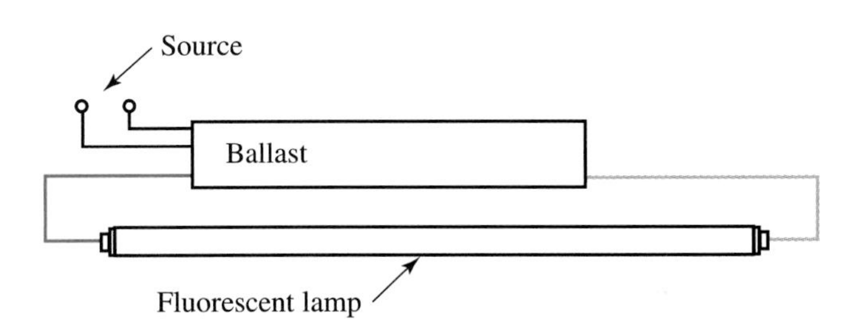

Figure 12-21. The fluorescent lamp ballast is another form of transformer. The ballast not only provides a high starting voltage to the lamp but also limits the current.

Coupling Transformer

The stages in an amplifier circuit are sometimes coupled together using transformers, **Figure 12-22.** The transformers are small in size and are not used to raise or lower voltage, but rather as a way of separating sections or stages of an amplifier system. The coupling transformer isolates each section of the amplifier circuit so that the design of each section and their individual resistance characteristics do not interfere with the other section. This process is referred to as impedance matching. The transformer is one way to ensure that each section will be compatible with the next. You will learn more about this in Chapter 18.

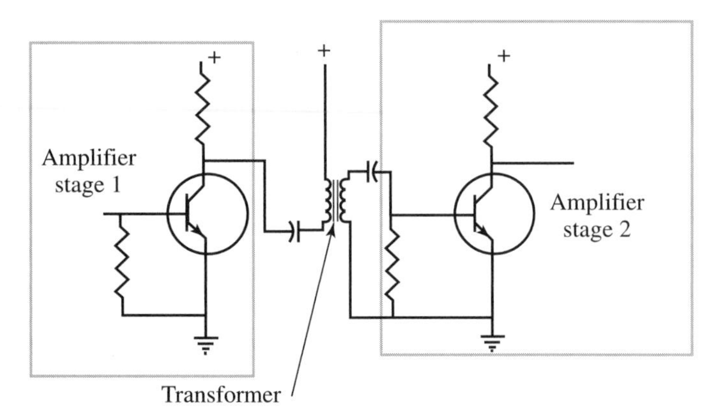

Figure 12-22. Transformers can be used to match impedance between different stages of electronic circuits.

Isolation Transformers

Isolation transformers are commonly used to separate commercial power systems from computer or other sensitive equipment power supply systems. Typically, they are in a one-to-one ratio, and they neither increase nor decrease voltage. By using an isolation transformer, the power supply is isolated from the rest of the electrical system. This isolation will prevent or suppress unwanted transient voltages. Transient voltages are voltage spikes that are higher than the typical supply voltage, examine **Figure 12-23.**

Transient voltage spikes are compared to typical sine waves. As you can see, the transient spike has a much higher voltage peak than the normal sine wave pattern. The high voltage peak can damage sensitive electronic equipment by exceeding its voltage rating, even if only for a brief period of time.

Transient voltages are caused by switching inductive circuits–such as lighting ballasts, motors, welders, or even the power company transformers–on and off. The isolation transformer suppresses the spikes in the voltage.

Automobile Ignition Coil

The high voltage spark in an automobile electrical system is produced by the ignition coil. The ignition coil is a form of autotransformer. It uses a high turns ratio to develop 30,000 volts or more across the spark plug gap. You may be wondering how the transformer principle is applied to a dc circuit. A car uses a 12-volt dc power supply, the battery. Look at **Figure 12-24** as we trace the ignition circuit. (Note that some of the devices found in a car ignition system have been eliminated to make the circuit easy and simple to follow.)

The 12-volt battery is connected in series with an ignition switch. The circuit connects to the ignition coil and then on to the distributor. The distributor turns, producing the opening and closing of the circuit. This action produces pulses of electrical energy flow to the coil circuit, turning the coil on and off. The pulses produce a rising an falling magnetic field across the winding in the coil, and thus, produces transformer action.

The 12 volts on the primary side of the autotransformer produces over 30,000 volts on the secondary side of the autotransformer. The electrical energy flows through the spark plug wire to the spark plug. At the spark plug, the circuit has an intentional open. The open in the circuit is at the spark plug gap. The 30,000 volts easily arcs across the open gap, completing the circuit to chassis ground. The arc across the spark plug gap ignites

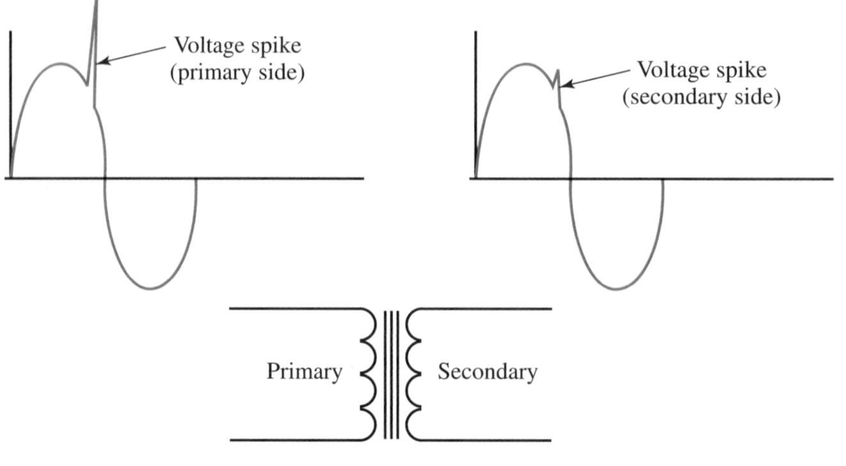

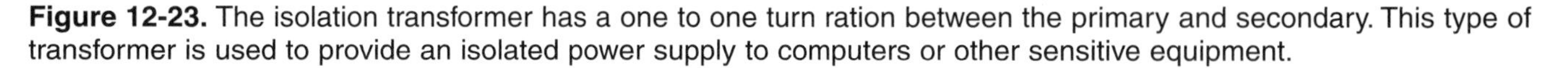

Figure 12-23. The isolation transformer has a one to one turn ration between the primary and secondary. This type of transformer is used to provide an isolated power supply to computers or other sensitive equipment.

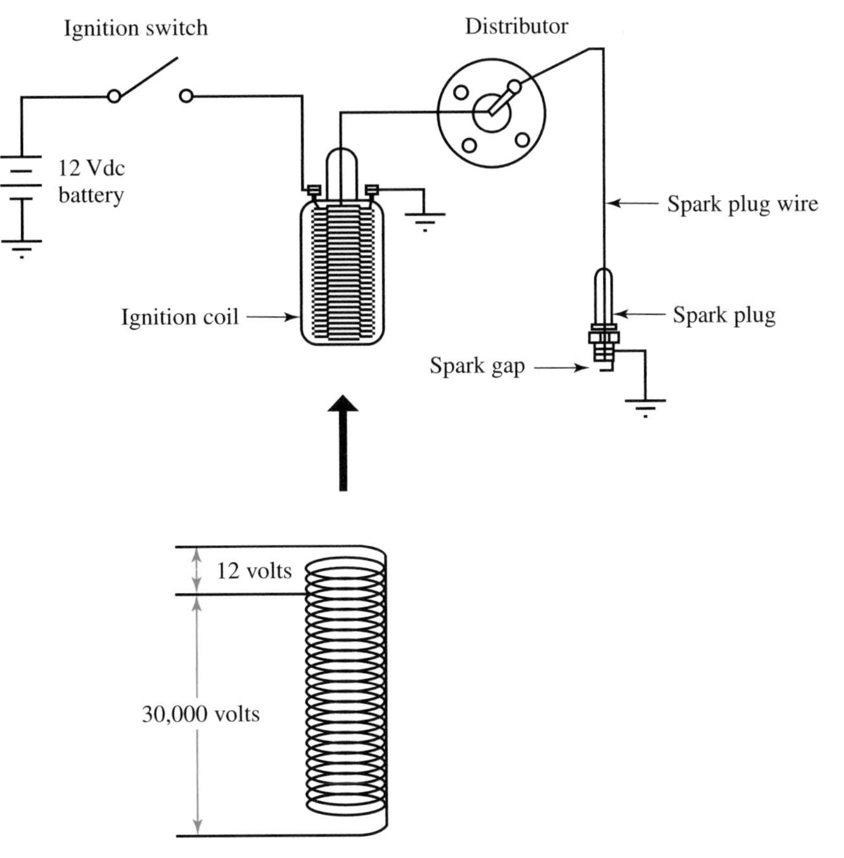

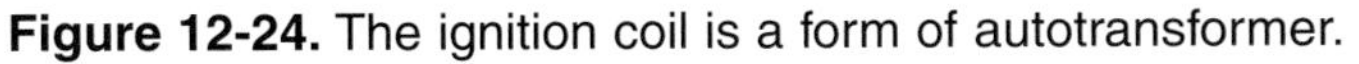
Figure 12-24. The ignition coil is a form of autotransformer.

the gasoline vapor causing an explosion. The explosion forces the piston to move, and thus, the engine runs.

Review Questions for Section 12.3

1. List four special transformer applications.
2. What is meant by electrical isolation?
3. What are the two jobs of the lighting ballast?

Summary

1. The input side of a transformer is called the primary.
2. The load side of a transformer is called the secondary.
3. The operation of the transformer is an application of induction.
4. The ability to produce electrical energy in a conductor without making physical contact with the conductor is referred to as induction.
5. Mutual induction exists between the primary windings and secondary windings of a transformer.
6. Self induction is emf in opposition to the emf producing the magnetic field of the coil.
7. The transformer can be used to step voltage up or down.
8. A step-up transformer raises the voltage in the secondary; the secondary side has less current than the primary.
9. A step-down transformer lowers the voltage in the secondary; the secondary side has greater current than the primary.
10. The transformer turns ratio, voltage ratio, and current ratio are related by the formula:

$$\frac{E_P}{E_S} = \frac{N_P}{N_S} = \frac{I_S}{I_P}$$

11. The power rating of a transformer is expressed in volt amps (VA) or kilovolt amps (kVA).
12. Transformer losses are due to copper losses, eddy current losses, and hysteresis losses. Eddy currents and hysteresis occur in the iron parts of the transformer. Copper losses are based upon the resistance of the conductor windings.
13. The total opposition to current in an ac circuit is referred to as impedance.

14. A center tap in a transformer allows the secondary windings to provide two equal voltages. One will be 180 degrees out of phase with the other.
15. When the transformer windings are connected in series they are referred to as delta connected. When windings are connected in parallel they are called wye connected.
16. Wye transformer banks produce 120/208 voltage and are used for power and lighting applications.
17. Delta transformer banks are used for 120/240 voltage and are used mainly for power applications.
18. Grounding an electrical system provides a safe path for fault current.
19. The purpose of isolation in a transformer is to separate the primary and secondary winding. There is no electrical connection between them. The secondary and primary are connected only by a magnetic field.

Test Your Knowledge

1. What is the difference between mutual induction and self induction?
2. The primary of a transformer has 200 turns of wire and the secondary has 800 turns of wire. If 117 volts ac is applied to the primary, what will the secondary voltage be?
3. The primary of a transformer has 50 turns and the secondary has 20 turns. If the current in the secondary is two amps, what is the current in the primary?
4. A transformer steps 120 volts down to 6 volts. A 30-ohm resistor is connected to the load side of the transformer. What is the current value in the primary caused by the secondary load? Assume 100 percent efficiency.
5. A current induced in the iron core of a transformer is called a(n) __________ __________.
6. Molecular friction caused by the atoms of the transformer core material resisting the changing magnetic polarization of the core is called __________.
7. Losses due to resistance of the conductor inside a transformer are called __________ losses.
8. The opposition to current in a transformer is called __________.
9. How is a typical home electrical system grounded?
10. Why would you use a coupling transformer between different stages in an amplifier?

For Discussion

1. Why do power companies use high voltage in cross-country lines?
2. Discuss the differences between a step-up and a step-down transformer.
3. How might electrical power be distributed in the future?
4. A transformer is used to step up 240 volts to 60,000 volts. Which side would have the bigger insulators, primary or secondary, and why?

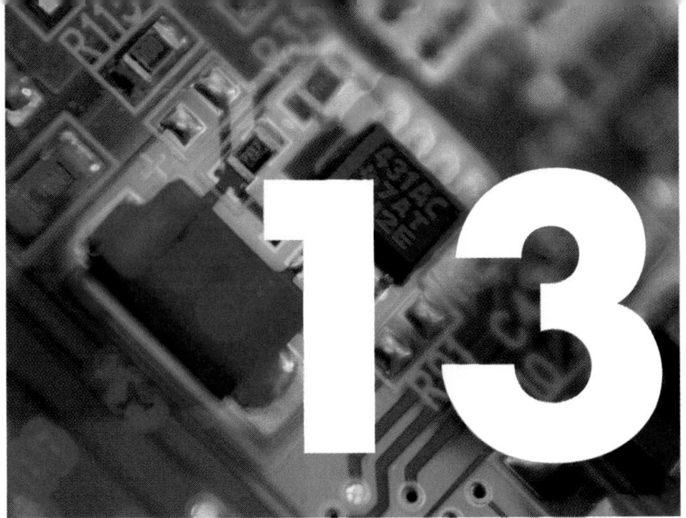

AC Motors

Objectives

After studying this chapter, you will be able to:

- Discuss the operation of an induction motor.
- Identify and explain the operation of various three-phase motors.
- Explain how a split-phase condition is created.
- Discuss the purpose of ac motor protection circuits.
- List the common causes of motor failure.
- Explain basic trouble shooting techniques for ac motors.

Key Words and Terms

The following words and terms will become important pieces of your electricity and electronics vocabulary. Look for them as you read this chapter.

branch circuit protection
dedicated line
locked rotor condition
rotor
run protection
running windings
selsyn unit
service factor
shading coil
shading the pole
single-phasing condition
slip
splitting phases
squirrel cage
starting windings
stator
synchro
synchronous speed

Industries use a tremendous number of ac motors to provide the power of automation. AC motors are found in factories, dairies, assembly lines, chemical plants, and many more locations. Your home also uses many types of ac motors. In this chapter you will learn about the main types of ac motors found in the home and industry.

13.1 INDUCTION MOTORS

The induction motor is a special class of motor. As its name implies, the motor operates on the theory of induction. In Chapter 12, the terms *mutual induction* and *self induction* were introduced. Both of these terms apply to the induction motor in a slightly different manner.

A motor converts electrical energy into mechanical energy. The mechanical energy takes the form of a rotating shaft. The rotation of the magnetic field inside a *stator* produces power in the same way as the revolving field stationary armature from the three-phase generator. The stator of a three-phase induction motor is a mirror image of the three-phase alternator. See **Figure 13-1**. The ***stator*** consists of stationary windings in a circular pattern inside a housing. When the three-phase power is applied to the stator, the coils in the stator produce electromagnetic fields in the same pattern produced by the revolving field in the alternator. Since the field rotates inside the alternator, the pattern of magnetic field poles inside the motor stator also produces a rotating pattern.

The ***rotor*** revolves inside the stator housing. The stator induces an emf on the rotor. This action is much like the operation of a transformer, **Figure 13-2.** Think of the rotor as the secondary of a transformer and the stator as the primary. The rotor is constructed of soft iron laminations with copper or aluminum bars running through the outside edges. These bars act like the secondary windings. The stator induces an emf on the rotor bars causing a current to flow through the bars. The bars connect at the ends of the rotor, completing the circuit and producing a magnetic field between the rotor bar pattern.

The magnetic field on the rotor follows the rotating magnetic field in the stator. As the rotor turns, the rotor bars cut across the magnetic field produced by the stator. This action adds to the current in the rotor bars. Note the similarities to the generator and transformer action.

The term ***squirrel cage*** rotor is used to describe the type of rotor in the common three-phase induction motor. If the rotor were made up of soft iron laminations only, it would simply vibrate when the stator was energized. To produce the rotation, either a squirrel cage or other specially constructed rotor must be used. The other special rotors will be discussed later.

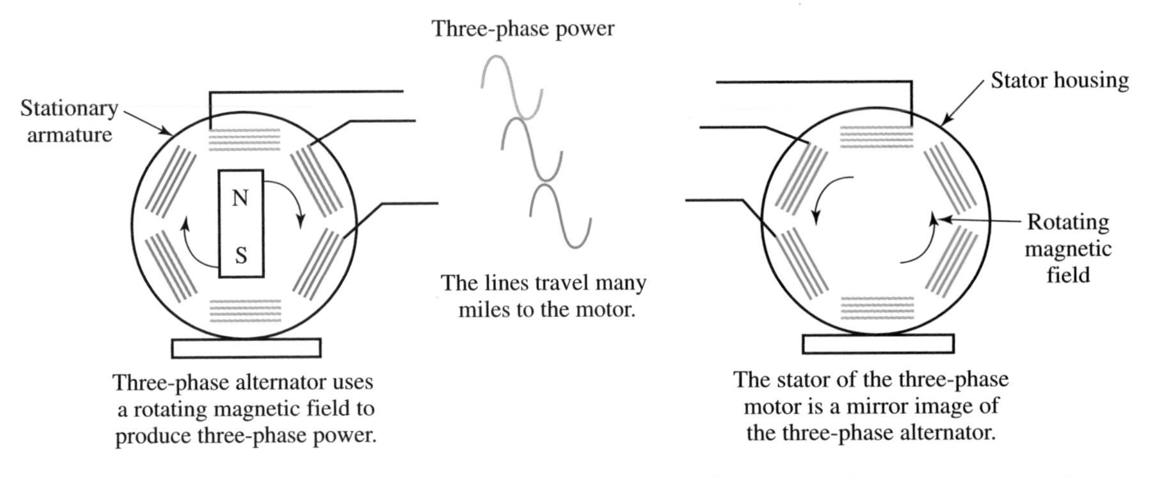

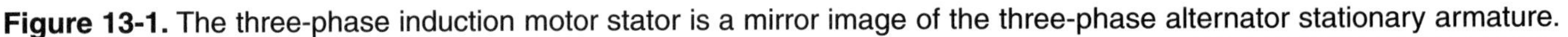

Figure 13-1. The three-phase induction motor stator is a mirror image of the three-phase alternator stationary armature.

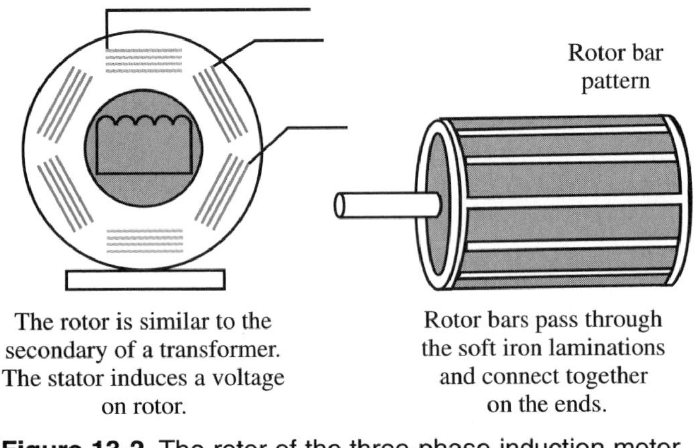

Rotor bar pattern
The rotor is similar to the secondary of a transformer. The stator induces a voltage on rotor.
Rotor bars pass through the soft iron laminations and connect together on the ends.

Figure 13-2. The rotor of the three-phase induction motor is constructed of copper or aluminum bars imbedded in soft iron laminations.

Figure 13-3 is a detailed illustration of a typical three-phase motor. The stator is inside the heavy cast iron frame and consists of coils made from moisture resistant, magnetic wire. The rotor bars are inside the soft iron laminations of the rotor. This type of rotor construction enhances the strength of the magnetic field of the rotor.

The rotor is mounted on a steel shaft with bearings. The bearings support the rotor as it turns inside the stator housing.

The rotation of a three-phase motor is reversed by reversing the connections of any two leads. Reversing any pair of motor lead connections causes the field of rotation inside the stator to reverse.

Synchronous Speed

Synchronous speed is used to describe the speed of the stator's rotating magnetic field. The rotating magnetic field speed is the product of the frequency of the three-phase electrical supply and the number of poles inside the stator housing. Because of this, the synchronous speed can be calculated from the following formula.

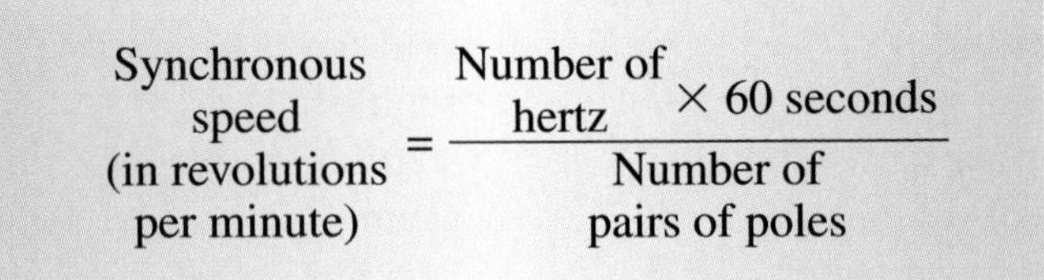

$$\text{Synchronous speed (in revolutions per minute)} = \frac{\text{Number of hertz} \times 60 \text{ seconds}}{\text{Number of pairs of poles}}$$

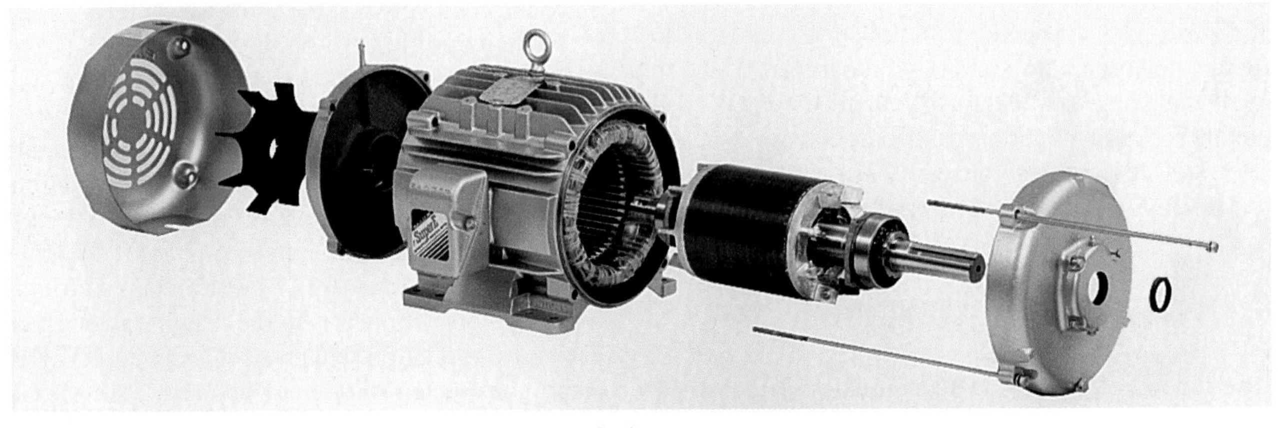

Figure 13-3. A typical three-phase motor. (Baldor Electric Co.)

Inserting the frequency coming from a standard outlet (60 hertz) into the formula along two poles (one pair) from a motor, the equation becomes:

$$\text{Synchronous speed} = \frac{60 \text{ Hz} \times 60 \text{ s}}{1} = 3600 \text{ rpm}$$

The synchronous speed of a three-phase induction motor with one set of wire coils per phase is 3600 rpm. If there are two sets of coils per phase then the synchronous speed would be 1800 rpm. *As the number of coils or poles per phase increases, the rpm of the motor decreases.*

Slip

Remember, rotor current is dependent upon the stator. As the rotor turns inside the stator housing, the rotor bars cut the magnetic field of the stator. The action of the rotor bars cutting the magnetic field produces a heavy current. If the rotor catches up to the synchronous speed of the magnetic field in the stator, the rotor bars would not cut across the magnetic field. Consequently, in most motors the rotor travels slower that the synchronous speed of the stator. The difference in these two speeds is described as ***slip***.

Generally, slip equals four to five percent of the motor's rpm. A four pole motor with a synchronous speed of 1800 rpm would have an actual speed of about 1725 rpm due to the slip. The exact rpm of the rotor is dependent on the number and size of the rotor bars embedded in the rotor. When the motor is loaded with a heavy mechanical load, the rotor slows down and the percentage of slip increases. As the slip increases, a greater current circulates in the rotor circuit, which then produces a greater torque.

Types of Induction Motors

Induction motors are used in a wide variety of situations. These motors are subject to different power sources as well as different loads. To supply these varied needs, induction motors have been created with a number of interesting features. Some important induction motors are:

- The three-phase synchronous motor.
- The wound rotor motor.
- The dual-voltage motor.
- The single-phase induction motor.
- The capacitor start, capacitor run motor.
- The repulsion-induction motor.
- The shaded-pole motor.
- The stepping motor.

Three-phase synchronous motor

Three-phase synchronous motor derives its name from the fact that it rotates at the same speed as the rotating magnetic field. This action is accomplished by using a squirrel cage rotor and embedding a set of windings inside the rotor. The rotor windings are constructed much like an electromagnet. The motor starts the same as any induction motor and reaches near synchronous speed. To create a synchronous state with the stator, the windings on the rotor are energized by a dc source, and the rotor becomes an electromagnet. The rotor develops its own north and south pole that is independent of the induction of the rotating magnetic field in the stator. There is no slip to maintain; thus, the rotor travels at the exact speed of the rotating magnetic field. This motor is used in applications where a constant speed must be maintained under a changing mechanical load condition.

Wound rotor motor

The wound rotor motor uses a stator constructed like the regular three-phase induction motor, however, its rotor is made up of three sets of windings wrapped around soft iron laminations. The three sets of windings in the rotor are connected to slip rings and brushes. The rotor circuit is connected to a rheostat. The windings are similar to the squirrel cage rotor. The motor is started and then brought up to near synchronous speed. The speed of the motor can be varied by adjusting the rheostat in the control circuit outside the motor housing. This allows the current induced in the rotor windings to be varied. By varying the current through the rotor circuit, the speed of the rotor can be adjusted. Although a full range of speeds cannot be accomplished this way, a great deal of variation in speed can be achieved.

Dual-voltage motor

Typical stator housing coils can be connected for more than one voltage. A dual-voltage motor is a motor that has the stator coils arranged in pairs. The pairs of coils can be connected in series or in parallel allowing the motor to be used with two different voltages. Three-phase motors are commonly dual-voltage motors. A typical three-phase motor is rated for 240/480 volts. This means that the motor can be operated by either a 240 volt or a 480 volt power supply. The dual-voltage motor is economical because fewer motors have to be stocked. The typical connection diagram of a dual-voltage motor is shown in **Figure 13-4**.

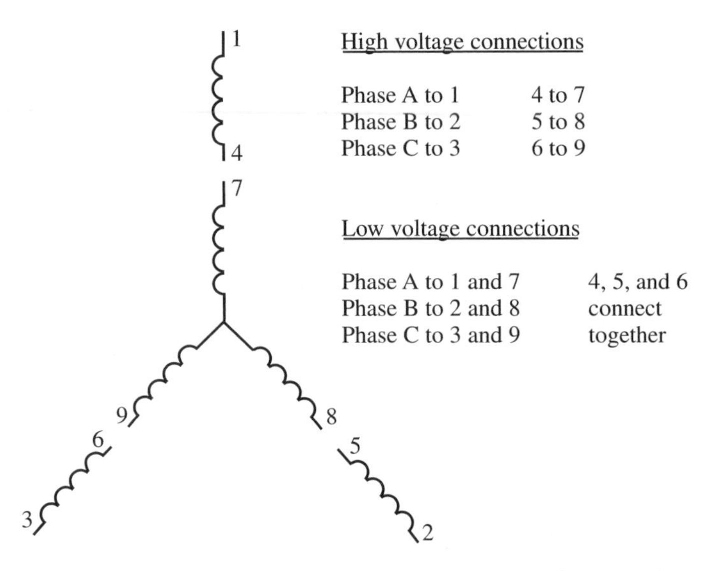

Figure 13-4. High and low voltage connections for a dual-voltage motor.

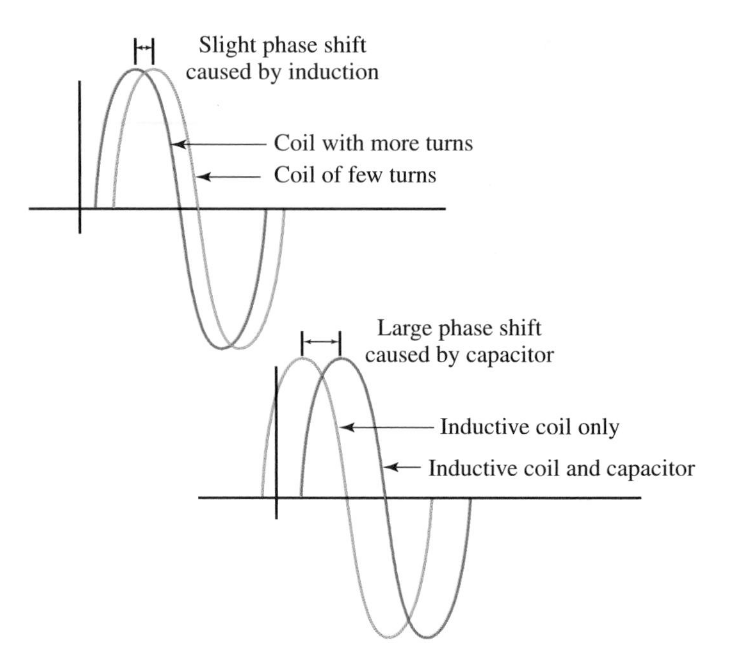

Figure 13-5. A phase can be split by using two coils with different inductive characteristics or by connecting a capacitor in series with a coil.

Single-phase induction motor

Single-phase motors are used in many home appliances, as three-phase electricity is not readily available. Most single-phase motors can be run in either a clockwise or counter clockwise direction. To reverse the direction of rotor rotation you must reverse the connections to the start windings. Not all single phase motors are designed for reversing rotation.

In industry, three-phase electricity is used to run large induction motors. Other means must be provided for starting single-phase induction motors. If, for instance, a single-phase current could be divided into a polyphase current, the starting problem would be solved. This action is possible, and it is called ***splitting phases***.

Two methods are used to produce a split phase. One method uses a capacitor, and the other method is based on induction and inductive reactance, **Figure 13-5**. Inductive reactance will be studied in the next chapter, and capacitance is discussed in Chapter 15, so only the basics are covered here.

Inductive reactance phase splitting. When wire is shaped into coils, the coil creates a certain amount of inductive reactance. Inductive reactance means that there is a force that opposes the current flowing through the coil. The amount of opposing force depends on the number of turns of wire in the coil. More turns of wire produce more inductive relative force. The induction causes the current flowing in the coil to slightly lag the applied voltage. Two coils having different number of turns of wire will have different amounts of inductance. Since the current from these coils occurs at two slightly different times, the magnetism in the coils also occurs at two different times. This action is the basis for splitting phases.

If the single-phase current flows through two parallel paths that contain unequal amounts of inductance, one current lags behind the other. A phase displacement will exist between them and a two-phase current has been created from a single-phase current.

The many turns of wire that comprise the field windings of the induction motor are excellent inductors. They can be used for the purpose of phase splitting. A schematic diagram of a split-phase motor is shown in **Figure 13-6**.

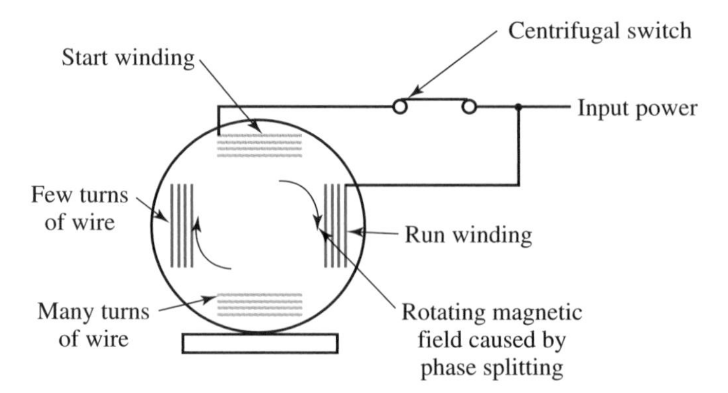

Figure 13-6. A single-phase motor constructed from two sets of coils with a different wire size and number of turns each. The characteristics of the two coils cause the magnetic field in each set of coils to occur at slightly different times, thus causing a rotating magnetic field.

The ***starting windings*** consist of many turns of relatively fine wire. The ***running windings*** have many turns of heavier wire. The windings are placed in slots around the inside of the stator.

When electricity is applied to the motor terminals, current flows in both windings because they are in parallel. The running winding has many turns of wire and a high inductance. This causes the current to lag almost 90 degrees. The starter windings of fine wire have much less inductance. Current does not lag by as great an angle. The phase displacement creates a rotating field similar to the three-phase motor, and the rotor starts to run. When the motor reaches speed, a centrifugal switch opens the starter winding circuit. The motor then runs as a single-phase inductor motor. The fine wire in the starting winding cannot withstand constant use, so it is cut out of the circuit after it has performed its function.

Capacitor phase splitting. The second method of phase splitting involves the use of a capacitor. If a capacitor is connected in series with the starting winding, a much larger phase displacement can be created. A larger phase displacement means that the motor will have a much higher starting torque than an induction start motor. The capacitor is switched out of the circuit when the motor approaches full speed, **Figure 13-7**.

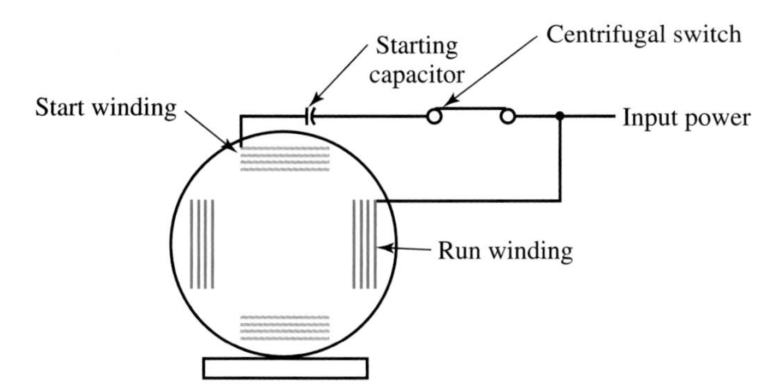

Figure 13-7. A capacitor start motor has a capacitor wired in series with the start winding and the centrifugal switch.

This motor is called a capacitor start induction motor. These motors are used for many jobs around the laboratory and home. They maintain a fairly constant speed under varying loads. They do not develop a strong starting torque as compared to the three-phase squirrel cage motor. A capacitor start motor is shown in **Figure 13-8**. Note the starting capacitor mounted on top.

Figure 13-8. A capacitor start motor. (Delco-Remy Div., General Motors Corp.)

Capacitor start, capacitor run motor

A capacitor start, capacitor run motor is designed for heavy mechanical load conditions when only single-phase power is available. It is similar to the capacitor start motor, but this motor uses the *same* size wire on the starting winding and the running winding. The two capacitors are connected to the motor windings, one capacitor to the start winding and the other to the run winding. See **Figure 13-9**. Both capacitors are energized while starting the motor. The starting capacitor is disconnected by the centrifugal switch when the running rpm is reached. The other capacitor is left continuously connected to the run winding. This motor is designed for high starting and good running torque under changing mechanical load conditions. Single-phase air-conditioning equipment is a major application of this type of motor.

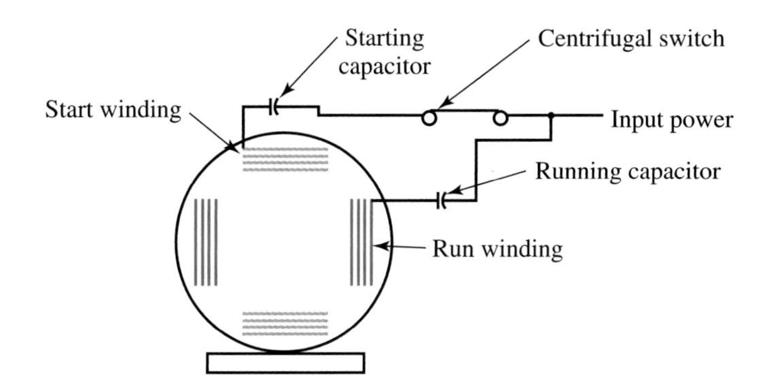

Figure 13-9. A capacitor start, capacitor run motor has two capacitors. One is in series with the start winding and the other is in series with the run winding.

Repulsion-induction motor

A repulsion-induction motor resembles a dc motor because it also has a commutator and brushes. However, in the repulsion-induction motor, the brushes are connected to each other, rather than to the source of power. The brushes are arranged so that only selected armature coils are closed to form complete circuits at any given time. The currents induced by transformer action in the rotor windings are displaced by the shorting brushes. Poles are created that oppose the stator poles.

Rotation is started by repulsion. When the motor reaches about 75 percent of its speed, a centrifugal shorting ring shorts out all of the commutator sections. The motor runs as an induction motor. The repulsion-induction motor has high starting torque, but does not quickly come up to speed under load. It is currently being replaced by the less costly, capacitor-start motor.

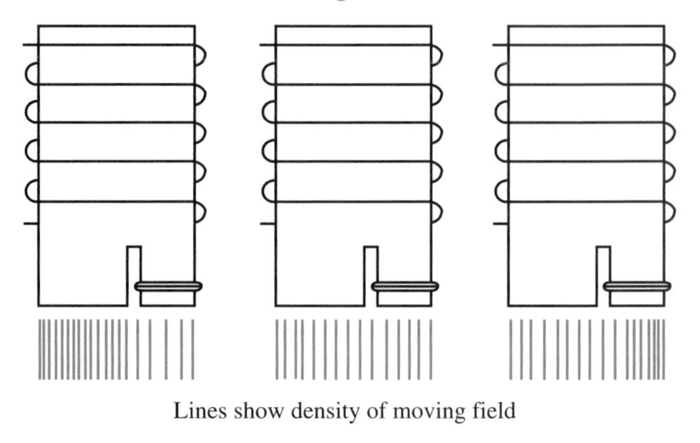

Figure 13-11. The magnetic field moves from left to right as the result of shading.

Shaded-pole motor

A two-pole motor having an uncommon method for starting rotation is the shaded-pole motor, shown in **Figure 13-10**. The rotor is a squirrel cage type. The stator resembles the common dc motor, except that a slot is cut in the face of each pole, and a single turn of heavy wire is wound in the slot. This is called ***shading the pole***. The single turn of wire is the ***shading coil***.

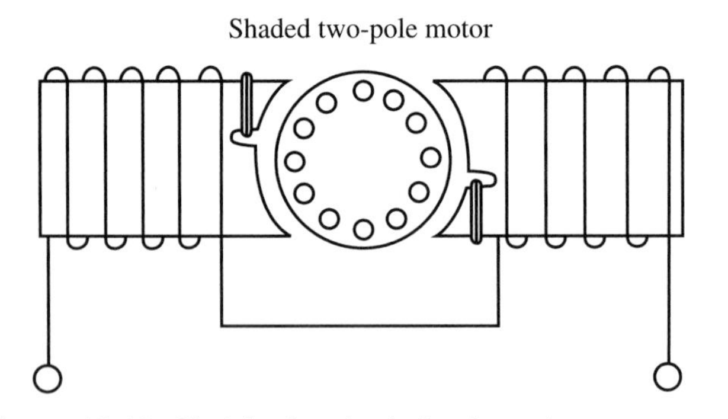

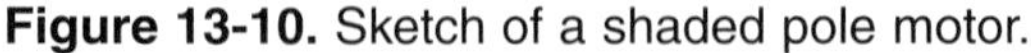

Figure 13-10. Sketch of a shaded pole motor.

The action of the shading coil can be seen in **Figure 13-11**. As the current rises in the first quarter cycle of the ac wave, a magnetic field is formed in the field winding. However, the expanding magnetic field cuts across the shading coil. This interaction induces a current (and thus a magnetic field) that *opposes change* in the magnetic field of the field coil.

When the wave is increasing, this condition tends to decrease, and weaken, the total field on the side of the shading coil. At the top of the ac wave very little current change occurs. The total magnetic field becomes equalized across the face of the pole. When the current starts to decrease in the next quarter cycle, the polarity of the induced current in the shading coil resists the decreasing current. So the pole has a strong magnetic field on the right side. Notice in Figure 13-11, during this half-cycle, that the magnetic field has moved from left to right. It is this movement of the field that causes the rotor to start rotation.

The starting torque of this motor is very weak. Therefore, it has many uses in items such as small fans and electric clocks.

Stepping motor

Stepping motors derive their name from the fact that they rotate in steps, or small angles, one by one. See **Figure 13-12**. Notice that the stepper rotor resembles a gear. It has teeth around its entire circumference, and it is usually magnetic. The stator consists of a gear-like structure with windings imbedded in a pattern of two phase, three phase, or four phase. The windings receive pulses of power rather than continuous application of power. The number of pulses applied to the stator windings determines the distance of travel the rotor makes. The rotor is a permanent magnet with a gear-like construction, but the number of teeth does not match that of the stator.

When one of the stator windings receives an electrical pulse, the rotor only turns its teeth far enough to line up with the energized teeth in the stator. The speed of the rotor is determined by how fast the power is pulsed. The direction of rotation is controlled by the polarity of the electrical pulse to the stator. Once a certain position is reached, the magnetic pull of the rotor to the stator holds the rotor in position. Stepping motors are used for positioning mechanical levers and linkages. They are found in

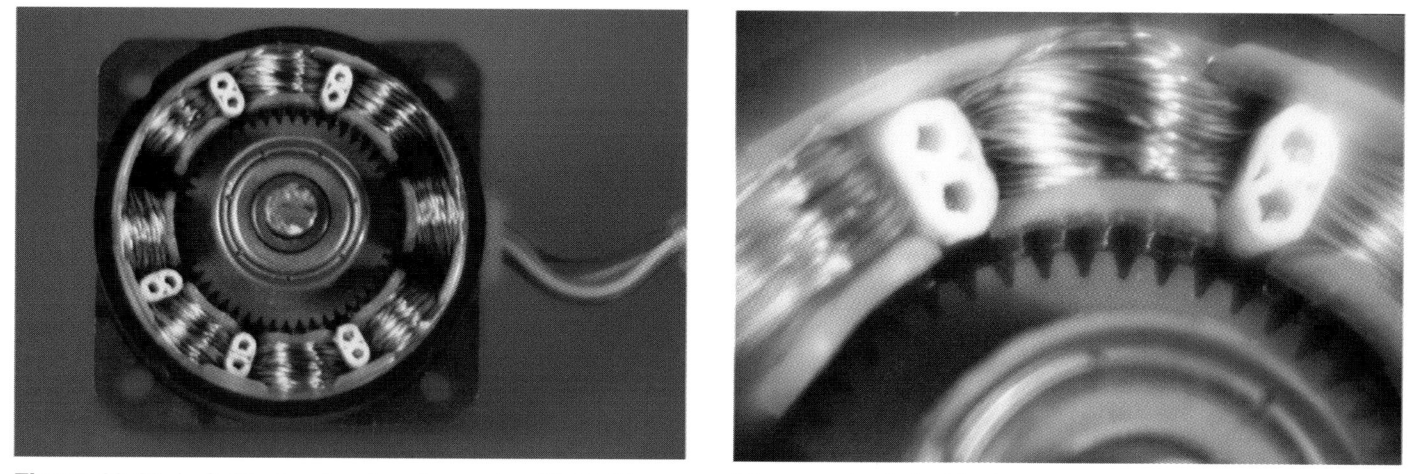

Figure 13-12. Left. Photo of a stepping motor. Right. Close up of the gear teeth inside the motor.

robots or disk drive systems, **Figure 13-13.** There are many varieties of stepping motor construction, however, their principles of operation are the same.

Figure 13-13. This stepping motor operates a computer's hard drive.

Selsyn Unit

A ***selsyn unit***, or ***synchro*** as it is sometimes called, is a unique transformer principle application designed to relay information through motor units, **Figure 13-14**. The selsyn unit consists of two motor units. Each unit is constructed with a typical stator and windings spaced equally around the stator housing. Each rotor is similar to the synchronous rotor. The rotors are electromagnets. Both the stator units and the rotors are connected to each other.

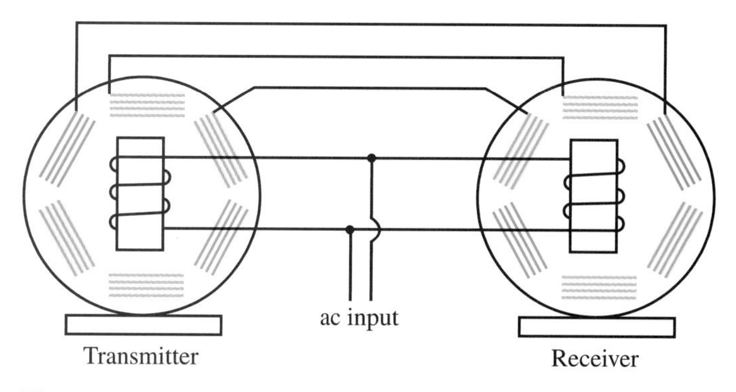

Figure 13-14. A typical selsyn unit consists of a transmitter and receiver. The receiver electromagnet will duplicate any angle the transmitter electromagnet assumes.

The rotors are energized by an ac power supply. When the rotors are energized, they induce an emf on the stator windings. When energized, the two rotors will be at exactly the same angle of rotation. One unit is usually referred to as the transmitter, and the other is called the receiver. When the transmitter is rotated by a mechanical force, the receiver duplicates the exact same turn. These units are used on ships to indicate rudder position.

They can be used to monitor the change in position of almost any mechanical item such as a weather vane or a gun turret.

Review Questions for Section 13.1

1. The stationary field poles and windings in an induction motor are called a __________.
2. The rotating member of a three phase induction motor is called a __________.
3. What is the synchronous speed of a motor if it has 6 poles and is operating on 50 Hz?
4. Define slip.
5. How are dual-voltage motors different than regular induction motor?
6. A split-phase induction motor has many turns of fine wire in the __________ windings, while __________ windings are made with many turns of heavier wire.
7. How does a repulsion induction motor differ from a standard ac motor?
8. Explain how a selsyn unit operates.

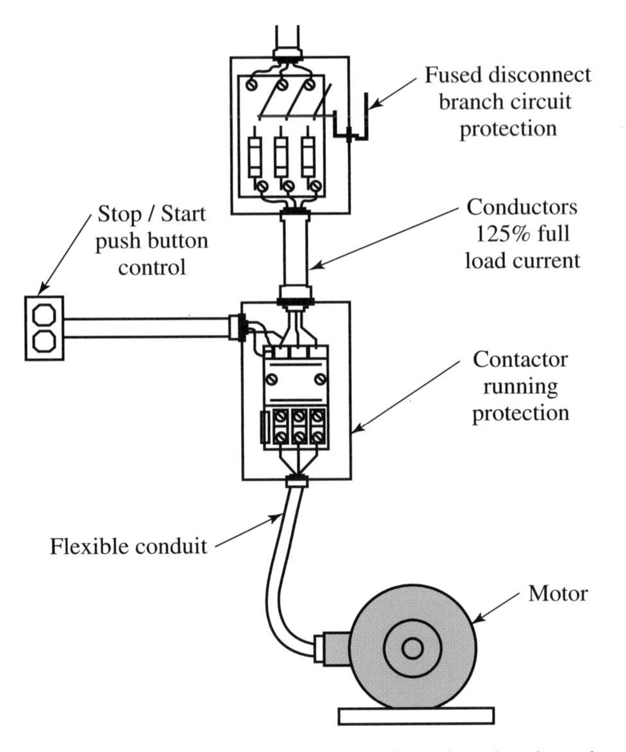

Figure 13-15. Typical motor installation showing locations of running protection and branch circuit protection.

13.2 MOTOR PROTECTION, FAILURE, AND TROUBLESHOOTING

Motor protection is important to any motor system. Most motor protection involves safety devices installed in the power lines that feed the motor. These devices protect more than just the motor. They protect the load and any other device connected in the motor system.

Motor Installation and Protection

Unless the motor being used is very small, motors are installed with dedicated lines. A ***dedicated line*** means that the circuit is used for only one purpose. For a proper installation of a motor, two forms of protection are needed in this dedicated line—branch circuit protection and running protection. See **Figure 13-15**.

Branch circuit protection means that a fuse or breaker is installed to protect the circuit feeding the motor. This fuse or breaker can be set to trip if the current exceeds anywhere from 150 percent to 500 percent of the motor's full load current. The full load current of a motor is the current the motor is expected to draw while it is under full mechanical load.

A table saw, for example, can be rated at 12 amps for full load condition. This means that when the saw is cutting the maximum size of wood it is designed for, it will draw 12 amps of current at the recommended applied voltage. When the saw is turning, but not cutting wood, the amperage will be small in comparison, maybe as little as 1.5 amps. However, the circuit protection must be selected at a high enough level to allow for a high starting current or a surge. A brief surge as high as 600 percent of full load current can be obtained when the motor is energized. If the circuit protection is set too low, the fuse will blow or the breaker will trip.

The second form of protection is called run protection. ***Run protection*** protects the motor from excessive heat damage or fire while it is in a normal mode of operation. Run protection is a necessity since only a slight mechanical overload to a motor will, over a period of time, cause damage or fire. Running protection must be closely sized to a motor's service factor. All motors have a service factor. The ***service factor*** of a motor is its ability to withstand or avoid damage from an overload condition. The service factor is based on a percentage of a motor's normal current under full load condition.

To interpret a service factor, subtract one from the number and multiply the remainder by 100. This gives you the percent overload a motor can handle. For example, if a motor has a service factor of 1.25, it can safely withstand a 25 percent overload. A motor with a 1.15 service factor can withstand a 15 percent overload, and a motor with a 1.0 service factor can withstand no overload.

Motor Failure

Most ac motors are of simple design and, unless brushes are used, require very little maintenance. However, motors will eventually fail. There are a few faults that account for most motor failure.

Excessive bearing wear

The most common cause of motor failure is excessive bearing wear. Some bearings are sealed and do not require maintenance, while others require periodic lubrication. Dry motor bearings cause friction, heat up, and wear down rapidly. In a motor, there is very little clearance between the stator and rotor. As the bearings wear down, the rotor will begin dragging on the stator. Usually this causes a loud roar or high, shrill whining sound as well as excessive vibration. The worn bearings cause an excessive load on the motor and cause it to go into an overload condition.

Mechanical overload

Another cause of motor failure for small ac motors is mechanical overload. An example of mechanical overload is when a power tool is used continuously under sever mechanical conditions. The tool heats up until the temperature exceeds the temperature rating of the stator's insulation. Once this occurs, the windings are irreversibly damaged. The damage can be so severe as to produce a short between the windings or to the iron laminations.

Single-phasing condition

Three-phase motors are often exposed to another damaging situation called a ***single-phasing condition***. A single-phasing condition occurs when the power supply to the three-phase motor loses one of its three lines due to an open or blown fuse. If the motor is running at that time the loss occurs, it will continue to run. However, there will be an increase in current in the two remaining motor windings. Unless the overload protection trips the relay coil in the starter, this current increase will damage the motor.

Small motors can have a thermo device inserted in the motor windings that will open when excess heating occurs. This device offers some protection for a motor experiencing mechanical overload or a single-phasing condition. If this type of running protection is included as a part of the motor, the data plate will indicate that the motor is thermo protected.

Locked rotor condition

The most severe damage occurs to a motor when it has a locked rotor condition. A ***locked rotor condition*** is when the motor is energized, but the rotor is not turning. A mechanical failure in the system to which the motor is connected can cause this condition. When locked rotor occurs, the current in the windings is extremely excessive and can severely damage the motor windings.

Low voltage

Another common cause of excessive motor current is low voltage. Under a low voltage condition, a motor will maintain its horsepower output rating until the low voltage becomes so severe that the rotor locks. This low voltage condition causes excessive current in the windings while the motor is under mechanical load. Remember, when checking for low voltage conditions, the motor must be under mechanical load. Otherwise, the volt meter readings will not be accurate. Review the voltage drop formula for conductors in Chapter 3. As current increases, so does the voltage drop along a conductor. As the mechanical load on a motor increases, so does the current. As the current increases, so does the voltage drop at the motor.

Look at the illustrations in **Figure 13-16**. Many burnt winding patterns can reveal why a motor failed.

Troubleshooting Motors

The most useful tool when troubleshooting motors is the clamp-on ammeter. The clamp-on ammeter operates using the transformer principle of induction. The ammeter clamps around the line to be read, **Figure 13-17**. The clamp is the primary coil and the secondary coil is inside the meter housing. The current in the line being measured produces a flux field that cuts across the clamp. This generates an emf that is measured by the meter movement. The clamp-on ammeter is used only for ac circuits. Its greatest advantage over other forms of ac ammeters is that the circuit does not have to be opened to connect the meter in series with the circuit.

As you have learned so far, a properly operating motor should not have a current reading higher than the motor's rating. A simple ammeter reading can tell you a lot about the condition of the motor. As a motor's windings break down, they will draw more current in the lines to the motor. This will show up on the ammeter.

A—Winding in good condition.

B—Winding single-phased.

C—Winding shorted turn-to-turn.

D—Winding grounded in the slot.

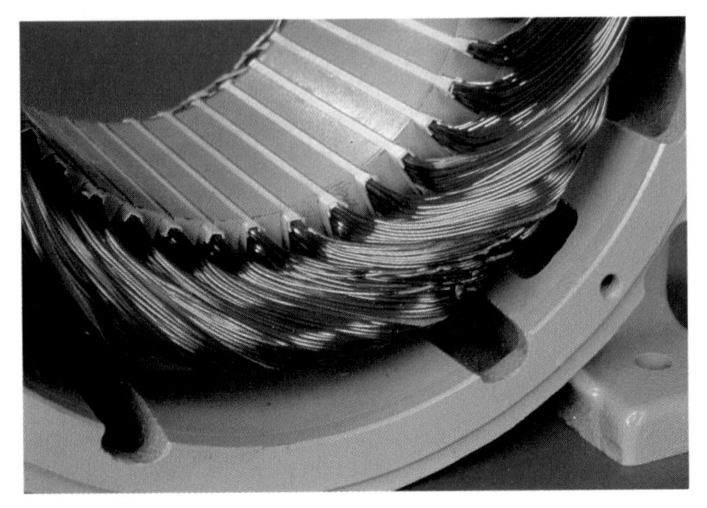

E—Phase damage do to unbalanced voltage.

F—Winding damaged due to overload.

G—Damage caused by locked rotor.

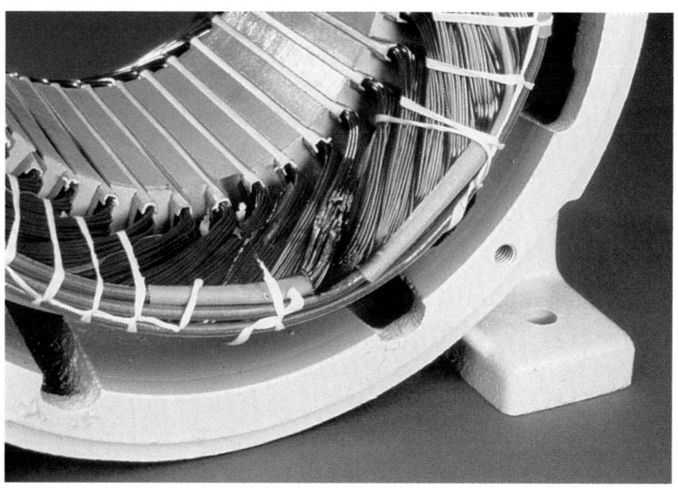

H—Winding damaged by voltage surge.

Figure 13-16. Figure A shows a motor with windings in good condition. Figures B through H show motor windings damaged through a variety of causes. Can you explain the patterns of the scorch marks by their causes? (Electrical Apparatus Service Association, Inc.)

Figure 13-17. Clamp-on ammeter in use. (Fluke Corp.)

Another quick check is to measure and compare all three lines of a three phase motor. The three lines should have *identical* current values. Any deviation from a near perfect match means the motor windings are breaking down.

Bearing wear in motors can be identified by sound, vibration, or excessive heat at the bearing housing. Large industrial motors use temperature monitors made from thermocouples to monitor bearing heat and windings. These monitors provide an accurate prediction of motor failure and allow the technicians to safely shut down the system and replace bearings or windings before failure occurs.

Another useful tool is the megohmmeter. As the name indicates, the megohmmeter is a type of ohmmeter that measures resistance in the millions of ohms range. This meter generally uses a much higher operational voltage than a standard ohmmeter. A standard ohmmeter uses only nine volts in the ohmmeter circuit. The standard type of meter would not have sufficient voltage to detect a short or ground in a motor that operates at a higher voltage such as 480 volts.

Review Questions for Section 13.2

1. What function does a dedicated line serve?
2. What is the difference between branch circuit protection and run protection?
3. A motor with a service factor of 1.9 can withstand a(n) __________ percent overload.
4. List four reasons for motor failure.
5. The current values on the three phases of a three-phase motor should be __________.
6. What is a megohmmeter?

Summary

1. An induction motor changes electrical energy to mechanical energy.
2. In an induction motor, the field poles and windings remain stationary.

3. Synchronous speed is used to describe the speed of the stator's rotating magnetic field. It can be calculated from the following formula.

$$\text{Synchronous speed (in revolutions per minute)} = \frac{\text{Number of hertz} \times 60 \text{ seconds}}{\text{Number of pairs of poles}}$$

4. In most motors the rotor travels slower that the synchronous speed of the stator. The difference in these two speeds is slip.
5. There are a number of different types of induction motors. They include: the three-phase synchronous motor, wound rotor motor, dual-voltage motor, single-phase induction motor, capacitor start, capacitor run motor, repulsion-induction motor, shaded-pole motor, and stepping motor.
6. Single-phase motors split the incoming single-phase current into a polyphase current.
7. Split-phase induction motors have starting and running windings.
8. A dedicated line for a motor will often have branch circuit protection and running protection.
9. Common causes of motor failure are: bearing wear, mechanical overload, single-phasing condition, locked rotor condition, and low supply voltage.
10. A single-phasing condition for a three-phase motor means that power has been lost on one of the three lines feeding the motor.
11. The clamp-on ammeter operates using the transformer principle of induction.

Test Your Knowledge

Please do not write in the text. Place your answers on a separate sheet of paper.

1. Name three different types of three-phase motors.
2. What electrical part do the motors in question one have in common.
3. How do you reverse the direction of rotation for a three-phase motor?
4. How do you reverse the rotation of a single-phase motor?
5. Explain the operation of a three-phase induction motor.
6. A(n) __________ __________ is a heavy wire wound into a slot cut on the face of a pole.
7. What is running protection?
8. What is service factor?

For Discussion

1. Why does an overloaded motor heat up?
2. Why are motor starters necessary on heavy-duty motors?
3. Compare an induction motor to a transformer.

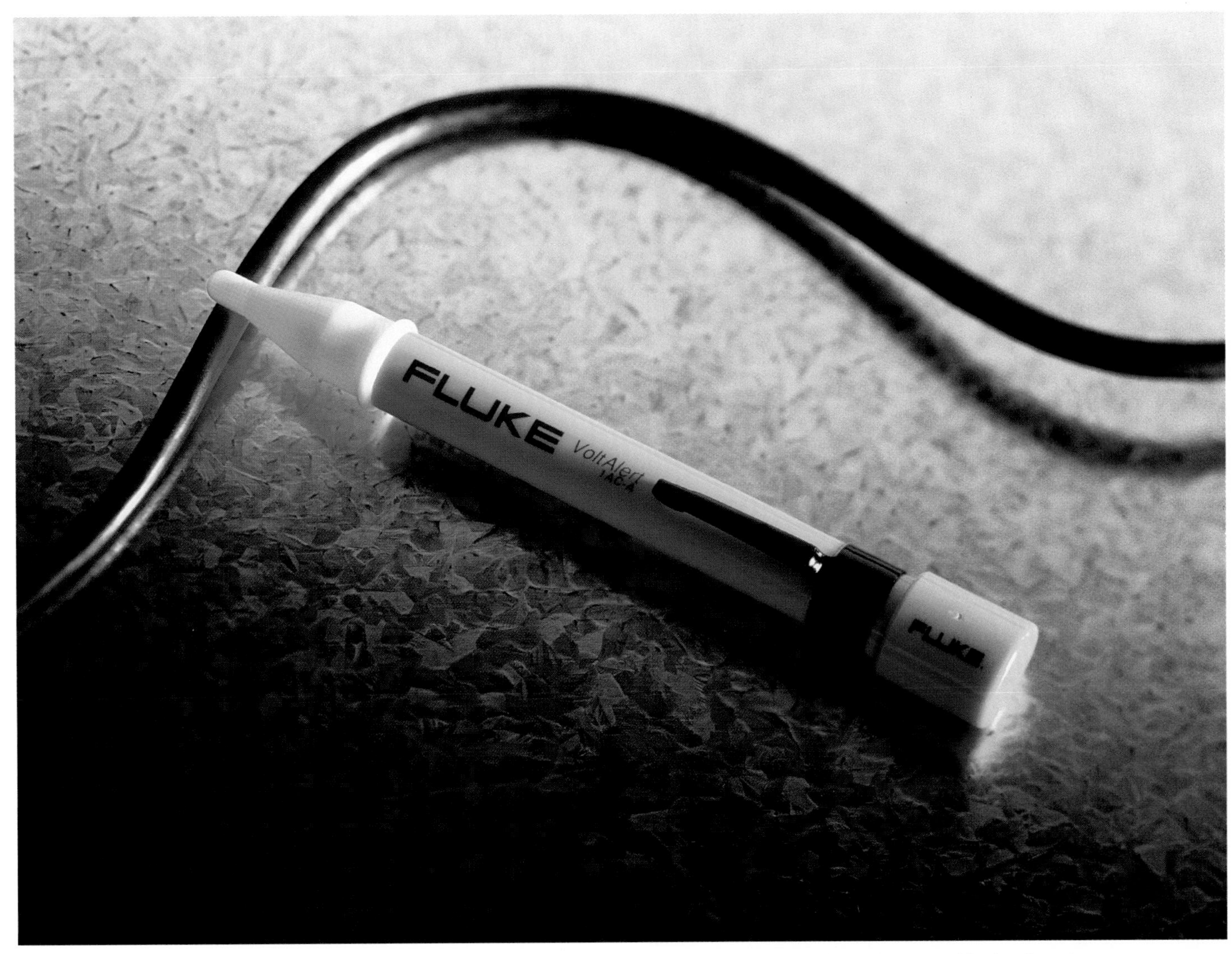

This test device takes advantage of the magnetic field that surrounds current carrying wires, and induction, to warn a user that a wire is live. (Fluke Corp.)

Advanced Electronic Circuits

The true heart of electronic fundamentals are explored in this section. Chapters 14, 15, and 16 are very important. They cover the operation of capacitors, inductors, and their combinations in series and parallel circuits. These three chapters lay the foundation for radio and television transmission and reception. Many electrical phenomena cannot be understood without first grasping these three chapters.

In Chapter 17, atomic structure is explored once more, but in more detail. Semiconductors are then introduced and explained using atomic theory. This is the foundation of solid-state electronics. Chapter 18 explains transistors, amplifiers, and electron tubes. You will see how each chapter is based on the foundation of the previous chapter.

The integrated circuit has revolutionized electronics. Chapter 19 discusses digital and analog devices. There is also expanded information on the operational amplifier and 555 timer. Chapter 20, on digital circuits details the fundamentals needed to keep up with the rapidly expanding field of electronics. Section 4 closes with a chapter on oscillators.

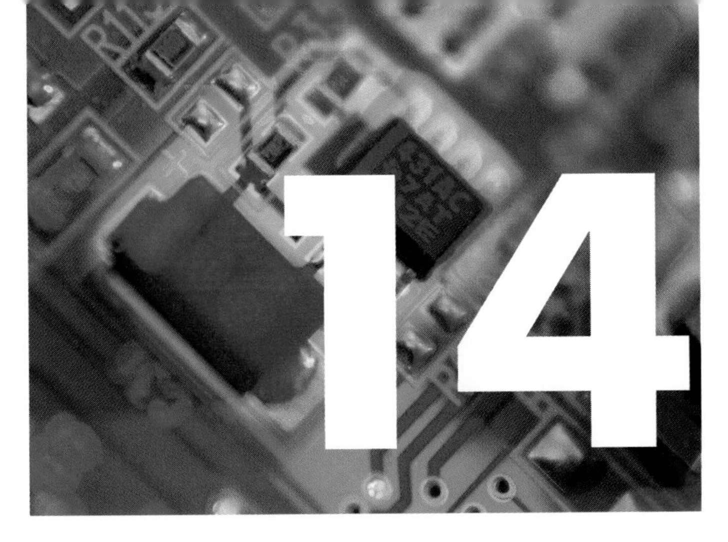

Inductance and RL Circuits

Objectives

After studying this chapter, you will be able to:

- Define the terms inductor and inductance.
- Explain how inductance affects a current.
- Describe an RL circuit's transient response.
- Define mutual inductance.
- Describe the effect of inductance in ac circuits.
- Explain and compare true power and apparent power.
- Use various measuring and computing methods to determine the values of currents and voltages in inductive circuits.

Key Words and Terms

The following words and terms will become important pieces of your electricity and electronics vocabulary. Look for them as you read this chapter.

apparent power
henry (H)
impedance (Z)
inductance
inductive reactance (X_L)
inductor
Lenz's law
mutual inductance
phase angle (θ)
power factor (PF)
Pythagorean theorem
reactance (X)
reactive power
RL circuit
self induction
time constant (τ)
transient response
true power
volt-ampere-reactive (VAR)
volt-amperes (VA)

The study of electricity and electronics revolves around inductance, capacitance, resistance, and combinations of these in series and parallel circuits. Resistance in a circuit was studied in Chapter 3. This chapter will help you answer these questions: What is an inductor and inductance? What is the effect of inductance in a circuit? What methods are used to measure and compute values of current and voltage in an inductive circuit?

14.1 INDUCTANCE

Inductance can be defined as the property in an electric circuit that resists a change in current. This resistance to a change in current is the result of the energy stored within the magnetic field of a coil. A coil of wire has inductance. Some electrical items that contain coils are relays, motors, transformers, radio transmitters, radio receivers, and speakers. An ***inductor*** is an electronic component that is used to produce inductance in a circuit. **Figure 14-1** shows the schematic symbol for an inductor with and without and iron core. There are a number of other names for the inductor. These names include ***coil, reactor,*** and ***choke.***

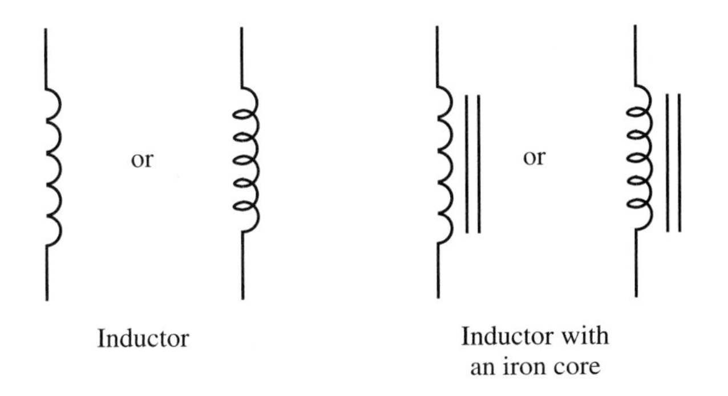

Figure 14-1. Schematic symbols for an inductor with and without a core.

Set up the experiment from Chapter 10, Figure 10-2, using the hollow coil connected to a galvanometer. Move the permanent magnet in and out of the coil and notice the deflection of the needle in the meter. When the magnet moves in, current flows in one direction. The current reverses direction when the magnet is withdrawn. No current flows unless the magnet is moved.

This principle was explained in Chapter 10 in relation to the theory of generators. By moving the magnet, a voltage is induced in the coil, which causes current to show on the meter. There is a distinct link between magnet movement and the current's direction. This is an application of ***Lenz's law:*** *The field created by induced current is of such a polarity that it opposes the field of the permanent magnet.*

A coil connected to a source of direct current builds up a magnetic field when the circuit is closed. The expanding magnetic field cutting across the coil windings induces a counter emf. This voltage opposes the source voltage and opposes the rise in current. When the current reaches its maximum value, there is no further change in the current. Consequently, there is no longer an induced counter emf. The current is now only limited by the ohmic resistance of the wire. If the source voltage is disconnected, the current in the circuit falls to zero. As the current falls to zero, the magnetic field collapses. As the magnetic field collapses, it again induces a counter emf. This counter emf is in the opposite direction of the counter emf created when the coil was energized. This counter emf now retards the reduction of current.

The inductance of the coil resists any change in current value. The symbol for inductance is L. Inductance is measured in a unit called a ***henry (H).*** A henry represents the inductance of a coil if one volt of induced emf is produced when the current is changing at the rate of one ampere per second. This is expressed mathematically as:

$$E = L \times \frac{\Delta I}{\Delta t}$$

where E equals the induced voltage, L equals the inductance in henrys, ΔI equals the change of current in amperes, and Δt equals the change of time in seconds. Note that the symbol Δ means "a change in."

Producing Stronger and Weaker Magnetic Fields

The strength of an induced voltage depends upon the strength of the field. It follows then that a stronger magnetic field produces a higher induced voltage. This magnetic field can be strengthened by inserting a core in the coil. An iron core has higher permeability than air and concentrates the lines of force. This concept was explained in the study of magnetism, review Chapter 9.

Large inductors are most often wound on laminated iron cores. Their inductance is measured in henrys. Smaller inductors, used in circuits that operate at higher frequencies, can have powdered iron or air cores. They have inductance measured in millihenrys (1/1000 of a henry) and microhenrys (1/1,000,000 of a henry). Radio frequency chokes using air cores are shown in **Figure 14-2.**

Figure 14-2. Radio frequency chokes.

The core material is just one factor that affects the size of the inductance produced by a coil. **Figure 14-3** is a chart that diagrams inductor traits that can strengthen or weaken a coils inductance.

Self Induction

A changing current through an inductor produces an expanding or collapsing magnetic field cutting across the wires of the coil. A counter emf is induced that opposes the change of current. This action is called ***self induction.*** The strength of the self induction depends on the number of turns of wire in the coil, the link between the length of the coil and its diameter, and the permeability of the core.

Transient Responses

The response of the current and voltage in a circuit after an instant change in applied voltage is known as a ***transient response***. A coil is connected to a dc voltage source. Refer to the diagram in **Figure 14-4.** When the switch is closed, the current builds up gradually. This effect is due to the inductance of the coil and the internal resistance of the battery. When the switch is opened, the current decays in a like manner. The rise and decay of the current in this circuit is shown in the graph in **Figure 14-5.**

It is important to understand that the opposition to the rise or decay of current occurs only when there is a change in applied voltage. When there is no change, the current remains at its steady-state value. This steady-state value depends only on the resistance of the coil.

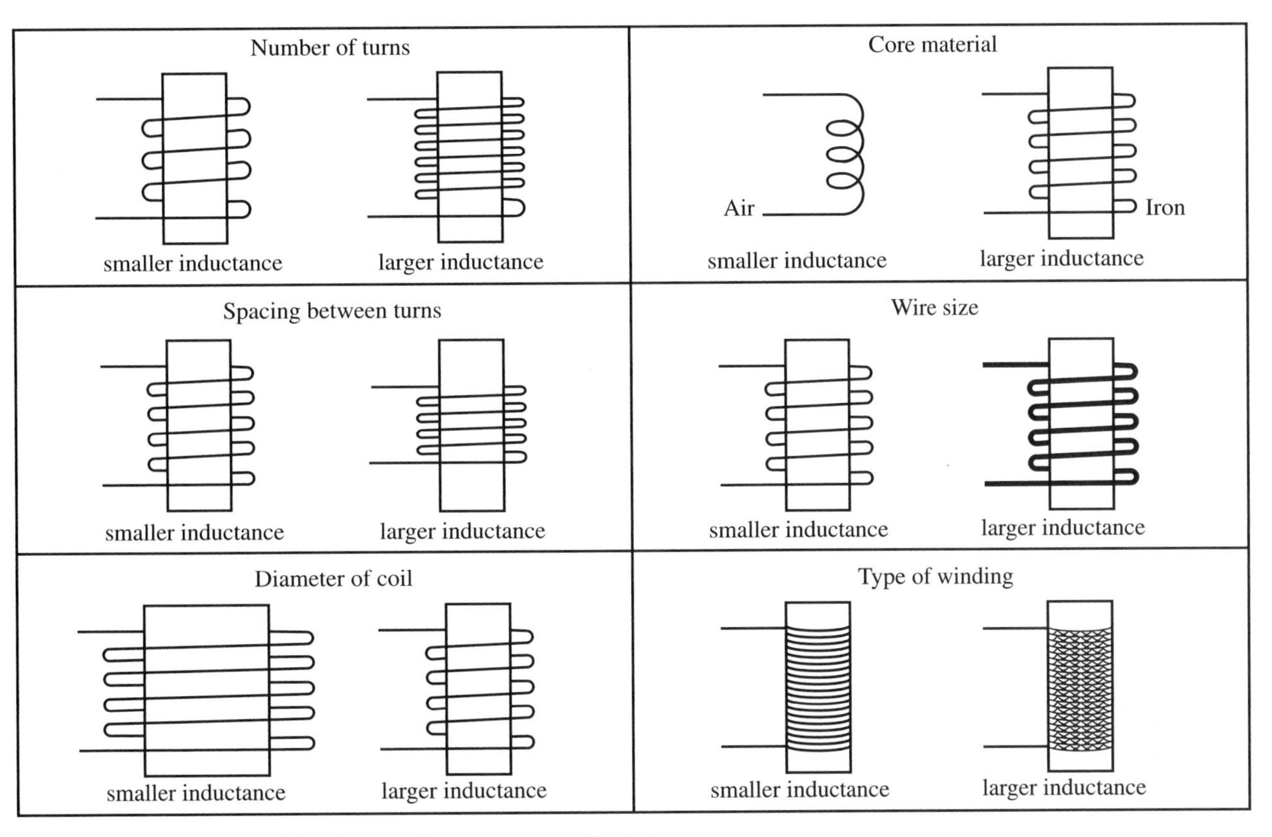

Figure 14-3. Chart of some of the factors that affect a coil's inductance.

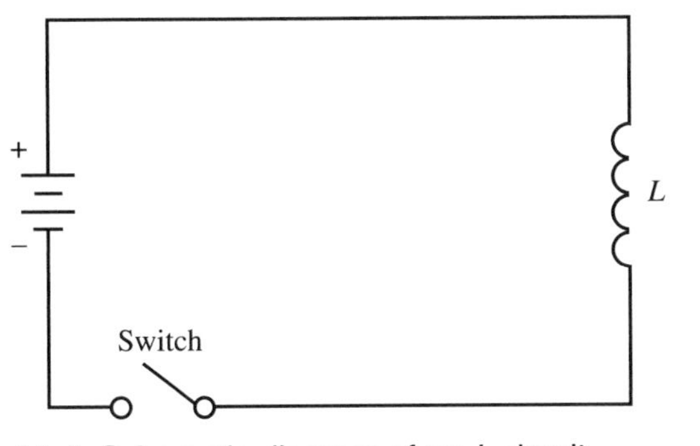

Figure 14-4. Schematic diagram of an *L* circuit.

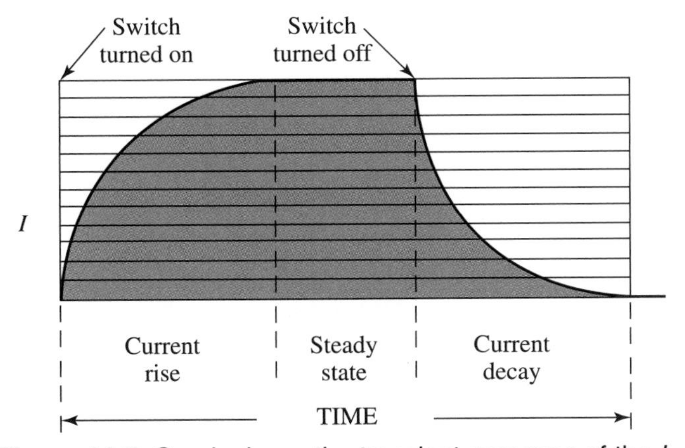

Figure 14-5. Graph shows the transient response of the *L* circuit as the switch is closed and opened.

A resistance is joined in series with an inductor in the circuit in **Figure 14-6,** view A. A circuit that contains resistance and inductance is called an ***RL circuit.*** The action of the voltage and current should be studied for increases and decreases. For example, when the switch is closed, the counter emf of the coil, *L*, equals the source voltage, view B. Since no current has started to flow, the IR drop across *R* equals zero. As the current gradually builds up, the voltage E_R increases. And counter emf E_L decreases until a steady state condition exists. All the voltage drop is E_R, and there is no drop across *L*, view C.

If the RL circuit is shorted by another switch, **Figure 14-7** view A, the stored energy in the field of *L* instantly develops a voltage, view B. This voltage is of the opposite polarity of the applied voltage. The circuit is discharged by current flowing through *R*. The graphs showing the charge and discharge of the circuit appear in **Figure 14-8.**

The baseline, or x-axis, of these graphs represents time. The transient response of an RL circuit does require a certain amount of time. The amount of time to complete the transient response depends upon the values of *R* and *L* in the circuit. This value is called the ***time constant (τ)*** of the circuit. For an RL circuit, it is found using the formula:

EXPERIMENT 14-1: Self Inductance

This experiment allows you to observe the effects of self inductance.

Materials

1–cylinder shaped magnet
–selection of coils (at least 2) with different numbers of windings
1–dc power supply
3–small light bulbs with sockets
1–SPST switch
–connecting wire

1. Set up the experiment shown in **Exhibit 14-1A.** Check with your instructor before applying power to the circuit.

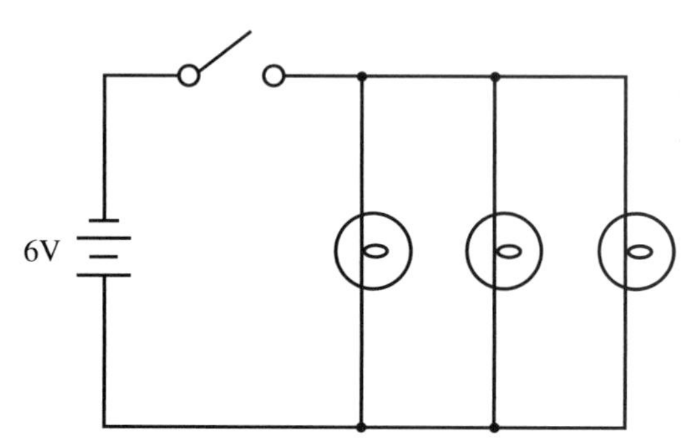

Exhibit 14-1A.

2. Close the switch. With the light connected directly to the six-volt power source, notice that the light burns at full brilliance instantly.
3. Open the switch and connect a coil in series with the light and close the switch, **Exhibit 14-1B.**

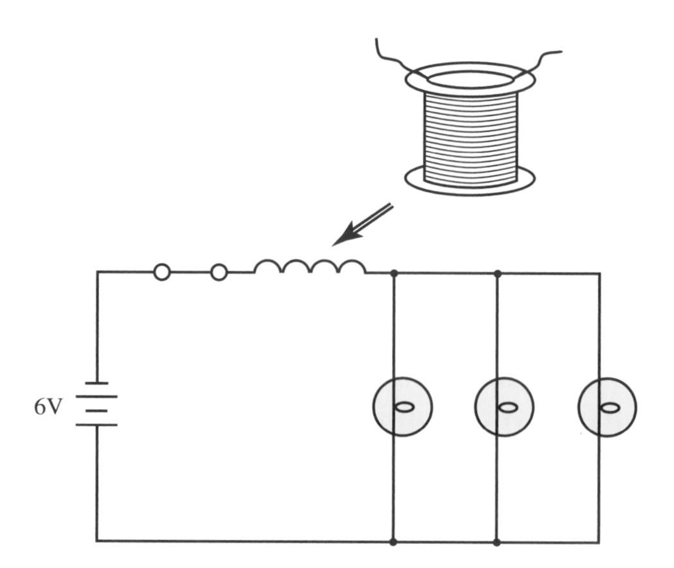

Exhibit 14-1B.

4. Close the switch. You now should notice a slight delay in the light approaching full brilliance. This delay is caused by the inductance of the coil resisting a change in the current.
5. Turn the switch off, and insert the iron core into the coil.
6. Once again, close the switch. The light now comes to brilliance rather slowly and never does reach full brilliance. Why does the core increase the delay in brilliance?
7. Move the core in and out of the coil. Note the change in brilliance. How does this relate to self inductance?
8. Repeat the above experiment using a coil with more windings. This should increase the self inductance. How does this affect your results?

$$\tau = \frac{L \text{ (measured in henrys)}}{R \text{ (measured in ohms)}}$$

where τ equals the time in seconds for the current to increase to 63.2 percent of its maximum value or to decrease to 36.7 percent of its value.

In most cases, a circuit is considered fully charged or fully discharged after a period of five time constants. The top of **Figure 14-9** is a graph showing the percentage of full charge against the number of time constants. From this graph, you can see why five time constants is considered full charge or discharge. The bottom of Figure 14-9 is a chart giving the voltage at the end of each time constant, assuming a source voltage of 100 volts.

Review Questions for Section 14.1

1. Define inductance.
2. What is the difference between inductance and resistance?
3. The unit for inductance is the __________.
4. Find the time constant in the circuit shown.

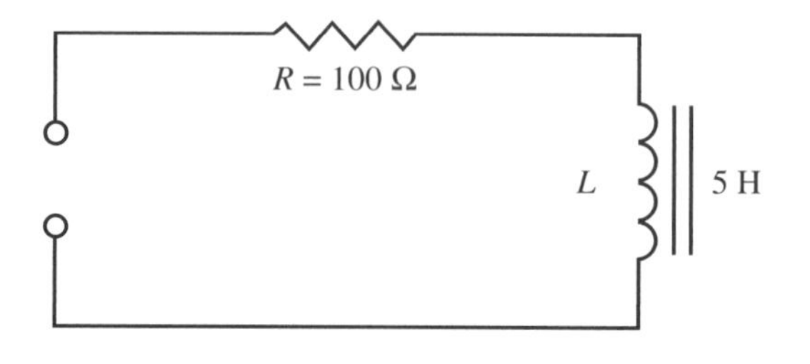

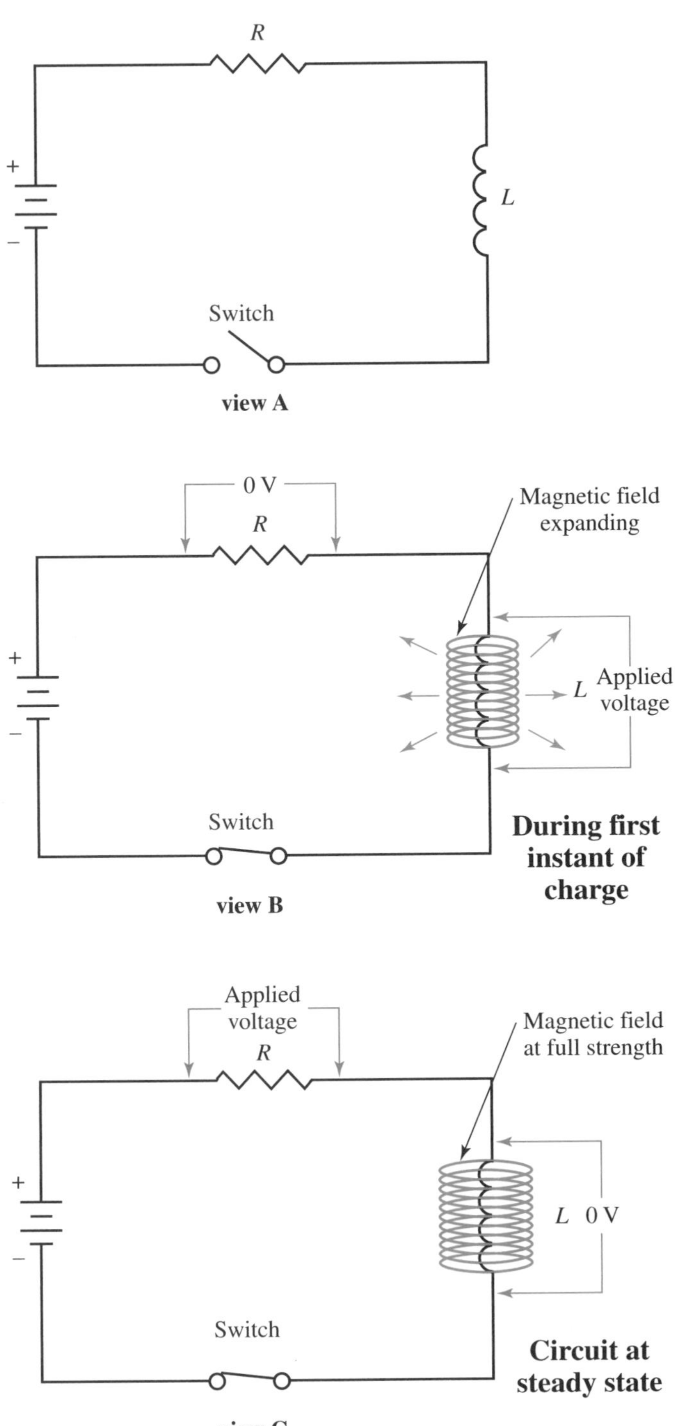

Figure 14-6. A diagram of the RL circuit showing the transient response.

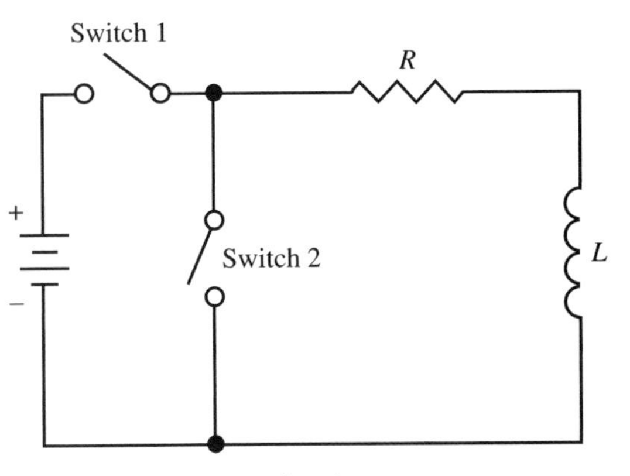

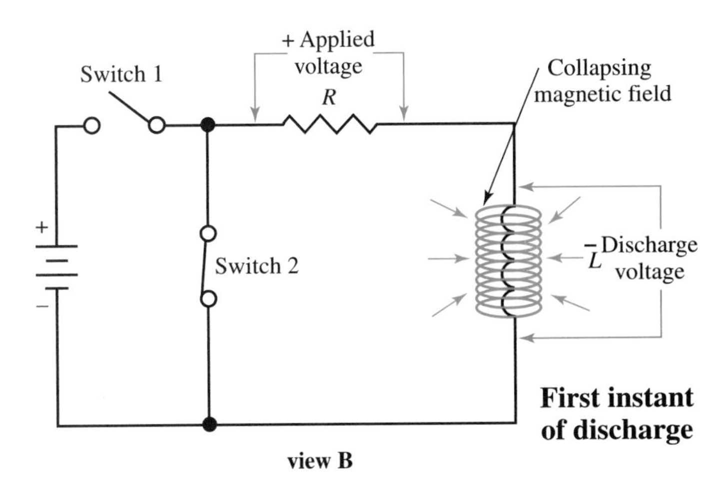

Figure 14-7. The coil L is shorted through switch 2. The magnetic field collapses.

14.2 MUTUAL INDUCTANCE

When two coils are within magnetic reach of each other so that the flux lines of one coil link with, or cut across, the other coil, they have ***mutual inductance***. If the coils are close to each other, many magnetic lines of flux from the first coil will link with the second coil. On the other hand, if the coils are a distance apart, there might be very little linkage.

The mutual inductance of two coils can be increased if a common iron core is used for both coils. The degree to which the lines of force of one coil link with the windings of the second coil is called ***coupling.*** If all lines of one cut across all the turns of the other, we have what is called ***unity coupling.*** Any number of percentages of coupling can exist due to the mechanical position of the coils. The amount of mutual inductance can be found using the formula:

$$L_M = k\sqrt{L_1 \times L_2}$$

where L_M is the mutual inductance in henrys, k is the coefficient or percentage of coupling, and L_1 and L_2 are the inductances of the respective coils.

Mutual inductance must be considered when two or more inductors are connected in series or parallel in a circuit.

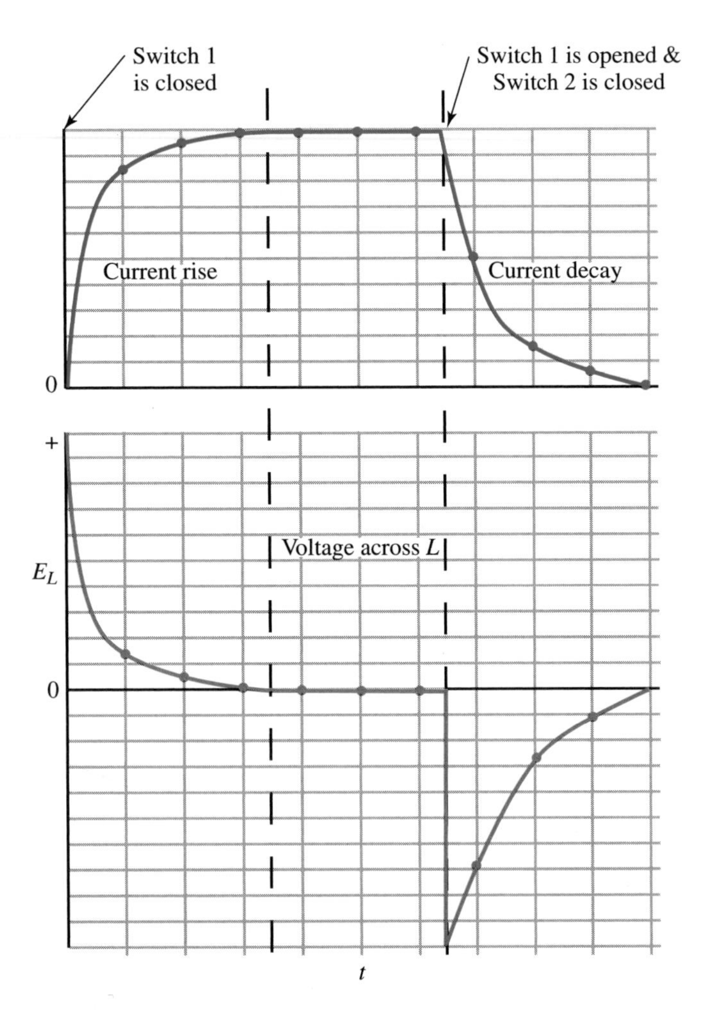

Figure 14-8. The transient response curves for current and inductive voltage of the RL circuit.

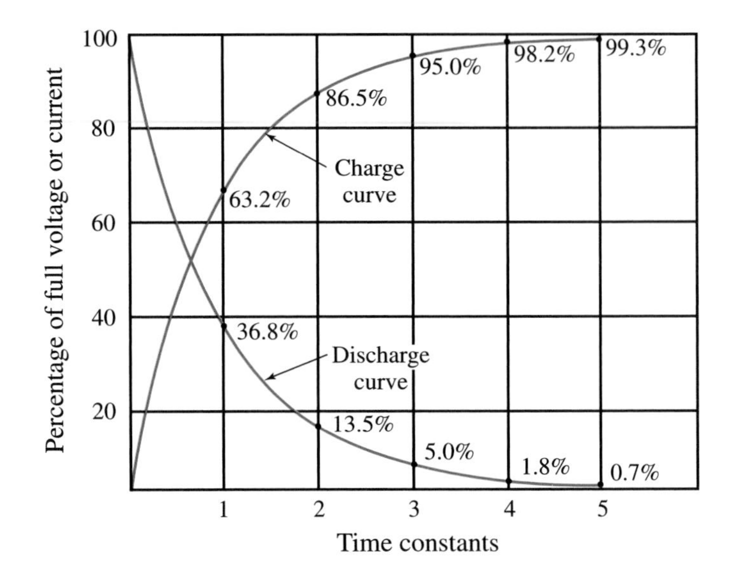

Time constant	Charging	Discharging
1	63.2 V	36.8 V
2	86.5 V	13.5 V
3	95 V	5 V
4	98 V	2 V
5	99 V	1 V

Figure 14-9. Top–Time constant curves. Bottom–Assuming the *E* source is 100 volts, these are the voltages at the end of each time constant.

Phase Relationship in Series and Parallel Inductance

When coils are joined in series, the total inductance is:

$$L_T = L_1 + L_2 + L_3 ... + L_N$$

Notice that the individual inductance's add together like series resistors. However, when there is mutual inductance the formula becomes:

$$L_T = L_1 + L_2 \pm 2L_M$$

The plus or minus sign (±) before the $2L_M$ means that the mutual inductance value might be added or subtracted depending on whether the magnetic fields are aiding or opposing each other. The plus (+) should be used when the coils aid each other. The minus (–) should be used when the coils oppose each other.

When inductors are connected in parallel without mutual inductance, the total inductance can be calculated using the formula:

$$\frac{1}{L_T} = \frac{1}{L_1} + \frac{1}{L_2} + \frac{1}{L_3} ... + \frac{1}{L_N}$$

The formula is structured the same as the parallel resistance formula.

When only two inductors are being added in parallel, the product over the sum formula can be used.

$$L_T = \frac{L_1 \times L_2}{L_1 + L_2}$$

Notice that here again, inductors in parallel add like resistors in parallel.

Mutual inductance is widely used in electricity and electronics. A skilled technician must have a complete understanding of the characteristics of circuits containing these components.

Review Questions for Section 14.2

1. The effect of two or more coils sharing the energy of one coil is called __________ __________.
2. As coils are connected in series, their total induction __________. (increases, decreases)
3. As coils are connected in parallel, their total induction __________. (increases, decreases)
4. A 1.5 H and a 500 mH coil are connected in series. What is their total inductance in henrys?
5. A 3 H and a 6 H coil are connected in parallel. What is their total induction in henrys?

14.3 INDUCTANCE IN AC CIRCUITS

In an ac circuit, the applied voltage constantly varies and reverses polarity. Any inductance in the circuit will generate a counter emf, which will oppose the source voltage.

This event can be observed by setting up the experiment in Experiment 14-1 again. Connect a light to a six-volt dc source. Note its brilliance. Now, connect the light to a six-volt ac transformer, and note the brilliance. Is it brighter or dimmer? The inductance has some indirect effect on the current.

Opposition to an alternating current due to inductance or capacitance is called ***reactance.*** The letter symbol for reactance is ***X***. When reactance is caused by an inductor, it is called ***inductive reactance.*** Inductive reactance is measured in ohms, and its symbol is X_L.

Induced Current and Voltage

Remember that the induced voltage in a coil is the counter emf. Counter emf *opposes* the source. Therefore, it is *180 degrees out of phase* with the source voltage. This counter emf opposes change, and the greatest counter emf is induced when the current change is at a maximum.

Study **Figure 14-10.** The greatest rate of change of current is at 90 degrees and 270 degrees. At these points, the applied voltage must be at a maximum. The induced voltage is also at a maximum. At 180 degrees and 360 degrees, the current change is minimum. The current is at its maximum value, ready to start its decline. As you can see in Figure 14-10, in a circuit containing pure inductance, the current is 90 degrees out of phase with the applied voltage. The current is lagging behind the voltage.

The magnitude of the reactive force opposing the flow of ac is measured in ohms. It can be expressed mathematically as:

$$X_L = 2\pi fL$$

where X_L equals inductive reactance in ohms, f equals frequency in hertz, L equals inductance in henrys, and π is approximately 3.1416. Remember, as frequency or inductance is increased, the inductive reactance increases. *They are in direct proportion.*

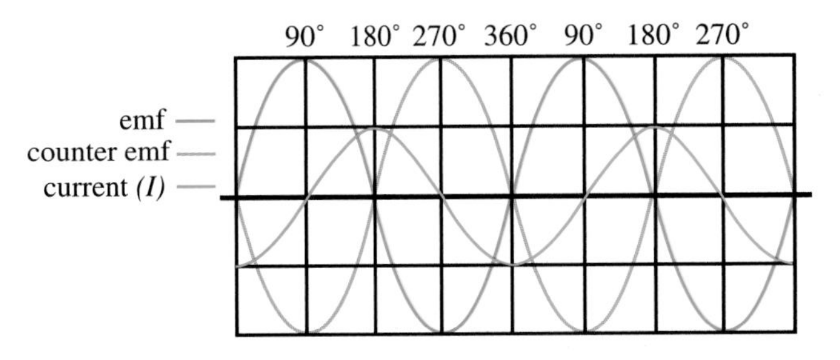

Figure 14-10. Comparison of emf, counter emf, and current. The greatest counter emf occurs when current is changing at its most rapid rate.

An inductance in a circuit behaves differently as various frequency signals are applied, **Figure 14-11.** An eight henry inductor connected to a dc ($f = 0$) signal has a reactance of zero. This means the eight-henry inductor does not impede the dc signal at all. Using the formula, $X_L = 2\pi fL$, the reactance can be computed for frequencies of 50, 100, 500, and 1000 Hz. A partial graph of the results shows the linear increase of reactance as the frequency is increased.

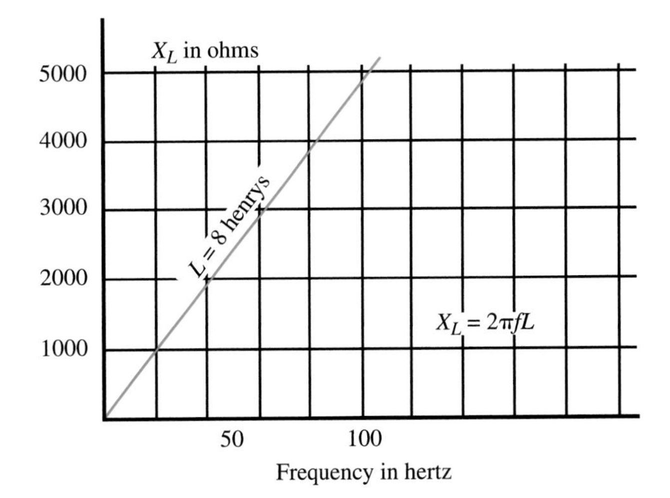

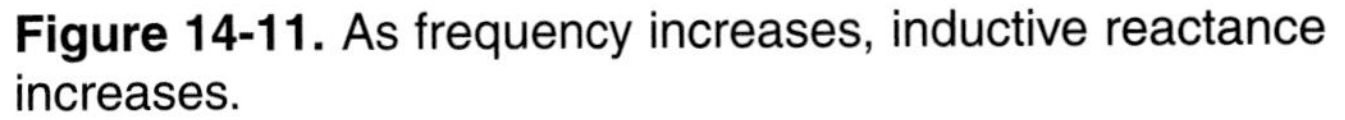
Figure 14-11. As frequency increases, inductive reactance increases.

In **Figure 14-12,** the frequency has been held constant at 1000 Hz. The reactance is plotted as the inductance is increased from 0 henrys to 0.8 henrys. Notice the linear increase in reactance as the inductance is increased.

These principles can be applied to filter and coupling circuits. However, you must first understand the traits of reactance and how they change as a result of inductance and frequency of the applied voltage.

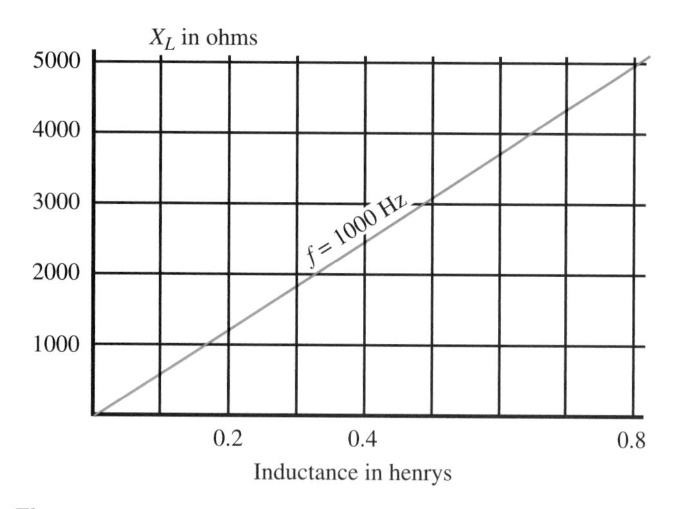

Figure 14-12. As inductance increases, inductive reactance also increases.

Power in Inductive Circuits

Power consumed in a purely resistive circuit is the product of the voltage and the current: $P = I \times E$. This is the actual power used by a circuit. It is called ***true power,*** and it is given in units of watts.

However, in an ac circuit containing inductance only, this formula does not hold true. In this type of circuit, a current will flow bound by the reactance of the circuit. The power used to build the magnetic field is returned to the source when the field collapses. This is ***reactive power.*** Reactive power is measured in ***VAR,*** which stands for ***volt-ampere-reactive.***

The combination of true power and reactive power gives rise to a measure of power that *appears* to be delivered to a load. This power is called ***apparent power.*** It is equal to the product of the effective voltage and the effective current. Apparent power is measured in ***VA,*** or ***volt-amperes.***

$$\text{Apparent power} = E_{eff} \times I_{eff}$$

For example, a 100 E_{eff} is applied voltage to a certain inductive circuit. This causes an I_{eff} of 10 amps. The apparent power equals:

Apparent power = 100 V × 10 A = 1000 VA

Remember, watts are not used for apparent power.

Power factor

The ratio of true power to apparent power in an ac circuit is called the ***power factor (PF).*** It has no units. It can be calculated using trigonometry. The power factor is equal to the cosine of the phase displacement between current and voltage. The formula is written:

$$\text{Power factor} = \cos\theta = \frac{\text{True power}}{\text{Apparent power}}$$

where θ is the phase displacement. Note that by rearranging this formula, we arrive at the formula for true power.

$$\text{True power} = \text{Apparent power} \times \cos\theta$$
$$\text{True power} = E_{eff} \times I_{eff} \times \cos\theta$$

As an example, assuming our previous circuit is purely inductive, determine the power factor and phase displacement.

$$\text{PF} = \cos\theta = \frac{0\text{ W}}{1000\text{ VA}} = 0$$

The power factor is 0. The angle whose cosine is 0 is 90 degrees. This tells us that current and voltage in the purely inductive circuit are 90 degrees out of phase, **Figure 14-13.**

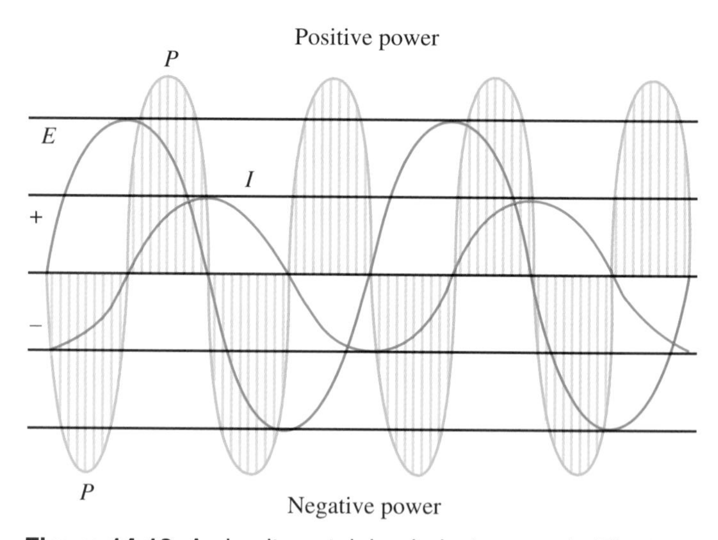

Figure 14-13. A circuit containing inductance only. The true power is zero and current lags the voltage by 90 degrees.

Resistance and Inductance in an AC Circuit

If an ac circuit contains only resistance, the current and voltage are in phase, **Figure 14-14.** The power consumed is equal to $I \times E$. Even though the polarity of the voltage changes and the current reverses, positive power is consumed. A resistor consumes the same amount of power no matter which direction the current is moving. The power factor in a circuit of this type equals one. The apparent power is equal to the true power.

The circuit traits change when an inductor is added in series with the resistor. The component does not have to be a resistive component. The wire from which the coil is wound will have a certain amount of resistance.

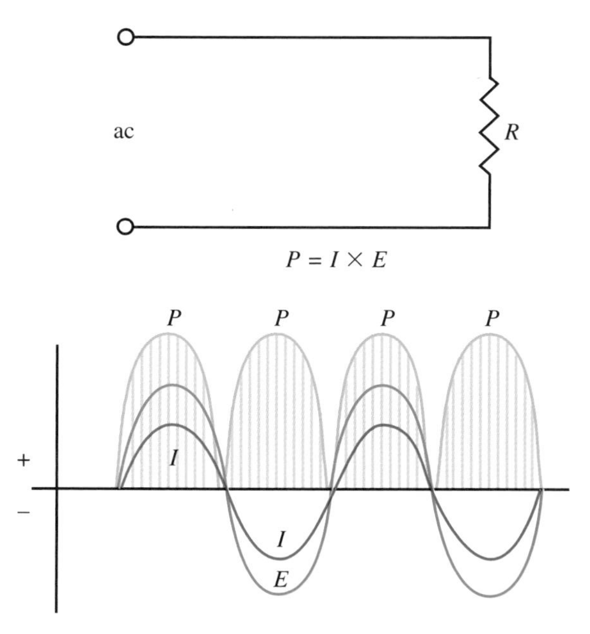

Figure 14-14. In a purely resistive circuit, true power and apparent power are the same. The current and voltage are in phase.

Refer to **Figure 14-15.** The series resistance equals 300 ohms. The inductive reactance equals 400 ohms. This reactive component will cause the current to lag by an angle of 90 degrees or less. The forces opposing the current can be thought of as the resistance and reactance. The resistance and reactance are 90 degrees out of phase. To find the final opposition, the combination of resistance and reactance, add the vectors of the two forces. Refer to **Figure 14-16.**

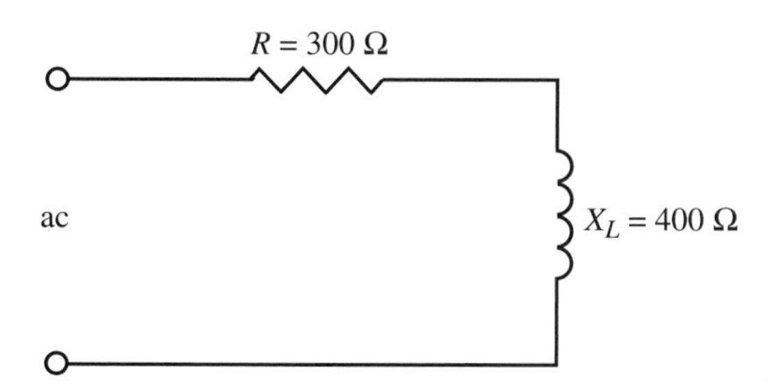

Figure 14-15. RL circuit having series resistance of 300 ohms and inductive reactance of 400 ohms.

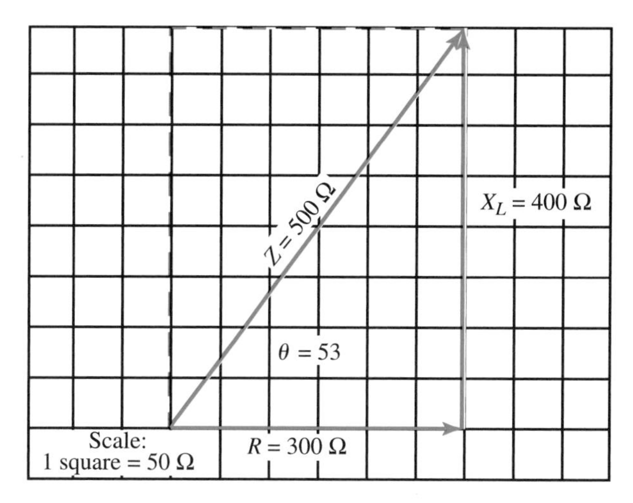

Figure 14-16. Vector addition of X_L and R, which are 90 degrees out of phase.

The vectors can be added using the graph. Place the tail of the X_L vector on the arrowhead of the R vector. Then draw a vector from starting point of the resistance vector to the head of the reactance vector, X_L. This new vector shows the magnitude and direction of the combined forces. The angle between this new vector and the resistance vector is θ, see Figure 14-16.

The total opposition to an alternating current in a circuit having resistance and reactance (the new vector you have just created) is the ***impedance.*** The symbol for impedance is ***Z,*** and it is measured in ohms.

Impedance problems are commonly solved using the Pythagorean theorem. The ***Pythagorean theorem*** states that the hypotenuse of a right triangle is equal to the square root of the sum of the squares of the two sides. It is commonly written:

$$a = \sqrt{b^2 + c^2}$$

or

$$a^2 = b^2 + c^2$$

where a is the hypotenuse and b and c are the other two sides of the right triangle.

In the problem of Figures 14-15 and 14-16, X_L equals 400 and R equals 300. Find the impedance, Z.

$$Z = \sqrt{R^2 + X_L{}^2}$$

$$Z = \sqrt{(300\ \Omega)^2 + (400\ \Omega)^2}$$

$$Z = \sqrt{(90{,}000\ \Omega) + (160{,}000\ \Omega)}$$

$$Z = 500\ \Omega$$

The impedance of the circuit is 500 ohms. The angle between vector Z and vector R is called the ***phase angle*** ***(θ).*** It represents the phase displacement between the current and voltage resulting from the reactive component. So, since cos θ equals the power factor:

$$\cos \theta = \text{PF}$$

$$\cos \theta = \frac{R}{Z}$$

$$\cos \theta = \frac{R}{Z} = \frac{300}{500} = 0.6$$

Since the cosine of θ equal 0.6, the angle whose cosine is 0.6 (53.1° approximately) is equal to the phase angle. Therefore, the current lags the voltage by an angle of 53.1 degrees.

The true power in this circuit can now be calculated. The true power in this circuit equals the apparent power times the power factor, or cos θ.

True power = Apparent power × cos θ

$$\cos \theta = \frac{\text{True power}}{\text{Apparent power}}$$

The waveforms for current, voltage, and power are drawn in **Figure 14-17.** Assuming an applied ac voltage of 100 V, the current in the circuit will equal:

$$I = \frac{E}{Z} = \frac{100 \text{ V}}{500\ \Omega} = 0.2 \text{ A}$$

The apparent power equals:

Apparent power = $E \times I$

Apparent power = 100 V × 0.2 A = 20 VA

The true power equals:

True power = $E \times I \times \cos \theta$

True power = 100 V × 0.2 A × cos 53.1°

True power = 20 W × 0.6

True power = 12 W

These figures can be checked by inserting them into the power factor formula.

$$\text{PF} = \frac{\text{True power}}{\text{Apparent power}}$$

$$\text{PF} = \frac{12 \text{ VA}}{20 \text{ W}} = \frac{3}{5} = 0.6$$

The power factor is not only considered in theory but in practice. The power factor must be considered whenever a power company connects power lines to a manufacturing plant. Industries must keep the power factor of their circuits and machinery within specified limits, or pay the power company a premium.

A reactive power is sometimes called ***wattless power.*** It is returned to the circuit. In Figure 14-17, the power in the shaded areas above the zero line is used. The power below the line is wattless power.

Ohm's Law for AC Circuits

In computing circuit values in ac circuits, Ohm's law is used with one exception. Z is used in place of R. Z represents the total resistive force opposing the current. Therefore:

$$I = \frac{E}{Z} \quad \text{and} \quad E = I \times Z \quad \text{and} \quad Z = \frac{E}{I}$$

Example: A series circuit contains an 8-henry choke and a 4000-ohm resistor. It is connected across a 200-volt, 60 Hz ac source. Find the impedance, circuit current, voltage drops across the resistor and inductor, phase angle (θ), apparent power, and true power. See **Figure 14-18.**

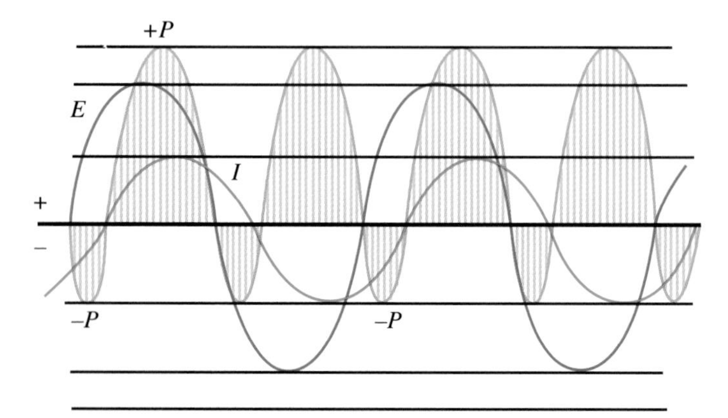

Figure 14-17. The relationship between voltage, current, and power in the circuit described in the text.

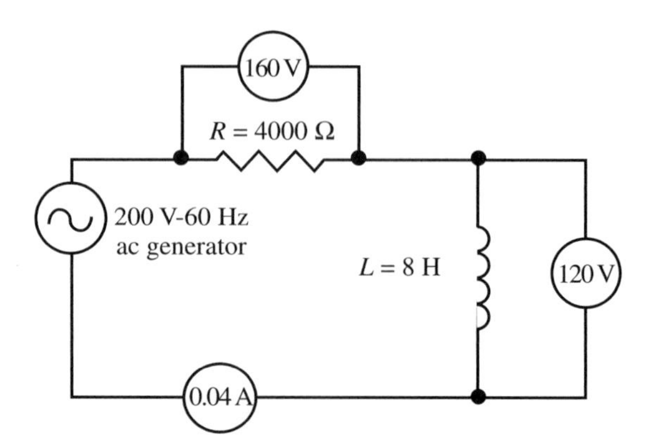

Figure 14-18. Circuit having an 8-henry choke, 4000-ohm resistor connected across a 200-volt, 60 Hz ac source.

First, find the reactance of L.

$X_L = 2\pi fL$
$X_L = 2 \times 3.14 \times 60 \text{ Hz} \times 8 \text{ H}$
$X_L = 3014\ \Omega$ (3000 Ω approximately)

Find the impedance of the circuit.

$$Z = \sqrt{(R^2 + X_L{}^2)}$$

$$Z = \sqrt{(4000\ )^2 + (3000\ )^2}$$

$$Z = \sqrt{(16{,}000{,}000\ ) + (9{,}000{,}000\ )}$$

$$Z = \sqrt{25{,}000{,}000}$$

$$Z = 5000$$

Find the current in the circuit.

$$I = \frac{E}{Z} = \frac{200 \text{ V}}{5000} = 0.04 \text{ A}$$

Find the voltage drop across R and X_L.

$E_R = I \times R = 0.04 \text{ A} \times 4000\ \Omega = 160 \text{ V}$
$E_{X_L} = I \times X_L = 0.04 \text{ A} \times 3000\ \Omega = 120 \text{ V}$

Notice that the sum of the voltage drops does not equal the applied voltage. This is because the two voltages are 90 degrees out of phase. This requires vector addition.

$$\text{Source voltage} = E_S = \sqrt{(160 \text{ V})^2 + (120 \text{ V})^2}$$

$$E_S = \sqrt{(25{,}600 \text{ V}) + (14{,}400 \text{ V})}$$

$$E_S = \sqrt{40{,}000 \text{ V}}$$

$$E_S = 200 \text{ V}$$

Find the phase angle θ between I and E.

$$\cos \theta = \frac{R}{Z}$$

$$\cos \theta = \frac{4000}{5000}$$

$$\cos \theta = 0.8$$

0.8 is the cosine of an angle of 37 degrees.

$$\theta = 37$$

Find the true power and apparent power.

Apparent power = $I \times E$
Apparent power = 0.04 A × 200 V
Apparent power = 8 VA
True power = $I \times E \times \cos \theta$
True power = 0.04 A × 200 V × 0.8
True power = 6.4 W

Note that some figures have been approximated to make the calculations simpler.

Parallel RL Circuit

In a circuit containing a resistance and an inductance in parallel, the voltage of each circuit element is the same as the source voltage. Further, there is no phase difference among the elements because they are all in parallel.

There will, however, be a phase difference among the total and branch currents. The current is in phase with voltage in the resistive branch. Current lags the voltage across the inductor by 90 degrees. The total current will lag the source voltage by some angle between 0 and 90 degrees. If the reactance of the inductor is larger than the resistance of the resistor, the angle will be closer to 0 degrees. If the resistance is much larger than the reactance the lag will be closer to 90 degrees.

In the parallel RL circuit, we do not find impedance through a vector sum. Instead, we apply Ohm's law.

$$Z = \frac{E_S}{I_T}$$

As discussed, voltage in the parallel RL circuit is in phase with I_R, and it leads I_L by 90 degrees. Thus, we can say that I_R leads I_L by 90 degrees. The parallel RL circuit, then, has two current components—I_R and I_L. Both of these can be represented by phasors. Since they are out of phase, we cannot simply add the two components together to figure the total circuit current. We must find their phasor sum. See **Figure 14-19.**

$$I_T = \sqrt{I_R{}^2 + I_L{}^2}$$

Note that the phasor for I_L is below the horizontal reference. This is because I_L lags voltage—the horizontal reference for the parallel RL circuit. *(Since a phasor, or rotating vector, rotates counterclockwise, a lagging phasor would be behind, or clockwise to, a leading phasor.)*

The phase angle represents the phase displacement between current and voltage resulting from a reactive circuit element. For the parallel RL circuit the phase angle

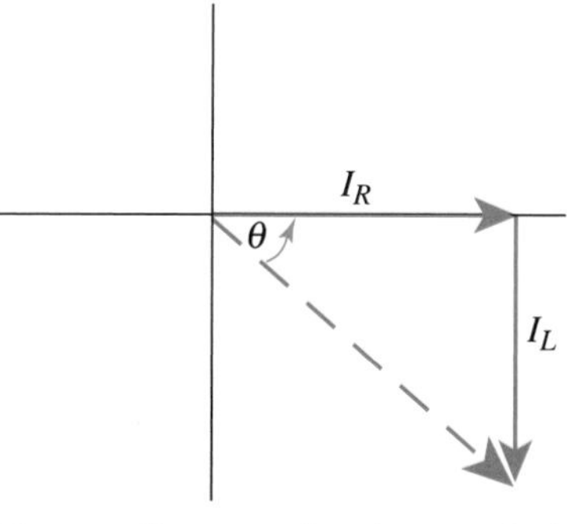

Figure 14-19. Phasor diagram showing θ relation with I_R and I_C.

is found on the current phasor diagram. The horizontal reference of this circuit is voltage, since it is common to all circuit elements.

In the current phasor diagram, we do not have a voltage phasor. We use I_R as the horizontal component since it is in phase with voltage, Figure 14-19. Using this diagram, the phase angle can be found from:

$$\theta = \arctan \frac{I_L}{I_R}$$

Example: Using the circuit with the assigned values in **Figure 14-20,** determine the phase angle between applied voltage and current. Draw the current phasor diagram and find the circuit impedance.

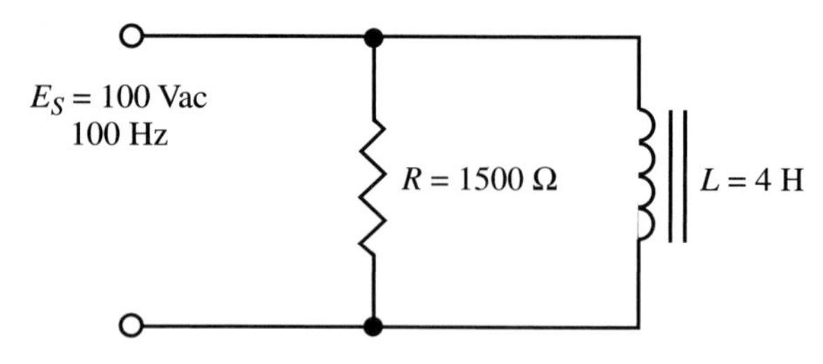

Figure 14-20. A parallel RL circuit.

Step 1. Compute the value of inductive reactance, X_L.

$$X_L = 2\pi fL$$
$$X_L = 6.28 \times 100 \text{ Hz} \times 4 \text{ H}$$
$$X_L = 2512\ \Omega$$

Step 2. Compute the branch currents. Using Ohm's law:

$$I_R = \frac{E_S}{R} = \frac{100 \text{ V}}{1500\ \Omega} = 0.067 \text{ A}$$

$$I_L = \frac{E_S}{X_L} = \frac{100 \text{ V}}{2512\ \Omega} = 0.04 \text{ A}$$

Step 3. Determine the phase angle to see how much the circuit current lags the voltage.

$$\theta = \arctan \frac{I_L}{I_R} = \arctan \frac{0.04 \text{ A}}{0.067 \text{ A}}$$

$$\theta = \arctan 0.597$$

$$\theta = 30.8°$$

Step 4. Draw the current phasor diagram. Use any convenient scale. Remember that I_R is drawn as the horizontal component. I_L is drawn downward at 90 degrees from I_R since it lags voltage—the horizontal reference. See **Figure 14-21.**

Step 5. Find the total circuit current.

$$I_T = \sqrt{I_{R_2} + I_{L_2}} = \sqrt{(0.067 \text{ A})^2 + (0.04 \text{ A})^2}$$

$$I_T = 0.078 \text{ A}$$

Step 6. Find the impedance of the circuit. Using Ohm's law:

$$Z = \frac{E_S}{I_T} = \frac{100 \text{ V}}{0.078 \text{ A}} = 1282\ \Omega$$

Once again, it is important to understand how the various quantities would be affected by a change in frequency. The effects of changing frequency are summarized in **Figure 14-22.** It shows how values change in a parallel RL circuit as frequency is changed and inductance is held at a constant value.

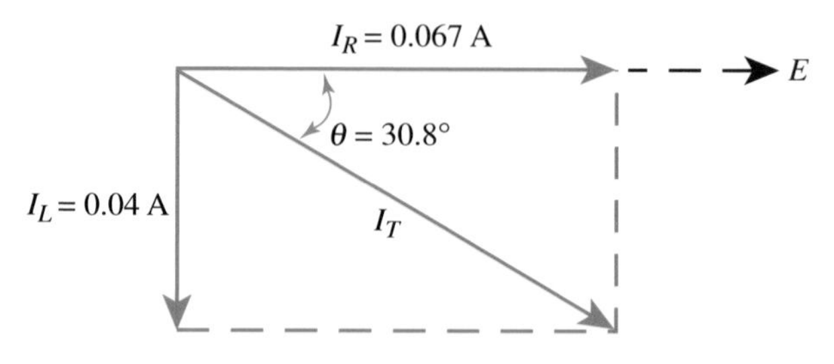

Figure 14-21. Current phasor diagram.

	X_L	θ	Z	I_R	I_L	I_T
f increases ↑	increases ↑	decreases ↓	increases ↑	remains constant	decreases ↓	decreases ↓
f decreases ↓	decreases ↓	increases ↑	decreases ↓	remains constant	increases ↑	increases ↑

Figure 14-22. This table shows the effect on various values in a parallel RL circuit as frequency is changed and the inductance value is held constant.

Review Questions for Section 14.3

1 What is reactance?
2. The actual opposition to a change in current offered by an inductor is __________ __________.
3. Find X_L in the following diagram.

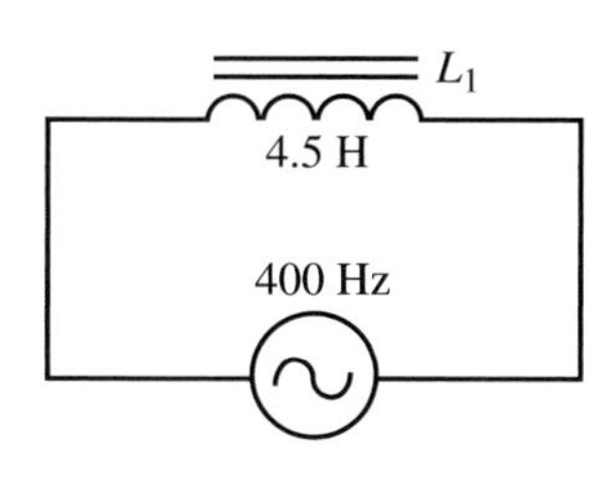

4. The actual power used by a circuit is the __________ power.
5. Find the values in the following figure.
   a. X_L.
   b. Z.
   c. I (lags, leads) E by __________ degrees.

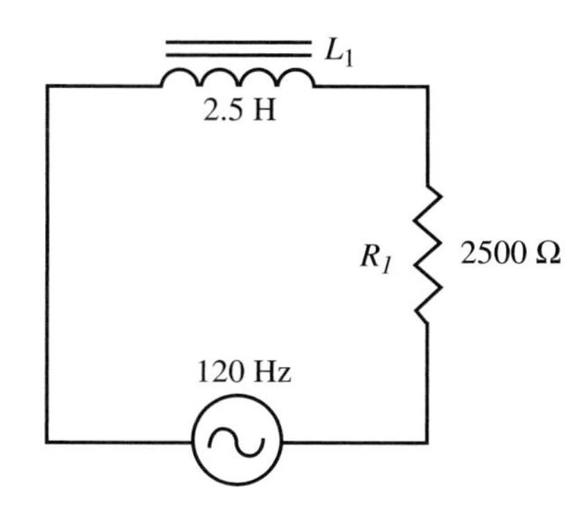

Summary

1. Inductance is that property of a circuit that opposes any change in current.
2. An inductor is a component that produces inductance in a circuit. Other names for an inductor are coil, reactor, and choke.
3. The unit for inductance is the henry (H).
4. Some factors that affect the amount inductance are:
   - Number of turns of coil.
   - Diameter of coil.
   - Core material of coil.
   - Type of winding.
   - Spacing between the windings. (length of the coil)
   - Size of wire.
5. The time constant of a coil is the amount of time it takes for the current to rise from 0 to 63.2 percent of its maximum value. The formula for time constant is:

$$\tau = \frac{L \text{ (in henrys)}}{R \text{ (in ohms)}}$$

6. Mutual inductance is two or more coils linked by magnetic lines of flux.
7. The following formulas are used to compute inductance in a circuit.
   Inductor in series: $L_T = L_1 + L_2 + L_3 ... + L_N$

   Inductors in parallel: $\frac{1}{L_T} = \frac{1}{L_1} + \frac{1}{L_2} + \frac{1}{L_3} ... + \frac{1}{L_N}$

   or a shortcut when only two inductors are in parallel:

$$L_T = \frac{L_1 \times L_2}{L_1 + L_2}$$

8. Inductive reactance (X_L) is the amount of opposition offered to a change in current by an inductor. It is measured in ohms.
9. The formula for inductive reactance is $X_L = 2\pi fL$, where π equals 3.14, f equals a frequency in hertz, and L equals inductance in henrys.
10. Reactive power is power reflected back to the source. It is measured in VAR. True power is power actually used in a circuit. It is measured in watts. Apparent power is the combination of reactive power and true power. It is measured in VA.
11. Power factor is the relationship between true power and apparent power.
12. The total opposition to a change in current in an inductive and resistive circuit is the impedance (Z). It is measured in ohms.
13. Impedance can be calculated using a graph of the resistance and reactance vector, using the Pythagorean theorem, or using the phase angle and trigonometry.
14. When a resistor and an inductor are in parallel, the branch voltages are the same. However, the branch currents will be 90 degrees out of phase with each other.

Test Your Knowledge

Please do not write in the text. Place your answers on a separate sheet of paper.

1. Draw a graph showing the rise and decay of current in a RL circuit.
2. What is the formula for determining the time constant?
3. Draw the symbol for an inductor with an iron core.

4. Reactance is measured in __________.
5. In a series RL circuit: $L = 2$ H, $R = 500\ \Omega$, $E_S = 100$ V at 60 Hz ac. Find:
   a. X_L
   b. Z
   c. I
   d. E_R
   e. E_{X_L}
   f. θ
   g. PF
   h. True power
   i. Apparent power
6. On graph paper, draw vectors and sine curves representing current and voltage of the circuit in Question 5. With a red pencil, draw the power curve.

For Discussion

1. What is the relationship between induction and frequency.
2. What is the difference between apparent power and true power?
3. Can any inductive circuit be completely 180° out of phase?

Inductive Heating

An iron or steel part inside an energized coil produces magnetism as well as heat. This heating effect is caused by *hysteresis* and *eddy currents* generated by the magnetic field. The amount of heat produced is directly related to the *intensity* of the magnetic field and the *frequency* of the applied ac. In normal electronic applications, this heating effect is a waste of power and is undesired. The effects can be reduced by using laminated metal.

Today there are many uses for the once undesired heating effect. One such product is found in a relatively new household stove top that is designed to incorporate the heating effect of induction. In this stove top, a cooking unit is designed with a coil located under a ceramic top.

When a metallic or some other induction-efficient cooking pot is placed on the cooking top, a magnetic field is induced by the coil hidden beneath the counter surface. The coil does not produce heat in the same manner as a conventional electric cook top. The conventional electric cook top uses resistance wire inside coils to produce heat. Instead, the magnetic field produces heat in the pot itself, not on the cook top. The top of the cooker remains relatively cool. This lowers the risk of burns. In addition, the surface of these stoves is completely sealed to allow for easy cleanup.

Another use of the induction heating principle is in the metalworking industry. When a piece of metal, such as a pipe, is inserted inside an energized coil, the metal pipe is heated. This heating allows for easy bending and shaping. The heat can be controlled to allow for heat treating processes such as annealing. *Annealing* is a process that adds strength to the metal by slowly and evenly heating a metal to a desired temperature and then systematically cooling it to produce a highly tempered metal.

Following is a drawing of the Ajax Magnethermic induction furnace.

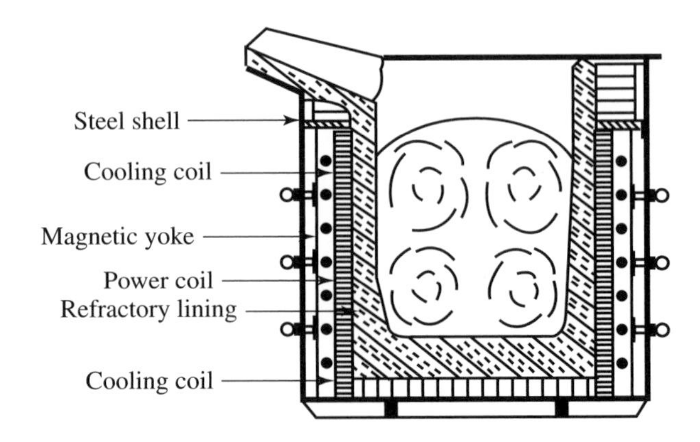

The furnace is rated at 3600 kW and can easily melt over 10 tons of steel.

Capacitance and RC Circuits

Objectives

After studying this chapter, you will be able to:

- Define capacitance and capacitor.
- Identify the many different types of capacitor.
- Describe the transient response of a capacitor.
- Explain how a capacitor behaves in a dc circuit.
- Discuss the effect of capacitance on an ac circuit.
- Describe the results of combining capacitance and resistance in a circuit.

Key Words and Terms

The following words and terms will become important pieces of your electricity and electronics vocabulary. Look for them as you read this chapter.

alignment tool
can type electrolytic capacitor
capacitance
capacitive reactance
capacitor
ceramic capacitor
dielectric
dielectric constant
farad (F)
fixed paper capacitor
mica capacitor
RC circuit
rectanglular oil filled capacitor
rotor
stator
tantalum capacitor
trimmer capacitor
tubular electrolytic capacitor
variable capacitor
working voltage (WV)

15.1 CAPACITANCE AND THE CAPACITOR

In Chapter 14, you learned that inductance is the property of a circuit that opposes any change in current. ***Capacitance,*** on the other hand, is a property that opposes any change in voltage. A ***capacitor*** is a device that temporarily stores an electric charge. A capacitor accepts or returns this charge in order to maintain a constant voltage. Schematic symbols that are used to represent a capacitor are shown in **Figure 15-1.**

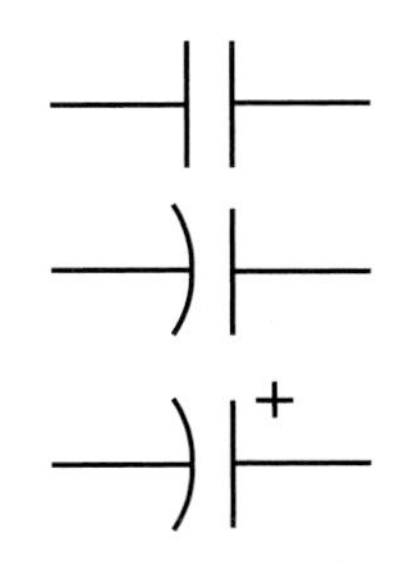

Figure 15-1. Schematic symbols for the capacitor.

The capacitor is made of two plates of conductive material, separated by insulation. This insulation is called a ***dielectric,*** **Figure 15-2.** In the figure the plates are connected to a dc voltage source. The circuit appears to be an open circuit because the plates do not contact each other. However, the meter in the circuit will show some current flow for a brief period after the switch is closed.

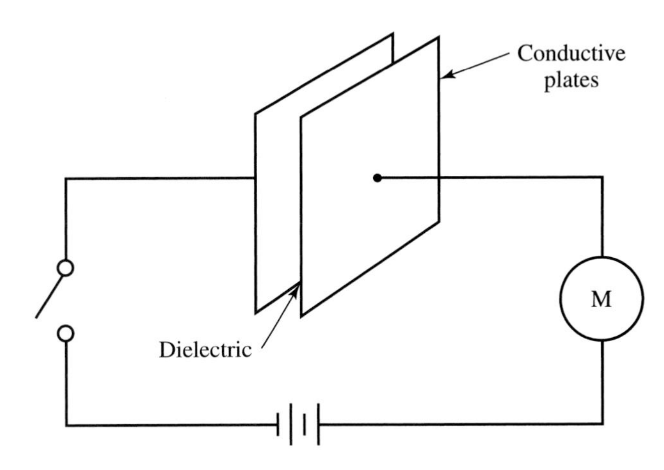

Figure 15-2. Basic form of a capacitor.

In **Figure 15-3,** as the switch is closed, electrons from the negative terminal of the source flow to one plate of the capacitor. These electrons repel electrons from the

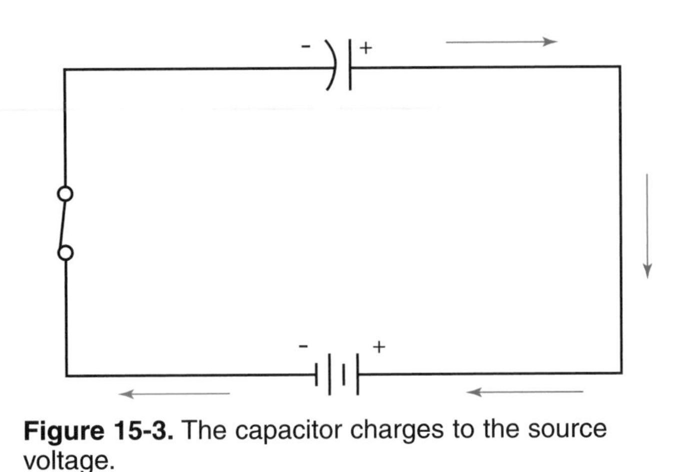

Figure 15-3. The capacitor charges to the source voltage.

second plate (like charges repel), which are then drawn to the positive terminal of the source. The capacitor is now charged to the same potential as the source and is opposing the source voltage. If the capacitor is removed from the circuit, it will remain charged. The energy is stored within the electric field of the capacitor. Once the capacitor is fully charged, current ceases to flow in the circuit.

It is important to remember that in the circuit in Figure 15-3, no electrons flowed through the capacitor. This is because a capacitor blocks direct current. However, one plate did become negatively charged and the other positively charged. A strong electric field exists between them.

Insulating or dielectric materials vary in their ability to support the electric field. This ability is known as the ***dielectric constant*** of the material. The constants of various materials are shown in **Figure 15-4.** These numbers are based on comparison with the dielectric constant of dry air. The constant for dry air has been assigned as 1.

The dielectrics used for capacitors can only withstand certain voltages. If this voltage is exceeded, the dielectric will break down and arcing will result. This maximum voltage is known as the ***working voltage (WV).***

Material	Dielectric Constant
air	1.0
oil	2.2
mica	5.0 – 8.5
porcelain	5.0 – 7.0
ceramic	5.0 – 8.0
glass	8.0
aluminum oxide electrolytic	8.4
tantalum oxide electrolytic	26
pure water	81

Figure 15-4. Dielectric constants. Larger numbers are better able to support electric fields.

Exceeding the working voltage can cause a short circuit and can ruin other parts of the circuit connected to the dielectric.

Increased voltage ratings require special materials and thicker dielectrics. When a capacitor is replaced, check its capacitance value and dc working voltage.

When a capacitor is used in an ac circuit, the working voltage should safely exceed the peak ac voltage. For example, a 120-volt effective ac voltage has a peak voltage of 120 V × 1.414 = 169.7 volts. Any capacitors used must be able to handle 169.7 volts.

Calculating Capacitance

Capacitance is determined by the number of electrons that can be stored in the capacitor for each volt of applied voltage. Capacitance is measured in ***farads (F).*** A farad represents a charge of one coulomb that raises the potential one volt. This equation is written:

$$C = \frac{Q}{E}$$

where C is the capacitance in farads, Q is the charge in coulombs, and E is the voltage in volts.

Capacitors used in electronic work have capacities measured in microfarads (1/1,000,000 F) and picofarads (1/1,000,000 of 1/1,000,000 F). Microfarad is commonly written as μF or sometimes written as mfd. Picofarad is written as pF. Nanofarad is not a common measurement of capacitance. A conversion chart for these units is shown in **Figure 15-5.**

Farad	1.0	1.0
Millifarad	0.001	10^{-3}
Microfarad	0.000 001	10^{-6}
Picofarad	0.000 000 000 001	10^{-12}

Figure 15-5. Prefixes used with the farad. Take special note that the prefix nano is missing. Nanofarad is not a standard rating size for a capacitor.

Capacitance is determined by:

- The material used as a dielectric. (The larger the dielectric constant, the greater the capacitance.)
- The area of the plates. (The larger the plate area, the greater the capacitance.)
- The distance between the plates. (The smaller the distance, the greater the capacitance.)
- These factors are related in the mathematical formula:

$$C = 0.225 \frac{KA(n-1)}{d}$$

where C is the capacitance in picofarads, K equals the dielectric constant, A equals the area of one side of one plate in square inches, d equals the distance between plates in inches, and n equals the number of plates.

This formula illustrates the following facts:

1. Capacity *increases* as the area of the plates *increase,* or as the dielectric constant *increases.*
2. Capacity *decreases* as the distance between plates *increases.*

LESSON IN SAFETY:

Many large capacitors in TVs and other electronic equipment retain their charge for a long time after power is turned off. Discharge these capacitors by shorting terminals to the equipment's chassis with an insulated screwdriver. If capacitors are not discharged, the voltages can destroy test equipment, and persons working on the equipment can receive a severe shock!

Types of Capacitors

Capacitors are made in hundreds of sizes and types. Several of these will be discussed in the following section.

Fixed paper capacitors are made of layers of tinfoil. The dielectric is made of waxed paper. Wires extending from the ends connect to the foil plates. The assembly is tightly rolled into a cylinder and sealed with special compounds. Some capacitors are enclosed in plastic for rigidity. These capacitors can withstand severe heat, moisture, and shock.

Rectangular oil filled capacitors are hermetically sealed in metal cans. They are oil filled and have very high insulation resistance. This type of capacitor is used in power supplies of radio transmitters and other electronic equipment.

Can type electrolytic capacitors use different methods of plate construction. **Figure 15-6** shows in detail how three separate can type electrolytic capacitors are put together. **Figure 15-7** shows several of the single-ended capacitors.

Some capacitors have aluminum plates and a wet or dry electrolyte of borax or carbonate. A dc voltage is applied during manufacturing. Electrolytic action creates a thin layer of aluminum oxide that deposits on the positive plate. This coating insulates the plate from the electrolyte. The negative plate is connected to the electrolyte. The electrolyte and positive plates form the capacitor. These capacitors are useful when a large amount of capacity is needed in a small space.

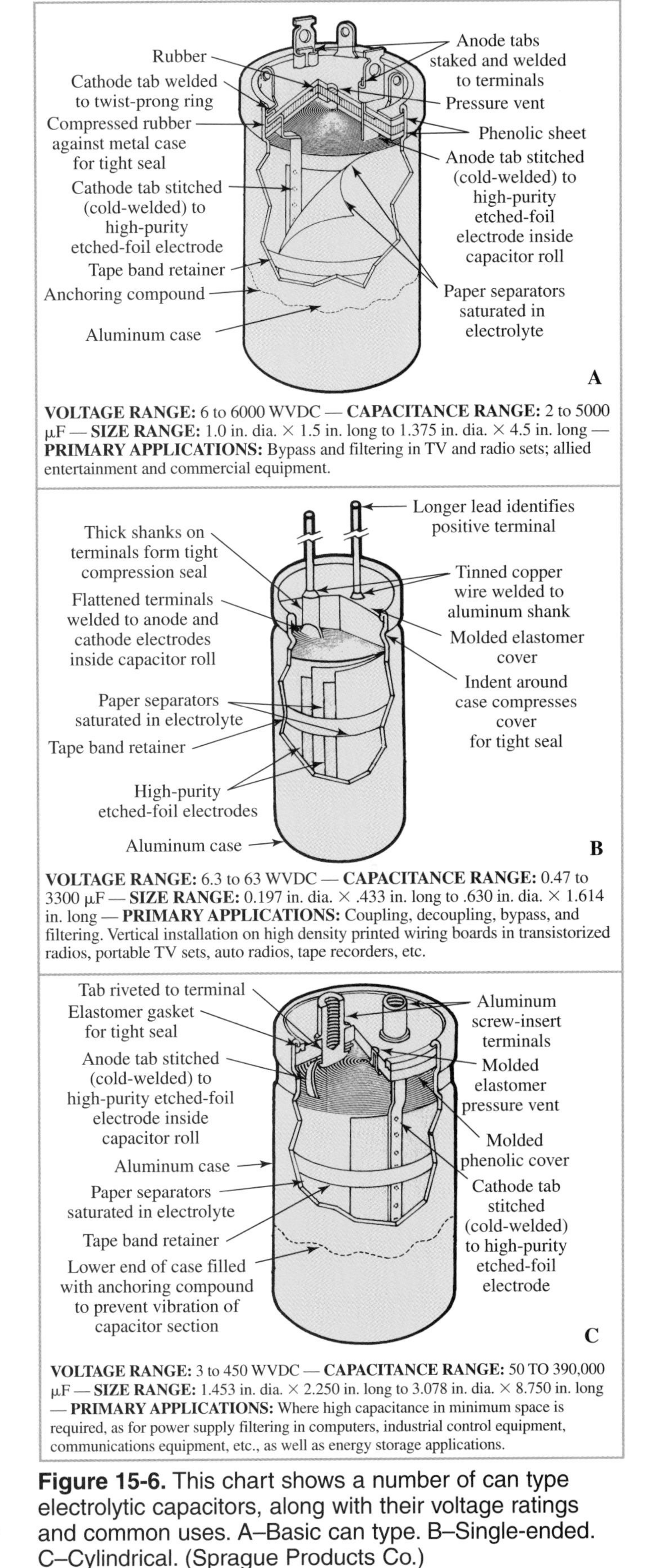

Figure 15-6. This chart shows a number of can type electrolytic capacitors, along with their voltage ratings and common uses. A–Basic can type. B–Single-ended. C–Cylindrical. (Sprague Products Co.)

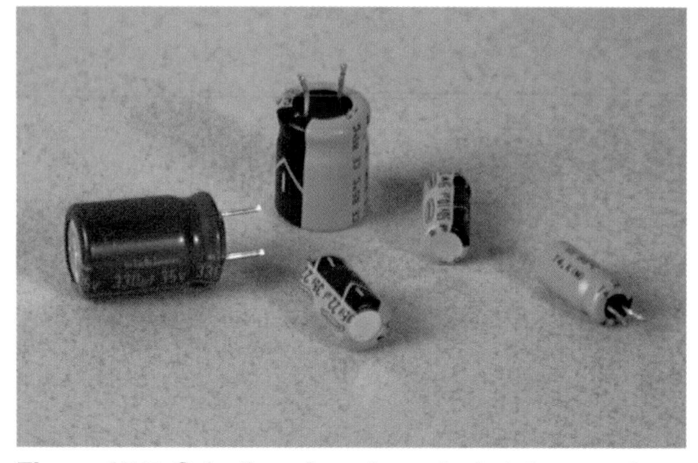

Figure 15-7. Selection of can type electrolytic capacitors.

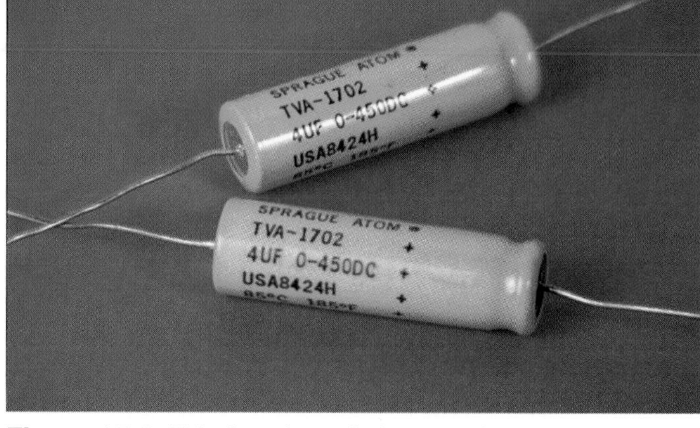

Figure 15-8. Tubular electorlytic capacitors.

Polarity of these capacitors is very important. A reverse connection can destroy them. The cans may contain from one to four different capacitors. The terminals are marked by △, ◠, □ and ▭ symbols. The metal can is usually the common negative terminal for all the capacitors. A special metal and fiber mounting plate is supplied for easy installation on a chassis.

Tubular electrolytic capacitor, **Figure 15-8,** construction is similar to the can type, **Figure 15-9.** The main advantage of these tubular capacitors is their smaller size.

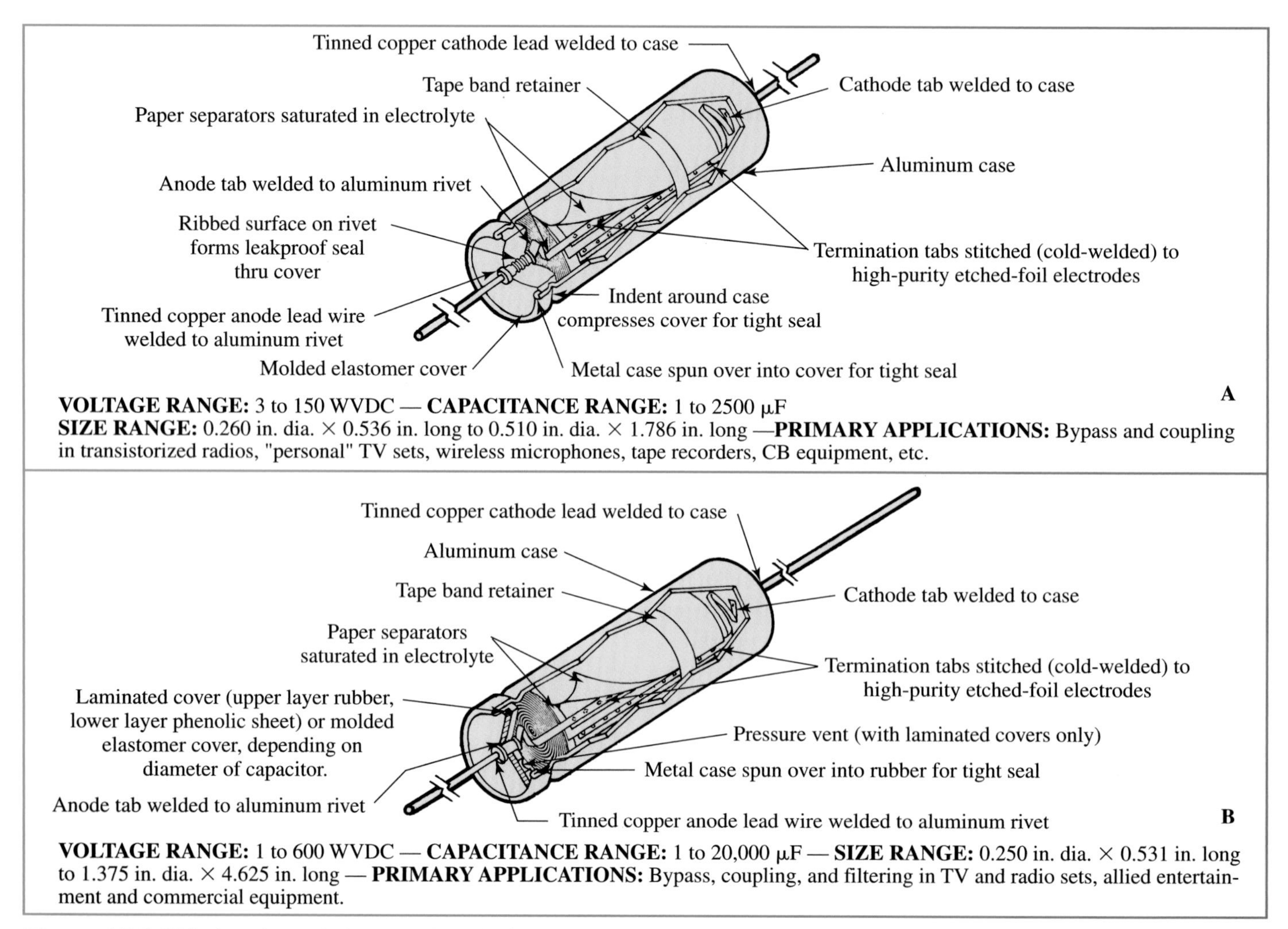

Figure 15-9. Tubular electrolytic capacitor. A–Standard tubular. B–Economy tubular. (Sprague Products Co.)

They have a metal case enclosed in an insulating tube. They are also made with two, three, or four units in one cylinder.

A very popular small capacitor used a great deal in radio and TV work is the ceramic capacitor, **Figures 15-10,** and **15-11.** The ***ceramic capacitor*** is made of a

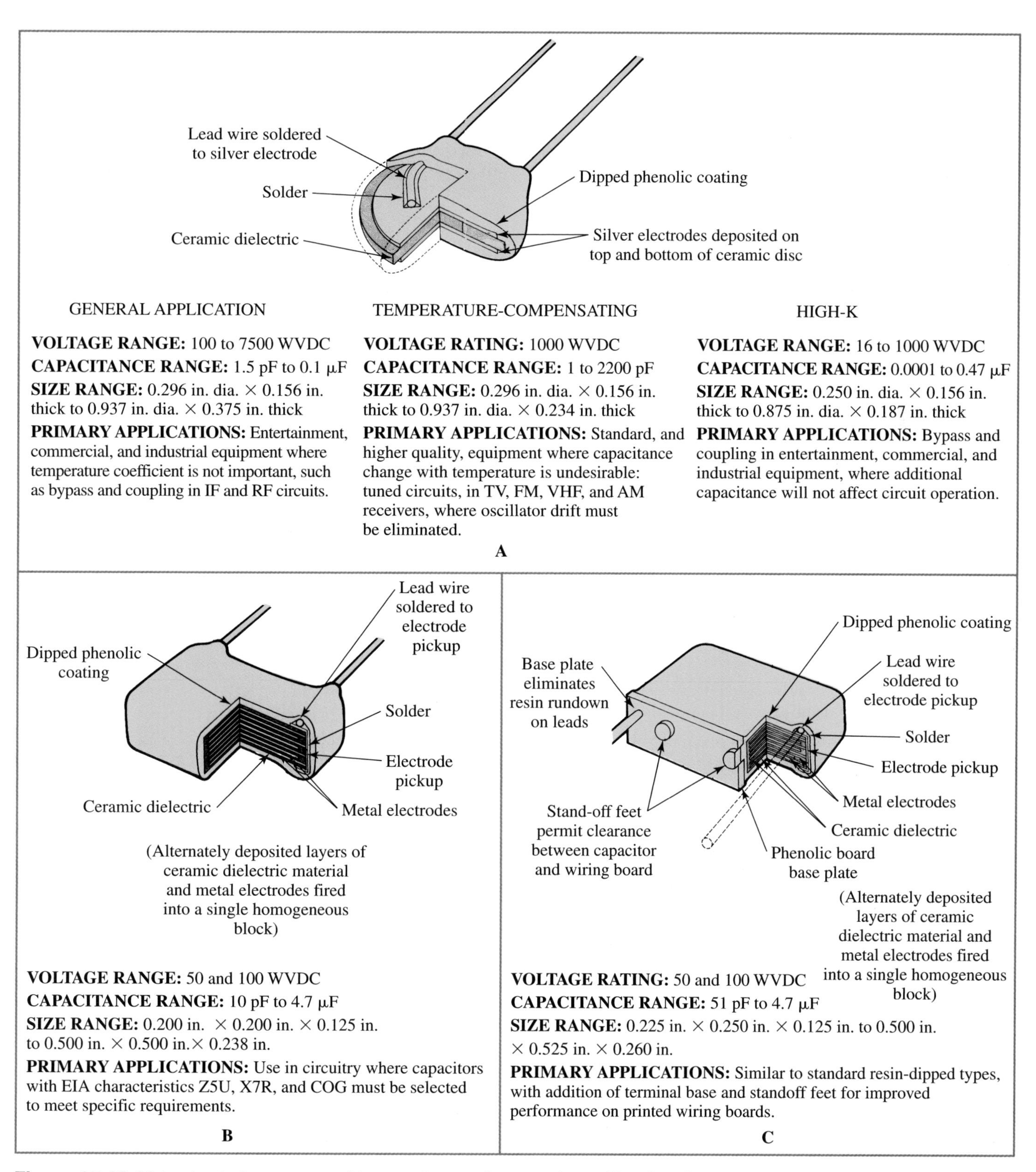

Figure 15-10. This chart shows several types of ceramic capacitors. Also listed are their voltage ratings and common uses. A–Disc type. B–Multilayer resin dipped. C–Multilayer terminal base. (Sprague Products Co.)

special ceramic dielectric. The silver plates of the capacitor are fixed on the dielectric. The entire component is treated with special insulation that can withstand heat and moisture.

Mica capacitors are small capacitors. They are made by stacking tinfoil plates together with thin sheets of mica as the dielectric. The assembly is then molded into a plastic case.

Figure 15-11. Typical ceramic capacitor.

Variable capacitors consist of metal plates that join together as the shaft turns, **Figure 15-12.** The stationary plate is called the ***stator.*** The rotating plate is called the ***rotor.***

When we adjust or turn the dial on a radio, we are actually adjusting a variable capacitor inside the radio. By changing the amount of capacitance inside the radio circuit, we are changing the radio frequency. This capacitor is at maximum capacity when the plates are fully meshed. The schematic symbol for a variable capacitor is shown in **Figure 15-13.**

A ***trimmer capacitor,*** **Figure 15-14,** is a type of variable capacitor. The adjustable screw compresses the plates and increases capacitance. Mica is used as a dielectric.

Trimmer capacitors are used where fine adjustments of capacitance are needed. They are used with larger capacitors and are connected in parallel with them. To adjust trimmer capacitors, turn the screw with a special fiber or plastic screwdriver called an ***alignment tool.*** A regular screwdriver should not be used for this purpose as the capacitance effect will cause an inaccurate adjustment.

Tantalum capacitors are similar to aluminum electrolytic capacitors, **Figure 15-15.** However, tantalum capacitors use tantalum, not aluminum, for the electrode.

Figure 15-12. Variable capacitors are made in many types and sizes. (Hammarlund Mfg. Co.)

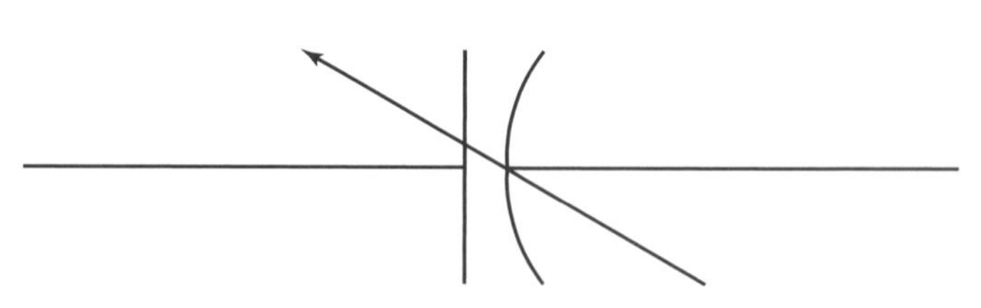

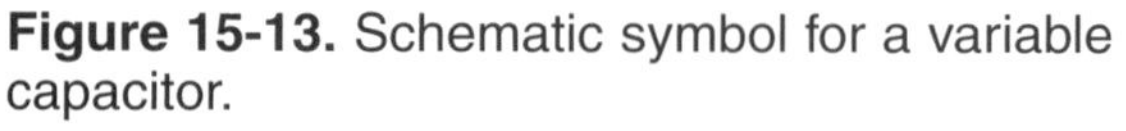

Figure 15-13. Schematic symbol for a variable capacitor.

Tantalum capacitors have three distinct advantages that make them quite useful.

- Tantalum capacitors have a larger capacitance over a smaller area, which makes them ideal for smaller circuits.
- Tantalum capacitors have a long shelf life.
- Tantalum resists most acids, consequently tantalum capacitors have less leakage current.

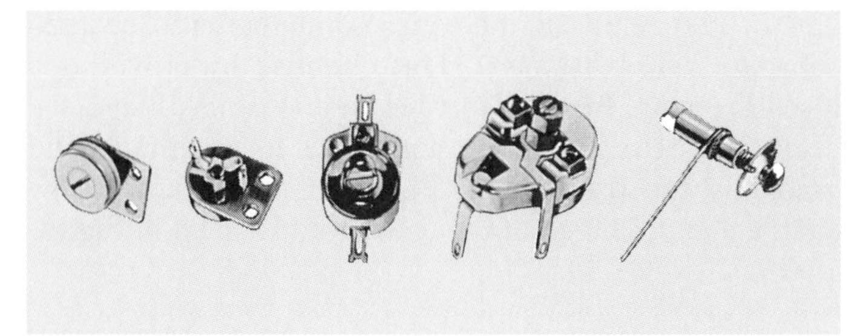

Figure 15-14. Several types of trimmer capacitors. (Centralab)

Review Questions for Section 15.1

1. A capacitor is made of two __________ of conductive material, separated by an insulator called the __________.
2. The ability of an insulator to support an electric field is known as the __________ __________.
3. Find the missing values for the following circuits.
   a. 20 volts, 1000 μF, __________ coulombs.
   b. 10 coulombs, 40 volts, __________ μF.
   c. 200 coulombs, 750 volts, __________ μF.
4. Convert the following units.
   a. 6240 pF = __________ μF.
   b. 0.05 μF = __________ pF.
   c. 150 μF = __________ F.
   d. 0.005 F = __________ μF.
5. Name the three items that determine capacitance.
6. List six types of capacitors.

15.2 TRANSIENT RESPONSE OF THE CAPACITOR

Recall that the response of current and voltage in a circuit immediately after a change in applied voltage is called the *transient response.*

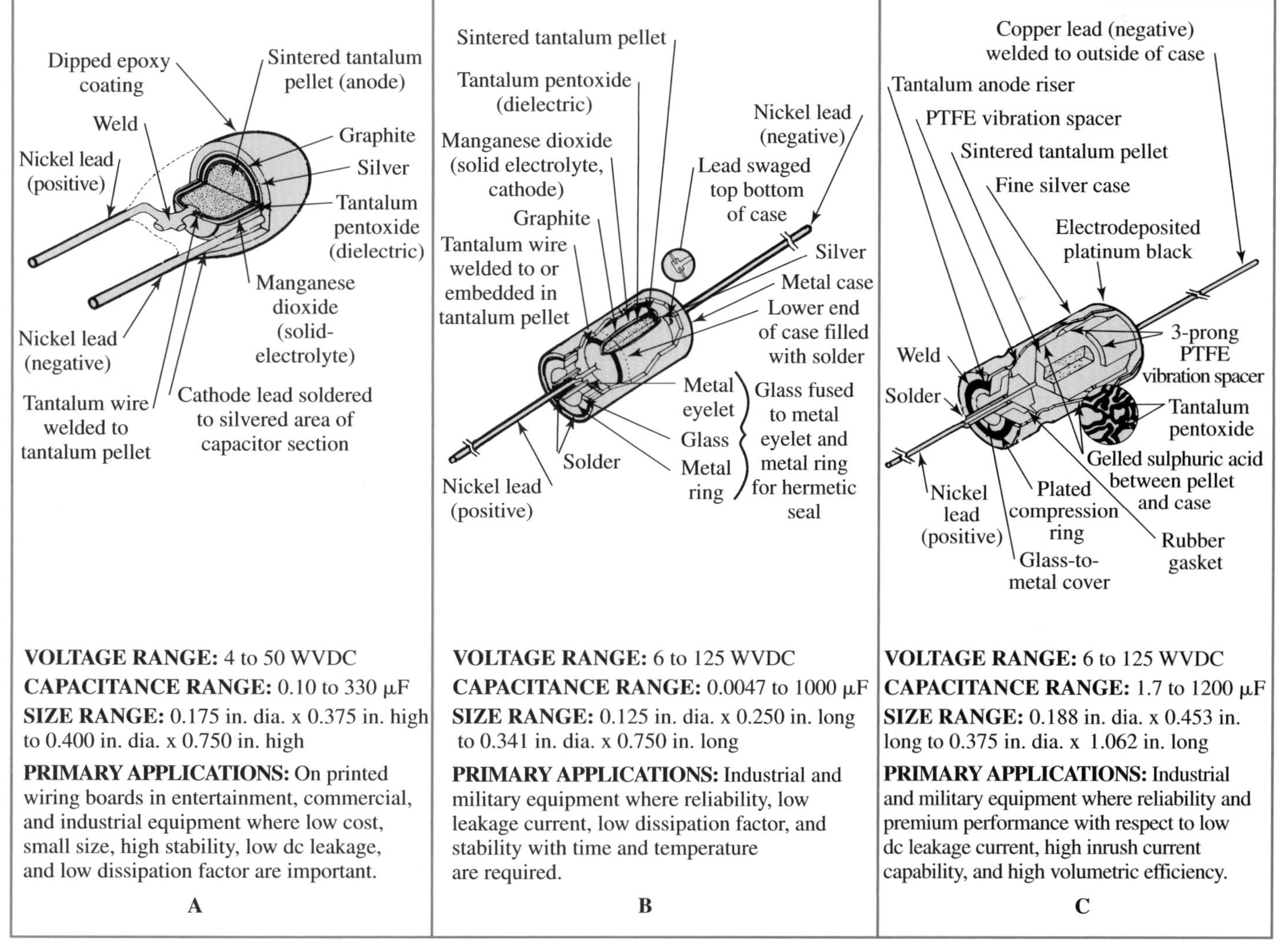

VOLTAGE RANGE: 4 to 50 WVDC
CAPACITANCE RANGE: 0.10 to 330 µF
SIZE RANGE: 0.175 in. dia. x 0.375 in. high to 0.400 in. dia. x 0.750 in. high
PRIMARY APPLICATIONS: On printed wiring boards in entertainment, commercial, and industrial equipment where low cost, small size, high stability, low dc leakage, and low dissipation factor are important.

A

VOLTAGE RANGE: 6 to 125 WVDC
CAPACITANCE RANGE: 0.0047 to 1000 µF
SIZE RANGE: 0.125 in. dia. x 0.250 in. long to 0.341 in. dia. x 0.750 in. long
PRIMARY APPLICATIONS: Industrial and military equipment where reliability, low leakage current, low dissipation factor, and stability with time and temperature are required.

B

VOLTAGE RANGE: 6 to 125 WVDC
CAPACITANCE RANGE: 1.7 to 1200 µF
SIZE RANGE: 0.188 in. dia. x 0.453 in. long to 0.375 in. dia. x 1.062 in. long
PRIMARY APPLICATIONS: Industrial and military equipment where reliability and premium performance with respect to low dc leakage current, high inrush current capability, and high volumetric efficiency.

C

Figure 15-15. This chart shows three types of tantalum capacitors. Voltage ratings and common uses are also listed. A–Epoxy dipped solid electrolyte. B–Hermetically sealed solid electrolyte. C–Hermetically sealed sintered-anode. (Sprague Products Co.)

Refer to **Figure 15-16.** A capacitor and a resistor are connected in series across a voltage source. A circuit that contains resistance and capacitance is called an ***RC circuit.*** When the switch is closed in this RC circuit, the maximum current will flow. The current gradually decreases until the capacitor has reached its full charge. The capacitor will charge to the level of the applied voltage.

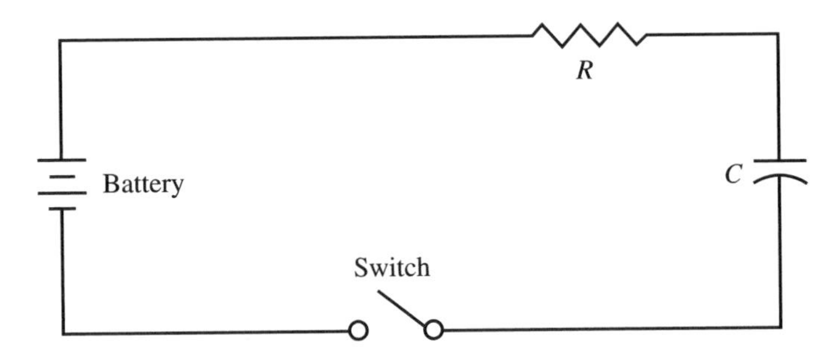

Figure 15-16. This series RC circuit demonstrates the transient response of a capacitor.

Initially, however, the voltage across the capacitor is zero. When the switch is closed, the voltage across the capacitor gradually builds up to the value of the source voltage. This charging of the capacitor is shown in **Figure 15-17.** The current in the RC circuit is also shown in this figure. Notice that when the switch is closed, the current rises to a maximum almost immediately. The current falls as the capacitor charges. When the capacitor reaches full charge, the current is zero.

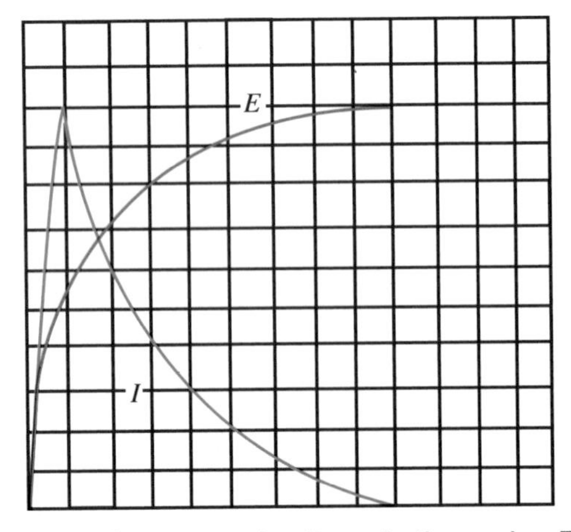

Figure 15-17. Current and voltage in the series RC circuit.

When the switch is opened, the capacitor remains charged. Theoretically, it would remain charged indefinitely, but there is always some leakage through the dielectric. After some period of time, the capacitor will discharge itself.

In **Figure 15-18,** the series combination of charged capacitor and resistor are short circuited by providing a discharge path. Because there is no opposing voltage, the discharge current will instantly rise to maximum and gradually fall off to zero. The combined graph of the charge and discharge of the capacitor is shown in **Figure 15-19.**

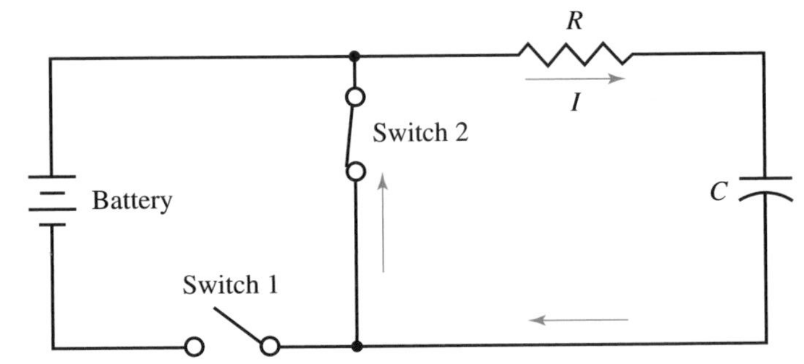

Figure 15-18. A short circuit occurs in the RC circuit when switch switch 2 is closed.

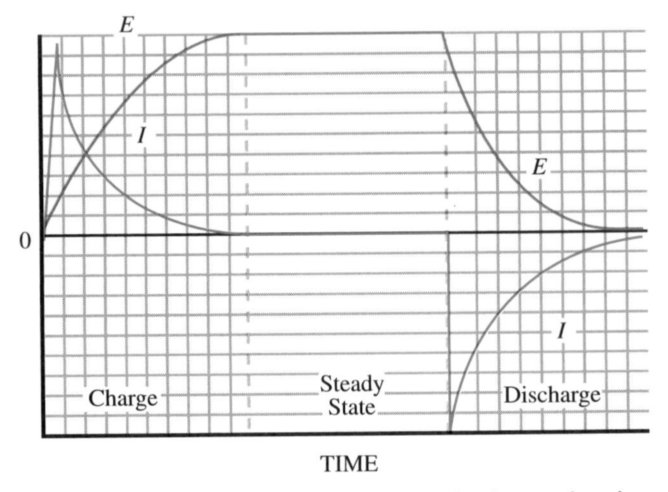

Figure 15-19. This combination graph shows the rise and decay of current and voltage in the series RC circuit.

Voltages appear across the resistor and capacitor in this circuit. The voltage across R is a result of the current, $E = IR$. Thus, the maximum voltage appears across R when maximum current is flowing. This condition exists immediately after the switch is closed in Figure 15-16, and after the discharge switch is closed in Figure 15-18. In both cases, the voltage across R drops off or decays as the capacitor approaches full charge or discharge. The graph of the voltage across R is drawn in **Figure 15-20.**

RC Time Constant

During the charge and discharge of the series RC network outlined above, a period of time elapsed. This

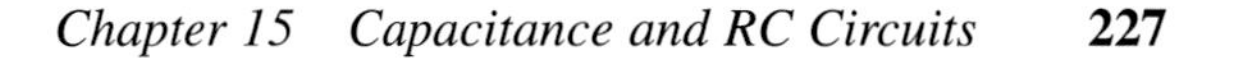

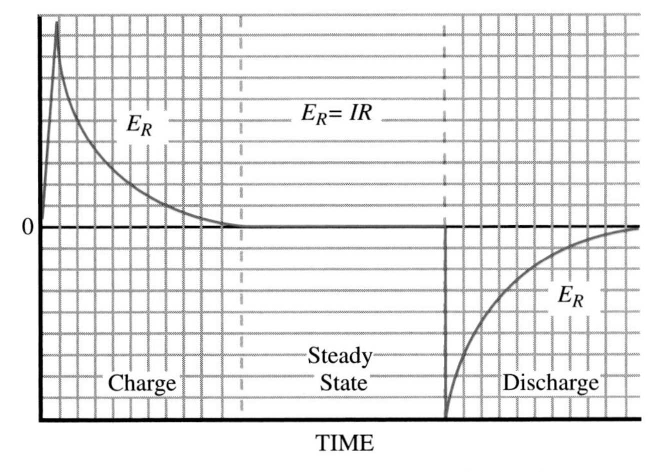

Figure 15-20. This graph shows the voltage drop across R as the capacitor is charged and discharged.

time is indicated along the base, or x-axis, of the graphs in Figure 15-19 and 15-20. The amount of time needed for the capacitor to charge or discharge 63.2 percent is known as the *time constant* of the circuit, as was discussed in Chapter 14. The formula to determine the time constant in RC circuits is:

$$\tau = R \times C$$

where τ is the time constant in seconds, R is the resistance in ohms, and C is the capacitance in farads. For complete charge or discharge, five time constant periods are required. Assuming a source voltage equal to 100 volts, **Figure 15-21** shows the time constant, percentage, and voltage.

Time constant	Percent of voltage	E charging	E discharging
1	63.2%	63.2 V	36.8 V
2	86.5%	86.5 V	13.5 V
3	95.0%	95.0 V	5 V
4	98.0%	98.0 V	2 V
5	99 +%	99 + V	1 V

Figure 15-21. A source voltage of 100 volts will create the time constant, percentage, and voltage shown.

Example: A 0.1 microfarad capacitor is connected in series with a one megohm (1,000,000 ohms) resistor across a 100 volt source. How much time will elapse during the charging of C? (Using powers of ten is explained in the Appendix.)

$$\tau = (0.1 \times 10^{-6}) \times (10^{6}) = 0.1 \text{ s}$$

One tenth of a second is the time constant. This means that C would charge to 63.2 volts during 0.1 seconds. At the end of five time constants, or $5 \times 0.1 = 0.5$ seconds, the capacitor is fully charged.

Set up the circuit in **Figure 15-22.** This neon lamp will remain unlit until a certain voltage level is reached. At the required voltage, called the *ignition,* or *firing voltage,* the lamp glows and offers little resistance in the circuit.

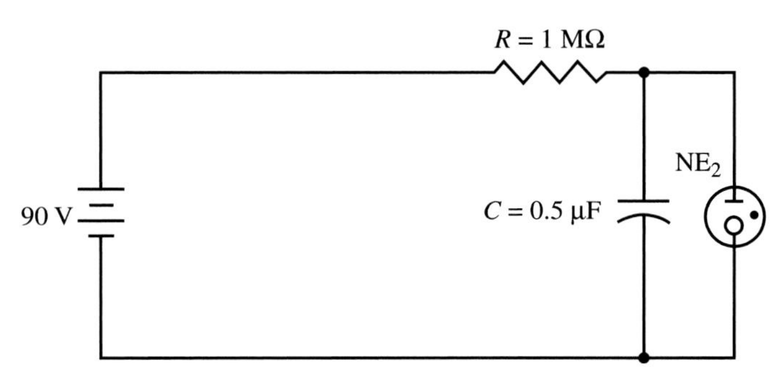

Figure 15-22. This flashing circuit is called a relaxation oscillator.

In this circuit, capacitor C charges through resistance R. When the voltage across C develops to the ignition voltage, the lamp glows. C is quickly discharged. This cycle repeats over and over, causing the neon light to flash. The frequency of the flashes can be changed by varying either the value of C or R. A telephone with a neon accent is shown in **Figure 15-23.**

Time constants have many uses in electronic circuits. Timing circuits are used in industry to control the sequence and duration of machine operations. A photographic enlarger uses a time delay circuit to control exposure time.

Figure 15-23. Neon lights are used in many new products. Here, a neon light highlights this designer phone. (Telean Technology Ltd.)

PROJECT 15-1: Flashing LED RC Circuit

Try building an RC blinker. This project is fairly easy to construct. See the schematic in **Exhibit 15-1A.**

Use the parts list to the right of Exhibit 15-1A. When you have a completed RC circuit, you can build an interesting case that covers all but the flashing LEDs.

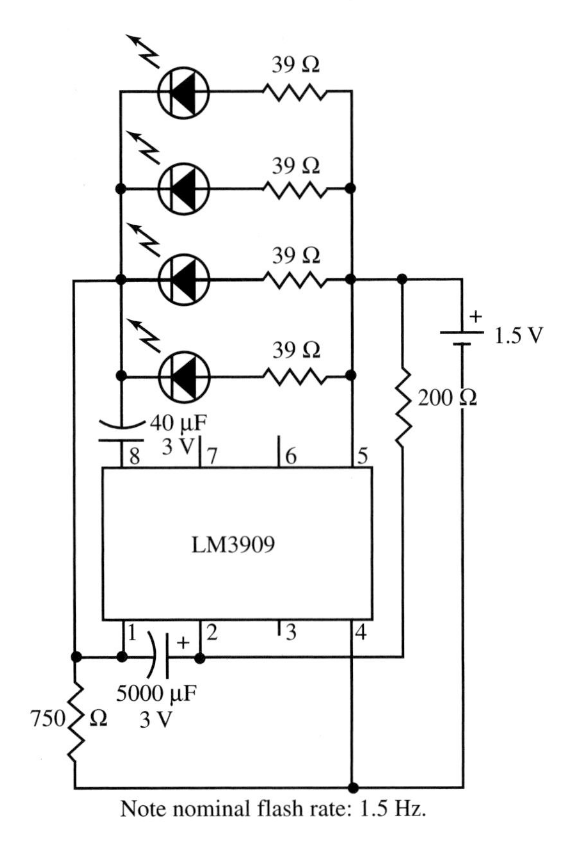

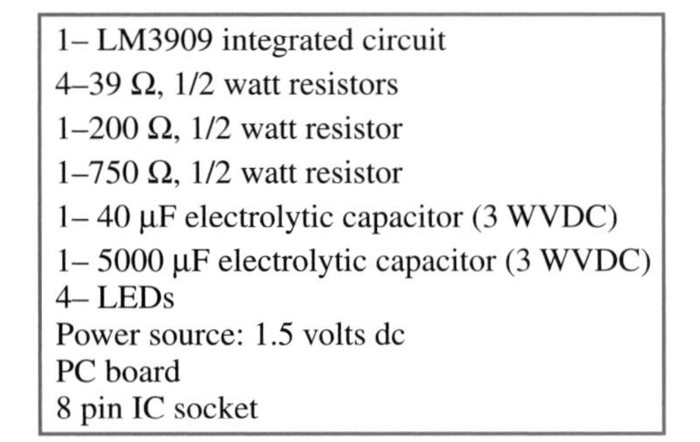

1– LM3909 integrated circuit
4–39 Ω, 1/2 watt resistors
1–200 Ω, 1/2 watt resistor
1–750 Ω, 1/2 watt resistor
1– 40 µF electrolytic capacitor (3 WVDC)
1– 5000 µF electrolytic capacitor (3 WVDC)
4– LEDs
Power source: 1.5 volts dc
PC board
8 pin IC socket

Exhibit 15-1A.

Capacitors in Series and Parallel

When two capacitors are connected in series, **Figure 15-24,** the total capacitance is:

$$C_T = \frac{C_1 \times C_2}{C_1 + C_2}$$

C_1 C_2

$$C_T = \frac{C_1 \times C_2}{C_1 + C_2}$$

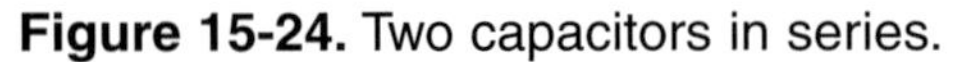

Figure 15-24. Two capacitors in series.

When two or more capacitors are connected in series, **Figure 15-25,** the total capacitance is:

$$\frac{1}{C_T} = \frac{1}{C_1} + \frac{1}{C_2} + \frac{1}{C_3} \ldots + \frac{1}{C_N}$$

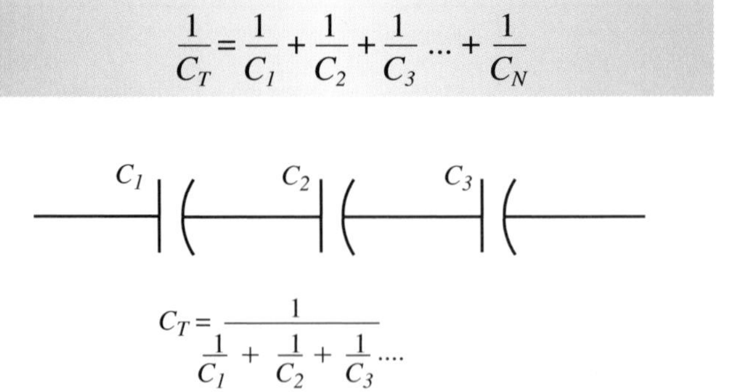

$$C_T = \frac{1}{\frac{1}{C_1} + \frac{1}{C_2} + \frac{1}{C_3} \ldots}$$

Figure 15-25. Two or more capacitors in series.

When capacitors are connected in parallel, **Figure 15-26,** the total capacitance is equal to the sum of the individual capacitances.

$$C_T = C_1 + C_2 + C_3 \ldots + C_N$$

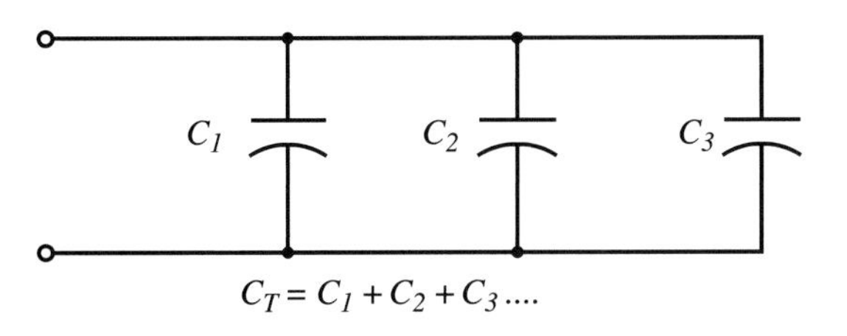

Figure 15-26. Capacitors in parallel.

Review Questions for Section 15.2

1. Can a capacitor that has a working dc voltage of 25 volts be used in a circuit that requires a working dc voltage of 10 volts? Explain your answer.
2. Give the total capacitance for the circuits below.

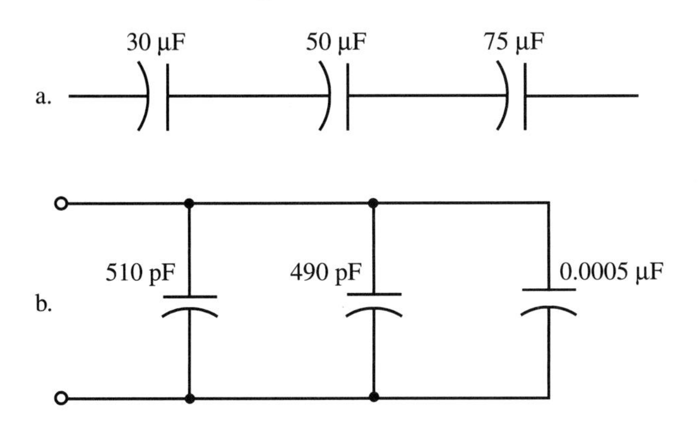

3. Find the time constants for the following circuits.
   a. 500 µF capacitor, 2500 ohms resistance.
   b. 225 µF capacitor, 3200 ohms resistance.
   c. 67 µF capacitor, 180 ohms resistance.

15.3 CAPACITANCE IN AC CIRCUITS

When an ac voltage is applied to a capacitor, the plates charge and discharge repeatedly. During the first half-cycle the plates charge up (one plate negative and one plate positive) and discharge back to zero. During the next half-cycle, the plates charge to the opposite polarities of the first half cycle and then discharge back to zero. An ac meter in the circuit shows a current flowing at all times.

To demonstrate this fact, connect a light and capacitor in series with a six-volt dc source, **Figure 15-27.** Does the light glow? Now connect the same circuit with a six-volt ac source. Notice that the light burns dimly. This experiment demonstrates that alternating current is flowing as a result of the alternate charging and discharging of the capacitor.

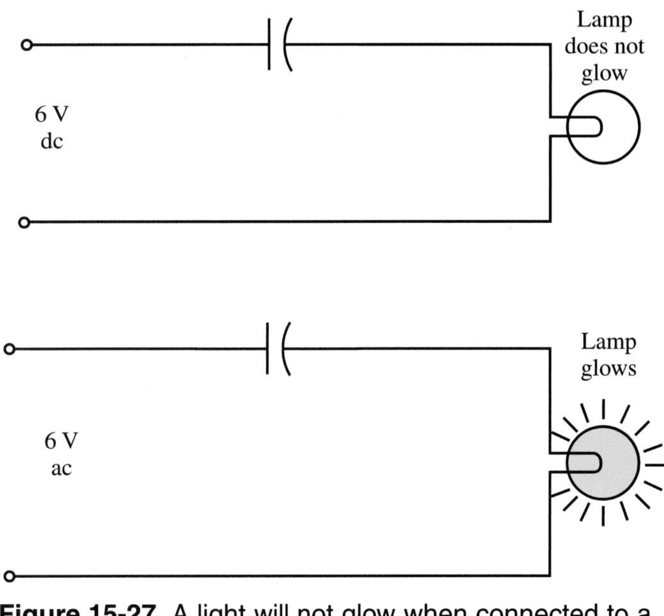

Figure 15-27. A light will not glow when connected to a dc source. The capacitor blocks direct current. When connected to an ac source, the light glows.

To review, refer to **Figure 15-28.** As the ac voltage starts to rise, current is at maximum because the capacitor, C, is in a discharged state. As C becomes charged to the peak ac voltage, the charging current drops to zero (point A). As the voltage begins to drop, the discharging current begins to rise in a negative direction. It reaches a maximum at the point of zero voltage (point B). This phase difference keeps going throughout each cycle. In a purely capacitive circuit, the current *leads* the voltage by an angle of 90 degrees.

The size of the current in the circuit depends upon the size of the capacitor. Larger capacitors (more capacitance) require a larger current to charge them. The frequency of the ac voltage also affects the current.

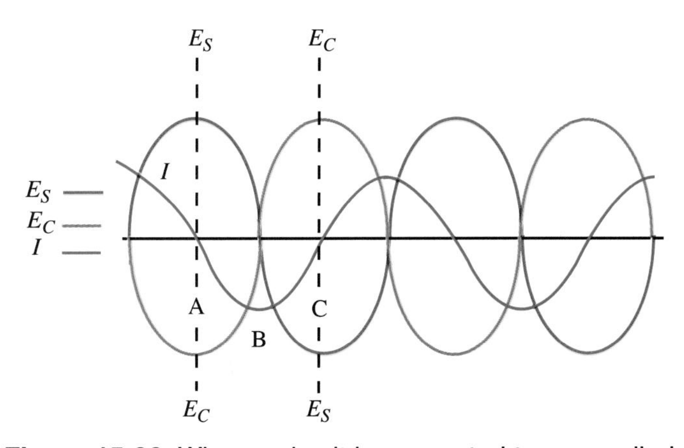

Figure 15-28. When a circuit is connected to ac, applied voltage, current, and voltage across C appear as shown.

The current depends upon the rate of charge and discharge of the capacitor. As the frequency of the ac is increased, current increases. These links are stated in the formula:

$$X_C = \frac{1}{2\pi fC}$$

where X_C equals the capacitive reactance in ohms, f equals the frequency in hertz, and C equals the capacitance in farads.

Like inductors, capacitors produce resistance to the flow of an alternating current. This resistance resulting from capacitance is called ***capacitive reactance.*** Capacitive reactance is measured in ohms, like dc resistance. As you can see from the formula:

- As the frequency increases, X_C decreases.
- As capacitance increases, X_C decreases.

Example: What is the reactance of a 10 μF capacitor working in a circuit at a frequency of 120 hertz?

$$X_C = \frac{1}{2\pi fC}$$

$$X_C = \frac{1}{2\pi \times (120\ \text{Hz}) \times (10 \times 10^{-6}\ \text{F})}$$

$$X_C = \frac{1}{6.28 \times (120\ \text{Hz}) \times (10 \times 10^{-6}\ \text{F})}$$

$$X_C = \frac{1}{0.007536} = 132.7\ \Omega$$

The reactance of a 0.1 μF capacitor as the frequency is varied can be seen in **Figure 15-29.** As frequency is changed to 50, 100, 1000, and 5000 Hz, each reactance is computed using the formula for capacitive reactance. Notice how the reactance on the graph approaches zero as the frequency heads toward infinity.

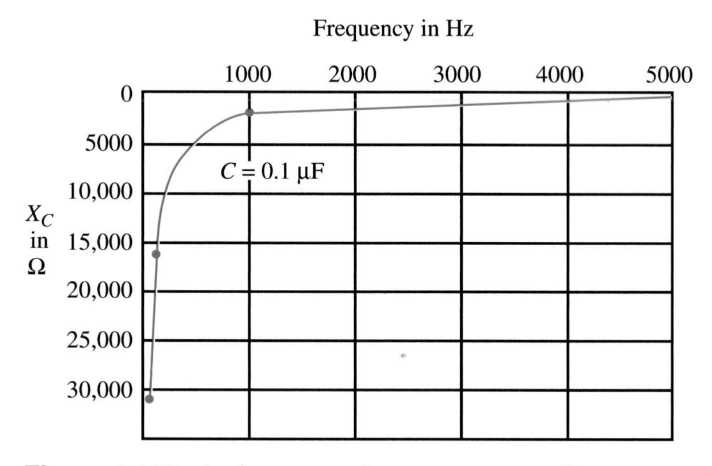

Figure 15-29. As frequency increases, capacitive reactance decreases.

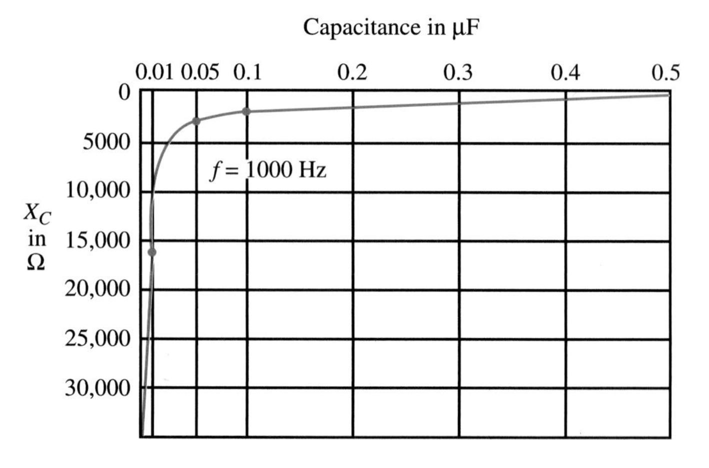

Figure 15-30. As capacitance increases, reactance decreases.

In **Figure 15-30,** the frequency is held constant at 1000 Hz. The reactance is plotted for capacitors of 0.01 μF, 0.05 μF, 0.1 μF, and 0.5 μF. These are common capacitor sizes used in electronic work. They are used in filtering, coupling, and bypassing networks.

Power in Capacitive Circuits

When a capacitor is discharged, the energy stored in the dielectric is returned to the circuit. This action is similar to the action of an inductor, which returns the energy stored in a magnetic field to the circuit. In both cases, electrical energy is used temporarily by the reactive circuit. This power in a capacitive circuit is also called *wattless power.*

In **Figure 15-31,** the voltage and current waveforms are drawn for a circuit containing pure capacitance. The power waveform results from plotting the products of the instantaneous voltage and current at selected points. The power waveform shows that equal amounts of positive power and negative power are used by the circuit. This condition results in zero power being used. The *true power,* or actual power used, then, is zero.

The *apparent power* is equal to the product of the effective voltage and the effective current. Look at **Figure 15-32.** An applied ac voltage to the capacitive circuit causes a 10 ampere current. The apparent power will equal:

Apparent power = $E_{eff} \times I_{eff}$

Apparent power = 100 V × 10 A = 1000 VA

As discussed with inductance, the ratio of true power to apparent power in an ac circuit is called the *power*

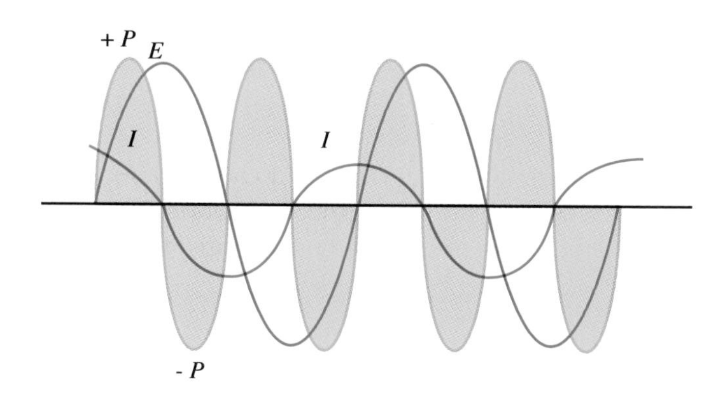

Figure 15-31. These waveforms show current, voltage, and power in a purely capacitive circuit.

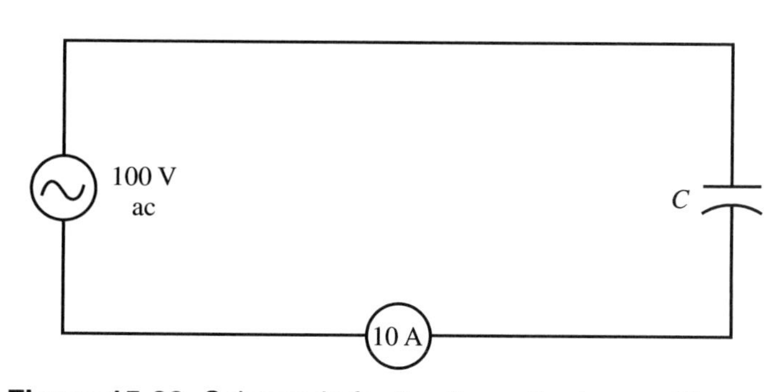

Figure 15-32. Schematic for the theoretical capacitive circuit.

factor (PF). It is found using trigonometry. It is the cosine of the phase angle between the current and voltage. Recall from Chapter 14:

$$\text{Power factor} = \cos\theta = \frac{\text{True power}}{\text{Apparent power}}$$

Example: Assuming our previous circuit is purely capacitive, determine the power factor and phase displacement.

$$\text{PF} = \cos\theta = \frac{0}{1000} = 0$$

The angle whose cosine is 0 is 90 degrees. This tells us that current and voltage in the purely capacitive circuit are 90 degrees out of phase.

Resistance and Capacitance in an AC Circuit

When resistance is present in a circuit, power is used. If a circuit contains only resistance, then the voltage and current are in phase. There is no phase angle θ, and the power factor is one ($\cos 0° = 1$). The apparent power equals the true power.

These circuit traits change when capacitance is added in series with the resistor. Capacitive reactance is also a force which resists the flow of an alternating current. Because the capacitive reactance causes a 90 degree phase displacement, the total resistance to an ac current must be the vector sum of X_C and R. These vectors are drawn in **Figure 15-33.** Assuming the circuit has a resistance of 300 ohms and a capacitive reactance of 400 ohms, the resulting opposition to current is 500 ohms. This opposition, like that generated with the RL circuit, is called the *impedance (Z),* of the circuit.

$$Z = \sqrt{R^2 + X_C^2}$$

$$Z = \sqrt{(300\ \Omega)^2 + (400\ \Omega)^2}$$

$$Z = \sqrt{(90{,}000\ \Omega) + (160{,}000\ \Omega)}$$

$$Z = 500\ \Omega$$

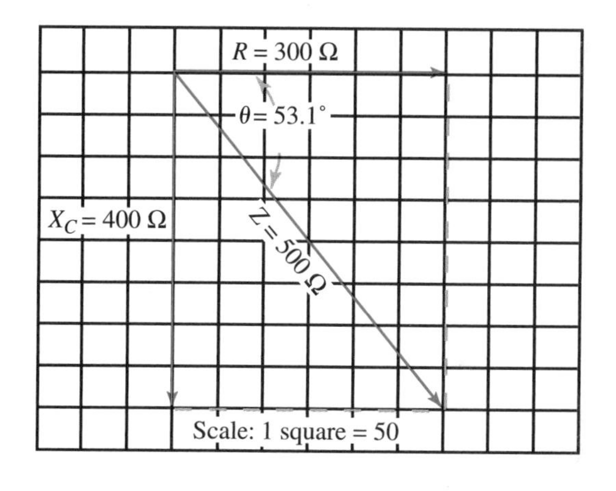

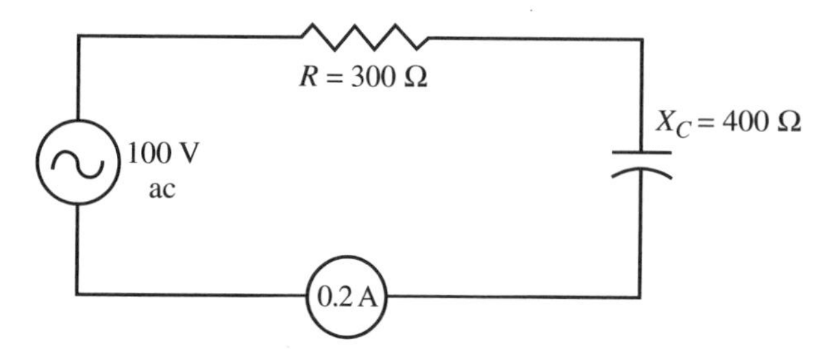

Figure 15-33. Top. The vector relationship between R and X_C and the resulting vector, Z, for impedance. Bottom. The circuit for the problem in the text.

The angle between vector Z and vector R represents the phase displacement between the current and the voltage as a result of the reactive component. This is angle θ. Since cosine θ is equal to the power factor, we can calculate θ using:

$$\cos\theta = \text{PF} = \frac{R}{Z}$$

$$\cos\theta = \frac{300\ \Omega}{500\ \Omega} = 0.6$$

An angle with a cosine of 0.6 is 53.1 degrees (approximately). Thus, the current leads the voltage by an angle of 53.1 degrees.

The true power in this circuit equals:

$$\text{True power} = \text{Apparent power} \times \cos\theta$$

The power factor is also the relationship between the true power and the apparent power. Since $\cos\theta$ equals the power factor, it follows that:

$$\text{True power} = \text{Apparent power} \times \text{Power factor}$$

or

$$\text{PF} = \frac{\text{True power}}{\text{Apparent power}}$$

Using Ohm's law for ac circuits, the current flowing in Figure 15-33 (100 volts ac applied) is equal to:

$$I = \frac{E}{Z} = \frac{100\ \text{V}}{500\ \Omega} = 0.2\ \text{A}$$

The apparent power is then:

$$\text{Apparent power} = I \times E = 0.2\ \text{A} \times 100\ \text{V} = 20\ \text{VA}$$

True power equals:

$$\text{True power} = I \times E \times \cos\theta$$

$$\text{True power} = 0.2\ \text{A} \times 100\ \text{V} \times \cos 53.1°$$

$$\text{True power} = 20\ \text{W} \times 0.6 = 12\ \text{W}$$

These values can be proved correct using this equation:

$$\text{PF} = \frac{\text{True power}}{\text{Apparent power}} = \frac{12\ \text{W}}{20\ \text{VA}} = 0.6$$

Parallel RC Circuit

In a circuit containing a resistance and a capacitance in parallel, the voltage of each circuit element will be the same as the source voltage. Further, there will be no phase difference among the voltages. This is because they are all in parallel.

There will, however, be a phase difference among the total and branch currents. The current is in phase with voltage in the resistive branch. The current leads the voltage across the capacitor by 90 degrees. The total current leads the source voltage by some angle between 0 and 90 degrees. If the reactance of the capacitor is larger than the resistance of the resistor, θ will be closer to 0 degrees. If the resistance is much larger than the reactance θ will be closer to 90 degrees.

In the parallel RC circuit, we do not find impedance through a vector sum of circuit resistances. Instead, we apply Ohm's law after finding the sum of the branch currents.

$$Z = \frac{E_S}{I_T}$$

As stated, voltage in the parallel RC circuit is in phase with I_R, and it lags I_C by 90 degrees. Thus, we can say that I_R lags I_C by 90 degrees. The parallel RC circuit, then, has two current components—I_R and I_C. Both of these can be represented by phasors. Since they are out of phase, we cannot simply add the two components together to figure the total circuit current. We must find their phasor sum.

$$I_T = \sqrt{I_R^2 + I_C^2}$$

Note that the phasor for I_C is above the horizontal reference. This is because I_C *leads* voltage—the horizontal reference for the parallel RC circuit.

For the parallel RC circuit, phase angle is found on the current phasor diagram. The horizontal reference of this circuit is voltage, since it is common to all circuit elements. On the current phasor diagram, the horizontal component is I_R since it is in phase with voltage. The phase angle, then, is the angle between I_R and the total current. This is the phase displacement resulting from the reactive element. In the parallel RC circuit, phase angle is:

$$\theta = \arctan\frac{I_C}{I_R}$$

Example: Using the circuit with assigned values in **Figure 15-34,** determine the phase angle between the applied voltage and current. Draw the current phasor diagram and find the circuit impedance.

Step 1. Compute the value of capacitive reactance, X_C.

$$X_C = \frac{1}{2\pi fC}$$

$$X_C = \frac{0.159}{fC} = \frac{0.159}{(1 \times 10^3\ \text{Hz})(2 \times 10^{-6}\ \text{F})} \cong 80\ \Omega$$

Step 2. Compute the branch currents. Using Ohm's law:

PROJECT 15-2: Building a Tunable Electronic Organ

This lab is an example of a two-stage amplifier connected as an astable mutivibrator oscillator circuit. You will learn about amplifiers and oscillator circuits later in the textbook. For now, just remember that an oscillator circuit causes a repetitive ac cycle. The repetitive ac cycle will cause a tone to be generated out of the speaker.

The part of the circuit that correlates to this chapter is the RC circuit. Look at the bank of variable resistors connected to a set of seven momentary push buttons. Each of the variable resistors is set to a different resistance value. When any push button is pressed, the corresponding variable resistor forms an RC circuit with capacitor C1. Since each of the variable resistors are set to a different resistance value, a different time constant is created for each push button. The various time constants each create a different tone in the speaker.

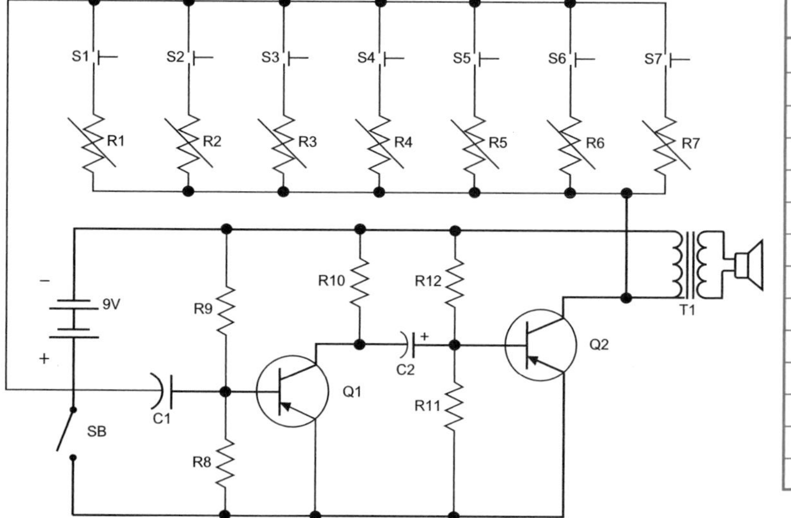

Greymark Parts List

Qty.	Part No.	Description
7	63558	R_1-R_7 Potentiometer, 25kΩ
1	61426	R_8, Resistor, 10kΩ
1	61374	R_9, Resitor, 100kΩ
1	61429	R_{10}, Resistor, 2.2kΩ
1	61853	R_{11}, Resistor, 15kΩ
1	61385	R_{12}, Resistor, 220kΩ
1	61231	C_1 Capacitor, disc. .022µF
1	62238	C_2, Capacitor, electrolytic, 10µF
2	62811	Q_1, Q_2, Transistor, PNP, 2N3906
1	62366	T_1, Transformer, Audio
7	62512	S_1-S_7, Switch, Push button, NO
1	61385	S_8, Switch, Slide, SPDT
1	62403	Battery Clip
1	61051	Speaker

Graymark

The electronic components and printed circuit board can be obtained from Graymark, Tunable Electronic Organ Model No. 117.

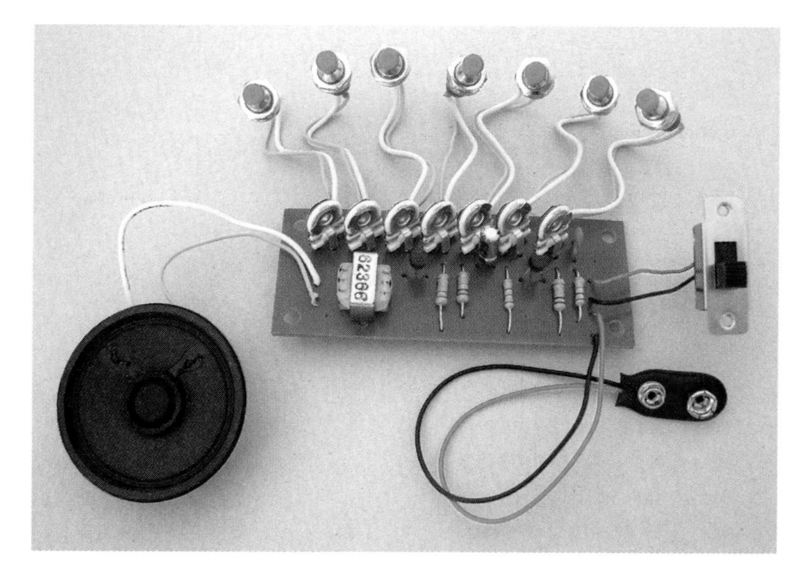

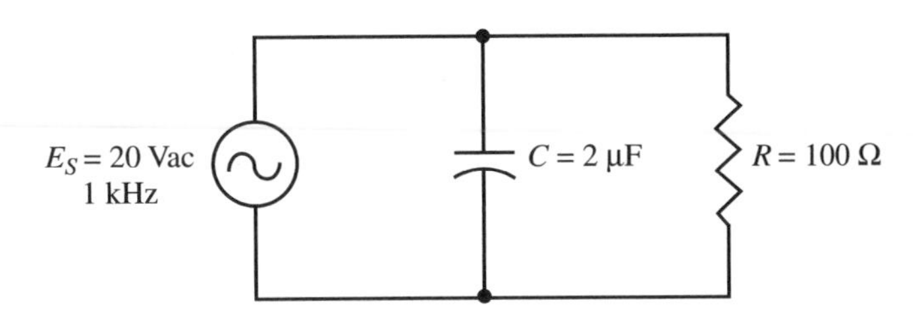

Figure 15-34. A parallel RC circuit.

$$I_R = \frac{E_S}{R} = \frac{20\text{ V}}{100\ \Omega} = 0.2\text{ A}$$

$$I_C = \frac{E_S}{X_C} = \frac{20\text{ V}}{80\ \Omega} = 0.25\text{ A}$$

Step 3. Determine the phase angle to see by how much the circuit current leads the voltage.

$$\theta = \arctan\frac{I_C}{I_R} = \arctan\frac{0.25\text{ A}}{0.2\text{ A}} = \arctan 1.25 = 51.3°$$

Step 4. Draw the current phasor diagram. Use any convenient scale. Remember that I_R is drawn as the horizontal component. I_C is drawn upward at 90 degrees from I_R since it leads voltage—the horizontal reference. See **Figure 15-35.**

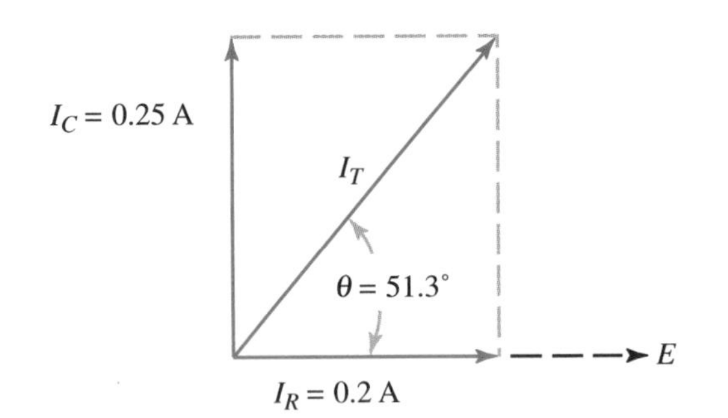

Figure 15-35. Current phasor diagram.

Step 5. Find the total circuit current.

$$I_T = \sqrt{I_R^2 + I_C^2} = \sqrt{(0.2\text{ A})^2 + (0.25\text{ A})^2} = 0.32\text{ A}$$

Step 6. Find the impedance of the circuit. Using Ohm's law:

$$Z = \frac{E_S}{I_T} = \frac{20\text{ V}}{0.32\text{ A}} = 62.5\ \Omega$$

All circuits have some combination of the three electrical properties: *R*, *L*, and *C*. These properties come in many arrangements, including R, RL, and RC networks. In Chapter 16, RCL circuits will be studied. Always review previous lessons if the theory covered is not clearly understood. The study of electricity and electronics requires a firm grasp of all previous lessons before progressing to more complex uses.

Review Questions for Section 15.3

1. Using the list of words given, fill in the spaces to make the statement true concerning the equation for capacitive reactance.
   - farads
   - X_C
   - C
   - hertz
   - f

   In the equation $X_C = \frac{1}{2\pi fC}$, the capacitive reactance is ____________, ____________ equals capacitance in ____________, and ____________ equals the frequency in ____________.

2. Define capacitive reactance.
3. Find X_C in the circuit shown.

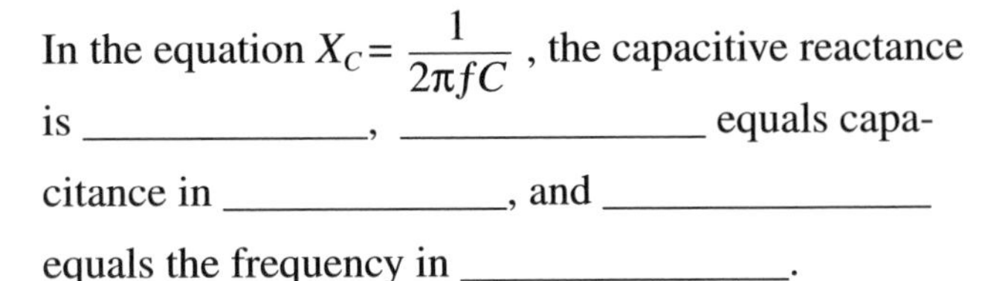

4. Find the following values using the circuit shown.
   a. *Z*.
   b. Cos *θ*.
   c. *I* ________ (leads, lags) *E* by ________°.

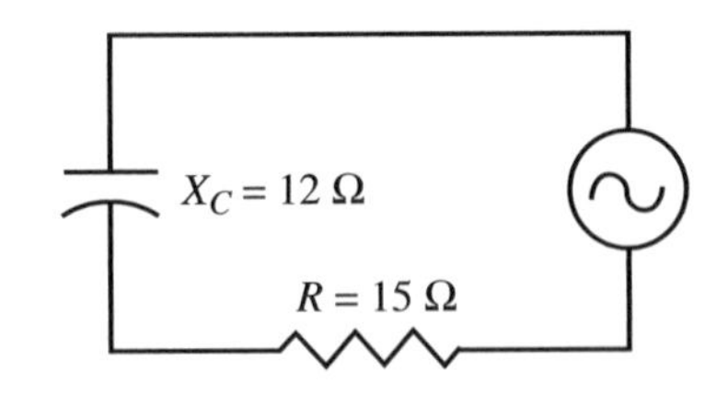

Summary

1. Capacitance is that property of a circuit that opposes any change in voltage.
2. A capacitor is a device that temporarily stores an electric charge. It is made up of two plates of conductive material separated by insulation, called the dielectric.

3. Capacitance is measured in farads. Commonly used units are the microfarad (μF) and picofarad (pF).
4. Factors affecting capacitance are:
   a. Distance between plates.
   b. Plate area.
   c. Dielectric material.
5. There are many types of fixed and variable capacitors.
6. The RC time constant can be found using the equation: τ (in seconds) = R (in ohms) × C (in farads)
7. The formulas for capacitors in series and parallel are:

   Capacitors in series: $\frac{1}{C_T} = \frac{1}{C_1} + \frac{1}{C_2} + \frac{1}{C_3} \ldots + \frac{1}{C_N}$

   or a shortcut when only two capacitors are in series:

$$C_T = \frac{C_1 \times C_2}{C_1 + C_2}$$

   Capacitors in parallel: $C_T = C_1 + C_2 + C_3 \ldots + C_N$
8. Working voltage is the maximum voltage that can be steadily applied to a capacitor without creating an arc.
9. Capacitive reactance, X_C, is opposition to the flow of ac resulting from capacitance.
10. The formula to find capacitive reactance is:

$$X_C = \frac{1}{2\pi f C}$$

11. True power can be found with the formula:

$$\text{True power} = I \times E \times \cos\theta$$

12. Apparent power can be found with the formula:

$$\text{Apparent power} = I \times E$$

13. Power factor can be found with the formula:

$$\text{Power factor} = \frac{\text{True power}}{\text{Apparent power}}$$

14. The impedance in a series RC circuit can be found using the formula:

$$Z = \sqrt{R^2 + X_C^2}$$

15. The phase angle can be found in an RC circuit using the formula:

$$\cos\theta = \frac{R}{Z}$$

16. When a resistor and a capacitor are in parallel, the branch voltages are the same. However, the branch currents will be 90 degrees out of phase with each other.

Test Your Knowledge

Please do not write in the text. Place your answers on a separate sheet of paper.

1. Define capacitance.
2. What safety precautions should be followed when working on radio and TV capacitors?
3. With respect to dielectric constants, the larger the number the __________ (better, worse) it supports an electric field.
4. Capacitance is determined by:
   a. dielectric material.
   b. plate area.
   c. distance between plates.
   d. All of the above.
5. The time constant of an RC circuit is equal to __________ × __________.
6. Determine total capacitance for:
   a. Two capacitors connected in series with values of 680 μF and 200 μF.
   b. Four capacitors connected in parallel with values of 25 μF, 60 μF, 2 μF, and 4 μF.
   c. Five capacitors connected in parallel with values of 1150 pF, 97 pF, 130 pF, 1240 pF, and 50 pF.
7. The symbol for capacitive reactance is __________. It is measured in __________ and can be found using the formula __________.
8. A capacitor has a value of 0.1 μF. Find its reactance value at:
   a. 50 Hz.
   b. 100 Hz.
   c. 1000 Hz.
   d. 10 Hz.
9. As capacitance increases, X_C __________. As frequency increases, X_C __________.

Matching Questions

Match each of the following terms with their correct definitions.

10. $\frac{R}{Z}$
11. $I \times E$
12. $I \times E \times \cos \theta$
13. $\sqrt{R^2 + X_C^{\,2}}$
14. $\frac{\text{True power}}{\text{Apparent power}}$

a. Apparent power.

b. Power factor.

c. True power.

d. Cos θ.

e. Impedance.

15. A series circuit has 300 ohms resistance and 0.1 µF capacitance. It is connected to a 50-volt, 400-hertz source. Find:
    a. X_C
    b. Z
    c. E_{X_C}
    d. E_R
    e. I
    f. θ
16. The frequency of the current in a circuit is 1000 Hz. What are the reactance values of the following capacitors?
    a. 0.5 µF.
    b. 0.1 µF.
    c. 0.05 µF.
    d. 0.01 µF.
    e. 0.001 µF.

For Discussion

1. Choose one type of capacitor and research it. Make a report to the class, addressing construction of the capacitor and its various applications.
2. Discuss the effect of using a common screwdriver to adjust a trimmer capacitor. Why is it necessary to use a plastic or fiber screwdriver?
3. Research the reason why a capacitor blocks dc, but passes ac.
4. Why does a theoretical circuit containing only capacitance consume no power?
5. Should a capacitor with a working voltage of 150 volts be used in circuit with 117 volts ac supplied from your house circuit? Explain your answer.

Tuned Circuits and RCL Networks

Objectives

After studying this chapter, you will be able to:

- Explain resonant frequency and how it affects various RCL circuits.
- Calculate a resonant frequency.
- Discuss the characteristics of a series RCL circuit at its resonant frequency.
- Discuss the characteristics of a parallel RCL circuit at its resonant frequency.
- Calculate circuit Q and bandwidth.
- Describe filtering action.
- List four types of filters and explain their action.

Key Words and Terms

The following words and terms will become important pieces of your electricity and electronics vocabulary. Look for them as you read this chapter.

acceptor circuit	low-pass filter
attenuation	oscillating
band-pass filter	quality factor, (Q)
band-reject filter	RCL circuit
bandwidth (BW)	RCL network
damping resistor	reject circuit
filter	resonant frequency (f_o)
flywheel action	selectivity
half-power points	tank circuit
high-pass filter	tuned circuit

16.1 RCL NETWORKS

RCL networks are ac circuits that have resistors, capacitors, and inductors placed in the circuit to pass, reject, or control current. A circuit containing all three factors—resistance, inductance, and capacitance—is called an ***RCL circuit.*** The resulting impedance in an RCL circuit is equal to the *vector addition* of R in ohms, X_L in ohms, and X_C in ohms.

The principles of resistance, capacitance, and inductance should be reviewed before studying these combination circuits.

- In an ac circuit containing resistance only, the applied voltage and current are in phase. There is no reactive power. The power consumed by the circuit is equal to the product of volts times amperes.
- In an ac circuit containing inductance only, the current lags the voltage by an angle of 90 degrees. They are not in phase. The power consumed by the circuit is zero.
- In an ac circuit containing resistance and inductance, the current lags the voltage by a phase angle of less than 90 degrees. The total resistive force is the vector sum of the resistance and the inductive reactance. This is the impedance of the circuit.
- In an ac circuit containing capacitance only, the current leads the voltage by an angle of 90 degrees. The power consumed is zero.
- In an ac circuit containing resistance and capacitance, the current leads the voltage by an angle of less than 90 degrees. The impedance is equal to the vector sum of the resistance and the capacitive reactance. This is the impedance of the circuit.

Resonance

A special condition exists in an RCL circuit when it is energized at a frequency at which the inductive reactance is equal to the capacitive reactance ($X_L = X_C$). Since X_L increases as frequency increases, and X_C decreases as frequency increases, there is one frequency at which both reactances are equal. This frequency is called the ***resonant frequency*** of the circuit, or f_o. A series or parallel RCL circuit at resonant frequency is known as a ***tuned circuit.***

In the vector diagram, **Figure 16-1,** X_L equals 100 Ω, X_C equals 100 Ω, and R equals 50 Ω. X_L and X_C are opposing each other because they are 180 degrees out of phase.

The algebraic sum of these vectors is zero, so only a resistance of 50 ohms remains. The current and voltage are in phase. This particular circuit frequency can be calculated using the resonant frequency formula. This formula is stated:

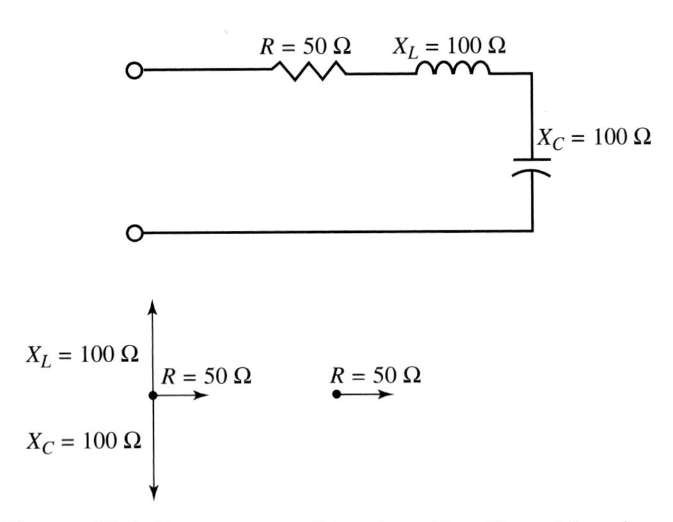

Figure 16-1. Resonance exists when $X_L = X_C$ and the circuit appears as a resistive circuit.

$$f_o = \frac{1}{2\pi\sqrt{LC}} = \frac{0.159}{\sqrt{LC}}$$

where f_o equals resonant frequency in hertz, L equals inductance in henrys, and C equals capacitance in farads.

This formula is arrived at using the following steps. At resonance, $X_L = X_C$, or:

$$2\pi f L = \frac{1}{2\pi f C}$$

Move the $2\pi L$ to the right side of the equation by dividing both sides by $2\pi L$. This leaves you with:

$$f = \frac{1}{(2\pi)^2 f LC}$$

Move the f in the denominator of the right side of the equation to the left side by multiplying both sides by f.

$$f^2 = \frac{1}{(2\pi)^2 LC}$$

Take square root of both sides of the equation.

$$f = \frac{1}{2\pi\sqrt{LC}}$$

The equation can be simplified further by:

$$\frac{1}{2\pi} = \frac{1}{6.28} = 0.159 \text{ (aproximately)}$$

Therefore:

$$f_o = \frac{0.159}{\sqrt{LC}} \text{ (approximately)}$$

Example: What is the resonant frequency of a circuit that has 200 μH inductance and 200 pF capacitance?

$$f_o = \frac{0.159}{\sqrt{LC}}$$

$$f_o = \frac{0.159}{\sqrt{200 \times 10^{-6} \times 200 \times 10^{-12}}}$$

$$f_o = \frac{0.159}{\sqrt{40{,}000 \times 10^{-18}}}$$

$$f_o = \frac{0.159}{\sqrt{4 \times 10^{4} \times 10^{-18}}} = \frac{0.159}{\sqrt{4 \times 10^{-14}}} = \frac{0.159}{2 \times 10^{-7}}$$

$$f_o = 0.08 \times 10^{7} = 800{,}000 \text{ Hz}$$

$$f_o = 800 \text{ kHz or } 0.8 \text{ MHz}$$

In electronic work, L and C usually have very small values. They are usually in the milli, micro, nano, and pico ranges. By restating the equation, you can create a shortcut for yourself. The equation can be stated as:

$$f_o = \frac{0.159}{\sqrt{LC}}$$

where f_o equals resonant frequency in *mega*hertz, L equals inductance in *milli*henrys, and C equals capacitance in *nano*farads.

or

where f_o equals resonant frequency in *mega*hertz, L equals inductance in *micro*henrys, and C equals capacitance in *micro*farads.

Acceptor Circuit

A series RCL circuit is drawn in **Figure 16-2.** At resonance $X_L = X_C$, so the impedance of the circuit equals:

$$Z = \sqrt{R^2 + (X_L - X_C)^2}$$

$$Z = \sqrt{R^2} = R$$

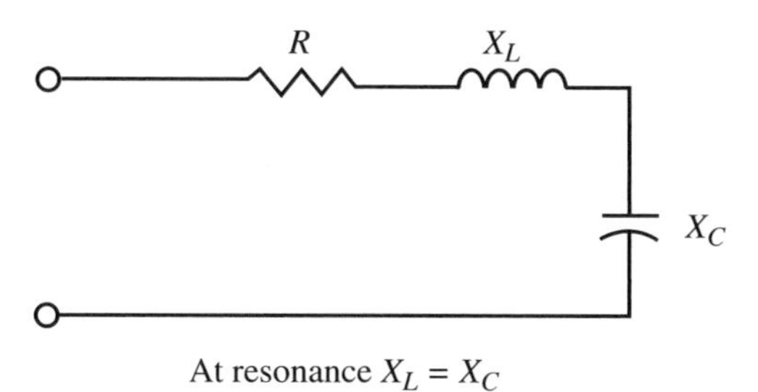

Figure 16-2. The series resonant circuit is an acceptor circuit.

Only the ohmic resistance impedes the current in the circuit. In a series resonant circuit the impedance is at a minimum. At frequencies above or below the resonant frequency, X_L is not equal to X_C and the reactive component *increases* the impedance of the circuit.

The response of a series tuned circuit appears as a bell-shaped curve, **Figure 16-3.** Notice, on the graph, that the impedance of the circuit is minimum at resonance. Note also that maximum current flows at resonance. The current rapidly falls off on either side of the resonant frequency due to the increased impedance, **Figure 16-4.** This circuit is called an acceptor circuit. An ***acceptor circuit*** provides maximum response to currents at its resonant frequency.

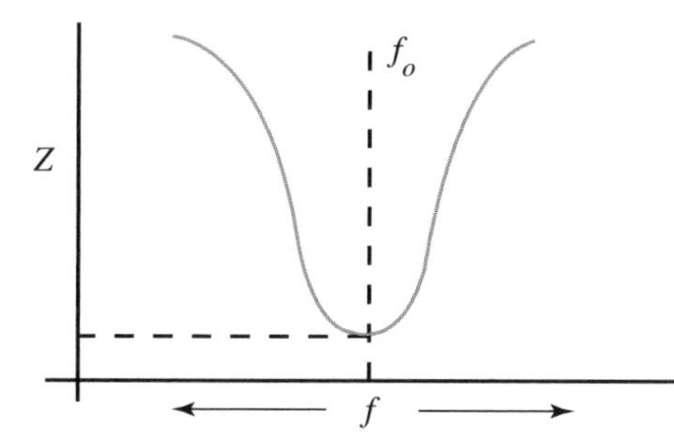

Figure 16-3. The curve shows the increase in impedance as the frequency is varied above or below resonance.

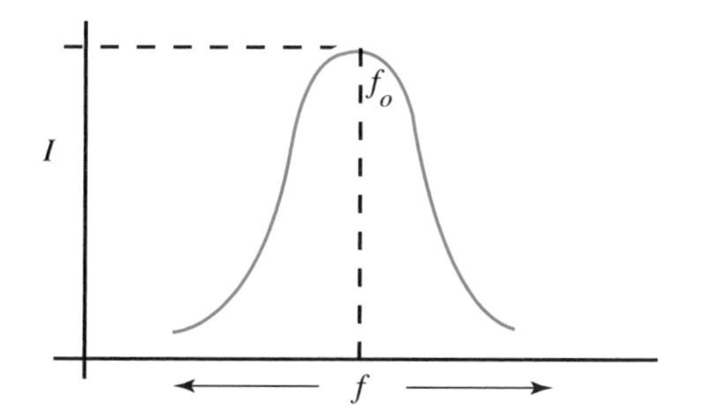

Figure 16-4. The response curve showing the falling off of the current in the circuit above or below resonance.

Example: Refer to **Figure 16-5.** This circuit has 10 Ω of resistance, 200 μH of inductance, and 200 pF of capacitance. These components are connected across a 500 μV radio frequency generator at 800 kilohertz. The resonant frequency of this RCL circuit is 800 kilohertz, so X_L equals X_C at 800 kHz. Z equals 10 ohms at this frequency and:

$$I = \frac{E}{Z} \text{ or } \frac{500 \times 10^{-6}\ \text{V}}{10\ \Omega}$$

$$I = 50\ \mu\text{A}$$

Also:

$$X_L = 2\pi \times 800 \times 10^3\ \text{Hz} \times 200 \times 10^{-6}\ \text{H}$$
$$X_L = 1000\ \Omega \text{ (approximately)}$$

and:

$$X_C = \frac{1}{2\pi \times 800 \times 10^3\ \text{Hz} \times 200 \times 10^{-12}\ \text{F}}$$

$$X_c = 1000\ \Omega \text{ (approximately)}$$

The voltage drops around the circuit equal:

$$E_R = 50\ \mu\text{A} \times 10\ \Omega = 500\ \mu\text{V}$$
$$E_{X_L} = 50\ \mu\text{A} \times 1000\ \Omega = 50{,}000\ \mu\text{V}$$
$$E_{X_C} = 50\ \mu\text{A} \times 1000\ \Omega = 50{,}000\ \mu\text{V}$$

It appears that the sum of these voltages would be 100,500 μV. However, E_{X_L} and E_{X_C} are 180 degrees out of phase. Therefore, a vector addition must be made,

$$E_{source} = \sqrt{(E_R)^2 + (E_{X_L} - E_{X_C})^2}$$

$$E_{source} = \sqrt{(500\ \mu\text{V})^2 + (50{,}000\ \mu\text{V} - 50{,}000\ \mu\text{V})^2}$$

$$E_{source} = \sqrt{(500\ \mu\text{V})^2} = 500\ \mu\text{V}$$

Note that, at resonance, the voltage drops across X_L and X_C are equal. Also, the voltage drop across R equals the source voltage.

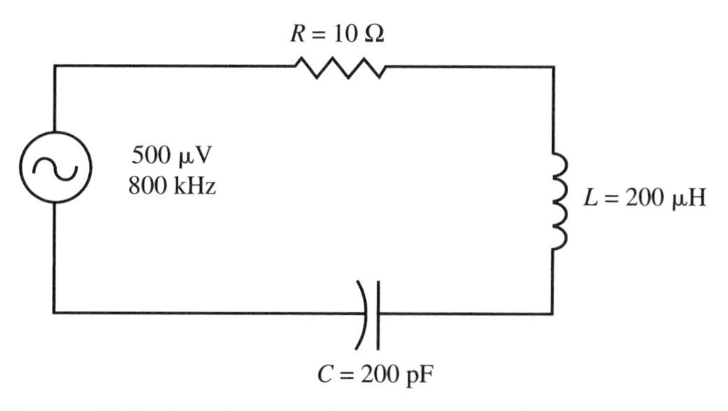

Figure 16-5. A series tuned acceptor circuit.

In a resonant circuit, an interchange of energy between the inductance and the capacitance builds up voltage that exceeds the supply voltage. In the first half cycle, the magnetic field of the inductor stores the energy of the discharging capacitor. In the next half cycle, the stored energy in the magnetic field of the inductor charges the capacitor. This action occurs back and forth, limited only by the series resistance between the two components. At resonance, the charging times for the inductor and capacitance must be the same, and they will have a canceling effect.

Summarizing the series tuned circuit:

- At resonance, the impedance is a minimum and the line current is a maximum.
- At resonance, the voltage drop, E_{X_L}, is equal to E_{X_C}, but is 180 degrees out of phase.
- The vector sum of all voltage drops equals the applied voltage.

Figure 16-6 shows how values change as circuit frequency changes.

Tank Circuit

Study **Figure 16-7.** It shows a parallel tuned circuit. This type of tuned circuit behaves very differently than a series tuned acceptor circuit.

In part A of Figure 16-7, switch 1 is closed and capacitor *C* is charged to the supply voltage. When switch 1 is opened, *C* remains charged. When switch 2 is closed (part B), the capacitor discharges through *L* in the direction of the arrows shown. As the current flows through *L*, a magnetic field builds around *L*. This field remains as long as the current flows. When the charges on the plates of *C* become equalized, current ceases to flow. The magnetic field around *L* then collapses. The energy stored in this field is returned to the circuit.

In Chapter 14, we learned that the induced emf opposes the current change. This is also true in the parallel RCL circuit. When the current drops to zero as a result of the discharged capacitor, a current is induced by the collapsing magnetic field. This current then drives a charge onto the capacitor, but opposite to the original polarity (part C). Later the capacitor discharges in the opposite direction. Once again the same cycle of events

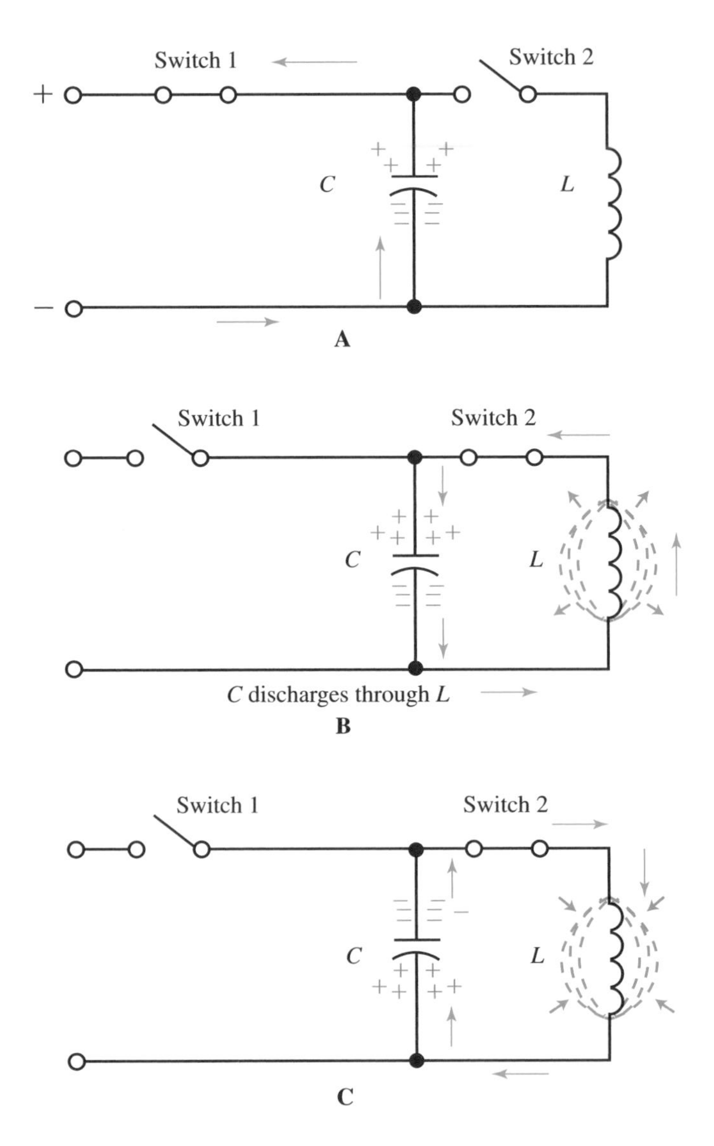

Figure 16-7. The charge and discharge of *C* through inductance *L* is similar to flywheel action. It is called a tank circuit. A–Capacitor charges. B–Capacitor discharges. C–The field around the inductor collapses, which charges the capacitor to the opposite polarity.

f (kHz)	X_L (Ω)	X_C (Ω)	*Z* (Ω)	*I* (μA)	E_{X_L} (μV)	E_{X_C} (μV)
200	250	4000	3750	0.13	33	520
400	500	2000	1500	0.33	165	660
600	750	1500	750	0.66	495	990
800	1000	1000	10	5.0	50,000	50,000
1000	1250	800	450	1.1	1375	880
1200	1500	666	833	0.6	900	400
1600	2000	500	1500	0.33	660	165

Figure 16-6. As circuit frequency changes, other values also change. The figures used are rounded to nearest whole number to ease understanding. Resistance of 10 ohms is insignificant. It is not included in computations, except at resonance.

occurs. The capacitor again becomes charged as in its original state. This discharge/charge cycle repeats over and over. The current periodically changes direction in the circuit. A circuit containing this periodic changing current is said to be ***oscillating.***

The periodic current changes in the circuit can be described as ***flywheel action.*** This parallel RCL circuit is called a ***tank circuit.*** If no energy were used during the cycles of oscillation, the circuit might oscillate indefinitely. But there is always some resistance due to coil windings and circuit connections. This resistance uses up the energy stored in the circuit and dampens out the oscillation. The amplitude, or size, of each successive oscillation decreases due to the resistance.

Compare the oscillators of a tank circuit to a child on a swing. The child can be swinging back and forth, or oscillating, but if no one adds a little push to the swing, the amplitude of the swing will decrease until it comes to rest. If not for friction or air resistance, the swing might continue swinging back and forth forever.

Now consider what would happen if the swing was pushed every time it reached its maximum backward position. This added energy would replace the energy lost to friction. The full swinging action would continue. A tank circuit is similar to this swing example. If pulses of energy are added to the oscillating tank circuit at the correct frequency, it will continue to oscillate.

What is the meaning of correct frequency? During the discharge and charge of the capacitor C in Figure 16-7, a set amount of time must pass. In other words, during one cycle of oscillation a set interval of time must pass. The number of cycles that occur in one second would be called the *frequency* of oscillation. As you learned in Chapter 10, frequency is measured in *cycles per second* (cps) or *hertz* (Hz).

Close study of this circuit reveals that if C or L were made larger (so that they require a longer time to charge) the frequency of the circuit decreases. This relationship can be shown using the resonance formula.

$$f_o = \frac{1}{2\pi\sqrt{LC}}$$

Notice that both L and C are inversely proportional to the f_o. If L or C are made large, f_o will become proportionally smaller.

Reject Circuit

Figure 16-8, shows a parallel tuned circuit connected across a variable frequency generator. Minimum line current flows at the resonant frequency of this tuned circuit. The resonant frequency can be found by observing the minimum value in the line current. This minimum value shows up as a dip that can be measured by a current meter in the line. Radio transmitter operators always "dip the final." This term means that the final tank circuit is tuned to resonance, which is indicated by a dip in current in the final circuit.

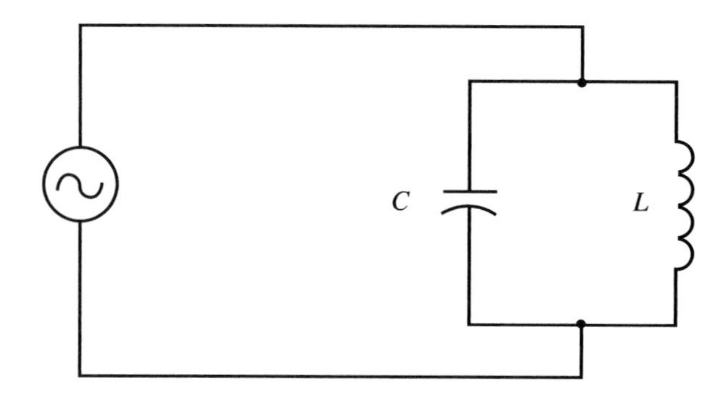

Figure 16-8. A parallel LC circuit connected to a variable frequency generator.

Since line current in the circuit at resonance is a minimum, a parallel tuned circuit has a maximum line impedance *(Z)*. At frequencies other than resonance, the impedance is much less. *So, the parallel tuned circuit rejects signals at or near its resonant frequency, and allows signals of frequencies other than resonance to pass.* This is why the parallel tuned circuit is called a ***reject circuit.*** These characteristics are shown in the response curves in **Figure 16-9.**

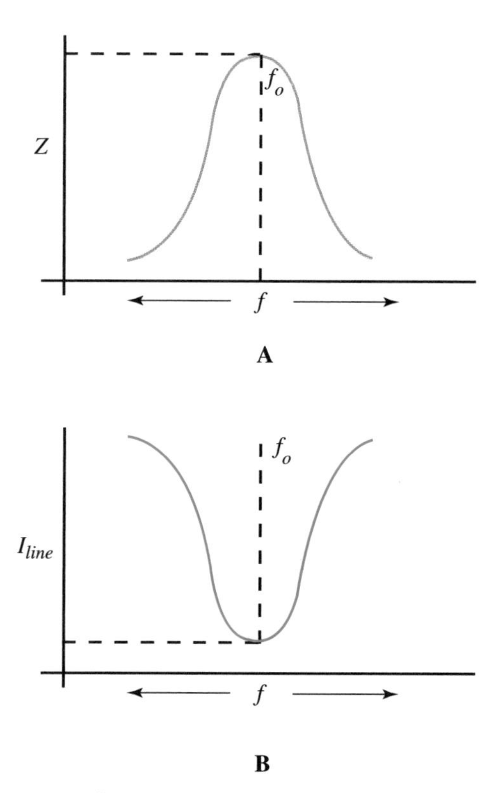

Figure 16-9. A–Curve shows maximum impedance at resonant frequency. B–Circuit response shows minimum line current at resonance.

Why does a parallel tuned tank present maximum impedance at resonance? At resonance $X_L = X_C$. Both paths have reactive values in parallel across the generator source. It would *appear* that these two reactive branches would combine to form a low reactive path for the line current. However, the current flowing in the X_L branch lags the applied voltage by 90 degrees. The current in the X_C branch leads the applied voltage by 90 degrees. The currents, therefore, are 180 degrees out of phase and cancel each other. The total line current is the sum of the branch currents. It is zero, except for a small amount of current that flows due to the resistance of the wire in the coil.

Example: Refer to **Figure 16-10.** A 200 μH inductor and a 200 pF capacitor are connected in parallel across a generator source of 500 μV. The resistance, 10 ohms, represents the lumped resistance of the wire of the inductor. These same components were used in the study of the series resonant circuit. Compare the results of the two circuits at resonance.

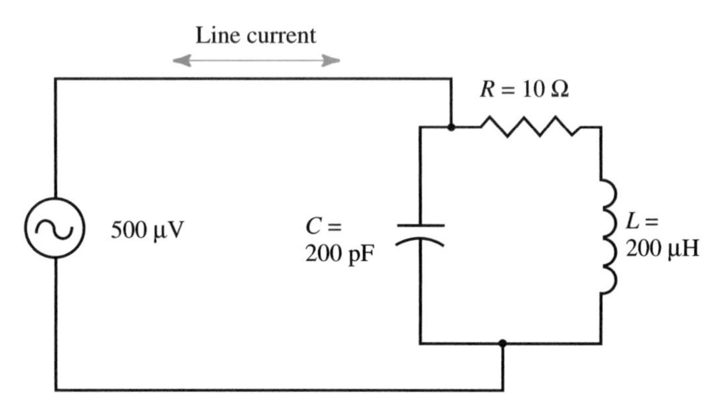

Figure 16-10. This parallel resonant circuit has a 200 μH inductor and a 200 pF capacitor. They are connected in parallel across a 500 μV generator source.

The resonant frequency of this tuned circuit is:

$$f_o = \frac{0.159}{\sqrt{LC}}$$

$$f_o = \frac{0.159}{\sqrt{200\ \mu\text{H} \times (0.2 \times 10^{-3})\ \mu\text{F}}}$$

$$f_o = \frac{0.159}{200} = 0.8\ \text{MHz (approximately)}$$

(Note use of convenient formula when L is in μH, C is in μF, and f_o is in megahertz.)

At resonance:

$$X_L = 2\pi fL$$

$$X_L = 2\pi \times (800 \times 10^3\ \text{Hz}) \times (200 \times 10^{-6}\ \text{H})$$

$$X_L = 1000\ \Omega$$

$$X_C = \frac{1}{2\pi fC}$$

$$X_C = \frac{1}{2\pi \times (800 \times 10^3\ \text{Hz}) \times (200 \times 10^{-12}\ \text{F})}$$

$$X_C = 1000\ \Omega$$

The voltage across both branches of the parallel circuit is the same as the applied voltage, or 500 μV. Therefore, the current in the X_L branch is:

$$I_{X_L} = \frac{E_L}{X_L} = \frac{500\ \mu\text{V}}{1000\ \Omega} = \frac{500 \times 10^{-6}\ \text{V}}{10^3\ \Omega}$$

$$I_{X_L} = 500 \times 10^{-9}\ \text{A} = 0.5\ \mu\text{A}$$

$$I_{X_L} = 500 \times 10^{-9}\ \text{A} = 0.5\ \mu\text{A}$$

The current in the X_C branch is:

$$I_{X_C} = \frac{E_C}{X_C} = \frac{500\ \mu\text{V}}{1000\ \Omega} = 0.5\ \mu\text{A}$$

Since these two currents are 180 degrees out of phase:

$$I_{X_L} = I_{X_C} = 0.5\ \mu\text{A} - 0.5\ \mu\text{A} = 0$$

In this exercise, *R* has been overlooked because of its small value. However, some current will flow as a result of *R*. For frequencies above and below resonance, the action of the circuit can be seen in **Figure 16-11.**

f (kHz)	X_L (Ω)	X_C (Ω)	I_{X_L} (μA)	I_{X_C} (μA)	I_{line} (μA)
200	250	4000	2.0	0.125	1.875
400	500	2000	1.0	0.25	1.75
600	750	1500	0.66	0.33	0.33
800	1000	1000	0.5	0.5	0
1000	1250	800	0.4	0.625	0.225
1200	1500	666	0.33	0.75	0.42
1600	2000	500	0.25	1.00	0.75

Figure 16-11. Circuit performance for frequency above and below resonance.

Note that the line current is the difference between the branch currents. Figure 16-11 could be carried a step further to show the decreasing impedance due to frequencies other than resonance for:

$$Z = \frac{E}{I}$$

As *Z* decreases, *I* must increase.

Q of Tuned Circuits

The ***Q, figure of merit,*** or ***quality factor,*** of the circuit is the link between inductive reactance and resistance in the circuit. Q has no units. The Q of a circuit can be found using the formula:

$$Q = \frac{X_L}{R}$$

where Q is the quality factor, X_L is the inductive reactance at resonance, and R is the resistance.

Example: In a series circuit, the inductive reactance is 1000 ohms at resonance, and the resistance of the wire of the coil is 10 ohms. The figure of merit, Q, would be:

$$Q = \frac{X_L}{R} = \frac{1000\ \Omega}{10\ \Omega} = 100$$

Acceptor circuit Q

The Q of a circuit indicates the sharpness of the reject or accept characteristics of the parallel or series RCL circuit. It is a quality factor. In the series acceptor circuit, an increase in resistance reduces the maximum current at resonance, **Figure 16-12.** The Q can also be used to determine the rise in voltage across L or C at resonance.

$$E_{X_L} = E_{X_C} = Q \times E_S$$

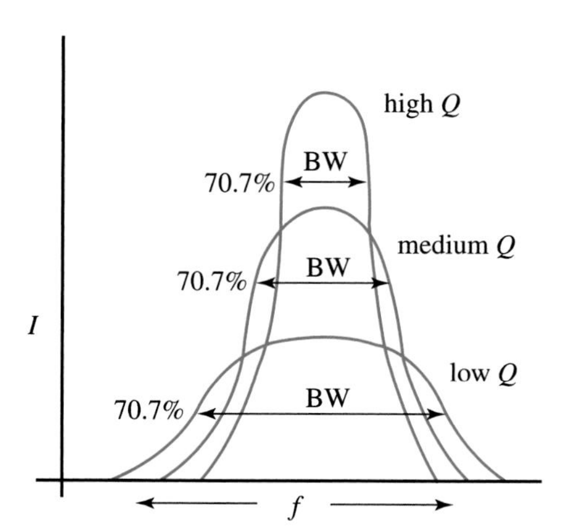

Figure 16-12. As the Q of a circuit is lowered, the curve flattens out. Its selectivity decreases and its bandwidth increases.

Refer back to Figure 16-5. The supply voltage is 500 µV and the Q of this circuit is 100. Therefore, the voltage rise across X_L or X_C at resonance is equal to:

$$E_{X_L} = E_{X_C} = 100 \times 500\ \mu V = 50{,}000\ \mu V$$

If the circuit had a lower Q, the magnified voltage at resonance would be much less. For example, increase R to 20 ohms. Now, the Q would equal only 50. The voltage rise would equal:

$$E_{X_L} = E_{X_C} = 50 \times 500\ \mu V = 25{,}000\ \mu V$$

High Q circuits are very useful in selective electronic circuits. Typical values for Q range from 50 to 250. The higher the value of Q, the greater response of the circuit at resonance. Also, a high Q circuit has increased ***selectivity.*** Selectivity is set by bandwidth. ***Bandwidth (BW)*** is the band of frequencies above and below resonance in which the circuit response does not fall below 70.7 percent of the response at resonance. Look again at Figure 16-12. The points at which the response falls to 70.7 percent are called the ***half-power points.*** As indicated by their name, at the half-power points the power is half that of the maximum.

The bandwidth of the tuned circuit can be found using the formula:

$$BW = \frac{f_o}{Q}$$

Continuing with the previous problem, the bandwidth equals:

$$BW = \frac{f_o}{Q} = \frac{800{,}000\ Hz}{100} = 8000\ Hz$$

The resonant frequency, 800 kHz, is at the maximum response point. The bandwidth extends 4000 Hz below resonance and 4000 Hz above resonance (for a total of 8000 Hz). This circuit can be considered, then, as passing all frequencies between 796 kHz and 804 kHz. Beyond either of these limits, the response will fall below the 70.7 percent value.

Reject circuit Q

Q can be computed for parallel tuned circuits using X_L at resonance and R, which is the resistance of the coil L. The Q of a tank circuit can be used to learn the maximum impedance of the circuit at resonance:

$$Z = Q \times X_L$$

where Z equals impedance at resonance, Q is the quality factor, and X_L equals reactance at resonance.

Referring back to Figure 16-10, the impedance at resonance would be:

$$Z = Q \times X_L = 100 \times 1000\ \Omega = 100{,}000\ \Omega$$

Damping resistor

One final example is shown in **Figure 16-13.** In the figure, a resistor is in parallel with a tank circuit. The resistor, R_S, is called a ***damping resistor.*** It broadens the frequency response of the circuit, because it carries a part

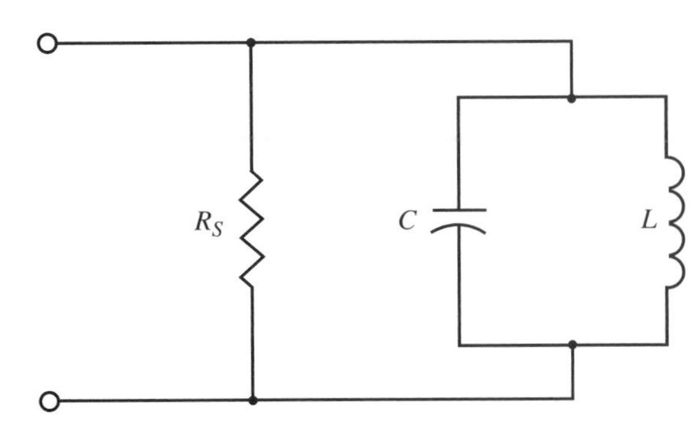

Figure 16-13. The resistor, R_S, is a damping resistor. It broadens the circuit response.

of the line current which cannot be canceled at resonance. Shunt damping lowers the Q of the circuit and makes it less selective.

Loading the Tank Circuit

The parallel tuned circuit is used when coupling energy from one circuit to another. Coupling transformers in radio and television sets use the tuned circuit to transfer signals from one stage to another.

A radio transmitter uses a coupling device attached to a tank circuit to feed energy to the antenna system. In this setup, a second coil is inductively coupled to the coil in the tank circuit. The varying magnetic field of the oscillating tank inductance induces a current to flow in the second coil.

Figure 16-14 shows a picture of, and the circuit for, an intermediate frequency transformer used in a radio receiver. Both the primary and secondary of this transformer are tuned circuits. Maximum radio frequency energy transfer is achieved by tuning the circuits to resonance. The intermediate frequency (IF) used in most radios is 455 kHz. (In television, a 10.7 MHz intermediate frequency is used.) The IF transformer is tuned for maximum response at this frequency. A band pass of 5 kHz above and below the center frequency is maintained. Signals between 450 kHz and 460 kHz can pass without attenuation. ***Attenuation*** of a signal is a decrease in its amplitude or intensity. The process of transformer adjustment is called ***alignment.***

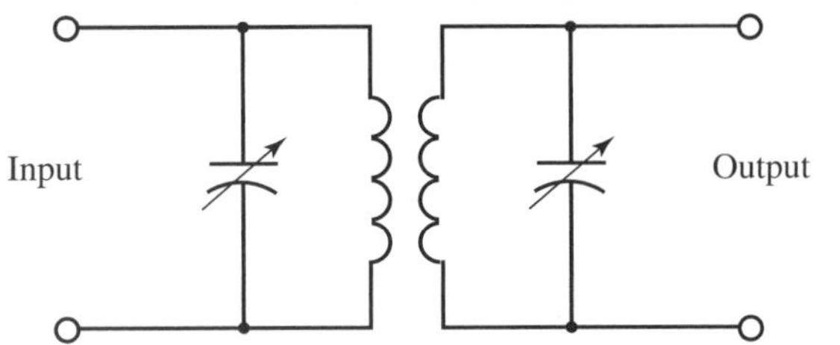

Figure 16-14. Top. An intermediate frequency transformer used to couple energy in a radio receiver. Bottom. Schematic for the transformer.

Review Questions for Section 16.1

1. Resonant frequency is a special condition of RCL circuits in which __________ __________ __________.
2. Find the resonant frequency using the values given in the following circuit.

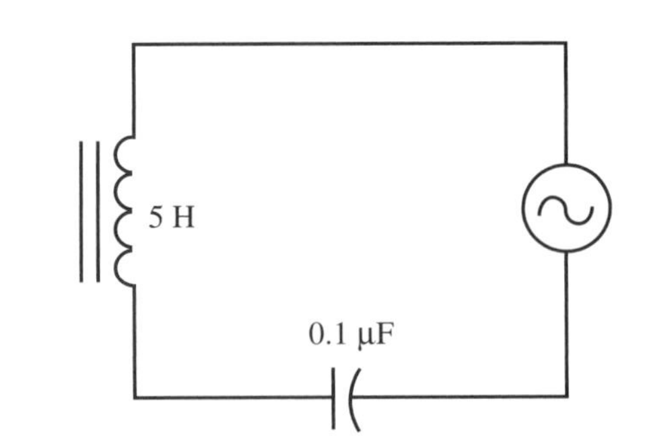

3. In a series resonant circuit, the impedance is a(n) __________.
4. Periodic current changes in a tank circuit are described as __________ __________.
5. In a parallel tuned circuit, the circuit __________ currents at the resonant frequency.
6. The relationship between X_L and R in a circuit is stated by the __________ of the circuit.
7. What causes the tank circuit to stop oscillating?
8. Define attenuation.

16.2 FILTERING CIRCUITS

Inductance and capacitance are quite useful in electronic circuitry. One example of this usefulness is the filter. A ***filter*** is a circuit that separates specific frequencies.

There are many filter designs. For example, filters can be designed to pass low frequencies and reject high frequencies; to reject low frequencies and pass high frequencies; to either pass or reject specified frequency bands.

Each filter type is named according to its function. There are four basic types of filters: low-pass, high-pass, band-pass, and band-reject.

Filters are also used to adjust a pulsating current coming from a rectifier. A *rectifier* is a device that converts ac into a pulsating dc. Because most electronic circuitry needs a pure direct current, the dc must be adjusted. This filter circuit levels out the peaks and valleys of the current. These filters are discussed in Chapter 17.

Filtering Action

In the study of filtering action, these points should be reviewed:

- A capacitor will block a direct current, but will pass an alternating current.
- A conductor can carry a current that has both a dc component and an ac component.

The graph in part A of **Figure 16-15** shows the flow of a steady 10-volt direct current. Part B of the figure shows an ac voltage whose peak value is 10 volts. Part C of the figure shows these two voltages combined. This wave depicts the sum of the two waves.

Note the new axis of the ac voltage. It now varies above and below the 10-volt dc level. Because the axis has been raised by the dc voltage, the ac voltage no longer reverses polarity. This current is a *varying direct current.*

The current's amplitude varies between zero and 20 volts. When this varying direct current is connected to a circuit, **Figure 16-16,** the capacitor *C* immediately charges up to the average dc level of voltage. In this case, that level is 10 volts. When the incoming voltage rises to 20 volts, current flows through *R* to charge capacitor *C.* In the next half cycle, the incoming voltage drops to zero, and capacitor *C* discharges to zero through *R.*

Notice that a voltage appears across *R* due to the charge and discharge of *C.* The output of this circuit, then, taken from across *R,* represents only the ac component of the incoming signal voltage. See **Figure 16-17.** The dc component is blocked by capacitor *C.* There is little phase shift between the input and output waves because the value of *R* is chosen as *ten times or more* the value of the reactance of *C* at the input voltage frequency. *When this ratio of resistance to reactance is maintained, the phase shift does not affect the waves.*

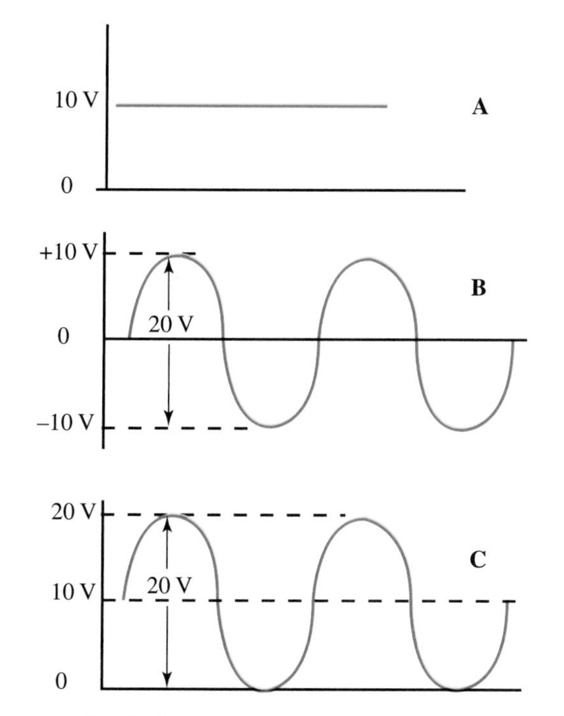

Figure 16-15. A–Steady direct current. B–10-volt peak alternating current. C–Combined dc and ac.

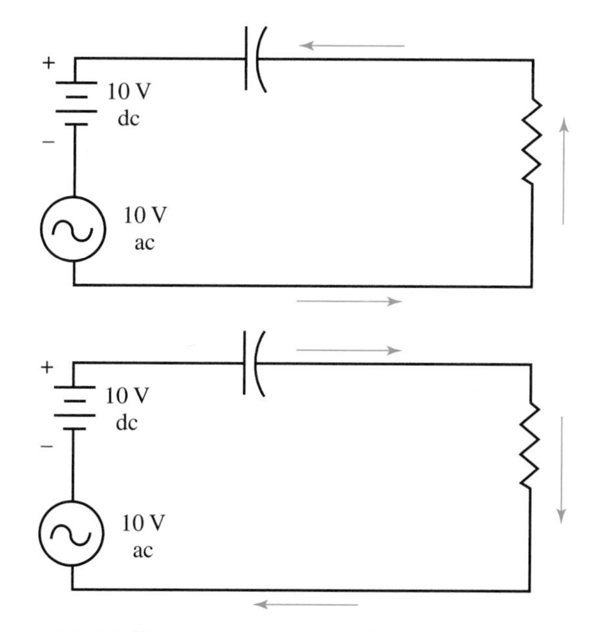

Figure 16-16. The ac generator and dc source are connected in series to the RC circuit.

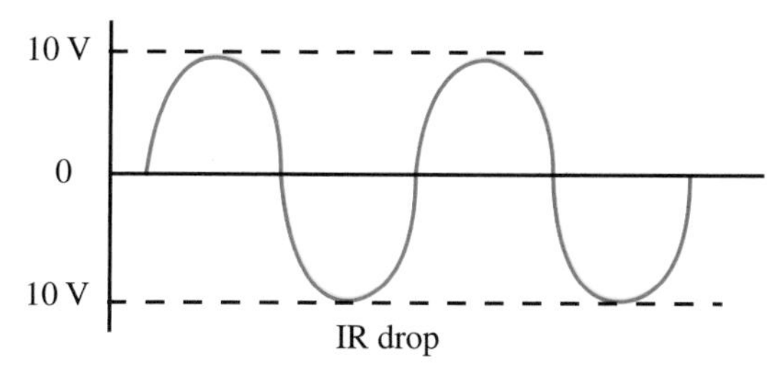

Figure 16-17. The voltage across *R* is the result of the charge-discharge current and represents the ac component of the signal.

EXPERIMENT 16-1: Demonstrate RC Coupling

During this experiment you will demonstrate *RC coupling.* You will construct a circuit upon which the dc component of the incoming signal appears across the capacitor and the ac component appears across the resistor. This important principle will be discussed in more detail in later chapters.

Materials

1–10-volt dc power supply
1–variable ac power supply
1–1 μF capacitor
1–25 kΩ resistor
1–multimeter

1. Examine the circuit shown in **Exhibit 16-1A.**

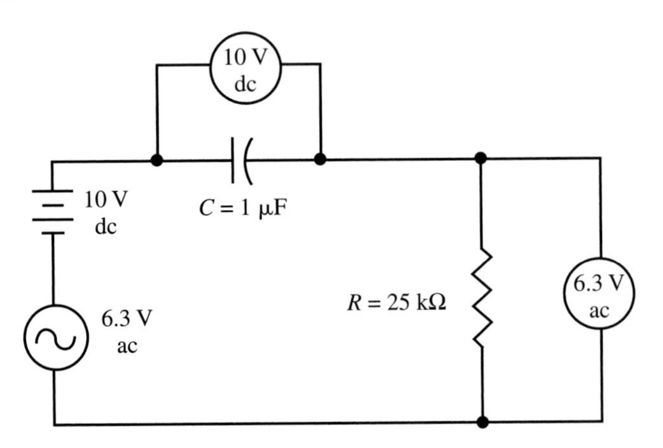

2. Compute the reactance of C as follows:

$$X_C = \frac{0.159}{fC} \text{ or } \frac{0.159}{60 \text{ Hz} \times 1 \times 10^{-6} \mu\text{F}}$$

$$X_C = 0.00265 \times 10^6 \,\Omega$$

$$X_C = 2650 \,\Omega$$

R should equal 10 times X_C, or 26,500 ohms. A 25,000-ohm resistor is suitable for this experiment.

2. Set up the circuit shown in Exhibit 16-1A. Set your power supply for 6.3 Vac at 60 Hz. Connect the 6.3-volt (rms) terminals of a power supply in series with a 10-volt dc supply. Connect the power supplies to the circuit. Have your instructor check your circuit before turning on the power supplies.
3. Take several measurements across C with a meter set to read dc volts. Measurements across C with a meter set for dc volts should read 10 volts. Do your readings agree?
4. Now measure for ac volts. A meter set for ac placed across the same points should read zero. Do your readings agree?
5. Next measure for ac voltages across the resistor R. An ac meter reading across R should read 6.3 volts. Do your readings agree?
6. Measure for dc voltages across the resistor. Your dc meter reading should show zero. Do your readings agree?

What conclusions can you draw from your results? How do your readings relate to RC coupling?

Bypassing

At times it is necessary to create a voltage drop across a resistor resulting from only the dc component of a signal voltage. This is done by bypassing an ac signal or voltage around a resistance. This bypassing can be accomplished using a capacitor.

Part A of **Figure 16-18** shows two 1500-ohm resistors joined to a source of 10 volts dc. Total resistance equals $R_1 + R_2$, or 3000 ohms. The current in the circuit is:

$$I = \frac{E}{R} \text{ or } \frac{10 \text{ V}}{3000 \,\Omega} = 0.0033 \text{ A or } 3.3 \text{ mA}$$

The voltage drop across each resistor is:

$$E_{R_1} = I \times R \text{ or } 0.0033 \text{ A} \times 1500 \,\Omega = 5 \text{ V}$$

$$E_{R_2} = I \times R \text{ or } 0.0033 \text{ A} \times 1500 \,\Omega = 5 \text{ V}$$

Connect a 10-volt ac source now and an ac voltage will appear across *both* resistors. But we wish to bypass this ac component around R_2. To do this, connect a capacitor that has low reactance to the ac voltage in parallel with R_2. If the reactance of C is one-tenth of resistance R, the greater part of the varying current will flow through C and not through R.

Use the values given in part B of Figure 16-18. C has a 150-ohm reactance to the ac frequency. This reactance is one-tenth the 1500-ohm value of R_2. For the most part, the impedance of $R_2 \parallel X_C$ (R_2 and X_C in parallel) can be given as a value of 150-ohms. Voltage distribution around this circuit can be measured or computed.

Magnetic Resonance Imaging (MRI)

A most interesting way to look at the human body is through a process known as MRI or *magnetic resonance imaging*. The MRI technique is based on the principles of magnetism and radio wave transmission.

Using MRI, physicians are able to identify abnormal tissue without performing surgery. Unlike *radiography* or *fluoroscopy*, MRI does not expose the patient to radiation. MRI is safe for most people but, because of the powerful magnet, it cannot be used on people with metal implants (such as pacemakers or artificial joints).

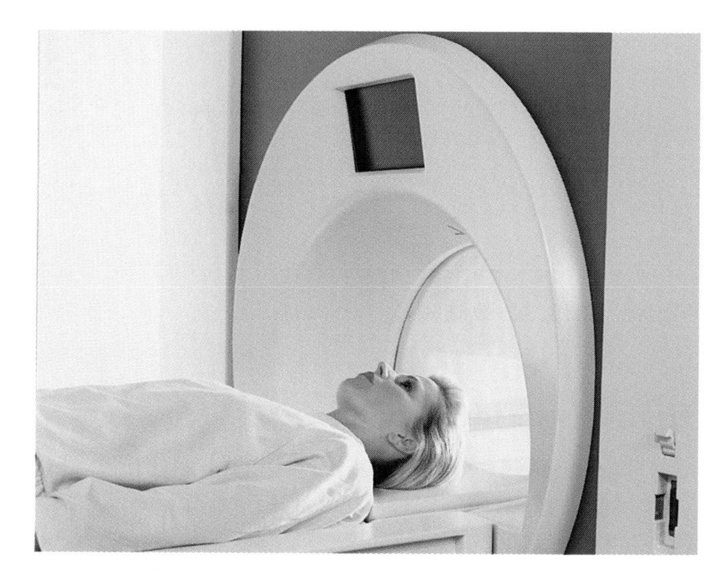

The patient is placed in a large circular magnet and a magnetic field is applied to the patient's body. The magnetic field causes nuclei in the hydrogen atoms inside the body to line up.

A radio wave is then transmitted through the patient and picked up by a receiver. The frequency of the radio waves sets up a *resonant condition* in the hydrogen atoms. This condition enables the nuclei to absorb the energy of the radio waves. When the radio-wave stimulation stops, the atoms return to their original state and emit energy in the form of weak radio signals.

The radio signals are then translated by a computer and the image is constructed based on the intensity of the radio waves interacting with the hydrogen atoms of the patient. Variances in the proportional content of hydrogen for each organ determines the type of image produced.

MRI is typically used to examine the head and spine but it is also used to examine organs such as heart and lungs, and joints. The detailed cross-sectional images allow physicians to see blood moving through veins and arteries, to see a swollen joint shrink in response to medication, and to see the reaction of cancerous tumors to treatment.

Photograph courtesy of Siemens Medical Systems, Inc.

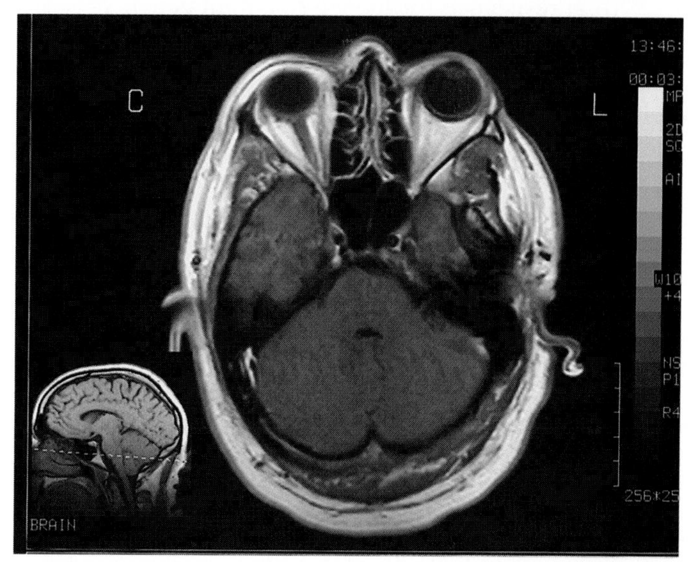

Some machines use the superconductor principle to produce the powerful magnetic field needed for MRI. The magnets in the ring are cooled to extremely low temperatures using liquid nitrogen or liquid helium. Once they have been cooled, electricity can be conducted with little to no measurable resistance. The lack of resistance in the coils is what enables such powerful magnetic fields to be produced.

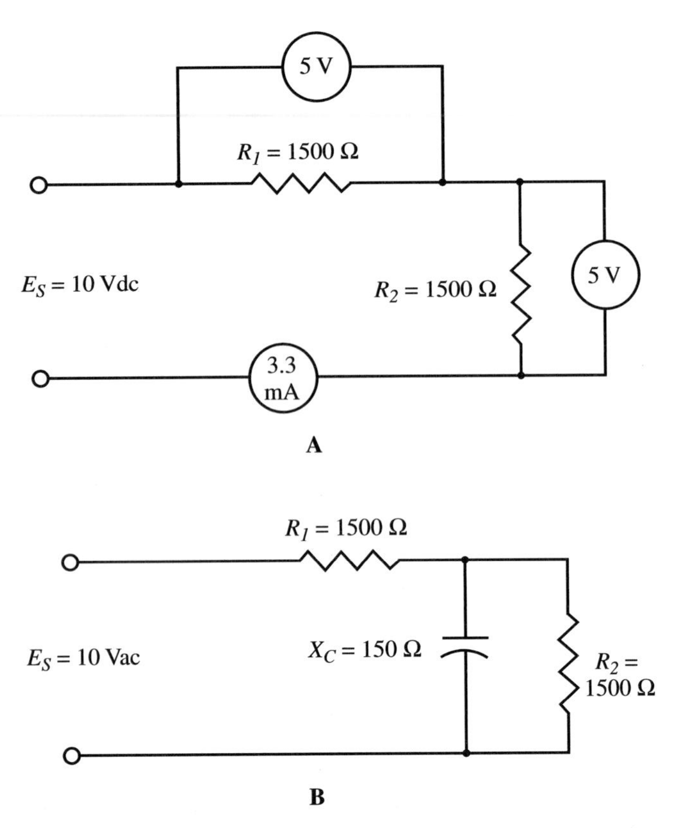

Figure 16-18. A–A dc voltage produces equal drops across R_1 and R_2. B–The voltage drop across R_2 is held constant by bypassing the ac component.

Total resistance for the entire circuit equals $R_1 + (R_2 \parallel X_C)$ or 1650 ohms. $R_2 \parallel X_C$ represents one-eleventh of the total resistance to the alternating current. Because voltage drop is a function of resistance, ten-elevenths of the voltage appears across R_1 and only one-eleventh across $R_2 \parallel X_C$. The applied ac voltage is 10 volts, so:

$$E_{R_1} = 9.1 \text{ V and } E_{R_2} = 0.9 \text{ V}$$

If a 10-volt dc voltage is connected in series with the 10-volt ac voltage for the combined input voltage, the dc voltage will divide equally between R_1 and R_2. The dc voltage will drop five volts across each resistor. C is an open circuit for a direct current.

The ac voltage divides in the ratio calculated earlier. Most of this voltage appears across R_1. The voltage across R_2 remains fairly constant due to the bypass capacitor action.

To summarize, choose a capacitor that will form a low reactance path around a resistor for currents of chosen frequencies. This produces a voltage across one resistor that is almost entirely dc.

Low-Pass Filters

At times, a filter is needed that will pass low frequencies, yet decrease the high frequency currents. This filter is called a ***low-pass filter.*** A low-pass filter circuit always has a resistance or an inductor in series with the incoming signal voltage. It also has a capacitor in shunt or across the line, **Figure 16-19.**

As the frequency is increased, the reactance of L increases so that a larger amount of the voltage appears across L. Also as the frequency increases, the reactance of C decreases. The capacitor thus provides a bypass for the higher frequency currents around the load resistance R.

Because of the increased reactance of the inductor and decreased reactance of the capacitor at high frequencies, the higher frequencies appear only in small amounts across the load. Low frequencies develop higher voltages across the load. Low frequencies are passed, high frequencies are rejected.

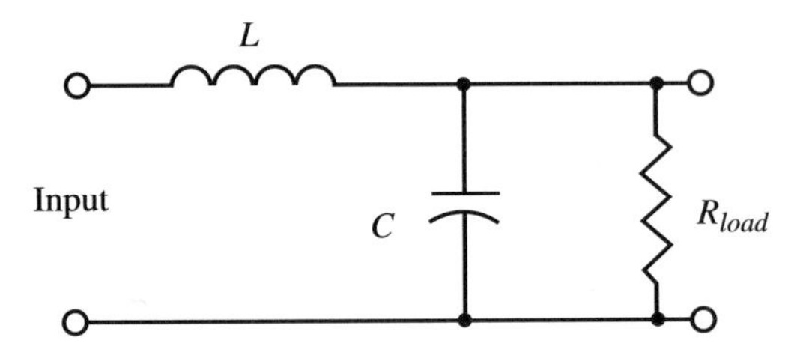

Figure 16-19. A low-pass filter circuit.

High-Pass Filters

The opposite to the low-pass filter is the high-pass filter. ***High-pass filters*** pass chosen high frequency current and reject low frequency currents. The filter circuit includes a capacitor in series with the incoming signal voltage and an inductance shunt across the line, **Figure 16-20.** As the frequency increases, X_L increases and a higher voltage is developed across L and R in parallel. As frequency increases, X_C decreases, providing a low reactance path for high frequency signals. Low frequencies are shunted or bypassed around the load R by the low reactance of L at low frequencies.

To improve the filtering action of both low- and high-pass filters, two or more sections are often joined. These sections are named according to the circuit makeup. **Figure 16-21** illustrates the schematic drawings and the names of several types of filters.

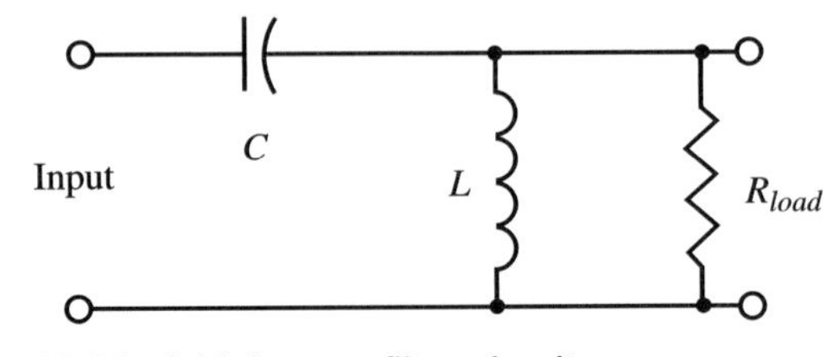

Figure 16-20. A high-pass filter circuit.

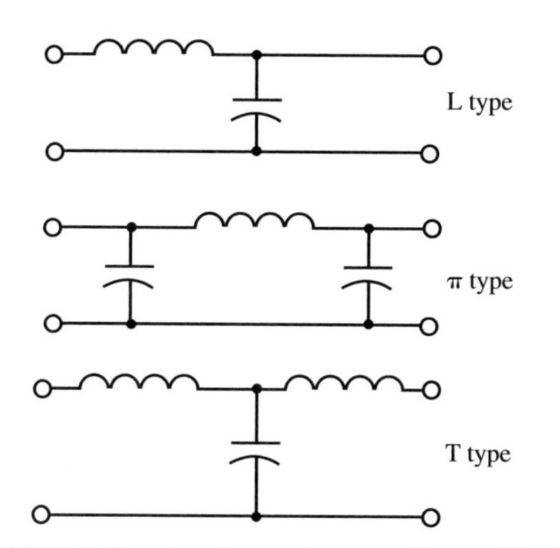

Figure 16-21. Filter circuits can be named according to circuit configuration.

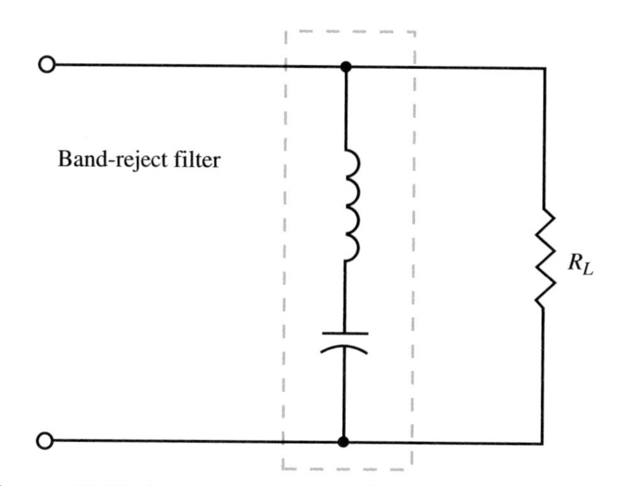

Figure 16-23. A series tuned circuit arranged as a band-reject filter.

Tuned Circuit Filters

Earlier in this chapter, series and parallel resonant circuits were said to be acceptor or reject circuits. These tuned circuits can be used as filters because they are able to give maximum response at their tuned resonant frequency.

In **Figure 16-22,** a series resonant circuit is used as a band-pass filter. A ***band-pass filter*** accepts only currents near its resonant frequency. In **Figure 16-23,** the series resonant circuit is shunted across, or parallel to, the load. This provides a low impedance path around currents at resonant frequencies (bypassing the load). In this case, the filter would respond as a band-reject filter. ***Band-reject filters*** reject a specific band of frequencies.

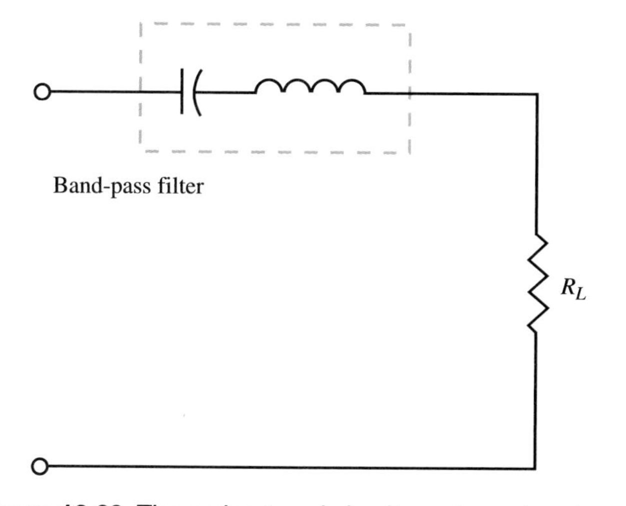

Figure 16-22. The series tuned circuit used as a band-pass filter.

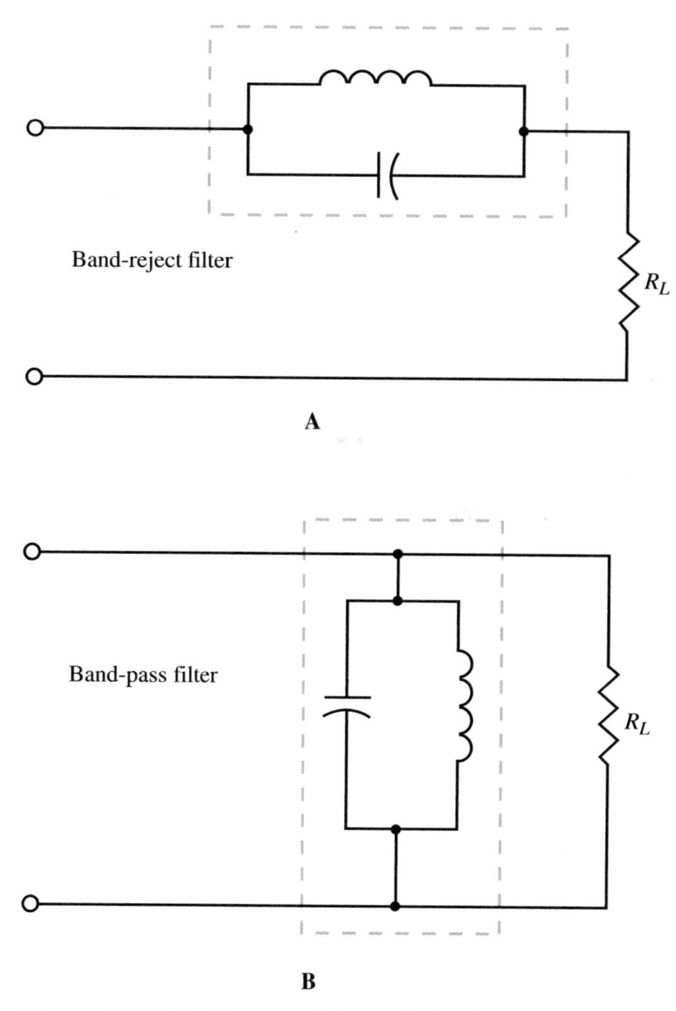

Figure 16-24. A–The parallel tuned circuit as a band-reject filter. B–The parallel tuned circuit as a band-pass filter.

The circuits in **Figure 16-24** use a parallel tuned circuit. The effect of the parallel tuned circuit is the opposite of the series tuned circuit. When the tank circuit is in series with the incoming currents, it provides maximum impedance and rejects the frequencies around its resonant frequency. When the tank circuit is shunted across the load, it causes the maximum response across the load at resonance. Frequencies other than resonance are bypassed because of the decreased impedance of the shunt tuned tank circuit.

Combinations of both series and parallel tuned circuits can be used to provide sharper cutoff and greater attenuation.

Nomograph

The nomograph, or alignment chart, shown in **Figure 16-25,** provides a rapid and handy method for solving problems involving X_L, X_C, and resonance. To use this nomograph, you need any two variables involved in finding resonance. The chart has lines for L, C, X_C, and f_0. To find the missing value, line a straightedge across the two variables that are known. Then, see where the straightedge crosses elsewhere on the chart.

The following examples will aid you in using and understanding this chart.

1. What is the reactance of a 0.01 μF capacitor at 10 kHz? Use the upper chart. Place a ruler or straight edge at 10 kHz on the frequency scale and 0.01 on the capacitance scale. The rule crosses the reactance scale at approximately 1600 ohms.
2. What is the reactance of an 8-henry choke coil at 60 Hz? Use the lower chart. Place the rule on 60 Hz and 8 H. The chart reads approximately 3000 ohms.
3. What is the resonant frequency of a tuned circuit when C equals 250 pF and L equals 200 mH? Use the upper chart. Place the rule on 200 mH and 250 pF and read 22.5 kHz on frequency scale.

Review Questions for Section 16.2

1. Circuits that separate specific frequencies are called __________.
2. A(n) __________ will pass ac and block dc.
3. In a low-pass filter, the __________ __________ are rejected.
4. When a series resonant circuit is used as a band-pass filter, it accepts only current near its __________ __________.
5. What is a nomograph?

Summary

1. Networks are ac circuits that have resistors, capacitors, and inductors placed in the circuit to pass, reject, or control current.
2. Resonant frequency of a circuit occurs when the inductive reactance (X_L) is equal to the capacitive reactance (X_C).
3. The formula for the frequency of resonance is:

$$f_o = \frac{1}{2\pi\sqrt{LC}}$$

It can be simplified to:

$$f_o = \frac{0.159}{\sqrt{LC}}$$

4. In a series resonant circuit, the current is maximum and the impedance is minimum at the resonant frequency.
5. Series resonant circuits pass currents at resonant frequency.
6. Tank circuits oscillate, providing an ac signal at a desired frequency.
7. In a parallel resonant circuit, the current is minimum and the impedance is maximum at the resonant frequency.
8. Parallel resonant circuits reject currents at resonant frequency.
9. The Q of a circuit describes the relationship between X_L and R. The formula is:

$$Q = \frac{X_L}{R}$$

10. Four types of filtering circuits are:
    a. low-pass filter.
    b. high-pass filter.
    c. band-pass filter.
    d. band-reject filter.
11. A nomograph is a chart that can be used to solve problems involving X_L, X_C, and resonance.

Test Your Knowledge

Please do not write in the text. Place your answers on a separate sheet of paper.

1. What is the formula for resonance?
2. Compute the resonant frequency of the following circuits:
    a. $L = 100$ μH, $C = 250$ pF.
    b. $L = 200$ μH, $C = 130$ pF.
    c. $L = 8$ μH, $C = 1$ μF.
3. A series resonant circuit is sometimes called a(n) __________ circuit.
4. Explain flywheel action of a tank circuit.
5. A parallel tuned circuit is sometimes called a(n) __________ circuit.

Reactance-frequency chart

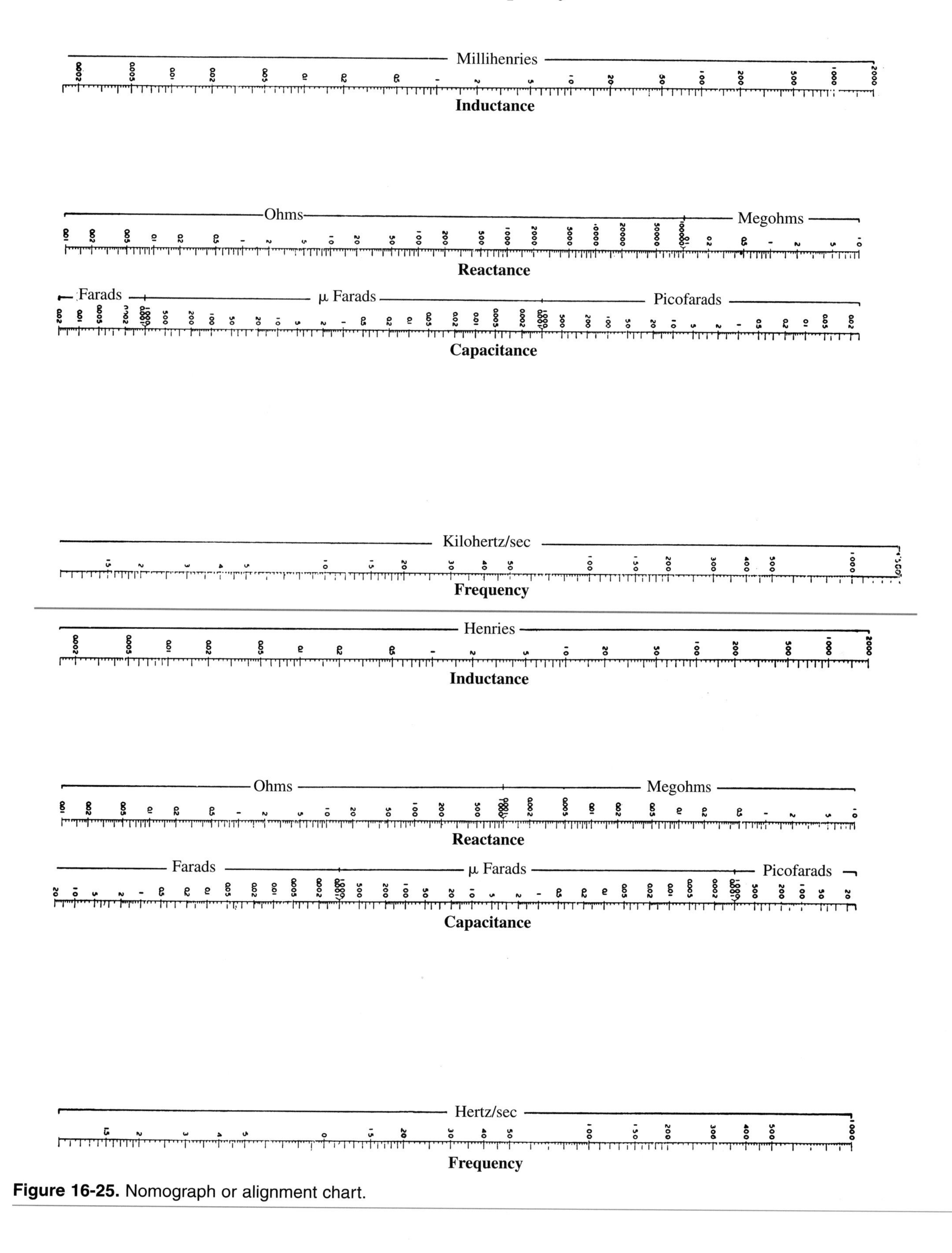

Figure 16-25. Nomograph or alignment chart.

6. At resonance in a parallel tuned circuit, line current is __________.
7. State the formula for determining the Q of a circuit.
8. A filter is a circuit that separates specific __________.
9. Name four general types of filters.
10. Distinguish between a high-pass and a low-pass filter.

For Discussion

1. How can a series resonant circuit develop voltages higher than the applied voltage?
2. Explain flywheel action and relate it to frequency.
3. Why is resonance indicated by a dip in the line current of a parallel RCL circuit?
4. What is the relationship between the Q and the selectivity of a tuned circuit?
5. List the similarities and opposites of capacitance circuits and inductive circuits.

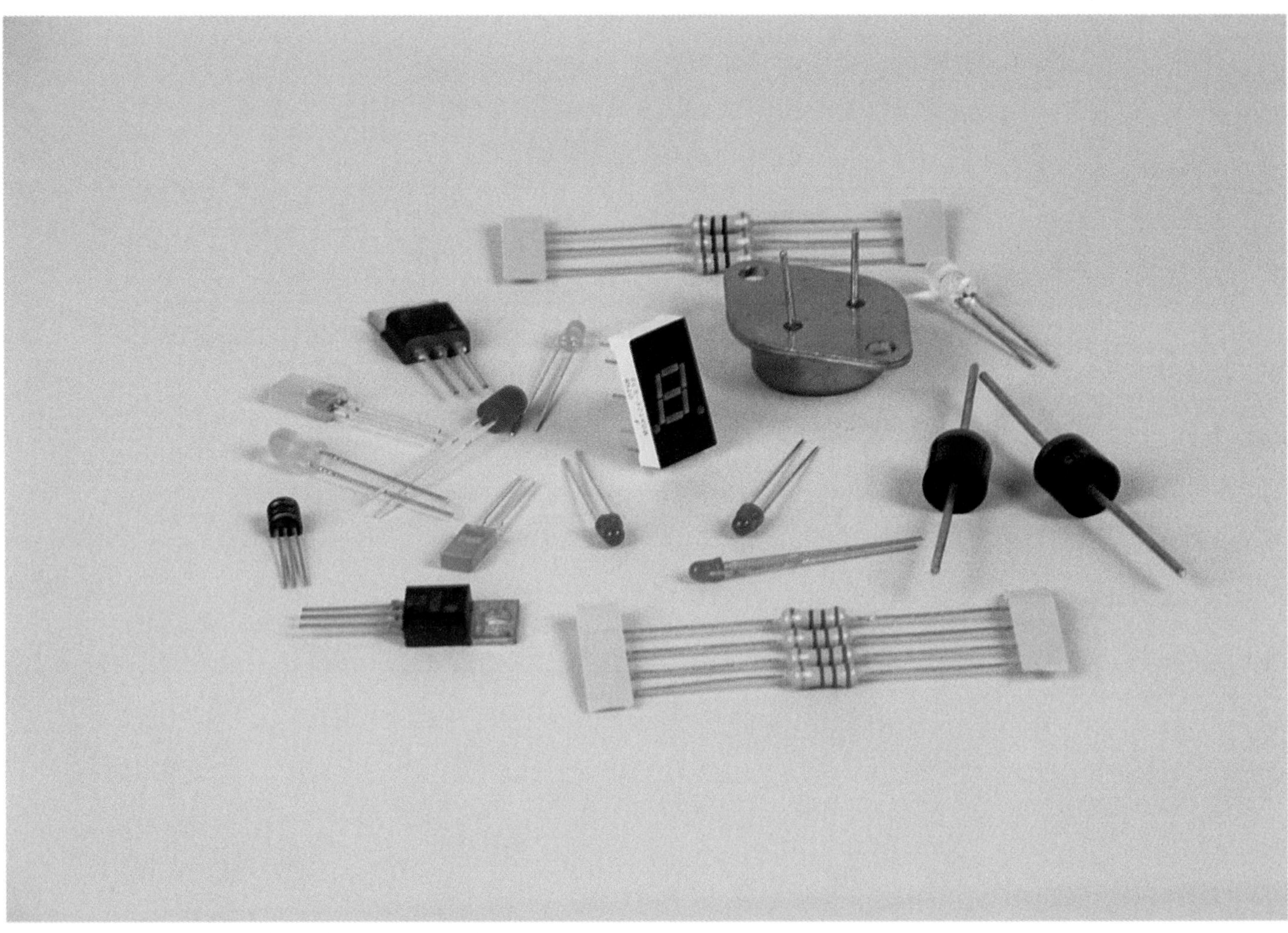

Resistors and a variety of semiconductor devices.

Introduction to Semiconductors and Power Supplies

Objectives

After studying this chapter, you will be able to:

- ❑ Define electronics.
- ❑ Explain the doping process. Explain how N-type and P-type materials are made.
- ❑ Discuss how N-type and P-type materials conduct electrical energy.
- ❑ Explain forward and reverse biasing.
- ❑ Discuss various types of semiconductor diodes.
- ❑ Explain the operation of a half-wave and full-wave rectifier.
- ❑ Explain power supply filtering.
- ❑ Explain power supply load characteristics.
- ❑ Outline various methods for regulating voltage.
- ❑ Discuss methods for raising voltages.
- ❑ Construct simple power supplies.

Key Words and Terms

The following words and terms will become important pieces of your electricity and electronics vocabulary. Look for them as you read this chapter.

anode	N-type
bridge rectifier	P-type
cathode	pentavalent
diode	percentage of voltage regulation
doping	potential barrier
electronics	potential hill
forward biased	rectification
full-wave rectifier	reverse biased
half-wave rectifier	ripple
hole	trivalent

Electronics can be defined as the study of the flow of electrons in active devices such as integrated circuits, transistors, and vacuum tubes. A time line showing the development of these three generations of electronic devices is shown in **Figure 17-1.**

During experiments in 1883 with the incandescent lamp, Thomas Edison noticed that electrons could be driven off the hot filament of a lamp. He called this the "Edison Effect." In 1904, J.A. Fleming built a diode (two element vacuum tube) that changed ac to dc. Changing an alternating current to a direct current is called ***rectification.*** That same year, Lee DeForest invented the first vacuum tube amplifier. It was called a triode. This material will be discussed in detail later in this chapter.

The first solid-state amplifier device was called the *transistor.* It was invented at Bell Laboratories in 1948. The technology used in transistors was key to the development of the third generation device, the integrated circuit. While it may seem that there have been more and more advances in electronics technology in the past few years, all recent advances are still based on the integrated circuit. Integrated circuitry, including the microchip, continues to get smaller and smaller, but the basic principles remain the same.

In the next three chapters, you will cover the three generations of electronic devices. The concepts of vacuum tubes, transistors, and integrated circuits will not be covered in chronological order. These systems will be presented in an order that will enhance your comprehension and mastery of this subject. The simple diode will be discussed first, followed by power supplies. These are the first systems encountered in almost every electronic system. Next, the vacuum tube will be introduced followed by the transistor and amplifier. The vacuum tube and transistor are presented together because they work on similar principles. The last concept in the series will be the integrated circuit. Integrated circuits are based on transistor technology. Your understanding of electronic systems will be greatly expanded upon completing these next three units.

17.1 SEMICONDUCTORS

The study of semiconductors, transistors and integrated circuits, requires an understanding of new

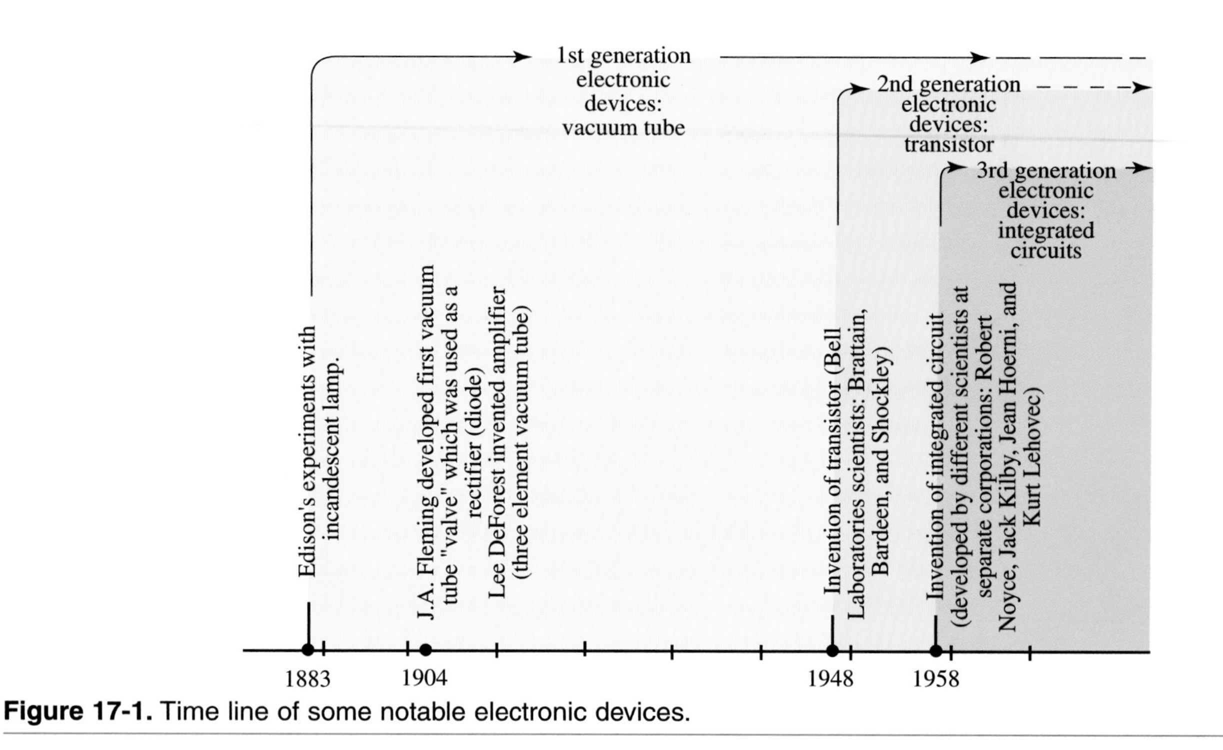

Figure 17-1. Time line of some notable electronic devices.

concepts in electronic theory. Let's review what have we learned about conductors in Chapter 1.

1. Copper and several other metals, such as silver and gold, make good conductors.
2. Certain materials, such as glass and rubber, are insulators.
3. Semiconductors are composed of materials that are neither good conductors nor good insulators.

Semiconductor diodes, transistors, and integrated circuits are all made of semiconductors. Their materials are neither good conductors nor good insulators. However, the ability of these devices to conduct electricity or to insulate can be greatly enhanced. It will be necessary to take a deeper look at atomic structure of materials to fully grasp the theory of semiconductor operation.

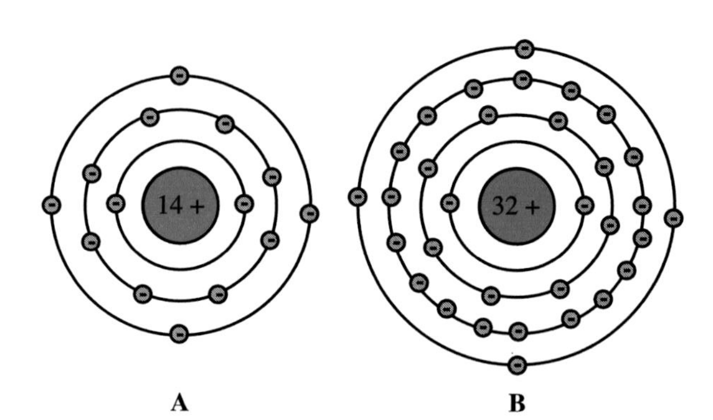

Figure 17-2. Atomic structures A–Silicon has an atomic number of 14. There are 14 protons within the nucleus and 14 electrons orbiting the nucleus. B–Germanium has an atomic number of 32. There are 32 protons in the nucleus and 32 electrons orbiting the nucleus.

Atomic Characteristics

Diagrams of the atomic structures of silicon (Si) and germanium (Ge) are shown in **Figure 17-2.** Silicon is in great supply. It makes up 28 percent of the world's surface. The atomic number for silicon is 14. The atomic number of germanium is 32. Recall that the *atomic number* is the number of protons or electrons that make up an atom of that element. Therefore, silicon has 14 electrons in its orbit. Germanium has 32 electrons in its orbit.

Note that both silicon and germanium have four electrons in their outer rings. These electrons on the outer rings are called ***valence electrons.*** In some atoms, valence electrons are able to bond with the valence electrons of another atom. This is called a ***covalent bond.*** The atoms are each bonded to their own nucleus and to each other. This bond creates a ***lattice crystalline structure.*** Germanium and silicon both lend themselves to this type of bond. **Figure 17-3** shows a silicon structure. It is a pure insulator. It will not conduct electricity.

Conduction of Electricity

Electrical current is best defined as the flow of electrons in a conductor. The flow of electrons results in energy. An example of this action is shown in **Figure 17-4.** An electron is added to one end of the conductor. Another electron leaves the opposite end. This chain reaction is the conduction of electricity by electrons.

Superconductivity

Superconductivity is a phenomenon in which certain metals, alloys, and ceramics conduct electricity *without* resistance. Superconductivity occurs in metals and alloys such as lead, mercury, and tin, at temperatures near *absolute zero* (–460°F [–273°C]). Some ceramics become superconductors at temperatures as high as –234°F (–148°C).

In 1911, Heike Kamerlingh Onnes discovered superconductivity while researching the effects of extremely cold temperatures on different metals. Onnes discovered that mercury lost all resistance to the flow of electricity when cooled to about 4° Kelvin.

Before Onnes' discovery, there wasn't any way to eliminate resistance. Although copper, aluminum, silver, and gold are excellent conductors, there is a loss of some electrical energy because of resistance.

The modern theory of superconductivity, the *BCS theory*, was developed by American physicists John Bardeen, Leon N. Cooper, and John Robert Schrieffer. According to the theory, a superconductor has no electrical resistance because of an attractive interaction between its electrons that results in the formation of pairs of electrons. These electron pairs are bound to one another and flow without resistance around impurities and other imperfections. In an ordinary conductor, resistance occurs because its unbound electrons collide with imperfections and then scatter.

Superconductivity is used in the field of *electromagnetics.* Powerful superconducting magnets use less electricity than ordinary magnets. These magnets are being used in many areas including medicine, transportation, and the study of physics.

In the medical field, *magnetic resonance imaging (MRI)* machines use a superconducting electromagnet to generate a powerful magnetic field. MRI technology allows physicians to identify abnormalities without performing surgery and without exposing the patient to radiation.

Maglev (magnetically levitated) trains use superconducting electromagnetic coils. The coils are mounted on the bottom of the train and aluminum coils or sheets are set in the guideway. When the train moves over the guideway, it induces a current. The guideway coils become electromagnets with the same polarity as the train coils. The two types of coils then repel each other. The repulsion force is strong enough to raise the train.

Superconducting magnets allow physicists to build more efficient *particle accelerators*. The ring of superconducting magnets produces a magnetic field that keeps the particles in the vacuum chamber by bending their paths into circular orbits.

The discovery of superconducting ceramic materials has made it possible to use superconductors in computers and other electronic devices. However, the discovery of a superconductor that could operate at room temperature would make it possible to use superconductors in an infinite number of applications. One of the most practical would be the generation and transmission of electricity. Generators would be smaller and more efficient and power lines could carry current from the fuel source over long distances without any loss of power.

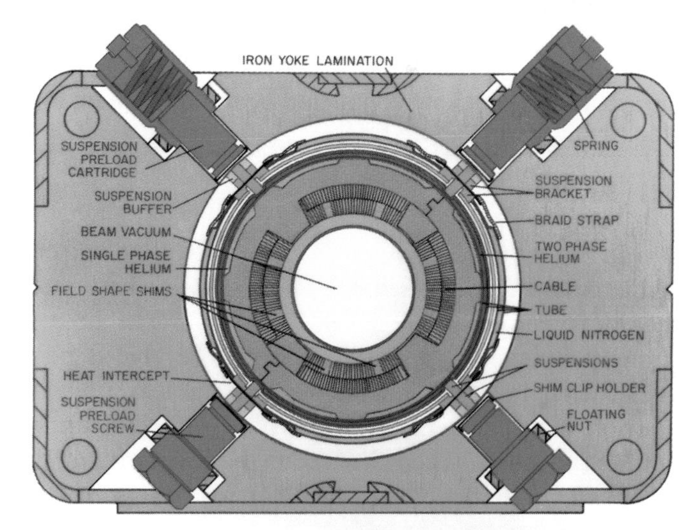

Scientists come from all over the world to study atomic particles at the Fermi National Accelerator Laboratory (Fermilab) in Batavia, Illinois, The particle accelerator at Fermilab lies in an underground tunnel that forms a circle 1.25 miles (2 km) in diameter. Protons are accelerated almost to the speed of light and reach an energy of up to 900 giga electronvolts (GeV). The bottom illustration is the cross section of a superconducting quadrupole magnet. (Fermilab Visual Media Services)

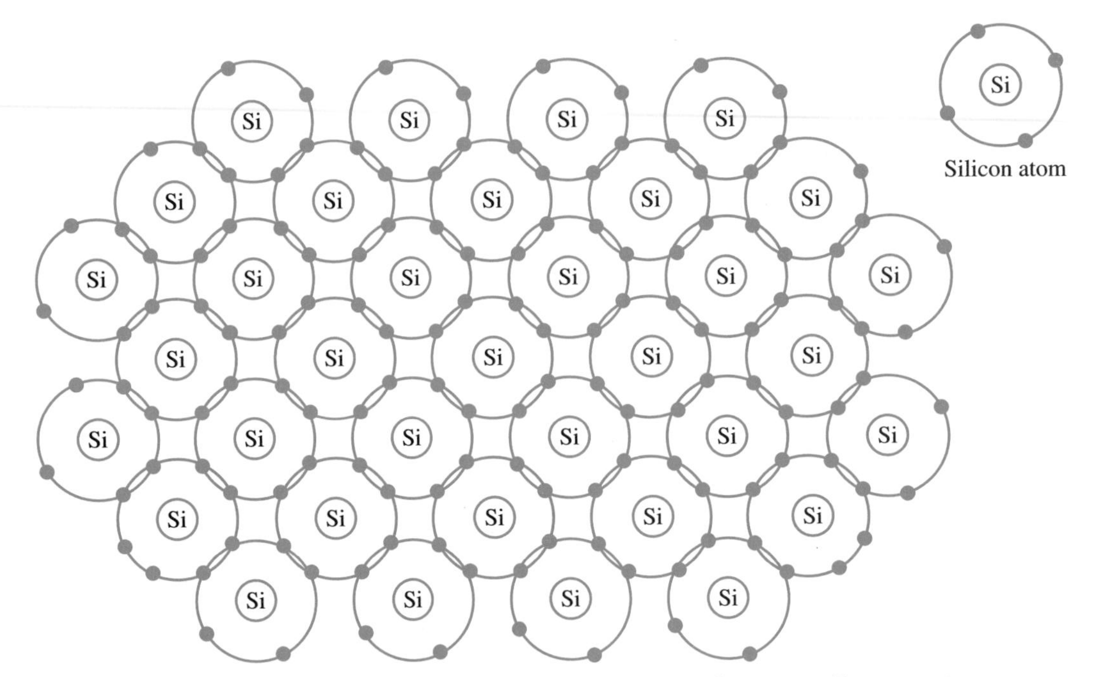

Figure 17-3. Silicon atoms form covalent bonds. This produces a lattice crystalline structure.

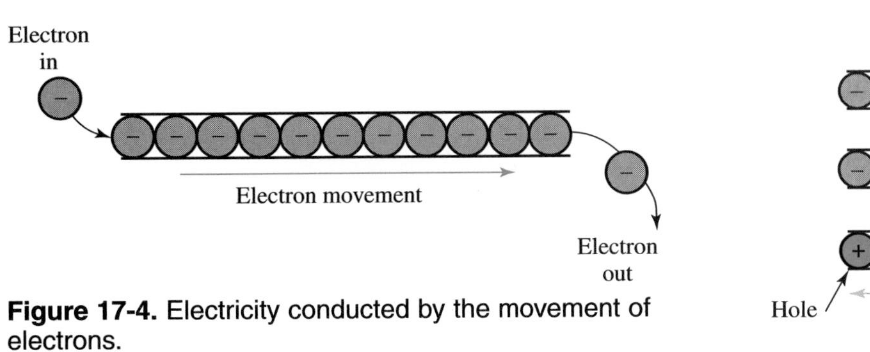

Figure 17-4. Electricity conducted by the movement of electrons.

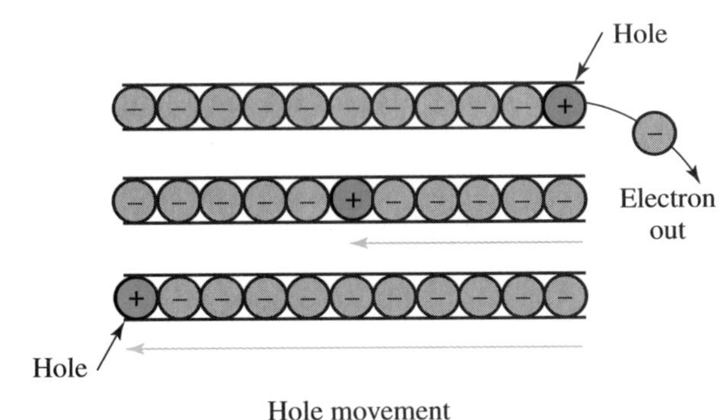

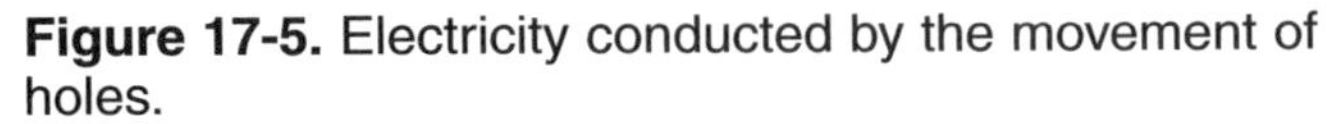

Figure 17-5. Electricity conducted by the movement of holes.

In **Figure 17-5,** an electron is removed from one end of the conductor. This action leaves a ***hole*** where there should be an electron, but there is not. The hole is strongly attracted to another electron. The hole is positive. The next electron fills up the hole and leaves a positive hole in its place. Each positive hole is filled in turn by the next electron. Then the vacant hole appears at the other end of the conductor. This chain reaction is the *conduction of electricity by holes.* The study of semiconductors is concerned with conduction by electrons and by holes.

Doping

Recall that *pure* silicon is a good insulator. Adding an impurity to the pure silicon changes the conduction traits of the material. The process of adding impurities to pure semiconductor material is called ***doping.*** The result of doping is an ***extrinsic semiconductor.*** This means the material is not in its pure form. A pure semiconductor is ***intrinsic.***

Trivalent and pentavalent impurities are most often used for doping in transistors. ***Trivalent*** impurities have three valence electrons. ***Pentavalent*** impurities have five valence electrons. **Figure 17-6** is a chart showing the quantity of valence electrons in several materials.

Adding doping elements to semiconductor material results in either an excess or shortage of electrons in the covalent bond. For example, arsenic, which is a pentavalent (five electrons in its outer shell), is added to silicon. Four valence electrons of arsenic form a covalent bond with the four valence electrons of silicon. One electron from the arsenic remains free, **Figure 17-7.** The arsenic donated a free electron to the structure. All pentavalent impurities donate free electrons. They are called ***donor***

	Element	Symbol	Atomic number	Valence electrons
Trivalent acceptor element produces P-type materials	boron aluminum gallium indium	B Al Ga In	5 13 31 49	3 3 3 3
Quadvalent pure semi-conductor material	silicon germanium	Si Ge	14 32	4 4
Pentavalent donor element produces N-type materials	phosphorus arsenic antimony	P As Sb	15 33 51	5 5 5

Figure 17-6. Comparison of trivalent, quadvalent, and pentavalent elements. Quadvalent elements are pure semiconductors.

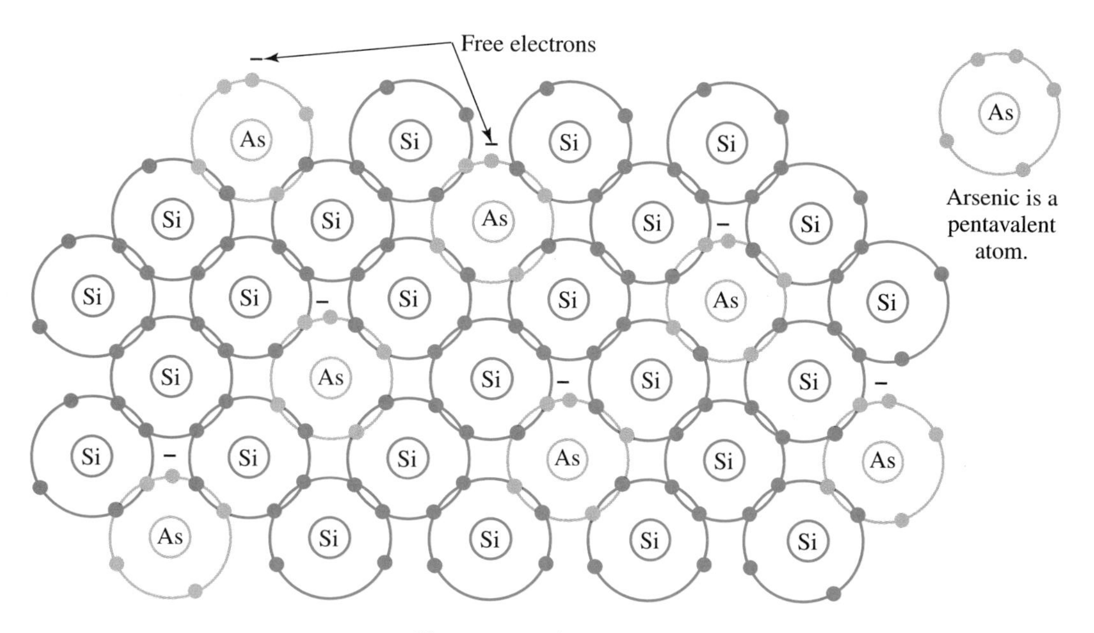

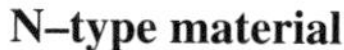

Figure 17-7. Arsenic (As) is a pentavalent impurity. When combined with a silicon crystal structure, an N-type material is formed.

impurities. In this structure, electricity is conducted by (negative) electrons. Therefore, the crystal becomes an ***N-type*** crystal.

What occurs when a trivalent impurity is added to a pure semiconductor material? It has only three electrons to join into a covalent bond. One more electron is needed to complete the lattice structure. A hole remains in the place of the missing electron, **Figure 17-8.** This hole will accept an electron. A trivalent impurity, then, is called an ***acceptor impurity.*** In this structure, electricity is conducted by (positive) holes. The crystal become a ***P-type*** crystal.

Figure 17-9 shows conduction through an N-type crystal. **Figure 17-10** shows conduction through a P-type crystal. Notice that the hole moves in the opposite direction of the electron flow.

Review Questions for Section 17.1

1. Define electronics.
2. Electrons on the outer ring of an atom are called __________ __________.
3. What is a covalent bond?

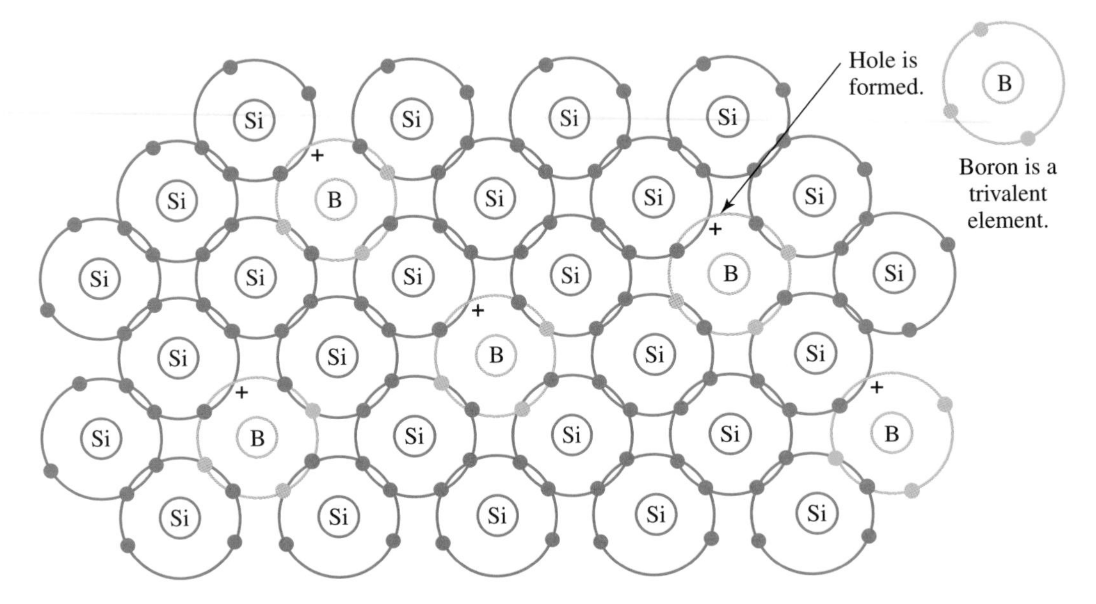

Figure 17-8. Boron is a trivalent impurity. When boron is combined with a silicon structure, positive holes are created.

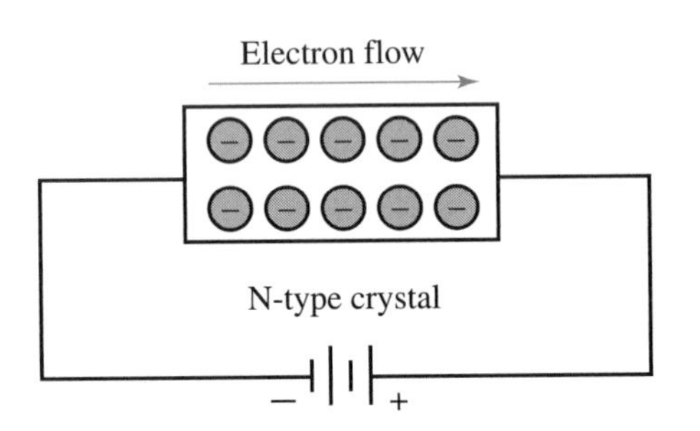

Figure 17-9. Conduction through an N-type crystal by electrons.

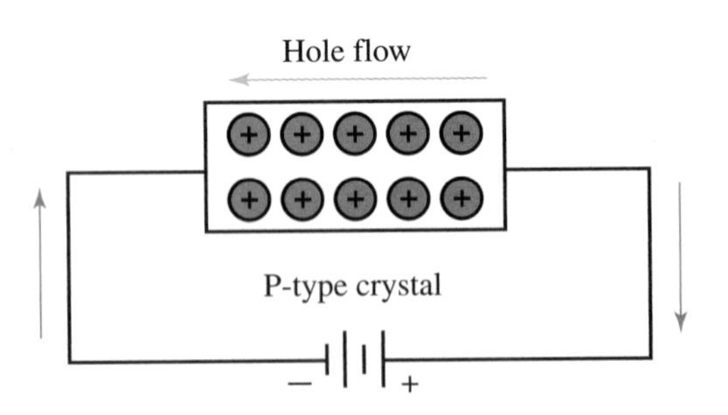

Figure 17-10. Conduction through a P-type crystal by holes.

4. Doping:
   a. is the process of adding impurities to pure semiconductor material.
   b. results in an extrinsic semiconductor.
   c. changes the conduction traits of a pure material.
   d. All of the above.
5. Pentavalent impurities are known as __________ impurities and they produce __________-type crystals.
6. Trivalent impurities are known as __________ impurities and they produce __________-type crystals.

17.2 SEMICONDUCTOR DIODES

A ***diode*** is a device designed to permit electron flow in one direction and block flow from the other direction. A diode consists of two electrodes: a cathode and an anode. A ***cathode*** is an electrode that emits (gives off) electrons. An ***anode*** collects the electrons and puts them to use.

A ***semiconductor diode*** is the result of fusion between a small N-type crystal and a P-type crystal, **Figure 17-11.** At the junction of the two crystals, the carriers (electrons and holes) tend to diffuse. Some electrons move across the barrier to join holes. Some holes move across the barrier to join electrons. Remember that unlike charges attract each other.

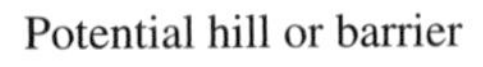

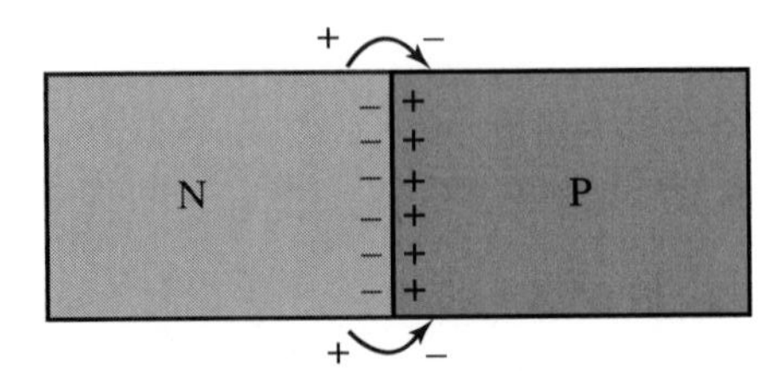

Figure 17-11. A semiconductor diode made of N-type and P-type crystals.

Due to this diffusion, a small voltage, or potential, exists between the regions near the junction. How does this happen? The region of the P crystal near the junction becomes negative. It has taken electrons from the N crystal. The region of the N crystal near the junction becomes positive. It has lost some electrons, but gained holes. This voltage or potential is called a ***potential hill*** or a ***potential barrier.*** The barrier prevents the other electrons and holes in the crystal from joining.

The symbol for a semiconductor diode is shown in **Figure 17-12.** The arrow side of the symbol denotes the anode portion of the diode. The anode contains the P-type material. The bar side of the symbol denotes the cathode portion of the diode. The cathode contains the N-type material.

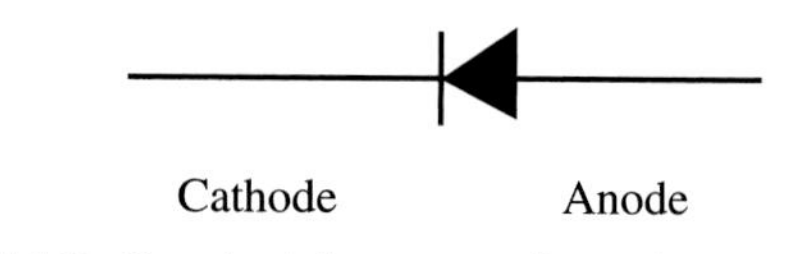

Figure 17-12. Symbol for a semiconductor diode.

Forward and Reverse Biasing

A voltage (potential) is connected across a diode in **Figure 17-13.** The positive terminal of the source is connected to the P crystal. The negative source is connected to the N crystal.

The negative electrons in the N crystal move toward the barrier. The positive holes in the P crystal move toward the barrier. The source voltage *opposes* the potential barrier and reduces its barrier effect. This allows the electrons and holes to join at the barrier. Therefore, current flows in the circuit. It flows in the P crystal by holes. It flows in the N crystal by electrons. The diode is *biased* in a *forward* direction, or ***forward biased.***

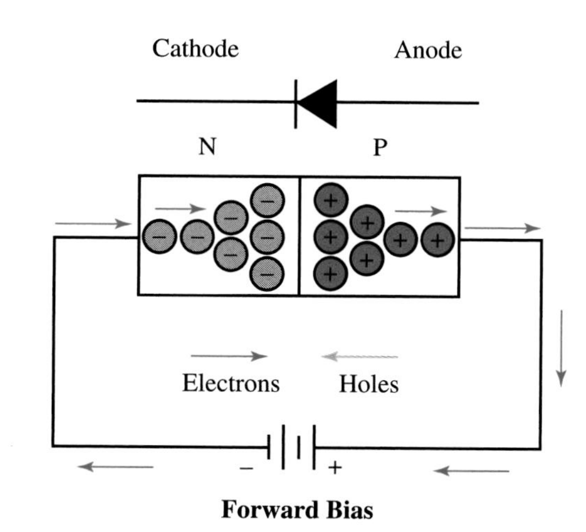

Figure 17-13. Conduction through a junction diode biased in the forward direction.

There is a minimum forward biased voltage needed to overcome the potential barrier for current to flow in the circuit. This minimum voltage depends on the type of semiconductor material used. It requires approximately 0.6 volts of potential to forward bias a silicon diode and 0.2 volts for a germanium diode. Once this barrier voltage is exceeded, current flows through the circuit.

Figure 17-14 the positive source is connected to the N crystal and the negative source is connected to the P crystal. This shows the same junction diode *biased* in a *reverse* direction, or ***reverse biased.***

In reverse bias, the source voltage aids the potential barrier. Electron/hole combinations are limited at the junction. The electrons in the N crystal are attracted to the positive source terminal. Very little current will flow in the circuit. A reverse voltage can be increased to a point where the diode will break down. The amount of reverse bias will vary according to the diode specifications.

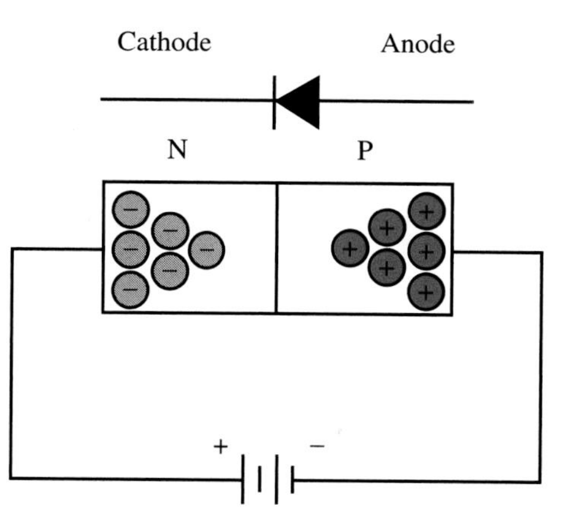

Figure 17-14. No conduction through a junction diode biased in a reverse direction.

Types of Semiconductor Diodes

Diodes are used for a number of purposes. Voltage rectification, voltage regulation, and even light production are some of their various uses. Following is a brief description of some diode types you might encounter.

Point contact diodes

The point contact diode is used for detection and rectification. It consists of a small piece of N-type germanium crystal. Against this crystal is pressed a fine phosphor bronze wire. While the diode is being made, a high current is run through the diode from wire to crystal. This forms a P-type region around the contact point in the germanium crystal. The point contact diode, therefore, has both the P- and N-type crystals. Its operation is like that of the junction diode.

Silicon rectifiers

Diodes conduct current more easily in one direction than in the other. This process is used to produce rectification. A key semiconductor diode is the ***silicon rectifier.*** **Figure 17-15** shows two typical rectifiers. However, silicon rectifiers come in a wide variety of shapes and sizes. **Figure 17-16** shows some common diode outlines.

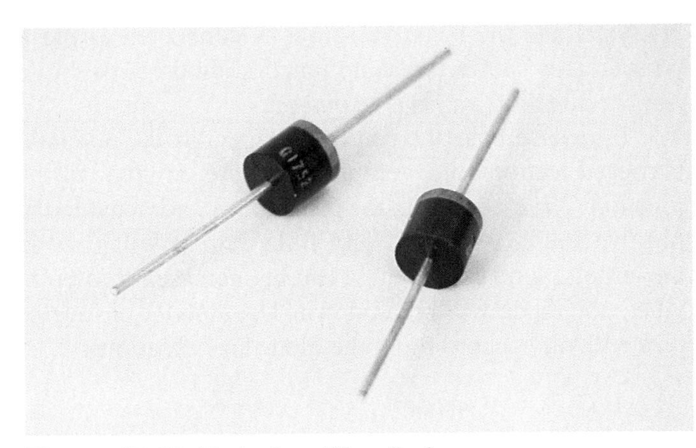

Figure 17-15. Typical rectifier diodes.

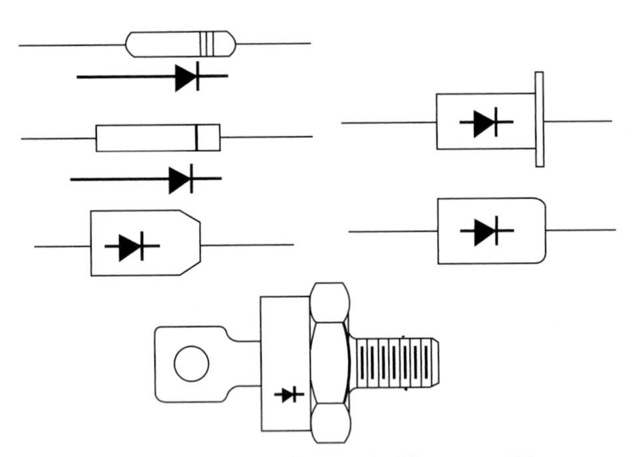

Figure 17-16. Some widely used silicon rectifier outlines.

Silicon rectifiers have high forward-to-reverse current ratios. This is the ratio of current allowed in the forward direction compared to the current allowed in the reverse direction. They can achieve rectification efficiencies of greater than 99 percent. These rectifiers are very small, light, and built to last a long time. In addition, they can be made resistant to shock and mishandling. Silicon rectifiers do not decay with age like a vacuum tube rectifier does.

Zener diodes

An electronic device that can be used as a voltage regulator is the ***zener diode,*** **Figure 17-17.**

Figure 17-18 shows a characteristic curve for a zener diode. When the diode is forward biased, it acts like a diode or a closed switch. However, the zener has unique reverse bias qualities that make it differ from the typical diode. The zener will go into reverse bias at various voltages. The amount of voltage required for reverse bias varies according to the zener diode selected. Some typical reverse bias voltages are 2.4 V, 5.1 V, 6.0 V, 9.1 V, 12.0 V, etc. At this point, when the applied voltage is increased, the forward current increases. This small reverse current flows until the diode reaches the zener breakdown point, V_2 in Figure 17-18. At the ***zener breakdown point,*** the zener diode is able to maintain a fairly constant voltage as the current varies over a certain range. Because of this attribute, the diode provides excellent voltage regulation. **Figure 17-19** shows a zener diode being used as a simple shunt regulator.

Light emitting diodes

The ***light emitting diodes (LED)*** is a special function diode, **Figure 17-20.** When connected in the forward bias direction, it emits light. Light emitting diodes are made from semiconductor compounds such as gallium arsenide, gallium arsenide phosphide, and gallium phosphide.

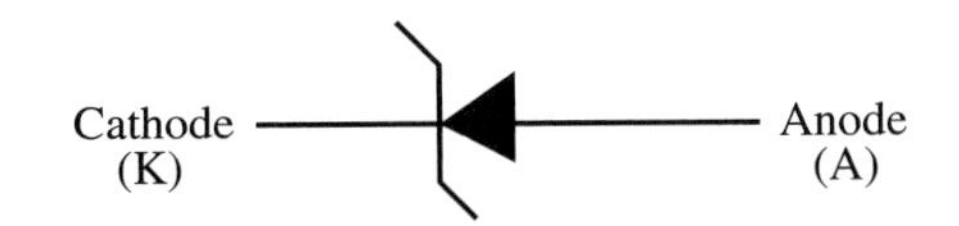

Figure 17-17. Schematic symbol for the zener diode.

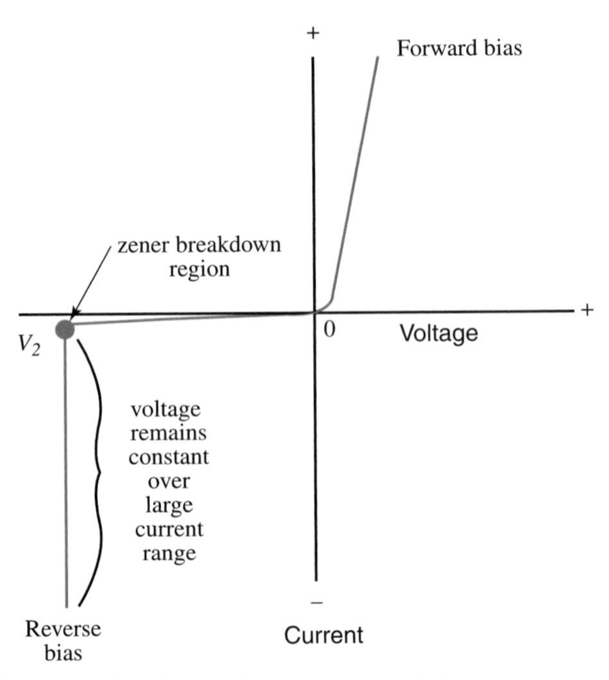

Figure 17-18. Zener diode characteristics.

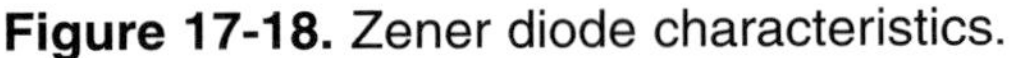

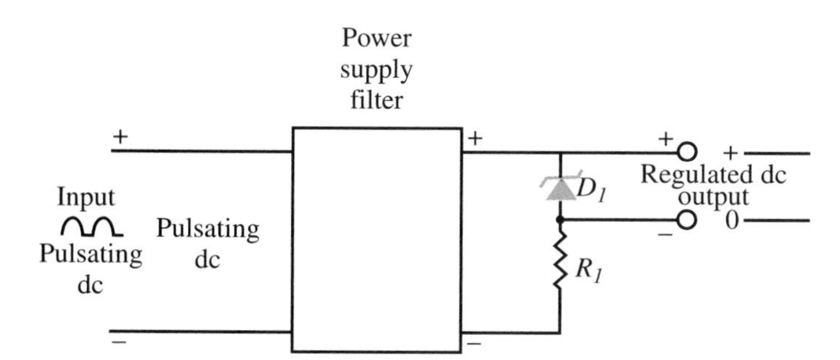

Figure 17-19. Zener diode as a shunt voltage regulator.

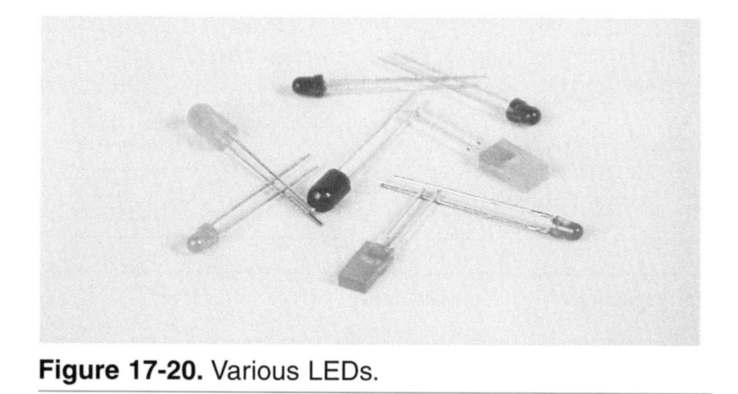

Figure 17-20. Various LEDs.

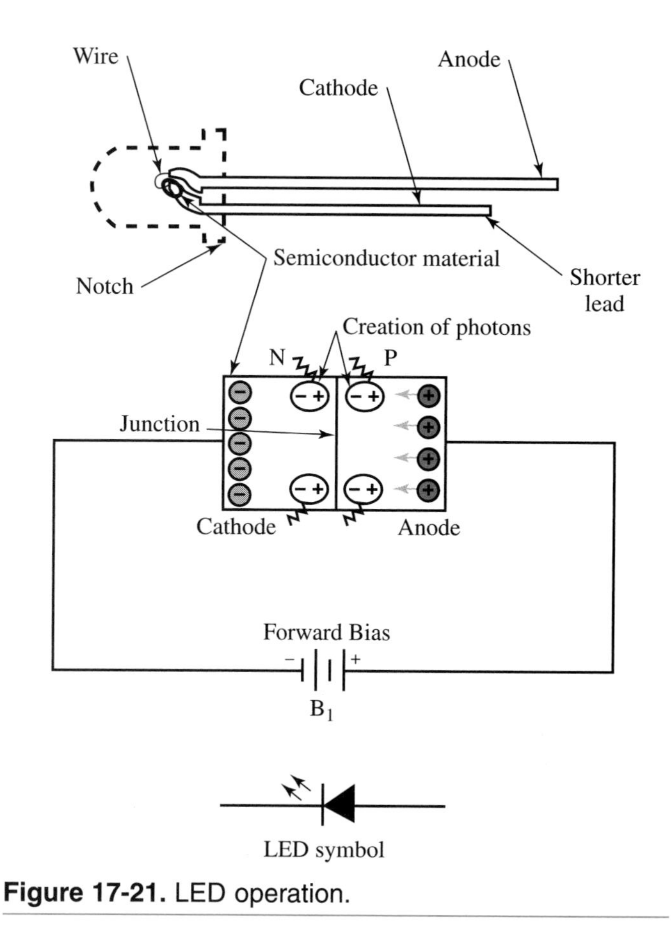

Figure 17-21. LED operation.

An LED receives energy from a dc power supply or a battery. This causes the electrons and holes to combine in the PN junction region of the diode. This combination creates ***photons,*** which are "particles" of light that can be seen. Each photon is produced from one hole (+ charge) and one electron (– charge). Different LED colors are achieved by using different materials in the manufacture of the LED. **Figure 17-21** shows the operating principle of an LED. The LED must be connected to a power source with some form of resistance in series with the LED. If the LED is connected directly to a source with forward bias, the LED will be destroyed.

Biasing diagrams for some common LEDs are shown in **Figure 17-22.** Note that the cathode lead is the shortest lead. It is also closest to the notch on certain diodes. LEDs have many uses. For example, they are used on the instrument panel of cars, clocks, watches, and many other electric devices.

Diode Characteristics and Ratings

Ratings of diodes and rectifiers are commonly based on current and voltage capabilities and peak inverse voltage. The ***peak inverse voltage (PIV)*** rating used by manufacturers defines the greatest reverse voltage that can be applied across a rectifier without damaging it. An example of this is one amp at 50 PIV for a 1N4001 silicon rectifier. This means that a 1N4001 silicon rectifier can withstand a maximum applied reverse voltage of 50 volts.

Many diodes are identified by a "1N." Examples of this are 1N4001 or 1N5400. Some manufacturers use their own labels: HEP 320, SK 3051, or 276-1102.

Series and Parallel Rectifier Arrangements

Diodes or rectifiers can be connected in series or parallel. This connection is made to improve the voltage or current capabilities of a single rectifier. Connecting diodes in *series* increases *voltage ratings* over the value of a single rectifier. Diodes can be connected in *parallel* to improve the *current handling* ability of the combination over that of only one diode. Care must be used when connecting diodes in a series or parallel arrangement. The electrical characteristics must match exactly. Otherwise, the current or voltage drop will not be exactly equal across both diodes—even when the diodes have the same part number.

Testing Diodes

You have learned that diodes with a forward bias direction have a low resistance to current flow. A reverse

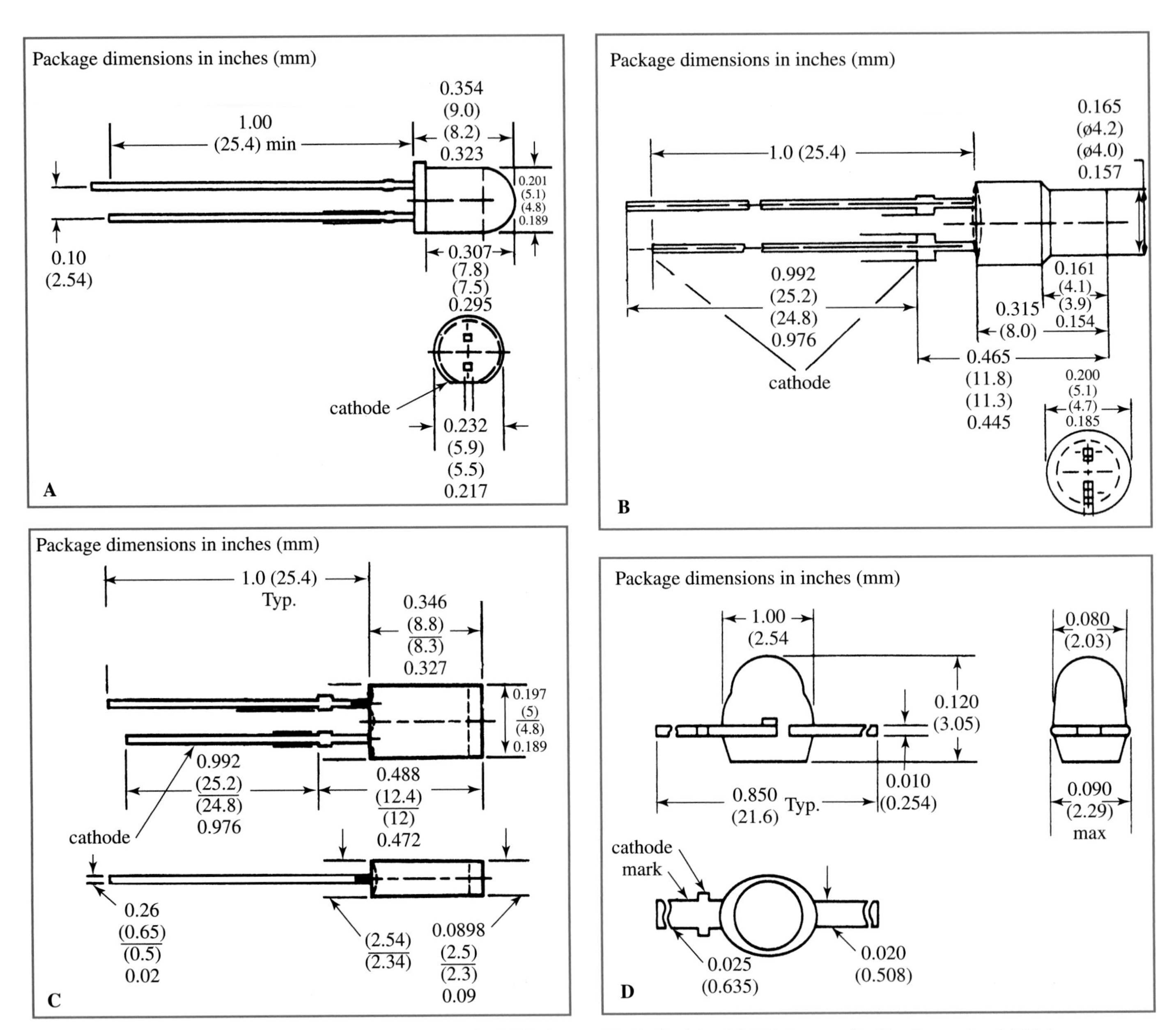

Figure 17-22. LED biasing diagrams. A–T 1 3/4 LED lamp. B–Cylindrical LED lamp. C–Rectangular LED lamp. D–Miniature axial lead LED lamp. (Siemens)

bias direction has a high resistance. This allows diodes to be tested using an ohmmeter.

To test a diode in the forward bias direction, connect the diode across the ohmmeter. Set the range switch on the low resistance setting (×1 or ×10). The reading should be low. A low reading is from 50 to 1000 ohms. See **Figure 17-23,** part A

To test a diode in the reverse bias direction, connect the diode across the ohmmeter leads. Set the range switch on a high resistance setting, such as ×10 k, ×100 k, or ×1 M. A good diode will show a very high resistance. See Figure 17-23, part B.

Many new digital meters have a diode test function built in. One model with a diode symbol on the selector switch is shown **Figure 17-24.** A properly connected diode can be tested by turning the switch to the diode function.

Review Questions for Section 17.2

1. A __________ is a device that will pass current in one direction and block it in the other direction.
2. What occurrence prevents electrons and holes in a crystal from joining?
3. Draw and label the symbol for a semiconductor diode and an LED.

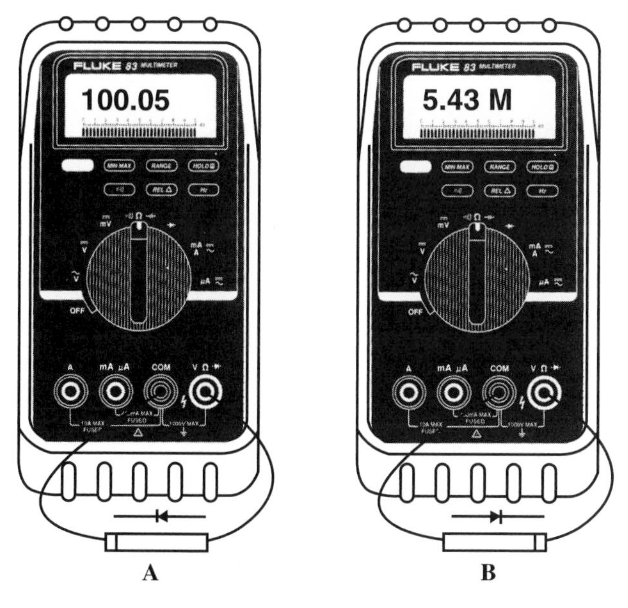

Figure 17-23. Testing a diode using an ohmmeter. A–Testing forward bias. B–Testing reverse bias.

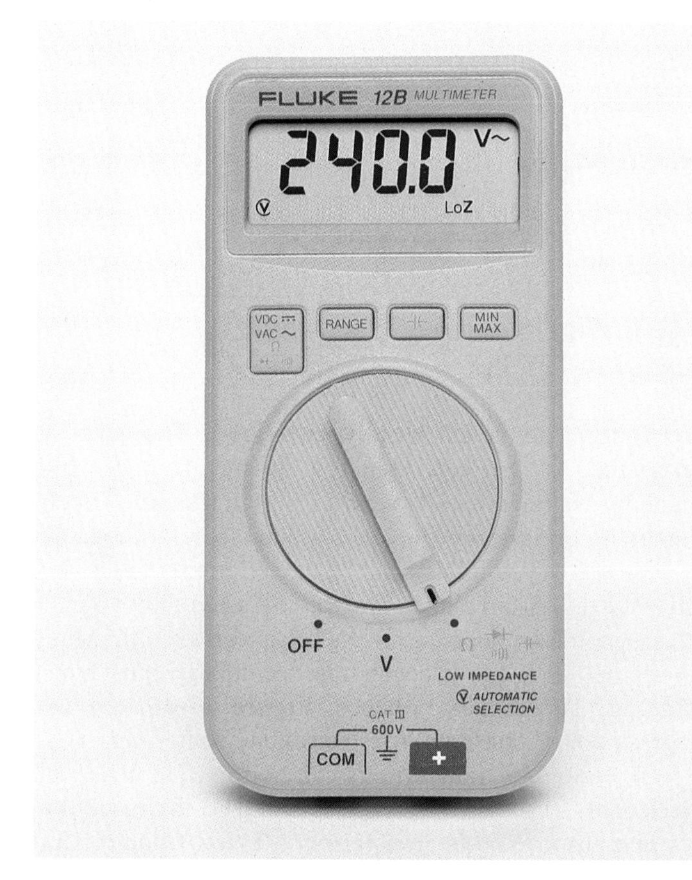

Figure 17-24. This digital meter can also be used to test diodes. (Fluke Corp.)

4. In __________ bias, the source voltage aids the potential barrier and resists electron flow. In __________ bias, the source voltage opposes the potential barrier and allows electron flow.
5. How much potential is required to forward bias a silicon diode and a germanium diode?
6. Look up a 1N4003 diode in a parts catalog, and determine the current rating and PIV rating.
7. A zener diode is an electronic device that acts as a(n) __________ __________.
8. A particle of light is called a(n) ____________.

17.3 POWER SUPPLIES

Proper operation of electronic equipment requires a number of source voltages. Low dc voltages are needed to operate ICs and transistors. High voltages are needed to operate CRTs and other devices. Batteries can provide all of these voltages. However, electricity for electrical and electronic devices are commonly supplied by the local power company. This power comes out of an outlet at 115 volt ac, with a frequency of 60 hertz. Different voltages are needed to operate some equipment.

A ***power supply*** is an electronic circuit designed to provide various ac and dc voltages for equipment operation.

Power Supply Functions

The complete power supply circuit can perform these functions:

1. Step voltages up or step voltages down, by transformer action, to the required ac line voltage.
2. Provide some method of voltage division to meet equipment needs.
3. Change ac voltage to pulsating dc voltage by either half-wave or full-wave rectification.
4. Filter pulsating dc voltage to a pure dc steady voltage for equipment use.
5. Regulate power supply output in proportion to the applied load.

A block diagram illustrating these functions is shown in **Figure 17-25.** Note that certain functions are not found in every power supply. See **Figure 17-26** for a typical commercial power supply.

Power Transformers

The first device in a power supply is the transformer. Its purpose is to step up or step down alternating source voltage to values needed for radio, TV, computer, or other electronic circuit use. Review transformer action, Chapter 12.

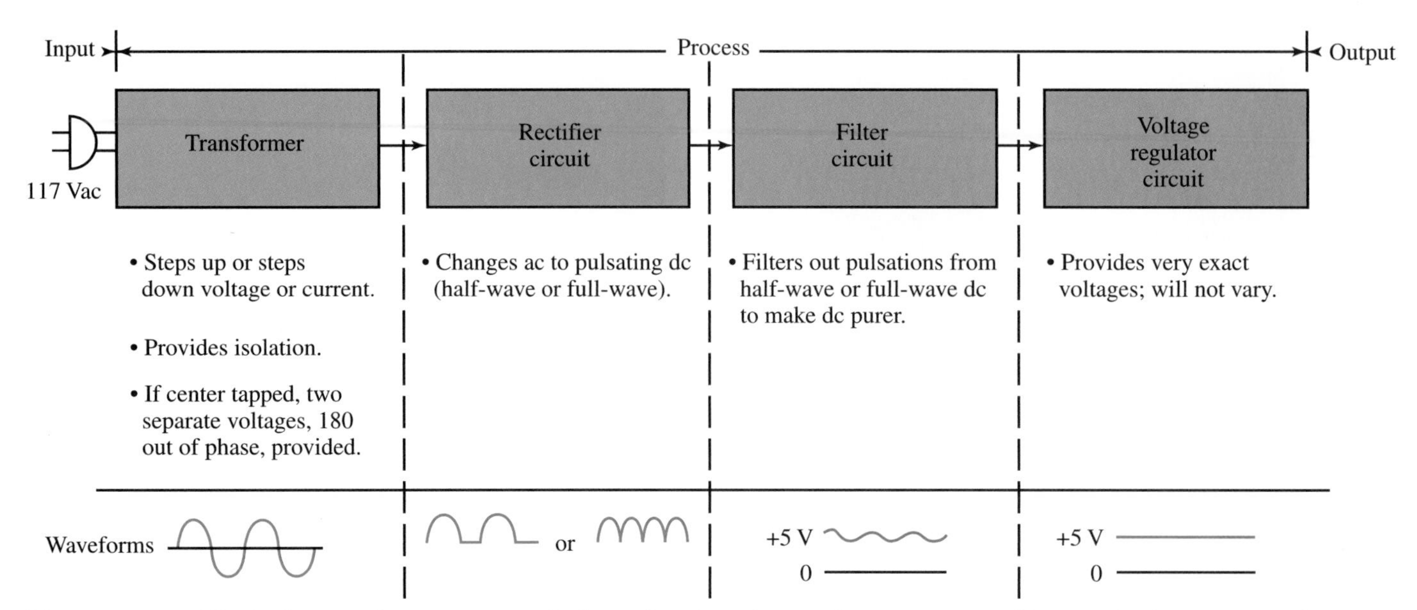

Figure 17-25. Block diagram for power supply. Input is 117 volts ac. Processes used in a typical power supply are shown below the blocks. Output of power supply can be dc or ac. The output of this supply is five volts dc.

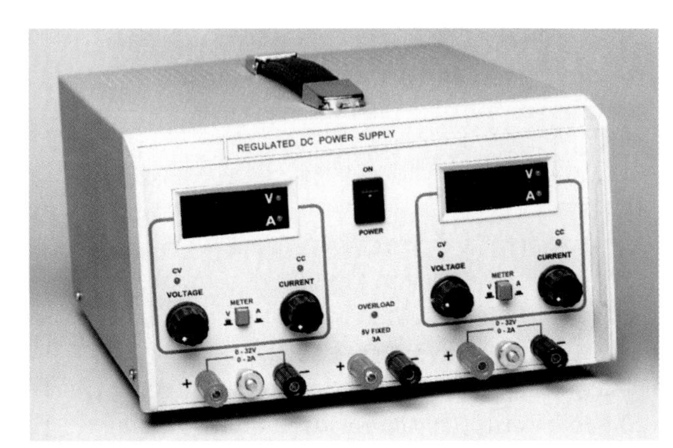

Figure 17-26. Regulated dc power supply. (Knight Electronics)

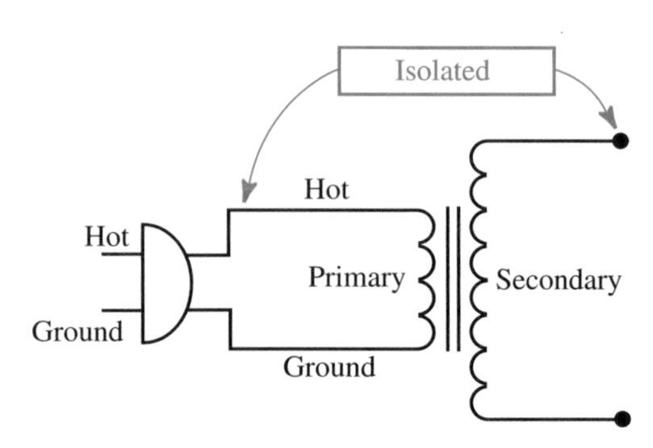

Figure 17-27. Isolation in a transformer.

Most transformers do not have any electrical connection between the secondary and primary windings. See **Figure 17-27.** This means that the transformer isolates the circuit connected to the primary from the circuit connected in the secondary. Isolation, defined in Chapter 12, is a term that means there are no electrical connections between the primary and secondary on the transformer.

An ***isolation transformer*** is a transformer that has the specific purpose of isolating the primary circuit from the secondary circuit. Using an isolation transformer is a safety feature because it helps prevent shocks in the secondary. Our body or hands must be joined across both leads of the secondary connections in order to receive a shock.

The safety condition described above does *not* hold true in the primary with commercial ac provided by the power company. One connection is ***hot,*** which means that the connection is electrically energized. The other is ***grounded,*** or ***neutral.*** Standing on the ground while touching the hot connection will result in a shock. Touching the ground connection alone will not result in a shock.

Secondary windings can be tapped to provide different voltages. A tap placed midway between the two ends of a secondary winding is called a ***center tap.*** Many power supplies use a center tap secondary transformer winding. The tapped voltages, **Figure 17-28,** are 180 degrees out of phase with respect to the center tap.

A variety of transformers can be found in nearly all electronic devices. You should understand the basic theory and purpose of the transformer. Review Chapter 12 if necessary.

LESSON IN SAFETY:

Transformers produce high voltages that can be very dangerous. Proper respect and extreme caution must be used at all times when working with, or measuring, high voltages.

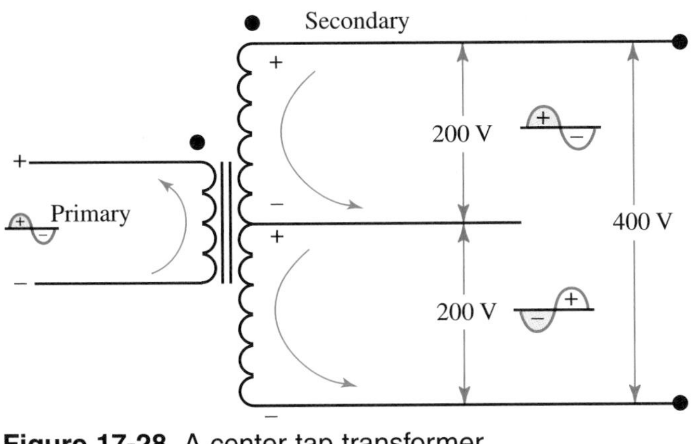

Figure 17-28. A center tap transformer.

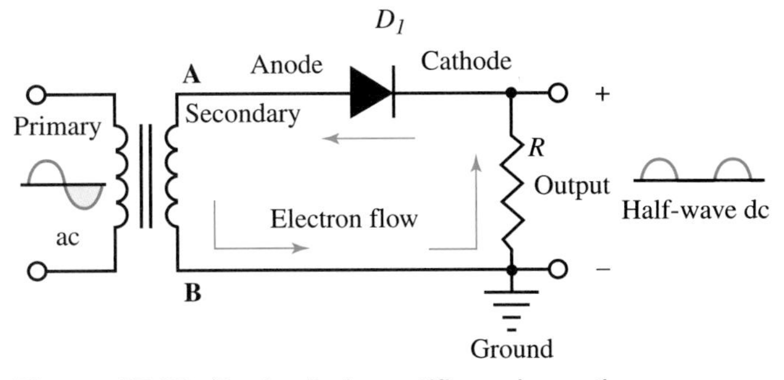

Figure 17-29. Basic diode rectifier schematic.

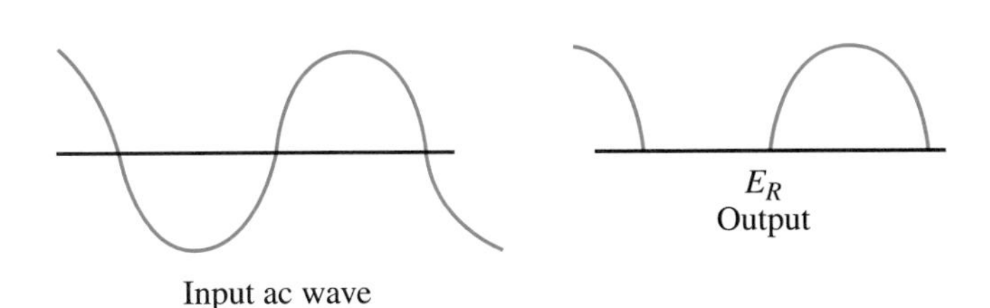

Figure 17-30. Input and output waveforms of a diode rectifier.

Half-Wave and Full-Wave Rectification

After a voltage has gone through a power supply's transformer, the next step is rectification. The process of changing an alternating current to a pulsating direct current is called rectification. When changing and ac signal to dc, there are two types of rectification: half-wave rectification and full-wave rectification. With the half-wave rectifier, only half of the input signal passes on through the rectifier. With the full-wave rectifier, the entire input wave is passed through.

Half-wave rectification

In **Figure 17-29,** the output of a transformer is connected to a diode and a load resistor that are in series. The input voltage to the transformer appears as a sine wave. The polarity of the wave reverses at the frequency of the applied voltage. The output voltage of the transformer secondary also appears as a sine wave. The magnitude of the wave depends on the turns ratio of the transformer. The output is 180 degrees out of phase with the primary. The top of the transformer (point A) is joined to the diode anode. Note that the B side of the transformer is connected to ground. During the first half cycle, point A is positive. The diode conducts, producing a voltage drop across resistor R equal to IR. During the second half cycle, point A is negative. The diode anode is also negative. No conduction takes place, and no IR drop appears across R.

An oscilloscope connected across R produces the wave form shown to the right in **Figure 17-30.** The output of this circuit consists of pulses of current flowing in only *one* direction and is at the *same* frequency as the input voltage. The output is a pulsating direct current.

Only one half of the ac input wave is used to produce the output voltage. This type of rectifier is called a ***half-wave rectifier.*** Look at the polarity of the output voltage in Figure 17-30. One end of the resistor R is connected to ground. The current flows from ground to the cathode. This connection makes the end of R connected to the cathode positive as shown in Figure 17-29.

A negative rectifier can be made by reversing the diode in the circuit, **Figure 17-31.** The diode conducts when the cathode becomes negative causing the anode to become positive. The current through R would be from anode to ground making the anode end of R negative and the ground end of R more positive. Voltages taken from across R, the output, would be negative with respect to ground. This circuit is called an ***inverted diode.*** It is used when a negative supply voltage is required.

It is possible to have a power supply that provides half-wave rectification without the use of a transformer. This circuit is not isolated. There is no step up or step

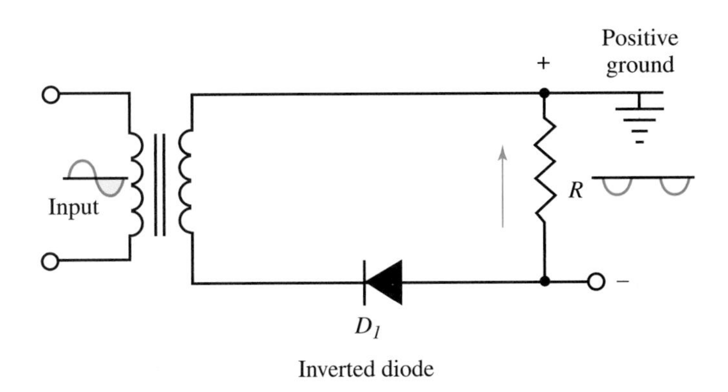

Figure 17-31. An inverted diode produces a negative voltage.

down of current voltages. This circuit is a simpler, less costly design, and since there is no transformer, it can be used in smaller spaces, **Figure 17-32.**

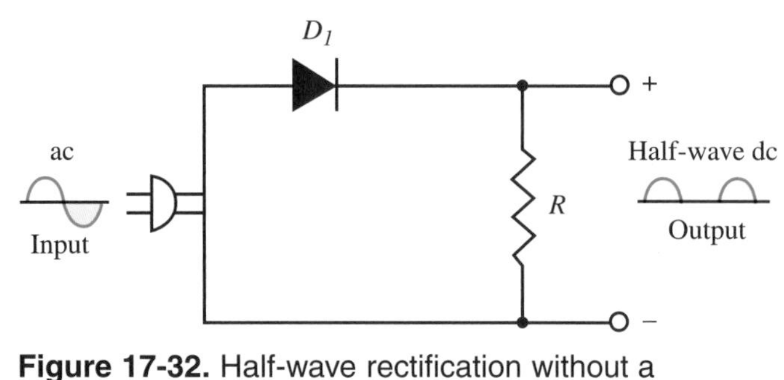

Figure 17-32. Half-wave rectification without a transformer.

Full-wave rectification

The pulsating direct voltage output of a half-wave rectifier can be filtered to a pure dc voltage. However, the half-wave rectifier uses only one half of the input ac wave. A better filtering action can be obtained by using two diodes. With this setup both half cycles of the input wave can be used. Both half cycles at the output have the same polarity in this ***full-wave rectifier.*** **Figure 17-33** follows the first half cycle. **Figure 17-34** follows the second half cycle.

To produce this full-wave rectification, a center tap is made on the secondary winding. This tap is attached to ground. In Figure 17-33, point A is positive and diode anode D_1 is positive. Electron flow is shown by the arrows. During the second half of the input cycle, point B is positive, diode anode D_2 is positive, and current flows as shown in Figure 17-34. No matter which diode is conducting, the current through load resistor *R* is always in the same direction. *Both positive and negative half cycles of the input voltage cause current through R in the same direction.*

Output voltage of this full-wave rectifier is taken from across *R*. It consists of direct current pulses at twice the frequency of input voltage, **Figure 17-35.** To produce this full-wave rectification in this circuit, secondary voltage was cut in half by the center tap.

The diodes, D_1 and D_2, used in Figures 17-33 and 17-34, are packaged both individually and in pairs. **Figure 17-36** shows a two rectifier package. The center lead, is used as the connection for the cathodes. The cathodes are wired together.

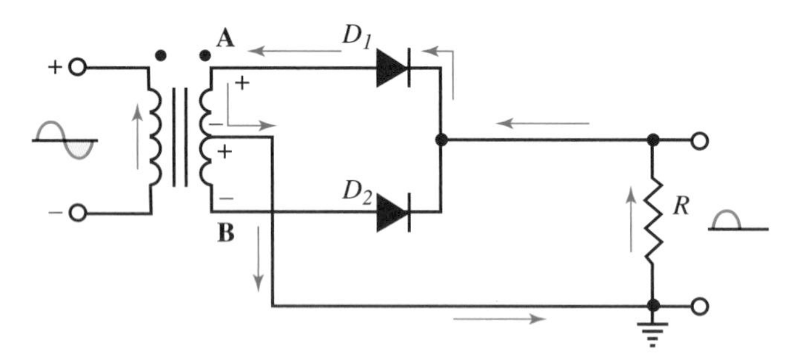

Figure 17-33. Arrows show current in full-wave rectifier during first half cycle.

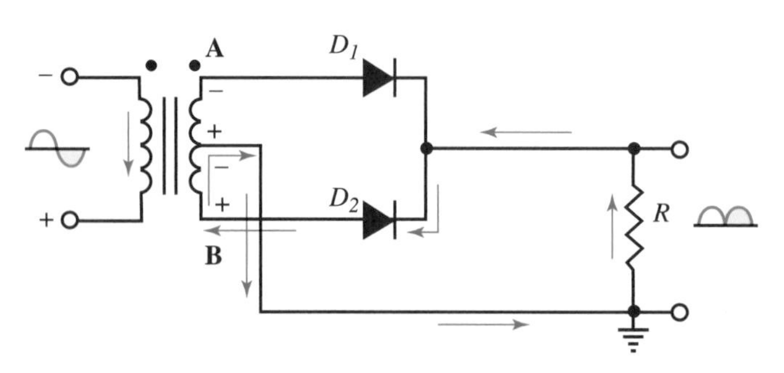

Figure 17-34. Direction of current during second half cycle.

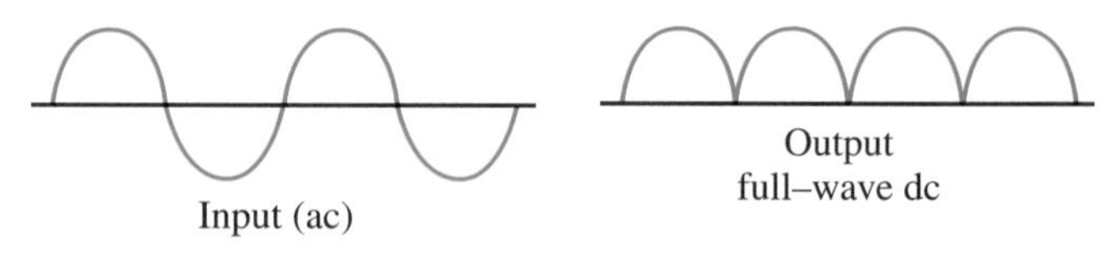

Figure 17-35. The wave forms of input and output of full-wave diode rectifier.

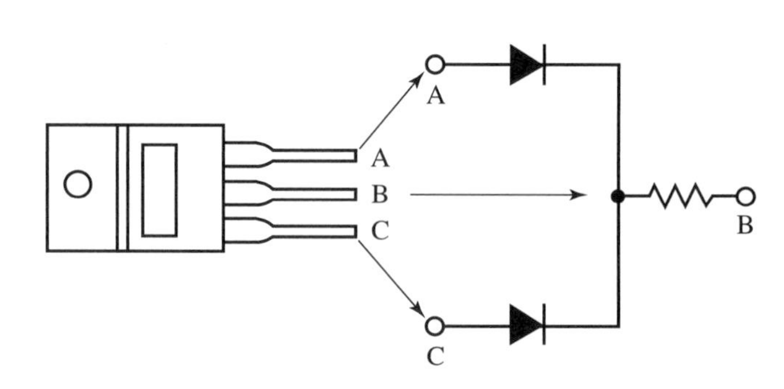

Figure 17-36. Dual diodes with center tap.

Bridge rectifiers

It is not always necessary to use a center tapped transformer for full-wave rectification. Full secondary voltage can be rectified by using four diodes in a circuit called a ***bridge rectifier,*** **Figure 17-37** and **17-38.** Two circuits are shown so that the current can be observed in each half cycle.

In Figure 17-37, point A of transformer secondary is positive. Current flows in the direction of the arrows. When point B is positive, current flows as in Figure 17-38. Again, notice that the current through *R* is always in one direction. Both halves of the input voltage are rectified and the full voltage of the transformer is used.

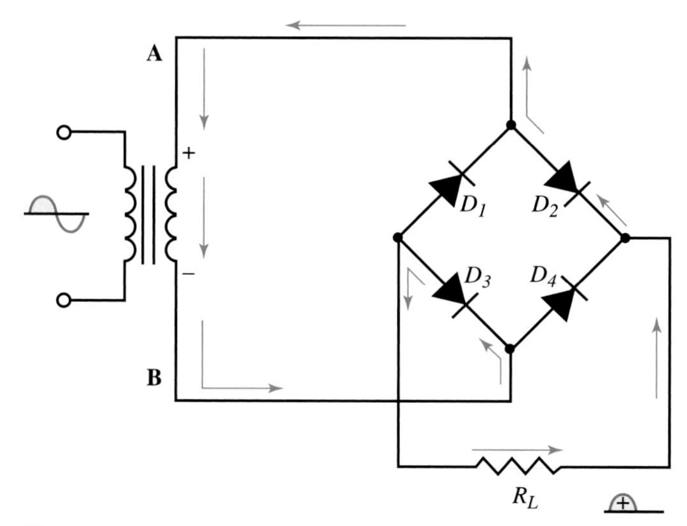

Figure 17-37. Current in bridge rectifier during the first half cycle.

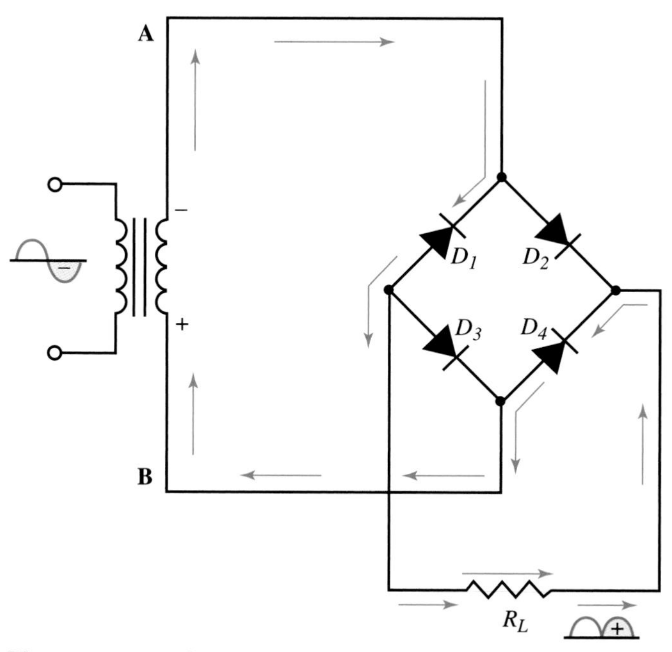

Figure 17-38. Current in bridge rectifier during the second half cycle.

Bridge rectifiers can be used in circuits without transformers. Without transformers, the voltage or current will not be stepped up or down. There will be no isolation. These circuits are also called ***line operated bridge circuits,*** **Figure 17-39.**

CAUTION:

Connecting an oscilloscope directly to a line operated bridge rectifier will result in a dead ground when the oscilloscope ground is connected to the line voltage bridge. An isolation transformer with a 1 to 1 ratio must be used to prevent the ground lead on the scope from being connected to the hot conductor.

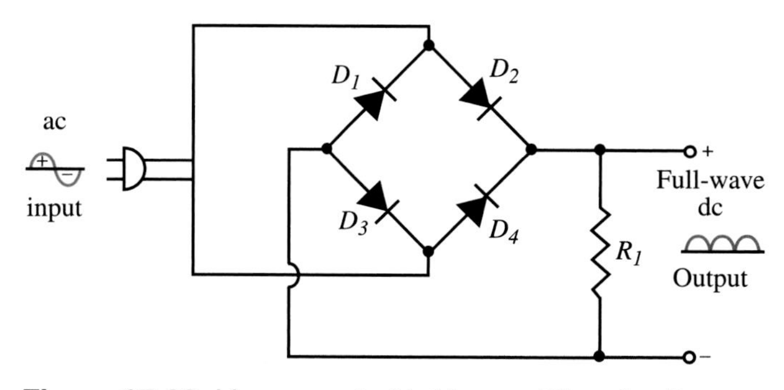

Figure 17-39. Line-operated bridge rectifier circuit.

Filters

The output of either the half-wave or full-wave rectifier is a pulsating voltage. Before it can be applied to other circuits, the pulsations must be reduced. A steadier dc is needed. It can be obtained using a ***filter network.***

In **Figure 17-40,** the line, E_{avg}, shows the average voltage of the pulsating dc wave. It is equal to 0.637 × peak voltage. The shaded portion of the wave above the average line is equal in area to the shaded portion below the line. Movement above and below the average voltage is called the ac ***ripple.*** It is this ripple that requires filtering. The percentage of ripple as compared to the output voltage must be kept to a small value. The ripple percentage can be found using the formula:

$$\% \text{ ripple} = \frac{E_{rms} \text{ of ripple voltage}}{E_{avg} \text{ of total output voltage}} \times 100\%$$

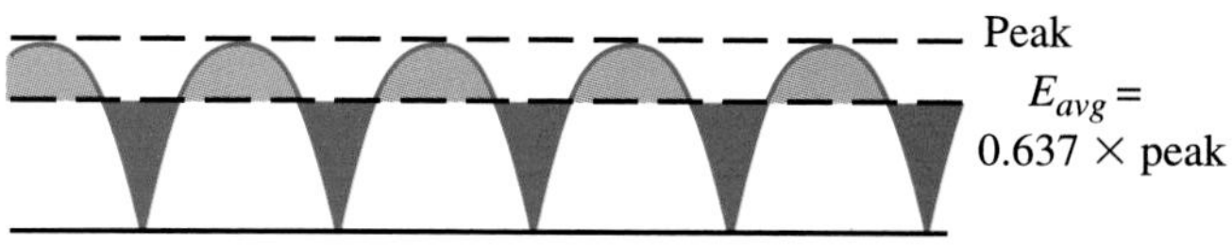

Peak

$E_{avg} =$ 0.637 × peak

Figure 17-40. Average value of full-wave rectifier output.

Capacitor filters

A capacitor connected across the rectifier output provides some filtering action, **Figure 17-41.** The capacitor is able to store electrons. When the diode or rectifier is conducting, the capacitor charges rapidly to near the peak voltage of the wave. It is limited only by the resistance of the rectifier and the reactance of the transformer windings. Between the pulsations in the wave, voltage from the rectifier drops. The capacitor then discharges through the resistance of the load. The capacitor, in effect, is a storage chamber for electrons. It stores electrons at peak voltage and then supplies electrons to the load when rectifier output is low. See **Figure 17-42.**

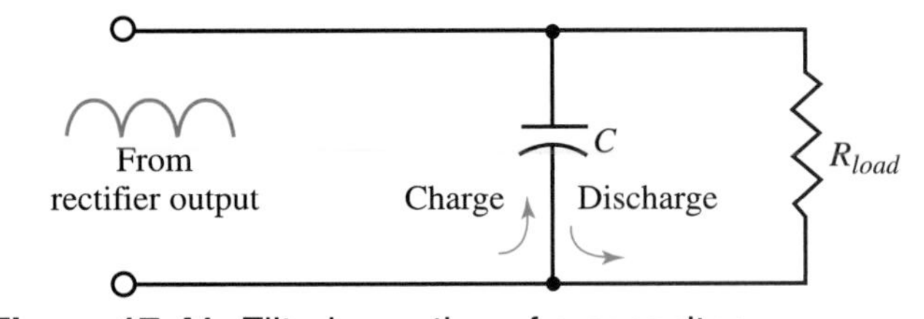

Figure 17-41. Filtering action of a capacitor.

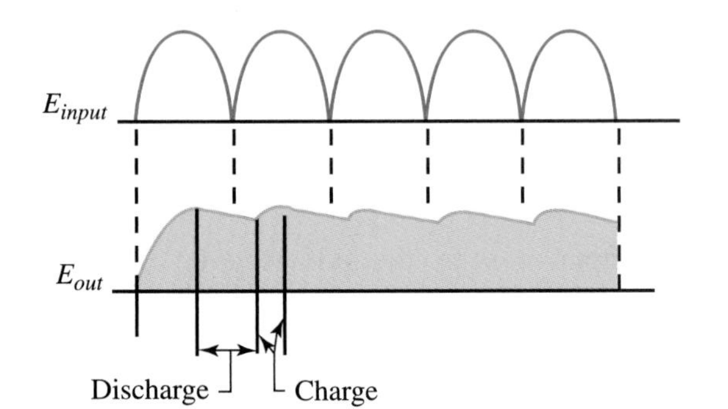

Figure 17-42. Input and output of capacitor filter showing the change in the waveform.

Capacitors used for this purpose are electrolytic types because large capacitances are needed in a limited space. Common values for the capacitors range from 4 to 2000 microfarads. Working voltages of capacitors should be in excess of the peak voltage from the rectifier.

LC filters

The filtering action can be improved by adding a choke in series with the load. This LC filter circuit appears in **Figure 17-43.** The filter choke consists of many turns of wire wound on a laminated iron core.

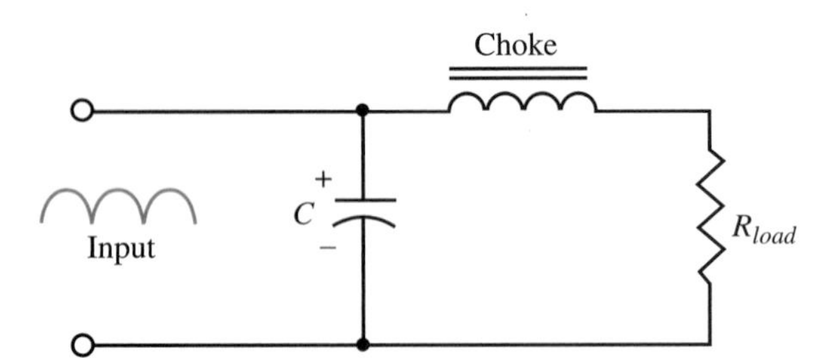

Figure 17-43. Further filtering is produced by the choke in series with the load.

Recall that inductance was that property of a circuit that resisted a change in current. A rise in current induced a counter emf that opposed the rise. A decrease in current induced a counter emf that opposed the decrease. As a result, the choke constantly opposes any change in current. Yet, it offers very little opposition to a direct current.

Chokes used in radios have values from 8 to 30 henrys. Current ratings range from 50 to 200 milliamperes. Larger chokes can be used in transmitters and other electronic devices. Filtering action as a result of the filter choke is shown in **Figure 17-44.**

A second capacitor can be used in the filter section after the choke, to provide more filter action. See **Figure 17-45.** Action of this capacitor is similar to the first capacitor. The circuit configuration appears as the Greek letter π. The filter is called a ***pi (π) section filter.***

When the first filtering component is a capacitor, the circuit is called a ***capacitor input filter.*** When the choke is the first filtering component, it is called a ***choke input filter,*** **Figure 17-46.** The choke input filter looks like an inverted L, so it is also called an ***L section filter.*** Several of these filter sections can be used in series to provide added filtering.

In the capacitor input filter, the capacitor charges to the peak voltage of the rectified wave. In the choke input, the charging current for the capacitor is limited by the choke. The capacitor does not charge to the peak voltage. As a result, the output voltage of the power supply using the capacitor input filter is higher than one using the choke input filter.

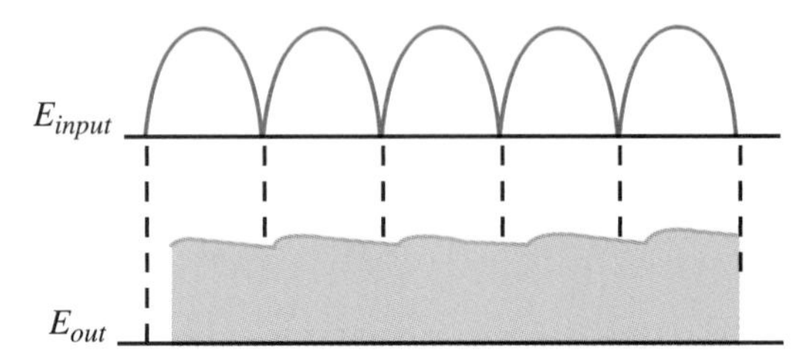

Figure 17-44. Waveforms show filtering action of the capacitor and choke together.

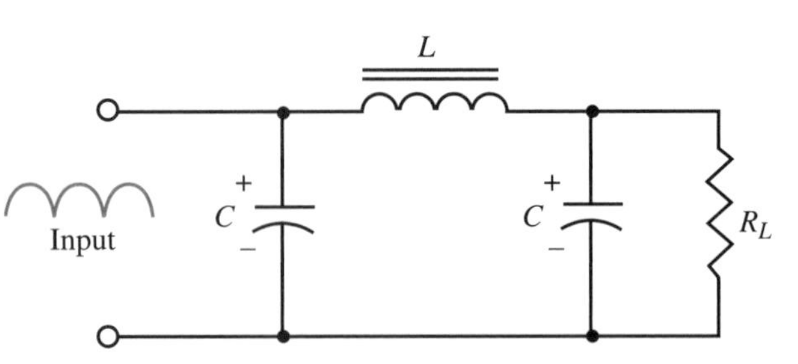

Figure 17-45. Pi (π) section filter.

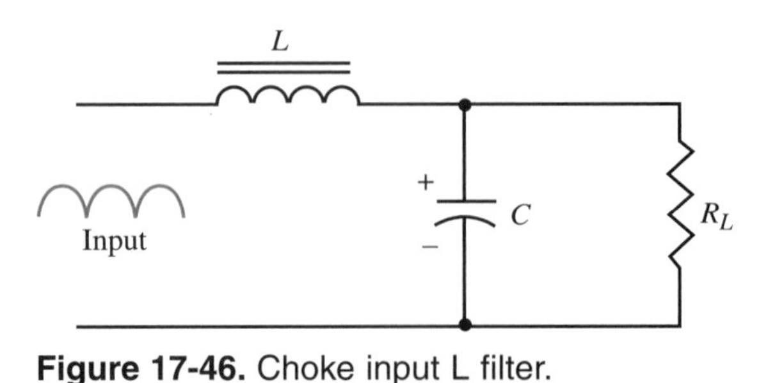

Figure 17-46. Choke input L filter.

Review Questions for Section 17.3

1. Define power supply.
2. List five functions provided by a complete power supply.
3. The first device in a power supply is the __________.
4. A circuit that changes alternating current to a pulsating direct current is a(n) __________.
5. Name two types of full-wave rectifier circuits.
6. What is ripple?
7. Name two types of filter circuit.

17.4 VOLTAGE REGULATION

The output voltage of a power supply will usually decrease when a load is applied. This decrease is not good and needs to be minimized. The size of this decrease is measured in comparison to the no-load voltage. The voltage decrease under load compared to the power supply voltage with no load is called ***percentage of voltage regulation.*** It is one factor used to determine the quality of a power supply. Expressed mathematically:

$$\% \text{ of voltage regulation} = \frac{E_{nl} - E_{fl}}{E_{fl}} \times 100$$

where E_{nl} equals voltage with no load, and E_{fl} equals voltage with full load.

Example: A power supply has a no load voltage of 30 volts. This voltage drops to 25 volts when a load is applied. What is its percentage of regulation?

$$\% \text{ of voltage regulation} = \frac{E_{nl} - E_{fl}}{E_{fl}} \times 100$$

$$\% \text{ of voltage regulation} = \frac{30 \text{ V} - 25 \text{ V}}{25 \text{ V}} \times 100$$

$$\% \text{ of voltage regulation} = \frac{5 \text{ V}}{25 \text{ V}} \times 100 = 20\%$$

Load Resistor

To complete the basic power supply circuit, a ***load resistor*** is connected across the supply, **Figure 17-47.** This resistor serves three important purposes.

First, the load resistor serves as a bleeder. A ***bleeder*** allows charged capacitors to drain. During operation of a power supply, peak voltages are stored in the capacitors of the filter sections. These capacitors remain charged after the equipment is turned off. These capacitors can be dangerous if accidentally touched by a technician. The load resistor allows these capacitors to discharge when

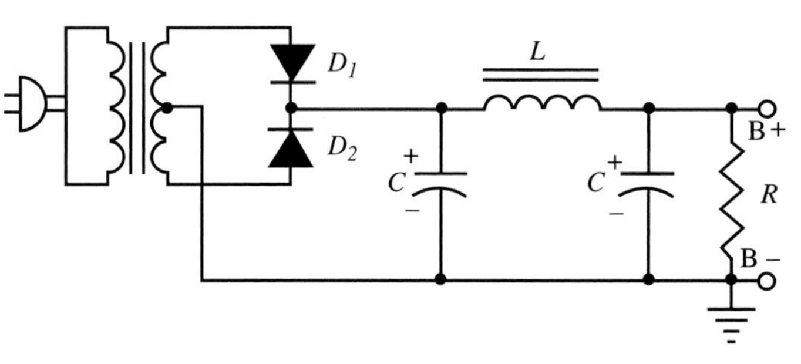

Figure 17-47. Complete power supply circuit with load resistor.

not in use. The wise technician will always take an added precaution and short capacitors to ground with an insulated screwdriver.

Second, the load resistor improves regulation. The load resistor acts as a preload on the power supply. It causes a voltage drop. When equipment is attached to the supply, the *added drop* is fairly small and the regulation is improved.

Example: Assume the terminal voltage of a power supply is 30 volts with no load resistor. No equipment is connected to it. When equipment is connected and turned on, the voltage drops to 25 volts. Regulation is 20 percent. (See previous example under voltage regulation.) If the resistor across the power supply produces an initial drop to 26 volts, then output voltage is considered 26 volts. If the equipment now connected to the supply causes the voltage to drop to 25 volts, then the power supply regulation is:

$$\% \text{ of voltage regulation} = \frac{E_{nl} - E_{fl}}{E_{fl}} \times 100$$

$$\% \text{ of voltage regulation} = \frac{26 \text{ V} - 25 \text{ V}}{25 \text{ V}} = \times 100$$

$$\% \text{ of voltage regulation} = \frac{1 \text{ V}}{25 \text{ V}} \times 100 = 4\%$$

The usable voltage of the supply has only varied four percent.

A further advantage of preloading the supply is an increase in the choke filtering action. The resistor allows current to flow in the supply at all times. A choke has better filtering action under this current condition than when current varies between a low value and zero.

Third, the load resistor acts as a voltage divider. The load resistor provides a way of obtaining several voltages from the supply. Replacing the single load resistor with separate resistors in series provides several fixed dc voltages, **Figure 17-48.** A sliding tap resistor can also be used to provide voltage adjustments.

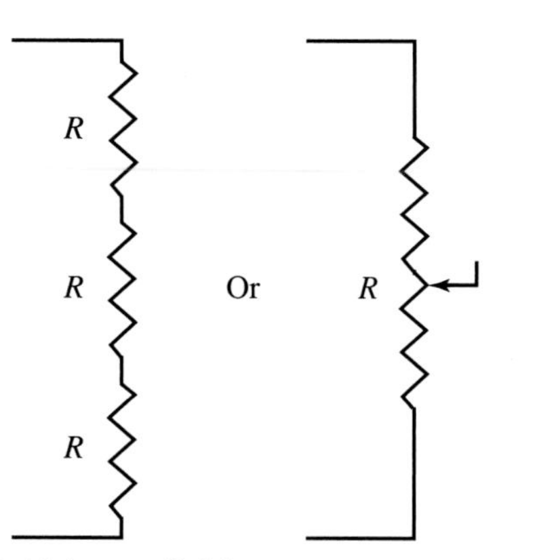

Figure 17-48. Voltage divider across power supply output.

This circuit is called a ***voltage divider.*** It takes advantage of Ohm's law (the voltage drop across a resistor equals current times resistance, or $E = I \times R$). In Figure 17-48, the three resistors are equal. The voltage drop divides equally between them, and their sum equals the applied voltage. If other voltages are required, resistors with other values can be used.

Do not neglect the effect of connecting a load to one of the taps on a voltage divider. This load is in parallel to the voltage divider resistor. It decreases total resistance and, therefore, changes the voltage at that tap.

Example: In part A of **Figure 17-49,** the voltage divider consists of three five-kilohm (kΩ) resistors. The supply of 30 volts divides to 10 volts, 20 volts, and 30 volts at terminals C, B, and A respectively. In part B, a load of five kilohms is connected to terminal C as shown. It is parallel with R_3 and resistance becomes:

PROJECT 17-1: Building a Voltage Regulator

You can build your own adjustable voltage regulator. Examine the schematic and its parts list. Be sure to use a proper heat sink on the voltage regulator case to stop overheating problems.

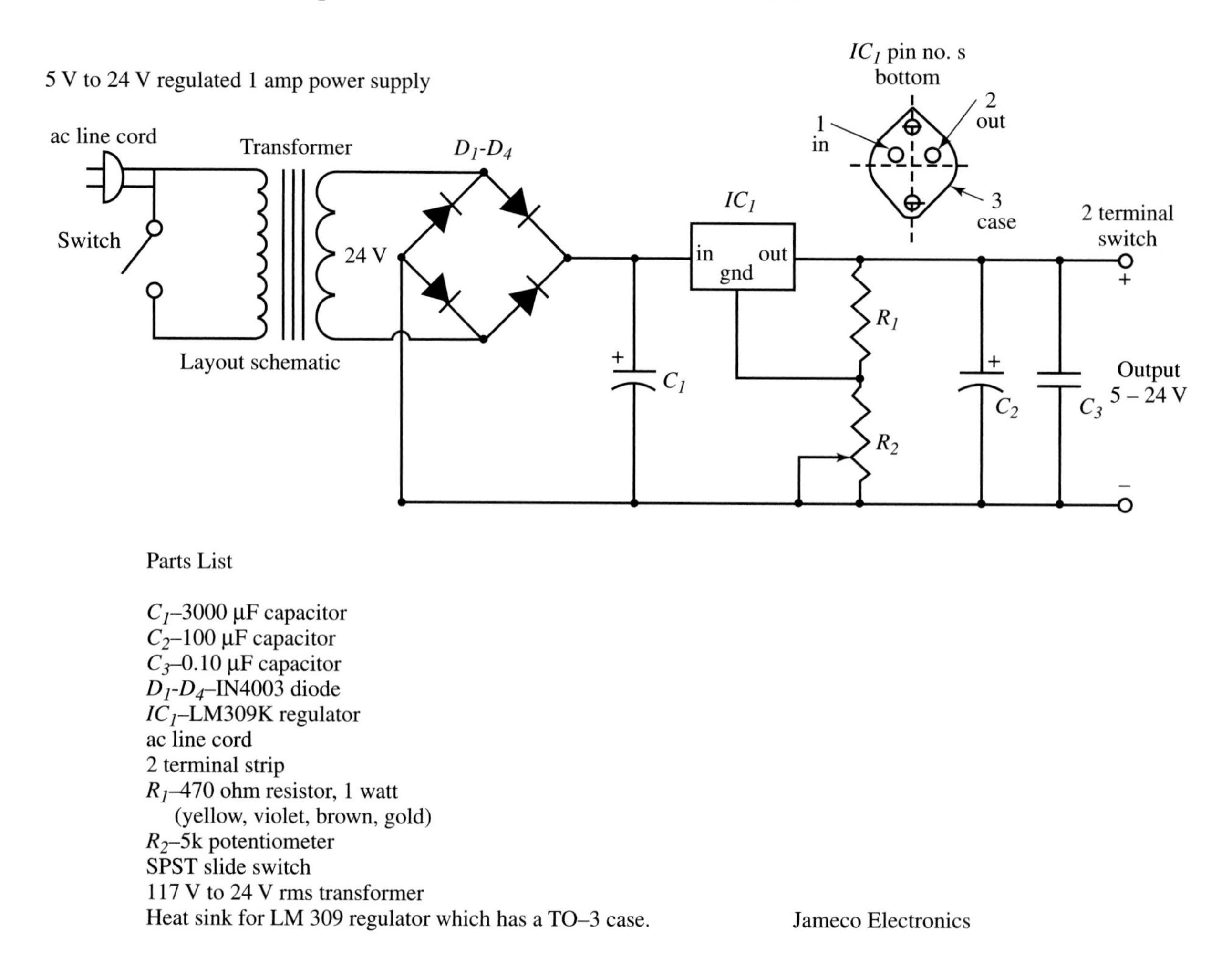

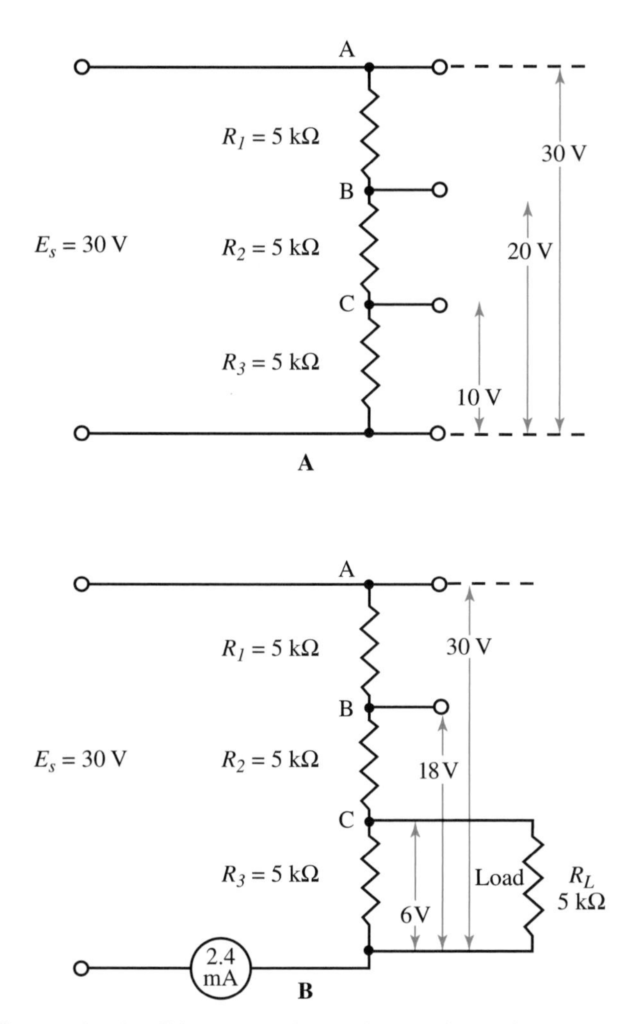

Figure 17-49. Diagrams show change in resistance in a voltage divider when a load is attached.

$$R_T = \frac{R_3 \times R_L}{R_3 + R_L} = \frac{5000\ \Omega \times 5000\ \Omega}{5000\ \Omega + 5000\ \Omega} = 2500$$

The total resistance across the power supply with R_L connected is 5000 Ω + 5000 Ω + 2500 Ω = 12,500 Ω. Now the current through the divider can be calculated.

$$I = \frac{E_{source}}{R}$$

$$I = \frac{30\ \text{V}}{12{,}500\ \Omega} = 0.0024\ \text{A} = 2.4\ \text{mA}$$

Using the total current, we can calculate the individual voltage drops. The voltage at point C is:

$$E_C = I \times R = 0.0024\ \text{A} \times 2500\ \Omega = 6\ \text{V}$$

The voltage at point B is 18 volts. If another load were connected to point B, a further change of voltage division would result.

Voltage Regulator Circuit

Some method for providing a constant voltage output at the power supply under varying load conditions is needed. This method would take into consideration the fact that a voltage drop across a resistor is equal to the product of current and resistance.

This method comes in the form of a circuit called a ***voltage regulator.*** It is shown in **Figures 17-50** and **17-51.** The total input of the power supply filter is applied to terminals A and B. Regulated output is across points C and B. The voltage regulator used in Figure 17-50 is often called a three terminal fixed voltage regulator. Common output regulated voltages can be 5, 6, 8, 12, 15, 18, 24 volts, etc. (Various current ratings are also available from manufacturers.)

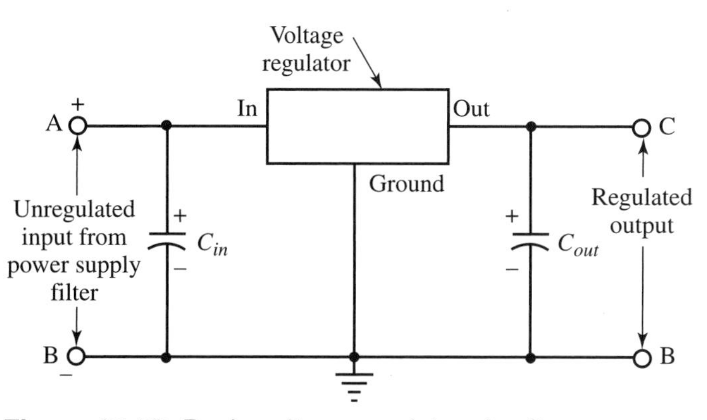

Figure 17-50. Basic voltage regulator circuit.

Regulators come in a number of common transistor package designs (TO-3, TO-39, TO-202, TO-220, etc.). These solid-state regulators are basically blowout proof. They require that a heat sink be used to remove excess heat from the device.

The internal circuits used for these voltage regulators are quite complex. They have a number of transistors, diodes, zener diodes, and resistors built into one small package. Figure 17-51 shows two voltage regulator schematics and their package designs.

An example of a use for a voltage regulator can be seen in an automobile. An automobile's voltage regulator controls the voltage level from the alternator.

Review Questions for Section 17.4

1. A power supply has a no load voltage of 65 volts. Voltage drops to 45 volts when a load is applied. What is the percentage of voltage regulation?

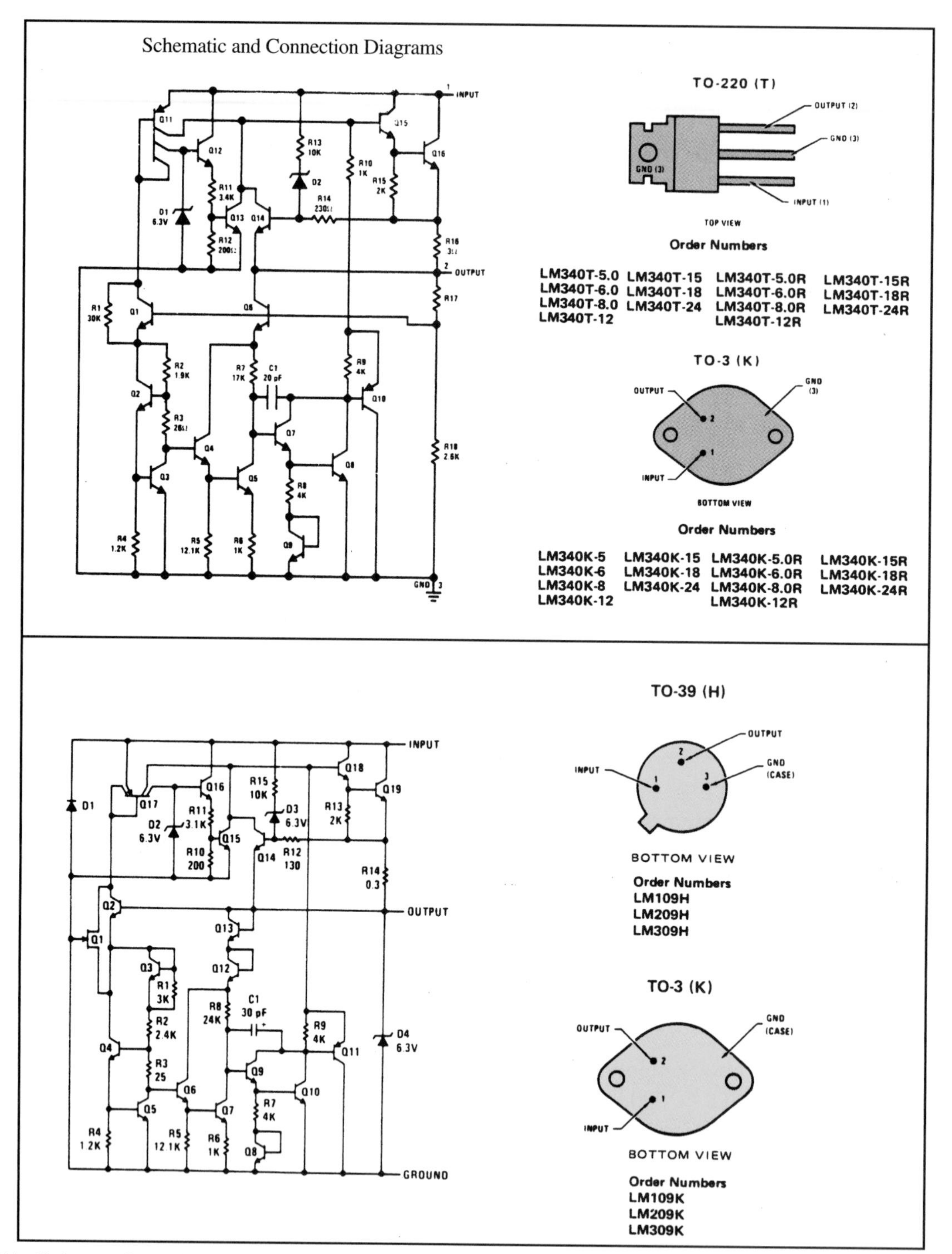

Figure 17-51. Schematic and connection diagrams for voltage regulators. (National Semiconductor Corp.)

2. List three purposes for using a load resistor in a power supply.
3. What is the purpose of a voltage regulator in a power supply?

17.5 VOLTAGE DOUBLERS

Up to this point, it has been assumed that the source of power was 117 volt ac found in homes and schools. The transformer was used to step up or step down the

PROJECT 17-2: Building a Trickle Charger

Shown here is a battery charger for an automobile. It is called a *trickle charger* because it provides a low level of charging current compared to service station chargers. This battery charger is excellent for recharging an automobile battery overnight, or for 12 to 24 hours.

The ammeter shows output current while charging. A minimum amount of output current means that the battery is reaching a maximum charge. The schematic shows a power supply that can be used as a battery charger. The wiring is simple. This project is fun to construct. The parts list is shown below.

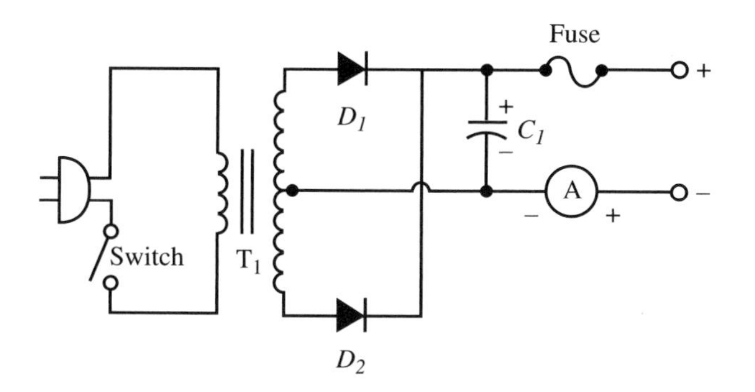

Switch	SPST switch
T	Step-down transformer, 117 V primary, 12 V CT secondary at 3A
D_1, D_2	Diodes, 50 PIV at 5A (RCA SK 3586)
*C_1	1000–5000 μF, 50 volt capacitor (electrolytic)
Fuse	3 amp fuse with holder
A	0–3 amp, ammeter

Also needed: line cord, chassis output leads with alligator clips.
*C_1 is only needed when operating 12 volt automotive equipment such as a radio, tape player, or compact disc player. It is not needed when using this power supply as a battery charger.

This power supply can also be used to operate 12 volt automotive radios by adding capacitor C_1, shown in the schematic.

voltages required for the electronic circuits. Because transformers are heavy and costly, voltage multiplying circuits have been devised to raise voltages without the use of transformers.

Study **Figure 17-52.** It shows the action in a half-wave voltage doubler circuit. In part A of Figure 17-52, the input ac voltage is on the negative half cycle. As a result, point A is negative. Current flows from point A, through the rectifier D_1, and charges capacitor C_1 to the polarity shown.

During the positive half cycle, point A is positive. The applied peak voltage of 165 volts is in series with the charged capacitor C_1. In the series connection the voltages add together. So, the output from the doubler is the applied voltage plus the voltage of C_1. Current cannot flow through the rectifier D_1 due to its one-sided conduction. The output waveform shows half-wave rectification with an amplitude of about twice the input voltage. Rectifier D_2 permits current to flow in only one direction to the load.

A full-wave voltage doubler is drawn in **Figure 17-53.** During the positive peak of the ac input, point A is positive. Current flows from point B, charging C_1 in the polarity shown, through D_1 to point A.

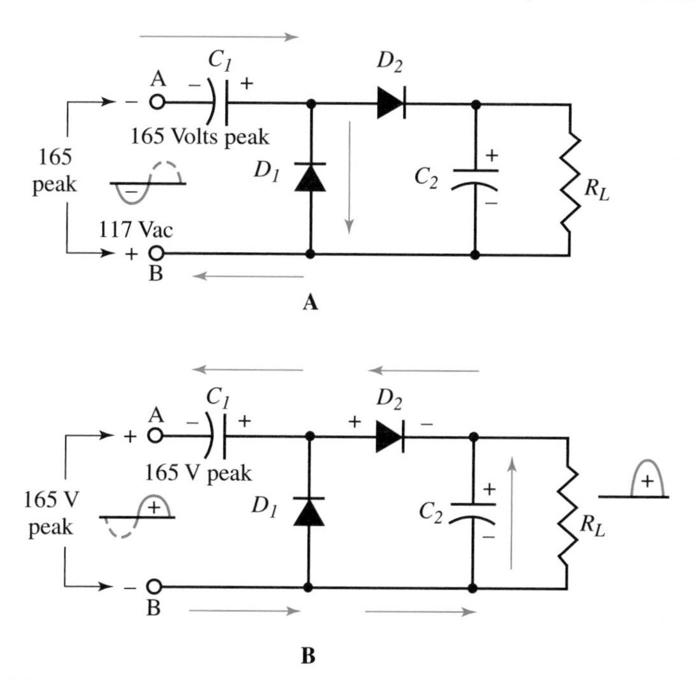

Figure 17-52. A–During first half cycle, C_1 charges through conduction of rectifier D_1. B–During second half cycle, applied line voltage is in series with charge on C_1. Current flows through D_2. C_2 gets the sum of line voltage and that from C_1.

PROJECT 17-3: Building a Small Power Supply

This is an excellent project for constructing a simple lab power supply for your home projects. The power supply can provide 300mA and 0-15 volts dc. Additional components can be added such as a plastic project box to serve as an enclosure, a dc voltmeter to indicate desired voltage, and banana jack plugs or terminals for connecting to circuits.

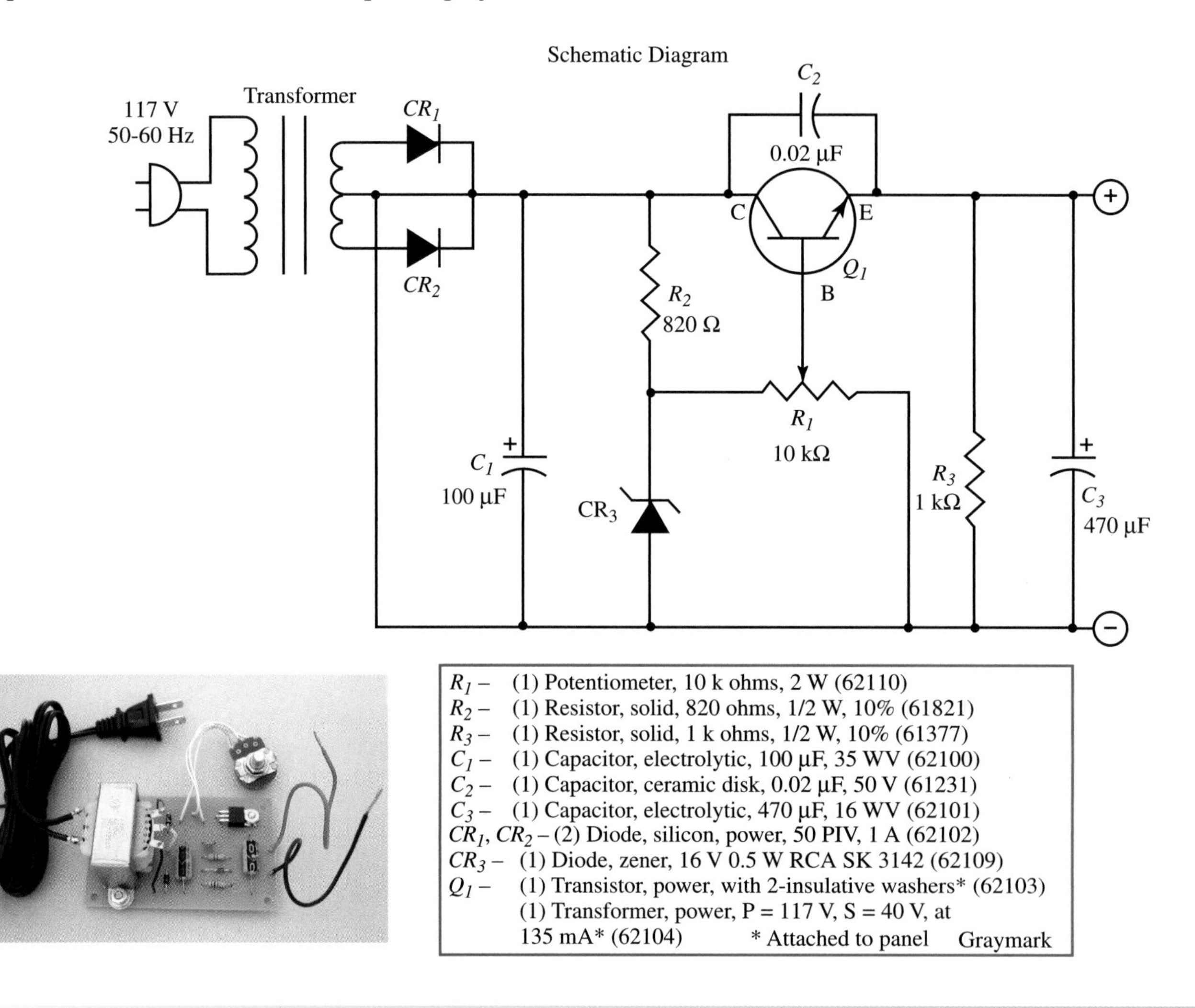

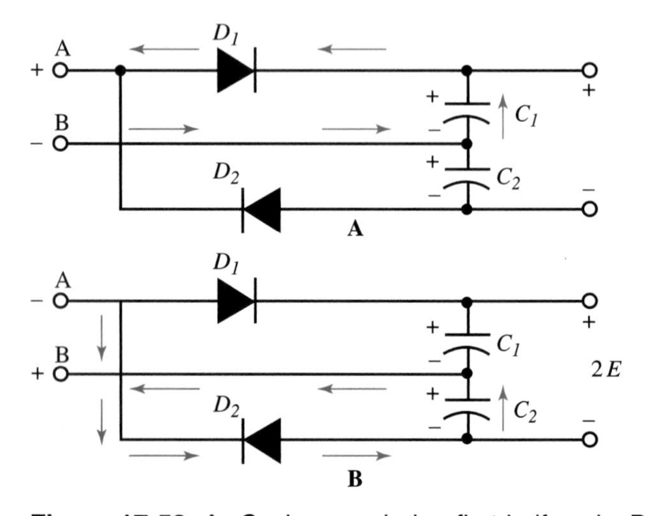

Figure 17-53. A–C_1 charges during first half cycle. B–C_1 + C_2 in series.

During the negative cycle of the input, point A is negative. Current flows through D_2 to C_2, charging it to the noted polarity, to point B.

Notice that during one cycle of ac input, capacitors C_1 and C_2 have been charged so that the voltages across C_1 and C_2 are in series. The output is taken from across these capacitors in series. The output voltage is the sum of both voltages or twice the input voltage.

Voltage doubler circuits provide useful high voltages for circuits needing low current. Because output voltage depends on charged capacitors, voltage regulation is poor. Conventional filter circuits are added to smooth out the voltage as in transformer rectifier circuits.

Review Questions for Section 17.5

1. What type of circuit is shown below? How does it operate?

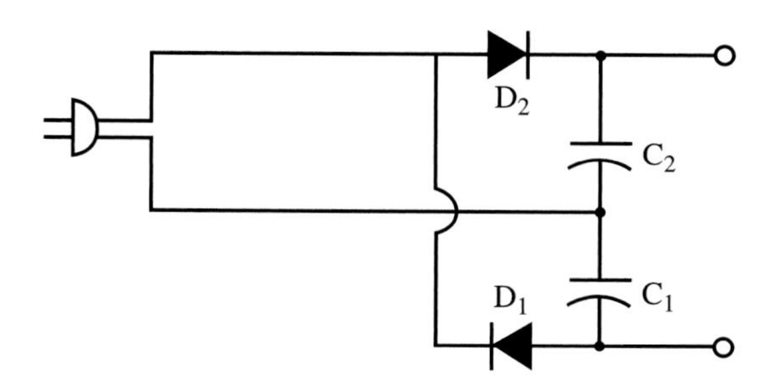

2. A line voltage of 117 volts ac produces a peak voltage of __________ volts when using a voltage doubler.
3. Why are voltage doubler circuits important for electronic equipment?

Summary

1. The three generations of electronic devices are: vacuum tubes, semiconductor materials, and integrated circuits.
2. Semiconductors are materials that fall between conductors and insulators in terms of resistance. Common materials used for semiconductors are silicon and germanium.
3. Trivalent and pentavalent elements are added to pure silicon or germanium to make N-type or P-type semiconductors. The process of adding impurities is called doping.
4. Rectification is the process of changing ac to dc. Diodes are PN semiconductor devices which can be used as rectifiers.
5. Reverse bias produces the effects of high resistance in a diode. Forward bias provides a low resistance path through the diode.
6. Diodes are labeled with a 1N prefix. They are rated by peak inverse voltage (PIV) and maximum current rating.
7. Diodes can be tested using an ohmmeter or a diode tester on a digital meter.
8. The light emitting diode (LED) is a special function diode that produces light when connected in the forward bias mode. Light consists of photons. Photons are created by joining of electrons and holes in the PN junction.
9. Power supplies are electronic circuits that provide various dc and ac voltages and currents for equipment.
10. The basic circuits in many power supplies include transformers, rectifiers, filters, and voltage regulators. Transformers and voltage regulators are not found in all power supplies.
11. The purpose of a transformer in a power supply can be to (1) increase voltage and decrease current, (2) increase current and decrease voltage, (3) provide isolation and/or (4) provide two equal voltages, one 180 degrees out of phase (using a center tap).
12. A rectifier changes ac to pulsating dc.
13. Following are schematics of some of the important basic line and transformer operated rectifier circuits.

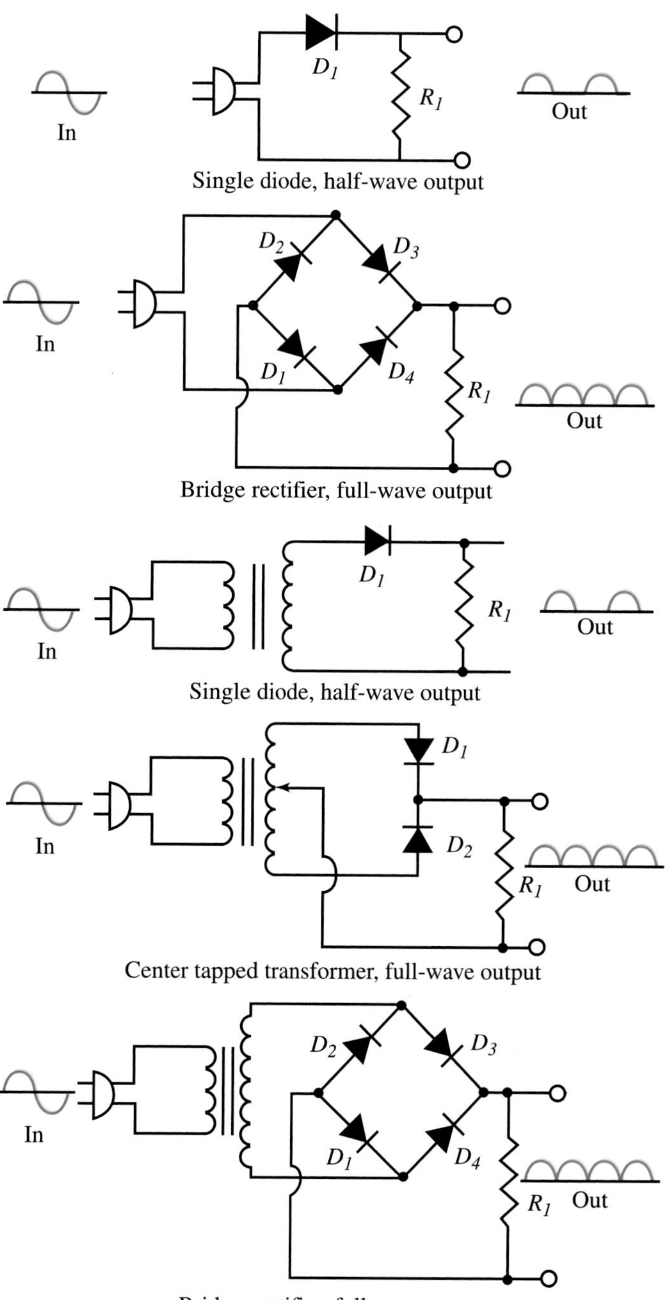

14. A power supply filter uses capacitors, inductors, and/or resistors to change pulsating dc to pure dc.
15. The percentage of voltage regulation can be used to determine the quality of a power supply. A smaller percentage means a higher quality.
16. A bleeder resistor is a resistor placed in parallel (shunt) across the output of a power supply. It provides a load, a discharge path on which charged current can flow, and a limited amount of regulation of voltage.
17. There are two basic types of voltage regulators: half-wave and full-wave. Both operate from the principle that capacitors, which always charge up to a peak value of a wave form, can be connected in series with each other (full-wave) or with a line voltage (half-wave) to provide an added voltage.
18. A floating ground is a common connection which is separated electrically from the chassis.

Test Your Knowledge

Please do not write in the text. Place your answers on a separate sheet of paper.

1. A semiconductor is neither a good __________ nor a good __________.
2. Define atomic number.
3. Electricity can be conducted in two ways. Name these ways.
4. Explain the difference between donor impurities and acceptor impurities.
5. A diode consists of two electrodes. Name these electrodes and state their functions.
6. A potential hill or barrier prevents electrons and holes from __________.
7. In forward bias, the barrier effect is __________. In reverse bias, the barrier effect is __________.
8. A point contact crystal has:
   a. P-type crystals.
   b. N-type crystals.
   c. both P- and N-type crystals.
   d. None of the above.
9. What is PIV?
10. Draw and label a typical diode. Label the anode, cathode, and polarity makings.
11. A power supply:
    a. provides various ac and dc voltages for equipment operation.
    b. can change ac voltage to pulsating dc voltage.
    c. can filter pulsating dc voltage to a pure dc voltage.
    d. All of the above.
12. List, in order, the devices a current goes through in a power supply.
13. A lack of electrical connection between secondary and primary windings is known as __________.
14. In order to rectify both half cycles of the input wave, a(n) __________ __________ is made on the secondary winding.
15. What is the purpose of a filter network?
16. State the equation for finding percentage of ripple.
17. Name two devices that can be used to improve filtering action.
18. Why is the output from a capacitor input filter higher than that of a choke input filter?
19. The terminal voltage of a power supply is 30 volts. When a load is applied, voltage drops to 28 volts. What is the percentage of regulation?
20. An electronic device that acts as a voltage regulator is a:
    a. diode.
    b. zener diode.
    c. bleeder.
    d. None of the above.
21. What is the purpose of a voltage regulator circuit?

For Discussion

1. Why is it important for electronic circuits to be made smaller and smaller?
2. What is the purpose of a power supply?
3. Do all power supplies need filters? Explain.
4. How does a transformer provide isolation?
5. What is the purpose of a center tap on the secondary winding of a transformer?
6. How does a zener diode provide voltage regulation?
7. Explain how a full-wave voltage doubler operates.

Power Supply Maintenance

- Use a dry, soft cloth or one dampened with isopropyl or denatured alcohol to remove dust from the exterior of the power supply. Do not use an abrasive cleaner or one with chemical agents as these may damage the plastic on the equipment.
- To prevent the power supply from overheating, keep the power supply in a well-ventilated area. Do not place other test equipment or any other items on top of or near its cooling vents.
- When the power supply is not in use, cover the power supply with a protective covering to prevent dust from getting inside the unit. A sufficient layer of dust inside the unit can cause overheating and component breakdown.

Tubes, Transistors, and Amplifiers

Objectives

After studying this chapter, you will be able to:

- Explain the operation of the vacuum tube.
- Describe the workings of a cathode ray tube.
- Explain the operation of the bipolar transistor.
- Explain the operation of field-effect transistors.
- Discuss different biasing techniques.
- Indentify various transistor circuit configurations.
- List the components of amplifier circuits and give the function for each component.
- Explain amplifier operation.
- Compute the gain of amplifier circuits.
- Perform dc load line analysis on a transistor circuit.
- Discuss the advantages and disadvantages of various mothods of amplifier coupling.
- Describe several common thyristors.

Key Words and Terms

The following words and terms will become important pieces of your electricity and electronics vocabulary. Look for them as you read this chapter.

amplifier
bipolar junction transistor (BJT)
cathode ray tube
common base (CB)
common collector (CC)
common emitter (CE)
cutoff
decibel (dB)
field-effect transistors (FET)
forward biased
junction field-effect transistor (JFET)
load line analysis
metal oxide semiconductor field-effect transistor (MOSFET)
NPN
PNP
quiescent point (Q-point)
reverse biased
saturation
thermal runaway
thyristors

Vacuum tubes led the way to the development of many electronic devices such as the radio receiver, radio transmitter, television, radar, and the first computers. Most of the vacuum tube applications have now been replaced by the transistor. Transistor circuits can operate more quickly, are more sensitive to small currents, and allow for much cheaper and more compact designs. However, there are still applications, some in the music industry, where vacuum tubes can still be found. Many musicians still prefer the rich mellow sounds produced by electronic tubes over those produced by transistor amplifiers. First, we will look at the design of the electronic tube. Then, we will explore the transistor and the operation of common amplifier circuits. Lastly, we will look at two common thyristors.

18.1 VACUUM TUBES

Vacuum tube use has been replaced in large part by the use of transistors and integrated circuits. The vacuum tube still finds some special uses, such as in cathode ray tubes (CRTs), which are found in most computer monitors. For the most part, however, vacuum tubes use more power, take up more space, and produce more heat than transistors and integrated circuits.

The vacuum tube is a direct result of Thomas Edison's experiment with the incandescent lamp. In this experiment, a metal plate was placed in the vacuum bulb containing the filament. A battery and series meter were connected between the light filament and the plate. At this point, Edison discovered that electric current would flow through the light if the positive terminal of the battery was connected to the plate. However, if the negative battery terminal was connected to the plate, no current would flow. See **Figure 18-1.** At that time, the electron theory had not yet been discovered, and the flow of current through the light was a mystery.

Today, we understand electron theory. Certain metals and metal oxides give up free electrons when heated. The heat supplies enough energy to the electrons to cause them to break away from the forces holding them in orbit. They become free electrons. This process is known as ***thermionic emission.***

In Edison's experiment, a cloud of free electrons was emitted from the filament of the light bulb. When a positive plate was placed in the bulb, the free electrons were attracted to the plate. (Unlike charges attract.) A flow of electrons means that a current is present. The meter attached to the plate circuit showed that electrons were passing from filament to plate. The vacuum tube uses the same principles discovered in the laboratory of Thomas Edison.

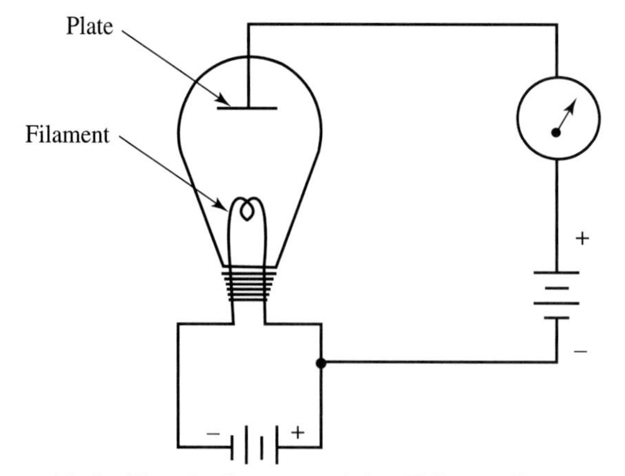

Figure 18-1. Circuit diagram of the Edison discovery.

Thermionic Emitters

Better emitting materials have been developed in the years since Edison's carbon filament. Many materials, when heated to the point of emission, will melt. For many years, tungsten seemed to be the best material. It required a great deal of heat for proper emission, yet it was strong and durable. Tungsten is still used in large, high-power vacuum tubes.

To lower the working temperatures and power usage, the thoriated-tungsten emitter was developed. It consisted of a thin layer of thorium placed on the tungsten emitter. Thorium is one of the heaviest metallic elements. Its symbol is Th, and its atomic weigh is 232.04. It is mined in Brazil, India, and South Africa. This type of emitter produces the proper emission at much lower temperatures than pure tungsten.

The most efficient emitter is the oxide-coated type. The emitter is a metal, such as nickel. On the emitter, a thin layer of barium or strontium oxide is formed. Because of its low power usage and high emission, this type of emitter was widely used in vacuum tubes for radios, televisions, and other electronic devices.

Cathodes

The emitter in the vacuum tube is called the ***cathode.*** Heat can be supplied either directly or indirectly. Both methods have certain advantages. Schematic diagrams of both types of tube are shown in **Figure 18-2.**

Equipment that uses batteries for power works best with directly heated cathodes. There is less heat loss, and the filament can be designed so that only a small amount of power is consumed during use.

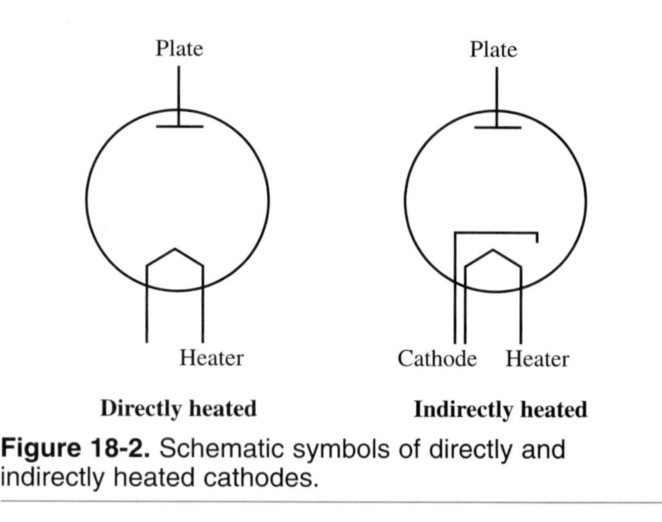

Figure 18-2. Schematic symbols of directly and indirectly heated cathodes.

When an ac source is handy, the indirectly heated cathode is more useful. There is little power loss, and the heater voltage source and cathode can be separated. This eliminates any humming in the circuit.

Diodes

Recall that a diode consists of two electrodes. In the case of a ***vacuum tube diode,*** the two electrodes are the cathode and plate. Again, the cathode can be heated directly or indirectly. The plate is a round piece of metal that surrounds all elements in the tube. The plate acts as the collector of electrons emitted from the cathode. See **Figure 18-3.** The heater connections are H_1 and H_2. These connections can be joined to an ac source.

The first number in a tube name is the approximate voltage that should be applied to the heaters. For example, a 6H6 tube needs 6.3 volts; a 12AX7 needs 12.6 volts; a 25Z6 needs 25 volts. When the tube in Figure 18-3 is turned on, the heaters indirectly heat the cathode. This causes thermionic emission of electrons. If the diode plate is joined to the positive terminal of the battery in the circuit, electrons will flow in the circuit from cathode to plate. If the connections are reversed, no electrons will flow. This electron tube acts as a one-way valve. It

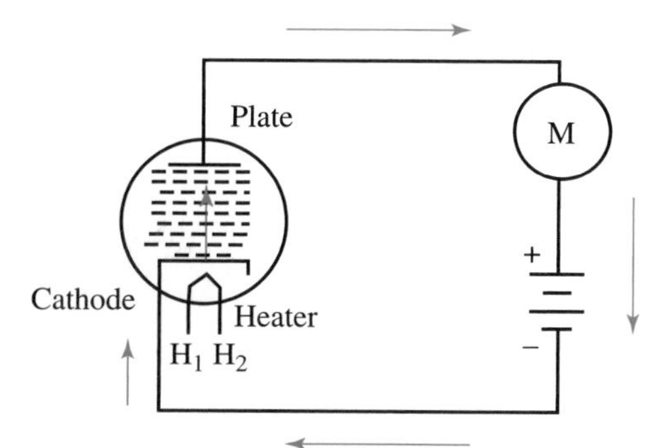

Figure 18-3. Diode circuit shows direction of electron flow.

permits electron flow in one direction only. This is just like the semiconductor diode discussed in Chapter 17.

At a certain temperature, the cathode emits the largest number of electrons. These electrons form a charge around the cathode. The cathode has been made slightly positive because of the emission of electrons. Some of these electrons are attracted back to the cathode. When the plate is made positive, many electrons are attracted to it. This is electron flow. The plate can be made *more* positive by applying a higher voltage. A greater number of electrons are attracted, and there is an increase in current.

Figure 18-4 shows the increase in current as a result of an increase in plate voltage. At some point, as voltage is increased, all the emitted electrons will be attracted to the plate. Any further increase in voltage will not increase the current. This is called the saturation point of the electron tube.

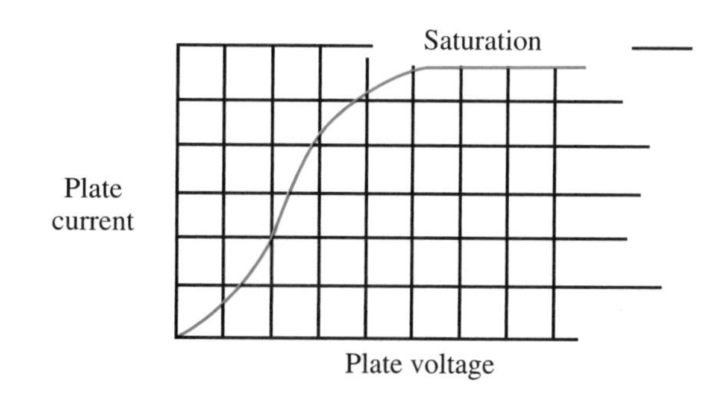

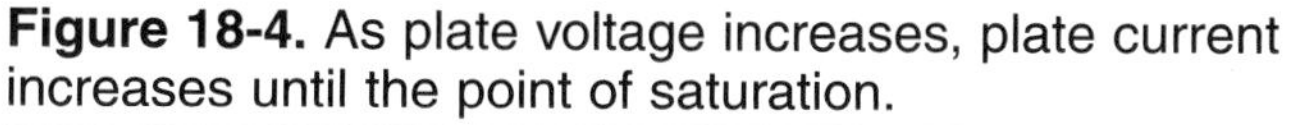

Figure 18-4. As plate voltage increases, plate current increases until the point of saturation.

Triodes

A ***triode*** is a three element tube made of a cathode, plate, and grid. The triode was developed by Dr. Lee DeForest. In his experiments, DeForest inserted a fine wire mesh between the cathode and the plate of the tube. In doing this, he was able to control electron flow through the tube. This wire mesh is the ***grid.***

The grid in the triode most often has a cylindrical shape and it surrounds the cathode. The space between the grid wires allows electrons room to pass through to the plate. The grid controls electron flow and is commonly called the ***control grid.***

Electron flow is controlled by changes in plate voltage. In the triode, the grid also affects electron flow. For example, a negative grid will repel many electrodes back to the cathode. This limits the number of electrons passing on to the plate. As the grid is made more and more negative, a point is reached where no electrons flow to the plate. This is the cutoff point of the tube. It is the negative voltage amount applied to the control grid that stops electron flow. The voltage applied to the control grid is called the ***bias voltage.*** At cutoff, it is called the ***cutoff bias.***

A triode with both plate and grid voltage is shown in **Figure 18-5.** Notice that the grid bias battery has its negative terminal connected to the grid. In electronic work, these voltages have specific names such as A, B, and C. The A voltage is for the heaters in the tube. The B voltage is for the plate of the tube. The C voltage is for the grid of the tube.

Figure 18-6 shows current through an electron tube as the grid bias is changed. The plate voltage is held at a constant value. The curve in this graph is plotted by measuring the value of current at each change of grid voltage. At a grid bias of negative two volts, the current is eight mA. At negative six volts, the current drops to three mA.

Tetrodes

Without neutralization circuits, the triode is limited as an amplifier due to the shunting effect of electrode capacitance at high frequencies, **Figure 18-7.** To overcome this drawback, another grid is inserted in the triode. This grid is called the ***screen grid.*** It is placed between

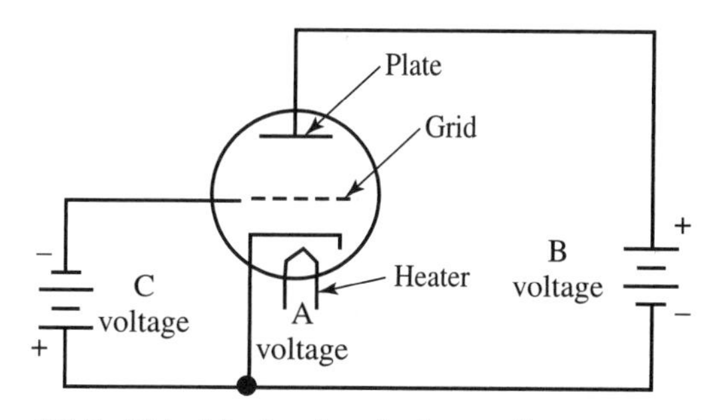

Figure 18-5. This triode circuit shows the connections for plate voltage and grid bias voltage.

These tubes have high amplification factors and high plate resistance. Interelectrode capacitance is at a minimum. Pentodes were once used as radio frequency (rf) amplifiers and as audio power amplifiers.

Cathode Ray Tubes

A ***cathode ray tube (CRT)*** is also a vacuum tube. A CRT works in the same manner. Common examples of CRTs include oscilloscope displays and computer monitors, **Figure 18-10.**

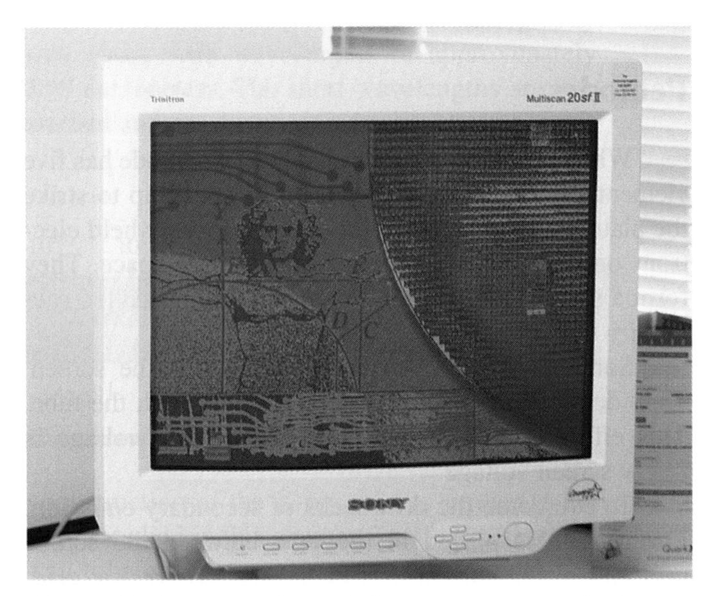

Figure 18-10. Cathode ray tubes are used as display screens in computer monitors.

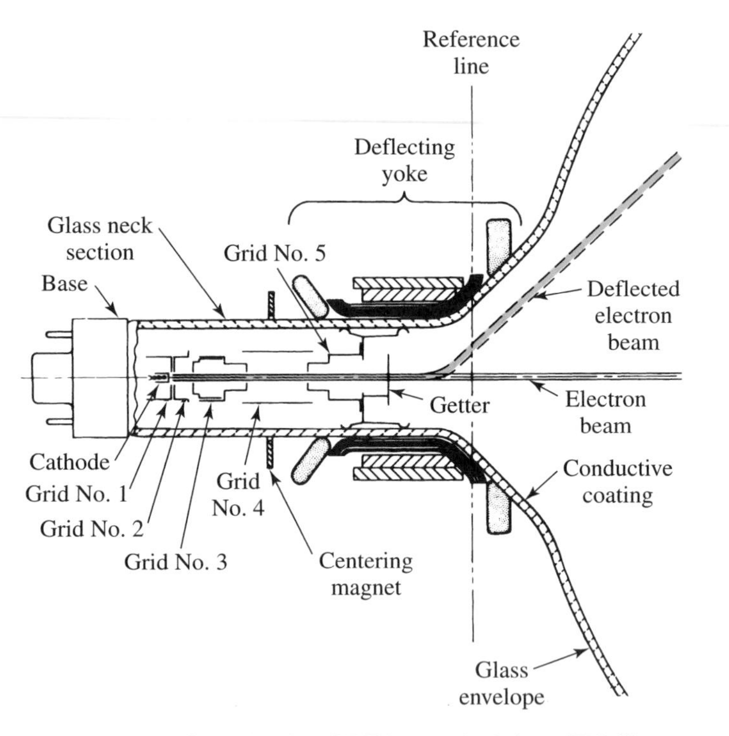

Figure 18-11. Sketch of a CRT in a television. (RCA)

The television picture tube is also a type of CRT. Refer to **Figure 18-11.** On the left is the cathode and heater. The cathode emits a cloud of electrons. Around the cathode is a metal cylinder. It is closed at one end except for a small hole (aperture) which controls the flow of electrons into the CRT. It is like the control grid. It also causes emitted electrons to move through the small hole in the shape of a beam.

Grid 2 works at a higher positive potential and is strongly attracted to the negative electrons. This grid causes the electrons to speed up. Grids 3, 4, and 5 further accelerate and focus the electrons. This results in a narrow beam of electrons.

Magnetic deflection coils around the neck of the tube move the electron beam. The electron beam is moved from left to right and top to bottom.

The inside of the picture tube face is treated with a special type of luminescent material called phosphor. The screen luminescence is a combination of fluorescence (the emission of light during a stimulus) and phosphorescence (the continuation of light after the stimulus has ceased). As the electrons strike the screen, the chemicals glow. The degree of brilliance on the screen is related to the strength of the beam. The picture is observed through the glass front of the tube.

The inside of the tube bulb contains a conductive coating. External connections to this coating are made from a high voltage power supply. When the electron beam strikes the CRT screen, electrons are knocked off. These electrons are called secondary emission and are collected by the inner coating, completing the circuit.

LESSON IN SAFETY:

Very high, dangerous voltages are present at the external connections to this anode (usually from 8000 to over 20,000 volts). Use caution when working close to these connections. Work with only one hand, to reduce the chance of death from accidental shock. Keep one hand in your pocket!

Review Questions for Section 18.1

1. Define thermionic emission.
2. The most efficient emitter is the __________-coated type.
3. The maximum limit of electron flow a vacuum tube will give off, regardless of increases in plate voltage, is the __________ __________.

4. The voltage applied to the control grid is the __________ voltage.
5. High amplification is possible with the:
   a. diode.
   b. triode.
   c. tetrode.
   d. None of the above.
6. Secondary emission is a problem connected with what tube?

18.2 TRANSISTORS

On July 1, 1948, the New York Herald Tribune announced:

"A tiny device that serves nearly all of the functions of a conventional vacuum tube, and holds wide promise for radio, telephone and electronics was demonstrated yesterday by the Bell Telephone Laboratories scientists who developed it. Known as the transistor..."

The scientists, John Bardeen and Walter Brattain invented the point contact transistor. It had two wires carefully fused on a crystal of germanium. William Shockly followed these inventions by creating the bipolar, or junction, transistor. These inventions were the beginning of microelectronics.

The transistor provided instant circuit operation and eliminated the warm-up time needed with the vacuum tube circuit. In addition, no large amounts of power were needed with the transistor. The transistor was, and still is, known for its small size, long life, and light weight.

Transistors are key devices in electronics for several reasons. They are able to amplify current. They can create ac signals at desired frequencies. They can also be used as switching devices. This makes them important in computer circuits.

The ***bipolar junction transistor (BJT)*** consists of three layers of impure semiconductor crystals. This transistor has two junctions. There are two types of bipolar transistors, NPN and PNP. Blocks and schematic symbols for these are shown in **Figure 18-12.**

The ***NPN*** bipolar transistor has a thin layer of P-type crystal placed between two N-type crystals, **Figure 18-13.** A ***PNP*** bipolar transistor has a thin layer of N-type crystals placed between two P-type crystals, **Figure 18-14.** In both types, the first crystal is called the ***emitter.*** The center section is called the ***base.*** The third crystal is called the ***collector.***

In the schematic symbols, notice the direction of the arrow. This indicates whether it is a NPN or PNP transistor. The arrow always points toward a N-type material. This will assist you in determining the proper polarity when making a connection in a circuit. The direction in which the emitter arrow points for an NPN transistor can easily be recalled by reciting "Never Points iN."

Figure 18-15 shows biasing diagrams for five transistors. Take special note of the fact that the base is not always the same pin location on the transistors. Never assume the proper connections. Always be sure by first checking the transistor part number in a catalog or product specification sheet.

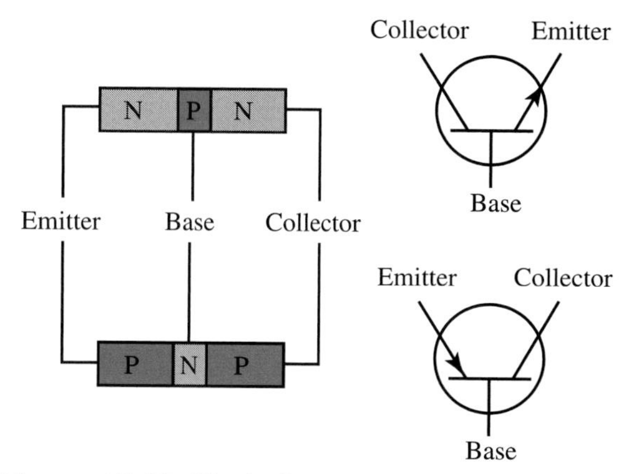

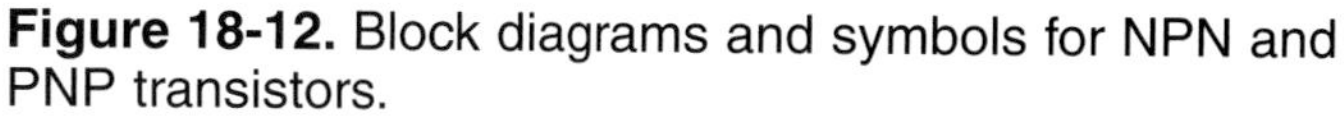
Figure 18-12. Block diagrams and symbols for NPN and PNP transistors.

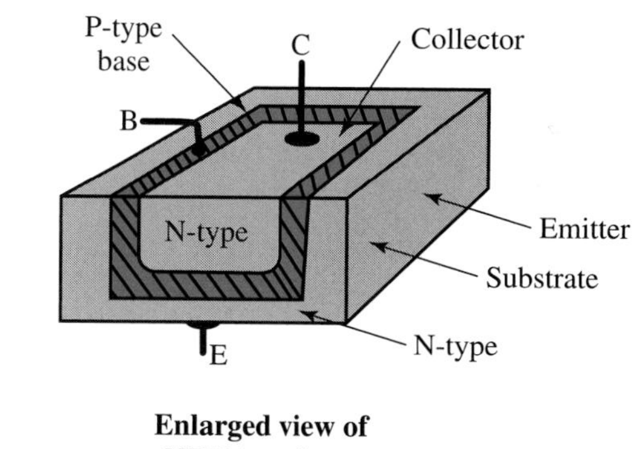

Enlarged view of NPN transistor construction

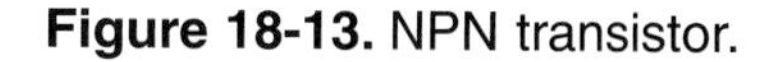
Figure 18-13. NPN transistor.

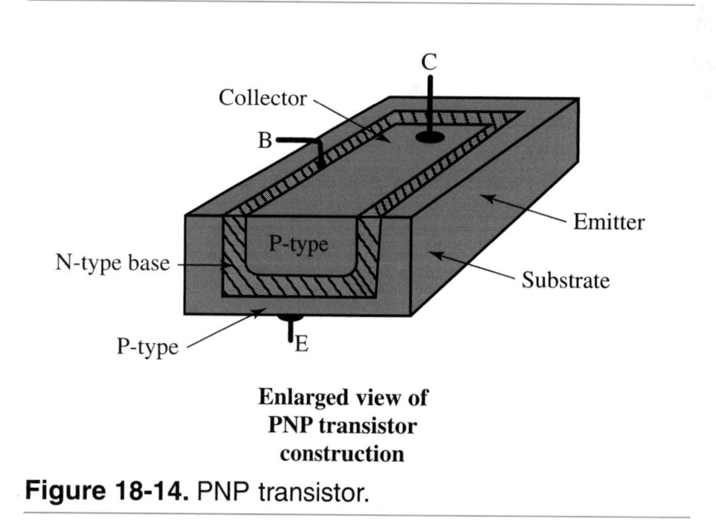

Figure 18-14. PNP transistor.

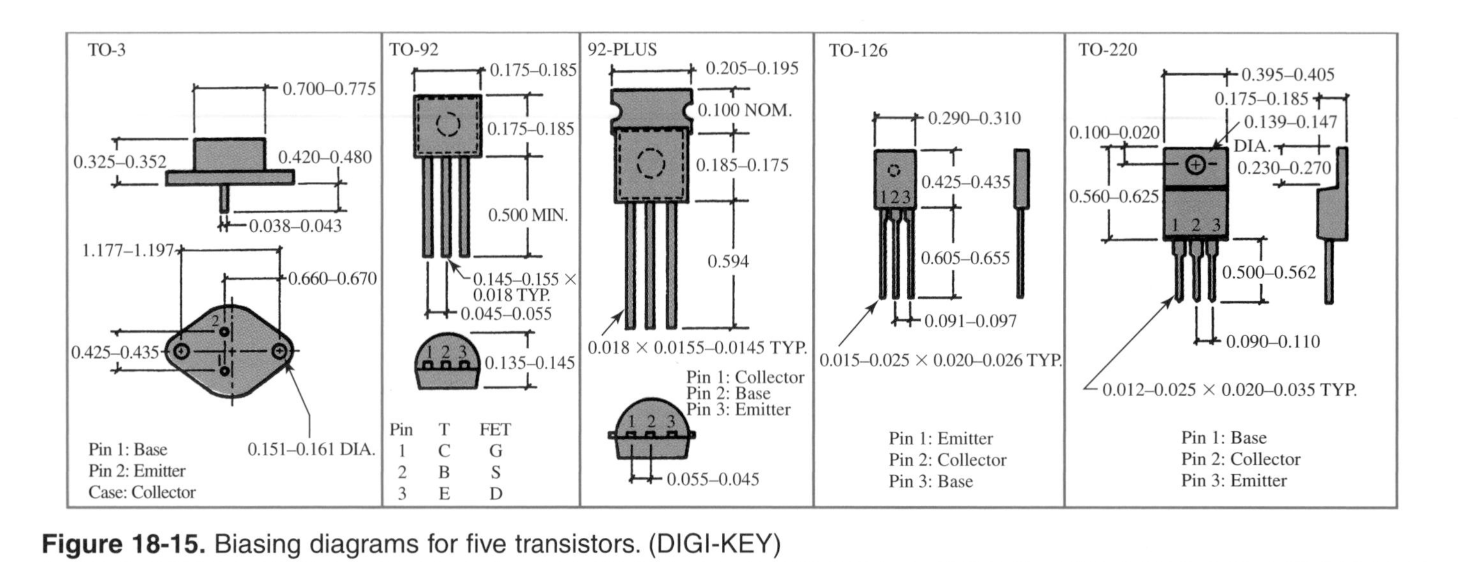

Figure 18-15. Biasing diagrams for five transistors. (DIGI-KEY)

The theory of NPN transistor operation is shown in **Figure 18-16.** Two batteries are used to simplify operation theory. Most applications require one voltage source. The negative terminal of the battery is connected to the N emitter. The positive terminal of the same battery is connected to the P-type base. Therefore, the emitter-base circuit is ***forward biased.***

In the collector circuit, the N collector is connected to, the positive battery terminal. The P base is connected to the negative terminal. The collector-base circuit is ***reverse biased.***

Electrons enter the emitter from the negative battery source and flow toward the junction. The forward bias has reduced the potential barrier of the first junction. The electrons then combine with the hole carriers in the base to complete the emitter-base circuit. However, the base is a very thin section, about 0.001 inch. Most of the electrons flow on through to the collector. This electron flow is aided by the low potential barrier of the second PN junction.

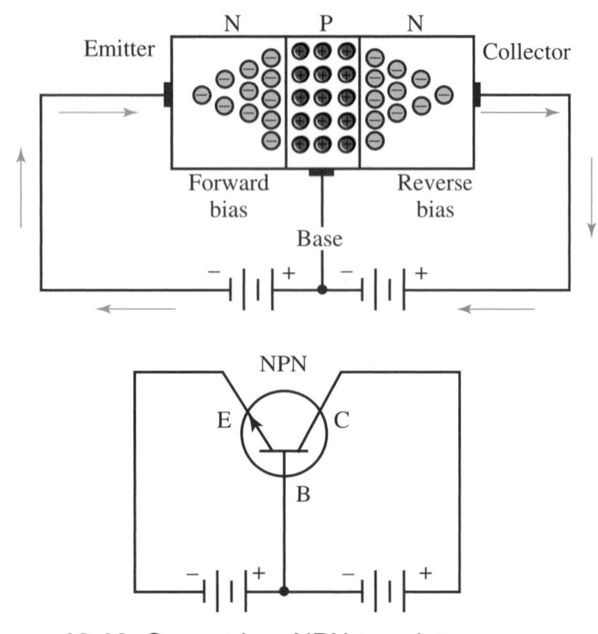

Figure 18-16. Current in a NPN transistor.

Approximately 95 to 98 percent of the current through the transistor is from emitter to collector. About two to five percent of the current moves between emitter and base. A small change in emitter to base bias voltage causes a somewhat larger change in emitter-collector current. This is what allows transistors to be used as amplifiers. The emitter-base current change, however, is quite small.

A PNP transistor has a P-type material for the emitter, an N-type material for the base, and a P-type material for the collector. See **Figure 18-17.** The power supply or battery must be connected in the opposite way as an NPN transistor. Like the NPN transistor, the emitter-to-base circuit has forward bias, and the collector-to-base circuit has reverse bias. In a PNP transistor, most carriers in the emitter-to-collector are holes.

Field-Effect Transistor (FET)

The major difference between bipolar junction transistors and ***field-effect transistors (FET)*** is that the BJT is a current device, and the FET is a voltage device. The current through the collector-emitter circuit of a BJT is controlled by the amount of current in the base-emitter circuit. A FET controls current in the source-drain circuit by the amount of potential applied to the gate. There are two types of FETs, the ***junction field-effect transistor (JFET)*** and the ***metal oxide semiconductor field-effect transistor (MOSFET).*** First we will look at the JFET.

The construction and symbols for the JFET are shown in **Figure 18-18.** The three main parts of the JFET

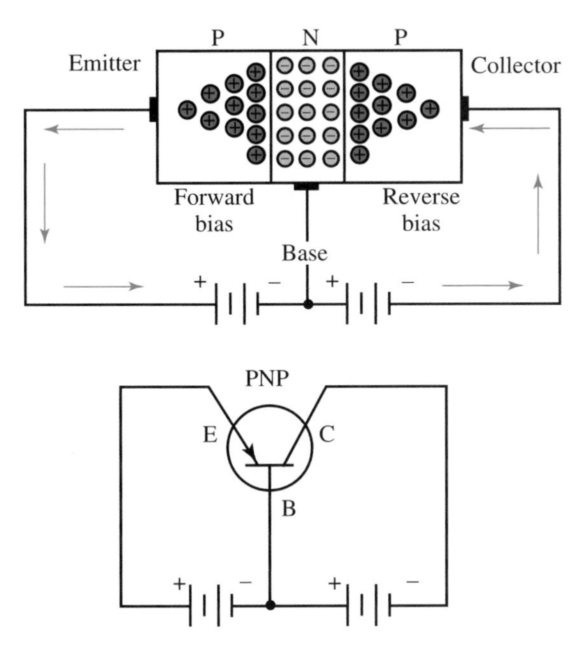

Figure 18-17. Current in a PNP transistor.

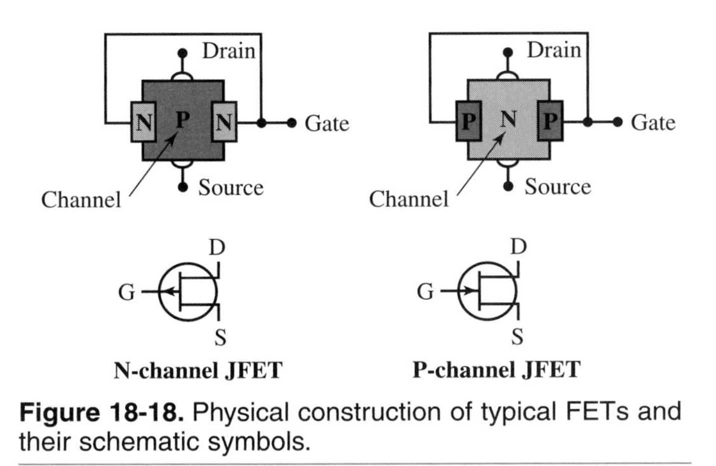

Figure 18-18. Physical construction of typical FETs and their schematic symbols.

are the source, drain, and gate. These three parts are similar to the three main parts of a typical bipolar transistor. The ***source*** compares to the emitter. The ***drain*** compares to the collector. The ***gate*** compares to the base. The gate is diffused into the channel material. The ***channel*** is a path from the source to the drain. The channel through the center of the device can be P- or N-type material.

The current through the device is controlled by the gate potential. See **Figure 18-19.** If a *small potential* is applied to the gate, a *large current* through the P channel will exist. If a *large potential* is applied to the gate, a *small current* will exist in the channel between the source and the drain. Notice how the potential applied at the gate *pinches off* the flow of electrical energy.

Another type of FET is the MOSFET. It is used extensively in digital circuits and memory circuits in computers. Look at **Figure 18-20.** The MOSFET is similar in construction to the JFET. The difference is that the MOSFET has a very thin film of insulation (silicon dioxide) between the gate and the channel area.

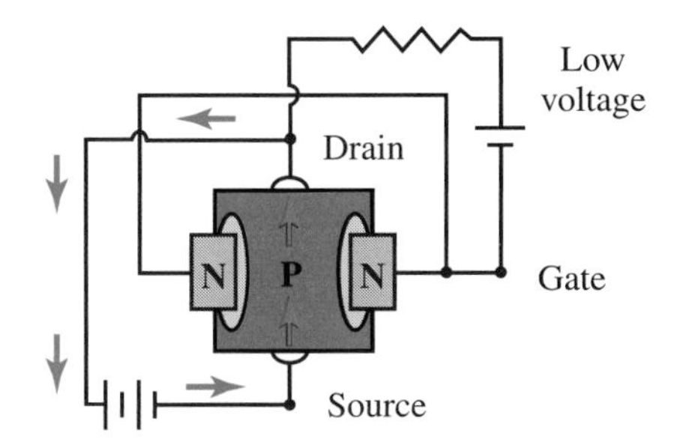

Low gate voltage means a large current through the P-channel.

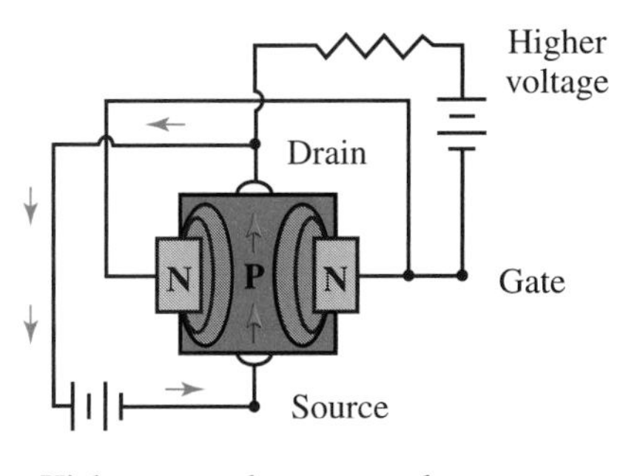

Higher gate voltage means less current through the P-channel.

Figure 18-19. The amount of voltage applied to the gate will determine the current value from the source to the drain.

CAUTION:

The layer of insulation is so thin that it can easily be damaged by static electricity. Care must be taken when handling MOSFET transistors and devices.

The thin layer of high resistance insulation prevents electron flow between the gate and the channel material. The high resistance between the gate and channel area makes for a very high impedance input device. A high impedance input device is very desirable in circuits such as amplifiers. Also, the channel consisting of the same material from the source to the drain provides a very low impedance path. A low impedance, or low resistance, path through the channel is also very desirable for devices such as amplifiers.

There are two main types of MOSFET, enhancement mode and depletion mode. In a ***depletion-mode MOSFET,*** current through the source drain circuit is reduced by gate voltage. In an ***enhancement-mode MOSFET,*** current through the source drain circuit is increased by the gate voltage. Look closely at the four symbols used for the MOSFET, Figure 18-20. The symbols differ for

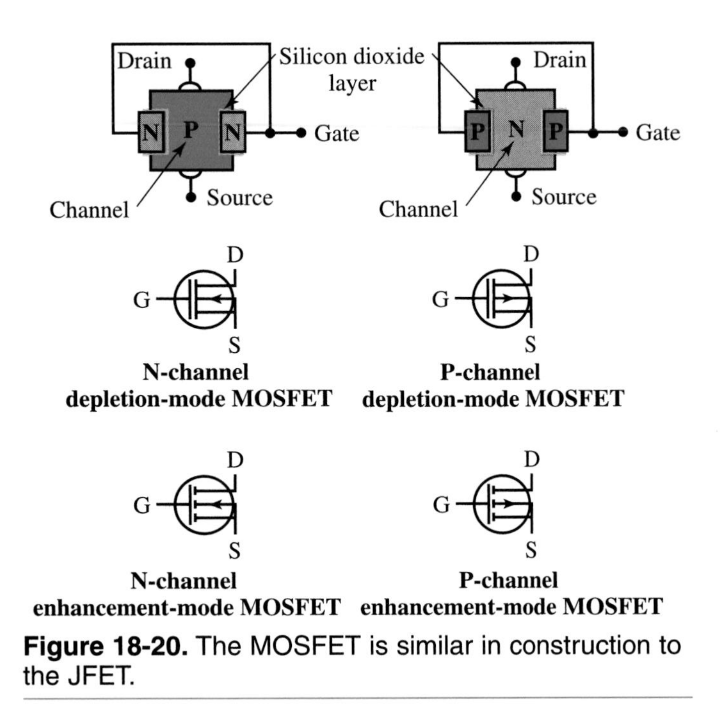

Figure 18-20. The MOSFET is similar in construction to the JFET.

N-type and P-type channel materials and for enhancement-mode and depletion-mode MOSFETs.

Review Questions for Section 18.2

1. Name the three parts of a bipolar transistor.
2. Draw the symbols for an NPN transistor and a PNP transistor. Label the parts.
3. Draw and label the parts of a JFET and MOSFET transistor.
4. Draw and label the symbol for an enhanced MOSFET.
5. Why is the MOSFET so easily damaged by static electricity?
6. What is the difference between a JFET and a MOSFET transistor?
7. What is the difference between depletion mode and enhancement mode?

18.3 TRANSISTORS AS AMPLIFIERS

An ***amplifier*** is an electronic circuit that uses a small input signal to control a larger output signal. In physics, weight can be amplified using levers. A small lever is used to lift a larger weight. In electronics, amplifiers have been used since the early twentieth century. Amplification can be accomplished using vacuum tubes or semiconductors devices such as transistors or integrated circuits.

The amount of amplification in a circuit is known as ***gain.*** The gain is the ratio between the strength of the output (current, voltage, or wattage) and the strength of the input (current, voltage, or wattage), **Figure 18-21.**

Amplifier circuits can be thought of as *control* circuits. A small amount of current or voltage can control a larger amount of voltage or current. These circuits produce outputs that vary, or are linear.

Amplifier devices such as transistors or ICs can also be used to *switch* current on or off, depending on how they are biased in the circuit.

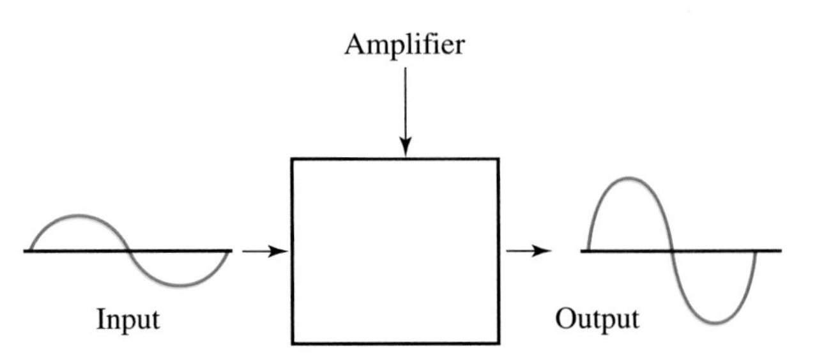

Figure 18-21. Block diagram of an amplifier.

Biasing

For amplifiers to operate properly, they must be correctly biased. ***Biasing*** means setting up the correct dc operating voltages between input leads of a transistor. In a transistor there are two junctions. One junction is between the emitter and the base and is called the ***emitter junction.*** The other is between the collector and the base and is commonly referred to as the ***collector junction.*** For proper transistor operation, energy from the internal power source (battery or power supply) is needed to overcome these junction resistances.

As discussed in section 18.2, a bias voltage must exist between emitter and base of an NPN transistor, **Figure 18-22.** Voltage applied to these elements with the correct polarity will create current flow. This is known as forward bias. Reverse bias is needed in the collector junction of an NPN transistor, **Figure 18-23.**

Both forward and reverse bias are needed for the operation of a transistor amplifier. **Figure 18-24** shows a complete NPN transistor circuit. Notice the forward bias in the emitter junction and reverse bias in the collector junction.

The same type of bias is needed for operation of a PNP transistor amplifier, **Figure 18-25**. The currents for each circuit are labeled. I_E is for emitter current, I_C for collector current, and I_B for base current. The emitter junction bias is provided by battery 1, and the collector junction bias is provided by battery 2.

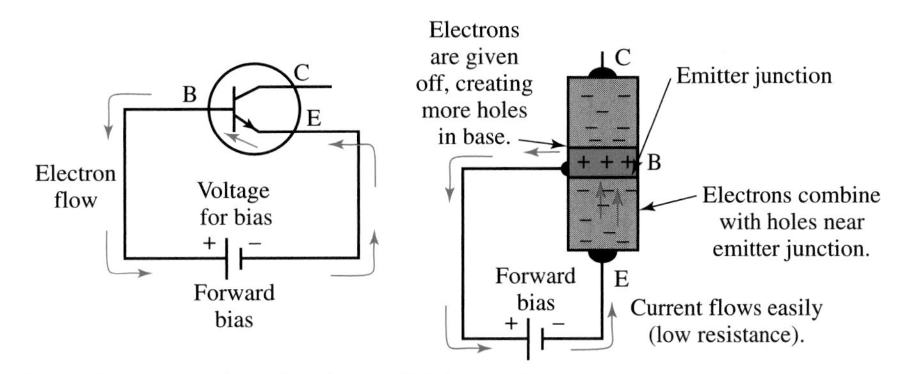

Figure 18-22. Forward bias for emitter junction in an NPN transistor.

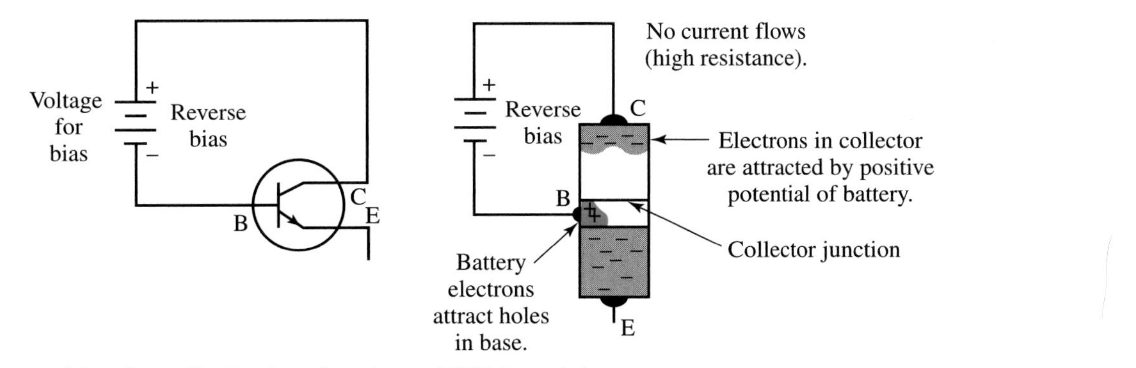

Figure 18-23. Reverse bias for collector junction in an NPN transistor.

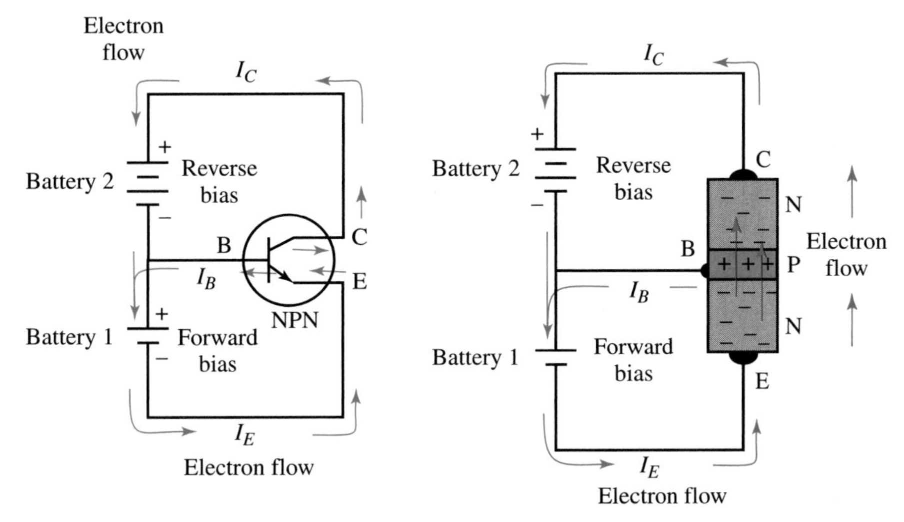

Figure 18-24. Forward and reverse bias in an NPN transistor amplifier circuit.

Single Battery Circuit

Two voltage sources have been used in all circuits discussed thus far. One source was used for the forward biasing of the emitter junction, and the other source was used for the reverse biasing of the collector junction. There is no need for two batteries. The amplifier in **Figure 18-26** uses only one battery.

There is no question about the reverse bias of the collector junction. The collector, C, is connected through R_C to the most positive point of the circuit, the positive supply terminal. The most negative point in the circuit is the ground, and it is connected directly to the negative terminal of V_{CC}.

Resistors R_F and R_B form a resistance voltage divider connected directly across V_{CC}. The voltage at the base, B,

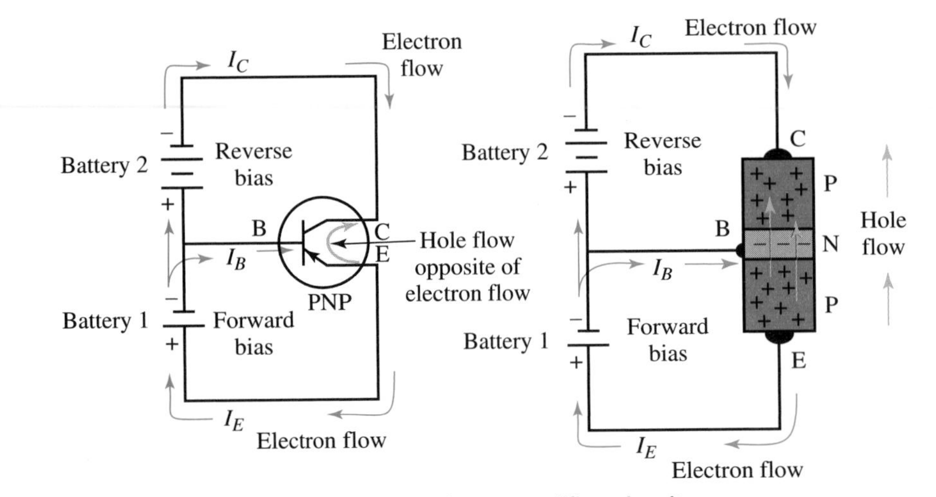

Figure 18-25. Forward and reverse bias in a PNP transistor amplifier circuit.

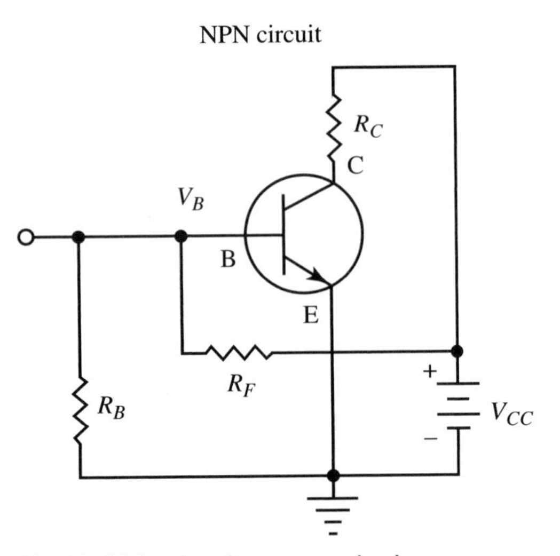

Figure 18-26. This circuit uses a single power source.

is *less* positive than the positive terminal of V_{CC} by the amount of the voltage drop across R_F. It is positive in respect to the emitter, E. E is at ground, which is the most negative point in the circuit.

By using the proper values for R_F and R_B, the desired forward bias voltage and current can be established in the emitter junction. The series resistance combination of R_F and R_B must be large enough so that current drain from the supply battery is small. This ensures long battery life.

Methods of Bias

The ***fixed bias*** method is shown in **Figure 18-27.** Notice that resistor R_B has been omitted. This circuit sets a constant base current. This bias is used for switching circuits. A switch is inserted in the base to control current through the emitter-collector. Proper selection of R (R_F and R_C) sets up the required forward bias voltages and base current.

The ***single battery bias*** scheme is another common method for biasing transistors, **Figure 18-28.**

Emitter biasing is a third method for setting forward bias of the emitter junction, **Figure 18-29.** In this case,

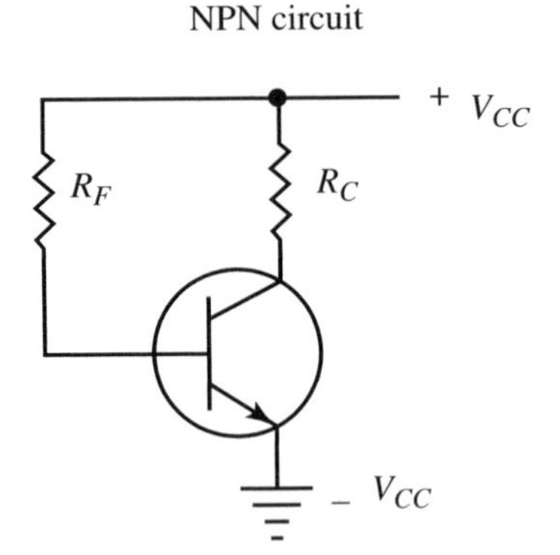

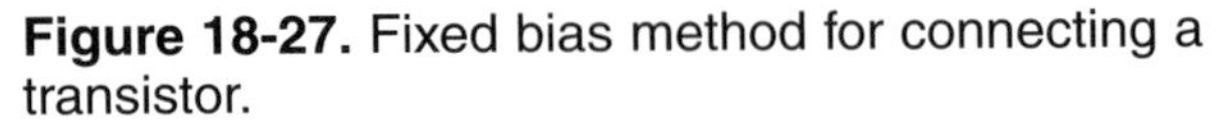
Figure 18-27. Fixed bias method for connecting a transistor.

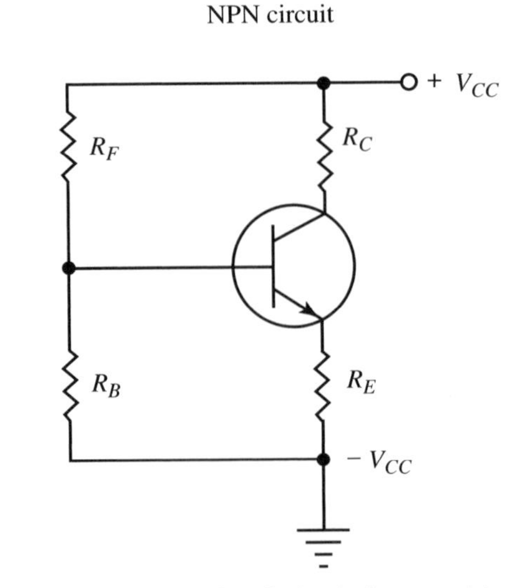

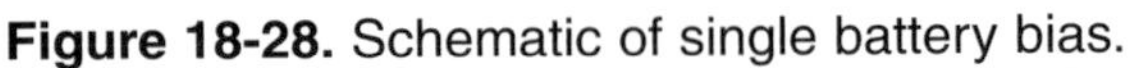
Figure 18-28. Schematic of single battery bias.

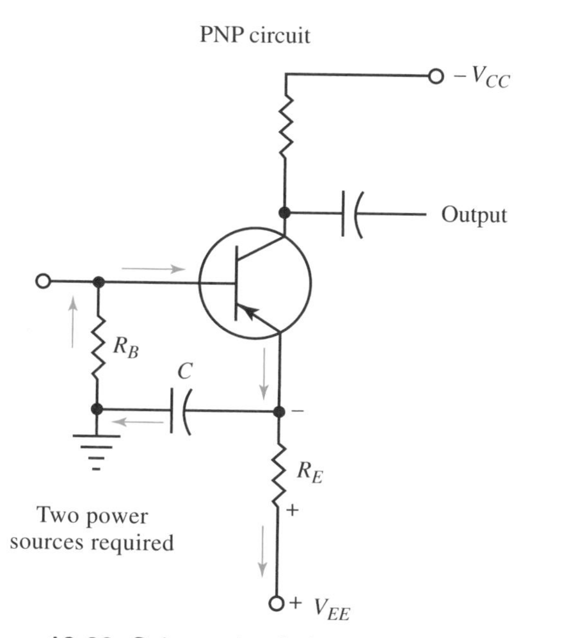

Figure 18-29. Schematic of circuit for emitter biasing.

the forward bias voltage V_{EE} will set a constant emitter current, I_E. This produces a voltage drop across R_E. R_E is chosen to provide the proper forward bias and R_B is the return to complete the emitter circuit. An ac signal applied to the amplifier base will produce a larger ac component in the collector current. This does not upset the emitter bias, because a low reactance path around R_E is provided by bypass capacitor *C*. This circuit ensures a stable operating point for the transistor. A disadvantage of this circuit, however, is the need for two power supplies or batteries.

A ***self bias*** schematic is shown in **Figure 18-30.** This circuit differs from the fixed bias method in that bias resistor R_F is connected to the collector rather than to V_{CC}.

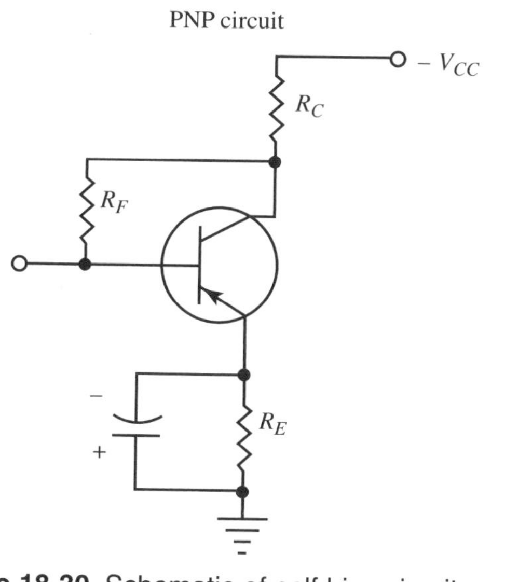

Figure 18-30. Schematic of self bias circuit.

This method provides a more stable operating point than fixed biasing and requires only one power source.

If a fixed collector current is assumed at some operating point, the collector voltage, V_C, will be constant. But it will be lower in value than V_{CC}, due to the voltage drop across R_C. Any change in I_C will also change the value of V_C. Since the base is connected to V_C through R_E, a certain amount of degeneration will result.

What is meant by degeneration? A positive signal at the input of the PNP amplifier makes the base more positive, decreases its forward bias, and decreases I_C. A reduction in I_C means a lesser voltage drop across R_C. Collector voltage V_C becomes more negative as it reaches the value of negative V_{CC}. This more negative voltage through R_F to the base tends to increase forward bias. It also opposes any increase caused by the input signal. In a phase relationship that opposes the input, a signal from the output of a device fed back to the input is called ***degeneration.*** The major disadvantage of self bias is the loss of amplifier gain due to degeneration.

Review Questions for Section 18.3

1. What is an amplifier?
2. Amplifiers can perform two functions. Name them.
3. The process of setting up the correct dc operating voltages between input leads of a transistor is called __________.
4. In a transistor there are two junctions. Name them.
5. Explain how a transistor needs both forward and reverse bias to operate?
6. Define degeneration.

18.4 AMPLIFIER OPERATION

As was described previously, an amplifier is an electronic device that uses a small input signal (voltage or current) to control a larger output signal. This principle works best when the output load, or impedance, of the circuit is greater than the input load.

Examine **Figure 18-31.** The input is a microphone in the emitter and base control circuit. The output is a loudspeaker connected to the collector and emitter circuit. The load of the loudspeaker is much greater than that of the microphone. The output current in the circuit is about the same as the input current. Recalling Ohm's law, if the input and output currents are about the same, and the resistance (load or impedance) is 50 times greater in the output than in the input, then the output voltage will be 50 times as great as the input signal. This process is signal amplification.

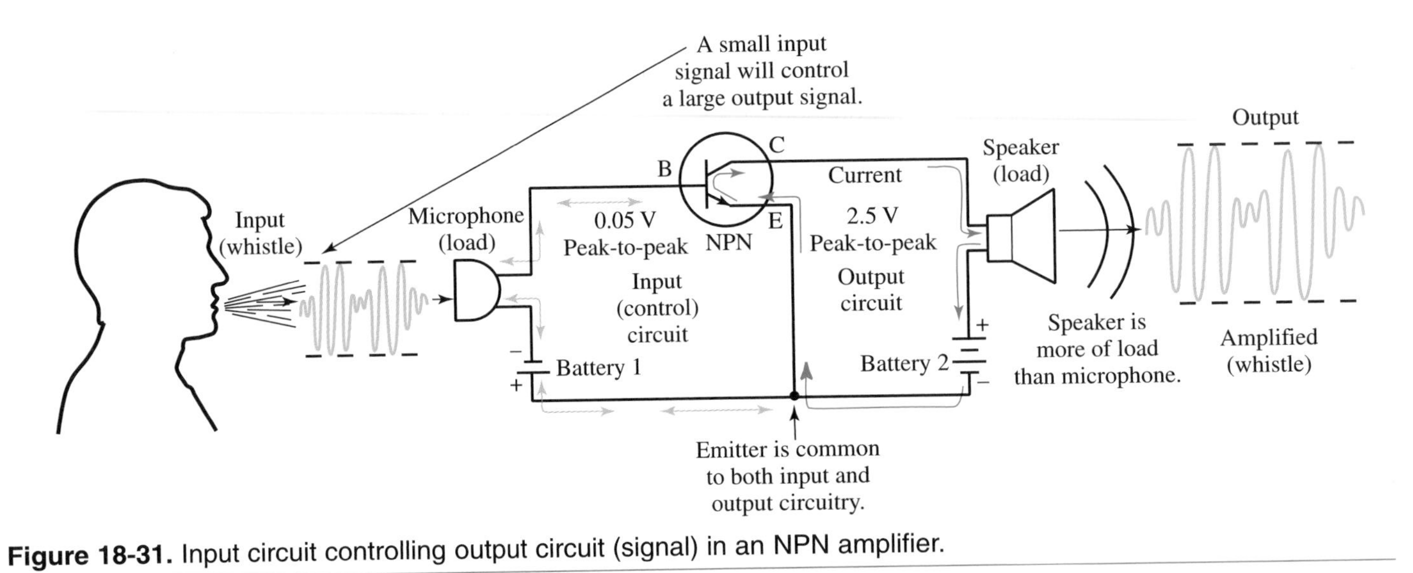

Figure 18-31. Input circuit controlling output circuit (signal) in an NPN amplifier.

Refer to **Figure 18-32.** It shows an NPN amplifier circuit with actual component values and specific voltages. The emitter is common to both the input and output circuits. The ac input signal is impressed across the emitter and base of the transistor Q_1. The output signal is taken across the collector and emitter of Q_1. A capacitor, C_1, is placed in the input signal to couple (or pass) the ac input signal to the base for control. A ground is added at a common connection in the battery 1, emitter, and the battery 2 circuit.

A 0.05 volt, peak-to-peak, ac input signal is applied across the emitter to base at points Y and Z. As the signal swings positive, the base-to-emitter forward bias voltage (signal plus bias) will increase. When the forward bias voltage increases, the base current will also increase resulting in a smaller resistance across the emitter-to-collector of transistor Q_1. The smaller resistance causes the current in the circuit made of battery 2, R_1, and the emitter/collector path of Q_1, to increase. This increase in current through a fixed value resistor, R_1, causes a greater voltage drop across the resistor. (See wave #1). The increase in voltage drop across R_1 causes a decrease in voltage across Q_1 because they are both in series with the 9-volt battery, battery 2. (See wave #2). Keep in mind the resulting voltage in a series circuit. The total voltage, E_{B2}, is equal to the sum of the voltage drops in that series circuit, including the transistor, Q_1, and the resistor, R_1.

During the negative half of the input signal, the bias voltage on the base is decreased or lowered. As a result, the resistance of the emitter-to-collector will increase

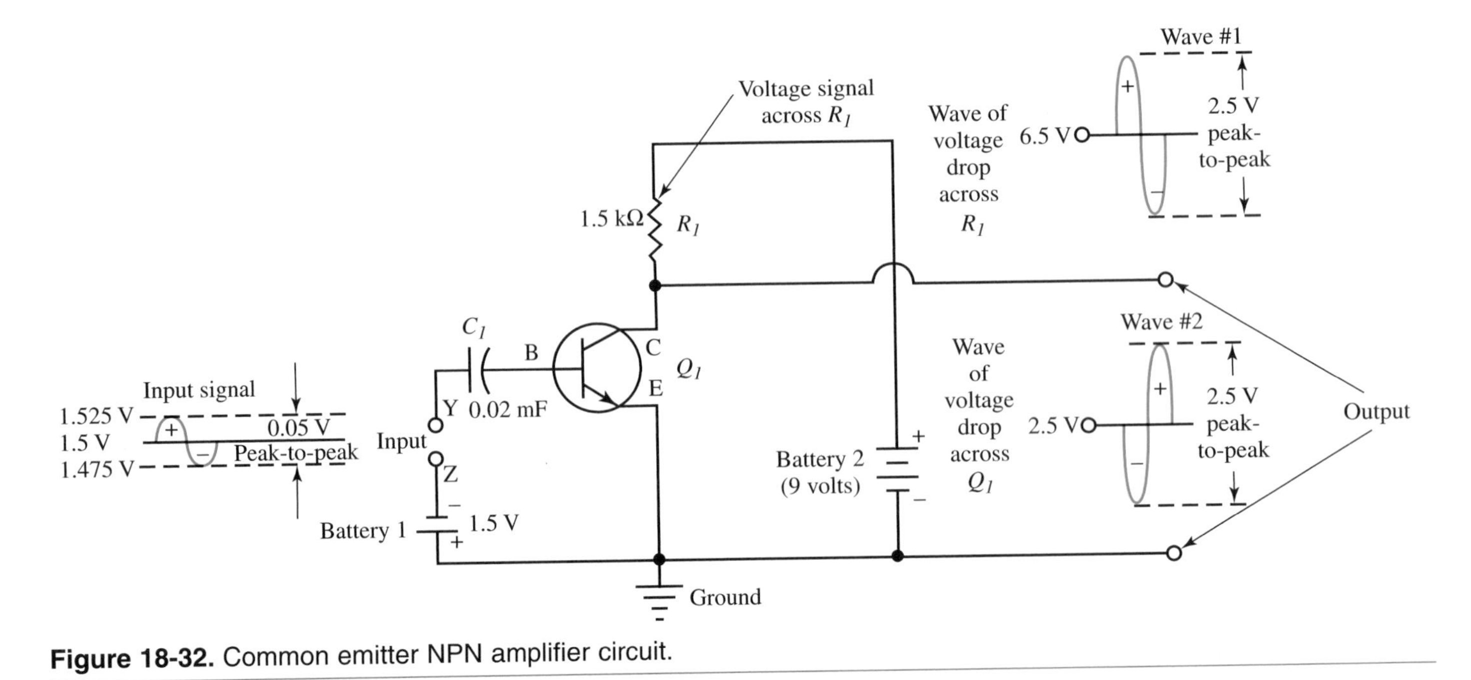

Figure 18-32. Common emitter NPN amplifier circuit.

causing current to be reduced. If the current flowing in the emitter-to-collector circuit is decreased, current will also decrease in R_1, since it is in series with the collector of Q_1. If the current through a fixed value resistor, R_1, is decreased, the voltage drop across it will also decrease. (See the negative half of wave #1). This decrease in voltage across R_1 causes an increase in voltage drop across the emitter-to-collector of Q_1 as the sine wave swings positive in wave #2. In any case, both wave #1 and wave #2 must add up to nine volts, **Figure 18-33.** Note that wave # 1 and wave #2 are 180 degrees out of phase.

A 0.05 volt signal in the input circuit (emitter-to-base in Q_1) has controlled a 2.5 volt signal in the output (emitter-to-collector in Q_1). This is amplification.

Computing Gain

There are a number of formulas that can be used to compute types of amplification in transistor amplifier circuits. When the formulas show a triangle, or delta (Δ), symbol, it means "the change in" a certain value or waveform.

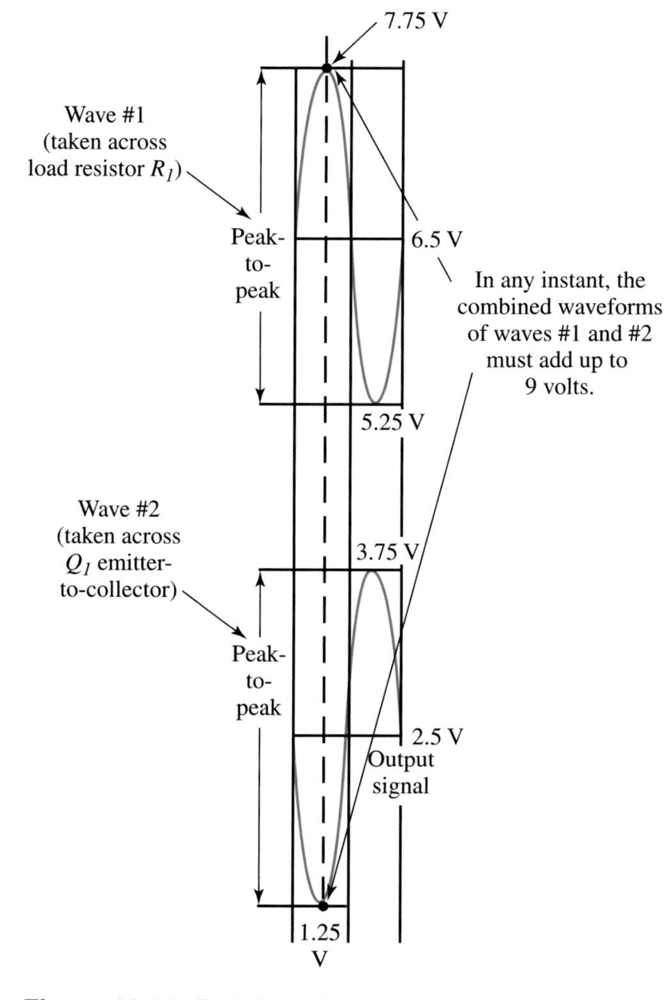

Figure 18-33. Relationship of output signal and ac signal across R_1.

Voltage gain

The ***voltage gain (A_V)*** is calculated by dividing the change in the output voltage by the change in the input voltage.

$$A_V = \frac{\Delta E_{out}}{\Delta E_{in}} = \frac{\Delta E_{out\ p\text{-}p}}{\Delta E_{in\ p\text{-}p}}$$

Current gain

The **current gain (A_I)** is calculated by dividing the change in the output current by the change in the input current.

$$A_I = \frac{\Delta I_{out}}{\Delta I_{in}}$$

Power gain

The ***power gain (A_P)*** is calculated like Watt's law. The voltage gain (A_V) is multiplied by the current gain (A_I).

$$A_P = A_V \times A_I$$

Gain is a ratio and, therefore, a dimensionless number. Gain simply tells how much a signal has been amplified. The value is a number, like 100, meaning the voltage gain is 100 times larger in the output than in the input.

Decibels

A ***decibel (dB)*** is a unit of *relative* measurement. When a measurement is definitive, it has a definite weight, size, volume etc. When a measurement is relative, it must be compared to a similar, known object.

Compare two different shades of blue, such as the blue sky and the blue water. There is no definite measurement as to just *how* blue sky is or *how* blue the water is. However, we can compare the shades of the sky and water to determine which is *more* blue.

The decibel was developed as a way to measure the relativity of sound or to compare the level of one sound to another. In the field of electronics, audio systems are expressed in decibels. Many times amplifiers are also classified in terms of decibels. Instead of having a definitive output, the output of the sound system or amplifier is expressed in decibels. This is a way of saying that the output is 10, 100, or 1000 or more times greater than the input. A decibel rating gives relative power level and not an actual power rating. To have a clear understanding of the decibel rating, we must understand logarithms of numbers. It is not within the scope of this textbook to cover logarithm tables.

Review Questions for Section 18.4

1. If the output ac signal voltage is 200 mV, and the input ac signal is 10 mV, what is the voltage gain?
2. Output ac current is 750 µA, and ac input current is 150 µA. Find the current gain.
3. Compute the power gain when the output voltage gain is 175 and the current gain is 5.

18.5 TRANSISTOR CIRCUIT CONFIGURATIONS

The transistor can be connected in three circuit configurations. The configuration used depends on the element that is common to both input and output circuits. Usually the common element is at ground potential.

In **Figure 18-34,** a ***common base (CB)*** amplifier circuit is shown. The ac input signal between emitter and base will vary the forward bias by alternately adding to, and subtracting from, the fixed bias. The increase and decrease in current in the emitter-collector circuit produces the amplified voltage across R_L. The input signal and the output signal are in phase. There is no signal inversion.

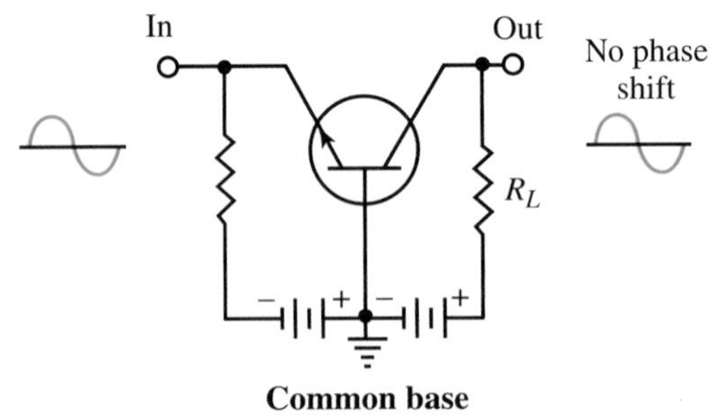

Figure 18-34. The common base (CB) or grounded base circuit with an NPN transistor.

In **Figure 18-35,** a ***common emitter (CE)*** circuit is drawn. This is the most common transistor circuit. In the common emitter circuit, a positive input signal makes the base more positive or less negative with respect to the emitter, thus reducing the forward bias of the circuit. A negative input signal makes the base more negative and increases the forward bias. As a result, the voltage across R_L in the output circuit is 180 degrees out of phase.

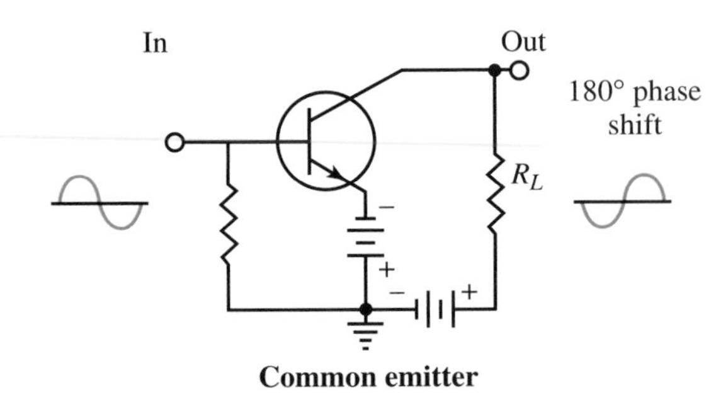

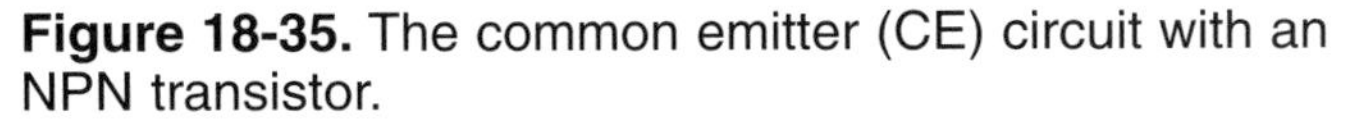

Figure 18-35. The common emitter (CE) circuit with an NPN transistor.

In the common emitter circuit, the gain is the ratio between the change in collector current and the change in base current. The current gain is represented by the Greek letter β (beta).

$$\beta = \frac{\Delta I_C}{\Delta I_B} \text{ where } V_C \text{ is constant}$$

In this equation, β (beta) equals current gain in common emitter circuit, ΔI_C equals the change in collector current, and ΔI_B equals the change in base current. V_C (collector voltage) is held constant.

Note: In transistor circuits, voltage is represented by a capital V instead of E as used elsewhere.

Figure 18-36 shows a ***common collector (CC)*** circuit. The signal is applied between the base and collector. The output is taken from the collector-emitter circuit. It has a very high input impedance and a low output impedance. This circuit is useful as an impedance matching circuit. The input and output signals are in phase. The voltage gain in this circuit must always be less than one.

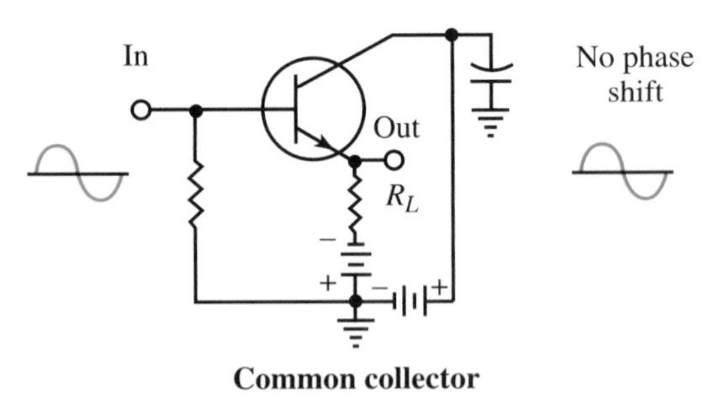

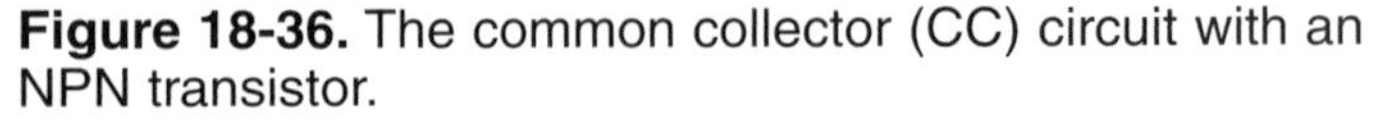

Figure 18-36. The common collector (CC) circuit with an NPN transistor.

Figure 18-37 shows common input and output impedance values for common base (CB), common emitter (CE), and common collector (CC) amplifier circuits. **Figure 18-38** shows the phase relationships between input signals and output signals in CB, CE, and CC amplifiers.

Type of circuit	Input impedance	Output impedance	A_V	A_I	A_P
Common base (CB)	Low 50–150 Ω	High 300 k–500 kΩ	High 500–1500 dB	Less than one dB	Medium 20–30 dB
Common emitter (CE)	Medium 500 Ω–1.5 kΩ	Medium 30 k–50 kΩ	Medium 300–1000 dB	Medium 25–50 dB	High 25–40 dB
Common collector (CC)	High 20 k–500 kΩ	Low 50 Ω–1 kΩ	Less than one dB	Medium 25–50 dB	Medium 10–20 dB

Figure 18-37. Common input and output impedance values of CB, CE, and CC circuits.

Type of amplifier	Input waveform	Output waveform	Status
Common base (CB)			In phase
Common emitter (CE)			180° out of phase
Common collector (CC)			In phase

Figure 18-38. Phase relationships of waveforms in CB, CE, and CC transistor amplifiers.

DC Load Line Analysis

A transistor ***load line analysis*** is a graphic display that plots the maximum current value of the collector against the maximum voltage across the emitter-collector. This graph is used to predict transistor characteristics under changing load conditions correlated to a particular voltage drop across the emitter collector. This graph is only an approximation. It does not take into consideration some influences that can affect transistor performance such as temperature and individual transistor characteristics.

Look at **Figure 18-39.** The bottom line of the graph represents voltage across the emitter-collector of the transistor. The maximum voltage is equal to the supply voltage. See **Figure 18-40.** When a transistor is not conducting, it is said to be at ***cutoff.*** In this condition, it is similar to an open switch. The full applied voltage can be read across the emitter collector circuit.

Again look at Figure 18-39 and compare it to Figure 18-40. See where the top of the slope line connects to the vertical axis, which indicates the current values in milliamps. This point represents the maximum current through the collector and is based on the current through the load resistor. Using Ohm's law, the collector current can be determined by dividing the supply voltage by the load resistor value in ohms. Since the load resistor is connected in series to the collector, the current values will be the same. Maximum current through the collector is the ***saturation*** point of the transistor.

Look at **Figure 18-41,** and you will see the ***Quiescent point,*** or ***Q-point,*** located at approximately the midpoint of the sloping line. The Q-point is where optimum performance characteristics for the transistor are displayed. For many amplifiers, the Q-point will be approximately one half of the supply voltage.

Rather than maximum gain, a small gain in an amplifier results in a better sound quality from the amplifier. If

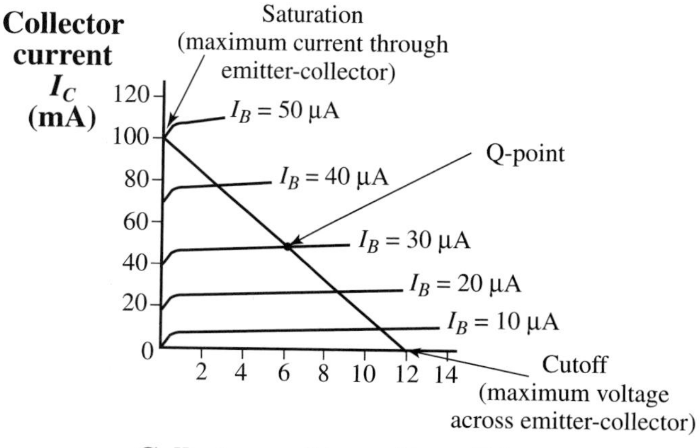

Figure 18-39. A line drawn from maximum collector current to maximum voltage across the emitter-collector can predict approximate transistor performance.

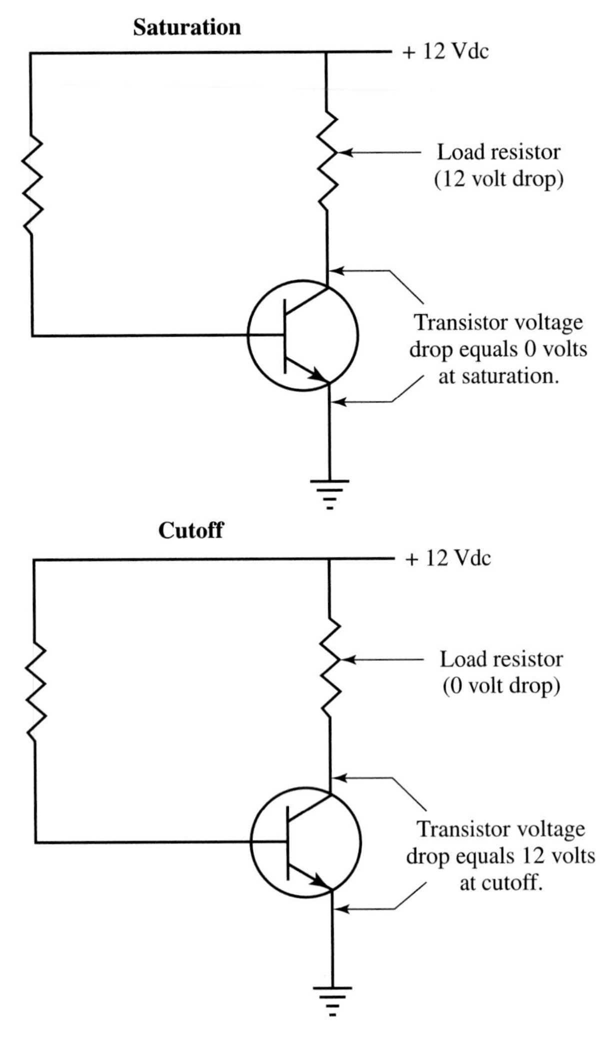

Figure 18-40. When a transistor is at saturation, there will be a zero-volt drop across the transistor and a 12-volt drop across the load resistor. Current is at a maximum through the collector circuit. When the transistor is at cutoff, the voltage drop will be equal to 12 volts across the transistor, and the current will be zero.

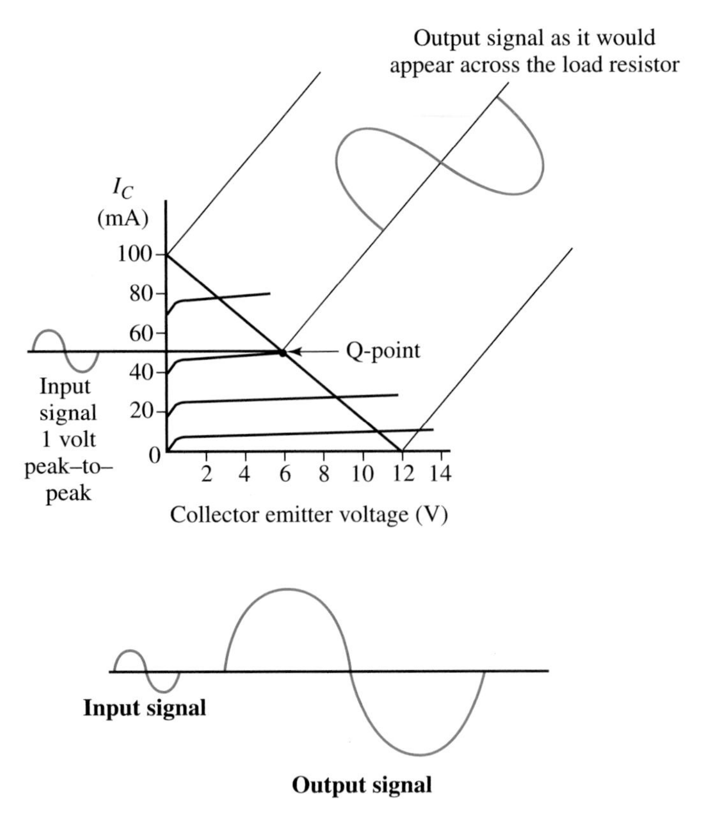

Figure 18-41. The output signal is a perfect replica of the input signal, only amplified.

the gain is too high, the transistor will be driven both into saturation and cutoff, which results in a distortion of the amplified signal.

Compare Figure 18-41 to **Figure 18-42.** A small gain results in a near perfect replica of the input signal. A large gain, or excessive amplification of an input signal, results in distortion. The amplified signal distortion is caused by the amplified signal exceeding the saturation and cutoff point, thus cropping off the top and bottom of the sine wave signal.

The same thing happens to an audio signal that is amplified too much. A multistage amplifier designed for small gains of 7 to 10 has far better sound quality than an amplifier with only a few transistors with large gains, of say, 50. This is one reason for the large variation in prices for amplifier systems. The amplifier with many stages will undoubtedly cost more and sound better.

Classes of Amplifiers

Amplifiers are classified according to their bias current. The bias current determines the output waveform compared to the input waveform.

Class A. The ***class A amplifier*** is biased so that an output signal is produced for 360 degrees of the input signal cycle. See **Figure 18-43.** Current flows at all times. The output of a class A amplifier is an amplified copy of the input signal. Its efficiency is low. This class is widely used in high fidelity sound systems.

Class B. A ***class B amplifier*** is biased so that an output signal is reproduced during 180 degrees of input signal. The class B amplifier is biased close to cutoff, **Figure 18-44.** Since collector current flows only one-half of the time, this amplifier has medium efficiency. It is similar to a half-wave rectifier. Its output is distorted.

Two amplifiers operated in a push-pull configuration can restore both halves of the signal in the output. The push-pull circuit has increased efficiency and greater power output. It will be used in many audio amplifiers as the final output stage. Push-pull coupling is covered in detail in section 18.6.

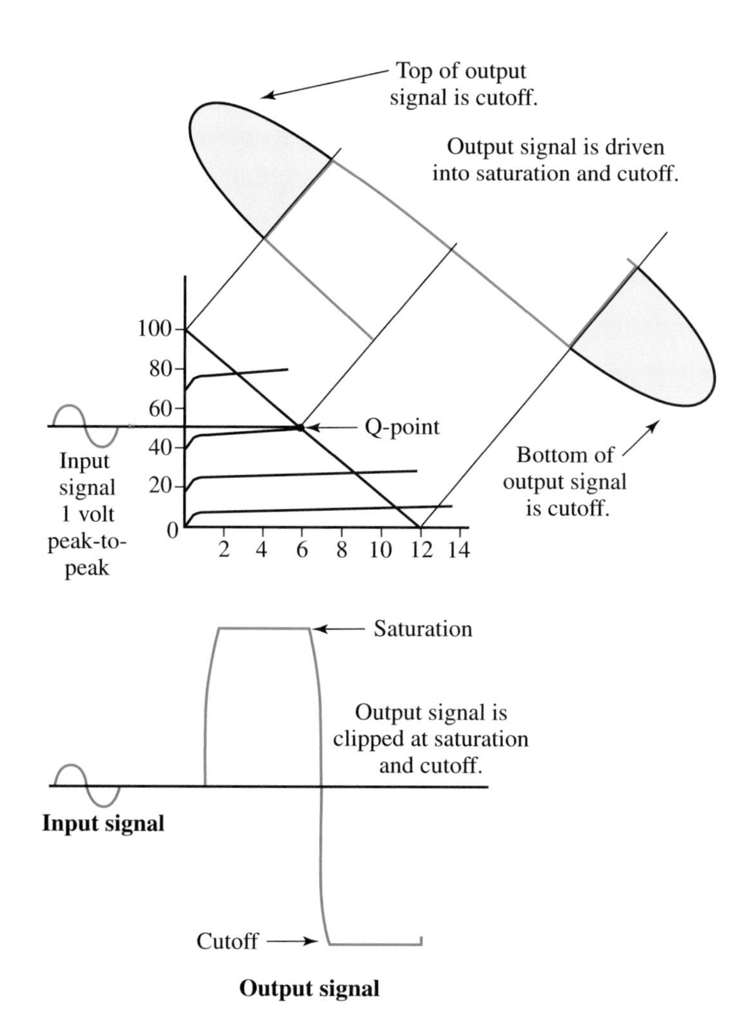

Figure 18-42. When amplified beyond the limits of saturation and cutoff, the amplified signal will be cropped on both ends. This results in severe distortion.

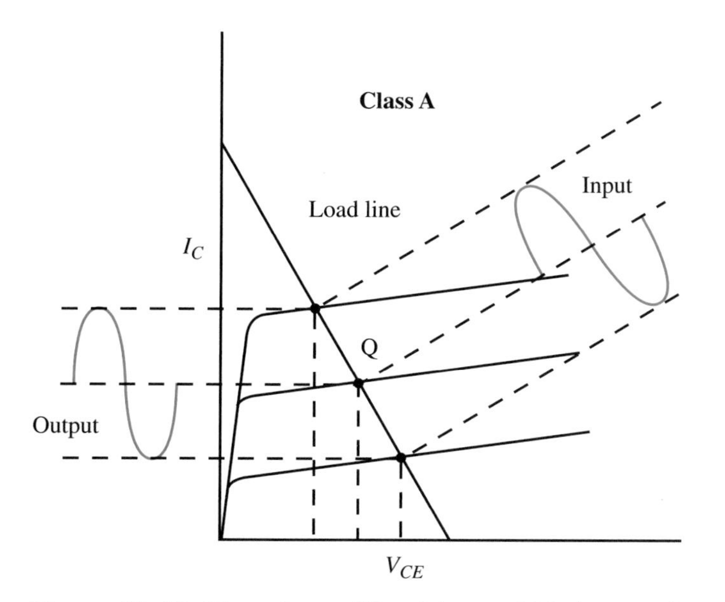

Figure 18-43. Class A amplifier. It has a 360 degree signal output for a 360 degree input signal.

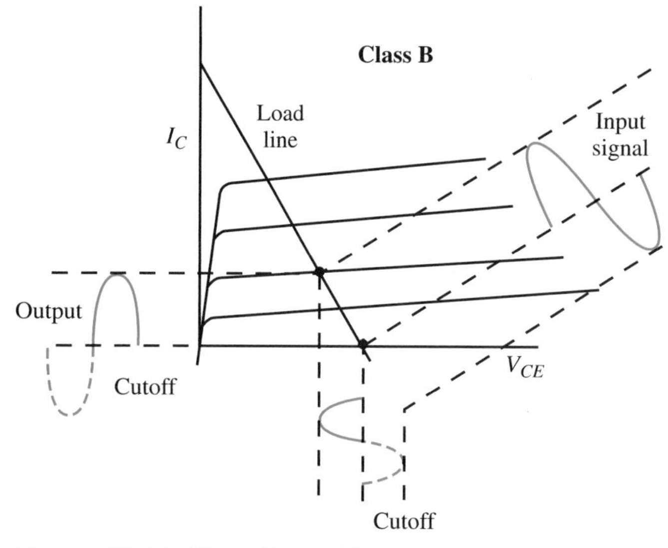

Figure 18-44. Class B amplifier. It has a 180 degree signal output for a 360 degree input signal.

Class C. A ***class C amplifier*** is biased so that less than 180 degrees of output signal is produced during 360 degrees of input signal. This class finds limited use in transistor circuits.

Thermal Considerations

Transistors are easily overheated. As operating temperature increases, there is greater thermal activity in the crystalline structure and a decrease in resistance. Repeated resistance decrease and current increase can cause destruction of the transistor. This condition is called ***thermal runaway.***

There are a few methods that prevent thermal damage to a transistor. One way of reducing overheating is through the use of special circuits. In **Figure 18-45,** the common emitter amplifier has a stabilizing resistor R_3.

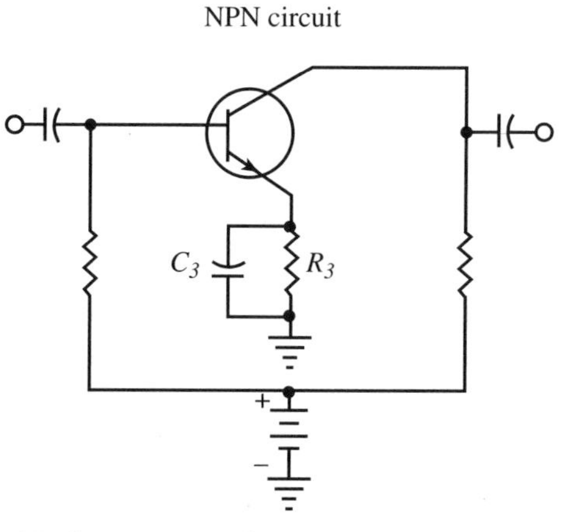

Figure 18-45. Common emitter stage with stabilizing resistor and bypass capacitor.

An increase in current makes the emitter end of R_3 more positive with respect to ground. The forward bias of the emitter-base circuit decreases. This decrease acts as a limiter to current increase resulting from thermal runaway. C_3 is a bypass capacitor that prevents degeneration and loss of gain.

Another method of preventing thermal overheating is through the use of heat sinks. ***Heat sinks*** are large pieces of metal, usually with fins, that absorb and disperse the heat that is generated in a transistor. See **Figure 18-46.**

Figure 18-46. Heat sink designs.

Transistor Precautions

Transistors are delicate electronic components. Following are a number of precautions you should take to protect transistors from damage.

- Do not remove or replace transistors in circuits when the power is on. Surge currents can destroy the transistor.
- Be very careful with close connections such as those found in miniature transistor circuits. A brief short circuit can burn out a transistor.
- When measuring resistances in a transistor circuit, remember that the ohmmeter contains a battery. If it is improperly applied, it will burn out a transistor. It is also important to observe the correct polarity when measuring ohms for accurate readings.
- Soldering the leads of transistors is a skill which must be developed. Heat will destroy the transistor. When soldering a transistor lead, grasp the lead between the transistor and the connection to be soldered with a pair of long nose pliers. These pliers will act as a heat sink and absorb the heat from soldering. Use a pencil type soldering iron of no more than 25 watts.

Review Questions for Section 18.5

1. Name three types of transistor configurations.
2. This class of amplifier has an output signal that is an amplified copy of the input signal.
   a. Class A.
   b. Class B.
   c. Class C.
   d. None of the above.
3. This class of amplifier has an output wave 180 degrees out of phase with the input signal.
   a. Class A.
   b. Class B.
   c. Class C.
   d. None of the above.
4. The total effect of uncontrolled decreasing resistance and increasing current in a transistor circuit is called __________ __________.
5. What is a heat sink?

18.6 COUPLING AMPLIFIERS

Transistor stages can be connected in series to produce the desired amplification. Connecting these stages in series is called ***cascading.*** The key challenge of cascading is matching the high impedance output of one transistor to the low input impedance of the next transistor without severe loss in gain.

Transistor stages can be connected using one of the four primary methods:

1. Transformer coupling.
2. RC coupling.
3. Direct coupling.
4. Push-pull coupling.

Transformer Coupling

The circuit in **Figure 18-47** shows two stages of transistor amplifier stages coupled with a transformer. Transistor Q_1 has an output impedance of 20 kΩ, and Q_2 has an input impedance of 1 kΩ. A severe mismatch and loss of gain results when they are directly connected.

A transformer can match these impedances. In this case, a step-down transformer is required. A low secondary voltage means a higher secondary current. This condition is acceptable for transistors because they are current operated. Many special sub-ouncer and sub-sub-ouncer transformers have been developed for this purpose, **Figure 18-48.**

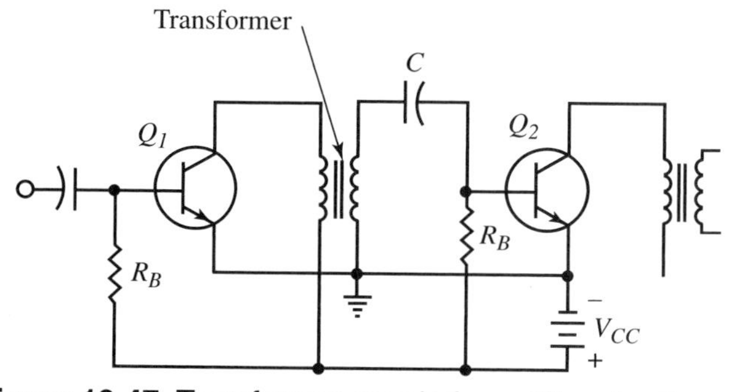

Figure 18-47. Transformer coupled amplifier circuit.

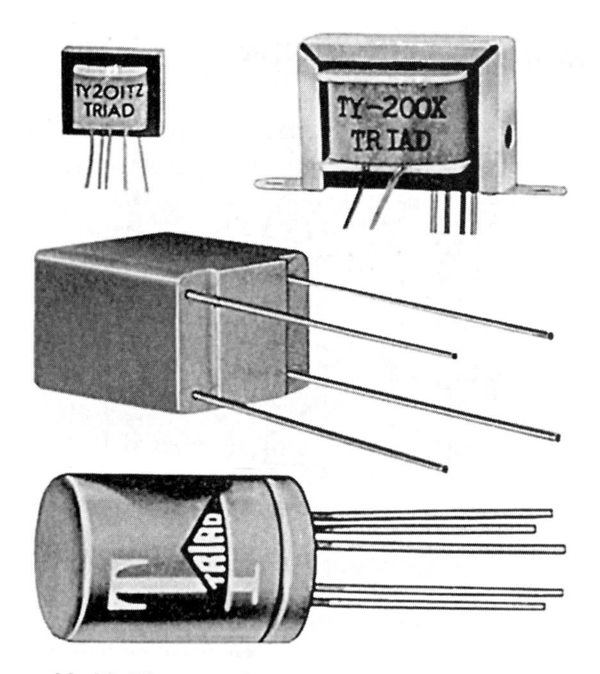

Figure 18-48. These sub-ouncer and sub-sub-ouncer transformers are used in transistor circuitry. (Triad)

The purpose of capacitor C in the transformer coupling circuit of Figure 18-47 is to block the dc bias voltage of the transistor from ground. Notice that if C is left out, the transistor base will be grounded directly through the transformer secondary.

The major drawback of transformer coupling, besides the cost, is the poor frequency response of transformers. They tend to saturate at high audio frequencies. At radio frequencies, inductance and winding capacitance presents problems.

A type of transformer coupling that avoids this problem uses a tapped transformer. These taps can be at medium, low, and high impedance points. Good impedance matching can be attained with taps, as well as good coupling and gain. Study the circuit in **Figure 18-49.** For radio frequency amplifier circuits, both the transformer primary and secondary windings can be tuned by variable capacitors for frequency choice.

There is one other key point about the transformer. The primary impedance of the transformer acts as a collector load for the transistor. This impedance only appears under signal conditions. The load is X_L of primary. From the dc point of view, the only load is the ohmic resistance of the wire used to wind the transformer primary. This information will be helpful when designing a power amplifier.

RC Coupling

A simple and less costly method of coupling transistor amplifier stages is to use resistor-capacitor (RC) coupling. See **Figure 18-50.**

In part A of Figure 18-50, capacitor C charges to 10 volts. Only during the charging of C will a current cause

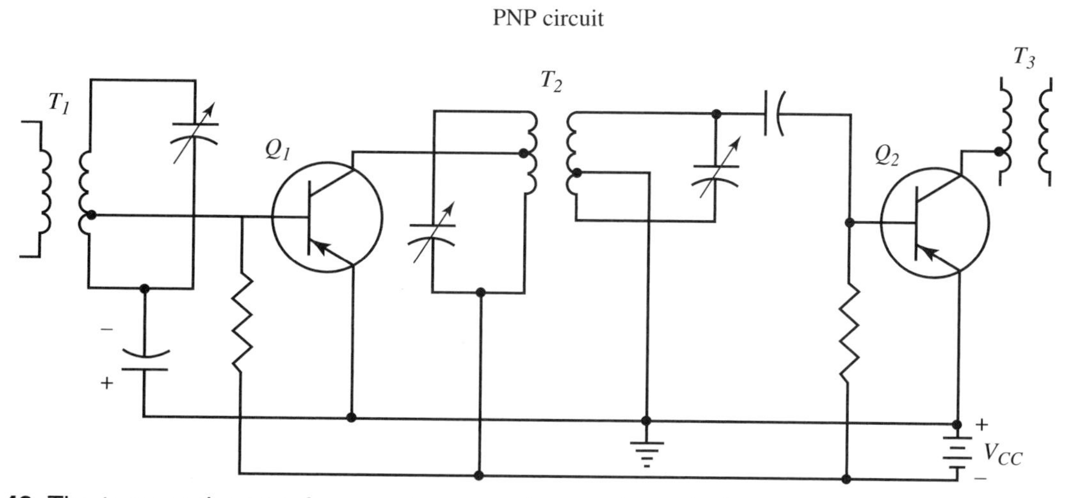

Figure 18-49. The taps on the transformer primary and secondary windings provide an ideal matching point. The transformer can be designed for good overall gain.

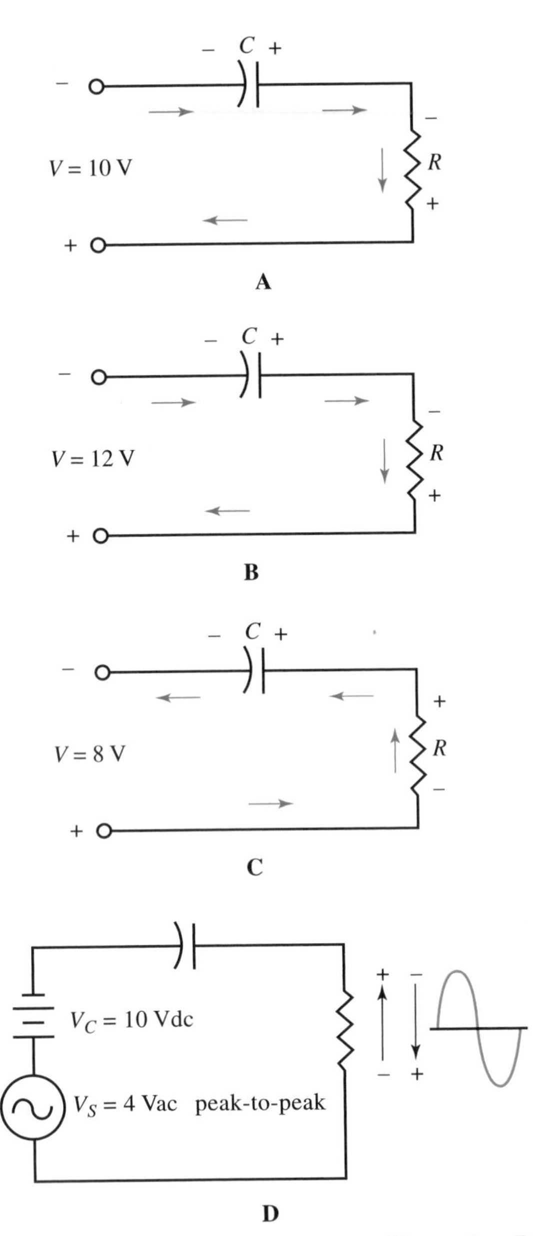

Figure 18-50. Coupling transistor amplifier using RC coupling.

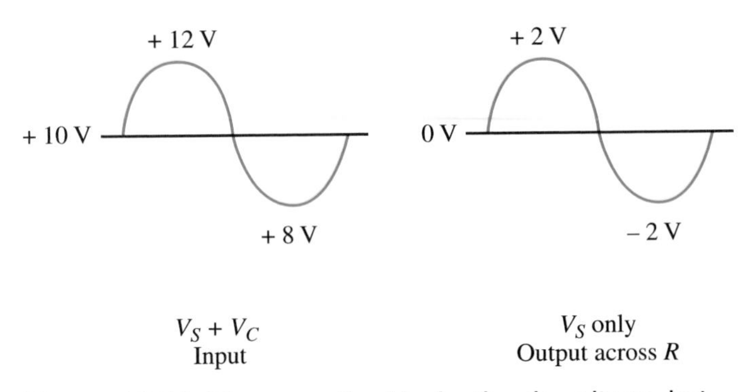

Figure 18-51. The capacitor blocks the dc voltage, but permits the ac signal to pass.

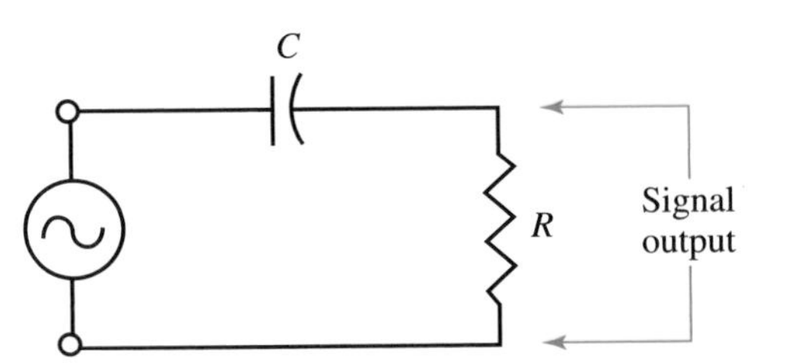

Figure 18-52. R and C form a voltage divider for the ac signal.

a voltage to appear across *R*. After *C* is charged, V_C will equal 10 volts, and V_R will equal zero.

If the source voltage is changed to 12 volts, V_C will also increase to 12 volts, part B of Figure 18-50. The charging current will produce a brief two-volt pulse across *R* in the polarity shown.

If the source voltage is changed to eight volts, *C* will discharge to eight volts, part *C* of Figure 18-50. The discharge current will produce a brief voltage pulse across *R* in the polarity shown.

In part D of Figure 18-50, both dc and ac voltages are connected to the RC circuit. The ac signal causes the total voltage to vary between 8 and 12 volts. Therefore, *C* charges and discharges at the ac generator frequency.

The voltage across *R* rises and falls at the same frequency as the generator voltage. Look at **Figure 18-51.** The input signal varies around a dc level of 10 volts. But, the output signal varies around the zero-volt level. The dc component has been removed. A capacitor blocks dc. A capacitor used to block dc is called a ***blocking capacitor.***

From a mathematical viewpoint, *C* and *R* in series form an ac voltage divider, **Figure 18-52.** Voltage division depends on the reactance of X_C at the frequency of the signal. Voltage output across *R* is a key consideration when the greatest amount of ac voltage must appear across *R*. With a 1000 Hz frequency signal and a *C* value of 0.01 µF:

$$X_C = \frac{1}{2\pi fC} = \frac{1}{6.28 \times 10^3\ \text{Hz} \times 1 \times 10^{-8}\ \text{F}}$$

$$X_C \cong 0.16 \times 10^5\ \Omega = 16{,}000\ \Omega$$

If the value of *R* is ten or more times greater than X_C, then most of the voltage will appear across *R*. Assuming *R* equals 160 kΩ, then:

$$V_R = \frac{R}{R + X_C} \times V_{in}$$

$$V_R = \frac{1.6 \times 10^5\ \Omega}{(1.6 \times 10^5\ \Omega) + (1.6 \times 10^4\ \Omega)} \times V_{in}$$

$$V_R = \frac{1.6 \times 10^5\ \Omega}{17.6 \times 10^4\ \Omega} \times V_{in} = 0.91 \times 10\ \text{V} = 9.1\ \text{V}$$

It seems that almost all of the voltage appears across R. Less than a volt has been lost as signal output. If R were made larger, then even more of the total output would be developed across R.

Look at the two stage amplifier using transistors in **Figure 18-53.** Consider the signal voltage at the collector of Q_1. There is more than one path for the voltage, and it will always take the easiest path. It can go through R_C or through the coupling network that is in parallel. If the network impedance is higher than R_C, signal currents will go through R_C instead of to the next stage. Therefore, the coupling capacitors used in transistor circuits must have values in the range of 8 to 10 μF.

To prove the point, assume Q_2 input impedance is 500 Ω. X_C would need to be 50 Ω or less. At 1000 Hz frequency:

$$C = \frac{1}{2\pi f X_C} = \frac{1}{6.28 \times 10^3 \text{ Hz} \times 50\ \Omega} \cong 3.2\ \mu\text{F}$$

At lower amplified frequencies, X_C would be higher. A capacitor of 8 to 10 μF would be required to prevent loss of amplifier gain.

Note also in Figure 18-53 that R_F and R_B are in parallel with the emitter-base circuit. These resistor values must be high enough so that the signal will not be bypassed around the emitter junction.

The values of R_B and R_F are based on the required bias and stability of the circuit. Current drain from the source is also considered.

Transistor circuit design can become quite complex and is often a matter of give and take. Compare the output and input impedances of the transistors in the CE configuration. The input can be in the 500 to 1.5 kΩ range. The output impedance is in the 30 kΩ to 50 kΩ range. This is a severe mismatch. With the RC coupling, the loss of power gain caused by the mismatch must be accepted. When cost is a factor, it may be cheaper to add another transistor stage. This addition will offset the loss due to mismatch and is less costly than purchasing a transformer for interstage matching.

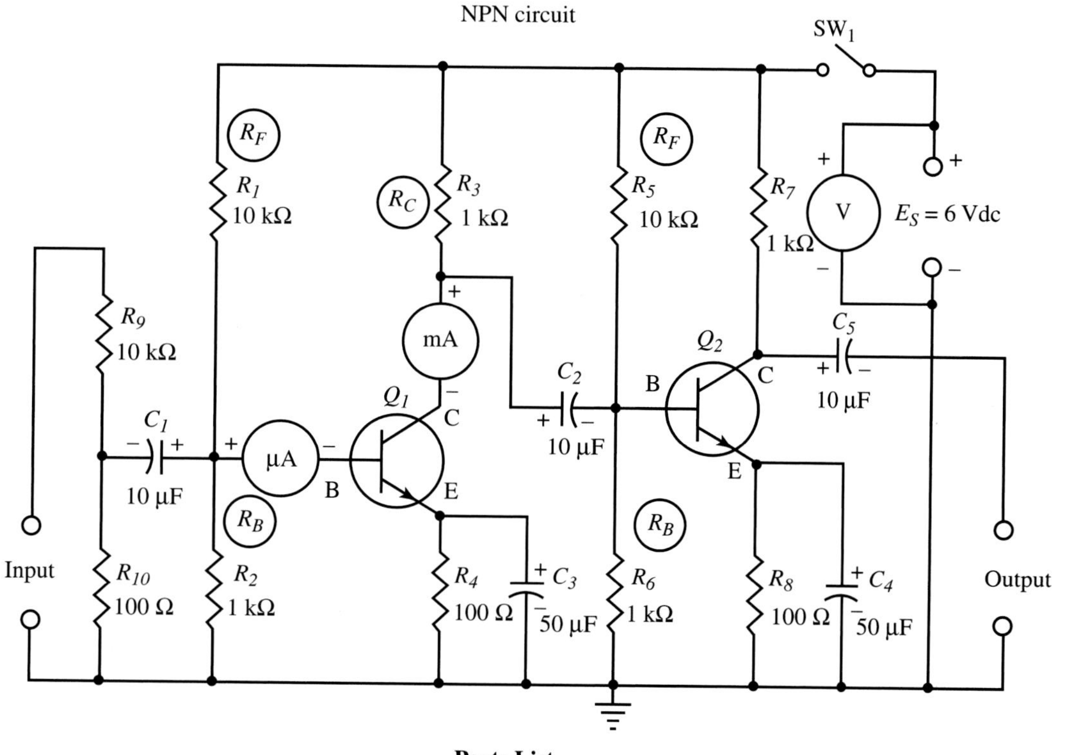

Parts List

Voltmeter 0–25 V dc.
Ammeter 0–10 mA dc.
Ammeter 0–0.1 mA dc.
Q_1, Q_2 – 2N649 transistor.
R_1, R_5, R_9 – 10 kΩ, 1/2W.
R_2, R_3, R_6, R_7 – 1 kΩ, 1/2W.

R_4, R_8 – 100 ohms, 1/2W.
R_{10} – 100 ohms, 1W.
C_1, C_2 – 10 μF electrolytic capacitor.
C_3, C_4 – 50 μF electrolytic capacitor.
C_5 – 10 μF electrolytic capacitor.
SW_1 – SPST switch.

Figure 18-53. Schematic and parts list for RC coupling.

Direct Coupling

In many circuits, very low frequency signals must be amplified or, perhaps, the dc value of a signal must be retained. Amplifier circuits using RC or transformer coupling block out the dc component. Direct coupling solves this problem, **Figure 18-54.**

In this circuit, the collector of Q_1 is joined directly to the base of Q_2. Collector load resistor R_C also acts as a bias resistor for Q_2.

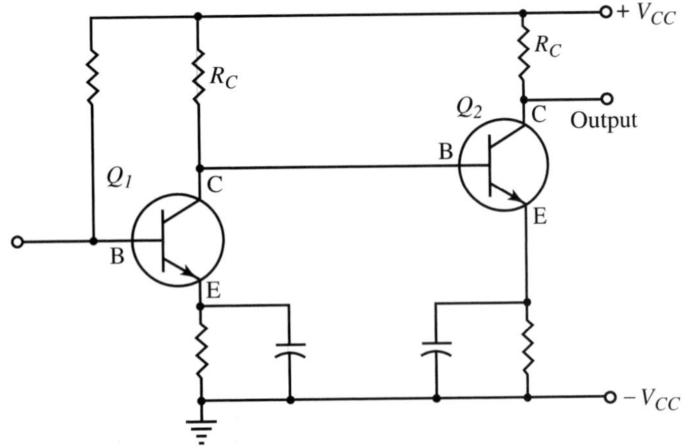

Figure 18-54. Directly coupled amplifier.

However, any change of bias current is amplified by the directly coupled circuit. This creates sensitivity to any temperature changes. This can be overcome with stabilizing circuits. Another drawback of direct coupling is that each stage needs a different bias voltage for proper operation. See **Figure 18-55.**

Push-Pull Coupling

Two transistors joined in parallel in a push-pull power amplifier circuit will achieve peak power output and efficiency from a power amplifier, **Figure 18-56.**

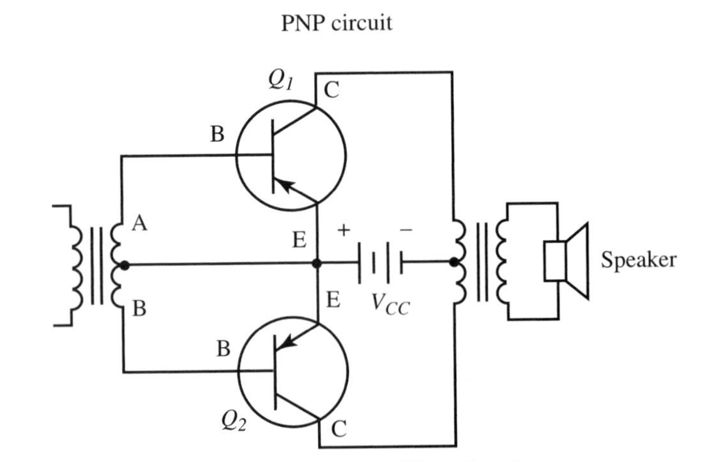

Figure 18-56. Push-pull amplifier circuit.

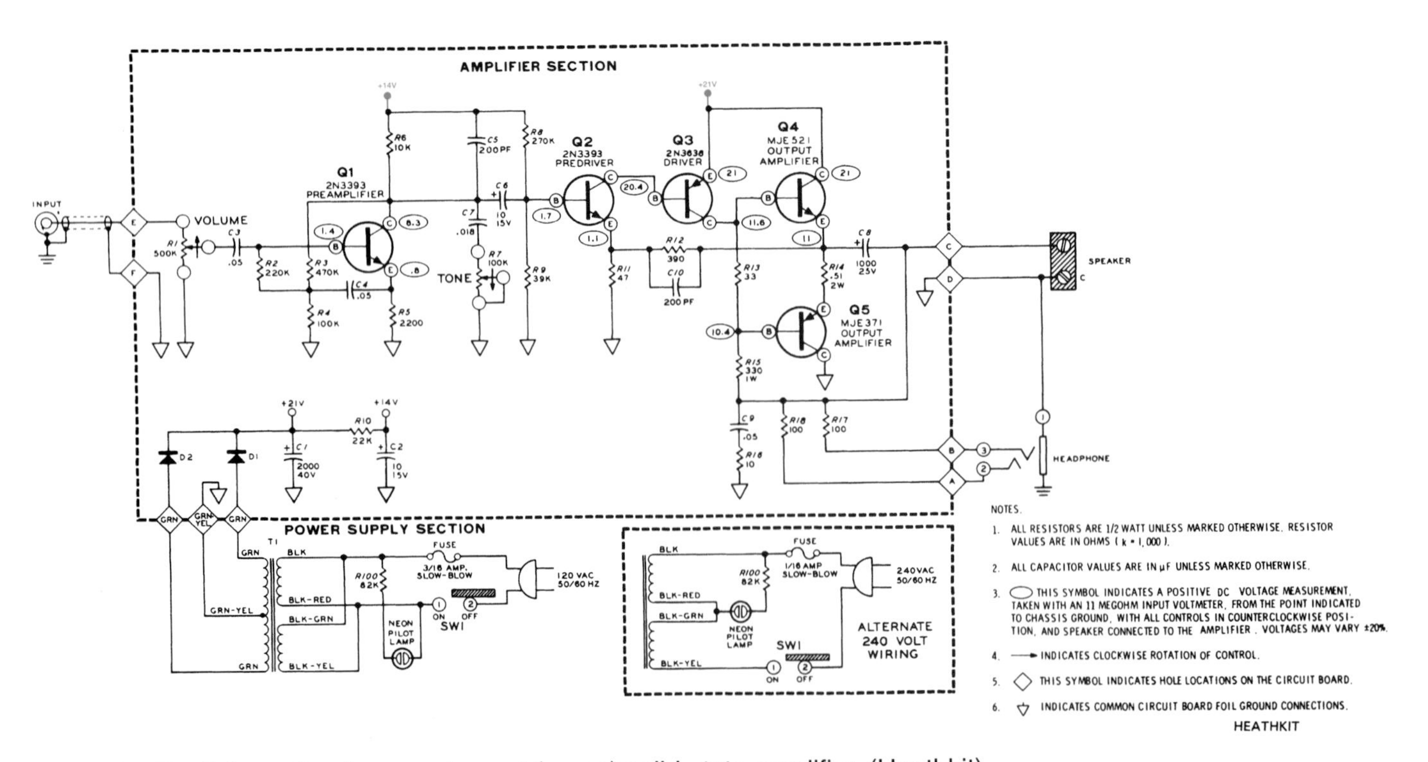

Figure 18-55. Schematic of a one channel (mono) solid-state amplifier. (Heathkit)

This basic circuit works at zero bias when no signal is applied. No voltage is applied across the EB junction, so the current I_B is zero. When point A of the input transformer becomes negative (result of a signal), then B of Q_1 is more negative than E. This is a forward bias and Q_1 conducts. At the same time, the base of Q_2 is being driven positive. This action increases the reverse bias on Q_2, and Q_2 does not conduct.

On the second half of the input signal, the reverse is true. Q_2 is driven into conduction when a negative signal is applied to its base. Q_1 is cut off as a reverse bias is applied by a positive signal to its base. So, Q_1 conducts first, then Q_2 conducts.

On one half cycle, current I_C flows in Q_1. On the second half cycle, current I_C flows in Q_2. These transistors are operating Class B. One half of the time each transistor is resting and cooling. This circuit definitely has increased efficiency. In fact, the maximum possible efficiency approaches 78.5 percent.

The two half signals, or waves of current, from the transistors are restored to their original input form by transformer action in the output transformer.

In **Figure 18-57,** when Q_1 conducts, current flows as shown by arrows. It creates a magnetic field of one polarity. This denotes a half-wave form in the secondary of the transformer. When Q_2 conducts, the polarity of the primary reverses. It induces the wave of opposite polarity in the secondary. Thus, the complete wave is restored.

Review Questions for Section 18.6

1. Connecting amplifier stages in series is called __________.
2. A(n) __________ can be used to couple amplifier stages together, as well as match input and output impedances.
3. Name two disadvantages of transformer coupling.
4. Refer to Figure 18-53. Q_1 is connected to Q_2 by the __________ coupling method.
5. If the collector of one amplifier is hooked to the base of the next amplifier, with no other components in between, the coupling method is called __________ __________.
6. Connecting two amplifier stages in parallel is referred to as __________-__________ coupling.

18.7 THYRISTORS

Thyristors are a special family of semiconductor devices. ***Thyristors*** are fast triggering semiconductor devices that are used as switches. Two common thyristor devices are the SCR and the triac.

SCR

The ***silicon controlled rectifier,*** or ***SCR,*** is used in many dc motor control applications. It is a four layer semiconductor consisting of P, N, P, N layers. The schematic symbol for the SCR is shown in **Figure 18-58.**

The SCR consists of an anode, a cathode, and a gate. When connected to a dc source and properly biased, the SCR does not conduct until it is triggered by a signal at the gate. After the SCR is triggered, it continues to conduct even after the gate is opened. Current flows until the

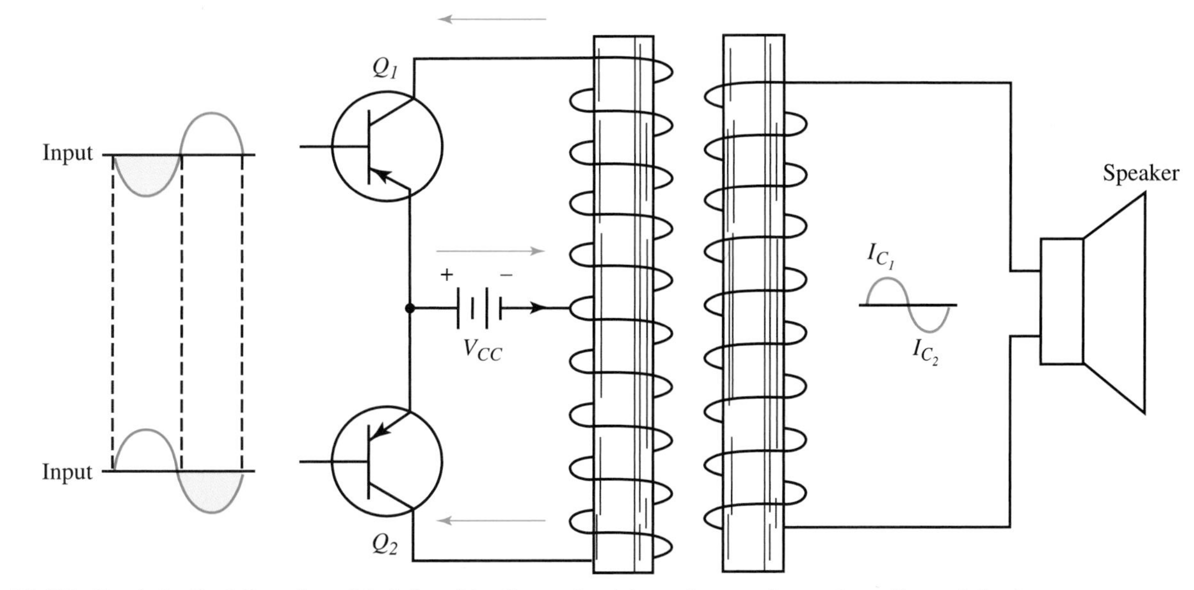

Figure 18-57. Each half of the signal is joined in the output transformer to restore the original wave.

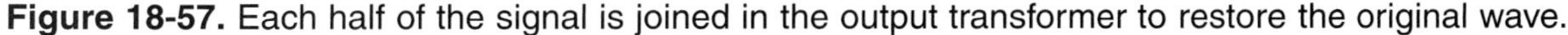

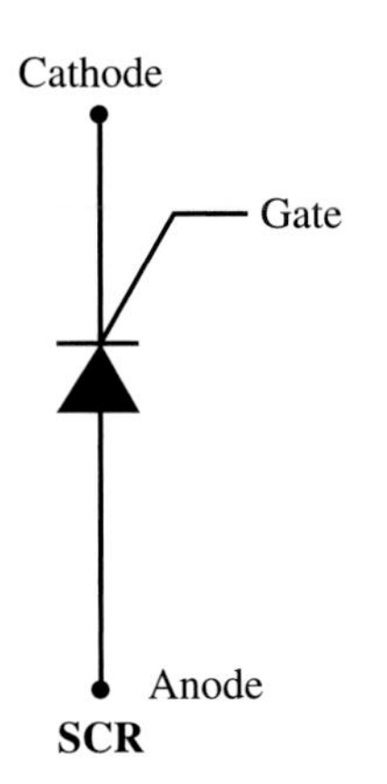

Figure 18-58. Schematic symbol for an SCR.

supply voltage to the anode is removed, reversed in polarity, or substantially reduced. The SCR usually utilizes a pulsating dc source of voltage for its applications. The gate is connected to some type of triggering circuitry that allows for control of the conduction of the SCR. The SCR is capable of handling large dc current values, **Figure 18-59.**

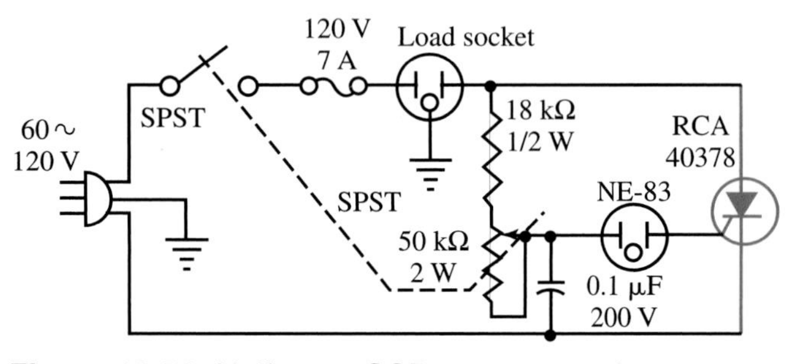

Figure 18-59. Half-wave SCR motor control (RCA circuit).

Triac

The triac is similar to the SCR in construction and is capable of handling large ac current values. **Figure 18-60.** The ***triac*** is constructed of five layers of semiconductor material such as N, P, N, P, N. The additional layer of semiconductor material allows the triac to control ac current. The triac is used extensively as a light dimmer control on ac circuits and in some ac motor applications such as hand drills.

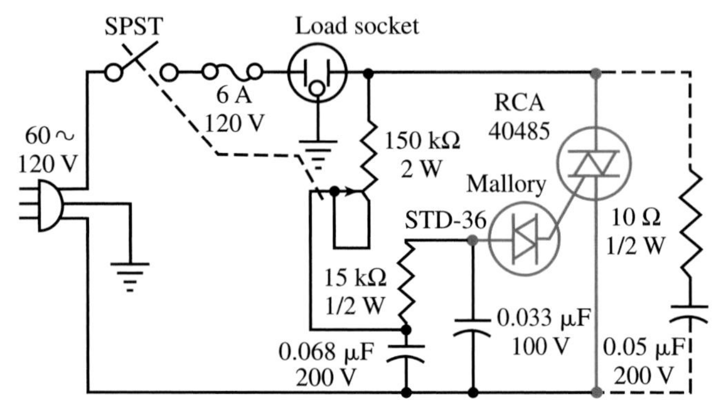

Figure 18-60. Triac motor control (RCA circuit).

Figure 18-61 shows the schematic symbol for the triac. The triac has three main leads: anode 1, the gate, and anode 2. The triac is designed as two SCRs facing opposite directions. It operates on the same principle as the SCR to control ac current.

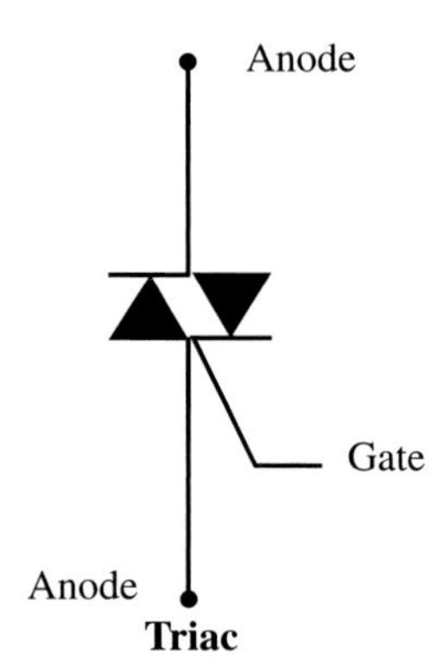

Figure 18-61. Schematic symbol for a triac.

Review Questions for Section 18.7

1. What are the three main parts of an SCR?
2. How must an SCR be turned off?
3. What are the three main parts of a triac?
4. Would you use a SCR or a triac to control a 100 W, 120 Vac lamp?

Summary

1. The three generations of electronic devices are:
   a. The vacuum tube—developed in the late 19th century and discovered in 1904 as a rectifier device.
   b. The transistor—invented in 1948.
   c. The integrated circuit—invented in 1958.
2. Vacuum tubes have been replaced in large part by the use of transistors and integrated circuits.
3. The vacuum tube still finds some special uses, such as in cathode ray tubes (CRTs). Cathode ray tubes are special vacuum tubes used in televisions, oscilloscopes, and computer screens.
4. Vacuum tubes operate on the principle that certain metals and metal oxides give up free electrons when heated. This process is known as thermionic emission.

PROJECT 18-1: Building a Sound Controlled Robot

The sound controlled robot model number 601A by Graymark is an excellent example of combining several different electronic system stages into one complete automated system. The robot combines a power supply, microphone, timer, inverter, dc motor control circuit, and dc motor into a complete automated system. When power is applied to the circuit, the robot moves in a forward direction until the microphone strikes an object or detects a load sound such as the sound produced by the clapping of hands.

Qty.	Part No.	Description	Qty.	Part No.	Description
1	61231	C_1 Capacitor, disc. .022μF	2	61418	R_5, R_{12}, Resistor, 1/4W, 5%, 22k
1	62695	C_2, Capacitor, electrolytic, 47μF	1	62594	R_6, Resistor, 1/4W, 5%, 22k
1	63235	C_3, Capacitor, electrolytic, 1μF	1	62151	R_7, Resistor, 1/4W, 5%, 1M
1	64234	Holder, battery 2xAA	1	62878	R_8, Resistor, 1/4W, 5%, 680
1	63242	ECM Microphone	1	61417	R_9, Resistor, 1/4W, 5%, 220
1	63249	M_1, dc motor	2	63803	R_{10}, R_{11}, Resistor, 1/4W, 5%, 15
1	63234	VR_1, Potentiometer, 100k	1	61411	R_{13}, Resistor, 1/4W, 5%, 100k
1	63247	Printed Circuit Board	1	63246	SW, Switch, SPDT, slide
1	61399	R_1, Resistor, 1/4W, 5%, 1k	2	63237	TR_4, TR_8, Transistor, 2SC2120Y
1	61410	R_2, Resistor, 1/4W, 5%, 2.2k	5	63037	TR_1, TR_2, TR_5, TR_6, TR_9, Transistor, 2SC1815Y
1	63802	R_3, Resistor, 1/4W, 5%, 2.7M	2	63236	TR_3, TR_7, Transistor, 2SA950
1	61416	R_4, Resistor, 1/4W, 5%, 47k			

Graymark project kit 601A Sound Controlled Robot contains all required electronic and mechanical components.

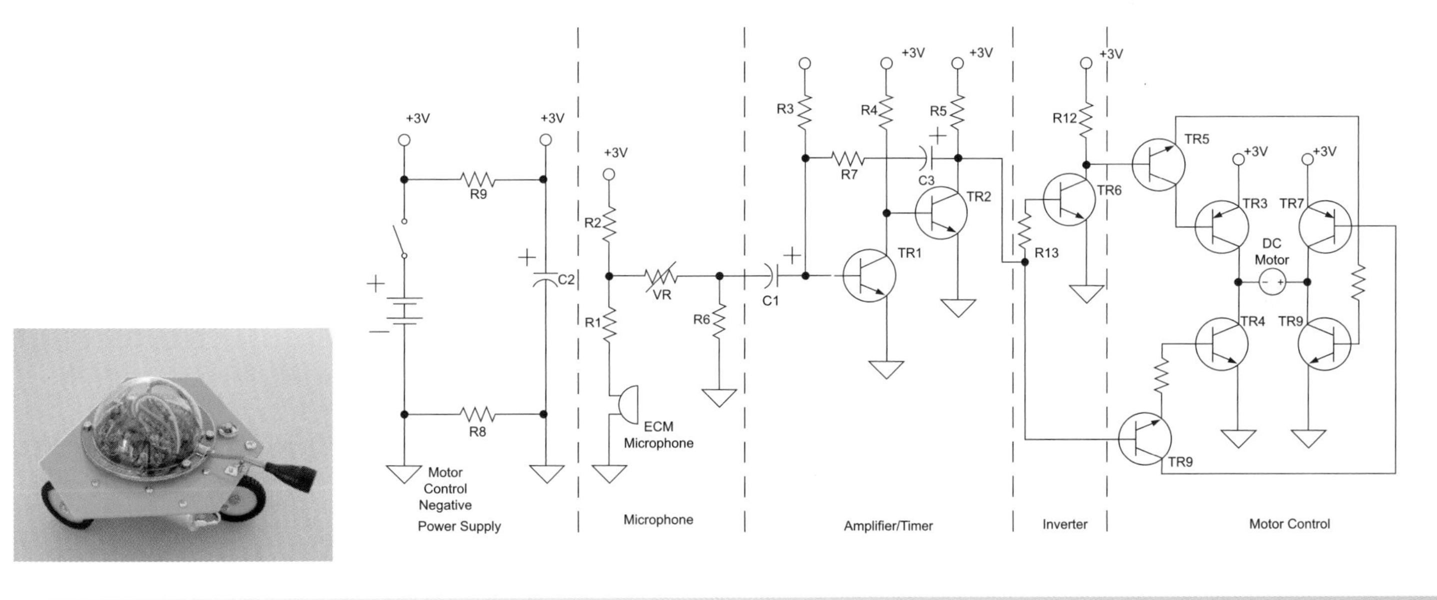

5. The transistor is a semiconductor device capable of amplification, oscillation, and switching.
6. The three basic leads of a transistor are the emitter, base, and collector.
7. Bias is the voltage difference between two input elements of a circuit.
8. In normal operation of a transistor amplifier, the input elements are usually forward biased while the output elements are reversed biased.
9. The three basic leads of an FET are the source, drain, and gate.
10. FETs are voltage control devices while transistors are current control devices.
11. There are two types of FET, the JFET and MOSFET.
12. The MOSFET can be operated in two modes, enhancement and depletion.
13. Amplifiers are electronic circuits that control output signals with input signals.

14. Formulas to compute gain are:

Voltage gain	$A_V = \frac{\Delta E_{out}}{\Delta E_{in}}$
Current gain	$A_I = \frac{\Delta I_{out}}{\Delta I_{in}}$
Power gain	$A_P = A_V \times A_I$

15. Three transistor circuit configurations are the common base amplifier (CB), common emitter amplifier (CE), and common collector amplifier (CC).
16. Amplifiers can be classified according to their bias current. Some classes are A, B, and C.
17. Transistors and other semiconductor materials are sensitive to heat. Heat sinks protect transistors and other semiconductor devices from excessive heat.
18. Some important methods of coupling amplifiers are:
    a. Transformer coupling.
    b. RC coupling.
    c. Direct coupling.
    d. Push-pull coupling.
19. Thyristors are special semiconductor devices capable of handling large currents. Two common thyristors are the SCR and the triac.

Test Your Knowledge

Please do not write in the text. Place your answers on a separate sheet of paper.

1. Name three disadvantages of the vacuum tube when compared to the transistor.
2. Draw and label the symbols for a diode, triode, tetrode, and pentode.
3. Determine the approximate voltage needed by the following tubes.
    a. 5Y3.
    b. 12AX7.
    c. 35W4.
4. What is the purpose of the grid in a triode?
5. In what tube is the suppressor grid found?
6. Draw an NPN transistor symbol and label the emitter, base, and collector.
7. How do the voltage connections from the source differ between an NPN and a PNP transistor?
8. Which part of an NPN transistor (common emitter configuration) has the largest amount of current present during amplification?
    a. Emitter.
    b. Base.
    c. Collector.
    d. Gate.
9. A field effect transistor is a(n) __________ controlled device.
10. Which part of an NPN transistor (common emitter configuration) normally has the least current present during normal operation?
    a. Emitter.
    b. Base.
    c. Collector.
    d. Gate.
11. Care must be taken to prevent __________ from coming into contact with a MOSFET.
12. The arrow pointing toward the gate in the MOSFET symbol means that the gate is connected to a(n) __________ polarity.
13. What are the two operating modes of the MOSFET?
14. An amplifier:
    a. is a control device.
    b. uses a small input signal to control a larger output signal.
    c. produces outputs that are linear.
    d. All the above.
15. Define gain.
16. Name three methods of transistor bias and draw a schematic of each.
17. Write the formula for computing voltage gain.
18. Write the formula for computing current gain.
19. Write the formula for computing power gain.
20. How is gain for a common emitter circuit computed?
21. The input and the output wave forms of a common collector amplifier are:
    a. in phase.
    b. 180 degrees out of phase.
    c. Both a and b.
    d. Neither a nor b.
22. Draw a direct-coupled amplifier using two NPN transistors.
23. What precautions should be observed in the use of and servicing of transistors circuits?

For Discussion

1. What types of devices still use vacuum tubes?
2. Why has the vacuum tube declined in use in favor of the transistor and IC in the field of electronics?
3. What is meant by solid-state?
4. What are some of the characteristics of a good sound system?
5. Discuss the relative merits of RC coupling verses transformer coupling.
6. Discuss several kinds of distortion that can be present in an amplifier.
7. Why should transistors be temperature controlled using a heat sink and ventilation fans?

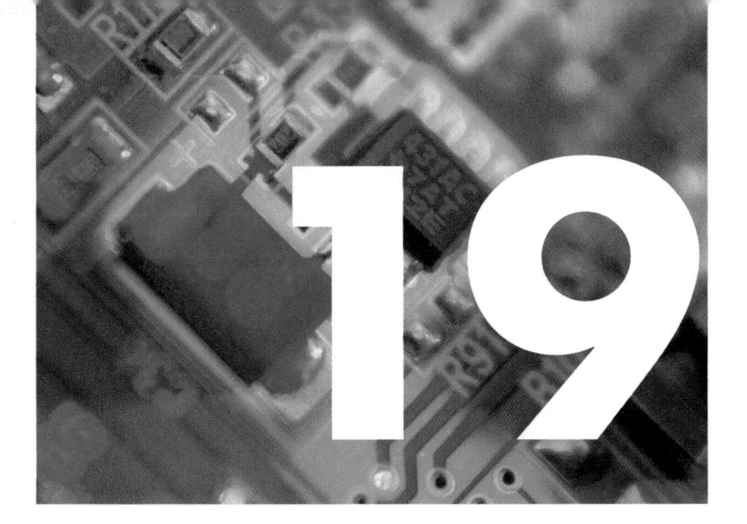

Integrated Circuits

Objectives

After studying this chapter, you will be able to:

- Define integrated circuit (IC).
- Give a brief history of the IC.
- Discuss the construction of an IC.
- Illustrate the steps in construction of an IC.
- Explain the operation of the operational amplifier.
- Explain the operation of the 555 timer

Key Words and Terms

The following words and terms will become important pieces of your electricity and electronics vocabulary. Look for them as you read this chapter.

analog	linear
astable	monostable
chip	offset null
circuit designer	operational amplifier (op amp)
comparator	photoengraving
diffusion	photomask
digital	photoresist
epitaxial	seed
555 timer	substrate
integrated circuit (IC)	
layout designer	

An ***integrated circuit,*** or ***IC,*** is a complete electronic circuit contained in one package, **Figure 19-1.** This package often includes transistors, diodes, resistors, and capacitors along with the connecting wiring and terminals. An IC is also called a ***chip.***

As discussed in Chapter 18, the transistor was invented in 1947 by Brattain, Bardeen, and Shockley of Bell Laboratories. The transistor served the same purpose as Lee DeForest's triode amplifier, but it did not need heat to operate. In addition, the transistor was solid (solid-state) and much smaller. Transistors were first used in small appliances such as hearing aids and small transistor radios. Small size and efficient operation also made transistors useful in defense items.

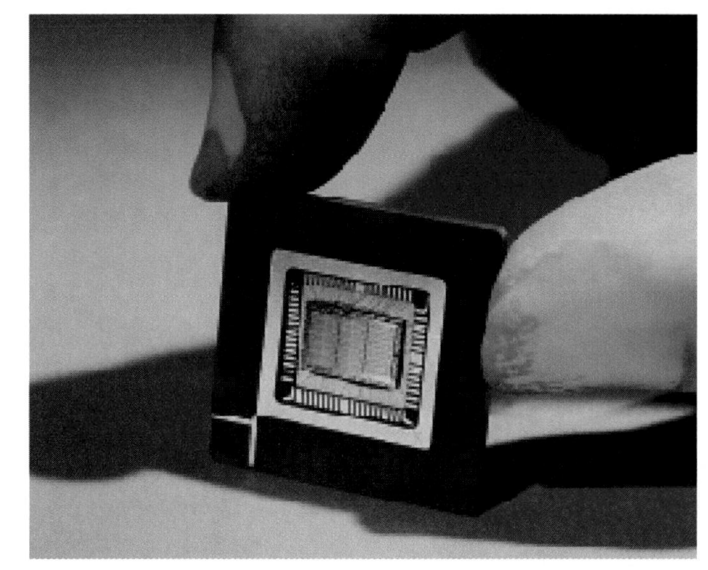

Figure 19-1. A single silicon chip. (Miller Electric Manufacturing Co.)

Transistors were also used in the newly developed electronic computer circuits of the 1950s and early 1960s. Computers used thousands of switching circuits, and transistors were able to quickly perform this switching function. But as computer circuits became larger and more powerful, electronic circuits needed to become smaller. Because the components of circuits have to be wired together, producing the smaller circuits was a complex task. Printed circuit boards helped, but wiring was still bulky. This problem was solved by integrating all these components into one solid piece of material, the integrated circuit. See **Figure 19-2.**

19.1 HISTORY OF THE INTEGRATED CIRCUIT

In 1952, G.W.A. Dummer of the Royal Radar Establishment in Great Britain had the idea for an integrated circuit. However, his ideas were not put to use at that time. In 1957, a new process for planar transistors was

Integrated Circuits

Advantages	• Low cost • Higher switching speed • Low power consumption • Small size • High component density
Disadvantages	• Some components cannot be fabricated • Large amounts of current and voltage cannot be handled

Figure 19-2. The advantages and disadvantages of ICs when compared with transistors.

Date	Type of component integration	Level of component integration
1964	Small scale integration	Up to 10 components or gates
1968–1969	Medium scale integration (MSI)	Up to 100 components or gates
1970	Large scale integration (LSI)	Up to 1000 components or gates
Early 1980s	Very large scale integration (VLSI)	1000 or more components or gates
Late 1980s	Mega integration	1 million or more components per chip

Figure 19-3. Levels of integration.

developed at Fairchild Semiconductors. This allowed semiconductor emitters, bases, and other parts to be made on the surface of a silicon wafer.

In early 1958, Jack Kilby of the Texas Instruments Corporation was developing micromodules. These were to be made by printing the components on a ceramic wafer. He realized that semiconductors and other components could be made on the same surface through a manufacturing process. The first commercially produced integrated circuit resulted from this work. It was made on a thin wafer of germanium. However, it still had wire connections, which caused a major problem when wiring together large numbers of transistors and other conductors.

About the same time, another process for making ICs was being studied at Fairchild Semiconductors. Using the principles of planar transistor manufacturing, Robert Noyce used silicon dioxide dopants to protect and insulate PN junctions.

The integrated circuit dramatically changed the electronics field. In 1965, about 30 components could be put on a silicon chip five millimeters (3/16 in.) square. By 1982, that number had increased to 1,000,000, **Figure 19-3.** While the IC has become smaller and smaller, even reaching microscopic sizes, the principles of operation remain the same.

Review Questions for Section 19.1

1. What is an integrated circuit?
2. Who invented the transistor? The IC?
3. List the advantages and disadvantages of ICs when compared to transistors.

19.2 IC CONSTRUCTION

An integrated circuit consists of many extremely thin layers of P and N type material arranged in configurations such as transistors, diodes, resistors, and capacitors. A single chip may contain millions of transistors and occupy less than one square inch of area.

The ***circuit designer*** begins the production process by designing the complete integrated circuit. One factor affecting design is the intended use of the IC. With this in mind, the designer plans the best IC for use. He or she submits the completed design in the form of a schematic diagram.

From this schematic diagram, the ***layout designer*** creates a detailed technical drawing. The circuit is drawn in a much larger scale than the final product so that when the drawing is reduced there will be enough space between parts. **Figure 19-4.** If any of the lines touch each another, the circuit will short out when tested.

Figure 19-4. This circuit diagram will be reduced hundreds of times before it is used to make circuits. (Martin Marietta)

Next, each circuit layout is ***photographically reduced.*** It is not unusual for the layout to be reduced over 1000 times or more. A reduced layout allows for thousands of circuits to be put on one wafer.

Working plates are made from the reduced layouts. These plates are called ***photomasks.*** Each photomask goes with a certain step in the production process. Each mask contains a large number of identical, actual size parts. The photomasks are now ready for use and production of the IC can now begin.

The structure of an IC is a pure silicon crystal. These pure crystals must first be produced. To make the crystal, liquid silicon is purified. A solid silicon particle, or ***seed,*** is dipped into the melted silicon. It is slowly withdrawn and placed in a cool area, **Figure 19-5.** The grown crystal is sliced into wafers about 0.5 mm thick. The wafers are then polished to rid the wafer of surface scratches and contaminants. Small portions of impurities are then added. The impurities give the silicon its electrical traits.

Figure 19-5. Pure silicon crystal. (Motorola)

On the thin wafers of doped silicon, the basic building process begins. The circuit is built, layer by layer, on the silicon wafer, or ***substrate.*** Each layer receives a pattern from the photomask.

In the example shown in **Figure 19-6,** the first layer on top of the silicon is a layer of N-type material. It is grown right on the wafer, and it is called the ***epitaxial*** layer. Epitaxy is a growth of one crystal on the surface of another crystal. This is the collector for a transistor or an element of a diode.

Next, a thin coat of silicon dioxide is grown over the N-type material by exposing the wafer to an oxygen atmosphere at about 1000°C. See **Figure 19-7.**

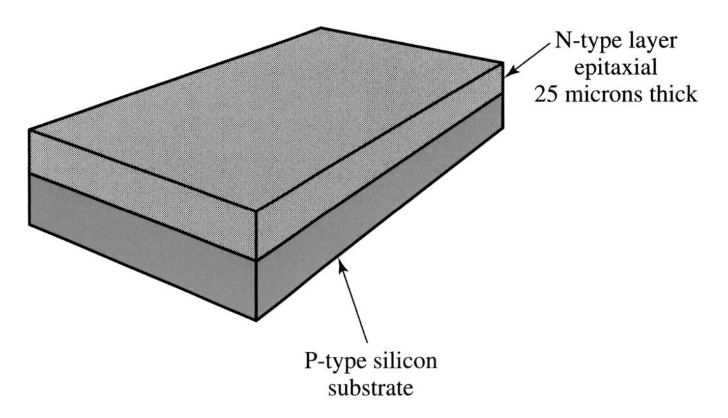

Figure 19-6. Growing N-type material on P-type substrate.

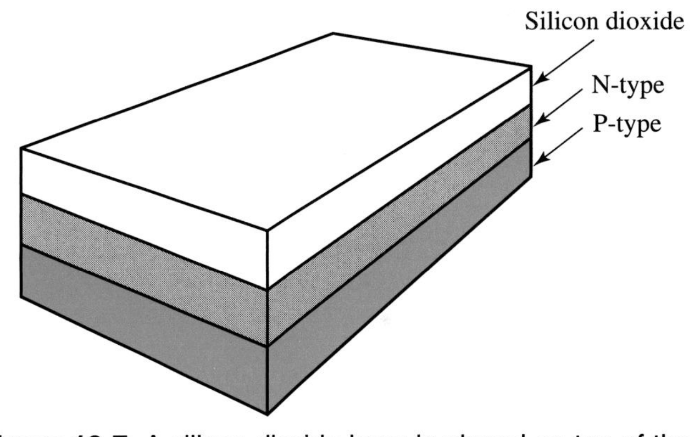

Figure 19-7. A silicon dioxide layer is placed on top of the N-type layer.

Next, a thin coat of a light-sensitive emulsion is placed over the N-type layer. The emulsion is called ***photoresist.*** In a process called ***photoengraving,*** the photomask is placed over the N-type layer. Then the entire wafer is exposed to ultraviolet light, **Figure 19-8.** The light causes the image of the photomask to transfer to the wafer.

The exposed photoresist hardens. The areas covered by the mask remain soft. Acids or solvents are used to etch away the unexposed (soft) area of the photoresist. This leaves the layer of the N-type silicon exposed. **Figure 19-9.**

The exposed N-type layer is further etched away by very hot gases. A chemical washes away any remaining hardened photoresist to expose all N-type silicon dioxide.

As parts of the IC are constructed, they must be isolated from each other. This is done by diffusion.

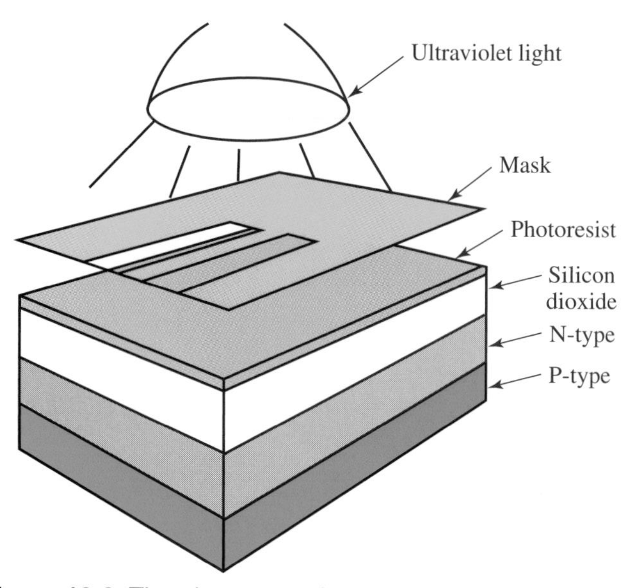

Figure 19-8. The photoengraving process.

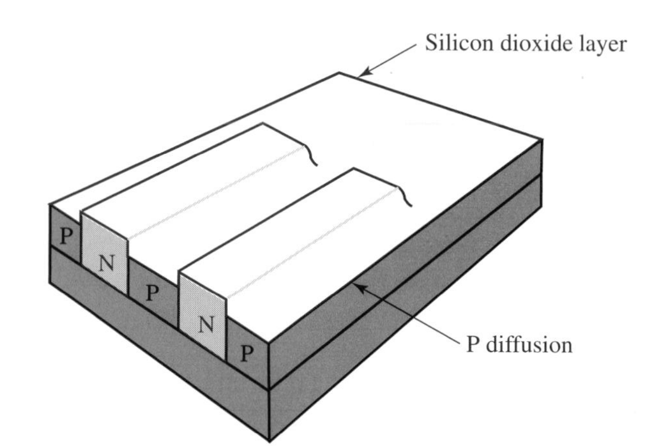

Figure 19-10. N-type material remains after P diffusion.

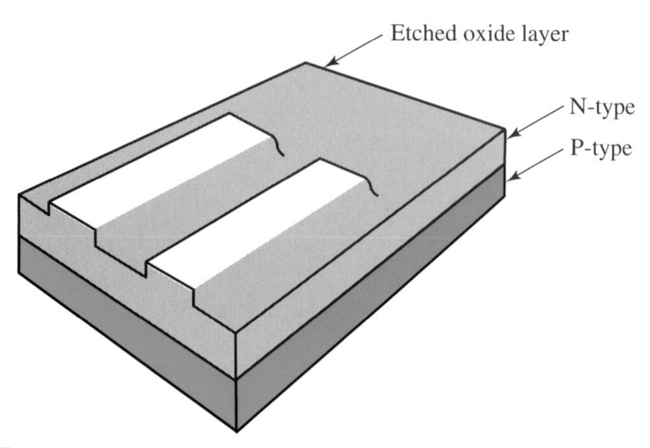

Figure 19-9. The first masking and etching isolates components.

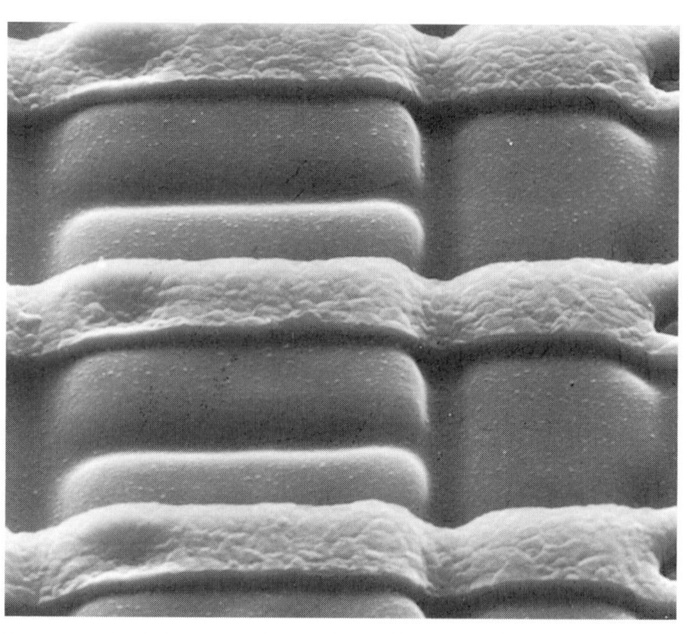

Figure 19-11. P-type diffusant on N-type silicon dioxide. (Lattice Semiconductor)

Diffusion is a process in which impurities are doped into the silicon wafer to form the needed junctions. Diffusion forms islands of N-type materials backing P-type materials.

The wafer is diffused using boron. The boron cuts into and forms a P-type material on all areas not protected by the silicon dioxide. The wafer has isolated islands of N-type material, **Figure 19-10.** NP junctions form around each island. There are back-to-back diodes between each N-type island.

During diffusion, a new layer of silicon dioxide forms over the diffused P-type areas, as well as on top of the islands, **Figure 19-11.**

The wafer is again coated with photoresist and exposed under a photomask. Areas in the N-type islands are etched away. Once again, the wafer is subjected to a P-type diffusant that forms areas for transistor base regions, resistors, or elements of diodes or capacitors. The wafer is then reoxidized, **Figure 19-12.**

The wafer is again masked and exposed to open windows in the P-type regions. A phosphorus diffusant is used to produce N-type regions for diodes and capacitors. Small windows are also etched through to the N-layer for electrical connections, **Figure 19-13.** The total wafer is again given an oxide coating.

The monolithic circuit is now complete except for the aluminum interconnections. The aluminum interconnections join the islands. They also join the circuit to other circuits and other devices.

A thin coat of aluminum is vacuum-deposited over the entire circuit. The aluminum coating is then sensitized

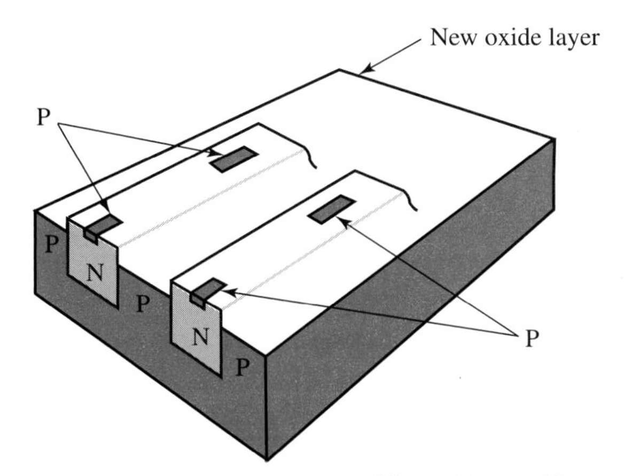

Figure 19-12. P-type regions are diffused in the N-type islands.

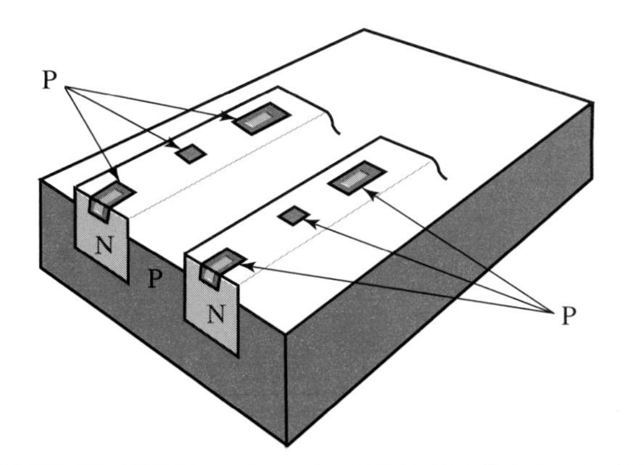

Figure 19-13. Windows are opened in the P-type regions.

and exposed through another special mask. After etching, only the interconnecting aluminum remains. It forms a pattern between transistors, diodes, and resistors, **Figure 19-14.**

Currently all ICs use aluminum interconnections. However, recent technology breakthroughs have made the use of copper connections possible. Soon the use of these copper connections will make ICs faster and cheaper.

The completed circuits are then tested, **Figure 19-15.** In a single test, the circuits are used to perform a series of electrical tasks, **Figure 19-16.**

After testing, the wafers are separated into individual chips, usually by scribing them with a diamond tipped tool. The chips are then mounted onto a small can or flat package, **Figure 19-17.** Leads are bonded, and the ICs are washed. The cavities that hold the ICs are sealed, and finally, the ICs are shipped to a distributor.

Figure 19-14. Aluminum interconnections. (Motorola)

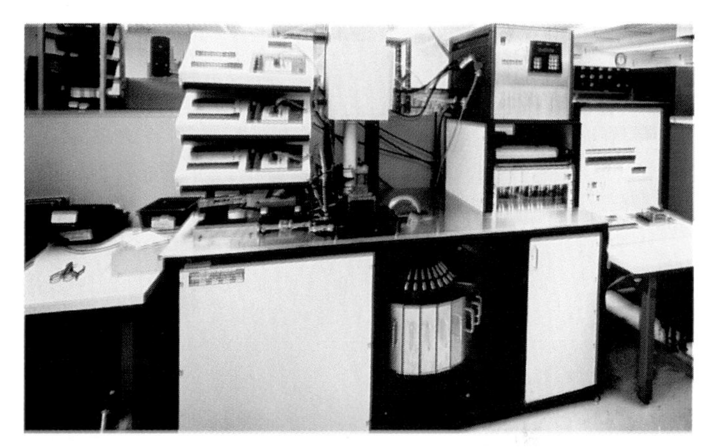

Figure 19-15. Wafer testing must take place in a sterile area. Even a small piece of dust can ruin a circuit. This machine automatically tests semiconductor parts for electrical specifications. (Motorola)

Figure 19-16. IC test clip.

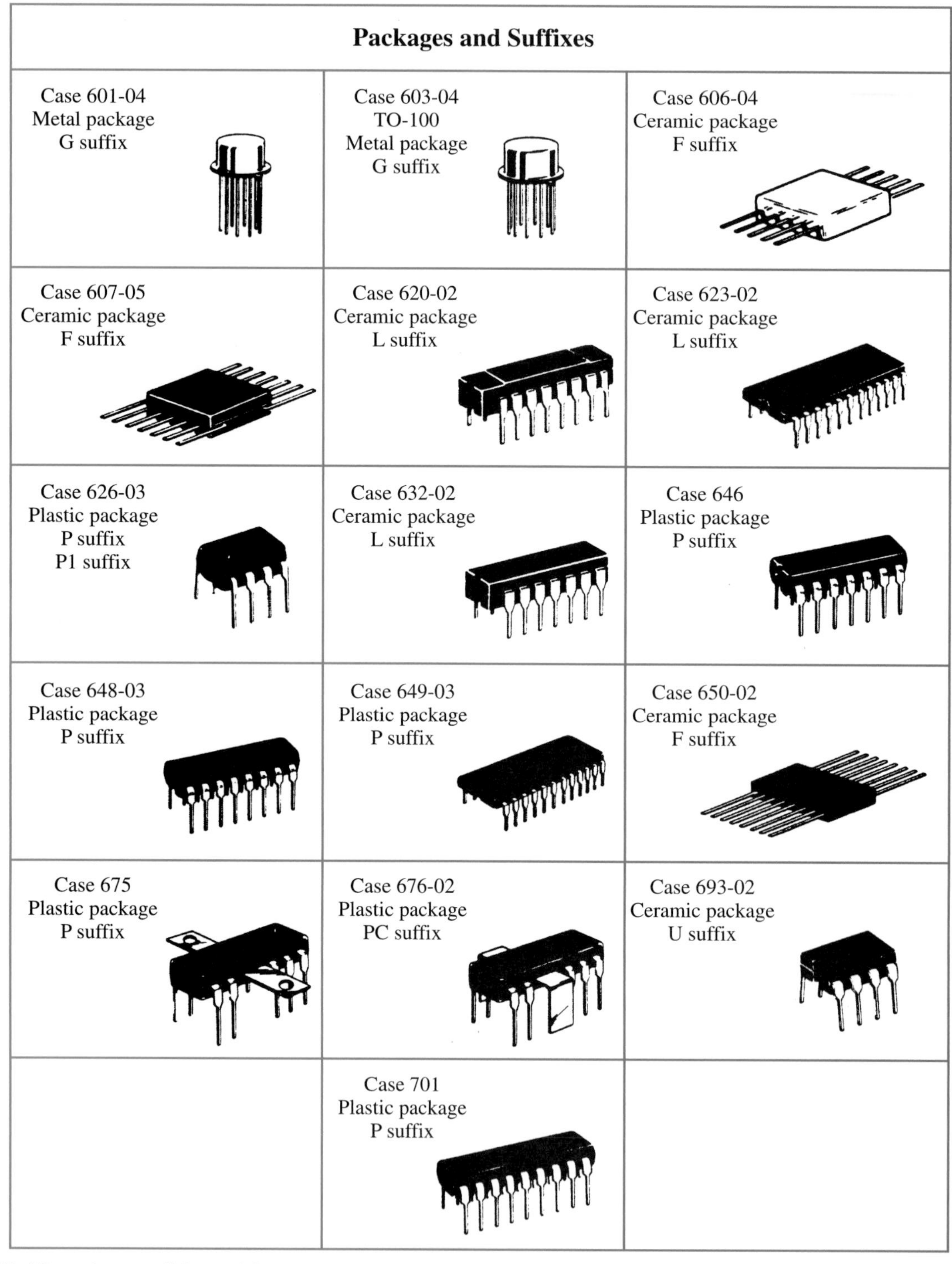

Figure 19-17. IC packages. (Motorola)

Resistors

The process just discussed is used to make semiconductor materials on ICs. This process can be used to make resistors, capacitors, and diodes.

Recall that N-type or P-type materials have certain resistances. Resistance depends on the physical size of the material (length or surface area) and the amount of dopants in the material. Semiconductors are made with very pure silicon. Through the doping process, impure trivalent or pentavalent atoms are added to produce the N-type or P-type substrate material.

For example, a P-type silicon material is used as the substrate. An N-type material is diffused into the surface of the chip, **Figure 19-18.** Then an other P-type material is added to the N-type material. Metal leads are fastened to the end of this P-type material. The P-type material and its two connections are used as a resistor.

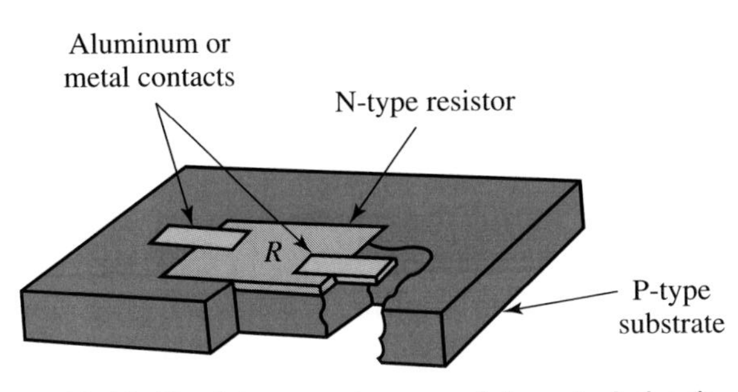

Figure 19-18. Resistors made on an integrated circuit.

Capacitors

Like resistors, capacitors can be made in an integrated circuit. Values for these capacitors are very small. However, they are still able to perform functions of coupling and storage. **Figure 19-19** shows how a capacitor can be made in an integrated circuit.

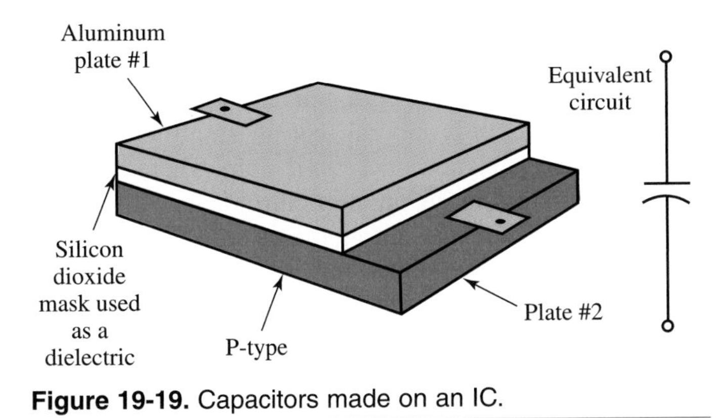

Figure 19-19. Capacitors made on an IC.

Putting It Together

An example of how a transistor, resistor, and capacitor can be integrated into one circuit is shown in **Figure 19-20.** Keep in mind that thousands of these circuits can be placed onto an area the size of the head of a pin.

Common Types of ICs

Integrated circuits come in two basic types. The type depends on their function. These types are linear and digital.

Linear ICs have variable outputs, controlled by variable inputs. These ICs are also called ***analog*** devices or circuits. Linear ICs are used as components in linear amplifiers, operational amplifiers, voltage regulators/buffers, voltage comparators, analog switches, and audio amplifier circuits.

Digital integrated circuits are used as switches. Their output operates in either on or off conditions. They are found in many logic and gate circuits in computers. These will be discussed in depth in Chapter 20.

Several IC designs, including pin numbering systems and dimensions, are shown in **Figure 19-21.**

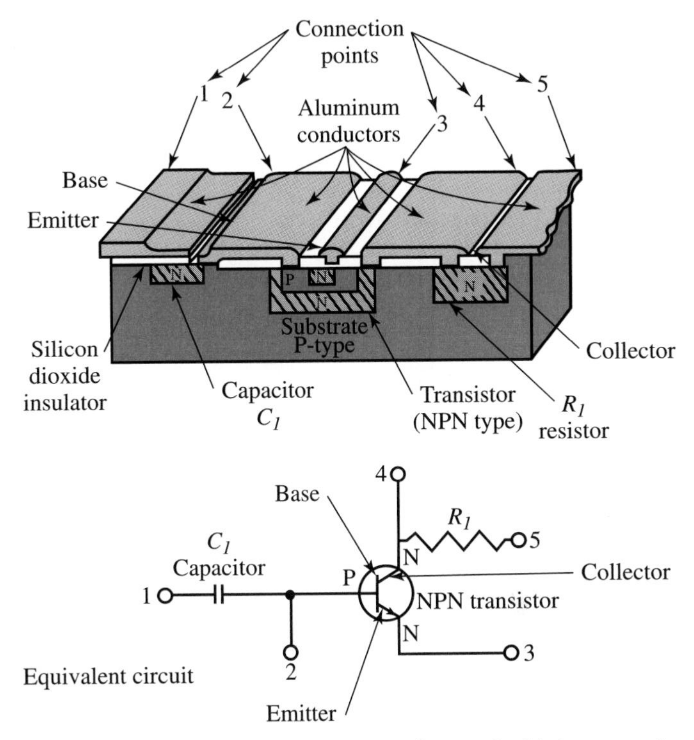

Figure 19-20. Various components integrated into a small circuit.

Review Questions for Section 19.2

1. What is the purpose of photomask?
2. The silicon wafer on which an IC is built, layer by layer, is called the __________.
3. A light-sensitive emulsion that accepts the transferred image of the photomask is called a:
   a. photoengraver.
   b. film.
   c. photoresist.
   d. None of the above.
4. What is diffusion?
5. What is the purpose of aluminum interconnections?
6. Name the two common types of ICs.

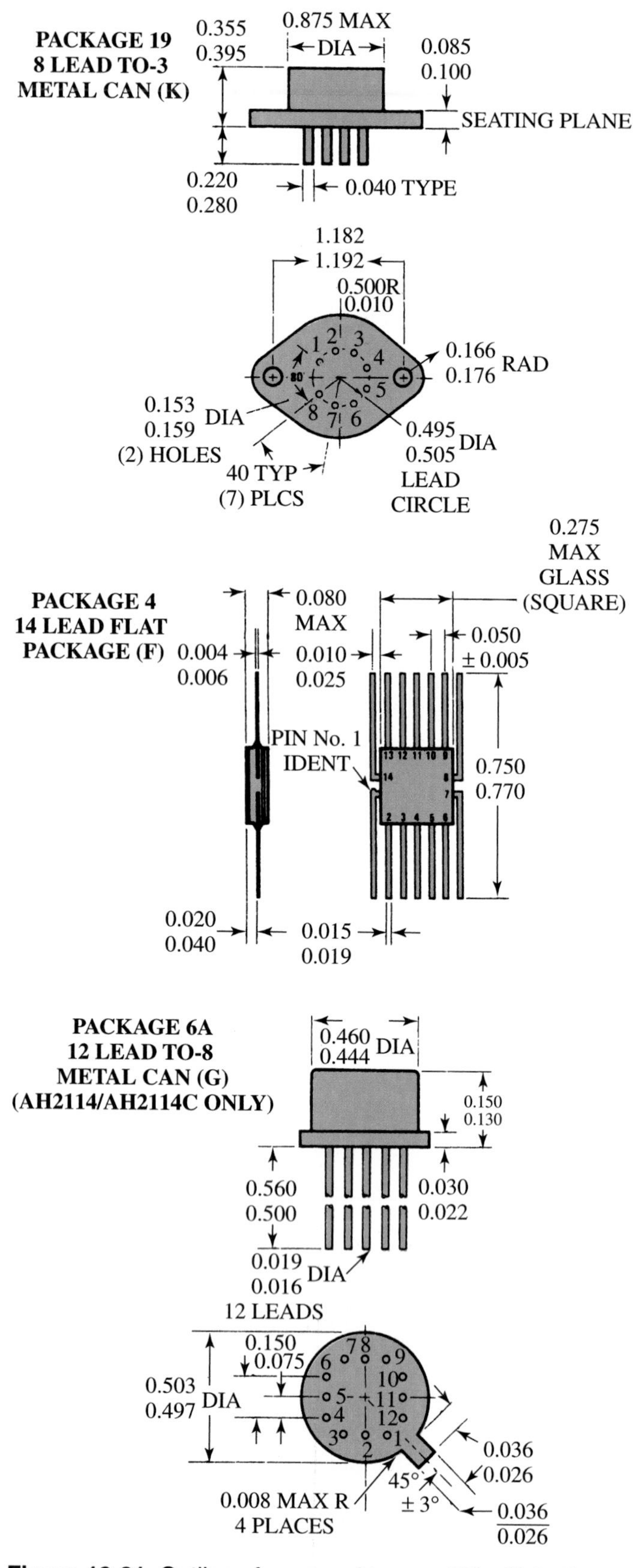

Figure 19-21. Outlines for several types of ICs. Note the manner in which the pins are numbered. (National Semiconductor)

19.3 OPERATIONAL AMPLIFIER (OP AMP)

One of the most commonly used integrated circuit chips today is the ***operational amplifier.*** The operational amplifier is often simply called the ***op amp.*** The 741 op amp is a near-perfect amplifier. It has a high input impedance and a low output impedance, which makes it an excellent amplifier. The op amp has a wide frequency response. This means the frequency of the signal being amplified has little effect on the operation of the amplifier as compared to other amplifiers. It has a high gain capability and can be adjusted for zero offset voltage. Having a zero offset voltage means it has features that allow the amplifier output to be adjusted to absolute zero voltage.

Many amplifiers have a slight output voltage due to temperature changes of the components. The op amp can be corrected easily to compensate for temperature changes.

The 741 op amp is a general purpose amplifier. It can be used to regulate power supplies, made into a simple signal generator, used as an oscillator, used as a radio or TV receiver, used as a timer, or used as a filter. It is also used extensively for instrumentation (metering current or voltage).

Originally, the term operational amplifier applied to any complete circuit designed from many discrete components that resulted in a high gain, high performance dc amplifier. It was only natural that an amplifier used so extensively be designed as a single chip. There are over twenty transistors in an op amp chip along with all the resistors needed for bias. Using the op amp IC saves time and money and also makes repair and troubleshooting much simpler. Op amps need only a few exterior components such as resistors and capacitors to create an amplifier or one of many other devices.

Figure 19-22 is an illustration of the pin configuration of a typical 741 general purpose op amp. Like most IC chips, the op amp does not have pin identification markings on the chip. The pins are identified by using specifications sheets and a reference point. The reference point for our chip is the notch at the top. The pins are numbered in a counterclockwise direction, starting at the notch.

The power for the op amp is provided through pins 4 and 7. Pin 4 is either connected to ground or a negative voltage value from 3 to 18 volts. Pin 7 is connected to the positive voltage of the power supply. The output of the amplifier is pin 6. There are two input pins, pin 2 and pin 3. Pin 2 is the inverting input. Any signal applied to pin 2 generates a signal of opposite polarity at the output. The noninverting input, pin 3, generates an output at pin 6 of

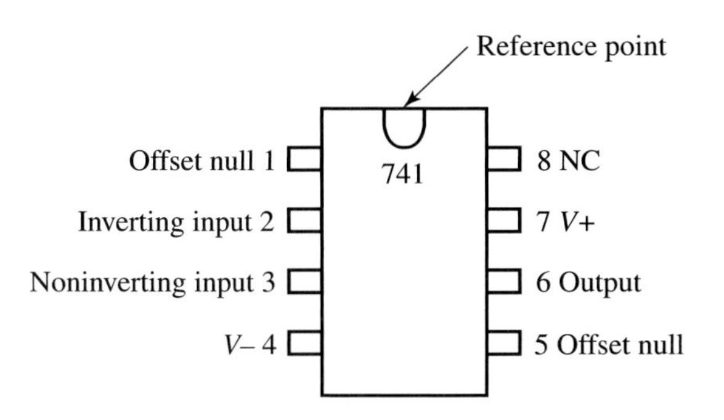

Figure 19-22. Pin identification of a typical dual-in-line package (dip) operational amplifier.

the same polarity. See **Figure 19-23.** Take special note that both inputs, pin 2 and pin 3, generate an output at pin 6, but they generate opposite polarities.

Offset null is a calibration feature of the op amp. The op amp is so sensitive to input voltage that at times the output will generate a signal even when there is no intentional input. To avoid this condition for certain applications, offset null pins, pin 1 and pin 5 are provided. They are usually connected to a variable resistance such as a potentiometer. The potentiometer can be adjusted to produce a zero output voltage from pin 6.

Look at pin 8 and take note of the NC identification. This pin is not used. The NC stands for no connection.

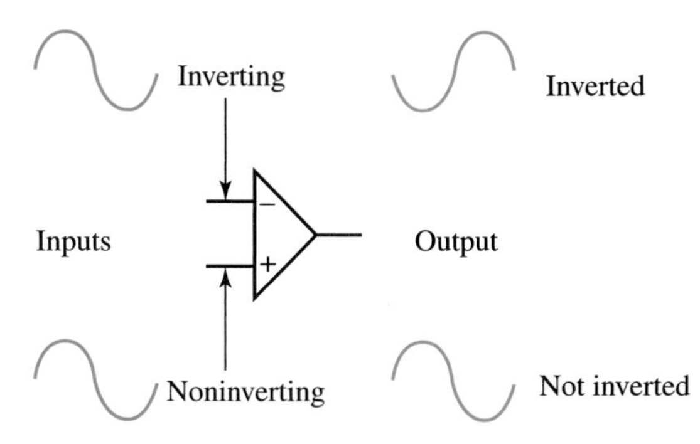

Figure 19-23. The polarity of the output signal of the op amp is determined by the input signal location. The inverting input generates an output signal of opposite polarity to the input signal. The noninverting input produces a signal of matching polarity at the output.

Op Amp Gain

The op amp gain is easily determined by the relationship of the feedback resistor and the input resistor. Look at **Figure 19-24.** The feedback resistor is labeled R_F and the input resistor is labeled R_I. To calculate the gain for the inverting op amp in Figure 19-24 we simply divide the feedback resistor value R_F (100 kΩ) by the input resistor value R_I (10 kΩ). The gain (A_V) for the op amp is 10. For a noninverting op amp the gain is equal to the feedback resistor value divided by the input resistor value plus one. The gain in the op amp circuit shown would be 11. In the form of an equation:

$$A_V \text{ (inverting)} = R_F \div R_I$$

$$A_V \text{ (noninverting)} = (R_F \div R_I) + 1$$

Some op amps can obtain a gain of 200,000 or higher.

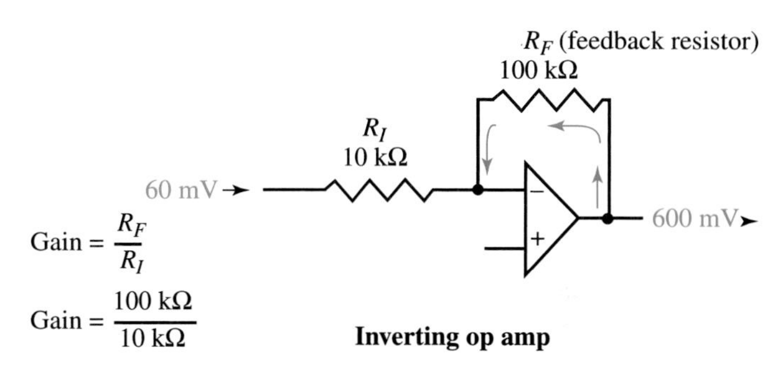

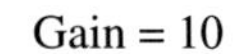

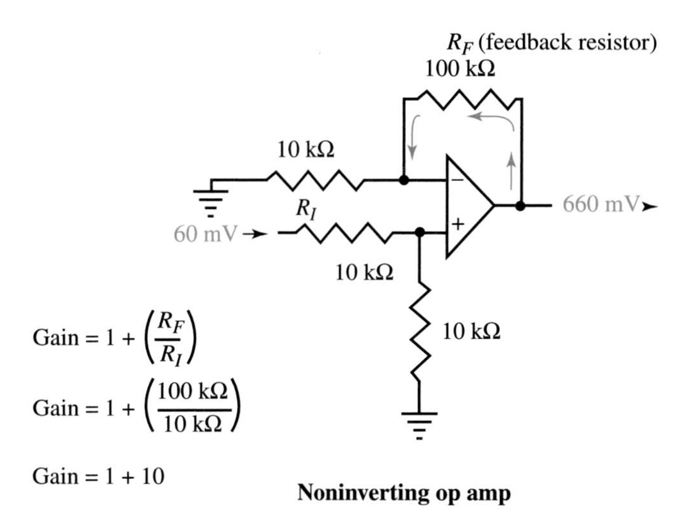

Figure 19-24. The gain is determined by the ratio of the input resistor to the feedback resistor value.

Comparator Circuit

Another mode of operation of the op amp is as a comparator circuit. When operating as a comparator, both inverting and noninverting inputs are used. As the name ***comparator*** implies, the two inputs are compared to each other. The output of the comparator is driven to maximum positive or negative, depending on which input is more dominant.

PROJECT 19-1: Building a Voice Recorder

Using an IC and a few other components, you can record a message of up to 10 seconds in length with this voice recorder project. Once recorded, the message can be played back at any time. Combined on this project's chip (of less than a square inch) are an oscillator circuit, microphone preamplifier, automatic gain control circuit, antialiasing filter, smoothing filter, and a speaker amplifier.

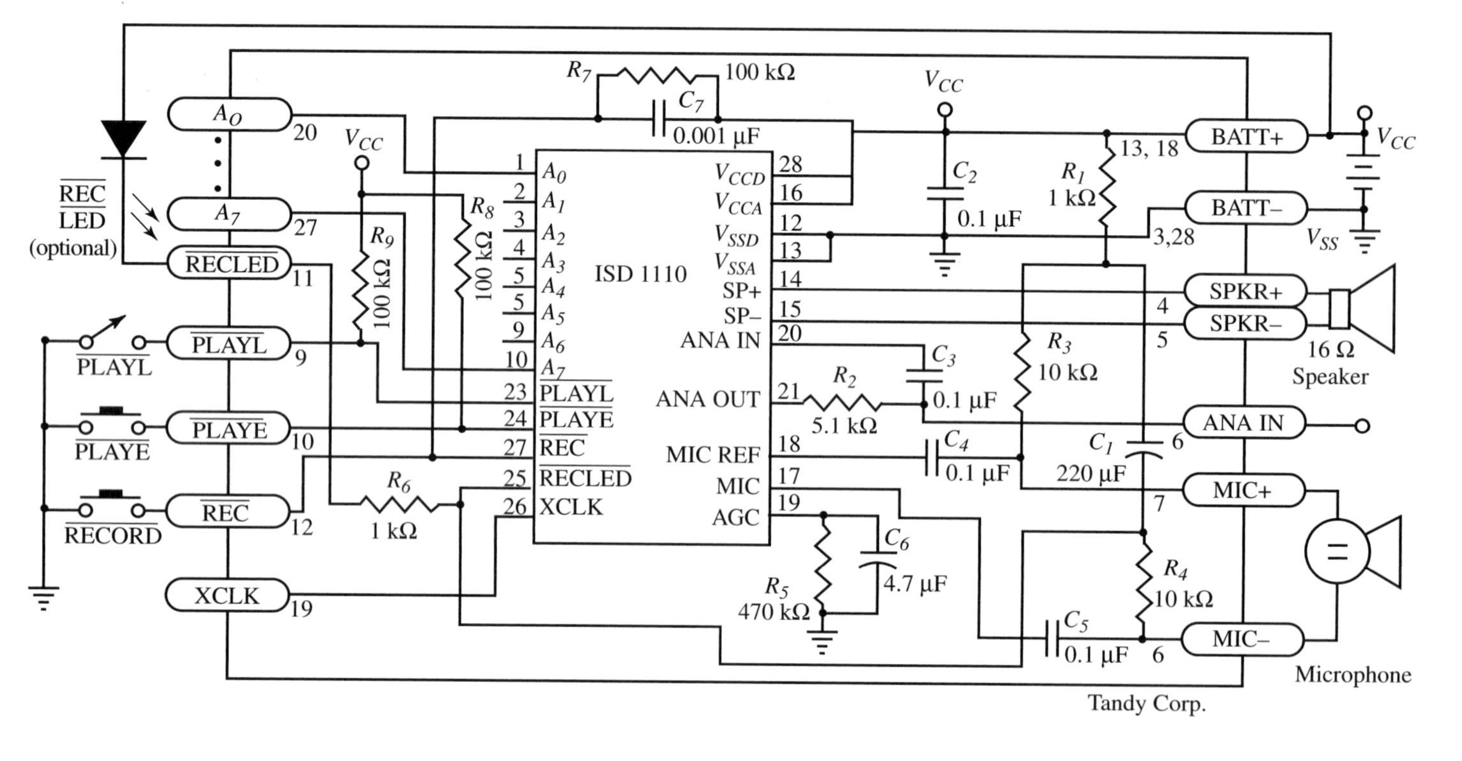

Parts List

Quantity	Description	Radio Shack Part Number
1	VM-1110A Voice Recorder/Playback Module	276-1324
2	SPST switch (normally open)	275-1571
1	SPST push-on, push-off	275-617
1	Electret microphone	270-090
1	Piezo speaker	273-091
1	4 AA battery holder	270-391
4	AA batteries	23-552
1	LED	276-044

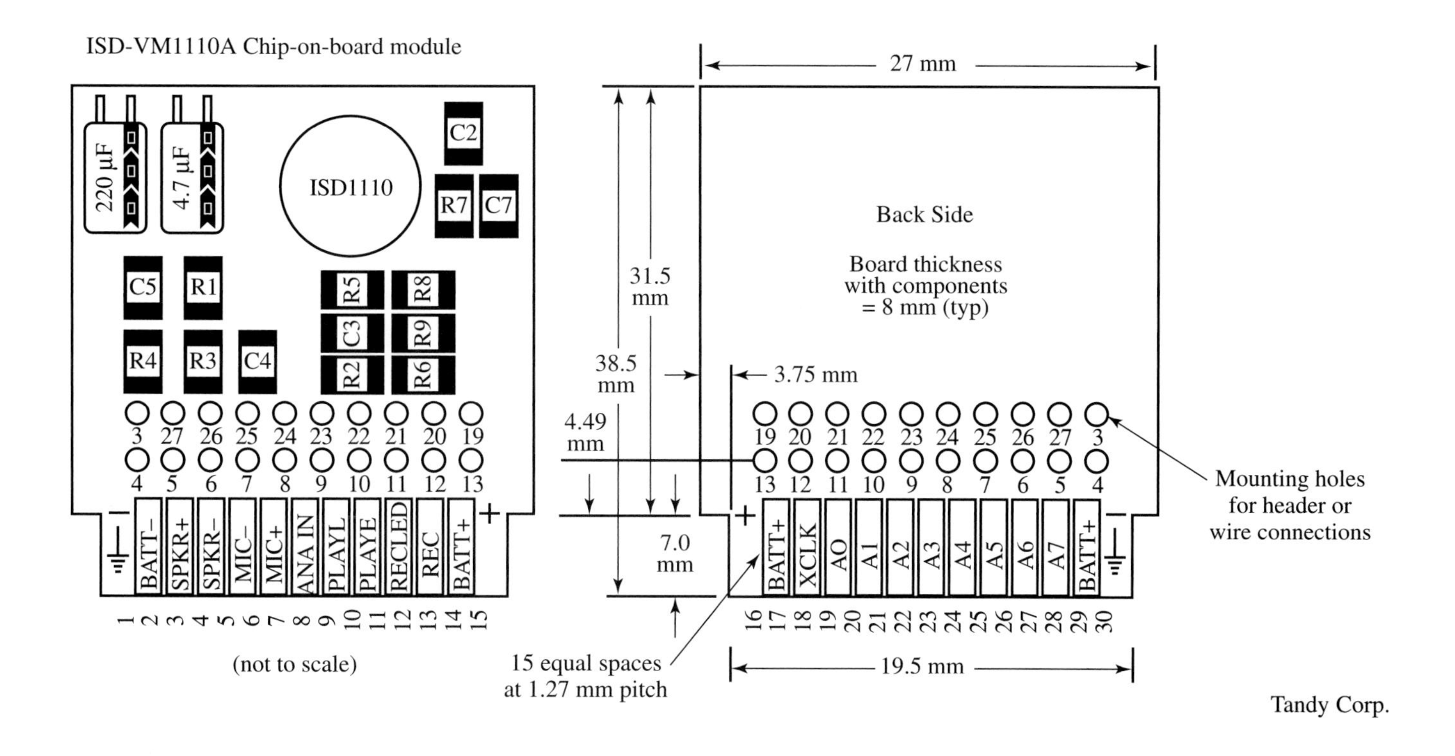
ISD-VM1110A Chip-on-board module
220 µF
4.7 µF
ISD1110
C2
R7
C7
C5
R1
R5
R8
C3
R9
R4
R3
C4
R2
R6
BATT–
SPKR+
SPKR–
MIC–
MIC+
ANA IN
PLAYL
PLAYE
RECLED
REC
BATT+
(not to scale)
27 mm
Back Side
Board thickness
with components
= 8 mm (typ)
31.5
mm
38.5
mm
3.75 mm
4.49
mm
7.0
mm
BATT+
XCLK
AO
A1
A2
A3
A4
A5
A6
A7
BATT+
Mounting holes
for header or
wire connections
15 equal spaces
at 1.27 mm pitch
19.5 mm
Tandy Corp.

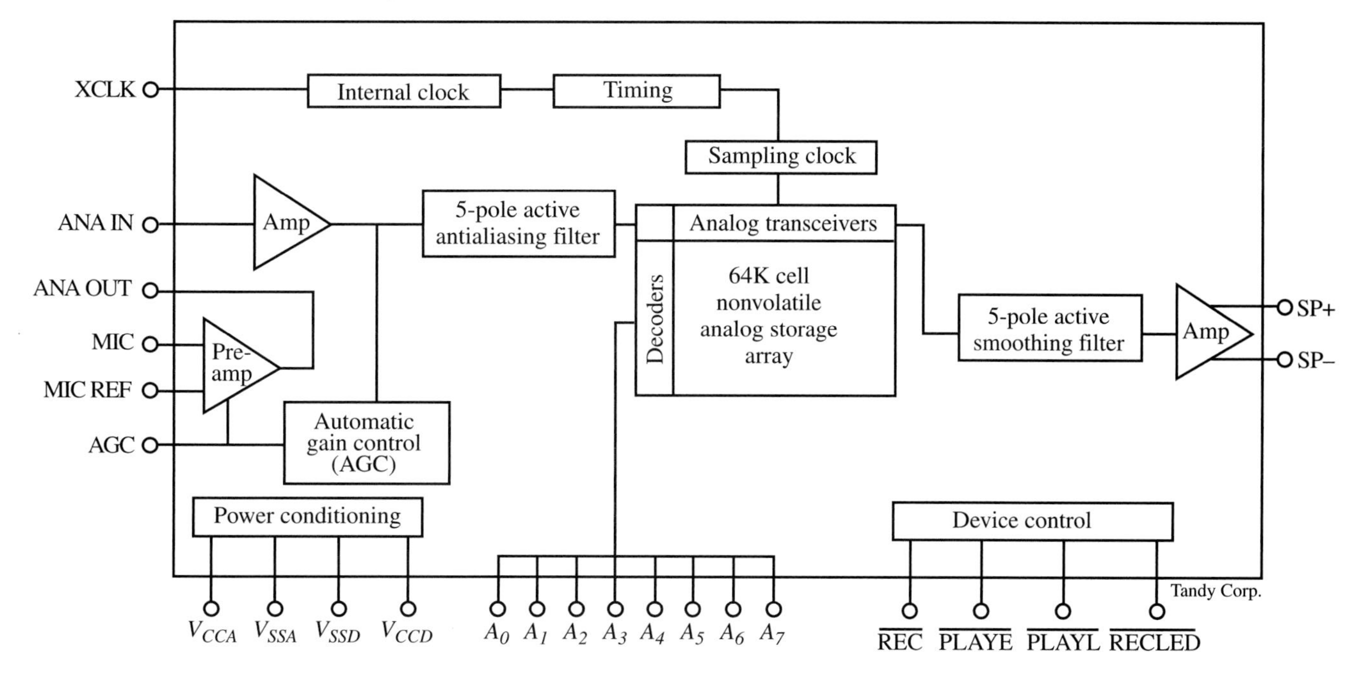
ISD1100 Series Block Diagram
XCLK
Internal clock
Timing
Sampling clock
ANA IN
Amp
5-pole active
antialiasing filter
Analog transceivers
ANA OUT
MIC
MIC REF
AGC
Pre-
amp
Decoders
64K cell
nonvolatile
analog storage
array
5-pole active
smoothing filter
Amp
SP+
SP–
Automatic
gain control
(AGC)
Power conditioning
Device control
Tandy Corp.
V_{CCA} V_{SSA} V_{SSD} V_{CCD}
A_0 A_1 A_2 A_3 A_4 A_5 A_6 A_7
$\overline{REC}$ $\overline{PLAYE}$ $\overline{PLAYL}$ $\overline{RECLED}$

PROJECT 19-2: Building an Audio Amplifier

The audio amplifier project 140 from Graymark is an excellent example of using op-amps to construct a simple audio amplifier circuit. Once completed, you will need an input device such as a microphone to make a simple PA system. With a coil of wire you can make an electrical line detector that will find electrical lines hidden in a wall behind sheet rock or plaster. Or you can connect a 100pf capacitor and a diode in series with the input to make a signal tracer.

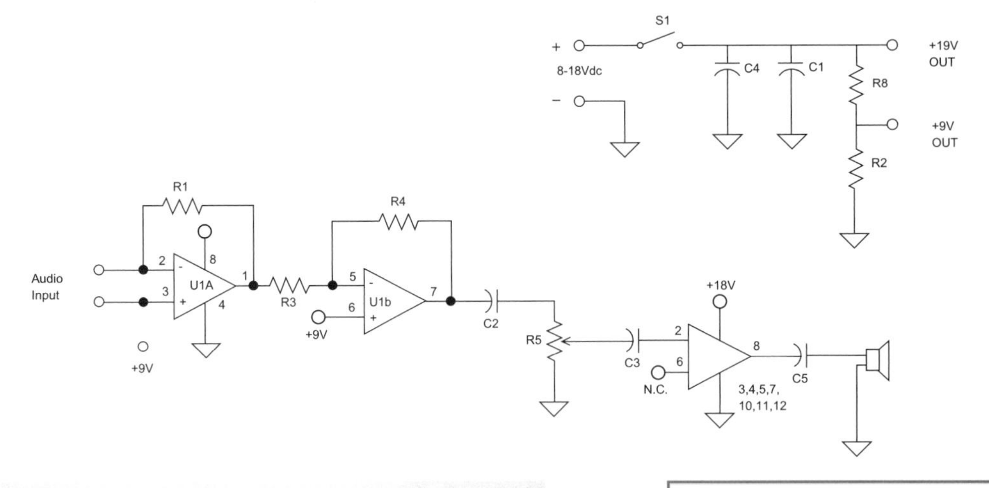

Qty.	Part No.	Description
1	63055	R_1 Resistor, 56kΩ
3	62650	R_2, R_3, R_6, Resistor, 10kΩ
1	61416	R_4, Resitor, 47kΩ
1	63054	R_5, Petentiometer, 5kΩ
1	61247	C_1 Capacitor, .001µF
2	62394	C_2, C_3, Capacitor, .1µF
2	62680	C_1 Capacitor, disc. .022µF
1	62238	C_4, C_5, Capacitor, Elect, 470µF
1	62840	U_1, IC, LM358, General Purpose Op Amp
1	62890	U_2, IC, LM380, Audio Amp
1	61358	S_1, Switch, Slide, SPDT
1	62578	IC Socket, 8-pin
1	62579	IC, Socket, 14-pin
1	61051	Speaker
1	61403	Battery Connector

Graymark

All parts, including the printed circuit board, are available as a kit through Graymark, Audio Amplifier Model No. 140

Look at **Figure 19-25.** The comparator circuit is designed as a motor control system. It brings a dc motor up to running rpm and then maintains it. One input is wired to a potentiometer, while the other is connected to a small generator on the dc motor shaft. As the motor turns, a dc voltage is generated. The op amp will continue to put out an inverted signal until the two inputs have a matching voltage amplitude. When the motor rpm is generating a slightly higher emf than the other input, the op amp switches to maximum opposite polarity. The op amp changes output condition to match the input conditions. This output switching occurs so rapidly that it maintains motor rpm under changing mechanical load conditions.

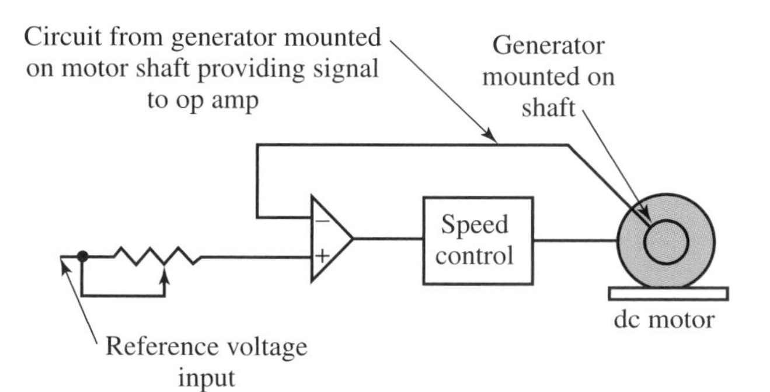

Figure 19-25. The comparator mode of operation can be used as a motor control system. The output of the op amp swings from and to maximum positive or negative polarity, depending on which input signal is greater.

Op amps come in various forms. There can be more than one op amp constructed on a single chip that takes the place of a multistage amplifier. They can be constructed from standard bipolar transistor construction or from MOSFET or JFET systems. The construction of an op amp is determined by the electrical characteristics desired.

Review Questions for Section 19.3

1. What is the difference between the inverting and noninverting op amp output signals?
2. What is the gain (inverting) for the op amp below?

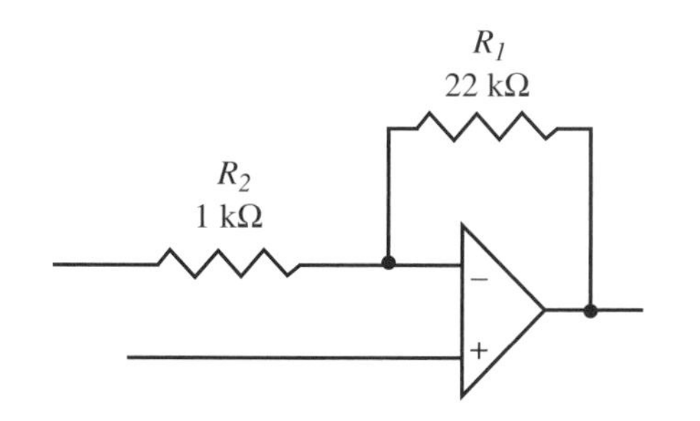

3. What does the offset null on the op amp do?
4. Power for energizing the op amp is connected to which two pins?
5. Which pin is used to input a signal when you desire the output polarity to match the input polarity?

19.4 555 TIMER

The 555 timer is another of the most popular integrated circuits in the industry today. It is very accurate and its construction is simple. Look at **Figure 19-26.** The ***555 timer*** can be used for highly accurate timing circuits with a range from a few microseconds to hours. Using more than one 555 timer, we can construct a timing device that will run for years. Other applications include sequencing operations for control systems, time delay operations, and repetitive pulses. The 555 IC becomes a timer with the simple addition of one resistor and capacitor.

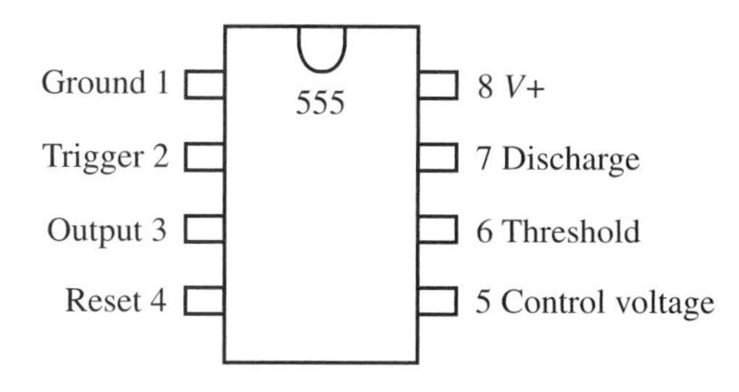

Figure 19-26. The 555 timer is used for many different timing circuits such as clocks, time delay circuits, sequential operation, and pulsing circuits.

The two main modes of operation for the 555 timer are referred to as monostable and astable. When operating in the ***monostable*** mode, the timer generates an output once for a predetermined period of time. This is also referred to as a "one shot" operational mode. In the ***astable*** mode, the timer puts out a repetitive signal from the output pin. The output in both cases is usually a square wave signal. See **Figure 19-27.** The top of the pulse wave is referred to as a high signal and the bottom of the pulse is referred to as a low signal. The width of the pulse and the distance between the pulses can be controlled by resistor and capacitor values that are connected to the chip.

Pin 1 is connected to ground and pin 8 is connected to a positive voltage. The number 2 pin is called the trigger. The trigger starts the timing process. The output of the 555 timer is located at pin 3. The timing of the output pulses is determined by the value of resistors and capacitors connected to pins 7 and 6. The reset, pin 4, can override the operation of the timer. When activated, the reset will return the timing operation back to the beginning of the timing cycle.

Review Questions for Section 19.4

1. Which two pins are used to power the 555 chip?
2. How are the timing cycles determined for the 555 timer?
3. The output timing signal is located at which pin?

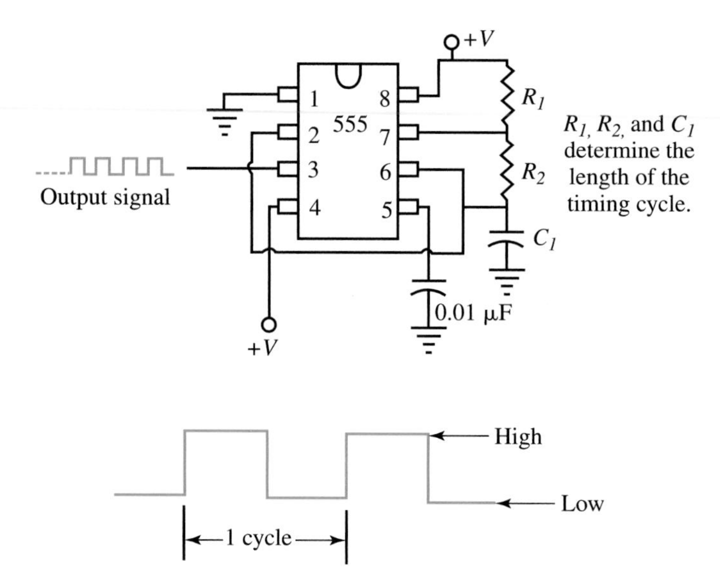

Figure 19-27. The output signal width between pulses is determined by the resistors and capacitor connected to pins 8, 7, and 6.

Summary

1. An integrated circuit, or IC, is a complete electronic circuit in a small package. It contains many transistors, diodes, resistors, and capacitors.
2. ICs are also called chips.
3. Advantages of the IC are its low cost, high component density, high switching speed, low power consumption, and small size. The disadvantages are that only certain parts can be built into an IC, and it is limited in the voltage and current capacity.
4. The IC production process is a detailed, involved one, in which one circuit is built on another circuit.
5. ICs come in many styles and sizes.
6. The two basic types of ICs are linear and digital.
7. The operational amplifier (op amp) is a common linear IC amplifier. It has a high gain capability and can be adjusted for zero offset voltage.
8. The gain in an op amp circuit can be calculated using the equations:

$$A_V \text{ (inverting)} = R_F \div R_I$$

$$A_V \text{ (noninverting)} = (R_F \div R_I) + 1$$

9. The 555 is a common linear IC timer.

Test Your Knowledge

Please do not write in the text. Place your answers on a separate sheet of paper.

1. An IC:
   a. is a complete electronic circuit.
   b. often contains transistors, diodes, resistors, and capacitors.
   c. is also called a chip.
   d. All of the above.
2. The __________ __________ begins the IC production process by illustrating the complete IC.
3. Why is the initial drawing of an IC done on such a large scale?
4. Why are silicon wafers polished?
5. __________ is the growth of one crystal on the surface of another crystal.
6. During photoengraving, the covered areas remain __________.
7. Label the layers in the following sketch.

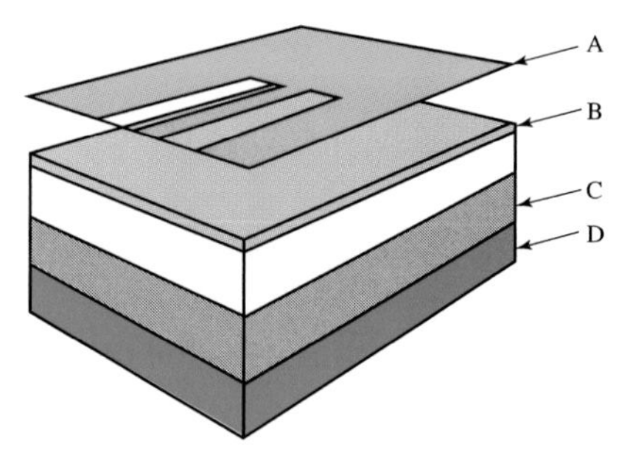

8. During diffusion, __________ are added to the silicon wafer to form needed junctions.
9. Why are completed ICs tested?
10. Linear ICs are also called __________ devices.
11. Digital integrated circuits are used as __________.

For Discussion

1. What are the two primary functions of an electronic amplifier device?
2. Explain how resistors are made on an IC chip.
3. Explain the difference between linear and digital ICs.
4. What has caused the drop in IC cost in the past 10 years?

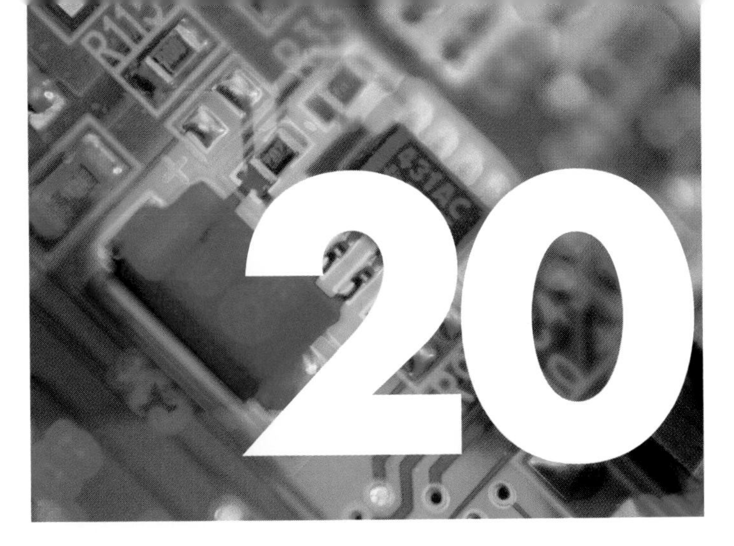

Digital Circuits

Objectives

After studying this chapter, you will be able to:

- Explain the difference between analog and digital systems.
- Convert decimal numbers to their binary equivalents and binary numbers to their decimal equivalents.
- Name seven types of logic gates.
- Explain the operation of various types of logic gates.
- Use truth tables to determine the output of a logic gate.
- Discuss two types of logic families.
- Explain the digital encoders and decoders.
- List three types of flip-flops and explain their truth tables.
- Explain analog to digital and digital to analog devices.

Key Words and Terms

The following words and terms will become important pieces of your electricity and electronics vocabulary. Look for them as you read this chapter.

AND gate
binary
bit
byte
complementary metal-oxide semiconductor logic (CMOS)
counter
decimal system
decoder
encoder
flip-flop
inverter
logic gate
NAND gate
NOR gate
NOT gate
OR gate
transistor-transistor logic (TTL)
truth table
XNOR gate
XOR gate

To review, an integrated circuit (IC) is a complete electronic circuit contained in one package. Usually this package includes transistors, diodes, resistors, and capacitors, along with the connecting wiring and terminals.

A major advantage of the IC is its size. High capacity computers that fit in the palm of a hand, or easily into a briefcase, are made possible by the small size of ICs. ICs are also very stable compared to individual components, and they are less costly to build and operate.

Figure 20-1 shows an enlarged IC. Millions of components make up this IC. This IC is used in computers.

As discussed in Chapter 19, there are two types of integrated circuits: linear and digital. *Linear* circuits are used as amplifiers and have variable outputs. *Digital* ICs are used as switches. They work in either the on or off state.

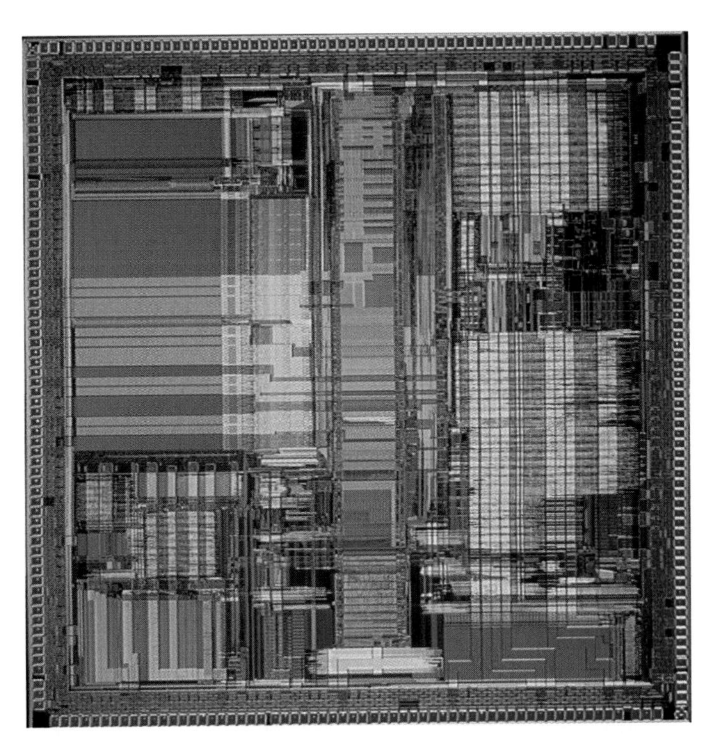

Figure 20-1. ICs can contain millions of components. (Intel)

Digital electronics have many uses in our lives. Digital electronics are used in communication systems such as television, radio, CD players, telephone systems, satellite systems, clocks, watches, fiber optics communications, calculators, computers, electronic meters, and much more. Digital systems are ideal for stepping motor drives, which make the digital electronic systems ideal for use in robotics. The CD player in **Figure 20-2** relies on digital electronics for its remarkable playback feature. **Figure 20-3** shows a small camcorder that records and displays both visual and audio information based on digital electronics.

Figure 20-2. Digital electronics make this portable CD player and its liquid crystal display possible. (Sony Electronics Corp.)

Digital electronics is a vast field of study. It is the basis of the modern computer system. Using digital, combined with analog systems, anything the mind can conceive can be created.

20.1 DIGITAL FUNDAMENTALS

There are two major fields of electronics, analog and digital. Analog is a system of continuous change without interruption. An example of an analog system would be a potentiometer used to control the intensity of an LED. As the potentiometer is rotated clockwise and counterclockwise, the LED will gradually brighten and then dim. On the other hand, a digital system is either on or off. A good example of a digital system is a single-pole switch. It will instantaneously bring a lamp to full brightness or darkness.

Figure 20-3. Camcorders record information in a digital format. (Sony Electronics Corp.)

At first, it may seem that the analog system is better than the digital. But, in fact both systems have their own advantages. Analog, digital, or both systems may be used for different applications. Below is a list of some of the major advantages of digital systems.

- Digital ICs are inexpensive as compared to similar analog systems.
- Information is much easier to store in digital systems.
- Digital systems are usually faster than analog systems.
- Digital systems are compatible with computer systems.
- Temperature has less effect on digital systems, which results in a much more stable operation than analog systems.

Through the use of very special integrated circuits, digital circuits are capable of replicating analog systems that will completely fool the human senses. A lamp dimmer constructed using special digital techniques would be indistinguishable from an analog system. Digital telephone technology can make it impossible for most people to tell if digital or analog electronic techniques are used for transmitting a conversation.

Digital integrated circuits handle information using switching circuits. They work as a result of the use of a combination of logic gate and flip-flop circuits. Let's look at gates and their operation using a system known as binary logic.

Binary Numbering System

Digital electronic circuits can be made to act in only two states: on and off. This two state system is called ***binary.*** This system can be compared to a single-pole, single-throw (SPST) switch, **Figure 20-4.** A switch in the off position represents a 0 in the binary numbering system. Likewise, a switch in the on position represents a 1. It is not practical to build large electronic logic circuits using manual switches. Manual switches do however, provide a good basis for understanding other switchable electronic components, **Figure 20-5.**

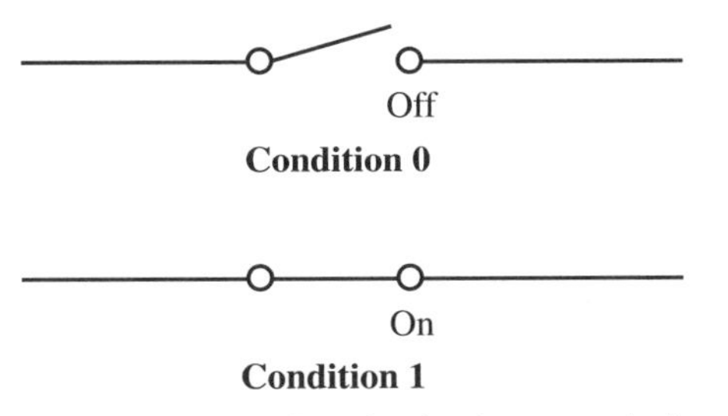

Figure 20-4. A single-pole, single-throw switch as a binary device.

Component	On or 1 state	Off or 0 state
SPST switch		
Magnetic core		
Digital pulse	+5 V 0 V	0 V
Relay		

Figure 20-5. Binary states of components.

The transistor is the most common electronic component that can be used as a switch. The transistor can allow current to flow. This is an on, or 1, state. Or it can stop current from flowing. This is the off, or 0, state. See **Figure 20-6.**

In Chapter 18, Class B amplifiers were discussed. These amplifiers are biased at the midpoint of the curve produced by the emitter voltage and collector current. Without any input signal (point A to B in Figure 20-6) the amplifiers are turned off. If the input signal changes to a negative level (B to C), the emitter-to-base bias will be

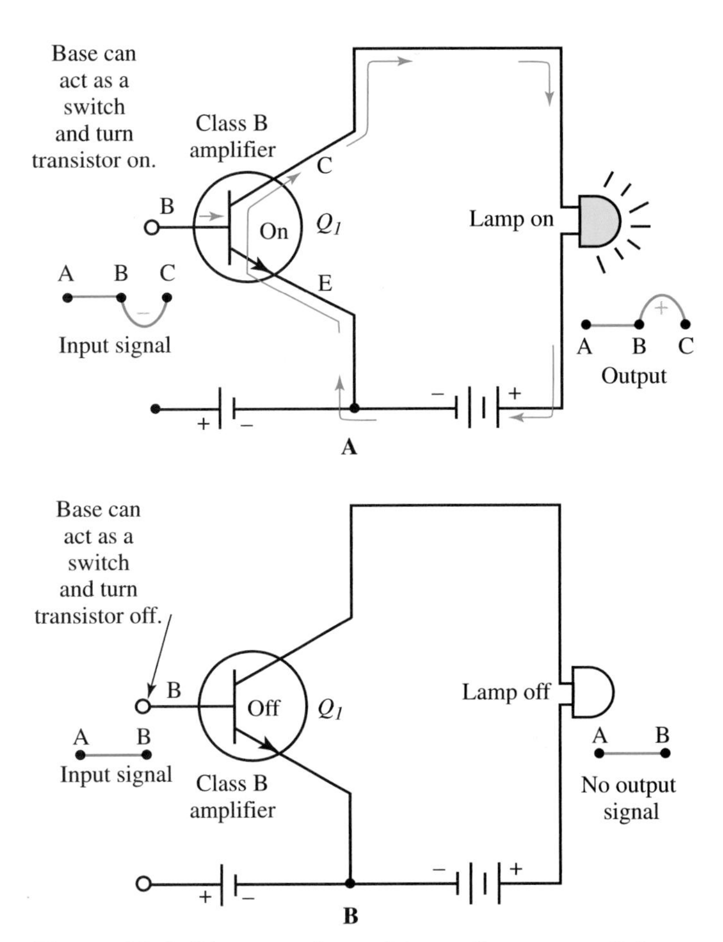

Figure 20-6. Diagram of transistor acting as a switch. A–On. B–Off.

decreased. This condition causes the output signal across the lamp to swing positive (input signal is 180 degrees out of phase with output signal in a CE circuit).

With only one SPST switch, we can produce only two outputs: 0 and 1. With additional switches we can count higher. To understand how to count to a higher number such as 45 or 79 with switches that are either on or off, we must learn the binary system.

There is a basic rule for counting in any system. Digits must be recorded one after the other for each counting unit. This continues until the count exceeds the total number of digits available. Then, a second column is started and counting continues. In the number system you are accustomed to, the ***decimal system,*** there are 10 digits available. Each time the count is raised from 0 you place the next highest digit (0, 1, 2, 3, ... 9). This takes you from 0 through 9. Nine is the last digit available in the decimal system. Thus, it is time start a second column and continue (10, 11, 12, ...).

The binary system works as shown in **Figure 20-7.** Since only two digits are available in the binary system, when you reach 2, you must start a new column. The digit 0 in the decimal system is 0 in the binary system. Like-

2^7	2^6	2^5	2^4	2^3	2^2	2^1	2^0
128s	64s	32s	16s	8s	4s	2s	1s
8th number	7th number	6th number	5th number	4th number	3rd number	2nd number	1st number

Figure 20-7. The binary numbering system.

wise, the digit 1 in the decimal system is 1 in the binary system. However, the number 2 in the decimal system is the number 10 in the binary system. There is no digit 2 in binary, so you have to move into the second column to create the number 2. The binary number for 6 is 110, **Figure 20-8. Figure 20-9** shows the decimal to binary conversion for the numbers 0 through 26. Try to extend this table to 50.

2^2	2^1	2^0
4s	2s	1s
1	1	0

Figure 20-8. Binary number for decimal number 6.

Decimal			Binary number				
Hundreds	Tens	Ones	2^4	2^3	2^2	2^1	2^0
			16s	8s	4s	2s	1s
		0					0
		1					1
		2				1	0
		3				1	1
		4			1	0	0
		5			1	0	1
		6			1	1	0
		7			1	1	1
		8		1	0	0	0
		9		1	0	0	1
	1	0		1	0	1	0
	1	1		1	0	1	1
	1	2		1	1	0	0
	1	3		1	1	0	1
	1	4		1	1	1	0
	1	5		1	1	1	1
	1	6	1	0	0	0	0
	1	7	1	0	0	0	1
	1	8	1	0	0	1	0
	1	9	1	0	0	1	1
	2	0	1	0	1	0	0
	2	1	1	0	1	0	1
	2	2	1	0	1	1	0
	2	3	1	0	1	1	1
	2	4	1	1	0	0	0
	2	5	1	1	0	0	1
	2	6	1	1	0	1	0

Figure 20-9. Decimal to binary conversion table.

If you are having trouble with binary numbers, try this activity. It may help you understand the binary numbering concept. Tear off five small pieces of tape. Number the tape as shown in **Figure 20-10.** Place one piece of tape on each finger of your left hand with palm up (in the order shown). Fingers pointing up represent 1s. Fingers folded down represent 0s. Position your hand as shown in **Figure 20-11.** What is the binary number? What is its decimal equivalent?

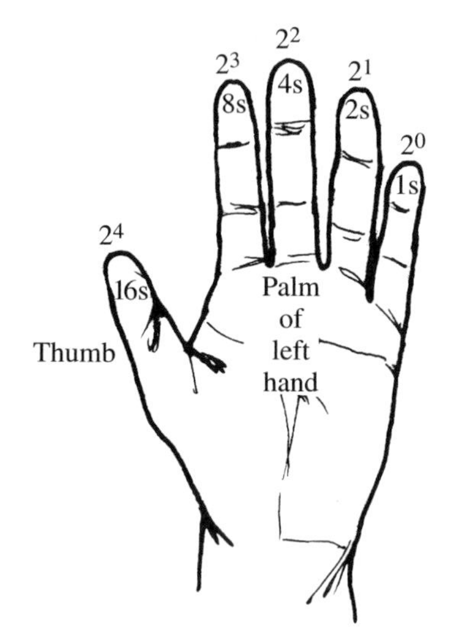

Figure 20-10. Learning the binary system.

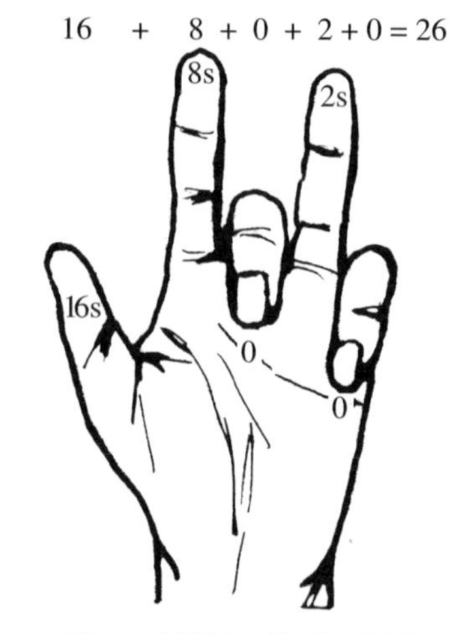

Figure 20-11. Converting 26 decimal to 11010 binary.

A larger decimal number, 79, is shown in binary form in **Figure 20-12.**

Binary addition is a simple process, **Figure 20-13.** When adding 0 and 0, the sum is still 0. When adding 1 and 0, the sum is 1. But adding 1 and 1 produces the sum is 10. The 0 is placed in the 1s column. The 1 is carried to the 2s column.

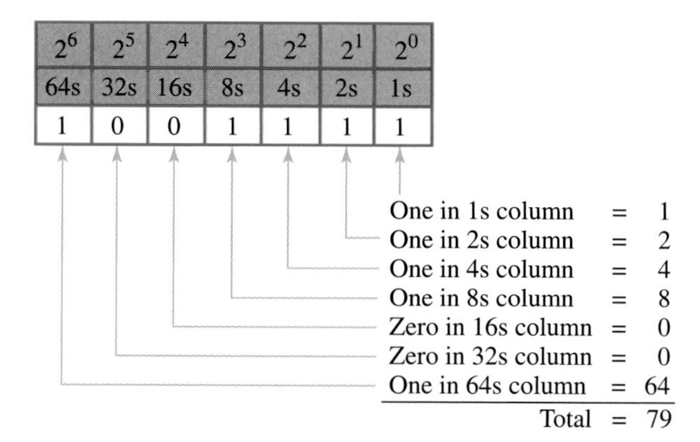

Figure 20-12. Binary number for decimal number 79.

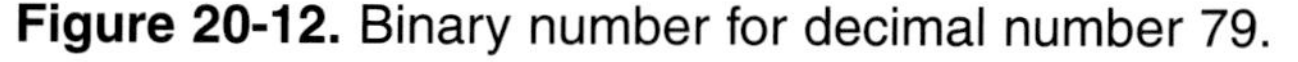

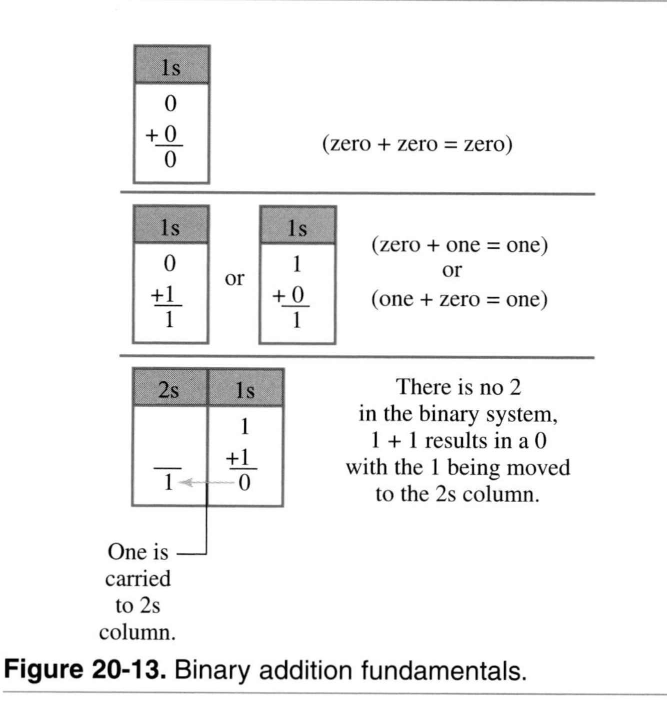

Figure 20-13. Binary addition fundamentals.

	Binary								Decimal	
	2^7	2^6	2^5	2^4	2^3	2^2	2^1	2^0		
	128s	64s	32s	16s	8s	4s	2s	1s	10s	1s
	0	0	1	1	1	0	0	1	5	7
+	0	0	0	1	1	0	0	0	+ 2	4
total	0	1	0	1	0	0	0	1	8	1

Figure 20-14. Binary addition of 57 and 24.

Decimal	Binary	Octal	Hex	BCD
0	0	0	0	0
1	1	1	1	1
2	10	2	2	10
3	11	3	3	11
4	100	4	4	100
5	101	5	5	101
6	110	6	6	110
7	111	7	7	111
8	1000	10	8	1000
9	1001	11	9	1001
10	1010	12	A	1 0000
11	1011	13	B	1 0001
12	1100	14	C	1 0010
13	1101	15	D	1 0011
14	1110	16	E	1 0100
15	1111	17	F	1 0101
16	10000	20	10	1 0110
17	10001	21	11	1 0111
18	10010	22	12	1 1000
19	10011	23	13	1 1001
20	10100	24	14	10 0000
21	10101	25	15	10 0001
22	10110	26	16	10 0010
23	10111	27	17	10 0011
24	11000	30	18	10 0100
25	11001	31	19	10 0101
26	11010	32	1A	10 0110
27	11011	33	1B	10 0111
28	11100	34	1C	10 1000
29	11101	35	1D	10 1001
30	11110	36	1E	11 0000
31	11111	37	1F	11 0001
32	100000	40	20	11 0010

Figure 20-15. Comparison of some common electronic numbering systems.

To add two larger binary numbers, the same steps are followed. **Figure 20-14** shows the addition of 57 (00111001) and 24 (00011000). **Figure 20-15** is a chart for comparison of some common electronic numbering systems.

Voltage Logic Levels in Digital Circuits

We know that digital circuits have only two states of 0 and 1. The operating voltages needed in a circuit for these two values depend on the type of logic circuitry or *family* used. Regardless of the logic family used, in positive logic, the high value of 1 is called the ***valid logic high*** range. This value most often varies from 2 to 3.5 volts to 5 volts. The low value, or 0, varies from 0 volts to 0.8 to 1.5 volts. This value is shown as the 0 to V_1 range in **Figure 20-16.** This is called the ***valid logic low.***

The area in between these two values acts as a buffer range. Any voltage in this range applied to the digital circuit causes confusion in the IC. The IC will not know whether to produce a 0 or a 1. This area is called the ***invalid value range,*** or the ***intermediate range.*** In Figure 20-16, this range is shown between V_1 and V_2.

Bits, Nibbles, and Bytes

In binary code, the smallest unit of information is called a ***bit.*** The word "bit" comes from joining the two

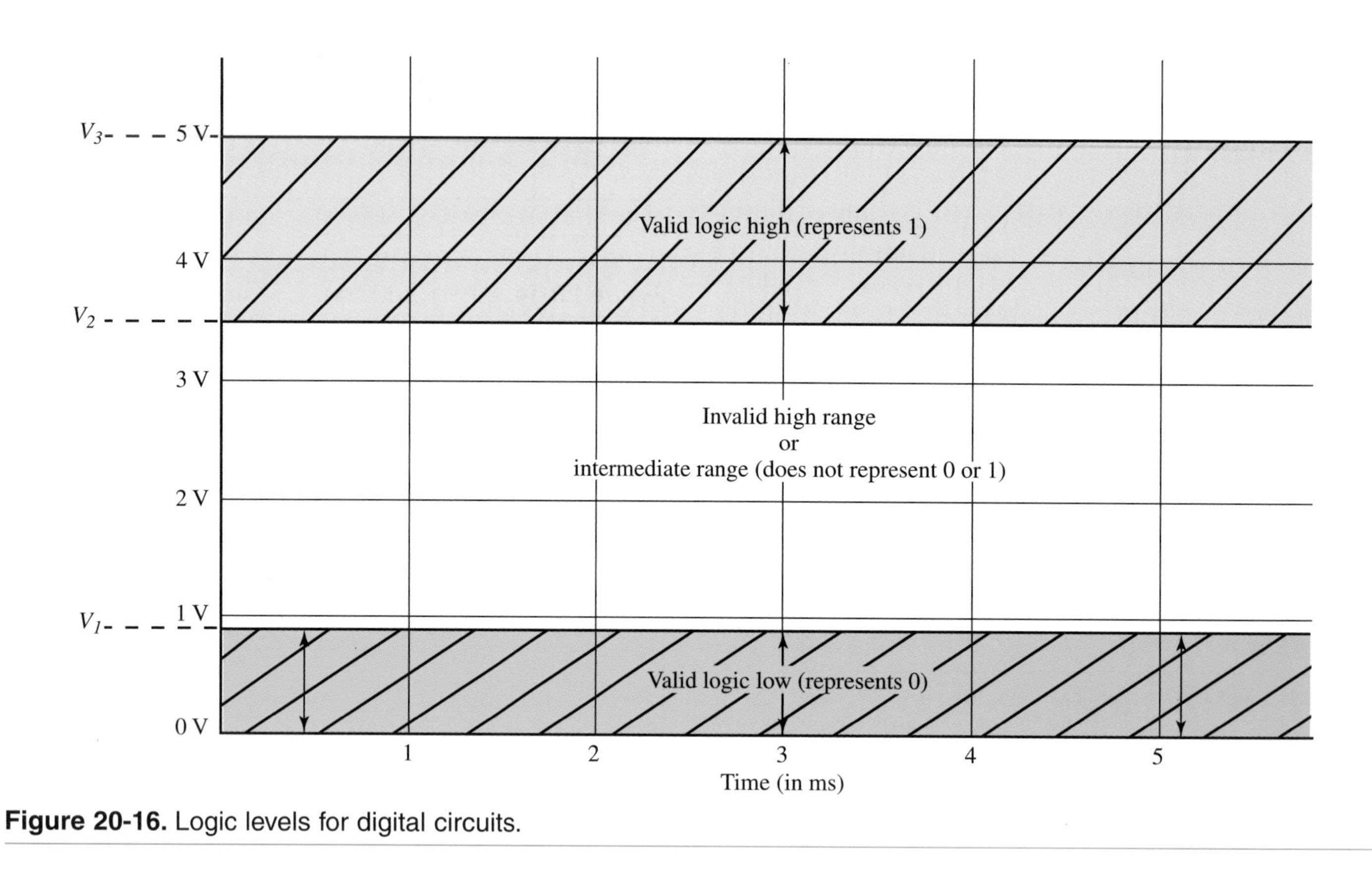

Figure 20-16. Logic levels for digital circuits.

words, **bi**nary digi**t**. A bit can be either 0 or 1. It is only one column or digit in a binary numbering system, **Figure 20-17.** Four bits of information make up a ***nibble.*** Two nibbles, or eight bits, make up a ***byte.*** A byte is a single unit of memory in a computer. For example, a computer with a 256 byte storage can hold 2048 bits of information. This is a small number, however. Most computer memory is given using the terms ***kilobytes, megabytes,*** or ***gigabytes.*** Computer storage abilities continue to grow.

Term	Symbol	Numeric Value
Kilobyte	K	1000
Megabyte	M	1,000,000
Gigabyte	G	1,000,000,000
Terabyte	T	1,000,000,000,000

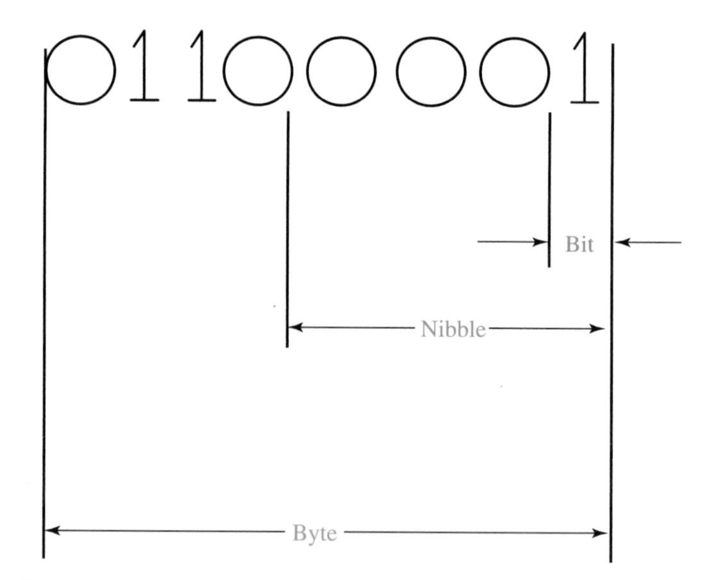

Figure 20-17. Bits versus nibbles versus bytes.

Review Questions for Section 20.1

1. Describe how a transistor biased as a Class B amplifier can act as a switch.
2. What is the binary number for 114?
3. What is the decimal number for 101101?
4. What is the binary sum of 01001010 and 01011001? What is the decimal equivalent of the sum?
5. A byte consists of __________ nibbles.
6. A logic value high is represented by the number __________, while the low logic level is represented by the number __________.
7. Between the valid high and valid low logic value is the __________ __________ range or __________ range, which does not produce a valid high or low.

Computerized Tomography (CT)

Computerized tomography (CT) is a diagnostic medical procedure that uses photomultiplier tubes (PMTs), computer scans, and X-rays to produce digitized images of the body. The technique is also called *computed tomography* or *computerized axial tomography (CAT).* The term *CAT scan* is still used today even though it is technically incorrect.

Let's look at the basic principle of CT scanning. An X-ray source is mounted opposite an array of PMTs. (A PMT is a sensitive vacuum tube that converts very weak light signals into measurable electric current.) The PMT uses a *photocathode* that gives off electrons when light strikes it. These electrons then strike the first of a series of plates called *dynodes.* The dynodes carry a high voltage charge that attracts the stream of electrons.

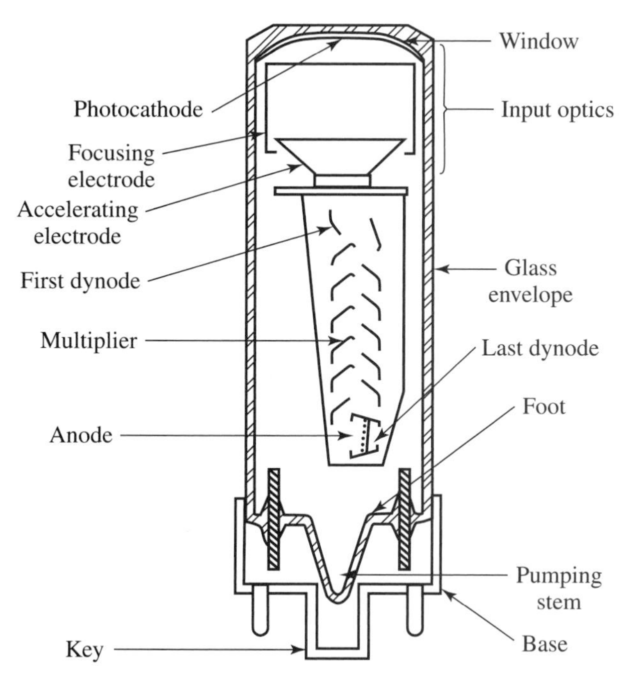

Main elements of a photomultiplier tube (PMT).

As the electrons bounce from plate to plate, they knock an increasing number of electrons from each plate. Millions of electrons may leave the tube for every electron given off by the photocathode. The tube thus multiplies the effect of the light that strikes it and enables the brightness of the light to be measured with extreme accuracy.

This procedure continues until the *anode grid* receives all the electrons. The electrons are then turned into a digital signal. The digital signal is stored in computer memory and then displayed graphically. Imaging systems can produce three-dimensional images at resolutions four times higher than normal television.

During the CT procedure, the patient lies on a table that passes through a circular scanning machine called a *gantry.* The table is positioned so that the area to be scanned lies in the center of the gantry. A tube on the gantry beams X-rays through the patient's body and into the photomultiplier tubes. The gantry rotates around the patient to obtain images from different angles. A computer processes the information to produce a cross-sectional image. Scans of the same organ or even the entire body are obtained by moving the table in the gantry.

CT scans are used to diagnose many conditions including tumors, infections, and blood clots. CT also assists in treating some conditions that might otherwise require surgery. For example, doctors can use a CT scan to guide small tubes to an abscess in the body to drain the infected area. Sometimes a *contrast agent* is injected into the body to make certain organs show up more clearly. For instance, to outline the inner surfaces of the stomach and bowel, the patient is given a barium mixture to drink.

The biggest advantage of the CT over conventional X-ray machines is that the patient is exposed to little radiation. CT scans can be saved as electronic files or even transmitted over modem lines to other doctors for consultation.

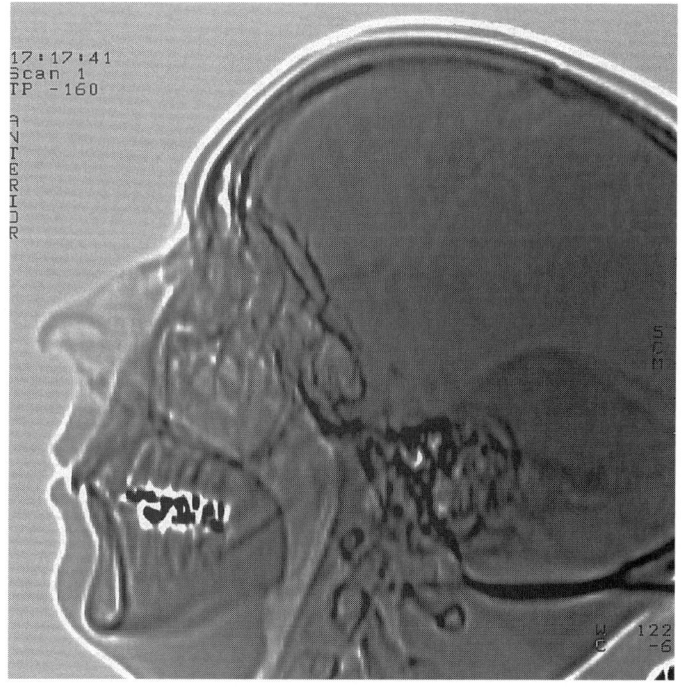

CT images are exposed on film in black and white and are very similar in appearance to regular X-rays.

20.2 LOGIC GATES

Electronic switching circuits that govern, or "decide," whether inputs will pass to output or be stopped are called ***logic gates.*** The logic gates discussed here are the building blocks for other logic gates. The basic logic gates are:

- AND gates.
- OR gates.
- NOT gates.
- NAND gates.
- NOR gates.
- XOR gates.
- XNOR gates.

All logic gates can be made from some combination of the first three gates in the list, the AND, OR, and NOT gates.

AND Gates

The ***AND gate*** accepts high and low inputs (1 and 0). Based on these inputs, the gate decides on the output. The outputs of AND gates are also high or low. The AND gate symbol is shown in **Figure 20-18.**

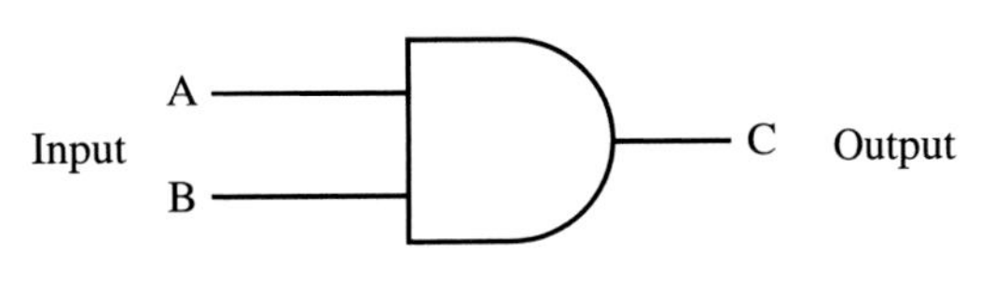

Figure 20-18. AND gate symbol.

The AND gate produce an output of 1 (high) if *all* inputs are 1 (high). Logic 1 or high also equals *on.* A simple circuit with switches that functions like an AND gate is shown in **Figure 20-19.** Both switches are off (0 or low). The lamp does not burn. When both switches are on, the lamp burns, **Figure 20-20.** If only one or the other gate is on, the lamp does not burn. This simple circuit using SPST switches illustrates AND gate operation, however, the actual AND gate used in a computer is very complex.

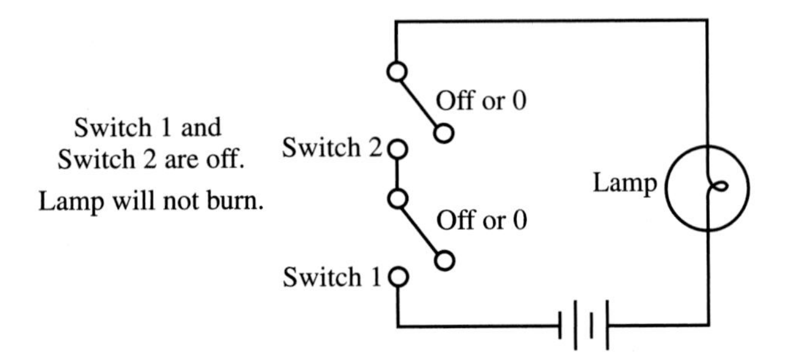

Figure 20-19. This simple switching circuit simulates the operation of an AND gate. The lamp will not burn because both switches are off.

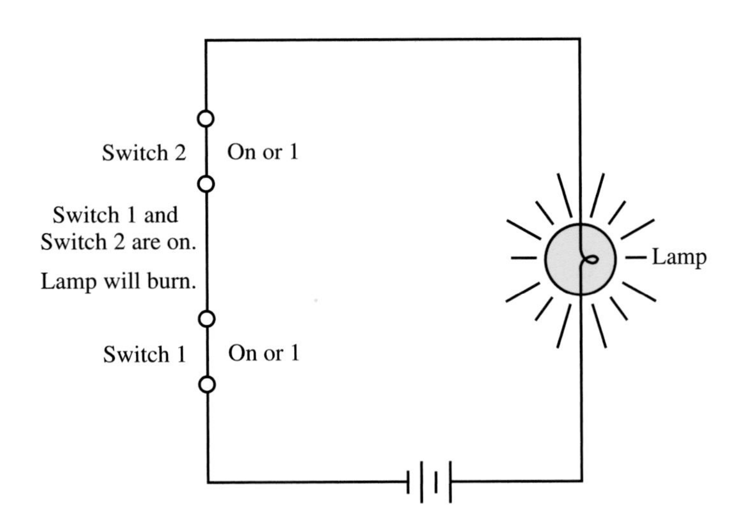

Figure 20-20. This is the same circuit as in Figure 20-19. However, the lamp will burn because both switches are on.

A binary table that explains the operation of the AND gate is shown in **Figure 20-21.** This table is called a ***truth table.*** The truth table shows the inputs on the left and outputs on the right.

Signal at inputs		Signal at output
A	B	C
0	0	0
0	1	0
1	0	0
1	1	1

Figure 20-21. AND gate truth table. Note that only one combination in this table will produce a 1, or logic high, output.

The AND gate is used to detect the presence of high signals, or ls, on both inputs A and B. If this occurs, the output signal will be 1. However, if even one input signal is 0, output will be 0. The output signal as affected by input signals in the AND gate is shown in **Figure 20-22.**

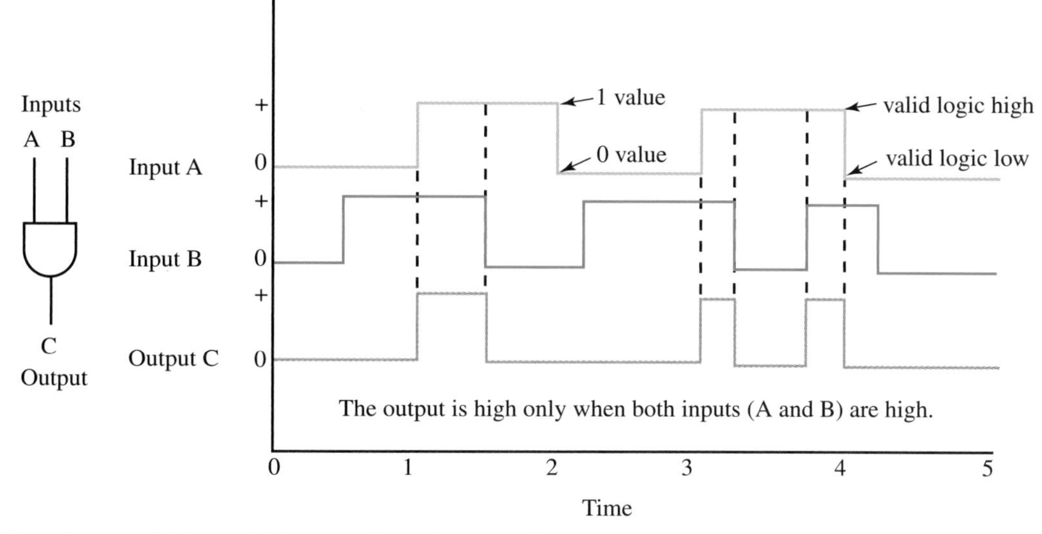

Figure 20-22. Resulting valid logic highs and lows in an AND gate as they appear on a three-channel oscilloscope.

OR Gates

The ***OR gate*** provides an output signal of 1 (high) when either one of its inputs or both of them is 1. If all inputs are 0 in an OR gate, the output is a 0 (low). The OR circuit detects the presence of any high input. The symbol for an OR gate is shown in **Figure 20-23.**

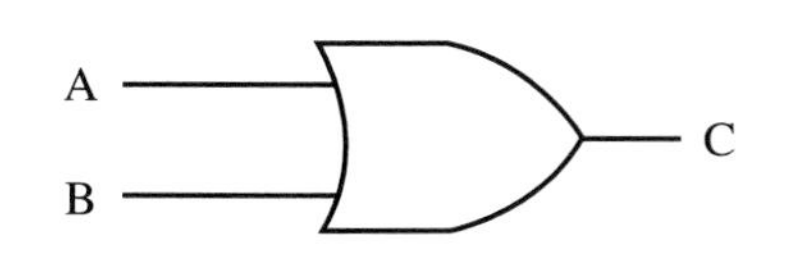

Figure 20-23. OR gate symbol.

The truth table for an OR gate is shown in **Figure 20-24.** The schematic for a OR gate created from switches is shown in **Figure 20-25.** The OR gate acts like two switches in parallel. The output signal, as it is affected by input signals in the OR gate, is shown in **Figure 20-26.**

Signal at inputs		Signal at output
A	B	C
0	0	0
0	1	1
1	0	1
1	1	1

Figure 20-24. OR gate truth table.

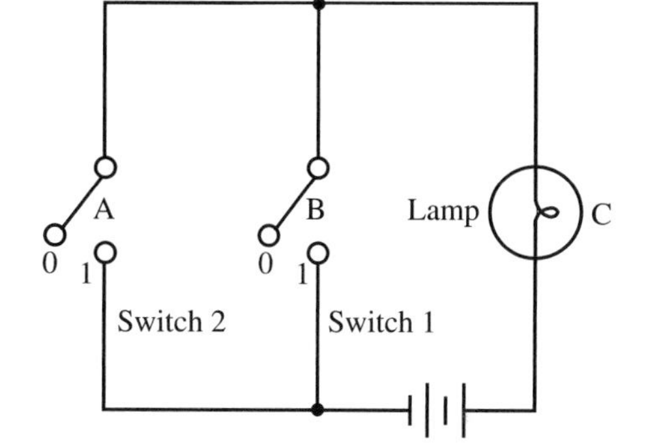

Figure 20-25. Schematic of a simulated OR gate. The lamp will not burn in this instance.

NOT Gates

The ***NOT gate*** is put into a circuit to invert the polarity of the input signal. The NOT gate is often called an ***inverter*** because the name describes its function. If the input signal is 1, the output signal will be 0. Likewise, if the input signal is 0, the output will be 1.

The symbol for the NOT gate is shown in **Figure 20-27.** Note that the NOT gate symbol has only one input lead. Notice that there is also a small circle at the end of the triangle in the symbol.

The schematic for a simulated NOT gate circuit is shown in **Figure 20-28.** The NOT gate truth table is shown in **Figure 20-29. Figure 20-30** shows the output signal in a NOT gate as affected by the input signals.

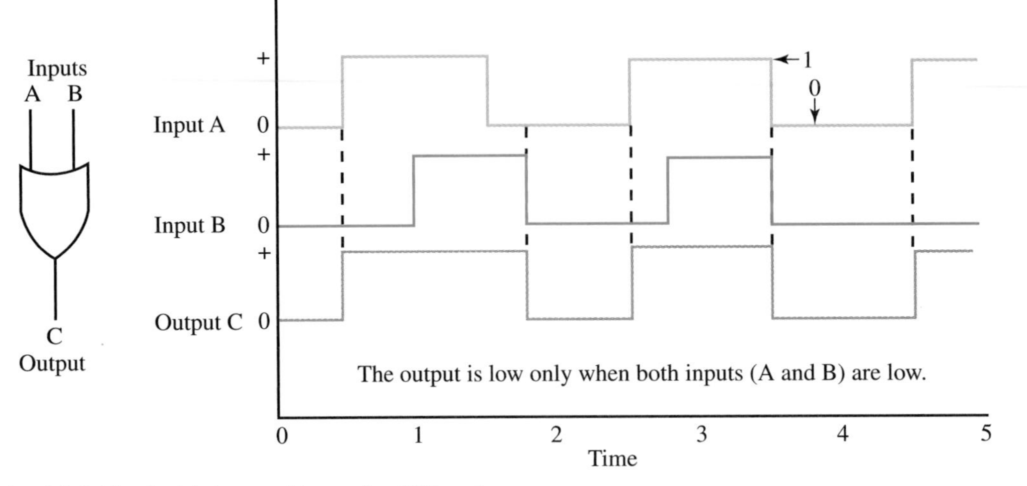

Figure 20-26. Valid logic highs and lows for OR gate.

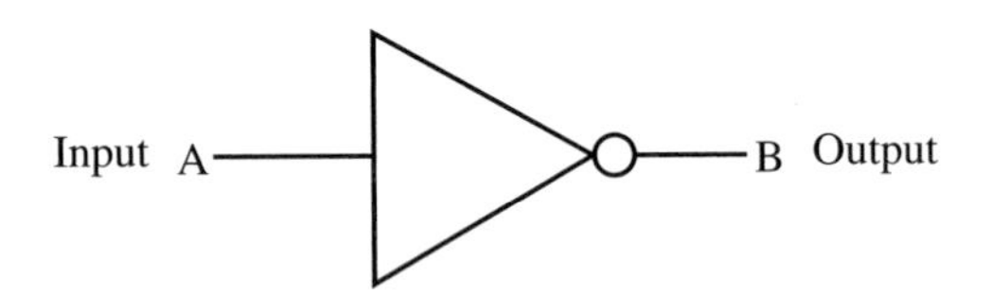

Figure 20-27. NOT gate symbol.

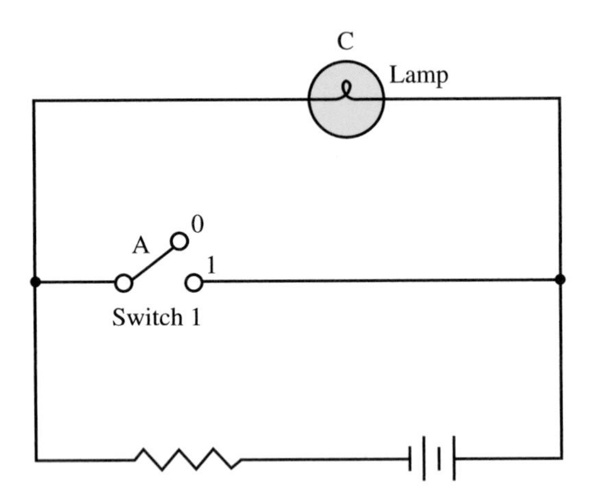

Figure 20-28. Schematic of simulated NOT gate. The input is 0, so the output is 1.

Signal at input	Signal at output
A	B
0 1	1 0

Figure 20-29. NOT gate truth table.

NAND Gates

All logic gates are combinations of the basic gates: AND, OR, and NOT. The ***NAND gate*** is a negative AND gate. It is made up of an AND gate and a NOT gate. It is also called a NOT AND gate.

The NAND gate symbol is like the AND gate symbol with a circle at the end, **Figure 20-31.** The schematic for a NAND gate circuit simulated from switches is shown in **Figure 20-32.** The NAND truth table is shown in **Figure 20-33.** Notice that this truth table is the reverse of the AND truth table. **Figure 20-34** shows the waveforms in a NAND gate circuit.

NOR Gates

The ***NOR gate*** gives the opposite (or negative) results of the OR gate. The NOR gate is made up of an OR gate and a NOT gate (inverter). The symbol for a NOR gate is shown in **Figure 20-35. Figure 20-36** shows the schematic for a NOR circuit simulated with switches.

The NOR circuit is used to test for any kind of input. If there is no input, output will be 1. If there is an input, output will be 0. This is shown in the truth table in **Figure 20-37.** Notice that the truth table for the NOR gate is the exact opposite of the truth table for the OR gate. The waveforms of a NOR gate circuit are shown in **Figure 20-38.**

XOR Gates

There is a special type of gate that provides a high output whenever *any,* but *not all,* inputs are logic high. It is called the ***exclusive OR gate*** or the ***XOR gate.*** Recall that in contrast, the OR gate provides a high output whenever *any* or *all* inputs are logic high.

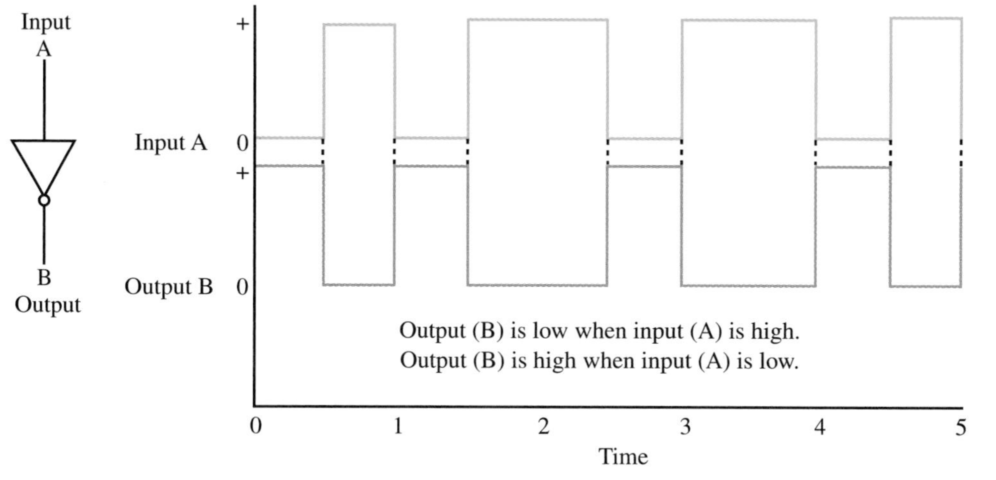

Figure 20-30. Valid logic highs and lows for NOT gate.

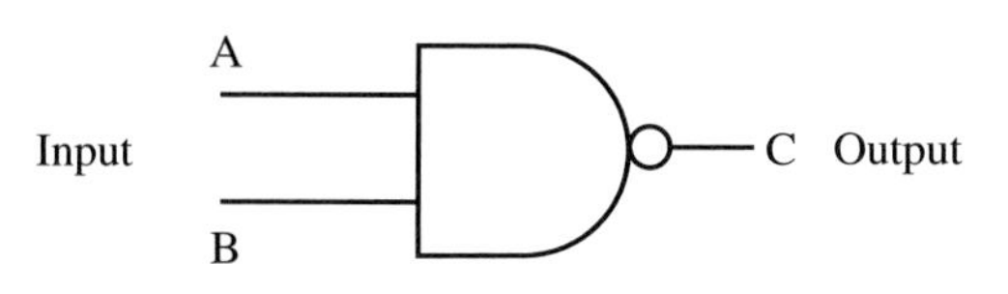

Figure 20-31. NAND gate symbol.

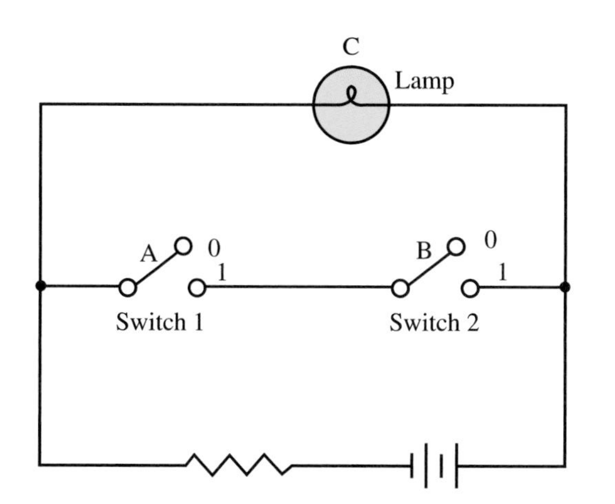

Figure 20-32. Schematic of a simulated NAND gate. The inputs are 0 and 0, so the output is 1.

Signal at inputs		Signal at output
A	B	C
0	0	1
0	1	1
1	0	1
1	1	0

Figure 20-33. NAND gate truth table.

The symbol for the XOR gate is shown in **Figure 20-39.** The circuit that simulates the XOR gate using two single-pole double-throw switches is shown in **Figure 20-40.** The XOR gate truth table is shown in **Figure 20-41.** The waveforms of a XOR gate circuit are shown in **Figure 20-42.**

XNOR Gates

A gate similar to the XOR gate is the ***exclusive NOR gate,*** or ***XNOR gate.*** This gate is the XOR gate with the output inverted. There is a high output only if all inputs are logic high or logic low. The symbol for the XNOR gate is shown in **Figure 20-43.** The truth table is shown in **Figure 20-44.** Notice that if only one input is high (column A or column B), there is a zero in column C. If column A and column B both have a 0s or both have 1s, then column C has a 1.

Review Questions for Section 20.2

1. Draw the symbols for three logic gates. Label the drawings.
2. In an AND gate, input A is 1 and input B is 0. What is output C?
3. The OR gate will provide a high output signal except:
   a. when both inputs are 1.
   b. when both inputs are 0.
   c. when either one of the inputs is 1.
   d. None of the above.
4. The __________ gate is often called the inverter.
5. What gate provides an output when any but not all of the inputs are 1?
6. Draw the truth table for the XNOR gate.

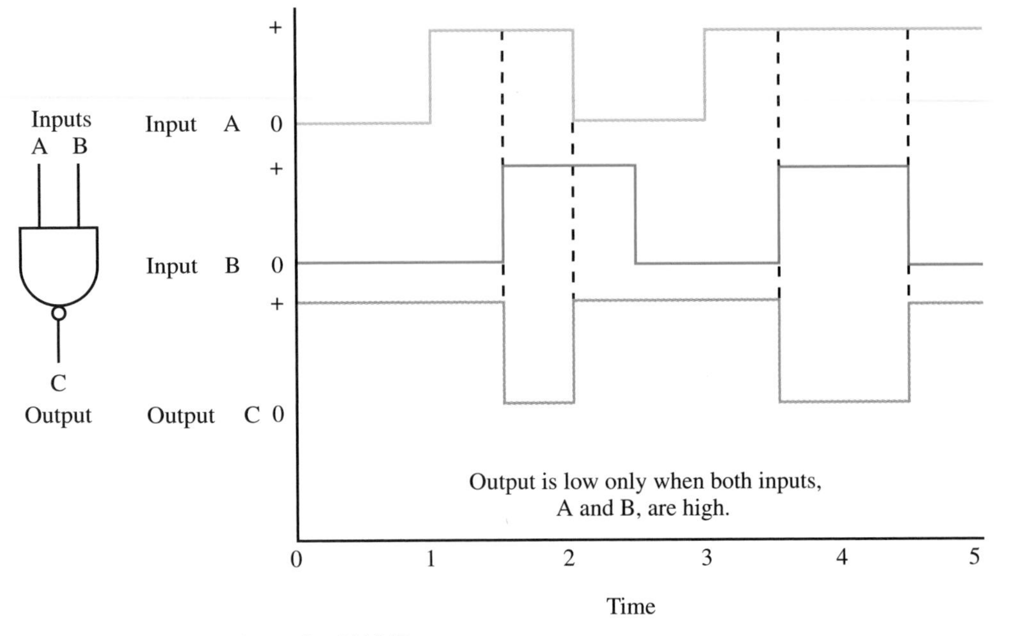

Figure 20-34. Valid logic highs and lows for NAND gate.

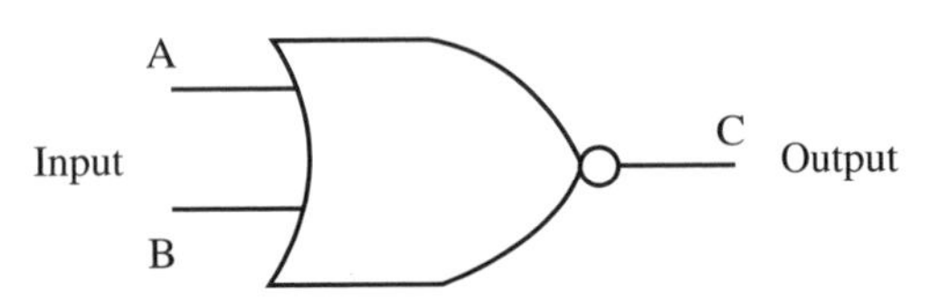

Figure 20-35. NOR gate symbol.

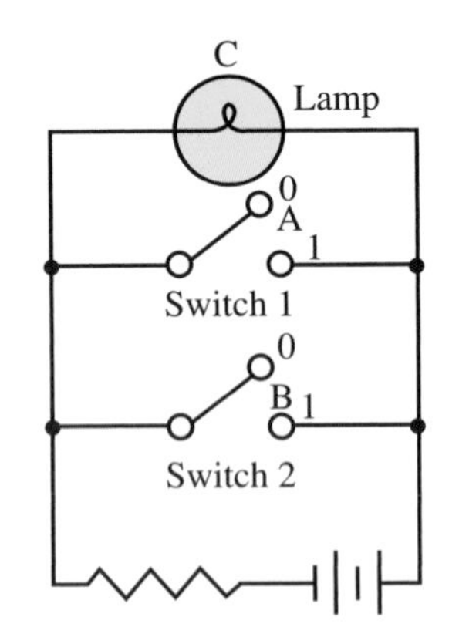

Figure 20-36. Schematic of simulated NOR gate. The inputs are 0 and 0, so the output is 1.

Signal at inputs		Signal at output
A	B	C
0	0	1
0	1	0
1	0	0
1	1	0

Figure 20-37. NOR gate truth table.

20.3 LOGIC FAMILIES

Manufacturing techniques have a major impact on the arrangement of digital circuits into groups or families. It is crucial that the traits of one logic family match the traits of another family when a number of digital ICs are used in one device. The devices may operate on different voltages, and timing can be affected.

In the time that has passed in the digital electronics field, certain logic families have evolved.

CMOS Logic

The ***complementary metal-oxide semiconductor logic (CMOS)*** family uses field effect transistors. CMOS integrated circuits have good resistance to noise. They require only small amounts of power. Voltage and current needs are not crucial.

CMOS ICs have one major problem, however. They can be damaged by static electricity. To prevent damage from static discharge, the worker and work surface should be grounded through a high resistance resistor (three to ten megohms at five watts or more).

CMOS ICs contain a manufacturer's identification number. One company uses a CD prefix. Another firm uses a C within the number, such as 74C30. High speed CMOS devices contain an H in the identification number, such as 74HC00 or 74HC190. CMOS ICs operate on voltages as high as 15 volts.

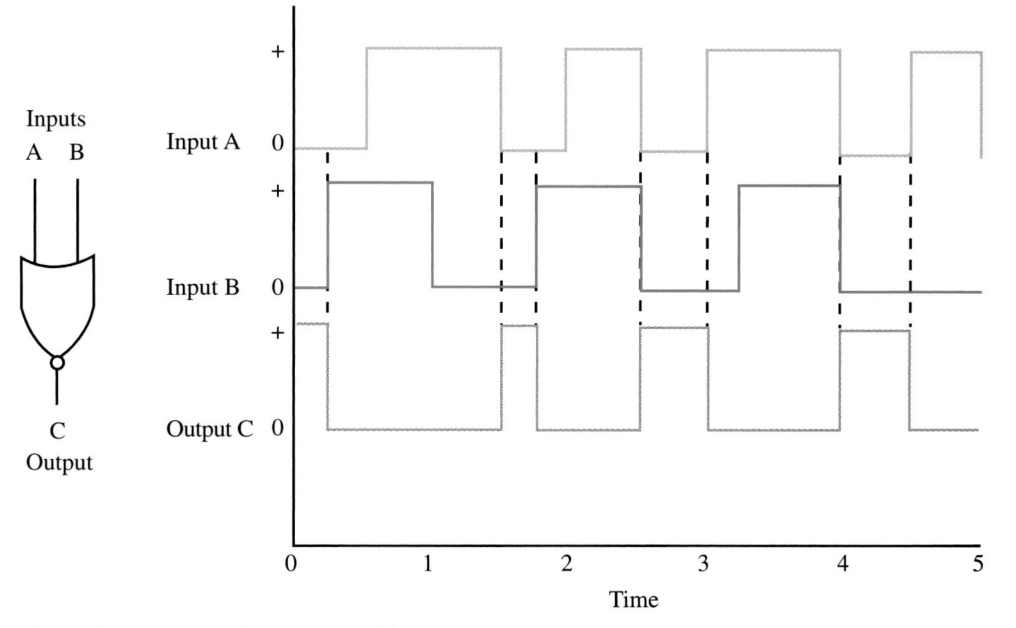

Figure 20-38. Valid logic highs and lows for NOR gate.

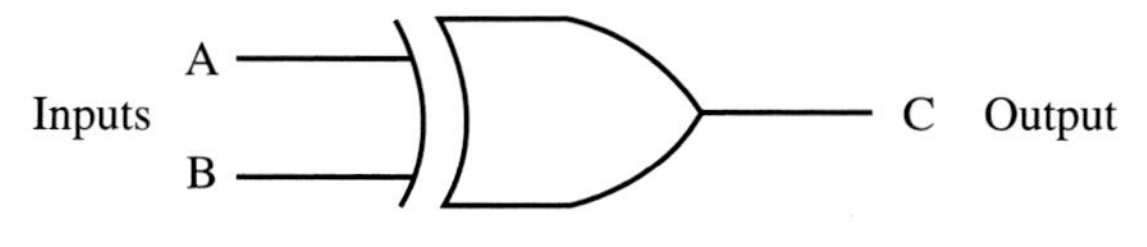

Figure 20-39. XOR gate symbol.

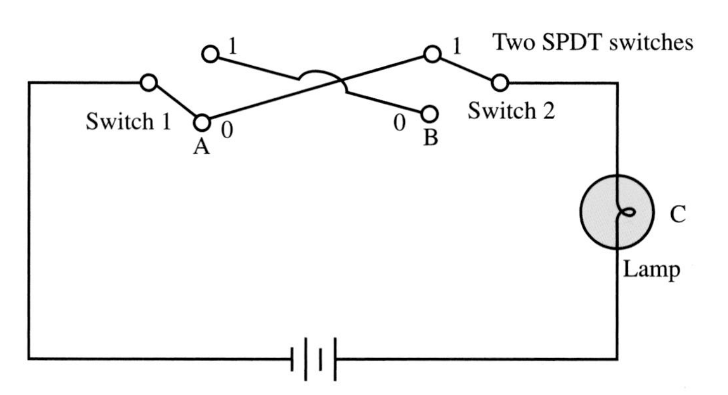

Figure 20-40. Schematic of simulated XOR gate. Note that it has two SPDT switches. One input is 1 and the other is 0. The output will be 1.

Signals at input		Signal at output
A	B	C
0	0	0
0	1	1
1	0	1
1	1	0

Figure 20-41. XOR gate truth table.

Transistor-Transistor Logic

The ***transistor-transistor logic (TTL)*** family is used widely in today's digital electronics. They work quickly. They perform logic functions from about 20 megahertz to 60 megahertz. TTL ICs work faster than CMOS ICs.

TTL ICs can be identified by their number. Their numbers begin with 74: 7400, 7402, 7404, 7408, 7432, and 7486. Unlike the CMOS logic, there is no C in the identification numbers for TTL logic.

Logic high for TTLs is two to five volts. Logic low is zero to 0.8 volts. The TTL requires high power dissipation and high current.

High power TTL ICs contain the part number 74H. They consume about twice the power of a regular TTL IC. Low power TTL ICs contain the part number 74L and consume one-tenth the power of a regular TTL IC.

A number of gates can be placed in one TTL IC. **Figure 20-45** shows some TTL ICs containing AND gates, NOT gates, NAND gates, NOR gates, and XOR gates.

Review Questions for Section 20.3

1. What two logic families are used widely today?
2. A 7408 IC is found in the __________ logic family.
3. List the advantages and major disadvantages of the CMOS ICs.
4. What type of transistor does the CMOS IC use?

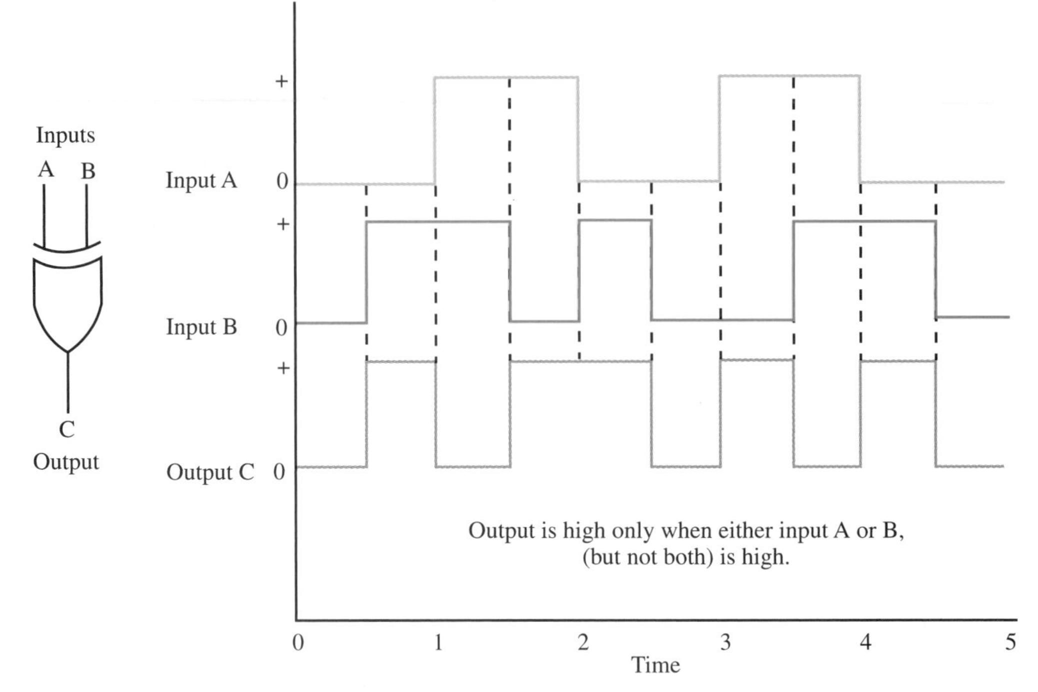

Figure 20-42. Valid logic highs and lows for XOR gate.

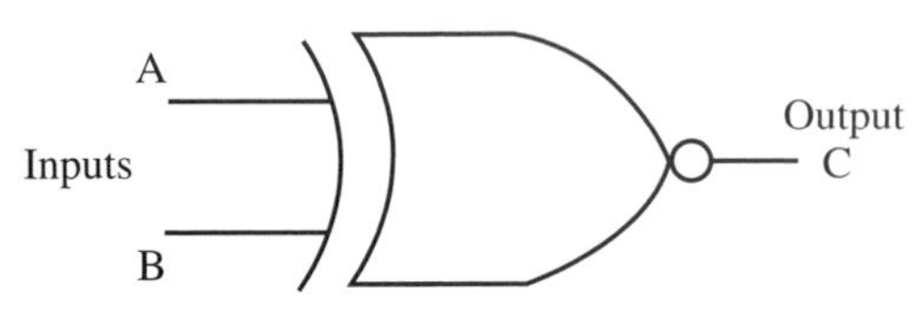

Figure 20-43. XNOR gate symbol.

Signals at input		Signals at output
A	B	C
0	0	1
0	1	0
1	0	0
1	1	1

Figure 20-44. XNOR gate truth table.

5. What number is used to signify a TTL IC?
6. For most TTL logic systems a low value is represented by a voltage of __________ to approximately __________, while a high is represented by a voltage of approximately __________ to a maximum of __________.
7. A CMOS IC can operate on a voltage as high as __________ volts.

20.4 DIGITAL APPLICATIONS

Digital circuits are used for many reasons. For example, digital circuits convert ac sine waves to digital pulses, **Figure 20-46.** The ac signal is sampled. Various points are given digital values. These are later converted to a binary value.

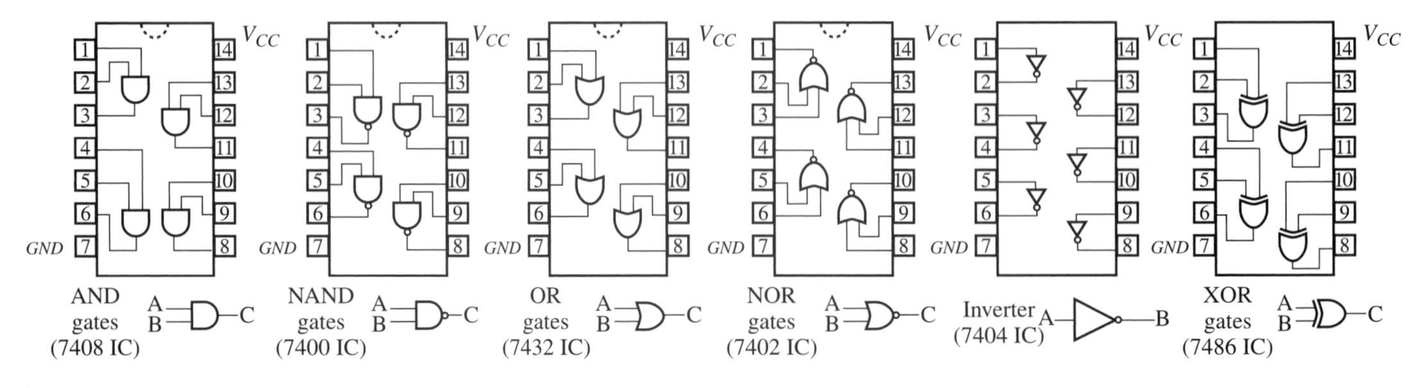

Figure 20-45. Logic gates in transistor-transistor logic ICs. Note the basing diagram for the ICs.

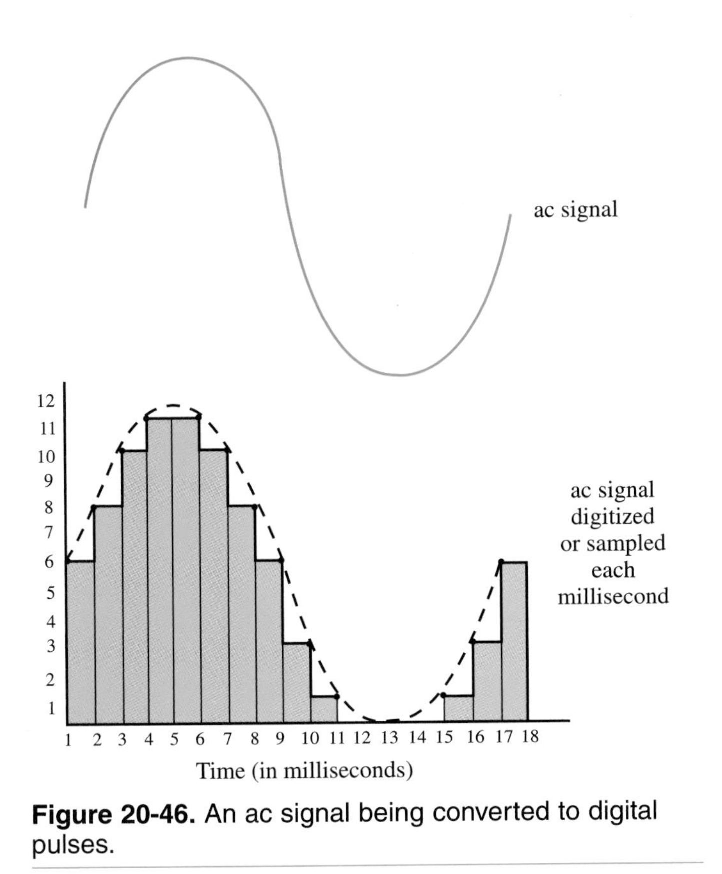

Figure 20-46. An ac signal being converted to digital pulses.

Figure 20-47 shows digital discs that have binary information recorded on their surface. A laser "reads" the information on the recording. Another example of digital electronics use is the Universal Product Code (UPC). It is found on most items purchased today, **Figure 20-48.** The UPC is read by a laser (light). The laser improves inventory (item movement) control. Some laser scanning systems contain a computer generated voice that tells the operator and the customer what has been purchased and the price of the item.

Figure 20-47. Digital discs.

Figure 20-48. Universal Product Code symbols. (IBM)

Logic Probe

The basic piece of test equipment used for digital circuits is the logic probe. The logic probe indicates either a *high* or *low* signal using LEDs. Look at **Figure 20-49.**

The logic probe connects to the power supply of the circuit being tested. The red lead connects to the positive side of the power supply while the black (negative) lead connects to ground. Most logic probes are equipped with a slide switch that allows you to select the logic family you will be testing, TTL or CMOS. The tip of the probe is touched to a pin on the chip or a connection point. The LED indicator lights to indicate a high or low logic value. When both LEDs light there is either an invalid logic level voltage or the probe is not making good contact with the circuit. Each logic probe manufacturer uses a unique design. Therefore, the instruction book of a logic probe should always be checked before use.

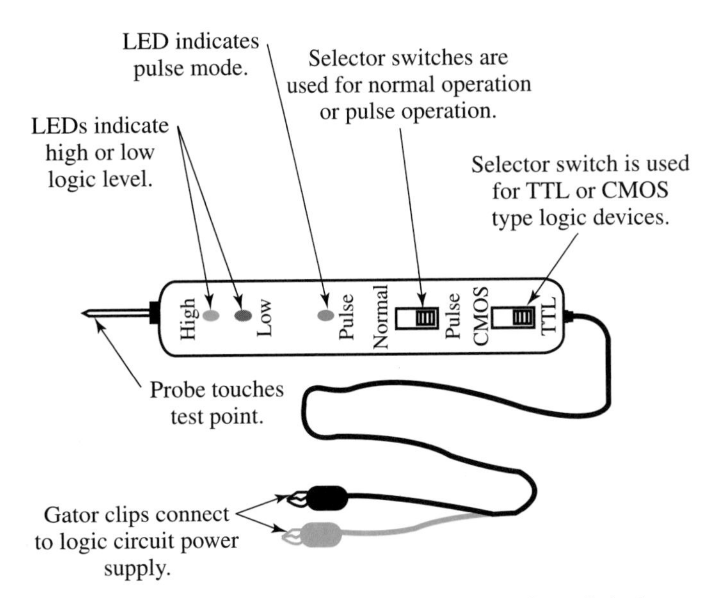

Figure 20-49. A typical logic probe for testing digital circuits.

Digital Encoders and Decoders

Most people have only learned the decimal number system. The digital system uses the binary numbering system, which is very different from the decimal system. To make the binary numbering easily understandable to us, we use an electronic system to translate, or decode, the binary system into the decimal system. A device that makes this translation is called a ***decoder.***

An example of a decoder is the 7448 chip. This chip converts a binary code to its decimal number equivalent on an LED seven segment display. This instrument allows us to easily read the binary number as a decimal number.

The conversion process also works in reverse. A decimal number can be converted to a binary number through an ***encoder.*** Once the decimal number has been encoded to a binary number, it can be utilized in a digital system such as a computer.

There are also analog to digital and digital to analog converters. These devices can change analog information such as temperature to a digital number equivalent. A digital multimeter is a good example of how analog information can be changed to digital information. When a voltage reading is taken from a circuit using a DMM, the input to the meter is an analog signal. Inside the meter, the encoder changes the analog signal into a binary code. This code is then turned into the decimal number value that appears on the meter's display.

Some circuits, such as the 7490 counter, are designed to convert electrical pulses into binary numbers. The counter advances by one each time the input receives an electrical pulse. The counter can be connected to a decoder and an LED driver such as the 7448 to display the pulse count as a decimal number. See **Figure 20-50.**

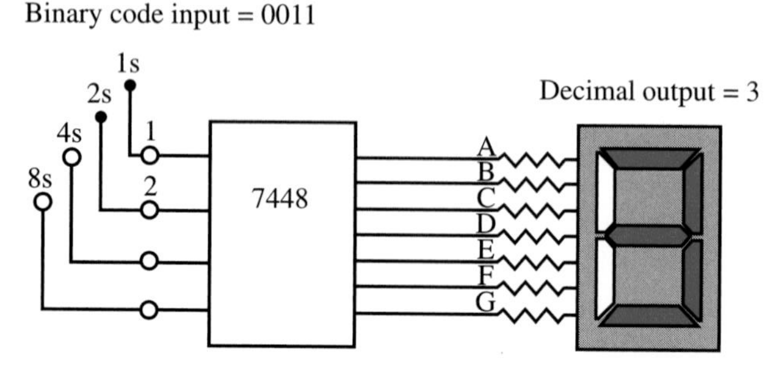

Figure 20-50. A 7448 chip will translate binary numbers to decimal numbers.

Digitized Analog Signals

A sound wave or linear voltage can be represented by a pattern of digital signals. There are many advantages to digitizing a sound wave pattern. A sound wave can be converted into a digital pattern using an analog-to-digital encoder. Each voltage level is represented by a binary code. **Figure 20-51** shows how the analog sound wave pattern might be represented by a series of binary numbers. The numbers can be transmitted by phone lines, satellites, microwave beams, etc. Once the sound pattern is encoded as a binary pattern, it can be transmitted faster than the time used to actually produce the sound. At its final destination, the digital pattern can be converted to an analog signal again.

The digital wave pattern will be in sharp steps and must be smoothed out. This smoothing of the digital pattern is accomplished using a filter similar to the filter used on a power supply. Once the pattern is smoothed out, it will be a perfect match of the original signal.

The digitized pattern shown in Figure 20-51 looks a bit crude because it is based on a four bit pattern. (Notice the bit patterns at the bottom of the graphs.) An actual digitized signal would use an 8, a 16, or a 32 bit encoder. The time increment between the binary codes would also be shortened. The combined effect of a larger bit pattern, say 32 bits, and a shorter time period between binary code creates a digital pattern *very close* to the original analog pattern. Only a slight filtering action would be needed to smooth the digitized pattern to match the original analog sound pattern.

The digitizing of sound has made it possible to store sound in computer memory. Once stored, it can be played back later, or even manipulated to distort or enhance the sound. For example, an echo can easily be made from any original sound pattern, or the volume and tone of the pattern can be changed.

Flip-Flops

Flip-flops are a rather unique digital device based upon the operation of combined logic gates. Flip-flops are an essential part of digital electronics. They are at the very heart of counters, timers, sequencing devices, and memories. ***Flip-flops*** are semiconductor devices that are capable of assuming one of two stable states. Two common flip-flops are the R-S and the J-K varieties.

R-S flip-flop

Look at **Figure 20-52.** Two NAND and two NOT gates have been configured to operate as a flip-flop. Two NOR gates have also been configured as a flip-flop. The standard pin markings are R, S, Q, and $\overline{Q}$. $\overline{Q}$ is pronounced "Q not". This flip-flop configuration is called a

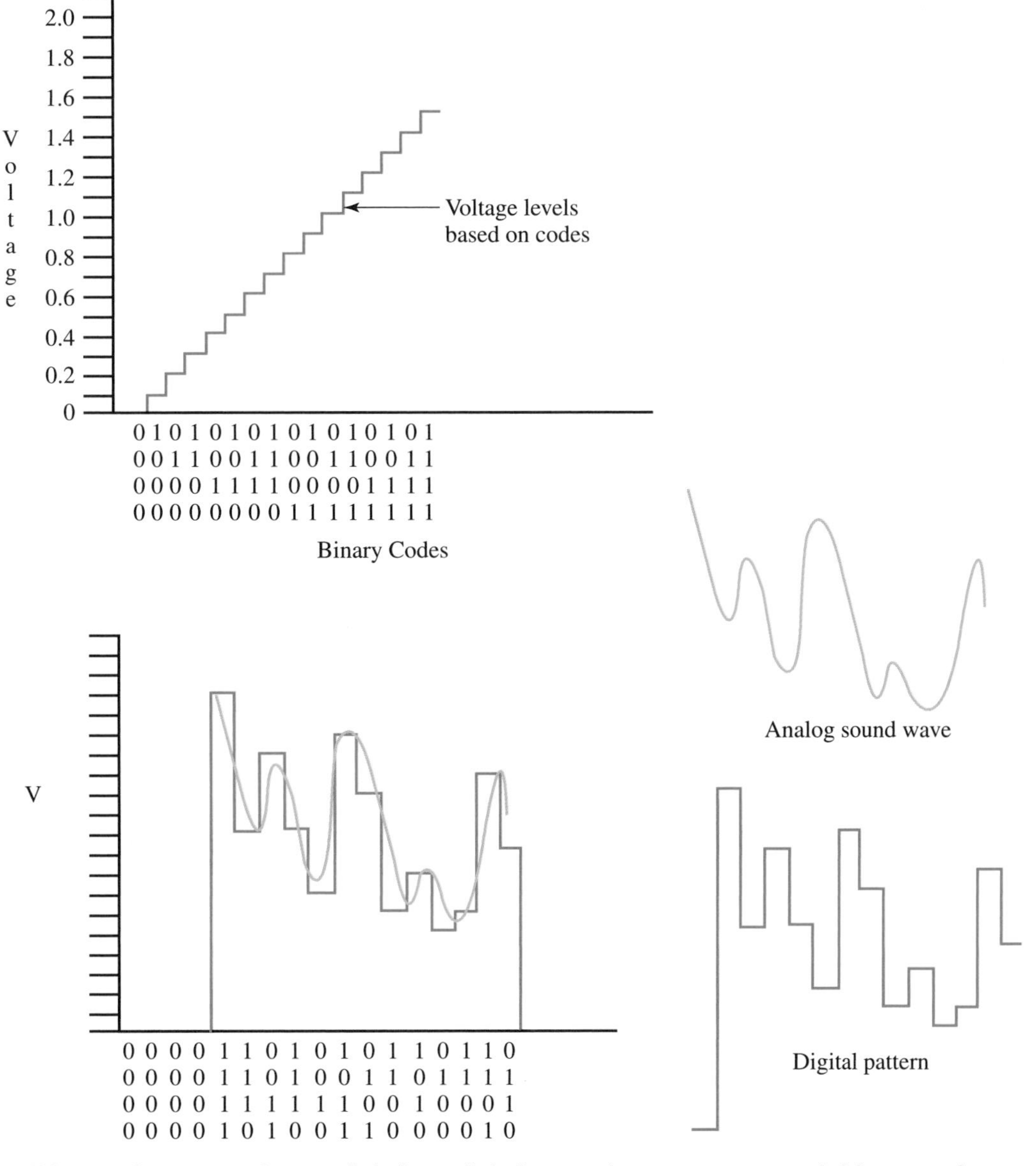

Figure 20-51. When using an analog-to-digital, or digital-to-analog converter, each binary code represents a different voltage value.

set-reset flip-flop, or ***R-S flip-flop.*** The S pin is called set and the R pin is called reset.

The operation of a R-S flip-flop is simple. When S is high, Q is also high. When R is high $\overline{Q}$ is high. When both inputs are low, the output will represent the *last* input setting. Both inputs being high is not a valid value. The output cannot be determined if both inputs are high.

Figure 20-52 also shows a truth table of the R-S flip-flop. The basic principle behind the operation of a flip-flop is that the outputs are complimentary. To say that two outputs are ***complimentary*** means that when one output is high, the other is low. Flip-Flops can retain their output condition.

A clocked R-S flip-flop uses a clock to synchronize the outputs. See **Figure 20-53.** With a ***clocked R-S flip-flop,*** the output changes when there is a change in the R or S input *and* a pulse appears at the clock input. By using devices that are clock driven, millions of parts can work together in unison to form an entire digital system. A digital clock is a string of pulses that varies continuously from high to low. The pulse train is the heartbeat of most digital systems.

The R-S flip-flop retains its output status even after the input is removed. This makes the clocked R-S flip-flop a good memory device.

J-K flip-flop

The ***J-K flip-flop*** is operationally similar to the R-S flip-flop. The J-K flip-flop is clock driven like the clocked R-S flip-flop. The difference is that the J-K flip-

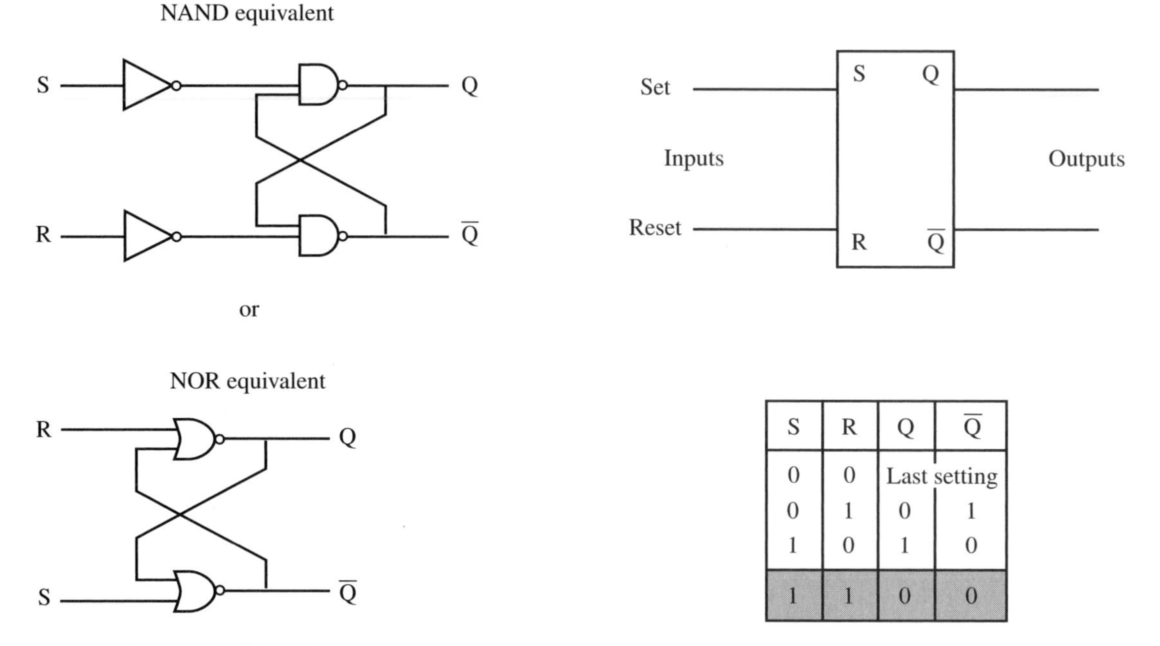

S	R	Q	$\overline{Q}$
0	0	Last setting	
0	1	0	1
1	0	1	0
1	1	0	0

Figure 20-52. Typical R-S flip-flop, truth table, and two equivalent flip-flops circuits. The shaded area in the truth table indicates a not-valid condition.

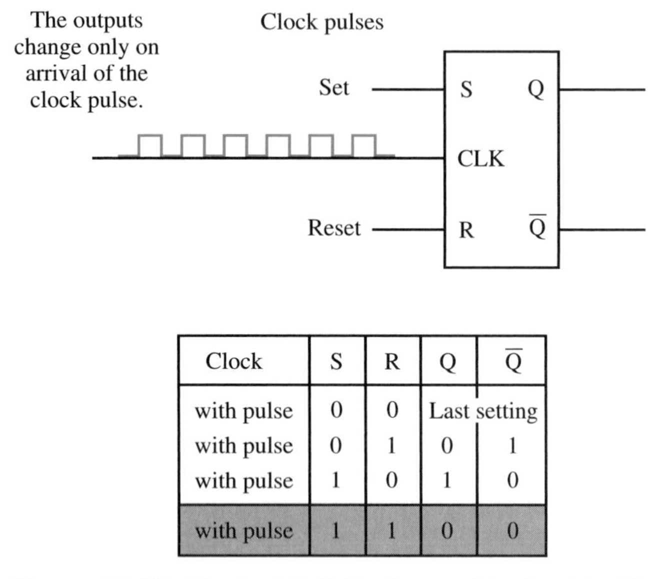

Clock	S	R	Q	$\overline{Q}$
with pulse	0	0	Last setting	
with pulse	0	1	0	1
with pulse	1	0	1	0
with pulse	1	1	0	0

Figure 20-53. Clocked R-S flip-flop and truth table. The shaded area in the truth table indicates a not-valid condition.

flop will retain its output status when two lows are present at its inputs. Also, when both inputs are high, the outputs will toggle on and off. See **Figure 20-54.**

D flip-flop

The ***D flip-flop,*** **Figure 20-55,** is similar to the J-K flip-flop except the D flip-flop does not require two inputs and the J-K does. When an input signal is received at the input, the Q outputs will toggle after a clock signal is applied. The output state of Q and $\overline{Q}$ will not change state until the clock signal is received.

By comparing the truth tables of the R-S, J-K, and D flip-flops you can see that the D flip-flop never has an unknown state, unlike the R-S and J-K. The R-S flip-flop has a not allowed state, and the J-K flip-flop has an output state that can not be determined unless the prior state of the flip-flop is known. D flip-flops do not have these problems.

D flip-flop output Qs are always complimentary. The J-K flip-flop can be made to simulate a D flip-flop by placing a not gate between its inputs.

Many flip-flops are used in binary counters. As you can see, the R-S, clocked R-S, D flip-flop, and the J-K flip-flop all have at least one of the two outputs high. They switch between these two states. Q is either high or low. Since the binary number system is composed of only 0s and 1s, you can see how the flip-flops might easily be used as the heart of a binary counting system.

Counters

A ***counter*** is a series of events. Events can be switches turning on and off or voltages rising and fallings. The events could be triggered by a pulse train or by some lever on a assembly line. The counter in **Figure 20-56** shows how a counter is fabricated from individual flip-flops. Counters are readily available as integrated

Clock	J	K	Q	$\overline{Q}$
with pulse	0	0	Last setting	
with pulse	0	1	0	1
with pulse	1	0	1	0
with pulse	1	1	T	T

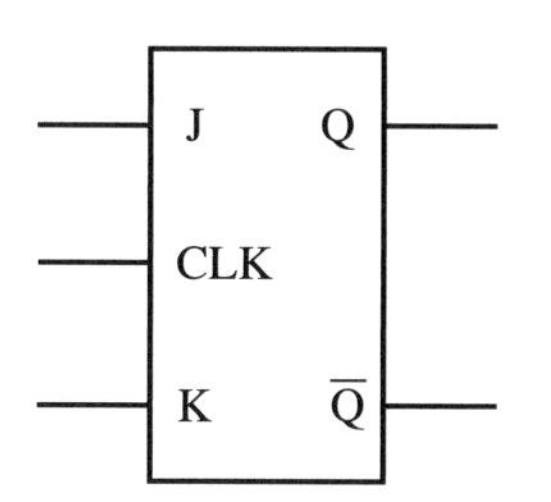

T = Toggle: outputs Q and $\overline{Q}$ both toggle on and off.

Figure 20-54. With the J-K flip-flop, if both data inputs are high, the outputs toggle on and off.

Clock	D	Q	$\overline{Q}$
with pulse	0	0	1
with pulse	1	1	0

Figure 20-55. The D flip-flop and truth table.

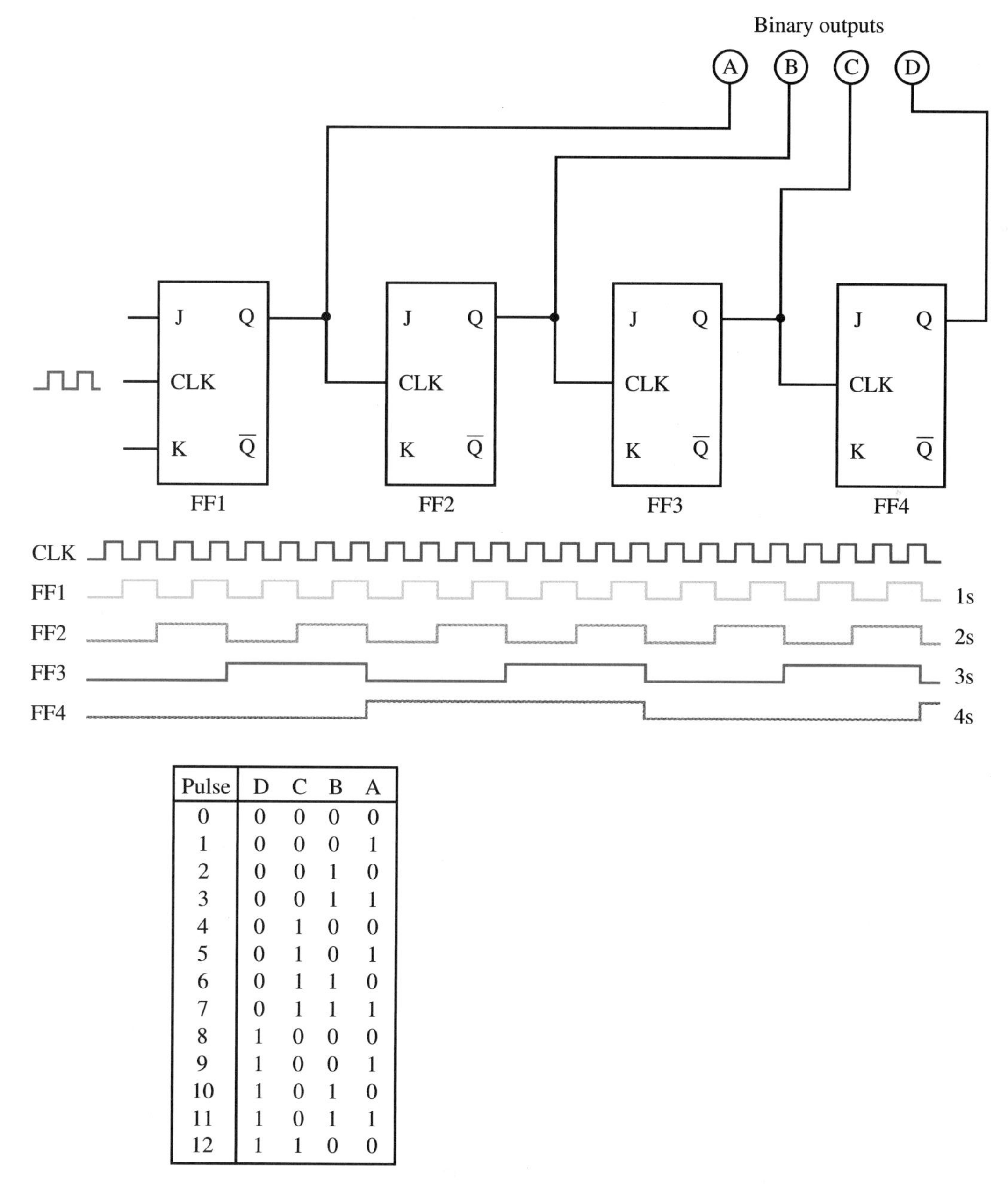

Pulse	D	C	B	A
0	0	0	0	0
1	0	0	0	1
2	0	0	1	0
3	0	0	1	1
4	0	1	0	0
5	0	1	0	1
6	0	1	1	0
7	0	1	1	1
8	1	0	0	0
9	1	0	0	1
10	1	0	1	0
11	1	0	1	1
12	1	1	0	0

Figure 20-56. Four J-K flip-flops connected as a ripple counter.

chips. One such counter is the 74HC193. It is an integrated circuit composed of flip-flops. Not only will it count up, (1, 2, 3, ...), but it will also count down (10, 9, 8, ...). The counter also has a *carry and borrow* functions. These principles are the basis of the common digital calculator.

Decade counter

The ***decade counter*** is a common digital counter. The term decade refers to its ability to count based on ten (a decade). A decade counter can also be used to divide by ten. Because of this feature, it is used often as a frequency divider for oscilloscopes or in a digital clock.

By connecting three decade counters together, a 60 Hz source can be used to make a digital clock to keep time. See **Figure 20-57.** The 60 Hz frequency is used to trigger a counter. Since the frequency is 60 Hz, the counter counts to 60 to equal one second. Sixty cycles of 60 seconds would signal one minute, and sixty minute cycles indicate one hour. Combine this with another counter that counts to 24 or 12, a decoder/driver, and four seven segment displays and you have constructed a time piece.

Review Questions for Section 20.4

1. The _____________ probe is the basic piece of test equipment for testing logic circuits.
2. A binary number can be converted to a decimal equivalent number using a(n) ___________.
3. What is the main difference between a J-K flip-flop and a D flip-flop?
4. What does the term complimentary mean in reference to a flip-flop?
5. In reference to flip-flops, the label CLK represents the term _____________.

Summary

1. There are two types of integrated circuits: linear and digital.
2. Digital circuits containing semiconductors use on/off or high and low pulse levels to activate them.
3. The binary numbering system has only two digits, 0 and 1.
4. Digital logic circuits are on (1) in the valid logic

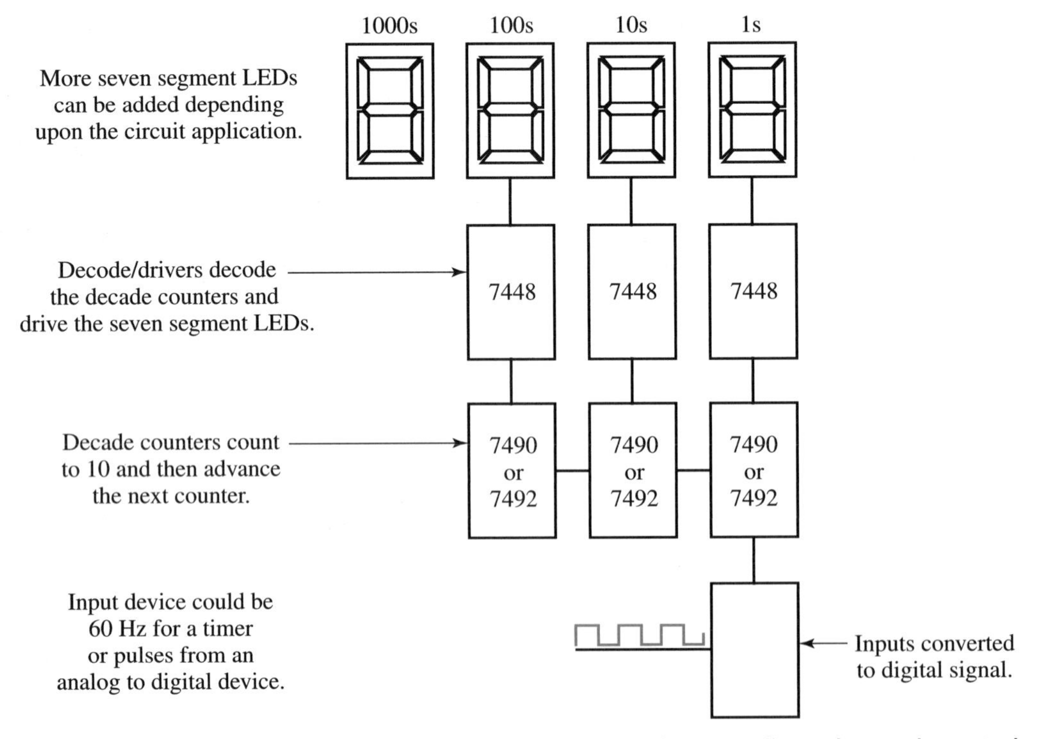

Figure 20-57. Decade counters can be used to create time clock circuits as well as electronic metering circuits. The seven segment displays can be used to indicate hours, minutes, seconds, or voltage values.

high voltage area and off (0) in the valid logic low voltage areas.
5. Basic logic gate circuits include:
   a. AND gate.
   b. OR gate.
   c. NOT gate.
   d. NAND gate.
   e. NOR gate.
   f. XOR gate.
   g. XNOR gate.
6. Digital ICs can perform all logic functions.
7. Truth tables show the condition of a gate's output, with varying input values.
8. The two main logic families for ICs are:
   a. Complimentary metal-oxide semiconductor (CMOS).
   b. Transistor-transistor logic (TTL).
9. Analog to digital devices convert analog signals to digital signals.
10. Digital to analog devices convert digital signals to analog signals.
11. Flip-flops have complementary outputs.
12. Some common types of flip-flops are R-S, J-K, and D.
13. Flip-flops can be made into counters and memory devices.

Test Your Knowledge

Please do not write in the text. Place your answers on a separate sheet of paper.

1. The two state system in which circuits are either on or off is the __________ system.
2. Give the binary equivalents for decimal numbers 1 through 10.
3. Complete the following binary addition problems.
   a. 1101 + 1001.
   b. 010101 + 100110.
   c. 11 + 111.
   d. 111001 + 100001.
4. The __________ __________ __________ is the high value of 1. The __________ __________ __________ is the low value of 0.
5. Electronic switching circuits that decide whether inputs will pass to output or be stopped are called __________ __________.
6. Name two common logic families.
7. Draw and label the schematic symbol for a J-K flip-flop.

Matching questions

Match the following terms with the correct definitions.

   a. AND gate.
   b. OR gate.
   c. NOT gate.
   d. NAND gate.
   e. NOR gate.
   f. XOR gate.
   g. XNOR gate.

8. A not AND gate.
9. Exclusive NOR gate.
10. Provides an output of 1 if all inputs are 1.
11. Inverts the polarity of the input signal.
12. Exclusive OR gate.
13. Provides an output of 1 when either input is 1.
14. A not OR gate.

For Discussion

1. What advantages do integrated circuits have over individual transistors?
2. Name the basic types of integrated circuits.
3. What is the difference between a NOT and a NOR logic gate?
4. What are the basic differences between the operational amplifier circuit symbol and the NOT gate?
5. What advances do you think will be made in the computer field in the next 25 years?

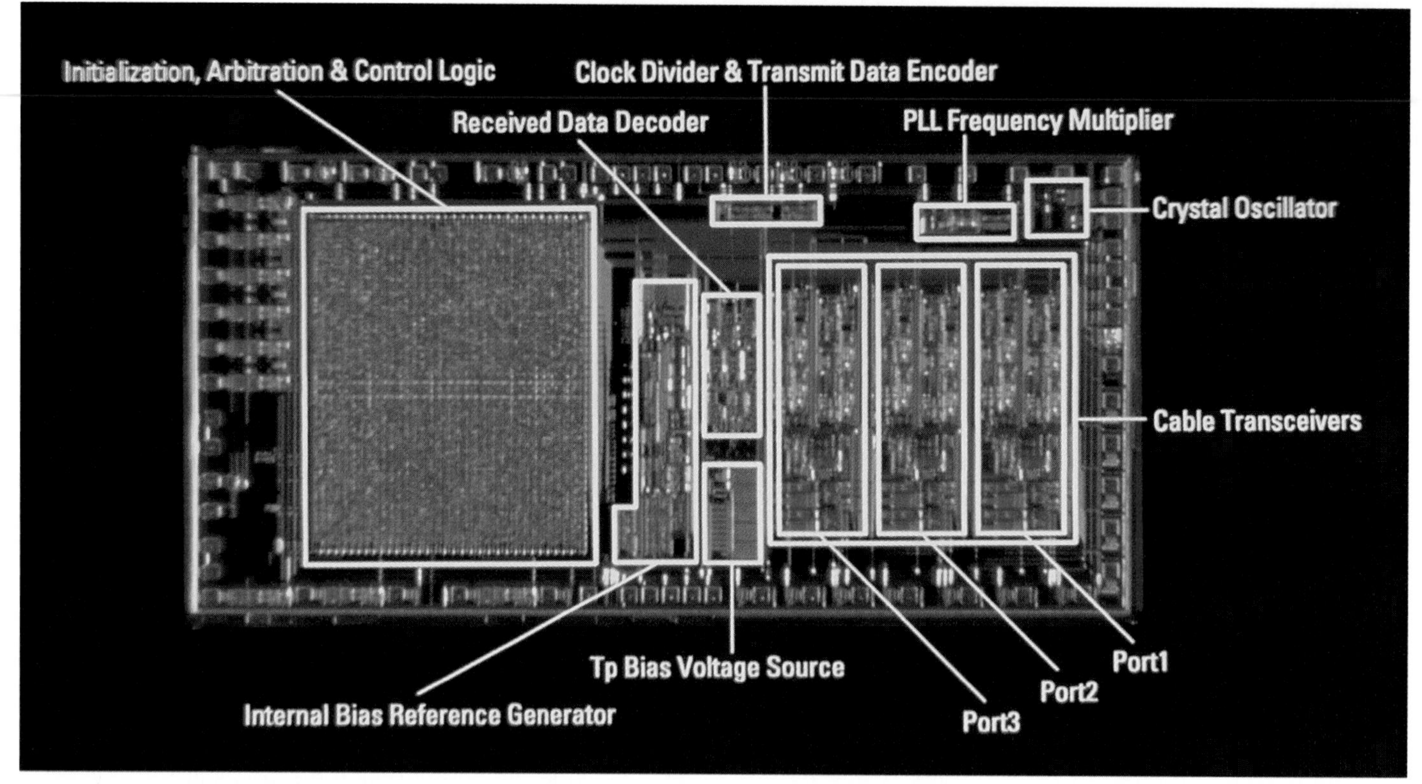

Oscillators are easily incorporated into integrated circuits. Note the crystal oscillator at the upper right in this high-performance bus. (Texas Instruments)

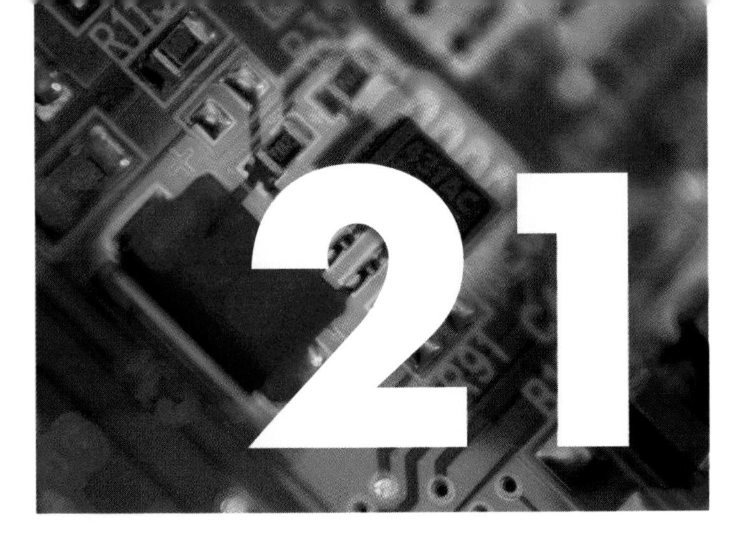

Oscillators

Objectives

After studying this chapter, you will be able to:

- Explain what occurs during an oscillation cycle.
- Indentify various oscillators.
- Discuss and compare the Armstrong oscillator and the Hartley oscillator.
- Outline the operation of the crystal oscillator and the power oscillator.

Key Words and Terms

The following words and terms will become important pieces of your electricity and electronics vocabulary. Look for them as you read this chapter.

Armstrong oscillator	Pierce oscillator
Colpitts oscillator	piezoelectric effect
crystal controlled oscillator	positive feedback
crystal oven	power oscillator
cycle	regenerative feedback
ganged capacitor	series fed oscillator
Hartley oscillator	shunt fed oscillator
oscillator	signal generator
period	

The pendulum on a grandfather's clock swings back and forth to keep time. It marks the time in seconds. The main spring that moves the pendulum is wound with a key. As the spring unwinds, the pendulum swings. How does the pendulum keep the correct time? Adjustments are made to the pendulum length so the time required for one complete swing matches one second.

The swinging pendulum can be thought of as an oscillator, **Figure 21-1.** An ***oscillator*** is an electronic circuit that generates an ac signal at a desired frequency. An oscillator circuit is usually made up of a wave producing circuit, amplifier, and feedback circuit. Variations in oscillator design are usually found in the feedback circuit. An oscillator is essential in many electronic applications.

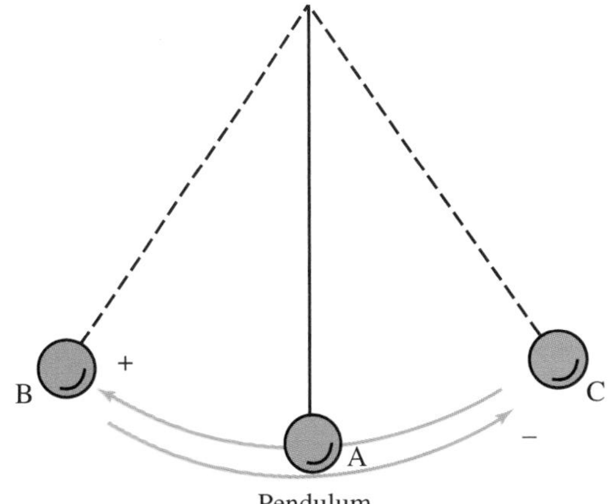

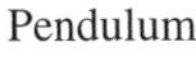

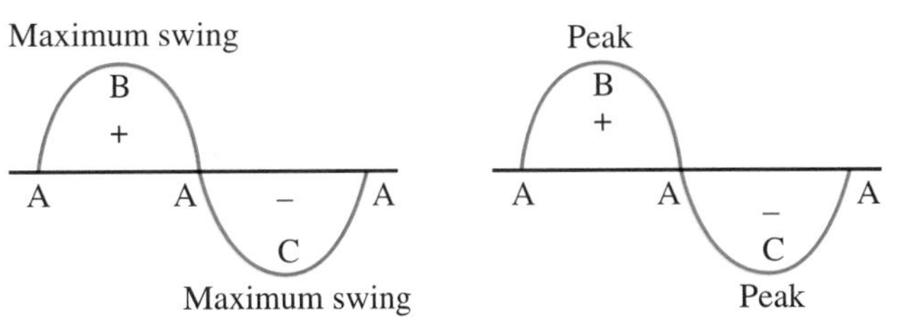

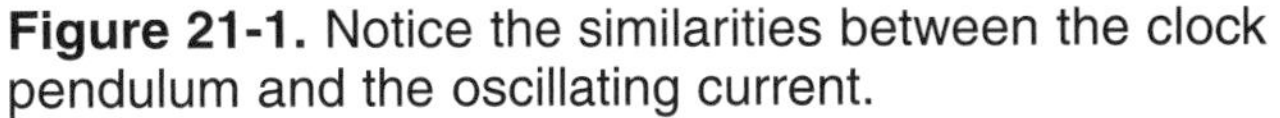

Figure 21-1. Notice the similarities between the clock pendulum and the oscillating current.

The oscillator is the very heart of radio transmission, microwave communications, radar, and much more.

21.1 BASIC OSCILLATORS

An oscillating current is one that flows back and forth. It moves first in one direction and then the other. You can compare the oscillator circuit output to a typical ac sine wave. The oscillator circuit changes dc from a power supply input into an ac waveform. The ac waveform maintains a steady frequency using a feedback circuit.

Follow the voltage amplitude in Figure 21-1. It starts from its reference line and rises to its peak in one direction and then falls to zero. Then it rises to its peak in the opposite direction and returns to zero. One ***cycle,*** a complete set of events in a repeated series, has been completed.

As oscillation continues, it repeats this cycle. The time that passes during one cycle is called the ***period*** of the cycle. The number of cycles occurring per second is measured and given as the frequency in hertz.

We learned earlier that the electricity used in homes and factories is an alternating, *or oscillating,* current. In the U.S. it alternates at a frequency of 60 hertz. The current is generated by dynamos driven by steam, water, or atomic power.

In our studies of electronics, voltages and currents of much higher frequencies are used. These are generated with semiconductor devices used as oscillators. These devices do not actually oscillate, but they act as valves. These valves feed energy to tuned circuits to maintain the oscillation. The basic block diagram of an oscillator is shown in **Figure 21-2.**

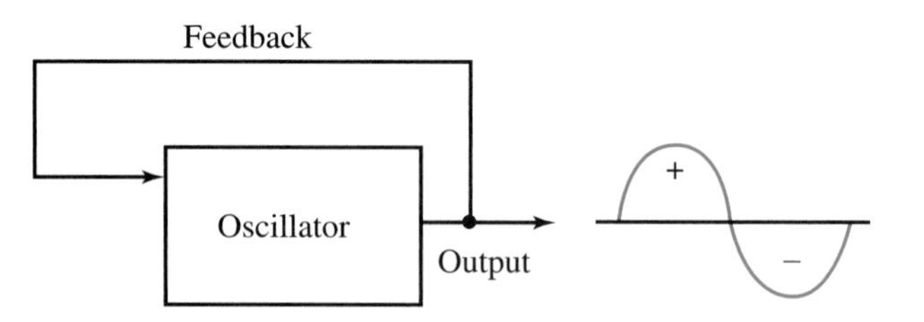

Figure 21-2. Block diagram of a sine wave oscillator.

Two conditions must exist to sustain oscillation in a tuned circuit:

- The energy fed back to the tuned circuit must be in phase with the first voltage. The oscillator depends upon this ***regenerative feedback,*** or ***positive feedback.***
- There must be enough feedback voltage amplitude to replace the energy lost by circuit resistance.

In Chapter 16, the tank circuit was discussed. This would be a good time to review Chapter 16, especially the tank circuit. The principles of the tank circuit operation are an essential part of the timing and feedback portion of many oscillator circuits.

Armstrong Oscillators

An ***Armstrong oscillator*** is shown in **Figure 21-3.** From this circuit, the basic theory of oscillators can be explained. Notice the tuned tank circuit $L_1 \| C_1$. This determines the frequency of the oscillator. Follow the sequence of events in this circuit.

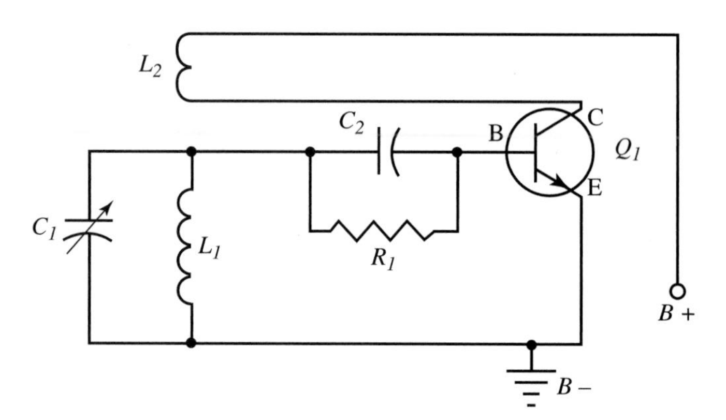

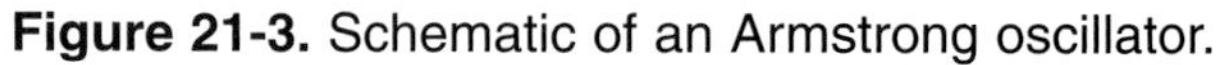

Figure 21-3. Schematic of an Armstrong oscillator.

Step 1. Examine **Figure 21-4.** When the voltage is applied to the circuit, current flows from $B-$, through the transistor and coil L_2, to $B+$. L_2 is sometimes called a tickler coil because it provides feedback to L_1. L_2 is closely coupled to L_1. The expanding magnetic field of L_2 makes the collector end of L_1 positive. C_1 charges to the polarity shown. The base of Q_1 also collects electrons. It charges C_2 in the polarity shown.

Step 2. Examine **Figure 21-5.** When Q_1 reaches its saturation point, there is no longer a change of current in L_2. Magnetic coupling to L_1 drops to zero. The negative charge on the base side of C_2 is no longer opposed by the induced voltage of L_1. The negative charge drives the transistor to cutoff. This rapid decrease in current through the transistor and L_2 causes the base end of L_1 to become negative. This *increases* the negative bias on Q_1. C_1 discharges through L_1 as the first half cycle of oscillation. C_2 bleeds off its charge through R_1.

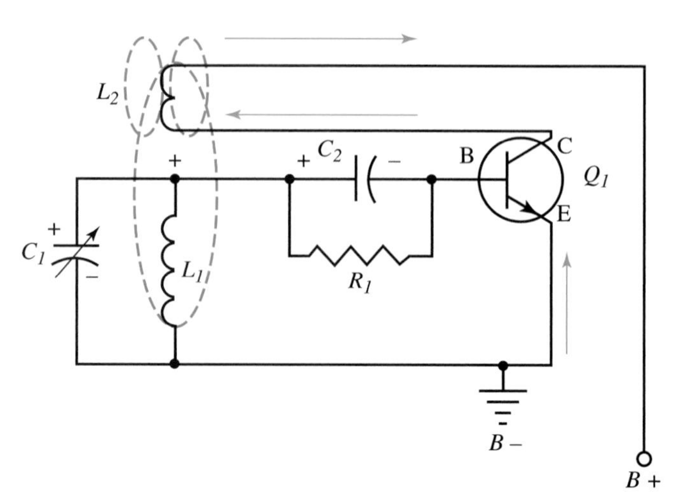

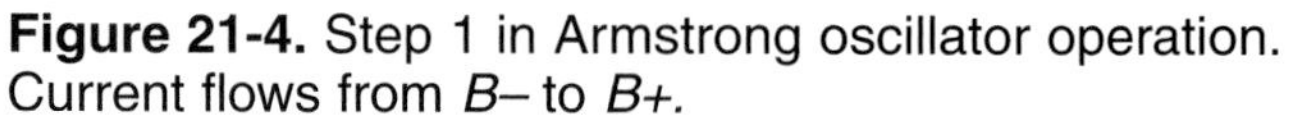

Figure 21-4. Step 1 in Armstrong oscillator operation. Current flows from $B-$ to $B+$.

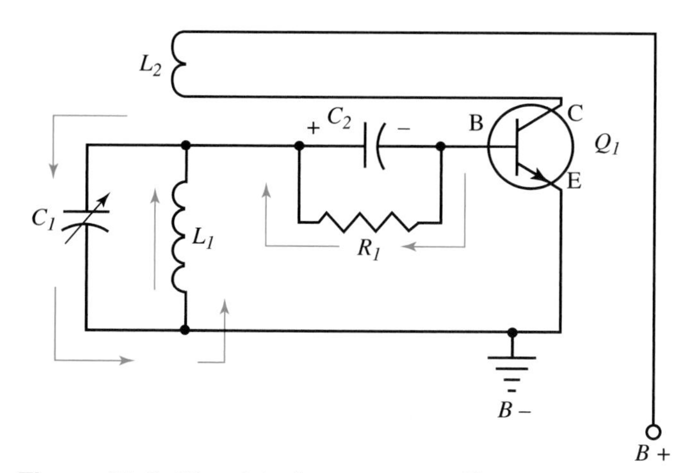

Figure 21-5. Step 2 in Armstrong oscillator operation. The first half-cycle of oscillation.

Step 3. The transistor, Q_1, is held at cutoff until the charge on C_2 is bled off to above cutoff. At that time the transistor starts conduction and the cycle is repeated.

There are a few points to remember in Armstrong oscillator operation.

- The voltage developed across L_1 first opposes, and then adds to, the bias developed by the $R_1||C_2$ combination.
- The energy added to the tuned tank circuit $L_1||C_1$ by the tickler coil L_2 is great enough to offset the energy lost in the circuit due to resistance. The coupling between L_1 and L_2 can be adjusted.
- The combination $R_1||C_2$ has a somewhat long time constant. It sets the operating bias for the transistor. Q_1 is operated Class C.

Study the voltage waveform on the base of Q_1 in **Figure 21-6.** The shaded portion is transistor conduction. At point B the bias is negative. This results from the charge on C_2 plus the induced voltage across L_1. The interval from B to C denotes the time that passes while C_2 discharges through R to the cutoff point, and conduction begins for the next cycle.

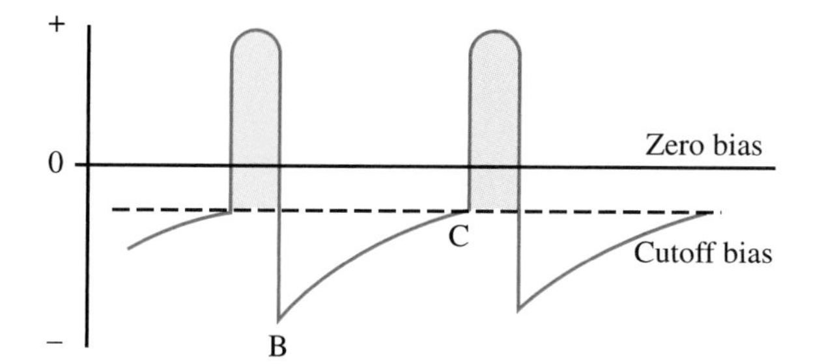

Figure 21-6. Voltage waveform on base of oscillator. Compare it to steps 1, 2, and 3.

Hartley Oscillators

An oscillator used commonly in radio receivers and transmitters is the Hartley oscillator. It is more stable than the Armstrong, but the theory of its operation is similar. The ***Hartley oscillator*** is set apart from other oscillators by the tapped coil, L_1 and L_2 in **Figure 21-7.**

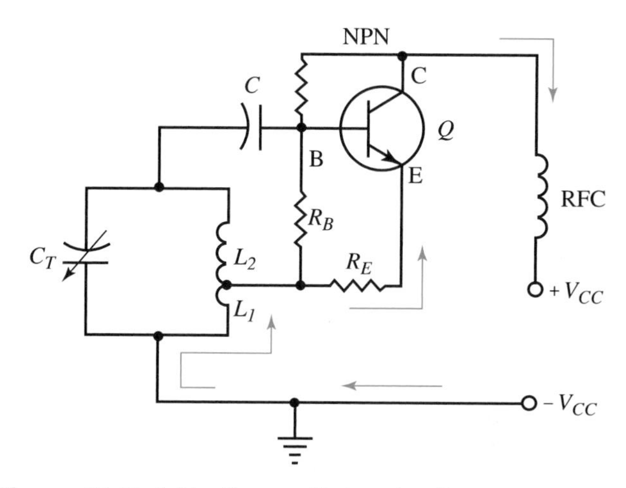

Figure 21-7. A Hartley oscillator circuit.

The Hartley parts in Figure 21-7 are labeled in a manner similar to the labeling used for the Armstrong oscillator in Figures 21-4 and 21-5. The L_1 section of the coil is in series with the emitter-collector circuit. It carries the total collector current.

The current I_E, which includes I_C, is shown by the arrows. When the circuit is turned on, current flow through L_1 induces a voltage on L_2. The voltage induced on L_2 makes the base of Q more positive and drives the transistor to saturation. Once the transistor is driven to saturation, the current ceases flowing. The magnetic coupling between coils L_1 and L_2 collapses to zero. The less positive voltage at the base of Q causes the transistor to decrease conduction. This decrease induces a negative voltage at the top end of L_2, which is reverse bias for the transistor. The transistor is quickly driven to cutoff and the cycle is then repeated. The tank circuit is oscillating, which causes the transistor to switch between saturation and cutoff. The switching action of the transistor is the frequency of the oscillator.

The resting bias condition of the transistor is set by resistors R_B and R_E. The radio frequency choke, RFC in the figure, blocks the RF signal from the power source. In this circuit, note that coil L_1 is in series with the transistor collector circuit. It is a ***series fed oscillator.***

In **Figure 21-8,** a ***shunt fed oscillator*** is shown. Operation is the same. Note that the dc path for the emitter-collector current is not through coil L_1. The ac signal path, however, is through C and L_1. At point A, the two current components are separated and required to take parallel paths. Both oscillators receive their feedback energy through magnetic coupling.

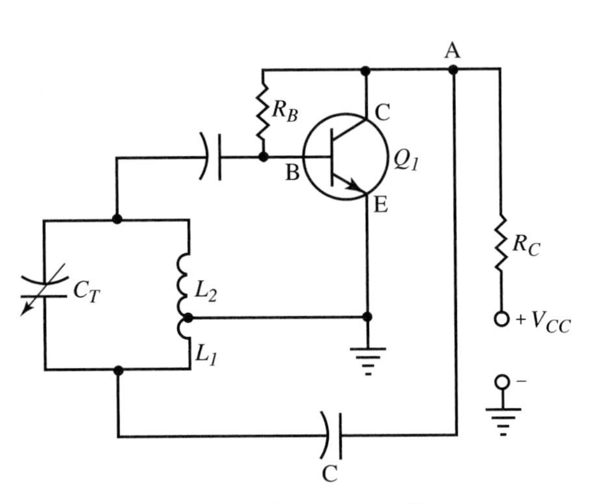

Figure 21-8. A shunt fed Hartley oscillator.

Review Questions for Section 21.1

1. What is the purpose of an oscillator?
2. The time that passes during one cycle of an oscillating current is called the __________ of the cycle.
3. What two conditions must exist to sustain oscillation in a tuned circuit?
4. The __________ oscillator has a tickler coil.
5. The __________ oscillator has a tapped coil.
6. The purpose of the tickler coil is to provide __________ to the tank circuit.

21.2 OTHER OSCILLATORS

There are quite a number of other types of oscillators you will encounter in the study of electronics. The following section discusses:

- Colpitts oscillators.
- Crystal controlled oscillators.
- Power oscillators.
- Operation amplifier oscillators.

As you study these oscillators, you will notice that they are similar in many ways to the two basic oscillators you have just studied.

Colpitts Oscillators

Feedback can also be created with an electrostatic field such as that found in a capacitor. Replace the tapped coil from the Hartley oscillator with a split stator capacitor. A voltage of proper polarity will be fed back causing the circuit to oscillate. This circuit is called a ***Colpitts oscillator,*** **Figure 21-9.**

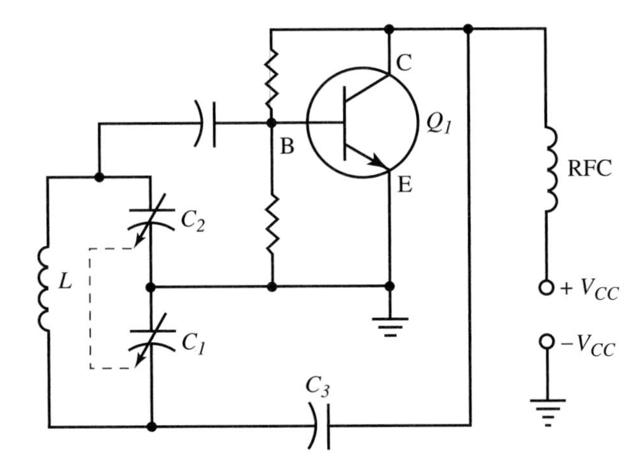

Figure 21-9. Schematic diagram of a Colpitts oscillator.

Operation of this oscillator is like that of the Hartley oscillator. However, the signal is coupled back to C_1 of the tank circuit through coupling capacitor C_3. A changing voltage at the collector appears as a voltage across the tank circuit $L||C_1C_2$.

The tank circuit must be in the proper phase with the transistor circuit to be a regenerative signal. The amount of feedback will depend on the ratio of C_1 to C_2. This ratio is most often fixed. Both capacitors C_1 and C_2 are controlled by a single shaft (called a ***ganged capacitor***). The frequency of the oscillator is set in the common manner. The tuned tank consists of L and C_1 and C_2 in series. The circuit is shunt fed. Series fed is not possible due to the blocking of dc by the capacitors.

Crystal Controlled Oscillators

A circuit with a stable high frequency is the ***crystal controlled oscillator.*** It is used in radio communications, broadcasting stations, and in equipment requiring a fixed frequency with little drift.

You learned earlier that an emf can be made with mechanical pressure and/or distortion of certain crystalline substances. The opposite is also true. A voltage applied to the surface of a crystal will cause distortion in the crystal. These effects are called ***piezoelectric effects.*** When electrical pressure is applied to a crystal, it will oscillate. The frequency of oscillation depends on the size, thickness, and kind of crystal used.

Look at **Figure 21-10.** The crystal is precisely cut and mounted on two connection leads using conductive

Radio Detection and Ranging (Radar)

The first major contribution to the development of radar dates back to the 1860s. James C. Maxwell predicted the existence of electromagnetic waves that travel at the speed of light. He also proposed the possibility of generating this type of wave. In the late 1880s, Heinrich Hertz proved Maxwell's ideas correct by producing radio waves and demonstrating that such electromagnetic waves could be reflected from solid objects.

Advances in radar technology continued throughout the early 1900s but the growing threat of war in the 1930s stimulated efforts to improve radar technology. A chain of radar stations were built along England's east and south coasts before World War II began, and by 1940 the U.S. was producing pulse-type radar for tracking planes and controlling antiaircraft guns. The radar units operated on the simple principle that microwaves directed toward an object will reflect off the object and back to the transmitter. The amount of time it takes the wave to travel to and from the object is in direct proportion to the distance the object is from the transmitter.

Although radar sets differ in size, most have similar parts including an oscillator, modulator, transmitter, duplexer, antenna, receiver, signal processor, display, and timer. The oscillator circuit of the radar unit produces microwaves. *Microwaves* are very short electromagnetic waves with frequencies of one GHz or higher. Like beams of light, microwaves are easily manipulated in terms of width and direction, and they will reflect off different materials. The intensity of the reflected wave depends on the type of material it strikes.

A microwave is transmitted out of the radar dish in short, but very high, energy pulses. When the wave strikes a reflective material, like the metallic skin of a helicopter, the wave is reflected in all directions. Some of the wave is reflected back to the radar dish.

The original radar units plotted only the distance of an object using an oscilloscope. Today's radar units can measure the angle of the object in relation to the radar unit itself as well as its distance. A servo unit, mounted in the base and arm of the radar unit, can be used to determine the exact elevation angle of the object and the angle of rotation of the radar dish when the echo returns. This information is fed into a computer and displayed on a screen showing the relative distance and altitude of the object. The strength of the reflected echo can also help identify the type of aircraft being tracked.

The most common type of radar is pulsar radar. *Pulsar radar* emits signals in powerful bursts that last only a few millionths of a second. *Continuous-wave radar* sends out a continuous signal rather than short bursts. Doppler and frequency modulated (FM) radar are both continuous-wave radar. *Doppler radar* operates on the basis of the Doppler effect and is used chiefly to make precise speed measurements. *FM radar* transmits a continuous wave while rapidly increasing or decreasing the frequency of the signal at regular intervals. As a result, FM radar can determine the distance to a moving or stationary object.

One of the first countermeasures devised for eluding radar detection was the dropping of *chafe* (long thin strips of aluminum foil). When dropped from a plane, chafe drifts through the air and circulates on the turbulence created by the aircraft. Each strip reflects radar signals and causes difficulty in recognizing echoes from the real planes. Other countermeasures include the use of high-powered radio transmitters to produce interference and the use of equipment that can modify pulses before they are returned. The U.S. Air Force has also designed two planes, the *stealth bomber* and the *stealth fighter,* that are almost "invisible" to radar.

The design of the stealth figher includes many perfectly flat planes or facets. The total reflective surface *facing* toward the radar unit is almost nonexistent. By comparison, the rounded shape of conventional aircraft provides an easily detectable reflective surface. Since the sharp and flat angles of the stealth aircraft are not in line with the radar transmitter unit, radio waves are not reflected in the direction of the radar unit. (U.S. Air Force)

cement. The entire crystal in then enclosed in a metallic can that is filled with dry nitrogen gas.

Figure 21-11 shows a circuit made of electrical components that is equivalent to a crystal. A crystal is placed between two metallic holders. This forms a capacitor C_H. The crystal itself is the dielectric. C_G denotes the series capacitance between the metal holding plates and the air gap between them (as a dielectric). *L*, *C*, and *R* denote the traits of the crystal. Note the likeness of the equivalent crystal circuit to a tuned circuit. Both, a tuned circuit and a crystal will have a resonant frequency.

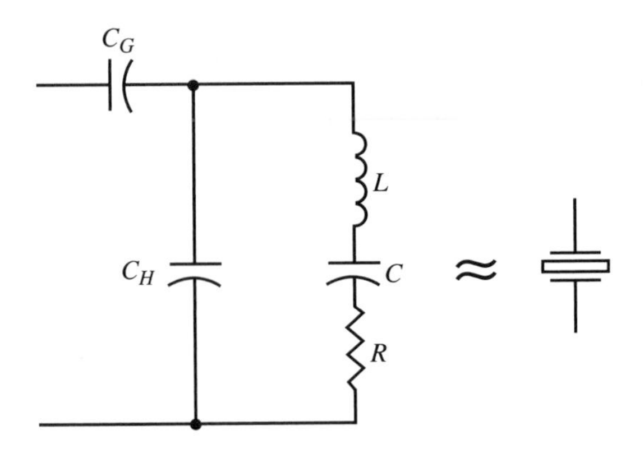

Figure 21-11. The electrical circuit on the left is roughly equivalent to a piezoelectric crystal (schematic symbol shown at right).

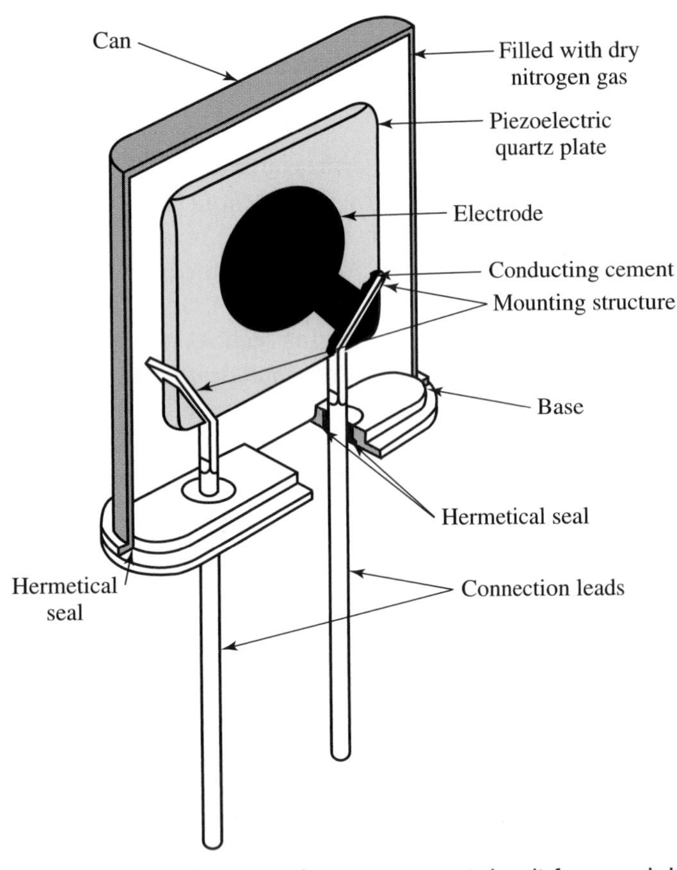

Figure 21-10. Section of a quartz crystal unit for special and industrial applications.

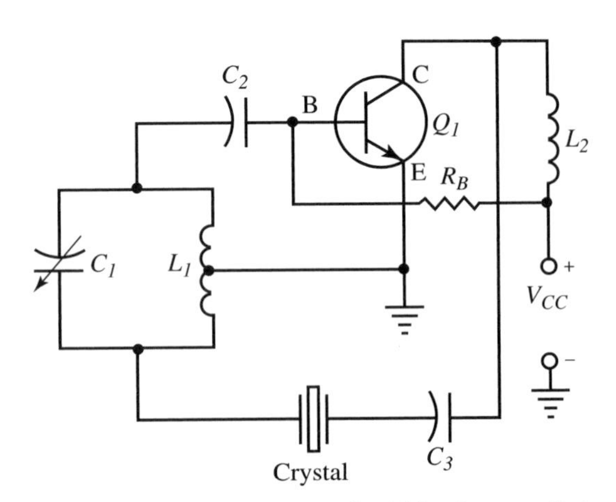

Figure 21-12. A crystal controlled Hartley oscillator circuit.

Crystals are used in amateur radio and commercial broadcasting stations to control the transmitter frequency. The frequency generated by the crystal is susceptible to temperature changes. In a commercial broadcasting station, crystals used to control transmitter frequency are placed in crystal ovens. The ***crystal oven*** maintains a constant temperature, which will in turn stabilize the frequency produced by the crystal.

A crystal oscillator circuit is shown in **Figure 21-12.** Compare this circuit to Figure 21-8. It is the same circuit with the crystal added to the *feedback circuit*. The crystal acts as a series resonant circuit. It sets the frequency of the feedback currents. The tank circuit must be tuned to this frequency.

Figure 21-13 the crystal is used in place of the inductor in the tank circuit of the Colpitts oscillator. This is the ***Pierce oscillator.*** Compare the circuit to Figure 21-9. The amount of feedback needed to energize the crystal, again, depends on the ratio of C_1 to C_2. These capacitors form a voltage divider across the emitter base of the transistor. The Pierce circuit is stable under changing circuit conditions.

Additional uses for crystals

A crystal changing frequency because of a change in temperature is undesirable in an oscillator circuit. However, this trait can be used to our advantage as well. The frequency developed by a quartz crystal varies in *direct proportion* to the temperature to which it is exposed.

Figure 21-14 shows a simple circuit for measuring temperature. Two crystals are used. One crystal is used as a reference, and the other crystal is exposed to the tem-

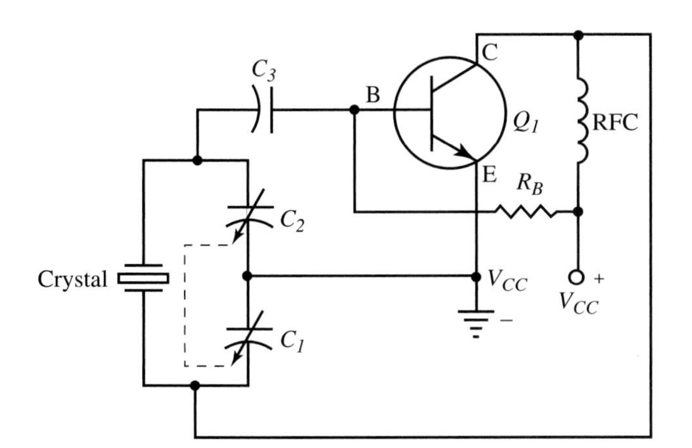

Figure 21-13. A crystal controlled Pierce oscillator circuit.

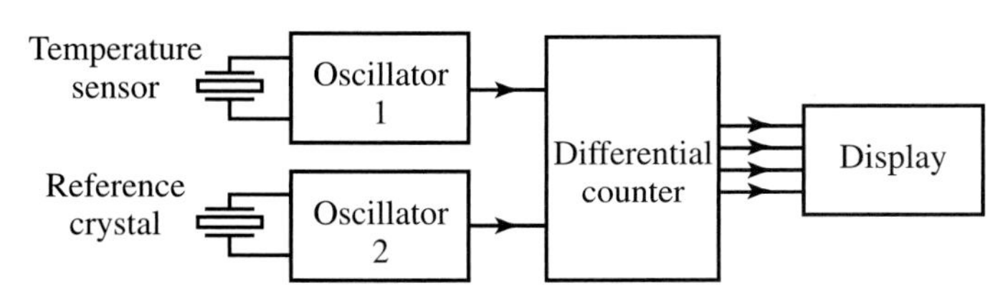

Figure 21-14. Quartz crystal used for temperature measurement.

perature being measured. The crystals will produce two different frequencies, which are input to the differential counter. The differential counter is a digital integrated circuit. The two frequencies are compared and then can be displayed on a number of different devices such as analog meter, digital meter, LED display, etc. Accuracy of this system can be up to ±0.001° Kelvin.

Quartz crystals are also used in timing devices because of their stability and accuracy. A ***signal generator*** is an electronic oscillator that generates various signals for testing, **Figure 21-15.**

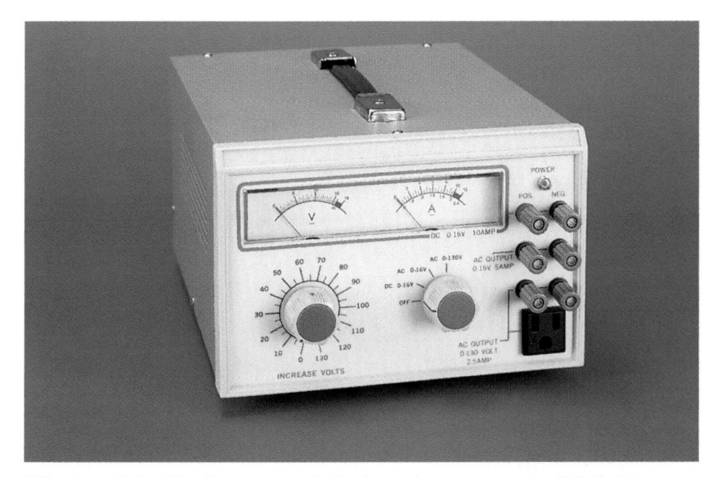

Figure 21-15. Commercial signal generator. (Knight Electronics)

Power Oscillators

Figure 21-16 shows a ***power oscillator*** circuit. This schematic is actually one of a push-pull oscillator. The collector load of each transistor is the primary of the transformer. The ac output is found at the secondary. A proper turns ratio is used if higher or lower voltages are needed.

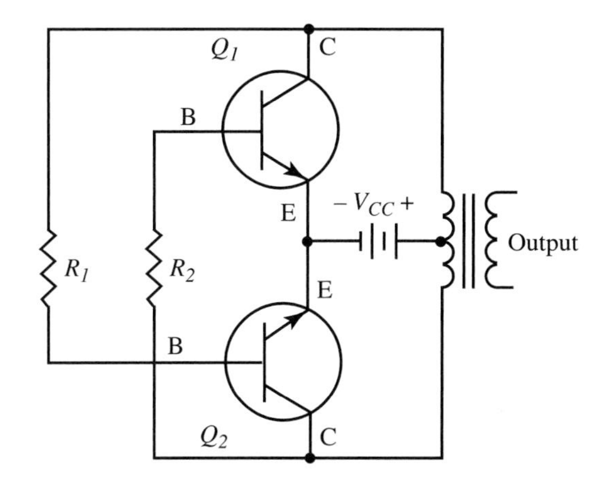

Figure 21-16. Power oscillator circuit.

A slight imbalance in conductivity between Q_1 and Q_2 will start oscillation. This imbalance is always present due to variances in transistor characteristics or temperature. This will send one of the transistors toward saturation and the other towards cutoff.

Assume Q_1 starts conducting. The voltage at C of Q_1 goes less positive. This makes the base of Q_2 less positive. Q_2 drives toward cutoff. A more positive voltage at C of Q_2 drives the Q_1 base more positive. Q_1 reaches saturation. When there is no change in current at saturation, transformer primary reactance drops to zero. Collector voltage toward the value of V_{CC} also increases. This more positive voltage coupled to the base of Q_2 through R_1 starts Q_2 toward conduction and saturation. The transistors conduct one at a time. The output is combined into a complete cycle at the transformer secondary output.

If the output from the transformer in Figure 21-16 is connected to a rectifier and some filter circuits, the output again becomes dc.

Operational Amplifier Oscillator Circuits

The operational amplifier (review Chapter 19) can be used in many oscillator circuits in place of a transistor in such circuits as the Hartley or Colpitts oscillator. See **Figure 21-17.** Notice that the crystal is used in the feedback circuit of the op amp. The charging circuit—R_1, R_2,

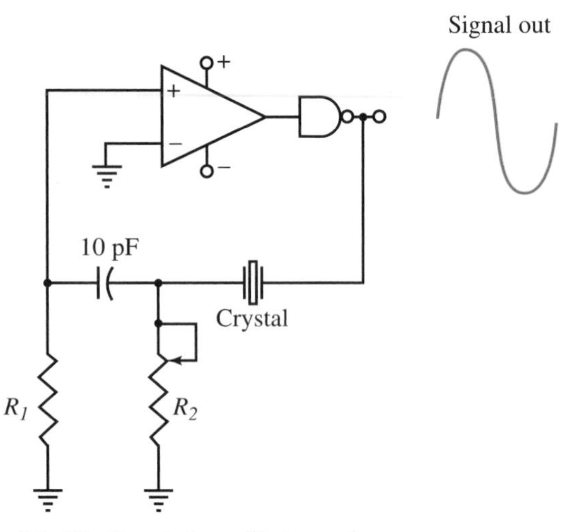

Figure 21-17. Crystal oscillator using an op amp.

and the capacitor—determine the amount of feedback and wave shaping. The NAND gate is used to start the output from the op amp. Without it, the circuit would need some exterior input to start the circuit.

Another oscillator that can utilize the op amp is the Wien bridge oscillator. See **Figure 21-18.** This circuit operates using both positive and negative feedback. The negative and positive feedback circuits must be balanced for the oscillator to work properly. The balancing is achieved by inserting the tungsten filament lamp in the circuit as part of the feedback system. The resistance of the lamp varies with the amount of current in the lamp filament. As current increases, temperature of the filament increases and so too the resistance value of the filament. The inverse is also true. As current decreases, temperature decreases and in turn the resistance value of the filament also decreases. This principle is the secret to balancing the amount of feedback in the circuit.

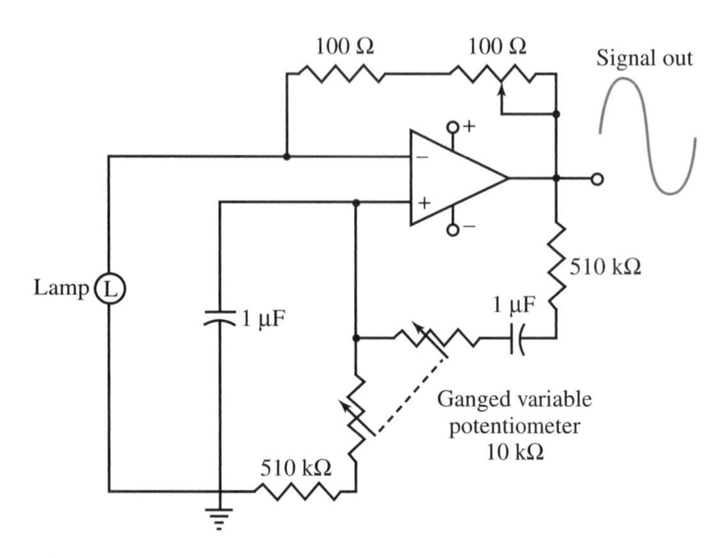

Figure 21-18. The Wien bridge oscillator

Many types of oscillators are formed on a single linear integrated chip. Only a few extra components are needed to produce a working oscillator. Some are voltage controlled oscillators (VCO), **Figure 21-19.** The frequency output of the oscillator is in direct proportion to the voltage level applied to one of the VCOs pins.

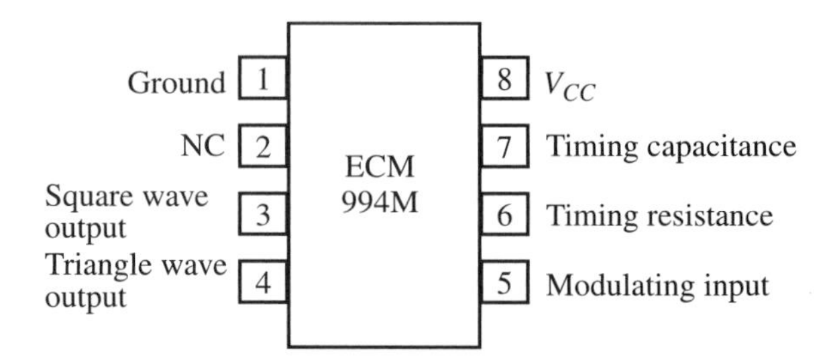

Figure 21-19. A general purpose voltage controlled oscillator (VCO) function generator. This chip can be used as a voltage controlled oscillator or as the wave shaping device for a function generator.

Review Questions for Section 21.2

1. Why is series feeding impossible in a Colpitts oscillator?
2. What is the piezoelectric effect?
3. A(n) __________ oscillator uses a crystal in place of the tuned circuit found in a Colpitts oscillator.
4. Draw the circuit for a power oscillator.

Summary

1. An oscillator is an electronic circuit that produces an ac signal at a desired frequency.
2. A cycle is a complete set of events in a repeated series for an ac signal.
3. Frequency is the number of cycles per second. It is measured in hertz.
4. An oscillator circuit is usually made up of a wave producing circuit, amplifier, and feedback circuit. The feedback circuit usually causes variations in oscillator designs.
5. The Armstrong oscillator has a tickler coil for feedback.
6. The Hartley oscillator has a tapped coil in the tank circuit.
7. A Colpitts oscillator has two capacitors in the tank circuit with a tap between them.
8. Crystal oscillators use the piezoelectric effect in crystals to maintain accurate frequencies.
9. Crystal oscillator circuits can be used for timing devices and temperature recording.

10. Operation amplifiers can be configured similarly to other oscillator circuits, taking the place of the transistor in most cases.

Test Your Knowledge

Please do not write in the text. Place your answers on a separate sheet of paper.

1. A(n) __________ is a complete set of events in a repeated series for an ac signal.
2. What is the period of a cycle?
3. The __________ __________ circuit determines the frequency in the Armstrong oscillator.
4. Outline the steps in the operation of the Armstrong oscillator.
5. What feedback method is used in the Hartley oscillator?
6. What feedback method is used in the Colpitts oscillator?
7. Draw the equivalent electrical circuit of a crystal.
8. What three types of electronic applications can the crystal oscillator circuit be used for?

For Discussion

1. How does an electronic oscillator compare to a clock pendulum?
2. Can a piezoelectric crystal be used to produce electricity if pressure is applied?
3. Explain the operation of a Hartley oscillator.
4. What is the difference between a series fed oscillator and a shunt fed oscillator?
5. Why are crystal oscillators used in many commercial transmitters?
6. How can a Colpitts oscillator be identified?

SECTION 5

Electronic Communication and Data Systems

Communication is the process of exchanging information. Electronics provides a pathway for the rapid exchange of information. The transmission of information can be through cables or across the airwaves. The various ways this task is accomplished are explored in this section.

In Chapter 22, AM and FM communication systems are explained. The transmission of electromagnetic waves are explored. You will also see how sound is converted into an electrical wave and then converted back to sound again. Chapter 23 explains the production, transmission, and reception of television signals. Both black-and-white and color systems are examined. You will also explore improvements to the conventional television. Included are satellite reception, the VCR, remote control, and high definition television.

Chapter 24 covers the rapidly expanding field of fiber optics. Fiber optics is steadily replacing conventional copper wire systems, and you will see why. Chapter 25 introduces you to the basic concepts and terminologies of the personal computer and its components. Chapter 26 covers one of the most powerful and widely used integrated circuit, the microcontroller. You will discover why the microcontroller is used in so many applications such as appliances, vehicles, and wherever a system must be controlled.

After you have sampled the world of electronics you may be ready to pursue further schooling or go directly to work. Chapter 27 is written to assist you with future choices in the electrical and electronics field.

AM and FM Radio Communications

Objectives

After studying this chapter, you will be able to:

- Discuss the history of radio wave receivers.
- Describe the different types of waves.
- Convert back and forth between frequency and wavelength.
- Describe several different types of microphone.
- Discuss amplitude modulation and frequency modulation.
- Calculate the percent of modulation.
- List the components and explain the operation of an AM receiver and transmitter.
- List the components and explain the operation of an FM receiver and transmitter.
- List the components and explain the operation of a superheterodyne receiver.
- Describe the use of transducers in radio wave receivers.

Key Words and Terms

The following words and terms will become important pieces of your electricity and electronics vocabulary. Look for them as you read this chapter.

amplitude modulation (AM)	heterodyning
broadcast band	intermediate frequency (IF)
carrier wave	lower sideband
center frequency	modulation
continuous wave (CW) transmitter	radio wave
demodulation	selectivity
detection	sensitivity
frequency deviation	significant sideband
frequency modulation (FM)	transducer
harmonic frequency	upper sideband

One of the greatest inventions of all time is the radio. Like many other inventions, the radio resulted from the work of many scientists. In 1864, James Maxwell theorized that electromagnetic waves existed. In 1887, Heinrich Rudolph Hertz confirmed this theory when he transmitted and received the first radio waves. The first continuous wave (CW) transmitter was developed in 1897 by Guglielmo Marconi.

Two key devices that furthered the development of the radio were the diode and the audion triode amplifier tube. The diode was invented in 1904 by Alexander Fleming. The amplifier tube was patented in 1907 by Lee DeForest.

A key scientist in radio history was Edwin H. Armstrong. He invented the regenerative radio circuit in 1913 and the superheterodyne radio circuit in 1918. Major Armstrong is also credited with the development of much of the FM radio theory.

In this chapter, the interesting concepts behind the broadcasting of sound will be explored. The concepts taught in this unit lay the foundations of many of today's communication systems. Microwave transmission, satellite communication, fiber optic systems, mobile telephones, electronic pagers, wireless computer modems, and television are just a few of these systems.

Many of the concepts and circuits explained in this section rely on concepts covered earlier in this text such as capacitance, inductance, RCL circuits, semiconductors, transistor amplifiers, and transformers. It is advisable to review any of these systems if difficulty is encountered in this chapter. This chapter will also lay all the foundation of the next chapter on television.

22.1 SIMPLE RADIO RECEIVER

For a basic understanding of radio and television operations, we will first look at a simple radio receiver. This radio receiver consists of very few parts, an antenna, a ground, a tank circuit, a diode, a filter, and a speaker or a set of headphones. In **Figure 22-1,** there are three radio stations each broadcasting at a different wavelength. Each station is broadcasting a radio signal consisting of a

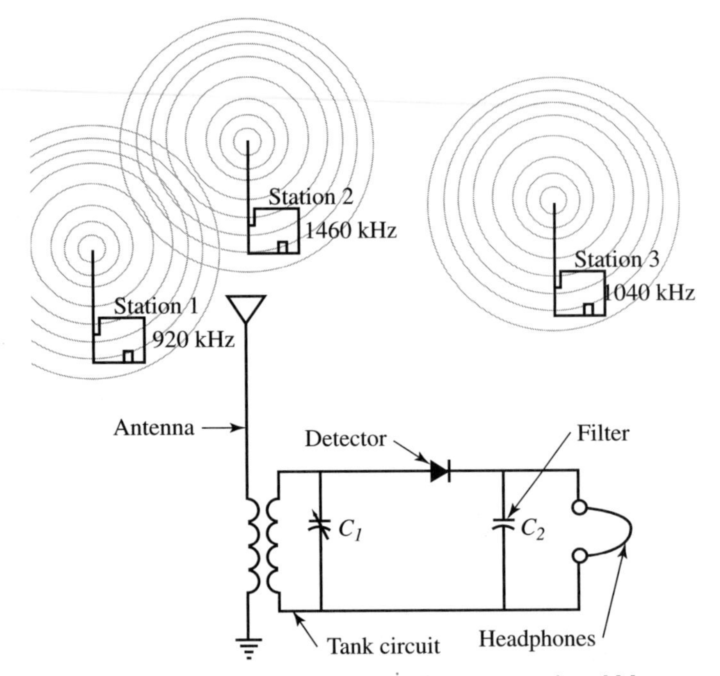

Figure 22-1. A simple crystal radio can receive AM radio signals and convert them to sound.

carrier wave and an audio signal. Station 1 is broadcasting at AM 920, station 2 at AM 1460, and station 3 at AM 1040.

The radio waves of all three stations come in contact with the radio receiver antenna. The antenna converts the radio signals to alternating current, which is conducted up and down the antenna to the ground. The antenna circuit is coupled to the tank circuit by mutual induction.

The tank circuit consists of an inductor and a variable capacitor connected in parallel. Recall from Chapter 16, that an inductor and capacitor connected in parallel will have a resonant frequency. By using a variable capacitor, you are able to vary the tank circuit resonant frequency until it matches the frequency of the desired station.

For example, if we wish to tune in station 1, the capacitor is varied until the resonant frequency of the tank circuit is equal to 920 kHz. Receiving a frequency of 920 kHz will cause the greatest voltage drop across the tank circuit. The other frequencies (1040 station 3, and 1460 station 2) will not produce a large voltage drop across the tank circuit.

The detector rectifies the radio signal to a pulsing dc signal. The filter capacitor smoothes the high frequency of the audio portion of the radio signal. The detector diode and filter capacitor are necessary to change the broadcast frequency and audio signal to a reproducible sound at the headphones.

The description above may sound simple, and that is because this is the simple operation of a radio receiver. The radio described is known as a ***crystal radio receiver.*** When constructed properly in the lab, you can actually receive and hear a few stations. The performance of this radio, however, is extremely poor by today's standards. Today's radios and televisions operate from the same *principles* just described, but they are a significant refinement of the crystal set. In this chapter, you will see how a better radio is developed by applying more of the electronic concepts you have learned thus far.

Take special note of the fact that there is no battery or other conventional power supply for this radio. First we will discuss the power source for this radio.

Radio Waves

A ***radio wave*** is electromagnetic radiation produced from current alternating through an antenna. A transmitting antenna is surrounded by electromagnetic radiation. In Chapter 9, in the study of electromagnetism, we learned that a conductor carrying an electric current is surrounded by a magnetic field. In a magnetic field created by an alternating current, the field expands, collapses, and changes polarity in step with the frequency. In Chapter 21, the oscillator was explained. An oscillator can produce high frequency alternating currents that produce a radio wave when connected to an antenna. In general, the radio wave is an electrostatic radiation of energy produced by an oscillator circuit.

The electrostatic field is perpendicular to the electromagnetic field. Both travel away from the antenna. As a result, a radio wave is made up of electromagnetic and electrostatic fields. See **Figure 22-2.** The direction these waves radiate, in respect to the earth, is called ***polarization.*** In **Figure 22-3** the waves are radiated from a vertical antenna. Note that the electrostatic, or E waves, are in the same plane as the antenna, yet perpendicular to direction of travel. The vertically polarized waves are perpendicular to the surface of the earth.

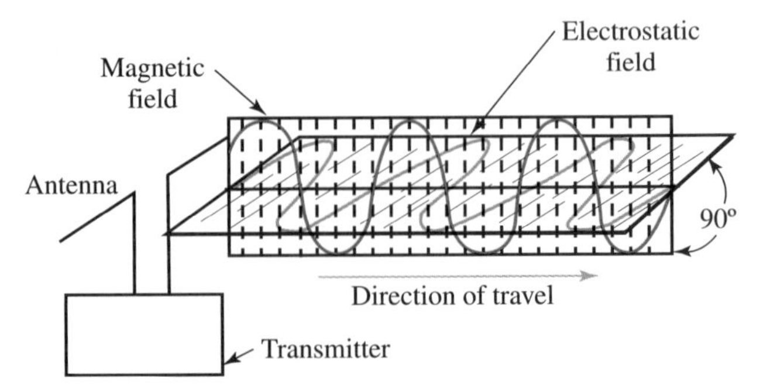

Figure 22-2. The relationship between electrostatic and electromagnetic waves. They are perpendicular to each other and both are perpendicular to the direction of travel.

PROJECT 22-1: Building a Crystal Radio

Crystal radios are often the first electronic project built at home or school. This project shows some of the principles we have studied, such as tuning and detection.

Commercial parts are used in this project. Its construction introduces more complex circuits. Being familiar with these parts will pay off in more advanced projects. The radio can be built on a plastic sheet or a wooden base. It will require a good antenna for proper operation.

Notes: The antenna coil consists of L_1 and L_2. It can be constructed from 10 feet of No. 30 AWG magnet wire. The wire is wound on four inches of 3/4 inch PVC. Use 30 turns for L_1. Use 70 turns for L_2.

If the variable capacitor is difficult to find, substitute a 360 pF fixed capacitor and vary the number of turns in L_2. Varying the number of turns in L_2 will change the resonant frequency in the tank circuit.

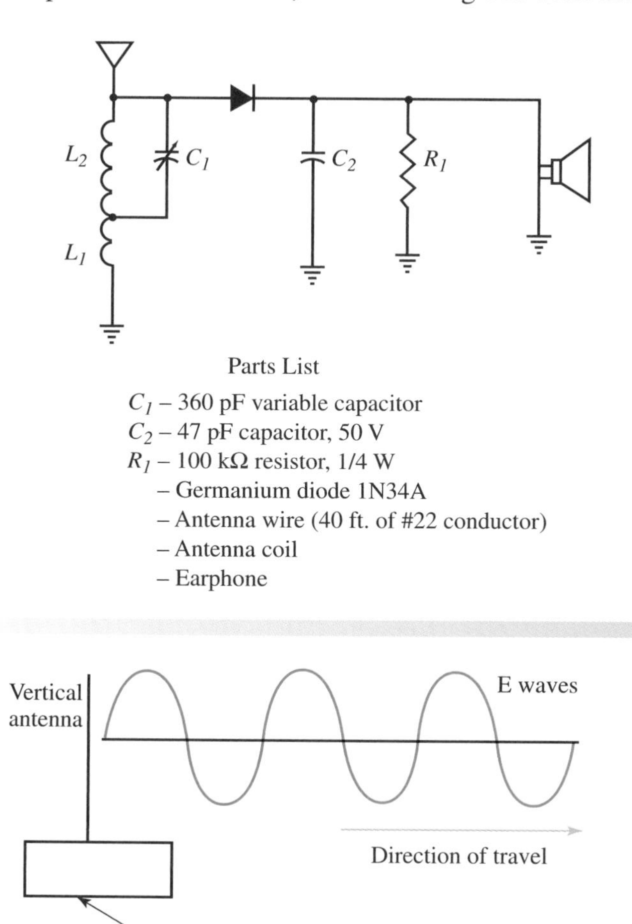

Parts List

C_1 – 360 pF variable capacitor
C_2 – 47 pF capacitor, 50 V
R_1 – 100 kΩ resistor, 1/4 W
– Germanium diode 1N34A
– Antenna wire (40 ft. of #22 conductor)
– Antenna coil
– Earphone

Figure 22-3. A vertical antenna radiates a vertically polarized wave.

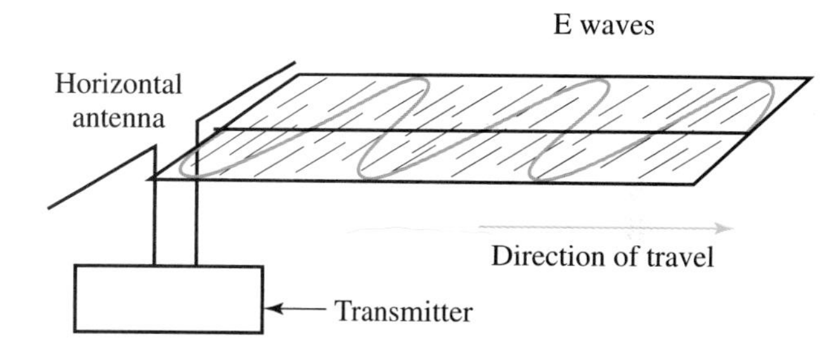

Figure 22-4. A horizontal antenna radiates a horizontally polarized wave.

In **Figure 22-4,** the wave is radiated from a horizontal antenna. It is still perpendicular to the direction of travel, but is parallel to the surface of the earth. Generally speaking, the antenna that receives these waves should be positioned in the same way as the transmitting antenna. At high frequencies, the polarization changes slightly as the wave moves.

Does all this mean the transmitting antenna radiates two waves? The answer is found in the fact that without one, there cannot be the other. A moving *electrostatic* field produces a moving *electromagnetic* field, and likewise, a moving *electromagnetic* field produces a moving *electrostatic* field. These conditions exist whether an actual conductor is present or not.

The radiated waves from an antenna can be divided into two groups. These are ground waves and sky waves.

Ground waves

A ***ground wave*** follows the surface of the earth to the radio receiver. The ground wave has three parts:

- The surface wave.
- The direct wave, which follows a direct path from the transmitter to the receiver.
- The ground reflected wave, which strikes the ground and is then reflected to the receiver.

The last two waves are combined and called a ***space wave.*** The waves that make up the space wave may or may not arrive at the receiver in proper order. They may join together or cancel each other, depending on distances traveled by each wave.

Broadcast stations depend on the surface wave for reliable communications. As the surface wave travels along the surface of the earth, it induces currents in the earth's surface. These currents use up the energy contained in the wave. The wave becomes weaker as the distance it travels increases.

An interesting side note is that salt water conducts surface waves about 5000 times better than the land. Overseas communication is very reliable when transmitters are near the coastline. These stations use high power and operate at lower frequencies than the normal broadcast band.

Sky waves

The second type of radiated wave is a ***sky wave.*** Sky waves use the ionized layer of the earth's atmosphere for transmission. This layer is called the ***ionosphere.*** It is located from 40 to 300 miles above the earth's surface. It is believed to consist of large numbers of positive and negative ions. As the sky wave radiates, it strikes the ionosphere. Some of the wave can be absorbed into the ionosphere. But some will bounce off the layer and be sent back to the earth's surface. See **Figure 22-5.**

Review Questions for Section 22.1

1. List the five main parts of a radio receiver and the function of each part.
2. What is polarization?
3. List the three parts of a ground wave.
4. A space wave is a combination of __________ wave and __________ __________ wave.
5. Sky waves use the __________ for transmission.

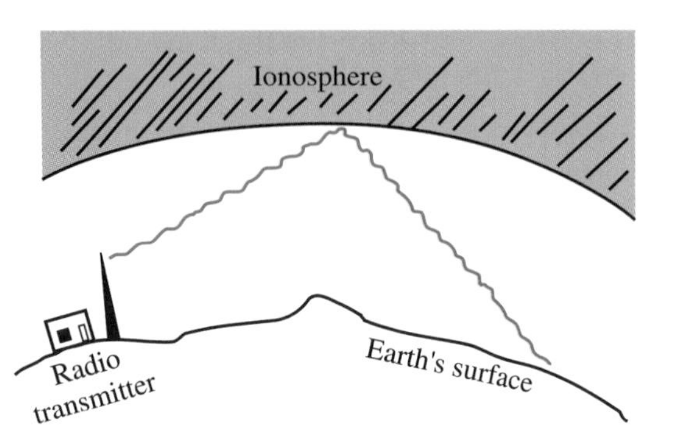

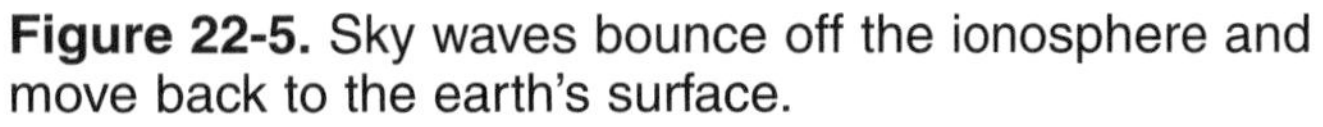
Figure 22-5. Sky waves bounce off the ionosphere and move back to the earth's surface.

22.2 FREQUENCY AND WAVELENGTH

Think of a radio wave moving through space as a wave on a pond rolling toward the beach. When a stone is thrown into water it creates a wave that moves outward. As the wave moves away from the center (where the stone was dropped), the amplitude of the wave decreases.

Does the water move along as the wave? No. The water moves *up* to a crest and *down* in a hollow, or trough, as the wave passes. The water has acted as the medium for transmitting the wave.

Radio waves travel through space at a speed of light, which is 186,000 miles per second (300,000,000 meters per second). This is the velocity of a radio wave.

Refer to **Figure 22-6.** The first radio wave has a frequency of one cycle per second (1 Hz). Starting at point A, the wave will move 186,000 miles by the time it completes one cycle and reaches point B. In the second wave, the frequency of the wave is 1000 Hz. The wave will move 186 miles by the time it completes one cycle and reaches point B. The second wave completes its one cycle in 1/1000th of a second. In the third wave, the frequency is increased to 1,000,000 Hz, or 1 megahertz. The wave will move only 0.186 miles by the time it completes one cycle and reaches point B.

A wave can be described not only by its frequency, then, but also by its length, **Figure 22-7.** The distance between the crests of the waves is the ***wavelength.*** As the frequency increases, the wavelength decreases. The Greek letter λ (lambda) stands for wavelength.

The length of a radio wave is important and must be taken into consideration when designing communication systems. The mathematical relationship of frequency and wavelength is:

$$\lambda = \frac{V}{f}$$

where λ is the wavelength given in miles, V is the velocity of the wave given in miles per second, and f is the frequency in hertz. Since the waves travel at the speed of light, the formula can be written:

$$\lambda = \frac{186{,}000 \text{ miles per second}}{f}$$

To solve for frequency, the equation is rearranged to read:

$$f = \frac{186{,}000 \text{ miles per second}}{\lambda}$$

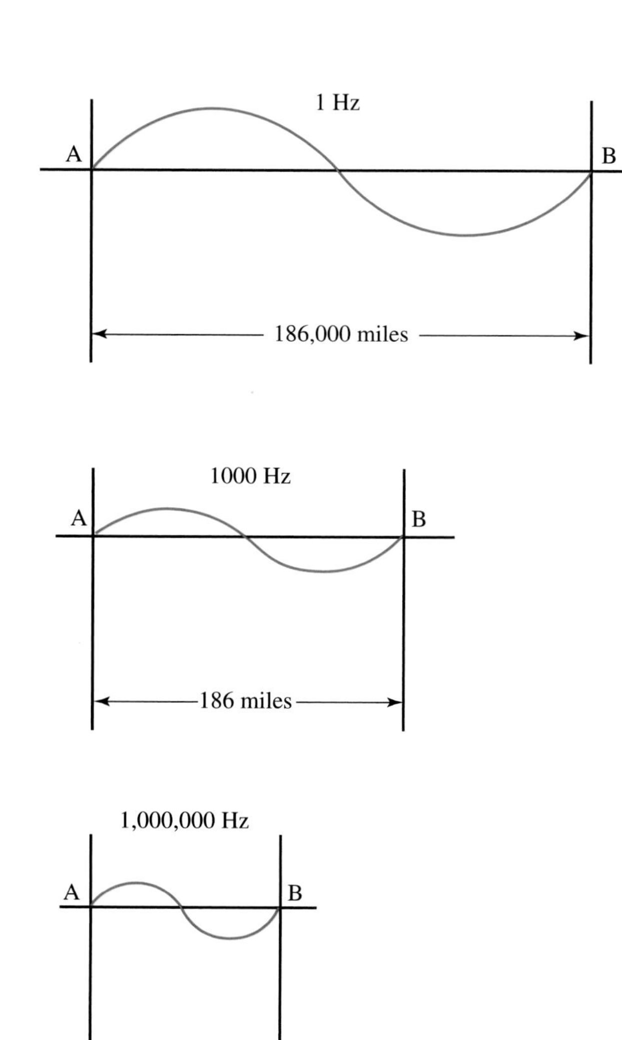

Figure 22-6. A comparison of the distances traveled in one cycle for 1 Hz, 1000 Hz, and 1,000,000 Hz waves.

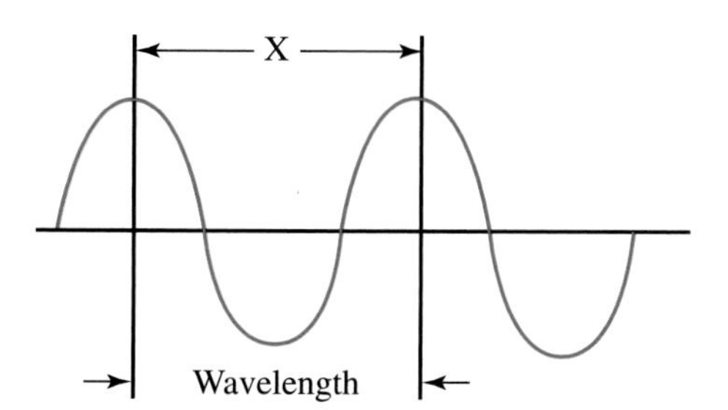

Figure 22-7. Radio waves are identified by their length and their frequency. X equals one wave.

When using metric measurements, the formulas are written:

$$\lambda = \frac{300{,}000{,}000 \text{ meters per second}}{f}$$

and,

$$f = \frac{300{,}000{,}000 \text{ meters per second}}{\lambda}$$

where λ is the wavelength given in meters and f is the frequency in hertz.

Example: An amateur radio broadcaster transmits on a frequency of 3.9 MHz. What is the wavelength of the radiated waves?

$$\lambda = \frac{186{,}000 \text{ miles per second}}{3.9 \text{ MHz}}$$

First, 3.9 MHz is converted to 3,900,000 Hz.

$$\lambda = \frac{186{,}000 \text{ miles per second}}{3{,}900{,}000 \text{ Hz}}$$

Observe how the math is simplified by using powers of ten.

$$\lambda = \frac{1.86 \times 10^5}{3.9 \times 10^6} = \frac{1.86 \times 10^{-1}}{3.9} = \frac{0.186}{3.9} \cong 0.05 \text{ miles}$$

Example: What is the frequency of a transmitter operating on 40 meters?

$$f = \frac{3 \times 10^8 \text{ meters per second}}{4 \times 10 \text{ meters}}$$

$$f = 0.75 \times 10^7 = 7{,}500{,}000 \text{ Hz} = 7.5 \text{ MHz}$$

A more useful form of this formula is available. Since the foot is the common unit of measure and radio frequencies are usually in megahertz:

$$\lambda = \frac{984}{f}$$

where λ is wavelength measured in feet and f is frequency measured in megahertz.

Memorize these formulas. They are used a great deal in the design of special frequency antennas. The antenna absorbs most of the radio wave when it matches the length of the radio wave.

Frequency Spectrum

Electromagnetic waves are produced at many different frequencies. Since the frequencies of radio transmitters could interfere with each other, a system of regulating the use of transmitters had to be developed. This system insures an orderly use of the air waves. See **Figure 22-8.** This system is known as a frequency spectrum. This chart shows the frequency bands used by

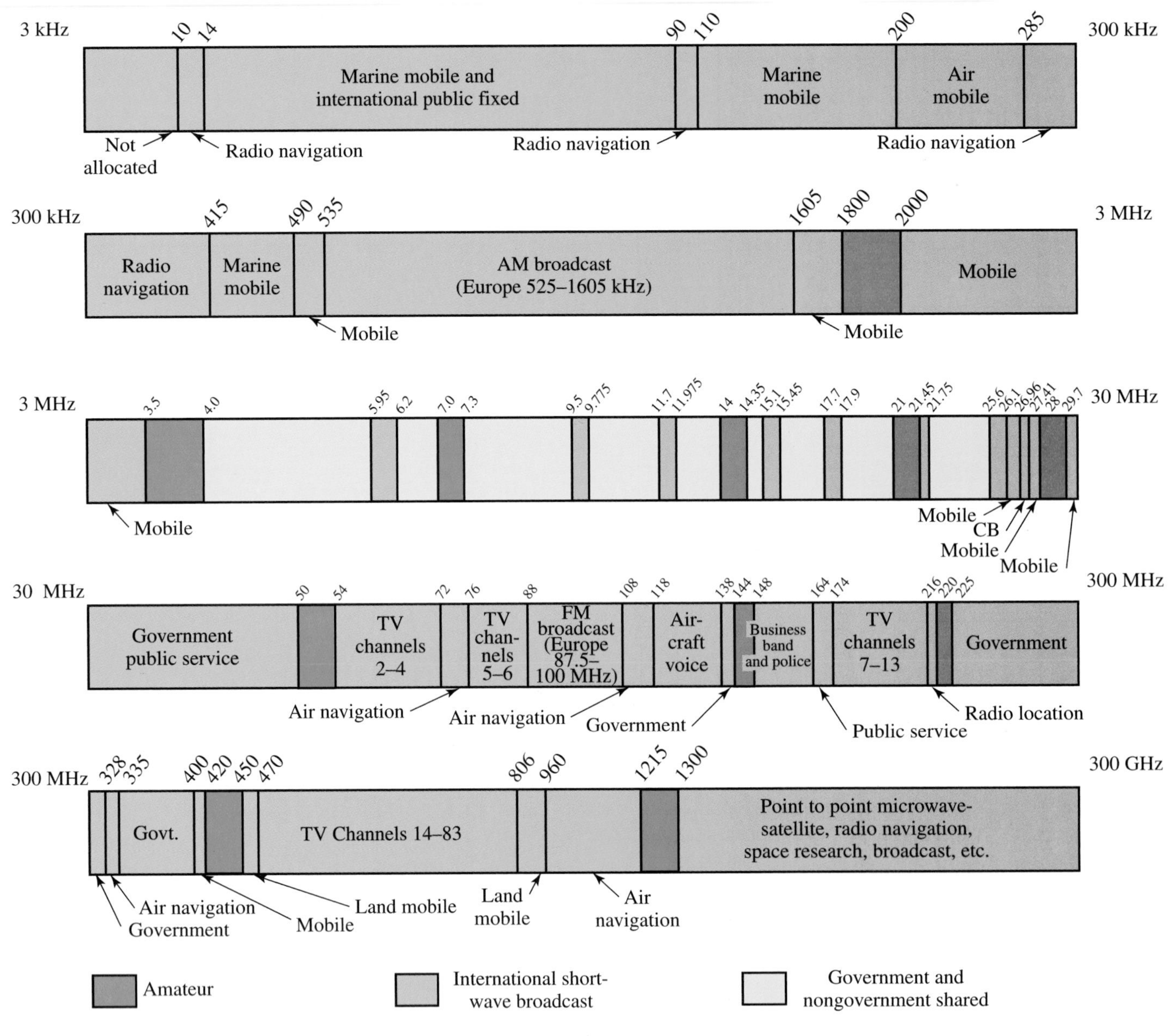

Figure 22-8. Frequency spectrum assignments. (Tektronix)

telecommunication areas such as radio, television, satellite transmission, aviation, radar, police, and many more.

The use of frequencies is strictly controlled. This is because there are many users of these systems. And each user requires a frequency on which to broadcast. Since there is only a limited number of frequencies available, the frequencies must be assigned to meet the needs of the public. At the same time, there must be a minimum of interference between the services. The Federal Communications Commission (FCC) controls frequency assignment in the United States.

Waves are also grouped according to frequency, **Figure 22-9.** Note the location of some services. AM

Frequency	Band
20–30,000 Hz	AF—Audio frequency
Below 30 kHz	VLF—Very low frequency
30–300 kHz	LF—Low frequency
300–3000 kHz	MF—Medium frequency
3000–30,000 kHz	HF—High frequency
30,000 kHz–300 MHz	VHF—Very high frequency
300–3000 MHz	UHF—Ultra high frequency

Figure 22-9. Study this chart of frequency ranges. Can you name one device that operates in each band?

broadcast stations use the MF range. Television uses the VHF and UHF ranges. A stereo receiver operates in the audio frequency range. Electricity for the home is 60 Hz. It operates at the audio frequency.

Review Questions for Section 22.2

1. At what speed does a radio wave travel?
2. Wavelength can be determined by dividing the __________ of the wave by the __________ of the wave.
3. The wavelength for a 93 megahertz signal is _________ meters.
4. What is the length of a 1040 AM radio wave?
5. Why are transmitted electromagnetic signals controlled by the FCC?
6. Radio waves are grouped according to their ____________.
7. The audio (or human hearing) range lies between __________ Hz and __________ Hz.
8. What are the frequencies of the following telecommunication systems?
   a. AM radio = __________ Hz to __________ Hz.
   b. CB radio = __________ Hz to __________ Hz.
   c. TV channels 2 through 4 = _________ Hz to __________ Hz.
   d. TV channels 5 through 6 = _________ Hz to __________ Hz.
   e. TV channels 7 through 13 = _________ Hz to __________ Hz.
   f. TV channels 14 through 83 = __________ Hz to __________ Hz.

22.3 RADIO TRANSMITTER

Any oscillator will produce radio frequency waves. When the oscillator is connected to an antenna system, it sends energy into the atmosphere. Amplification will increase the amplitude of the oscillator wave so that it will drive a final power amplifier.

Continuous Wave Transmitter

A block diagram of a simple ***continuous wave (CW) transmitter*** is shown in **Figure 22-10.** The first block is the conventional crystal oscillator and then the final power amplifier. A power supply is provided for the oscillator and the final power amplifier.

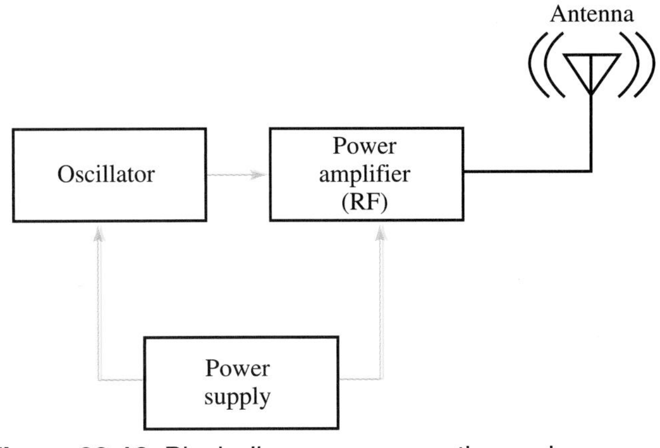

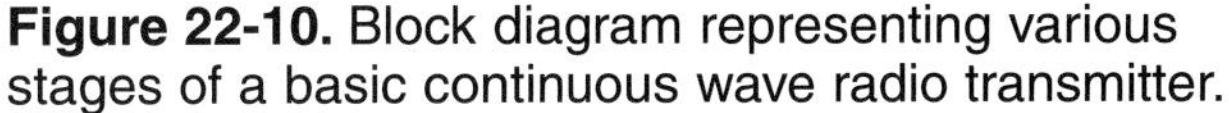
Figure 22-10. Block diagram representing various stages of a basic continuous wave radio transmitter.

Following the action in Figure 22-10, the oscillator creates an ac sine wave at a desired frequency. This signal is called the ***carrier wave.*** The carrier wave is then amplified by the radio frequency (RF) power amplifier to the desired output wattage. A power supply is required to provide the voltages and current needed to operate the oscillator and the RF power amplifier. The output is then fed to an antenna. From there, the energy is sent into the air as electromagnetic waves.

Notice that a CW transmitter sends energy that has no audio or video message. The CW transmitter has only two states, on or off. How can this type of transmitter be useful? By adding a switch, the transmitter can be turned on and off following a code. For example, such a transmitter could be used to send Morse code messages, **Figure 22-11. Figure 22-12** lists the character set for sending Morse code messages.

The basic switched, or keyed, CW transmitter can be improved by placing a buffer amplifier between the oscillator and the RF amplifier. The buffer amplifier isolates

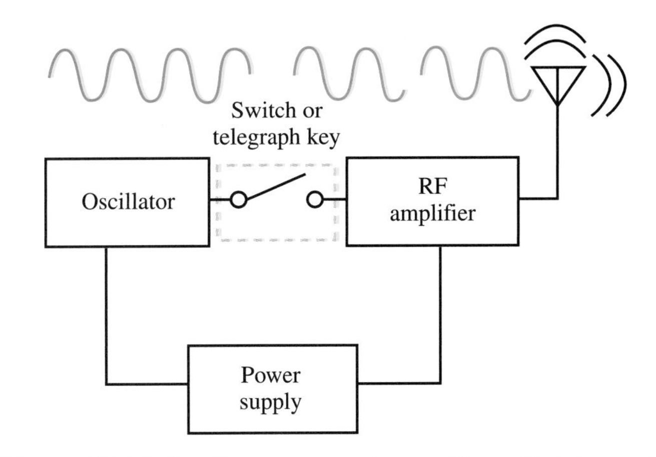

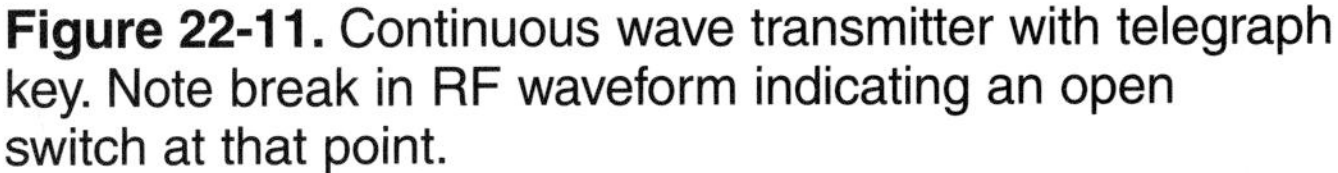
Figure 22-11. Continuous wave transmitter with telegraph key. Note break in RF waveform indicating an open switch at that point.

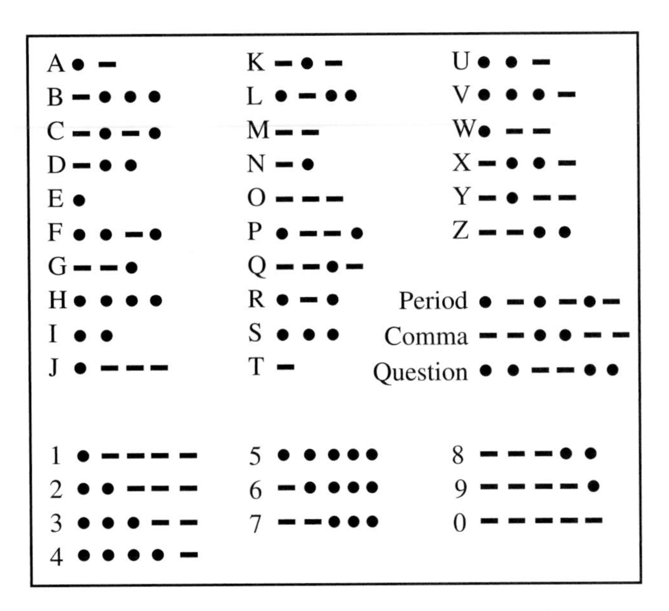

A • –	K – • –	U • • –
B – • • •	L • – • •	V • • • –
C – • – •	M – –	W • – –
D – • •	N – •	X – • • –
E •	O – – –	Y – • – –
F • • – •	P • – – •	Z – – • •
G – – •	Q – – • –	
H • • • •	R • – •	Period • – • – • –
I • •	S • • •	Comma – – • • – –
J • – – –	T –	Question • • – – • •
1 • – – – –	5 • • • • •	8 – – – • •
2 • • – – –	6 – • • • •	9 – – – – •
3 • • • – –	7 – – • • •	0 – – – – –
4 • • • • –		

Figure 22-12. The character set for Morse code.

Figure 22-13. This video camera is equipped with a stereo microphone. (Sony Electronics Corp.)

the oscillator from the RF amplifier and keeps it from shifting off of the desired frequency. It also provides some amplification to the carrier wave.

Many CW transmitters use frequency multipliers to increase the frequency produced by the basic oscillator. These circuits multiply the carrier wave by two (doubler) or three (tripler). These circuits operate on the principle of harmonics in the fundamental carrier frequency created by the oscillator. A ***fundamental frequency*** is the basic frequency produced by the oscillator. A ***harmonic frequency*** is a multiple of the fundamental frequency.

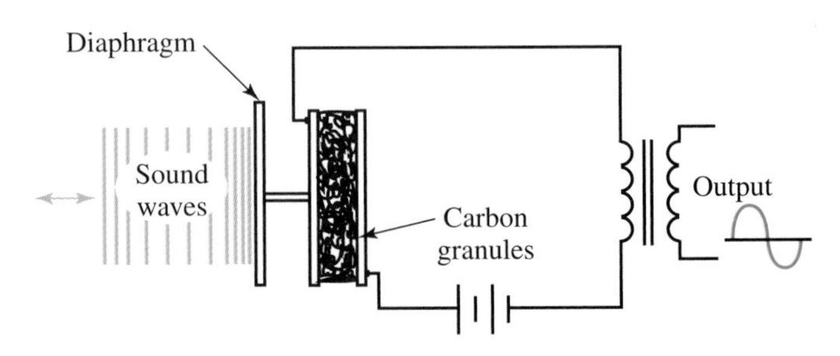

Figure 22-14. In a carbon microphone, sound waves change the resistance of the circuit.

Microphones

How is a sound wave converted to an electrical wave?

Your vocal cords send vibrations in the air. These waves move out to all persons within hearing range. A ***microphone*** will convert these sound waves to electrical audio waves of the same frequency and relative amplitude. Microphones are sometimes called transducers. This is because they transform one form of energy (vibrating air, or mechanical) to electrical energy. **Figure 22-13** shows a microphone built into a video camera.

Carbon microphone

A diagram of a ***carbon microphone*** is shown in **Figure 22-14.** Granules of carbon are packed in a small container. Electrical connections are made to each side. A transformer and a small battery are joined in series with the carbon. A diaphragm is attached to one side of the container. This diaphragm is sometimes called a *button.* Sound waves strike the diaphragm (button) and cause the carbon granules to be compressed or pushed together. This varies the resistance of the carbon. Varying resistance causes a varying current to flow through the carbon button and the transformer primary. The output is a current that varies at the same frequency as the sound waves acting on the diaphragm.

The carbon microphone is a very sensitive device. It has a frequency response up to about 4000 Hz. This is useful for voice communication, but not for reproduction of music. It provides a good response for its intended frequencies. A carbon microphone is *nondirectional,* which means it will pick up sound from all directions.

Crystal microphone

A second type of microphone uses the piezoelectric effect of certain crystals. It is called a ***crystal microphone.*** When sound waves strike a diaphragm, mechanical pressure is transferred to the crystal. The flexing or bending of the crystal creates a small voltage between its surfaces. This voltage is the same frequency and relative amplitude as the sound wave, **Figure 22-15.**

Crystal microphones have a frequency response up to 10,000 Hz. They are sensitive to shock and vibration. They should be handled with care.

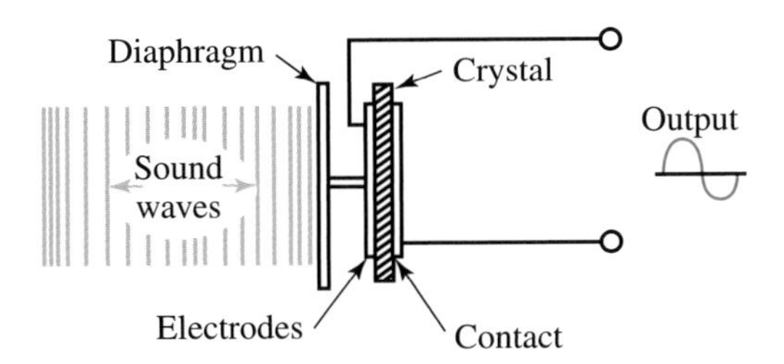

Figure 22-15. Mechanical pressure is used to produce electrical energy. The crystal microphone takes advantage of the piezoelectric effect.

Dynamic microphone

A ***dynamic microphone,*** or moving coil microphone, is sketched in **Figure 22-16.** As sound waves strike the diaphragm, they cause the voice coil to move in and out. The voice coil is surrounded by a fixed magnetic field. When the coil moves, a voltage is induced in the coil (Faraday's discovery). This induced voltage causes current to flow at a frequency and amplitude similar to the sound wave causing the motion. It has a frequency response up to 9000 Hz. It is directional and requires no outside voltage for operation.

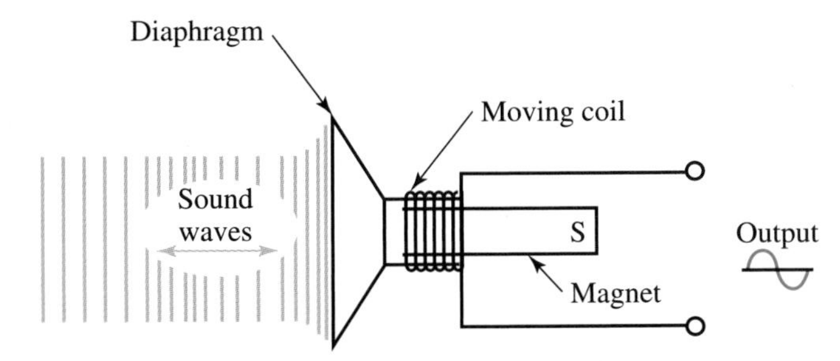

Figure 22-16. Dynamic microphone. Electrical audio waves are produced by a coil moving in a magnetic field.

Condenser microphone

A condenser microphone operates on the principle of capacitance. It is similar in construction to a capacitor, consisting of two plates separated by air. One plate is rigid while the other is moveable. As sound waves strike the moveable plate, the distance between the two plates will vary, varying the capacitance of the microphone. The varying capacitance of the microphone causes a reproduction of the audio signal similar in frequency and amplitude. The condenser microphone is very sensitive when compared to other types of microphones.

Velocity microphone

A high quality microphone, called a ***velocity microphone,*** is made by suspending a corrugated ribbon of metal in a magnetic field. Sound waves directly striking the ribbon cause the ribbon to vibrate. As the ribbon cuts the magnetic field, a voltage is induced. Proper connections at the ends of the ribbon bring the voltage out to terminals. This voltage varies according to the frequency and amplitude of the incoming sound waves.

The velocity microphone is a somewhat delicate microphone with a response above 12,000 Hz. When using this microphone, the speaker must speak across its face or stand about 18 inches away. Otherwise, a "booming" effect is created.

Modulation

When you turn on the radio or TV, you expect to hear music and voices you understand. The signals of the CW transmitter mean nothing to the average person. To make an understandable message, an audio wave is combined, or ***superimposed,*** on a carrier wave. See **Figure 22-17.** The process of combining an audio wave with a carrier wave is called ***modulation.*** Sound waves are converted by microphones into electrical waves, amplified, and then combined with the CW radio wave.

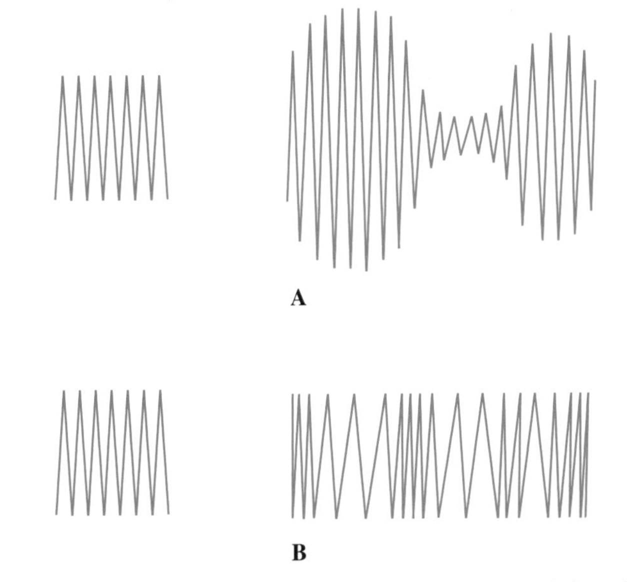

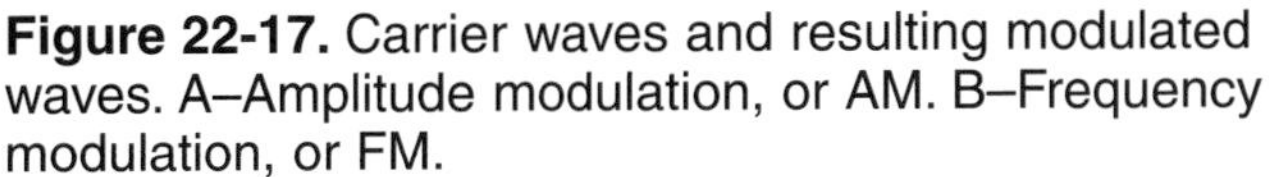

Figure 22-17. Carrier waves and resulting modulated waves. A–Amplitude modulation, or AM. B–Frequency modulation, or FM.

PROJECT 22-2: Building an AM Transmitter

In this project, you will make a working AM transmitter. Using it, you will be able to hear yourself over the radio. Examine the block diagram of the transmitter. The components in this circuit can be purchased individually from the parts list or as a package from Graymark Enterprises, Inc.

The RF carrier range is 550–1500 kHz with an output power of less than 100 milliwatts. Transmitting distance will depend on the environment. However, this distance should be less than 100 feet. Any AM receiver can be used to pick up the transmitted signal.

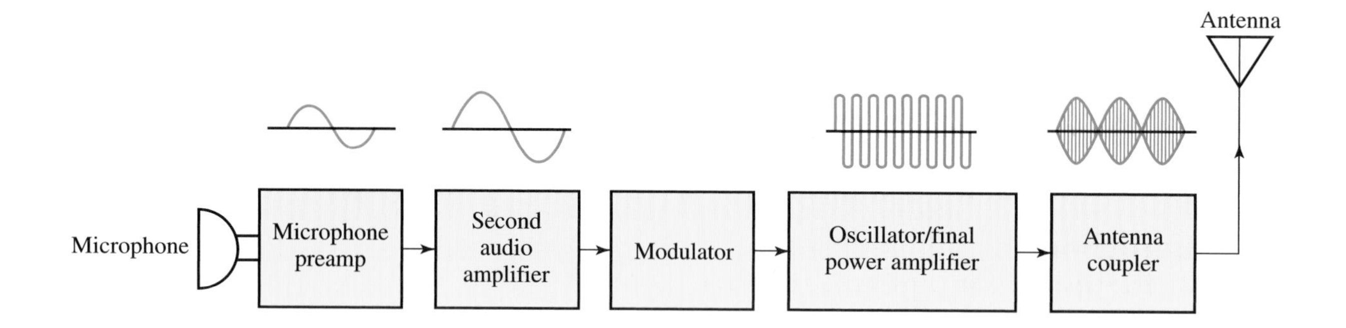

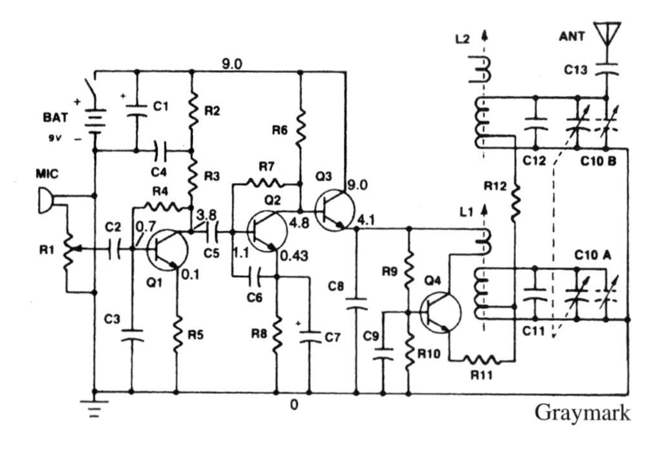

Graymark

Qty.	Symbol	Description
1	R_1	Potentiometer, with switch, 50 kΩ
2	R_2, R_8	Resistor, 470 Ω, 1/4-W, 10%
2	R_3, R_{10}	Resistor, 4.7 kΩ, 1/4-W, 10%
1	R_4	Resistor, 470 kΩ, 1/4-W, 10%
1	R_5	Resistor, 100 Ω, 1/4-W, 10%
1	R_6	Resistor, 3.9 kΩ, 1/4-W, 10%
1	R_7	Resistor, 560 kΩ, 1/4-W, 10%
1	R_9	Resistor, 3.3 kΩ, 1/4-W, 10%
1	R_{11}	Resistor, 22 Ω, 1/4-W, 10%
1	R_{12}	Resistor, 10 Ω, 1/4-W, 10%
2	C_1, C_4	Capacitor, electrolytic, 33 μF, 16 WV
2	C_2, C_5	Capacitor, disc, 0.1 μF, 50 WV
1	C_3	Capacitor, disc, 0.0047 μF, 50 WV
1	C_6	Capacitor, disc, 0.01 μF, 50 WV
1	C_7	Capacitor, electrolytic, 4.7 μF, 16 WV
2	C_8, C_9	Capacitor, disc, 0.047 μF, 50 WV
1	C_{10}	Capacitor, variable, (tuning), 266 pF
2	C_{11}, C_{12}	Capacitor, disc, 22 pF, 50 WV
1	C_{13}	Capacitor, disc, 220 pF, 50 WV
2	L_1, L_2	Coil, oscillator type
4	Q_1–Q_4	Transistor, NPN silicon, 2SC372-Y
1		Microphone
1		Antenna, telescopic
1	J_1	Jack, microphone, 3.5 mm type, NC
1		Battery clip, for 9-volt battery
1 set		Hookup wire set, stranded, 1-red, 1-yellow
1		Bare wire (buswire), single strand, #22
1		Solder, rosin core
1		Printed circuit board
1		Breadboard

Graymark

Amplitude modulation occurs when the *amplitude* of the CW radio wave is made to vary at an audio frequency rate. Amplitude modulation is referred to as ***AM.*** In a second method, the radio wave *frequency* is made to vary at an audio frequency rate. This is called ***frequency modulation*** or ***FM.*** Part A shows the modulation of the amplitude of a carrier wave. Part B show the modulation of the frequency of a carrier wave.

Review Questions for Section 22.3

1. Sketch a carbon microphone.
2. What is a fundamental frequency? What is a harmonic frequency?
3. Combining a continuous wave with an audio wave is called __________.

22.4 AMPLITUDE MODULATION (AM)

Modulation is a process in which an audio wave is combined or superimposed on a carrier wave. Assume a radio transmitter is operating on a frequency of 1000 kHz. A musical tone of 1000 Hz is to be used for modulation. Refer to **Figure 22-18.** Using a modulation circuit, the amplitude of the carrier wave is made to vary at the audio signal rate.

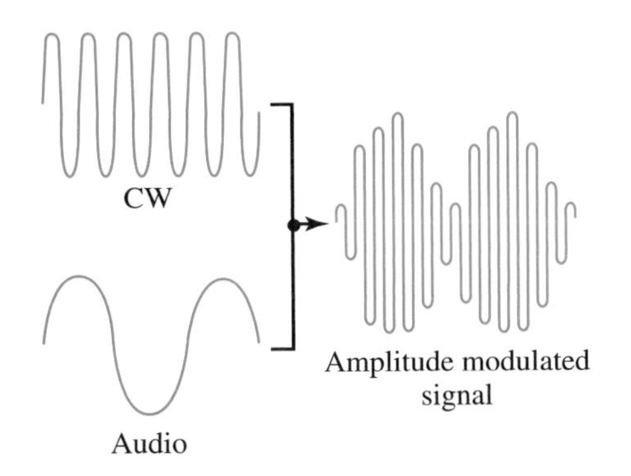

Figure 22-18. A CW wave, an audio wave, and the resulting amplitude modulated wave.

Let's look at this process another way. Mixing a 1000 Hz wave with a 1000 kHz wave produces a sum wave and a difference wave, which are also in the radio frequency range. These two waves will be 1001 kHz and 999 kHz. They are known as ***sideband frequencies.*** The ***upper sideband*** is the higher number and the ***lower sideband*** is the lower number, **Figure 22-19.**

The sum of the carrier wave and its sidebands is an amplitude modulated wave. The audio tone is present in both sidebands, as either sideband results from modulating a 1000 kHz signal with a 1000 Hz tone.

The location of the waves on a frequency base are shown in **Figure 22-20.** If a 2000 Hz tone was used for modulation, then sidebands would appear at 998 kHz and 1002 kHz. In order to transmit a 5000 Hz tone of a violin using AM, sidebands of 995 kHz and 1005 kHz would be required. The frequency bandwidth to transmit the 5000 Hz musical tone will be 10 kHz.

There is not enough space in the spectrum for all broadcasters to transmit. And, if all broadcasts contained the same message or operated on the same frequency, the effect would be confusing. Therefore, the ***broadcast band*** for AM radio extends from 535 kHz to 1605 kHz. It is divided into 106 channels, each 10 kHz wide.

Each radio station in a geographic area is licensed to operate at a frequency in one of these 106 channels. The channels are spaced far enough from each other to prevent interference.

In order to improve the fidelity and quality of music within these limitations, a vestigial sideband filter is used. A ***vestigial sideband filter*** removes a large portion of one sideband. *Recall that both sidebands contain the same information.* This way, frequencies higher than 5 kHz can be used for modulation, and the fidelity is improved.

Modulation Patterns

A radio transmitter is not permitted by law to exceed 100 percent modulation. This means that the modulation signal cannot cause the carrier signal to vary over 100 percent of its unmodulated value. Look at the patterns in **Figure 22-21.** Notice the amplitude of the modulated waves.

The ***100 percent modulation*** wave variation is from *zero to two times the peak value* of the carrier wave. ***Overmodulation*** is caused when modulation increases the carrier wave to *over two times its peak value.* At negative peaks the waves cancel each other and leave a straight line of zero value. Overmodulation causes distortion and interference called ***splatter.***

Percent of modulation can be computed using the following formula:

$$\% \text{ modulation} = \frac{e_{max} - e_{min}}{2e_c} \times 100$$

where e_{max} is the maximum amplitude of modulated wave, e_{min} is the minimum amplitude of modulated wave, and e_c is the amplitude of unmodulated wave.

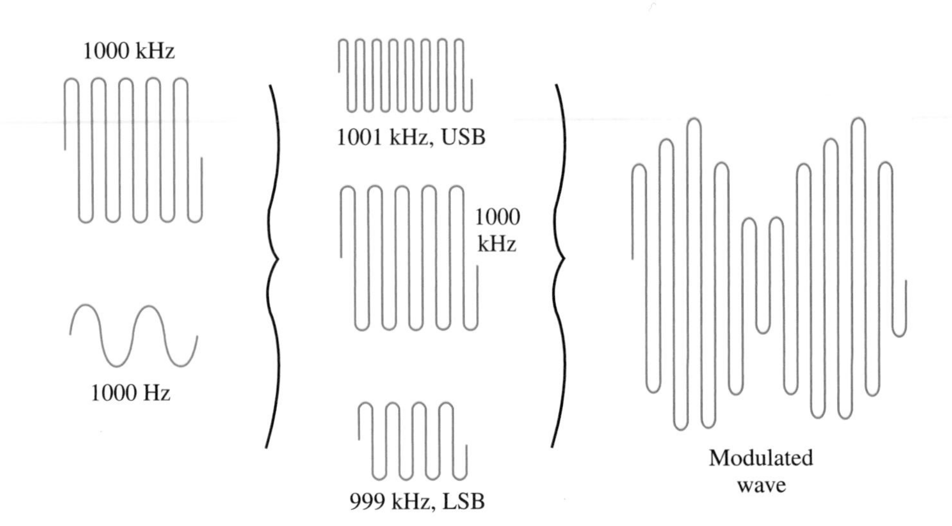

Figure 22-19. Wave mixing showing the formation of sidebands and modulation envelope.

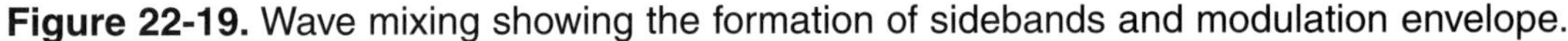

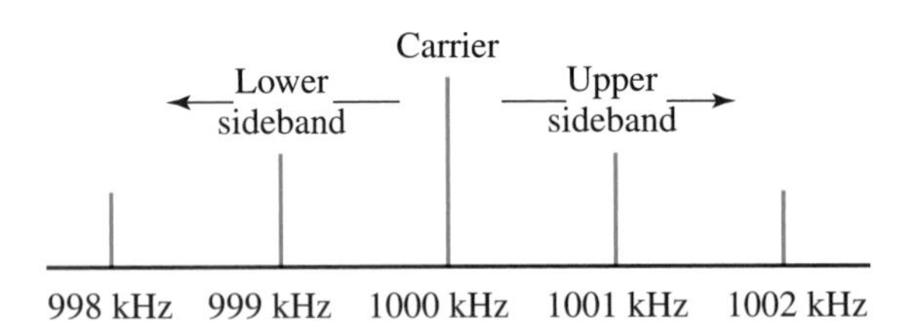

Figure 22-20. Carrier and sideband locations for modulation tone of 1 kHz and 2 kHz.

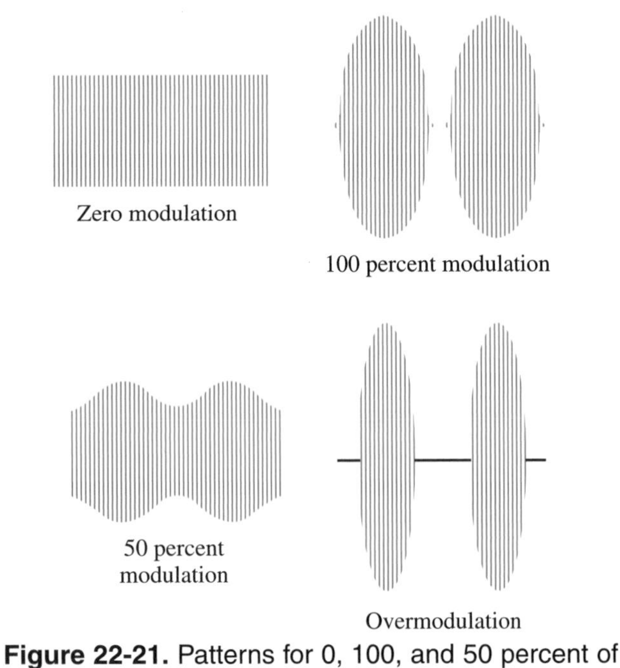

Figure 22-21. Patterns for 0, 100, and 50 percent of modulation and for overmodulation.

Sideband Power

The dc input power to the final amplifier of a transmitter is the product of voltage and current. To find the power required by a modulator, the following formula can be used.

$$P_{audio} = \frac{m^2 \times P_{dc}}{2}$$

where P_{audio} is the power of the modulator, m is the percentage of modulation (expressed as decimal), and P_{dc} is the input power to the final amplifier.

Example: What power is required to modulate a transmitter having a dc power input of 500 watts to 100 percent?

$$P_{audio} = \frac{(1)^2 \times 500 \text{ W}}{2} = 250 \text{ W}$$

This represents a total input power of 750 watts (250 W + 500 W). Notice what happens under 50 percent modulation.

$$P_{audio} = \frac{(0.5)^2 \, 500 \text{ W}}{2} = 62.5 \text{ W}$$

The total input power is only 562.5 watts (62.5 W + 500 W). Where the modulation percentage is reduced to 50 percent, the power is reduced to 25 percent. This is a severe drop in power that decreases the broadcasting range of the transmitter. It is wise to maintain transmitter modulation close to, but not exceeding, 100 percent.

The term *input power* has been used because any final amplifier is far from 100 percent efficient.

$$\% \text{ efficiency} = \frac{P_{out}}{P_{in}} \times 100$$

If a power amplifier had a 60 percent efficiency and a P_{dc} input of 500 watts, its output power would approach:

$$P_{out} = \% \text{ efficiency} \times P_{in} = 0.6 \times 500 \text{ W} = 300 \text{ W}$$

A transmitter has 100 percent modulation and power of 750 watts. 500 watts of this power is in the carrier wave and 250 watts is added to produce the sidebands. Therefore, there are 125 watts of power in each sideband or one-sixth of the total power in each sideband. Recall that each sideband contains the same information and each is a radio frequency wave that will radiate as well as the carrier wave. So why waste all this power?

In ***single sideband transmission,*** this power is saved. The carrier and one sideband are suppressed. Only one sideband is radiated. At the receiver end the carrier is put back in. The difference signal (the audio signal) is then detected and reproduced.

We will not cover the methods of sideband transmission and reception. But, you may wish to study this communication system on your own.

Review Questions for Section 22.4

1. Define modulation.
2. The sum of the carrier wave and its sidebands results in the __________ __________ wave.
3. A carrier wave has a peak value of 500 volts. A modulating signal causes amplitude variation from 250 volts to 750 volts. What is the percent of modulation?
4. Draw an amplitude modulated wave that has 50 percent modulation.

22.5 FREQUENCY MODULATION (FM)

In frequency modulation, a constant amplitude continuous wave (the radio wave) is made to vary in frequency at the audio frequency rate, **Figure 22-22.** FM radio is a popular method of electronic communication. Frequency modulation allows a high audio sound to be transmitted while still remaining within the space legally assigned to the broadcast station. Also, FM transmits dual channels of sound (stereo) by multiplex systems. The FM band is from 88 MHz to 108 MHz. A block diagram of an FM transmitter is shown in **Figure 22-23.**

Each FM station is assigned a ***center frequency*** in the FM band. This is the frequency to which a radio is tuned, **Figure 22-24.**

The amount of frequency variation from each side of the center frequency is called the ***frequency deviation.*** Frequency deviation is set by the amplitude, or strength, of the audio modulating wave.

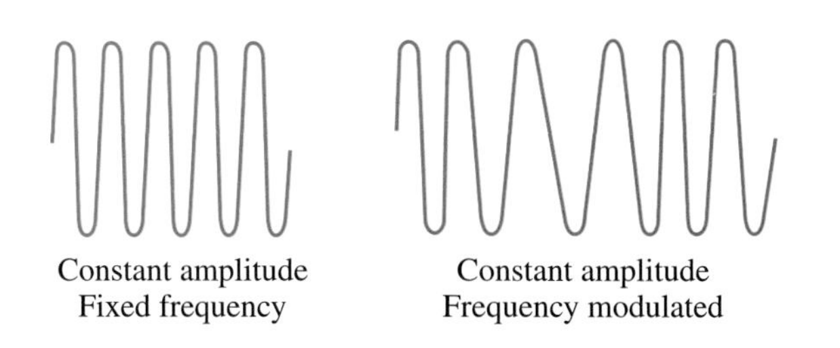

Figure 22-22. For FM, the frequency of the wave is varied at an audio rate.

In part A of Figure 22-24, a weak audio signal causes the frequency of the carrier wave to vary between 100.01 MHz and 99.99 MHz. The deviation is ±10 kHz. In part B of the figure, a stronger audio signal causes a frequency swing between 100.05 MHz and 99.95 MHz or a deviation of ±50 kHz. The stronger the modulation signal, the greater the frequency departure, and the more the band is filled.

The *rate* of frequency deviation depends on the *frequency* of the audio modulating signal. See **Figure 22-25.** If the audio signal is 1000 Hz, the carrier wave goes through its greatest deviation 1000 times per second. If the audio signal is 100 Hz, the frequency changes at a rate of 100 times per second. Notice that the modulating frequency does not change the amplitude of the carrier wave.

An FM signal forms sidebands. The number of sidebands produced depends on the frequency and amplitude of the modulating signal. Each sideband is separated from the center frequency by the amount of the frequency of the modulating signal, **Figure 22-26.**

The power of the carrier frequency is reduced a great deal by the formation of sidebands, which take power from the carrier. The amount of power taken from the carrier depends on the maximum deviation and the modulating frequency.

Although a station is assigned a center frequency and stays within its maximum deviation, the formation of sidebands determines the bandwidth required for transmission. In FM, the bandwidth is specified by the frequency range between the upper and lower significant sidebands. A ***significant sideband*** has an amplitude of one percent or more of the unmodulated carrier.

Narrow Band FM

Maximum deviation of a carrier wave can be limited so that the FM wave occupies the same space as an AM wave carrying the same message. This is called ***narrow band FM.*** Some distortion occurs in the received signal. This is satisfactory for voice communication but not for quality music sound systems.

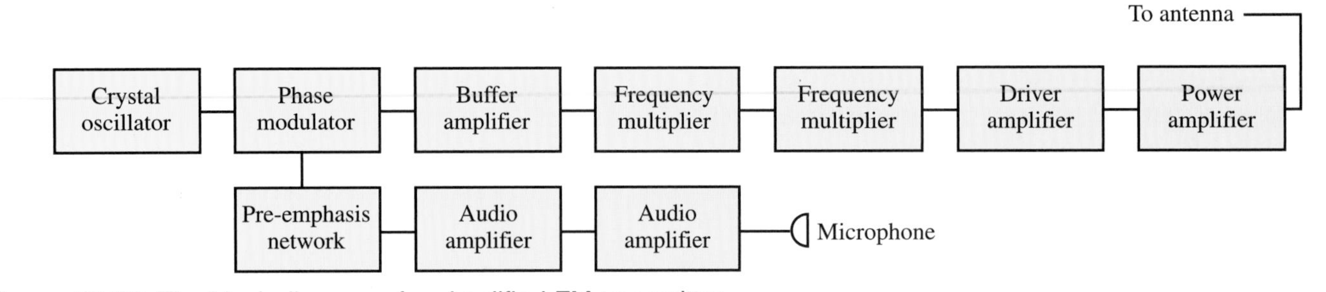

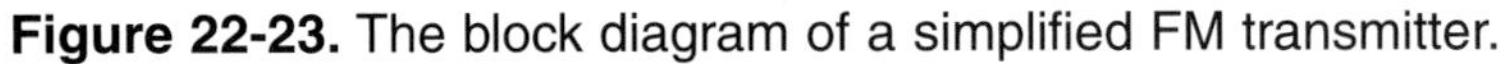
Figure 22-23. The block diagram of a simplified FM transmitter.

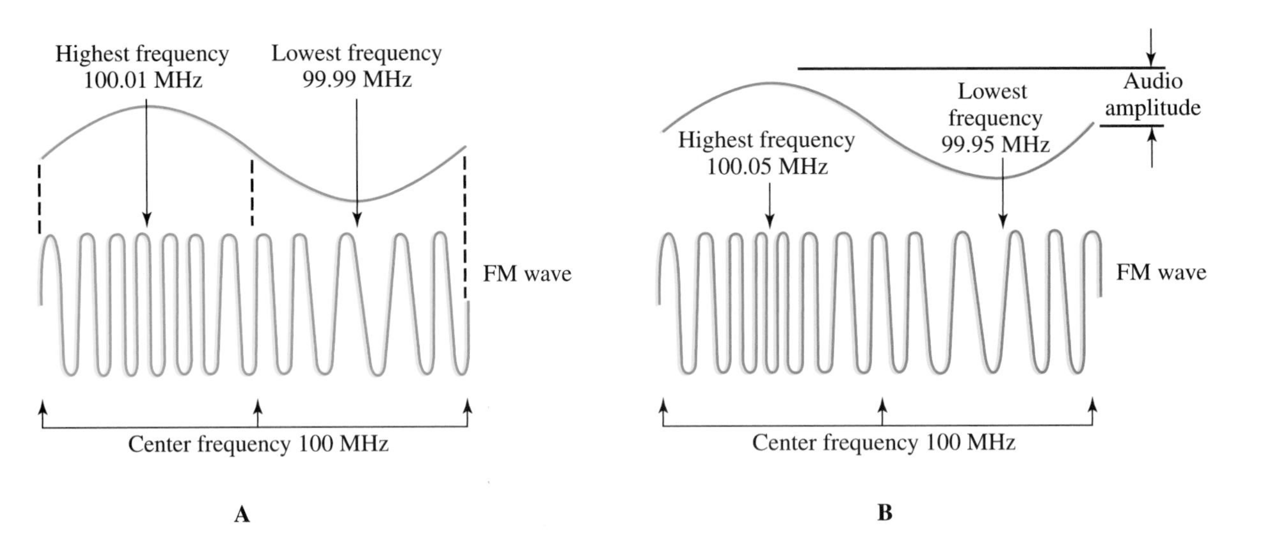

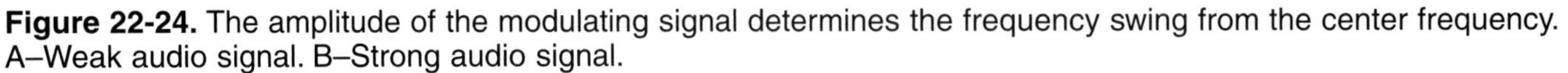
Figure 22-24. The amplitude of the modulating signal determines the frequency swing from the center frequency. A–Weak audio signal. B–Strong audio signal.

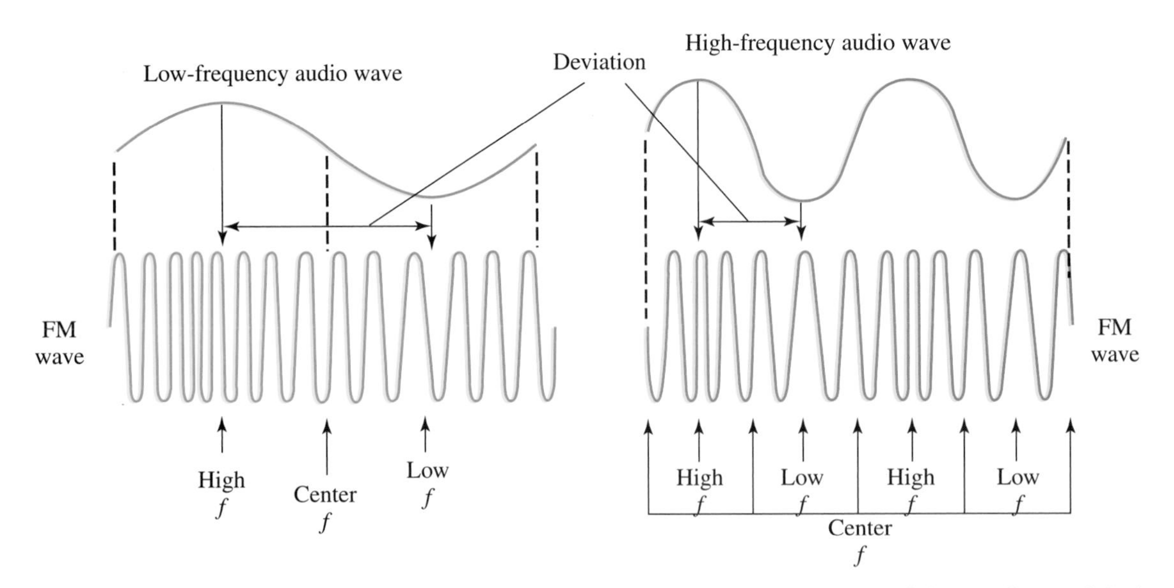

Figure 22-25. The rate of frequency variation depends on the frequency of the audio modulating signal.

Modulation Index

Modulation index is the relationship between the maximum carrier deviation and the maximum modulating frequency:

$$\text{Modulation index} = \frac{\text{Maximum carrier deviation}}{\text{Maximum modulating frequency}}$$

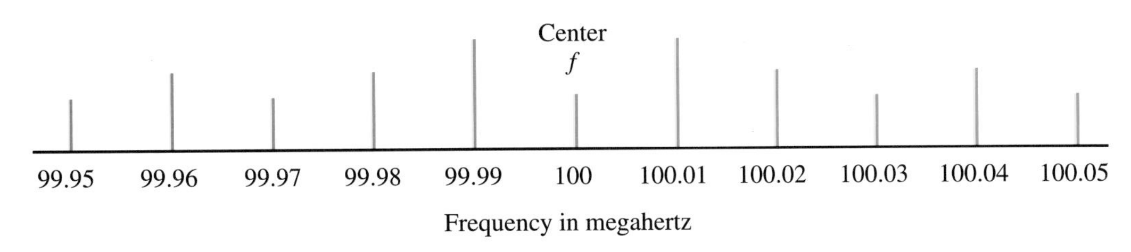

Figure 22-26. Sidebands generated by a 10 kHz modulating signal on a 100 MHz carrier wave.

Using this index, the number of significant sidebands and the bandwidth of the FM signal can be figured. The complete index can be found in more advanced texts. Examples of the use of the modulation index are given in **Figure 22-27.**

If the amplitude of a modulating signal causes a maximum deviation of 10 kHz and the frequency of the modulating signal was 1000 Hz, the index would be:

$$\text{Modulation index} = \frac{10{,}000}{1000} = 10$$

If you examine Figure 22-27, you can see that this FM signal would have 14 significant sidebands and occupy a bandwidth of 28 kHz.

Modulation index	Number of sidebands	Bandwidth
0.5	2	$4 \times f$
1	3	$6 \times f$
5	8	$16 \times f$
10	14	$28 \times f$

Figure 22-27. Examples of modulation index use. When calculating bandwidth, *f* is the modulating frequency.

Percent of Modulation

The percent of modulation has been set at a maximum deviation of ± 75 kHz for FM radio. The FM sound transmission in television is limited to ± 25 kHz.

Review Questions for Section 22.5

1. The __________ frequency is the frequency to which a radio is tuned.
2. Define frequency deviation.
3. On what does the rate of frequency deviation depend?
4. A(n) __________ __________ has an amplitude of at least one percent of the unmodulated carrier.
5. How is the modulation index for FM found?

22.6 AMPLITUDE MODULATED (AM) RECEIVER

A radio wave transmitted through space carries audio information, such as voices and music. This information has been combined with the carrier wave at the transmitter by the amplitude modulation process.

These radiated electromagnetic waves cut across and induce a small voltage in the receiving antenna. The small radio frequency voltages are coupled to a tuned circuit in the receiver which selects the signal to be heard. After selection, the voltages can be amplified, then demodulated. ***Demodulation,*** or ***detection,*** is the process of removing the audio portion of a signal from the carrier wave. It is a form of rectification. The audio signal is then amplified until it can drive an audio speaker.

A block diagram of this receiver is shown in **Figure 22-28.** The modulated RF wave is shown as it passes the antenna. The audio wave is shown as the output of the detector. The increase in amplitude between the blocks is the result of the amplification stages. This type of receiver was once quite popular. It was called the tuned radio frequency (TRF) receiver.

For satisfactory operation of the circuit, each stage had to be tuned to the correct incoming frequency. Some early radios had a series of tuning dials on the front panel. Adjusting these to receive a signal required skill and patience. The development of the superheterodyne receiver overcame these obstacles.

The Tuning Circuit

One function of any radio receiver is selection of the desired radio signal. **Figure 22-29** shows a schematic of the first stage of the TRF receiver.

Follow the signal of the antenna to the output of this circuit. Radio signals of many frequencies from many radio transmitters pass by the antenna. The induced voltage in the antenna causes small currents to alternate from antenna to ground and from ground to antenna through coil L_1. The magnetic field created by the antenna current in L_1 transfers energy to L_2. The combination coils L_1 and

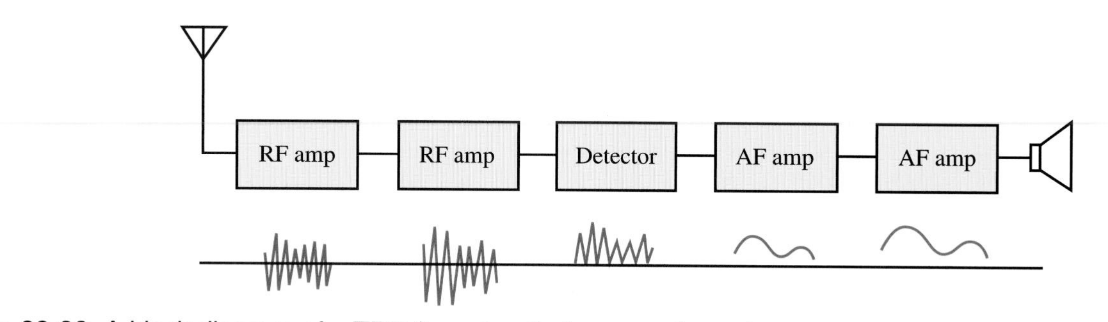

Figure 22-28. A block diagram of a TRF (tuned radio frequency) receiver.

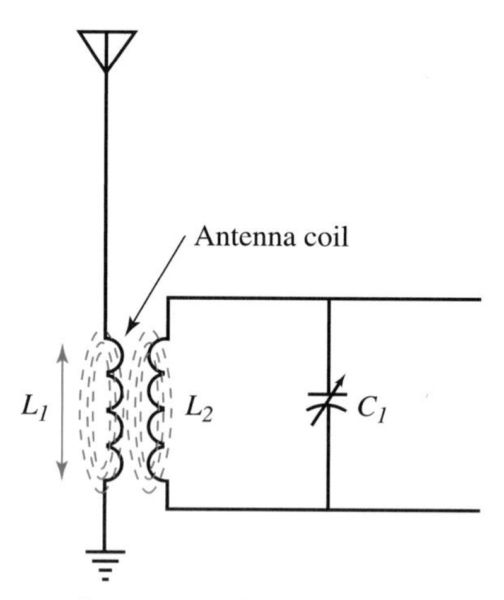

Figure 22-29. The tuning, or station selector, section of the radio receiver.

L_2 are called the ***antenna coil.*** Examples of coils are drawn in **Figure 22-30.**

A tank circuit is formed from L_2 and variable capacitor C_1. The resonance of the tank circuit is changed by adjusting C_1. When the resonant frequency of the tank circuit is the same as the incoming signal, high circulating currents develop in the tank. In other words, the tank circuit can be adjusted to give peak response for only a single frequency.

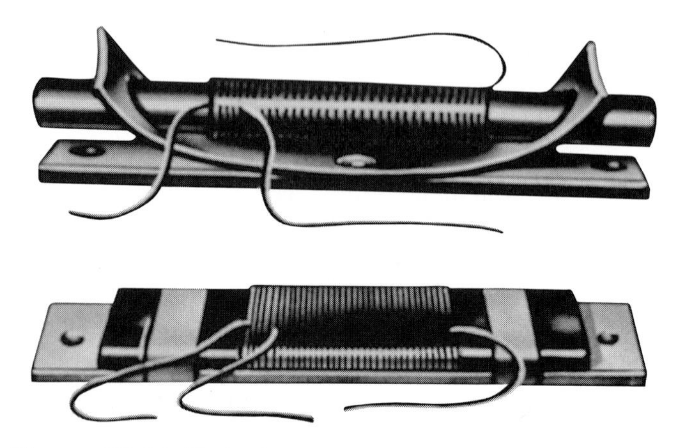

Figure 22-30. Typical antenna coils used with transistor receivers. (J.W. Miller Co.)

The L_2C_1 combination can cover only a certain range or band of frequencies. Most of our home radios cover only the broadcast band. For other groups of frequencies, such as shortwave (λ between 10 and 100 meters), another coil is switched in place of L_2. This coil changes the resonant frequency range of the circuit. This is called a ***band switching*** circuit.

The ability of a radio receiver to select a single frequency is called ***selectivity.*** The ability of a receiver to respond to weak incoming signals is called ***sensitivity.*** Both of these traits are desirable. Special circuits and components have been devised to improve selectivity and sensitivity.

RF Amplification

The radio frequency amplifier is most often the first stage to receive the signal from the antenna. The RF amplifier is tuned to incoming signal's frequency. It amplifies the signal to provide gain.

Generally, RF amplifiers are narrow band amplifiers that can amplify only the band or frequencies to be picked up by that receiver. An example is the RF amplifier in an AM broadcast band receiver. It can amplify frequencies from 550 to 1500 kHz. An RF amplifier improves the selectivity of a receiver.

The schematic for an RF amplifier is shown in **Figure 22-31.** The tuning circuit is made up of C_1 and L_2 formed into a tank circuit. The tuned signal is fed into the secondary winding of L_2. The amplifier is the 2N1637 transistor. Some less costly radio receivers do not have RF amplifier stages. High quality radio receivers will often have two or more RF amplifier stages.

Detection

Detection is needed to recover the audio signal from the modulated RF carrier wave. Assume that a modulated radio frequency signal has been amplified by several

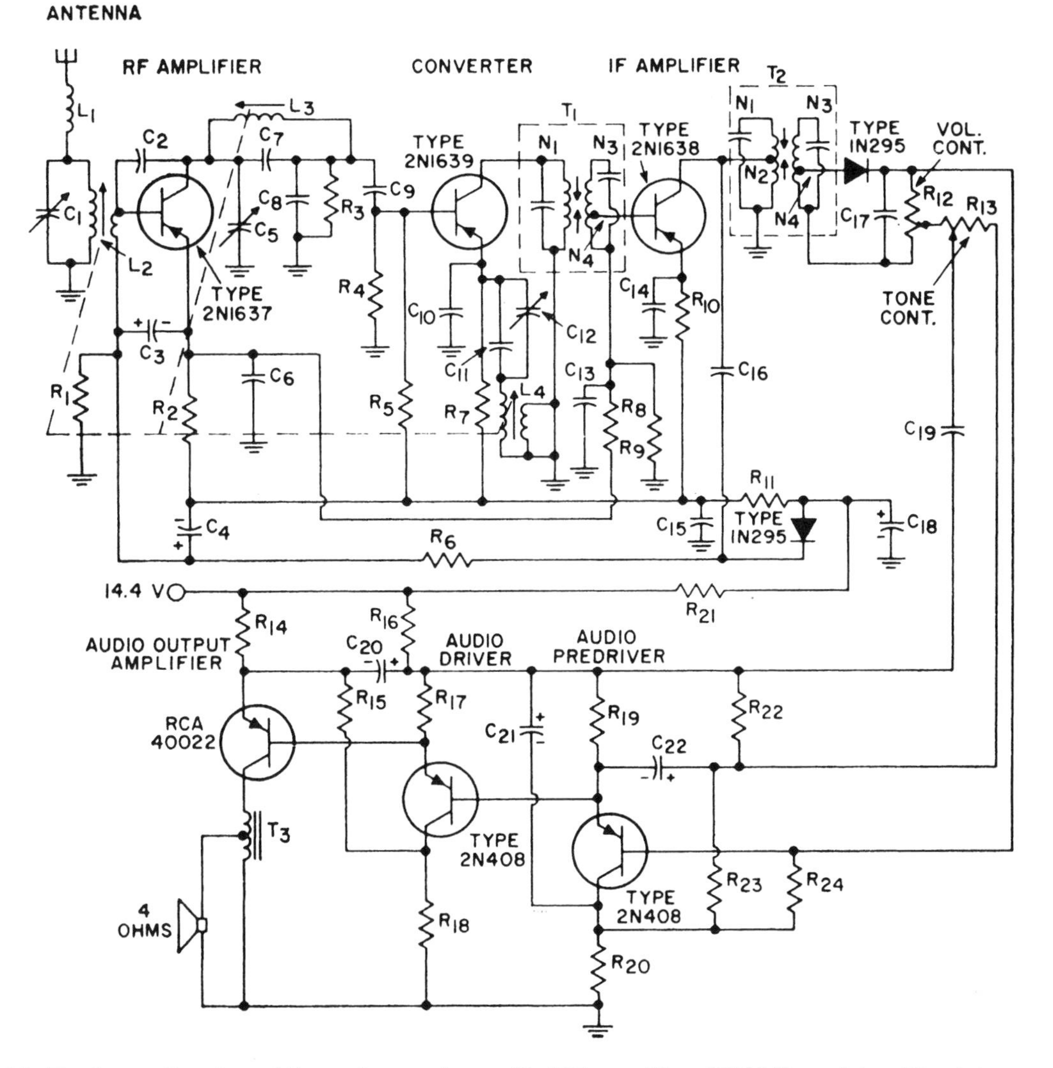

Figure 22-31. Twelve-volt automobile radio receiver with RF amplifier. (RCA Transistor, Thyristor, and Diode Manual SC-15)

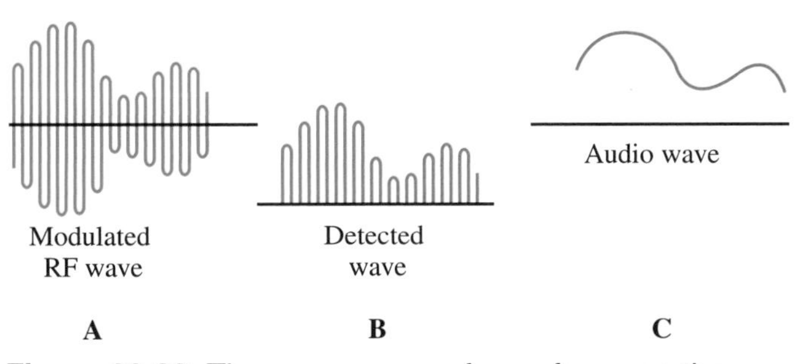

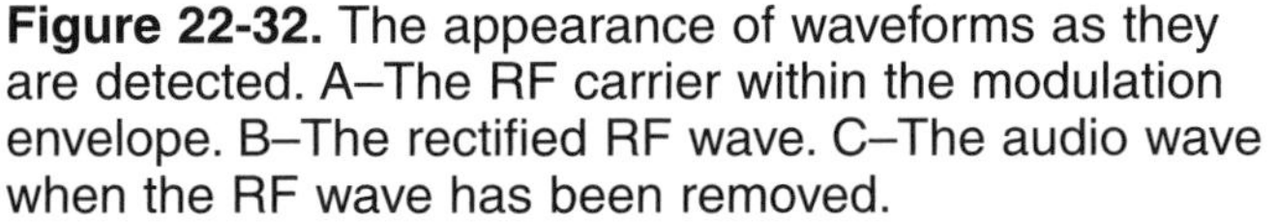

Figure 22-32. The appearance of waveforms as they are detected. A–The RF carrier within the modulation envelope. B–The rectified RF wave. C–The audio wave when the RF wave has been removed.

stages of RF amplification. It now has enough amplitude for detection. Recall that detection is a form of rectification. The RF wave is removed, leaving only the AF wave. See **Figure 22-32.**

The diode detector

One method of detection uses a diode as a unilateral (one direction) conductor. When the anode of the diode is driven positive, electrons flow through the diode from cathode to anode. The diode conducts. When the anode is driven negative, the diode does not conduct.

Figure 22-33 shows the basic diode detector. The amplified modulated signal is supplied to the detector by the previous amplifier stage. When the input signal is positive, the diode conducts. A voltage develops across the diode load resistor R_l.

When the incoming signal is negative, there is no current in the diode circuit. The diode has rectified the modulated RF signal into pulses, or waves, of dc voltage. These waves have a frequency and amplitude of the audio wave.

To understand half-wave rectification, look at the input and output wave forms in **Figure 22-34.** The black

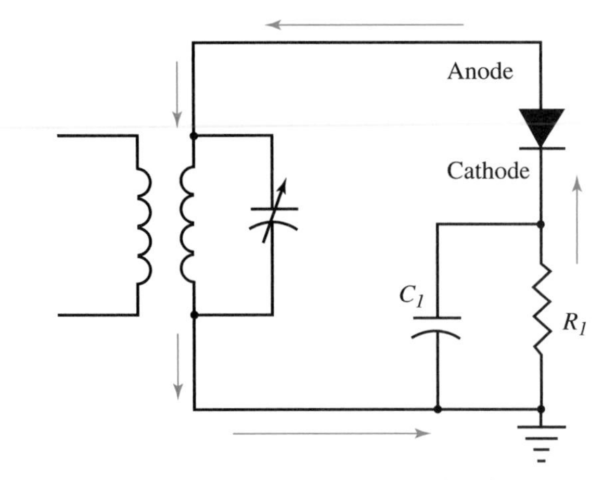

Figure 22-33. Basic diode detector circuit.

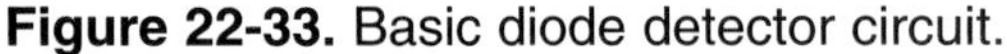

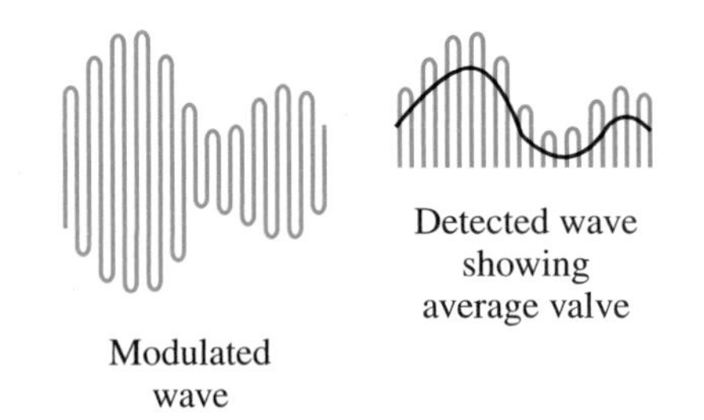

Figure 22-34. The average value of the dc component is represented by the black line.

curve in the output of the diode denotes the *average* dc value of the rectified voltage. Even though the signal has been rectified, the frequency is still too high to convert to a recognizable audio sound. The signal is rising and falling too fast for a speaker or headphone to convert. Raising the average dc output voltage will cause it to more closely reproduce the original audio input signal.

A filter capacitor of the correct value C_1, is shunted across R_1. The capacitor charges to the peak value of the signal and then slowly discharges slightly. The addition of the capacitor smoothes the signal so that it will be changed to a recognizable audio sound similar to the original sound mixed with the carrier wave and transmitted. The improved output waveform resulting from filtering is shown in **Figure 22-35.**

The time constant of R_1C_1 should be long when compared to the RF cycle and will deter C_1 from discharging to a low value. Likewise, this time constant should be

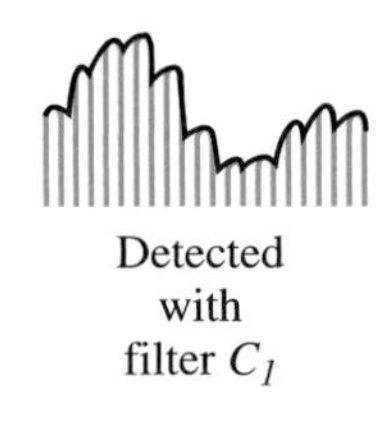

Figure 22-35. The dc average of the wave from Figure 22-34 is raised by adding the filter capacitor.

short when compared to the AF cycle. A properly selected R_1C_1 network will cause the voltage variations to follow the audio frequency cycle.

Review Questions for Section 22.6

1. Define demodulation?
2. Demodulation is a form of:
   a. rectification.
   b. detection.
   c. amplification.
   d. None of the above.
3. What was the drawback of the TRF receiver?
4. The ability of a receiver to choose a single frequency is __________.
5. The ability of a receiver to respond to weak incoming signals is __________.
6. __________ __________ are narrow band amplifiers that can amplify only the band, or frequencies, picked up by that receiver.

22.7 THE SUPERHETERODYNE RECEIVER

The superheterodyne circuit was developed as a solution to the problems of the TRF receiver. The term ***heterodyning*** means mixing of signals. Heterodyning converts all incoming signals to a *single intermediate frequency.* This signal can be then amplified with little loss and distortion.

See **Figure 22-36.** The signal picked up by the antenna is fed first to a stage of radio frequency amplification (RF amp). The output of the RF amp is then fed to the ***mixer*** or ***converter stage.***

The output of a ***local oscillator*** is also fed to the converter. When two signals are mixed together, four signals appear in the output. These signals include the two original signals (the RF signal and the signal from the local oscillator), the sum of the two signals, and the difference between the two signals. For example, if a 1000 kHz signal is mixed with a 1455 kHz local oscillator signal, appearing in the output is the 1000 kHz original signal, the 1455 kHz oscillator frequency, a 2455 kHz frequency sum, and a 455 kHz frequency difference.

The ***beat frequency*** is the name given to the combination of the two frequencies. The beat frequency of 455 kHz is key to the study of the superheterodyne receiver. A radio station is selected using the tuning circuits of the receiver. Turning the tuning knob on the front panel varies the capacitance of the tuning circuit. Attached to the shaft

PROJECT 22-3: Building a Four Transistor Radio

A four-transistor radio has good reception on local stations, and it is not very costly to build. Using the project schematic and parts list shown, you can construct your own radio receiver.

Station selection is done by tuned circuit L_1C_1. The signal is detected by diode CR_1, and amplified by four transistor stages to drive the speaker and to increase the sensitivity of the detector. The grounded emitter configuration is used by Q_2, Q_3, and Q_4. Interstage coupling is provided by C_4, C_5, and C_6.

It is best to build the radio on a breadboard first. After tests have been completed and your receiver is working, you can solder your components to a printed circuit board and design a case to hold the radio.

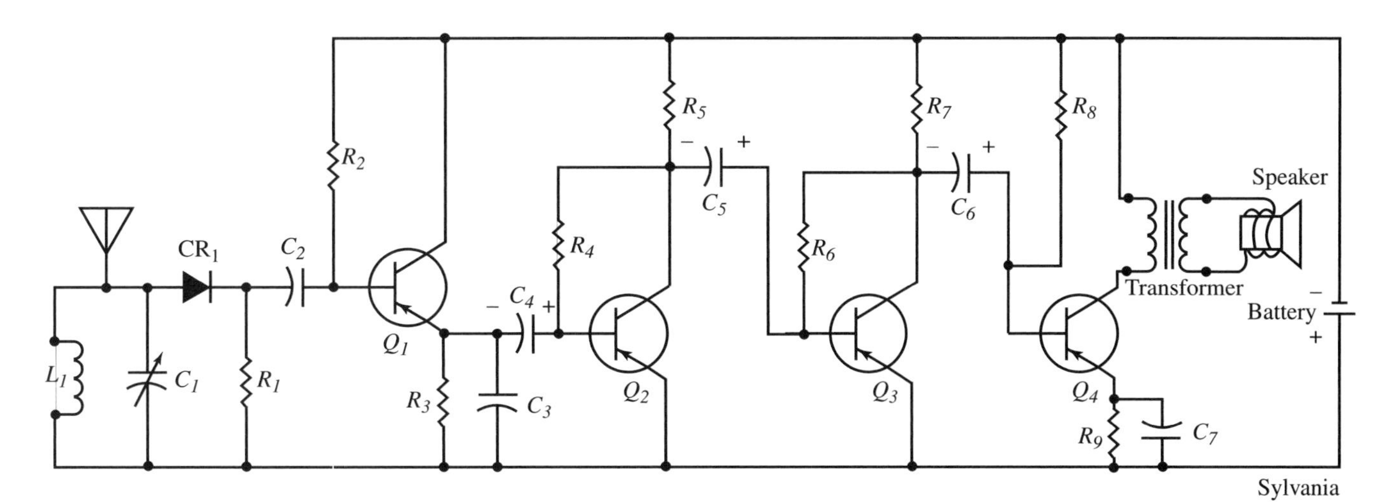

Parts List

– 6 volt battery	– 3-6 ohm speaker
C_1 – 0.365 pF tuning capacitor	Q_1, Q_2, Q_3, Q_4 – Sylvania 2N 1265, RCA SK 3003
C_2, C_3 – 0.02 µF ceramic disc or paper	R_1, R_8 – 470 kΩ, 1/2 W
C_4, C_5, C_6 – 1.0 µF electrolytic, 15 V	R_2, R_4, R_6 – 220 kΩ, 1/2 W
C_7 – 25.0 µF electrolytic, 15 V	R_3, R_5, R_7 – 2.2 kΩ, 1/2 W
CR_1 – Sylvania 1N64 or 1N34 diode or RCA SK 3087	R_9 – 100 Ω, 1 W
L_1 – Ferri-loopstick antenna coil	– Argonne AR-133 (pri. = 10 kΩ , sec. = 3.2 Ω, 100 mW)

of the tuning capacitor is another tuning capacitor that adjusts the frequency of the local oscillator, **Figure 22-37.**

These capacitors operate in step with each other. They provide a change in oscillator frequency as the

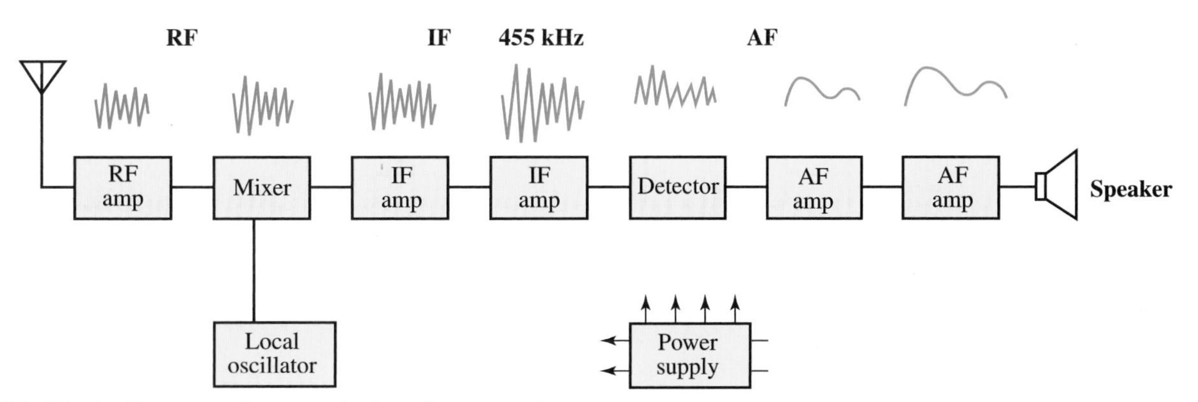

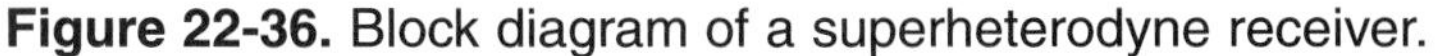
Figure 22-36. Block diagram of a superheterodyne receiver.

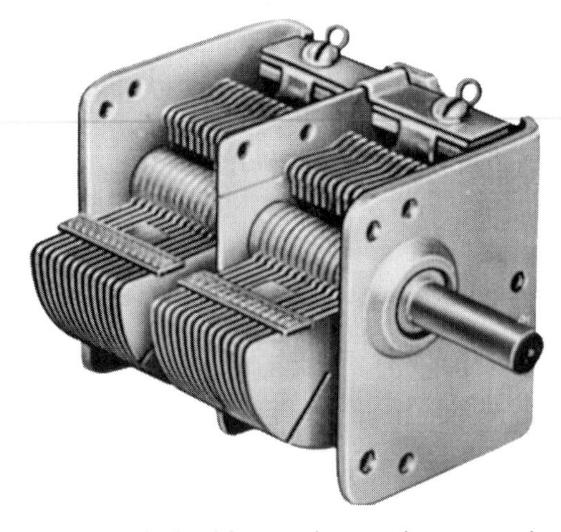

Figure 22-37. A double section tuning capacitor. (J.W. Miller Co.)

tuned frequency is changed. They always maintain a fixed difference, or beat frequency, of 455 kHz. This frequency is called the ***intermediate frequency,*** or ***IF.*** The IF output is then amplified by two stages of voltage amplification and fed to the detector. The ***detector*** output is an audio frequency voltage. It is amplified enough to operate the power amplifier and speaker. The waveforms at each stage are shown in Figure 22-36.

An eight-transistor AM superheterodyne receiver is shown in **Figure 22-38.** This circuit is part of the Graymark AM transistor radio. Each part of this superheterodyne receiver will be discussed in detail.

Mixer

The mixer, or frequency converter, performs three functions, **Figure 22-39.** Mixing is just one of the functions. The mixer also converts the incoming signal to a new frequency, called the intermediate frequency. This is done by mixing it with a signal produced by a local oscillator. At the same time, it provides amplification.

The mixer is a nonlinear circuit. Signals are combined to produce the sum and the difference frequencies of the original signals. The properties of a mixer can be used to produce a signal of the correct frequency for the IF amplifier. A signal must be produced locally (within the receiver) by an oscillator. This signal and the signal of interest are mixed. The original two signals, the sum signal, and the difference signal appear at the output of the mixer. An example is shown in **Figure 22-40.**

Notice that the signals of many stations could be mixed to a new frequency. However, the IF amplifier will

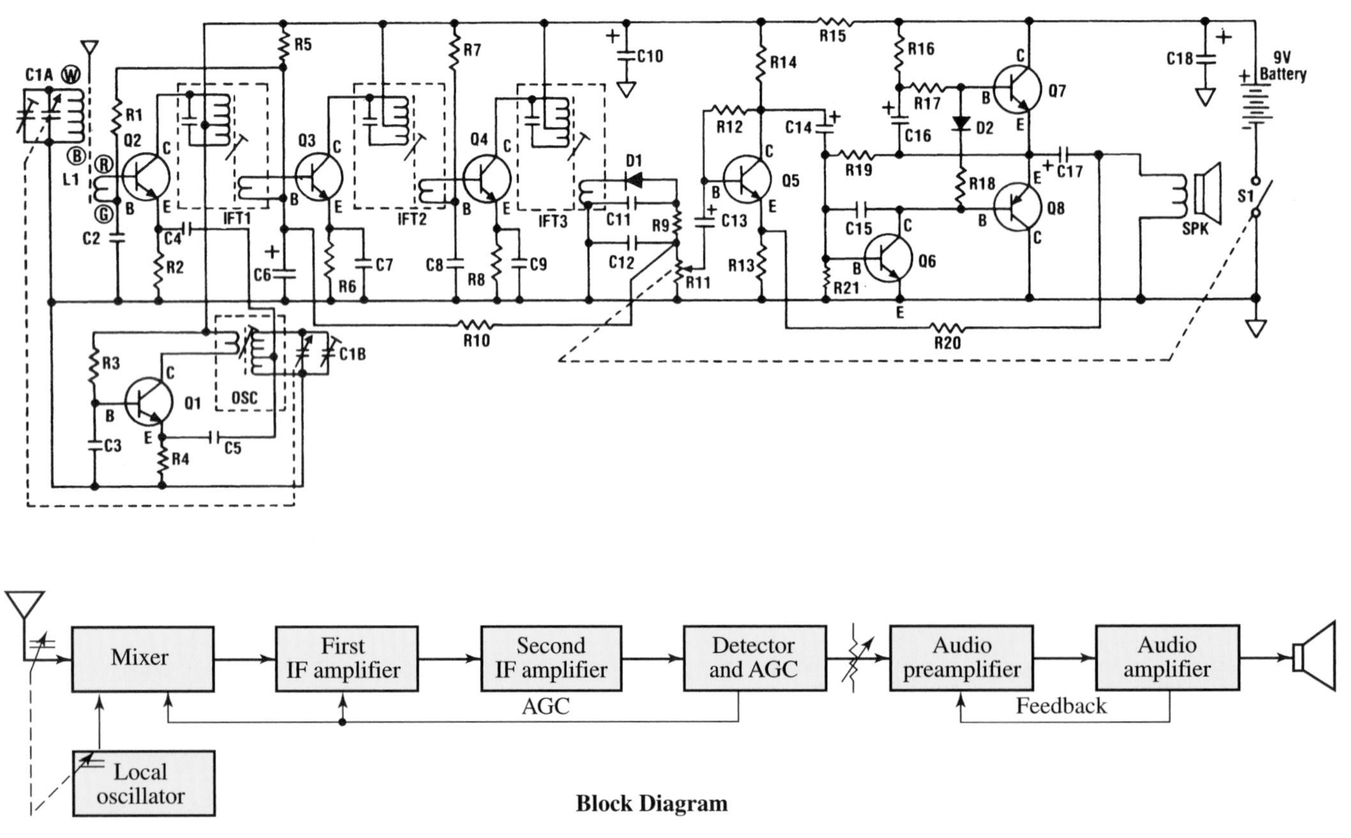

Figure 22-38. Model 536, eight-transistor radio. (Graymark)

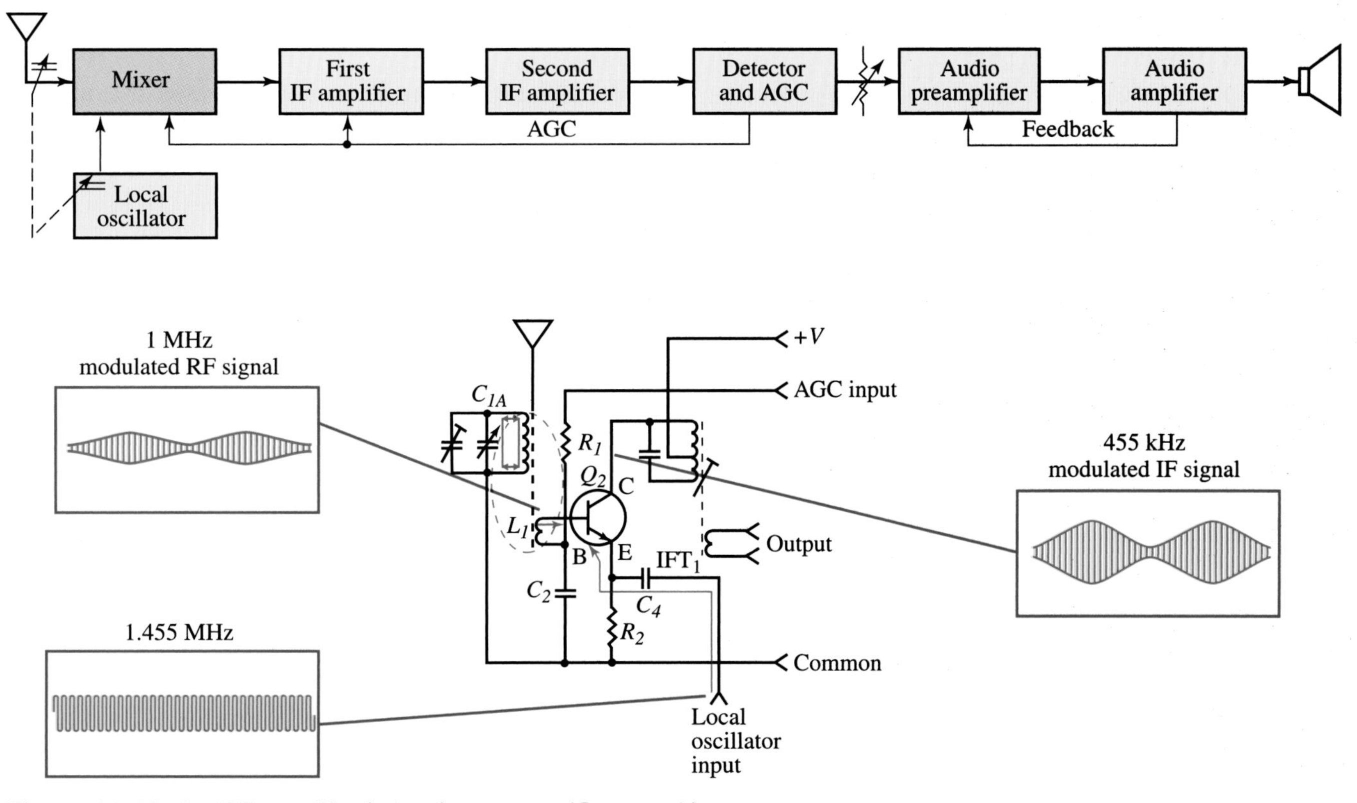

Figure 22-39. An RF amplifier/mixer/converter. (Graymark)

	Frequency (MHz)	Abbreviation
Incoming	1.000	RF
Local oscillator	1.455	LO
Local oscillator *plus* incoming	2.455	LO + RF
Local oscillator *minus* incoming	0.455	LO – RF = IF

Figure 22-40. Frequencies of signals being mixed. Their abbreviations are also shown.

	Frequency (MHz)
Desired RF	1.000
LO	1.455
LO – RF = IF	0.455
Image (LO + IF)	1.910
Image – LO = IF	0.455

Figure 22-41. Frequencies of desired and image signals.

not amplify these because they are not within the IF passband. The local oscillator is designed to be 455 kHz above the desired signal. The incoming signal must pass through a tuned antenna circuit to get to the mixer. This is because there are two possible signals that can be mixed to result in the IF. One is the desired signal, the other is called an ***image.*** See **Figure 22-41.**

Images are suppressed in front of the mixer by the tuned antenna circuit. The more selectivity in front of the mixer, the less of a problem images will be.

The incoming signal frequency, the local oscillator frequency, the sum, and the difference of the two are all presented to the input of the first IF amplifier. The IF amplifier is designed to amplify only the difference frequency (455 kHz). It will reject the other three frequencies because they fall outside of the passband of the IF amplifier.

The partial schematic shown in Figure 22-39 is the mixer used in the receiver of Figure 22-38. The input circuit is composed of tuning and trimmer capacitors C_{1A}, and coil L_1 (also serving as the antenna). This is a high Q resonant circuit. It is designed to select the incoming signal and to reject images. L_1 has a winding joined to the antenna circuit by an inductive couple. This winding will couple the selected signal to the base of Q_2. Bias current for Q_2 is provided by R_1. C_1 bypasses one end of the coupling winding to ground. The resonant circuit connected to the collector of Q_2 is tuned to the intermediate frequency. The emitter resistor of Q_2, R_2, provides bias voltage. The Q_2 emitter is the point at which the local oscillator output is put into the mixer. It moves through coupling capacitor C_4. Mixer gain is controlled by automatic gain control (AGC) feedback through the resistors R_{10} and R_1.

Local Oscillator

Oscillator pulling is caused when the local oscillator is inserted as part of the mixer. There is also an increase in distortion products. Consequently, although it uses more components, a separate local oscillator creates more stable design. **Figure 22-42** shows a radio with a separate local oscillator.

Transistor Q_1 is connected as a variable oscillator. Variable capacitor C_{1B} is mechanically coupled to the mixer input circuit. Capacitor C_{1B} is coupled to the tuning capacitor C_{1A}. The frequency of the signal made by the local oscillator will be adjusted to 455 kHz higher than the received signal. The positive feedback needed for oscillation is returned to the Q_1 emitter through C_5.

IF Amplifiers

IF amplifiers set selectivity and provide most of the voltage gain in the superheterodyne circuit. IF amplifiers are nontunable (except for alignment) radio frequency amplifiers. Because they amplify a fixed frequency, they can be quite useful. Gain and bandwidth can be tailored to meet the needs of the receiver. Some limits to this include:

- The lower the frequency the easier it is to obtain a narrow bandwidth.
- The IF frequency will set the image frequency (see mixer discussion). Thus, the IF frequency should be as high as practical. In AM broadcast receivers, 455 kHz has been chosen to be the best frequency. If the intermediate frequency is lower, the image response would suffer. And if higher, it would be harder to reach the proper bandwidth with just two IF stages.

A lower intermediate frequency permits more narrow bandwidth because of ***arithmetical selectivity.*** To explain this concept, consider two signals, one at 1.00 MHz and the other at 1.01 MHz. The separation of these two signals is 10 kHz which is a difference of 1%. But when both signals are converted to their intermediate frequencies in the mixer, the 1.00 MHz signal becomes 455 kHz. The 1.01 MHz signal becomes 465 kHz. The difference of 10 kHz is now 2.2%, or more than twice the percentage difference, of the intermediate frequency. The signals can now be easily separated by the IF amplifier.

The standard intermediate frequencies for receivers are:

AM receivers	455 kHz
FM receivers	10.7 MHz
Television receivers	41– 46 MHz

There are two IF amplifiers used in the AM receiver shown in **Figure 22-43.** Two transistor stages and three

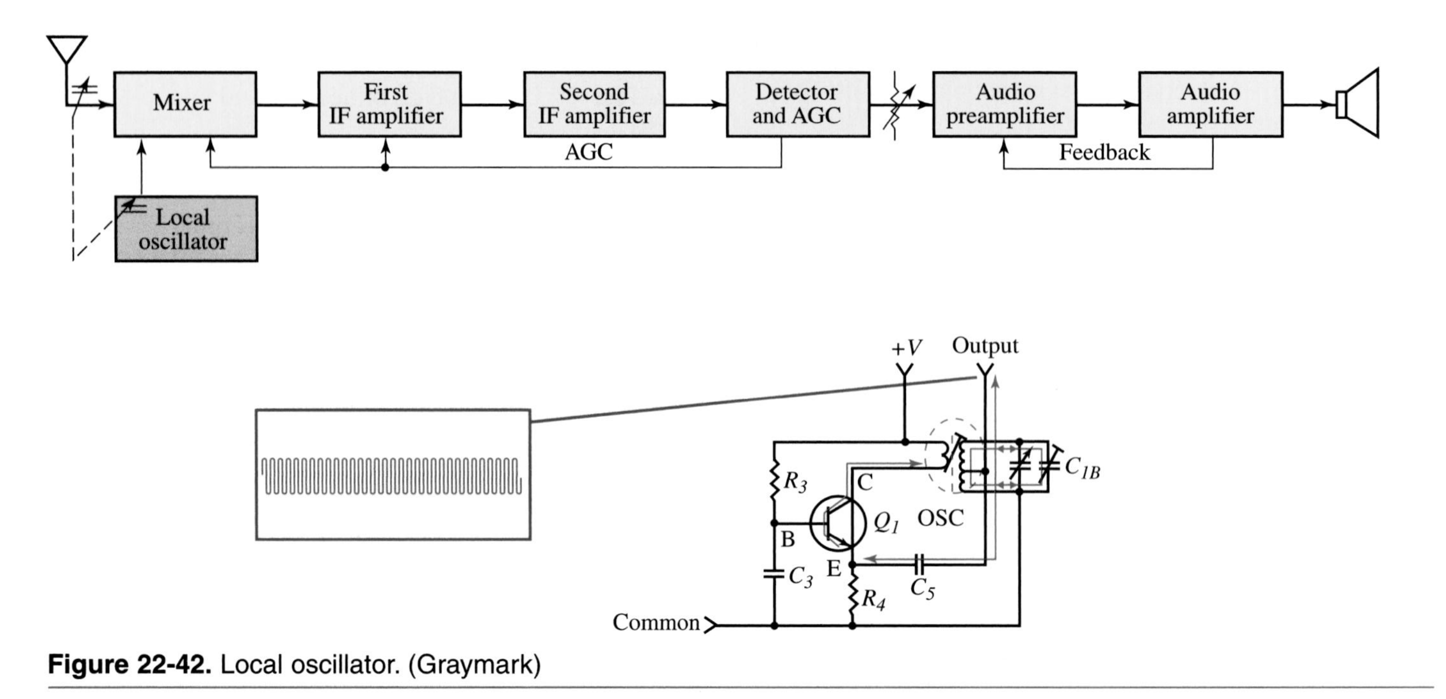

Figure 22-42. Local oscillator. (Graymark)

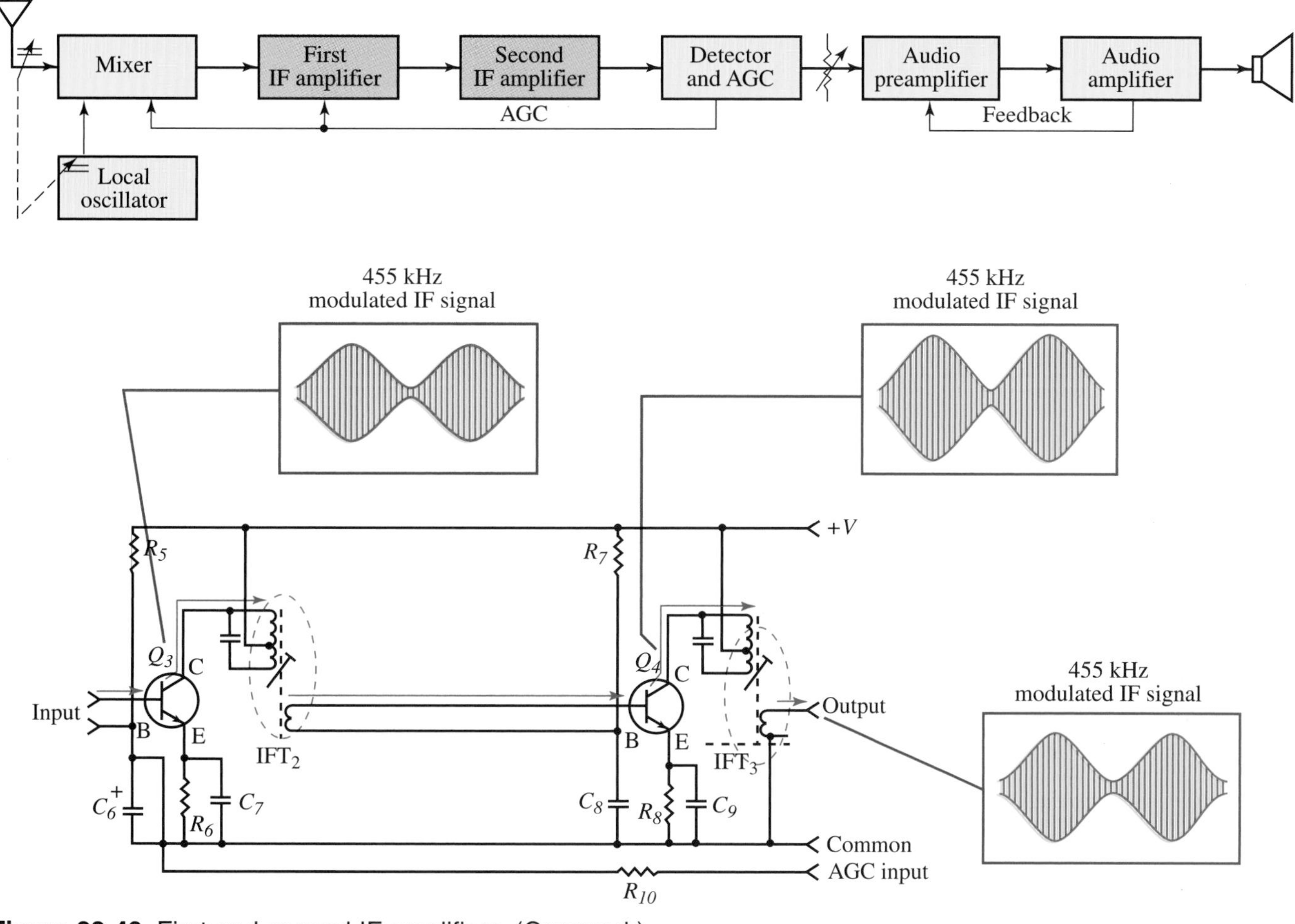

Figure 22-43. First and second IF amplifiers. (Graymark)

resonant circuits are provided to achieve the gain and passband required.

Transistors Q_3 and Q_4, each amplify the IF signal. The transformers IFT_1 (Figure 22-38), IFT_2, and IFT_3, and their internal capacitors, each have a narrow passband. This keeps out unwanted frequencies. The transistor Q_3 is connected to the output of the mixer transistor through a winding with an inductive coupling, transformer IFT_1. The operating bias for transistor Q_3 is developed by resistor R_5 and emitter resistor R_6. The bias current, and thus the collector current, are controlled by the AGC voltage. This voltage is developed at the output of the third IF transformer, IFT_3. Each of three resonant circuits are located in the IF transformers. All are tuned (user adjustable) to 455 kHz, the intermediate frequency. The Q and coupling coefficient are designed to provide the proper bandpass.

These factors are crucial. They are an integral part of the design and are not adjustable. Gain is fixed and maximized by proper biasing. The emitter resistor is RF bypassed. If this was not done, the negative feedback developed across the emitter resistor would reduce the gain a great deal.

The second IF stage is like the first IF stage. The gain is fixed and maximized by proper biasing with resistors R_7 and R_6. The emitter resistor is RF bypassed so that the negative feedback developed across the emitter resistor will not reduce the gain.

Detector

The detector circuit shown in **Figure 22-44** is much the same as is used in a simple crystal receiver. However, because there are two stages of IF amplification before the detector, and more gain is provided by the mixer stage, weak signals can be detected.

The gain in front of the detector allows the signal to overcome the threshold voltage of the detector diode. The detector detects the modulation information from the received signal. This is done by rectifying the amplified signal, then filtering the remaining RF from the signal. The detector diode (D_1) rectifies the RF modulated signal. Note that the envelope of the RF waveform contains the audio information first used to modulate the RF carrier at the transmitter.

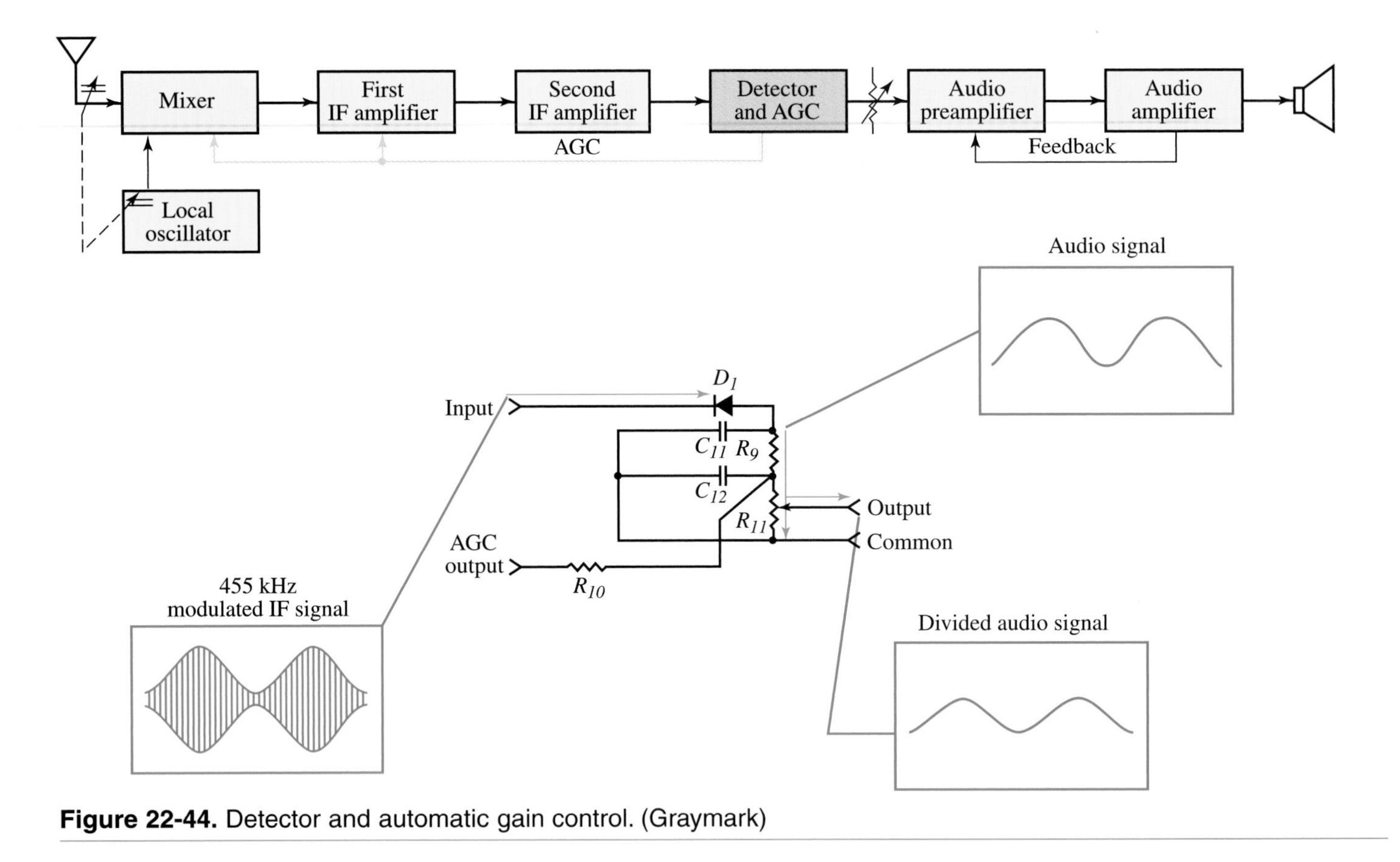

Figure 22-44. Detector and automatic gain control. (Graymark)

The RF is removed (bypassed to ground) by C_{12}. This occurs because the 0.02 µF capacitor presents a low impedance path for the RF signal, but very high impedance for audio frequencies.

Automatic Gain Control

Automatic gain control, or ***AGC,*** keeps the audio output level constant despite the varying strengths of the signals. AGC rectifies and filters the output of the IF stages. This signal is used to control the gain of the preceding stages. In this manner, input signals with a strength difference of 40 decibels (dB) can cause as little as 3 dB change in audio output level. *Without AGC, the volume would have to be adjusted as each station was tuned in.* AGC is also known as AVC (automatic volume control).

Refer again to Figure 22-44. The AGC voltage is taken from the junction of C_{12} and R_{10}. Resistor R_{10} keeps the audio from the further filtering by C_6, Figure 22-43. The AGC voltage appears across C_6. This AGC voltage controls the current into the bases of Q_2 and Q_3, Figure 22-38. The gain of transistors Q_2 and Q_3 are directly proportional to their collector currents. Detector diode D_1 is connected to supply a negative voltage that increases as signal strength increases. This negative voltage is applied to the bases of Q_2 and Q_3. It controls the gain. Therefore, as the AGC voltage increases, the gain of Q_2 and Q_3 goes down. Note that the detector diode is used both as an audio detector and AGC rectifier.

Audio Preamplifier

The audio preamplifier has a high impedance input. This provides a minimum load on the detector output. Refer to **Figure 22-45.** Variable resistor R_{11} is used to select the detector signal at different voltage levels. The preamp gain varies. It is controlled by the negative feedback from the audio amplifier through R_{20}, **Figure 22-46.**

Audio Amplifier

The audio output from the detector will have an amplitude of several volts. The impedance level will be fairly high. This means that little power will be available to drive the loudspeaker. In order to match the low impedance loudspeaker to the high impedance detector output, an audio amplifier with a low impedance output is required. In Figure 22-46, transistors Q_6, Q_7, and Q_8 do this.

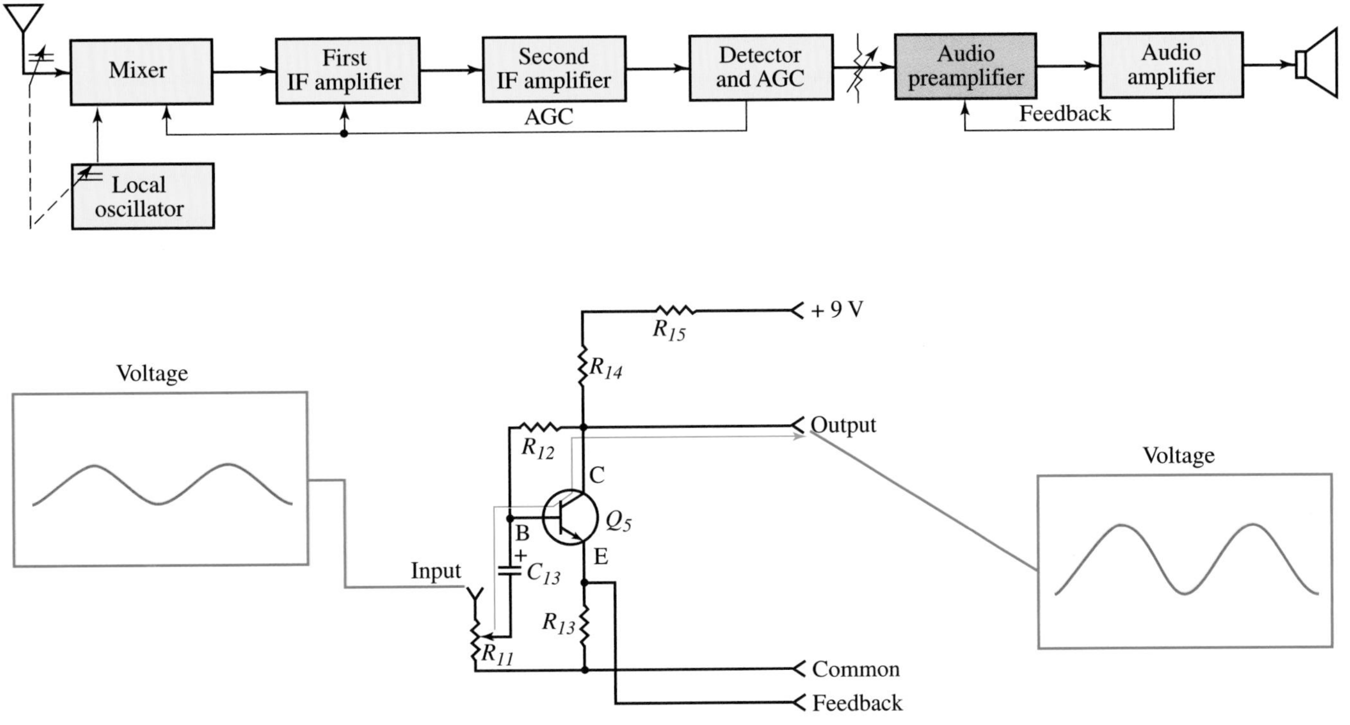

Figure 22-45. Audio preamplifier (Graymark).

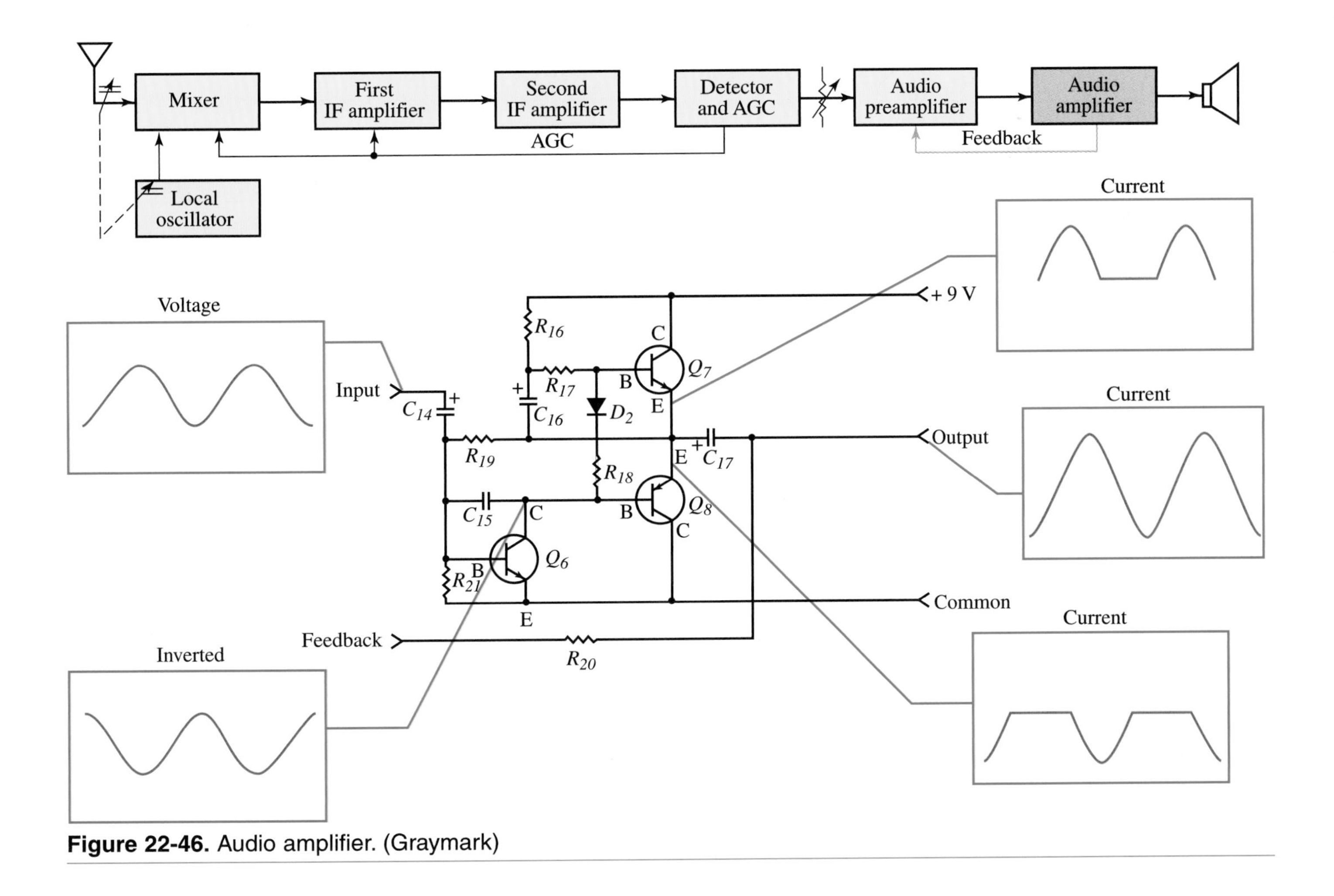

Figure 22-46. Audio amplifier. (Graymark)

In the previous stage, transistor Q_5 is biased Class A. It provides enough voltage and power gain to drive the output transistors, Q_7 and Q_8. The output stage is composed of Q_7 (NPN) and Q_6 (PNP). During the positive swing, Q_7 will conduct. During the negative swing, Q_8 will conduct. It is crucial that Q_8 and Q_7 not be on at the same time as this would cause a large current to flow through both transistors. Diode D_2 prevents this event from occurring.

It is best to use all the voltage produced by the nine-volt battery. To accomplish this, the junction of the Q_7 and Q_8 emitter must not be allowed to idle at 4.5 volts (9 V ÷ 2) under no signal conditions. This error would make a peak-to-peak audio signal of 9 volts ready for use before the signal is clipped.

If the no signal voltage is not centered at the halfway point, one side of the signal swing will clip at a lower voltage. This action causes uneven clipping distortion, which limits the ready-to-use power output. The audio amplifier in Figure 22-46 achieves halfway idle bias voltage as follows.

Transistor Q_7 is an NPN device. A positive voltage applied to its base-emitter junction will cause it to conduct. Resistors R_{16}, and R_{17} provide bias current that causes Q_7 to turn on. If no other circuit elements are present, the bias current will cause the Q_7 emitter to rise to 9 V minus the collector emitter voltage drop.

To prevent this, the emitter of Q_7 is connected via R_{19} to the base of Q_6. Q_6 is also an NPN transistor. As a higher base current is applied to Q_7, the collector emitter resistance becomes lower.

The collector of Q_6 is connected via R_{18} and D_2 to the base of Q_7. As Q_6 becomes lower in resistance, it steals base current from Q_7. This increases the collector emitter resistance of Q_7. The network then becomes a voltage divider. To complete the circuit, think of the effect of the PNP transistor Q_8. If the base-emitter junction of Q_8 becomes forward biased, it will conduct. As stated earlier, it must not be allowed to conduct at the same time as Q_7.

Under working conditions, capacitor C_{14} couples the audio signal to the base of driver/inverter Q_6. When the signal swings positive, Q_6 will conduct more. As a result, Q_8 will also conduct more. While this is happening, more current is stolen from the base of Q_7, and Q_7 conducts less.

On the negative swing of the signal, Q_6 will conduct less. This allows Q_8 to reverse bias. More bias current is then applied to Q_7, causing it to conduct more. As in many circuits, there are two conditions. These conditions are dc, or no signal conditions, and ac, or dynamic conditions. In order to study the circuit, both states should be understood and then combined.

Capacitor C_{17} is used to block dc current from flowing through the speaker.

Alignment

The superheterodyne receiver contains several tuned circuits. These include the primary and secondary windings of the IF transformers. These must be tuned to resonance, or maximum response, for those signals to be passed. Variable trimmer capacitors are connected in parallel with these coils to provide the adjustment, or ***alignment.***

Receivers often need adjustment as they age or after parts have been replaced in the service shop. "Peaking" a receiver to adjust it is also known as aligning the receiver. Tools needed for peaking include a signal generator, and an output indicator, such as an ac voltmeter.

A signal generator produces either a modulated or unmodulated RF wave. This can be selected by the controls on the panel. When an alignment job is needed, consult the technical manual for that radio or TV.

IC Radios

Most electronic parts of an AM/FM radio can be contained on one IC chip. See **Figure 22-47.** This is an excellent example of how microcircuits can reduce the size and cost, and provide improved performance over a transistor radio. The AM/FM IC contains most of the radio circuits active components. One of the interesting features of this integrated circuit is its ability to operate over a wide range of voltage supplies. Typically, it can operate up to 13 volts dc and down to two volts dc.

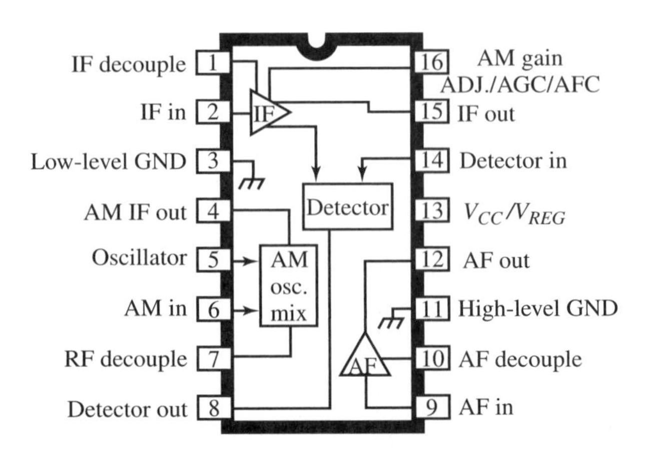

Figure 22-47. Pin diagram of an AM/FM integrated circuit. (Sprague Electric Co.)

Global Positioning Systems

Global Positioning Systems (GPS) are space-based radio positioning systems that provide 24-hour three-dimensional position, velocity, and time information to equipped users. The NAVSTAR system, operated by the U.S. Department of Defense, is the first global positioning system widely available to commercial enterprises and private citizens.

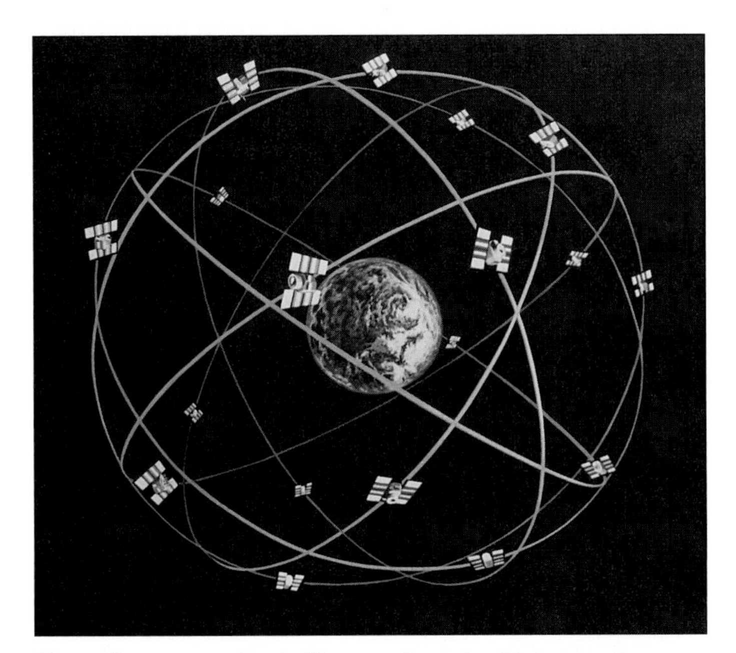

The solar-powered satellites are launched into nearly 11,000 mile circular orbits. At present, there are 21 operational satellites plus three on-orbit spares circling the globe in six orbital planes. Each of the satellites has an estimated life span of seven years and replacement satellites must be launched periodically. (NASA #S90-50865)

GPS satellites orbit the earth every 12 hours emitting continuous navigation signals on two different L-band frequencies. With a GPS receiver, these signals can be used to calculate time to within a millionth of a second, velocity within a fraction of mile per hour, and location within a few feet. The user does not need to transmit anything to the satellite, and the satellite is not aware of the user's presence. There also is no limit to the number of users that can be using the system at any one time.

For greater security, multiple satellites are used simultaneously to receive data. An encoded military signal is first received from a minimum of four satellites. The time required for the signal to travel from the satellites to the designated location is then calculated by a computer system. After the distance from each satellite is calculated, the information is converted to latitude and longitude coordinates. Coordinates can also be converted into an image on a computer screen. This computer screen illustrates the local regional map and then displays the exact location on the screen display.

In order to prevent the deciphering of secured information, the military limits public access of GPS signals to the *Coarse Acquisition code (C/A-code).* The C/A-code is less accurate, easier to jam, and easier to acquire. The *Precision code (P-code)*, is designed for authorized military users. To ensure that unauthorized users do not acquire the P-code, an encryption segment called *anti-spoofing (AS)*, can be implemented on the P-code.

GPS can be used to locate other satellites and for location systems of vehicles such as a fleet of trucks, ships, or airplanes. It can be used by weather bureaus to measure moisture content in the atmosphere and by geologists to measure *plate tectonics* (movements of the earth's crust).

GPS receivers have been developed for use in aircraft, ships, land vehicles, and for hand carrying.

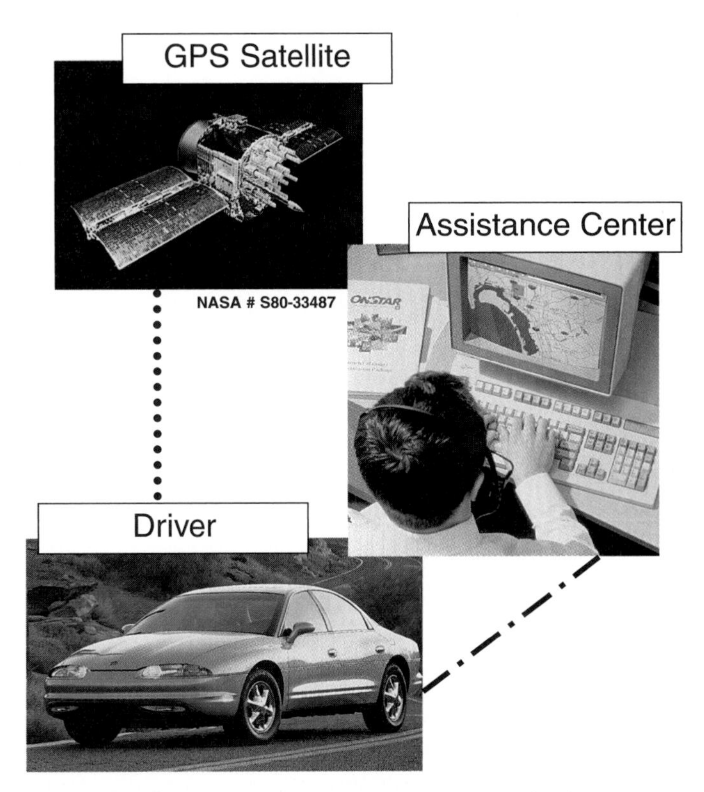

The OnStar® system by General Motors uses GPS technology and a cellular telephone to link equipped drivers to its 24-hour assistance center. The OnStar system can be used to dispatch emergency help, track stolen vehicles, perform diagnostics, unlock vehicle doors, offer alternate routes, and direct users to nearby lodging, food, ATMs, police/fire departments, and a wide range of other locations. (Used with permission of General Motors Corporation)

PROJECT 22-4: Building a Superheterodyne Receiver

This superheterodyne receiver (first shown in Figure 22-38) receives AM broadcast stations transmitting on frequencies between 550 and 1600 kHz. As with Project 22-2, the parts can be purchased individually from the parts list or as a package from Graymark Enterprises, Inc.

The RF amplifier-converter stage receives the modulated RF signal from the broadcast stations in your area. By tuning the radio, you select the modulated RF signal from one of the broadcast stations. In this stage, it is amplified and combined with the RF signal from the local

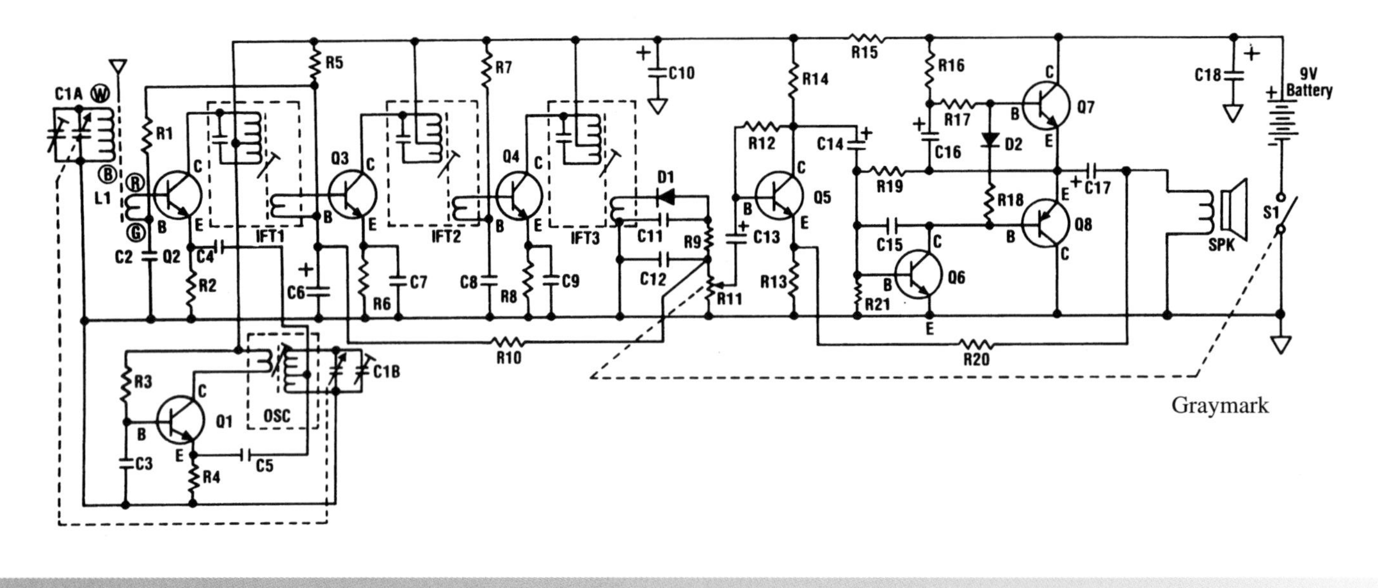

Graymark

Today, ***variable capacitance diodes*** are used in place of the large, bulky, variable capacitors to tune radio circuits. A diode has a natural capacitance caused by the space between the anode and cathode. The construction is similar to the typical capacitor constructed with two plates separated by an insulator. The amount of voltage applied to the variable capacitance diode determines the amount of capacitance. The variable capacitance diode is also incorporated into certain ICs.

Review Questions for Section 22.7

1. Define heterodyning.
2. How does the superheterodyne receiver differ from the TRF receiver?
3. The intermediate frequency in a superheterodyne radio is:
   a. 455 Hz.
   b. 45 kHz.
   c. 455 kHz.
   d. None of the above.
4. What three functions are performed by the mixer?
5. A local oscillator produces an unmodulated signal __________ kHz above the incoming frequency signal.
6. What is the purpose of the AGC circuit in a superheterodyne receiver?
7. What tools are needed to align a receiver?
8. The __________ __________ __________ is replacing the variable capacitor that utilizes large plates separated by air.

22.8 FM RECEIVER

A block diagram of a complete FM receiver is shown in **Figure 22-48.** Each block is labeled according to its function in the system.

The FM receiver is similar to the superheterodyne AM receiver, with three exceptions.

- The incoming signals to be tuned are from 88 to 108 megahertz.
- The IF frequency used in the FM radio is 10.7 megahertz. (Though, the same heterodyne principles apply as with the AM receiver.)
- The detection method used in an FM receiver is different.

Qty.	Symbol	Description
2	R_1, R_3	Resistor, 1/4 W, 5%, 470 kΩ
1	R_2	Resistor, 1/4 W, 5%, 1.5 kΩ
1	R_4	Resistor, 1/4 W, 5%, 5.6 kΩ
1	R_5	Resistor, 1/4 W, 5%, 56 kΩ
1	R_6	Resistor, 1/4 W, 5%, 1.2 kΩ
2	R_7, R_{12}	Resistor, 1/4 W, 5%, 560 kΩ
3	R_8, R_{17}, R_{20}	Resistor, 1/4 W, 5%, 1 kΩ
1	R_9	Resistor, 1/4 W, 5%, 680 Ω
1	R_{10}	Resistor, 1/4 W, 5%, 27 kΩ
1	R_{13}	Resistor, 1/4 W, 5%, 10 Ω
1	R_{14}	Resistor, 1/4 W, 5%, 3.9 kΩ
1	R_{15}	Resistor, 1/4 W, 5%, 150 Ω
2	R_{16}, R_{18}	Resistor, 1/4 W, 5%, 82 Ω
1	R_{19}	Resistor, 1/4 W, 5%, 12 kΩ
1	R_{21}	Resistor, 1/4 W, 5%, 2.7 kΩ
1	R_{11}	Petentiometer, 5 kΩ, with switch
1	C_1	Capacitor, Tuning
3	C_2, C_3, C_5,	Capacitor, ceramic, 0.01 mF, marked
1	C_4	Capacitor, ceramic, 0.005 mF, marked
1	C_6, C_8, C_9, C_{11}, C_{12}	Capacitor, electrolytic, 10 mF, 10 volt or 16 volt
5	C_7	Capacitor, ceramic, 0.02 mF, marked
3	C_{10}	Capacitor, electrolytic, 100 mF, 10 volt
1	C_{13}, C_{17},C_{18},	Capacitor, electrolytic, 0.47 mF, 50 volt
1	C_{14}	Capacitor, electrolytic, 4.7 mF, 25 volt
1	C_{15}	Capacitor, ceramic, 0.001 mF, marked
1	C_{16}	Capacitor, electrolytic, 47 mF, 10 volt
6	Q_1, Q_2, Q_3, Q_4, Q_6, Q_7	Transistor, 2SC1815Y, Silicon, NPN
1	Q_5	Transistor, 2SC1815GR, Silicon, NPN
1	Q_8	Transistor, 2SA1015Y, Silicon, PNP
1	D_1	Diode, 1N60
1	D_2	Diode
1	L_1	Antenna core with coil
1		Antenna holder (mounted to Printed Circut Board)
2	OSC	Coil, oscillator, red core
2	IFT_1	Transformer, IF, yellow core
2	IFT_2	Transformer, IF, white core
2	IFT_3	Transformer, IF, black core
1	SPK	Speaker, 8 Ω
1		Battery connector
1		Printed Circuit Board
2		Speaker holder
3		Sponge, double-sided adhesive type, 15 × 10 mm
1		Solder, rosin core, 600 mm
1		Wire, bare, solid, 22 gauge, 1100 mm
1		Wire stranded, black, 26 gauge, 1100 mm
1		Breadboard

oscillator in your radio to form an intermediate frequency signal. The IF signal is then amplified by the IF amplifier stage and passed on to the detector stage. The detector stage removes the carrier frequency and leaves only the AF signal that was originally combined with the carrier frequency at the broadcast station. This process is demodulation. The small AF signal is amplified by the first AF amplifier stage and then by the audio power amplifier stage. It is now strong enough to operate the speaker.

FM Detection

In the AM radio, the detector is sensitive to amplitude variations. An FM detector must be sensitive to frequency variations and remove them from the FM wave. The FM detector must produce a varying amplitude audio signal from the frequency variations contained in the FM wave.

Refer to **Figure 22-49.** Assume that a circuit has a peak response at its resonant frequency. All frequencies,

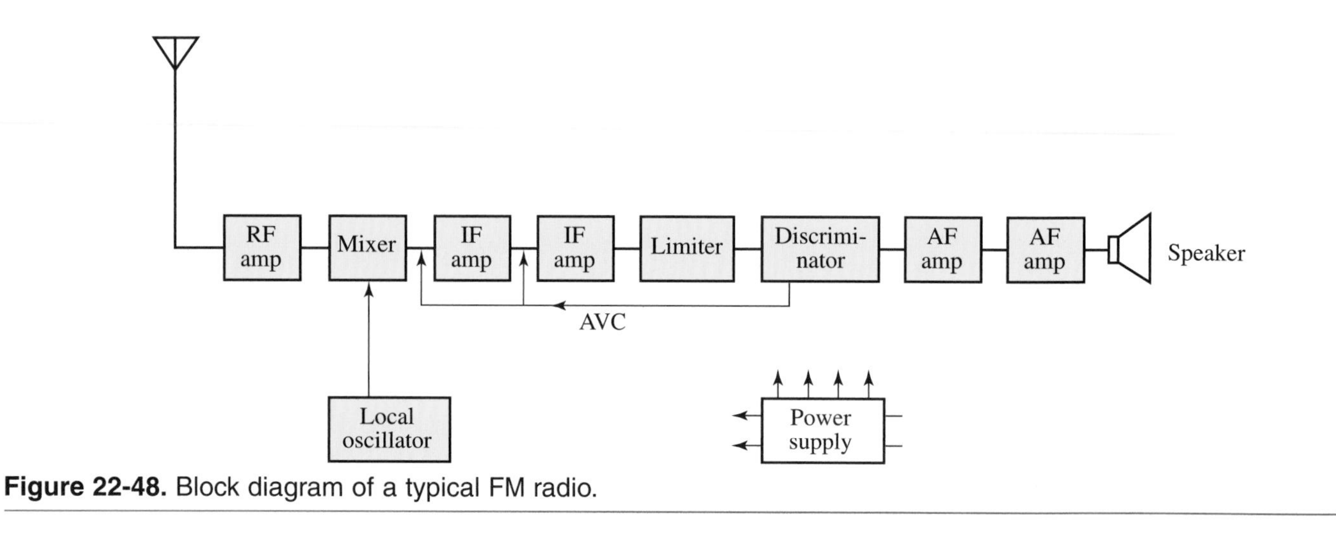

Figure 22-48. Block diagram of a typical FM radio.

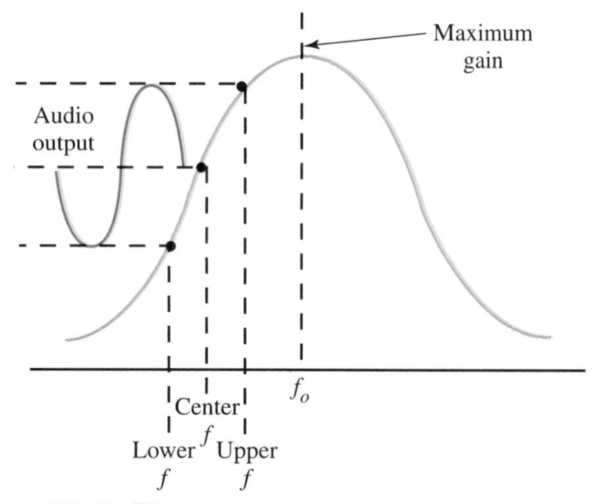

Figure 22-49. These curves demonstrate slope detection.

other than resonance, will have a lesser response. If the center frequency of an FM wave is on the slope of the resonant response curve, a higher frequency will produce a higher response in voltage. A lower frequency will produce a lower voltage response. The curve in Figure 22-49 reveals that the amplitude of the output wave is the result of the maximum deviation of the FM signal. The frequency of the audio output depends on the rate of FM signal frequency change.

Discriminator

The ***discriminator*** in the FM receiver takes the place of the detector in the AM receiver. It turns the frequency variations of the incoming wave into an audio signal. The discriminator in **Figure 22-50** uses three tuned circuits. In this circuit, L_1C_1 is tuned to the center frequency. L_2C_2 is tuned to above center frequency. L_3C_3 is tuned to below center frequency by an equal amount.

At center frequency, equal voltages are developed across the tuned circuits. D_1 and D_2 conduct equally. The voltages across R_1 and R_2 are equal and opposite in polarity. The circuit output is zero. If the input frequency increases above center, L_2C_2 develops a higher voltage. Then D_1 conducts more than D_2 and unequal voltages develop across R_1 and R_2.

The difference between these voltage drops will be the audio signal. The output, therefore, is a voltage wave varying at the rate of frequency change at the input. Its amplitude depends upon the maximum deviation. The capacitors across the output of the discriminator filter out any remaining radio frequencies.

The discriminator in **Figure 22-51** is a circuit encountered in some FM receivers. L_1 and C_1 are tuned to center frequency. At frequencies above resonance, the tuned circuit becomes more inductive. At frequencies below resonance, the circuit becomes more capacitive. The out of phase conditions produce voltages that determine which diode will conduct. The output is an audio wave. In advanced courses, you will study this type of detection in more detail.

Note that each diode in the discriminator must have equal conduction capabilities. This means that the semiconductor diodes used must be in ***matched pairs.***

Ratio detector

Another type of FM detector is drawn in **Figure 22-52.** It is called the ***ratio detector.*** The diodes are connected in series with the tuned circuit. At center frequencies, both diodes conduct during half cycles. The voltage across R_1 and R_2 charges C_1 to the output voltage. Capacitor C_1 remains charged because the time constant of C_1R_1 and R_2 is no longer than the period of the incoming waves. C_2 and C_3 also charge to the voltage of C_1. When both D_1 and D_2 are conducting equally, the charge of C_2 equals C_3. They form a voltage divider. At the center point between C_2 and C_3 the voltage is zero.

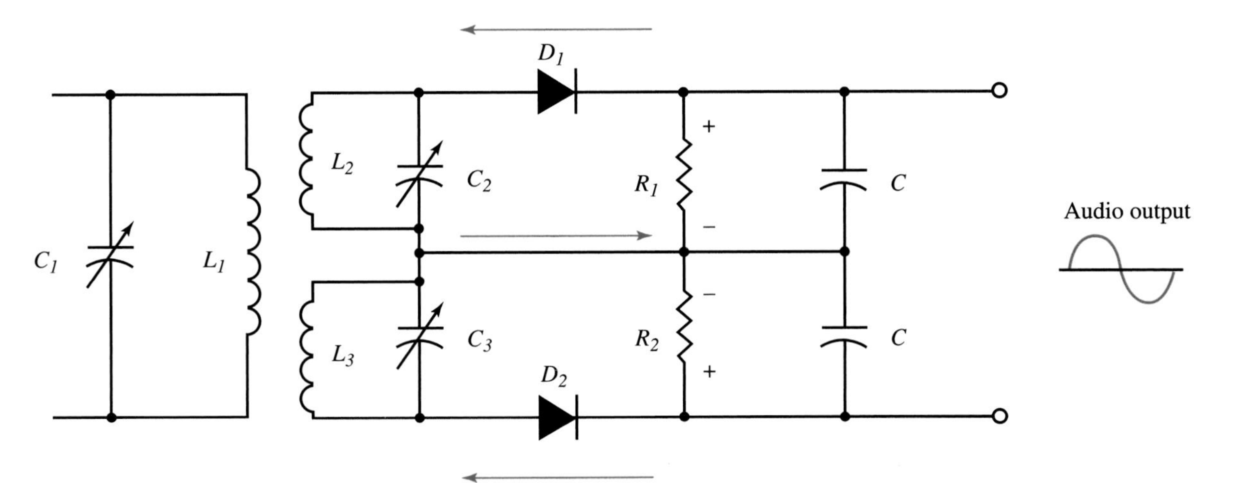

Figure 22-50. FM discriminator circuit using semiconductor diodes.

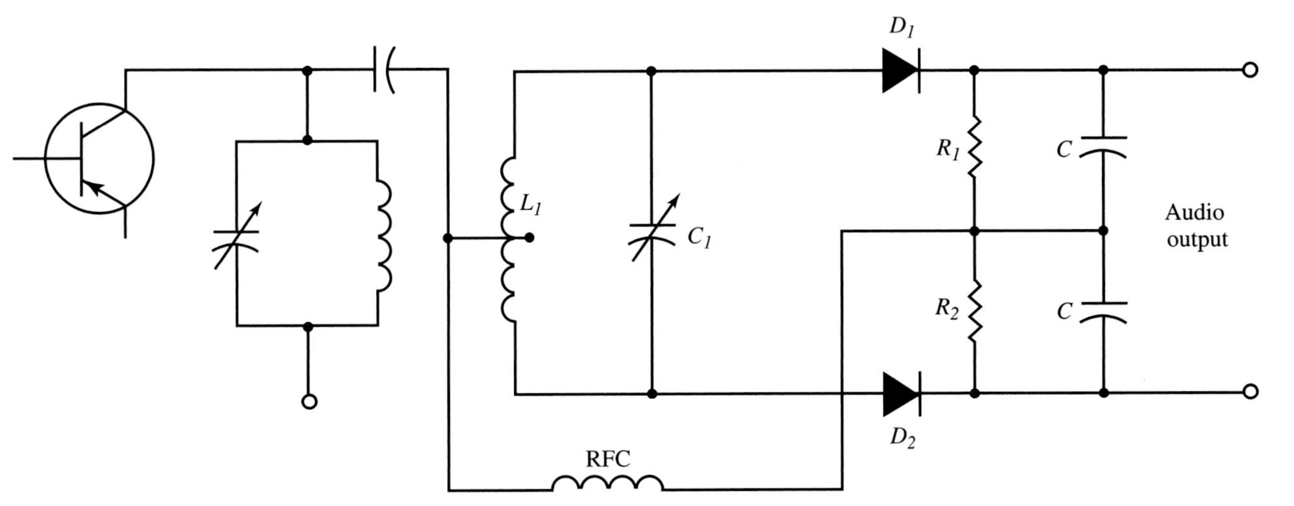

Figure 22-51. Foster Seeley discriminator using a special transformer designed for this purpose.

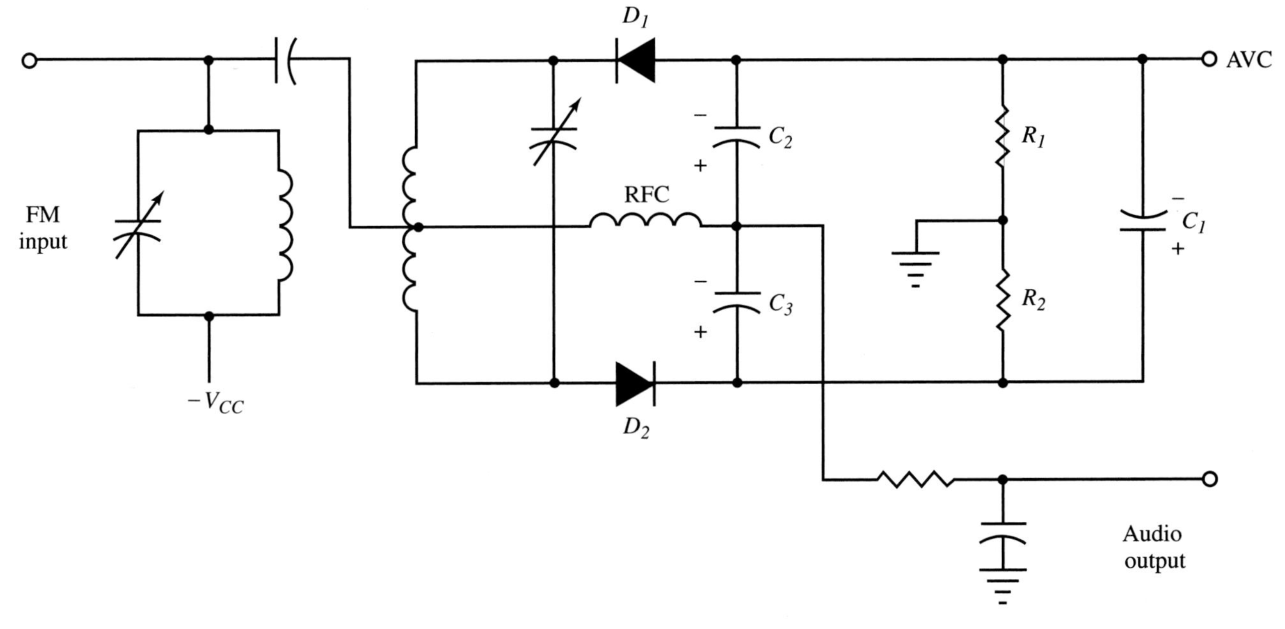

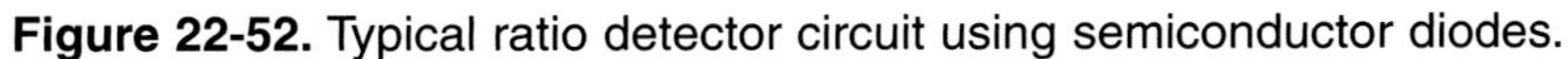
Figure 22-52. Typical ratio detector circuit using semiconductor diodes.

A frequency shift either below or above center frequency causes one diode to conduct more than the other. As a result, the voltages of C_2 and C_3 become unequal, but they will always total the voltage of C_1. This change of voltage at the junction of C_2 and C_3 is the result of the *ratio* of the unequal division of charges between C_2 and C_3. This will vary at the same audio rate as the rate of change of the FM signal.

Look again at the charge on C_1. It is the result of the carrier wave amplitude or signal strength. It is charged by half-wave rectification of the FM signal. It is, therefore, a fine point to pick off an automatic volume control voltage to feedback to previous stages to regulate stage gain.

Noise Limiting

FM radio receivers are sensitive to, and detect, frequency variations, not amplitude variations. Most noise and interference in radio reception are amplitude variations. They are called ***noise spikes.*** Noise spikes have little effect on the FM detector. Therefore, FM reception is mostly free of noise and disturbances.

The FM signal is held at a constant amplitude before detection in a discriminator circuit using a limiter. A schematic of a limiter stage is shown in **Figure 22-53.**

A limiter is an *overdriven* amplifier stage. If the incoming signal reaches a certain amplitude in voltage, it drives the transistor to cutoff (or saturation when the voltage is opposite in polarity). At either of these points, gain cannot increase. The output is confined within these limits. Any noise spikes will be clipped off.

Review Questions for Section 22.8

1. Draw the block diagram of an FM receiver.
2. What is the difference between the detection methods in an AM receiver and an FM receiver?
3. Most interferences in radio reception are amplitude variations called __________ __________.
4. The __________ circuit in an FM receiver does not allow signals to go over a certain amplitude after they leave the IF transformer and reach the discriminator circuit.

22.9 TRANSDUCERS

The final stage of the radio receiver is a transducer. The ***transducer*** converts the electrical energy of audio frequencies into sound energy (or vice-versa). Common transducer devices are microphones (discussed earlier in the chapter), speakers, and headphones.

Speakers

A common type of speaker is shown in **Figure 22-54.** The drawing diagramming the parts of this speaker appears in **Figure 22-55.** It is made with a permanent magnet. It is called a PM speaker.

In a PM speaker, a strong magnetic field is produced between the poles of a fixed permanent magnet. A small voice coil is hung in the air gap. It is attached to the speaker cone. The audio alternating currents are joined to

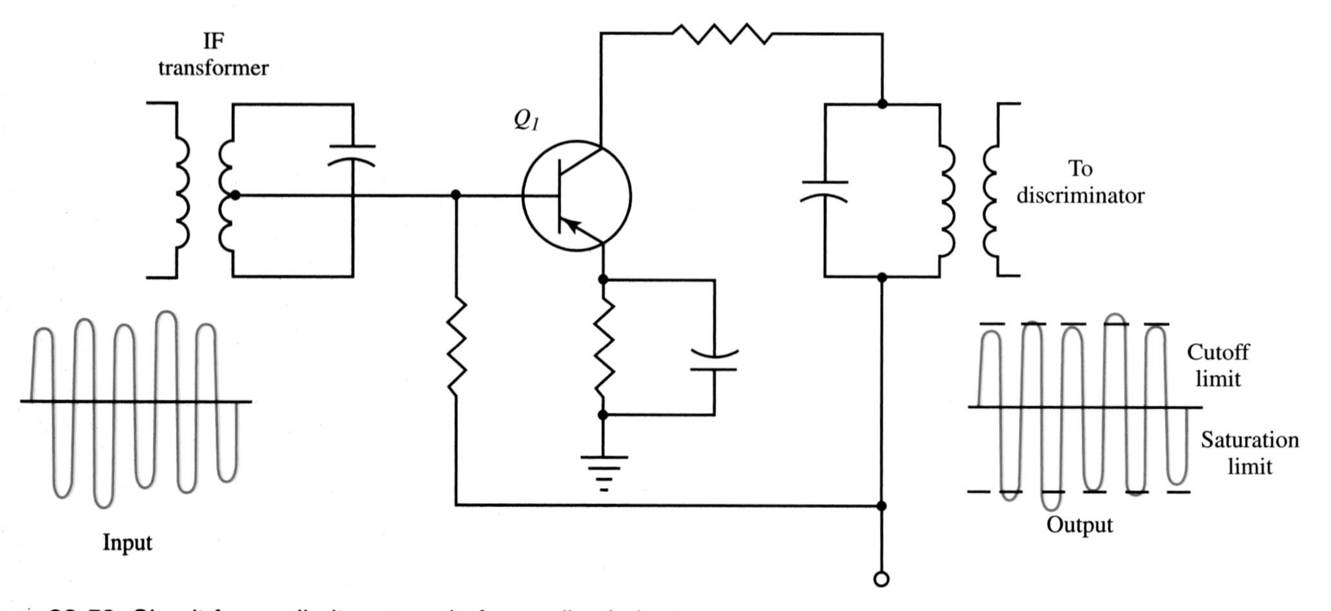

Figure 22-53. Circuit from a limiter stage before a discriminator.

Figure 22-54. Typical radio speaker.

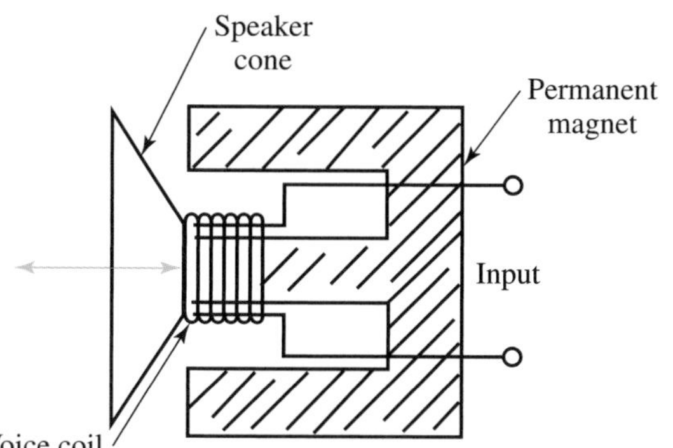

Figure 22-55. This sketch diagrams the parts of the PM speaker.

the voice coil. The action between the fixed field and the moving field causes the voice coil to move back and forth. This motion also causes the speaker cone to move back and forth. The air pressure, in the form of sound energy, changes back and forth, from highest to lowest pressure.

An electrodynamic speaker replaces the permanent magnet with an electromagnet, **Figure 22-56.** It works like the PM type. A strong source of direct current must be supplied to the electromagnet. This can come from the power supply. A common practice is to use the field coil of the speaker as the filter choke in the supply.

Special sizes and types of speakers have been developed. These provide the best response for certain bands of audio frequencies.

Low frequency speakers are ***woofers.*** High frequency speakers are called ***tweeters.*** A ***midrange*** speaker can be used to reproduce intermediate frequencies. Special filters and ***crossover networks*** allow signals of set

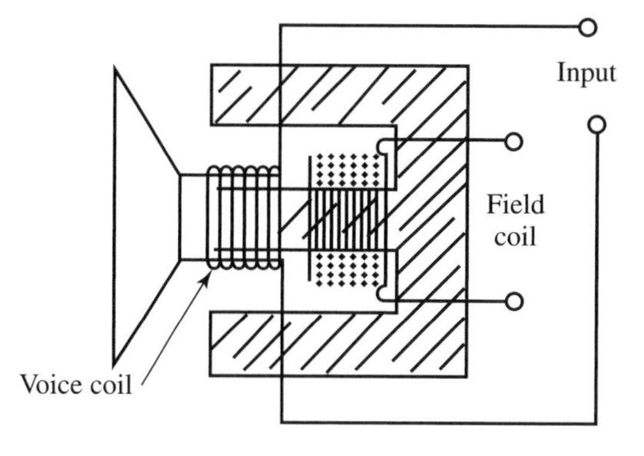

Figure 22-56. An electrodynamic speaker. The permanent field is replaced with an electromagnetic field.

frequency ranges to be channeled to the speaker that best reproduces the sound.

A crossover network is shown in **Figure 22-57.** Coil L is connected to the woofer. As the frequency of the sound increases, the reactance of L increases at the rate of $X_L = 2\pi fL$. The tweeter is connected through C. As the frequency increases, the reactance of C decreases at the rate of $X_C = 1/(2\pi fC)$. Values of L and C can be selected for the desired crossover point. This is usually between 400 and 1200 Hz. Note that at one frequency, X_L equals X_C. The response is equal for both speakers at this crossover frequency.

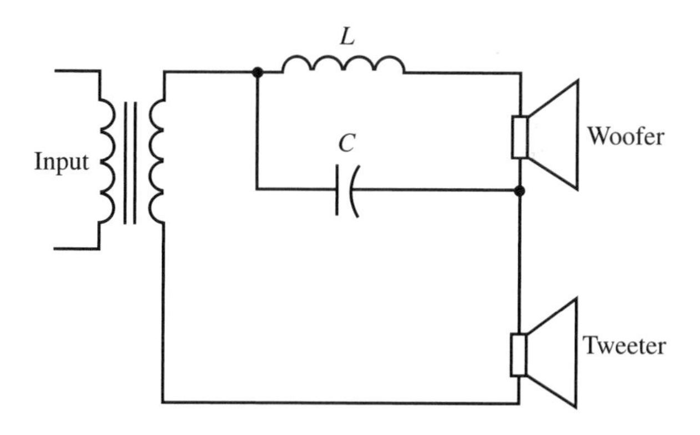

Figure 22-57. A basic crossover network design.

Headphones

Headphones provide excellent sound reproduction and allow for private listening. Like loudspeakers, there are two basic types of headphones: dynamic and electrostatic. See **Figure 22-58.**

Dynamic

Dynamic headphones are divided into two groups. Pressure-type dynamic headphones require an air seal

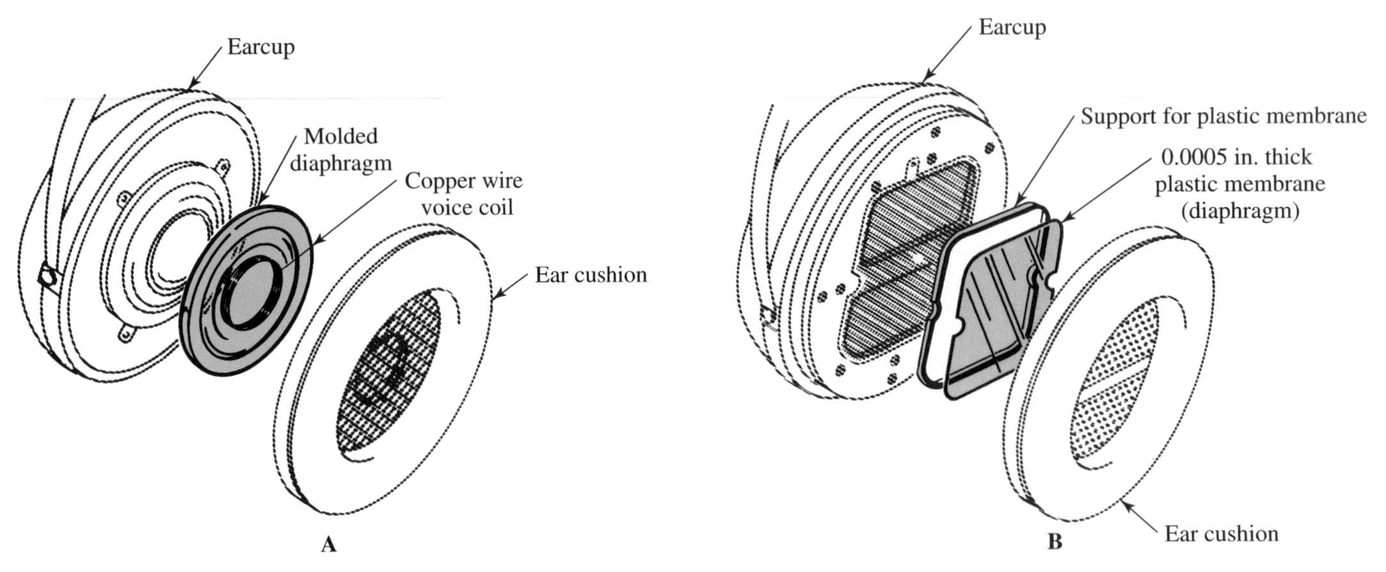

Figure 22-58. Headphone designs. A–The dynamic headphone uses a voice coil and molded diaphragm. B–With electrostatic headphones, a very thin membrane provides the diaphragm within the driver of the headphone.

around the ears for proper bass response, **Figure 22-59.** Hear-through or velocity headphones allow the listener to hear outside sounds (telephone, door bell, etc.) with the headphones in place.

Pressure and hear-through headphones use the same basic transducer. Some pressure headphones use a dynamic woofer and a separate tweeter. The tweeter can be electrostatic, ceramic, or dynamic in design. Correct crossover circuits are included in the headphone.

Pressure and hear-through headphones can have two or four channels. A copper voice coil is attached to one side of miniature loudspeaker cone or diaphragm. In quad headphones, there are two driver elements in each earcup.

The coil is hung in a magnetic field. It acts like a pulsating motor when the electrical form of the music energy flows through its windings. This causes the diaphragm to move the air in a manner similar to the original sound waves made by the musical instruments. See **Figure 22-60.**

Sound waves are made in headphones with either cone-type or element-type drivers. They both perform the same task. Cone-type drivers are loudspeakers, like those found in small radios. They are not designed for headphone use. But, they work well when used with the pressure-type cushions.

Figure 22-59. Pressure-type dynamic headphones.

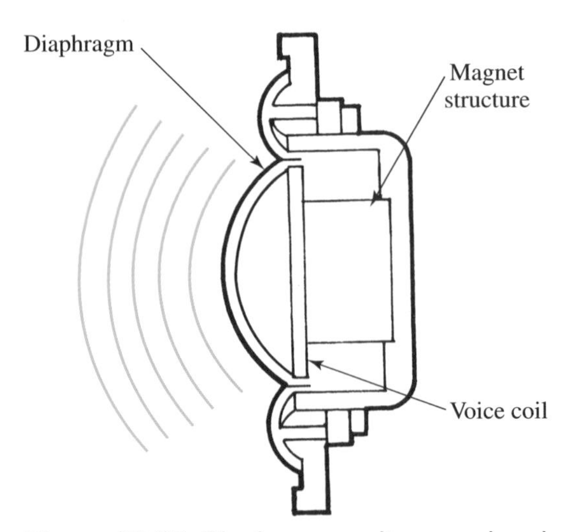

Figure 22-60. Diaphragm as it moves in a headphone. (Society of Audio Consultants, Koss Corp.)

PROJECT 22-5: Building an FM Wireless Mic

The Graymark wireless FM microphone project is designed to transmit a voice or audio signal to a typical FM receiver (FM radio). This project illustrates the principles of FM radio transmitters. The circuitry consists of RCL circuits, oscillators, a mixing circuit, and two general purpose transistors.

When the circuit construction is complete, 1½ volts is applied from a dc source. An FM radio is placed nearby, and the tuner is adjusted until a steady hum is heard coming from the radio receiver. To be sure that the hum is being generated from the wireless microphone you can either speak into the microphone or tap it with the tip of your finger.

The microphone works best if the frequency of the wireless microphone project oscillator does not match any existing radio stations in the area. If it does, the radio station may drown out the hum from the wireless microphone. If you suspect that the oscillator circuit matches a nearby radio station, simply adjust the depth of the tuning slug inside the inductor coil. It only takes a half turn or so to drastically changes the oscillator frequency.

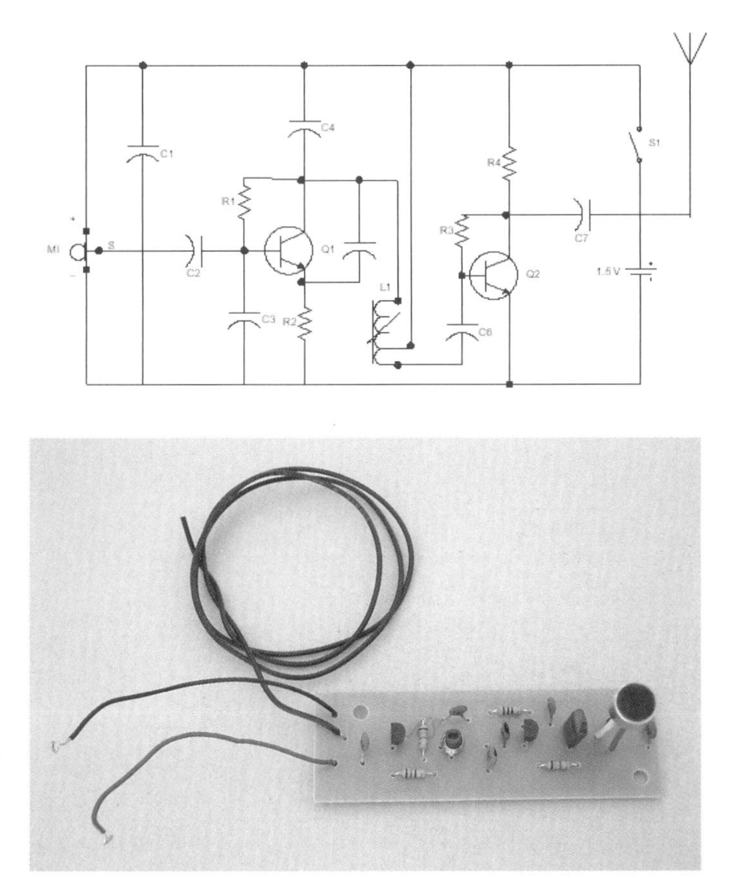

Qty.	Symbol	Description
2	R_1, R_3	Resistor, 18kΩ
1	R_2	Resistor, 100Ω
1	R_4	Resistor, 150Ω
2	C_1, C_3	Capacitor, 0.0047μF
1	C_2	Capacitor, 0.047μF
1	C_4	Capacitor, 5 PF
1	C_5	Capacitor, 10 PF
1	C_6, C_7	Capacitor, 100 PF
2	Q_1, Q_2	Transistor, 2N3904
1	S_1	Switch, Slide SPDT
1		Microphone, Condenser
1	L_1	Tuning Slub
1		Wire, #24, 6"
1		Wire, #22, 25"

Graymark

Element-type drivers are special loudspeaker structures. They are designed for headphone use. A special element is designed to work with a certain headphone housing. Grouping the driver element and housing into one package results in a driver that performs better and has better quality than the cone-type drivers.

Hear-through headphones are much different from the pressure-type. They are light weight and have porous foam ear cushions. Recall that the pressure-type headphone requires a closed volume of air. Hear-through headphones, however, vent the back sound waves through the rear of the cup.

The porous cushions provide some acoustic resistance. They help control the sound emitted by the headphone diaphragm. They also provide some acoustic openness. This allows the listener to hear outside sounds when the headphones are in place. A key advantage of this headphone is its weight. The drawback, however, is a loss in sound quality.

Dynamic headphones use a fairly heavy copper voice coil. It is attached to one side of a miniature loudspeaker cone. In theory, the moving parts act like a piston to reproduce the recorded sound. But the heavy voice coil lags behind the electrical energy. Therefore, the plastic or parchment cone also loses true piston action and distorts the sound wave.

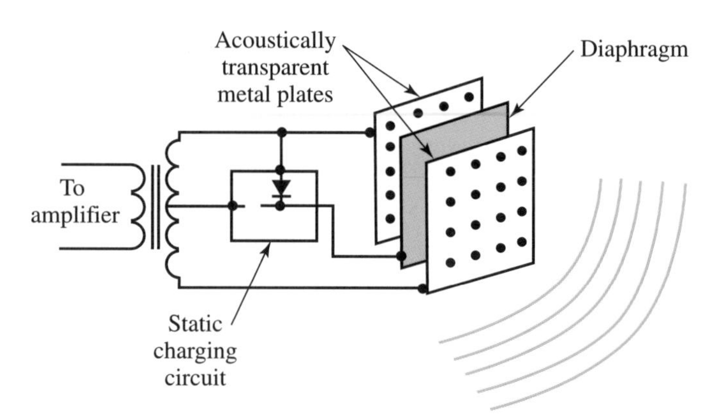

Figure 22-61. Electrostatic headphones circuit and operation. (Society of Audio Consultants, Koss Corp.)

Electrostatic

Electrostatic headphones look like pressure-type headphones. However, their mechanics are very different.

The best electrostatic headphones use light diaphragms of plastic instead of heavy plastic or parchment cones. The diaphragm in electrostatic headphones may be only 1/1000 of an inch thick, and weigh less than the surrounding air.

The diaphragm moves back and forth by controlled charges of static electricity. Lightness and control of the high quality electrostatic allows for excellent sound reproduction, **Figure 22-61.**

Frequency response of electrostatics is wider and flatter than dynamics. Using electrostatic headphones makes the drawbacks of disc and tape recordings, power source equipment, and broadcast stations more apparent. The noise and hiss from these sources are more noticeable to the listener.

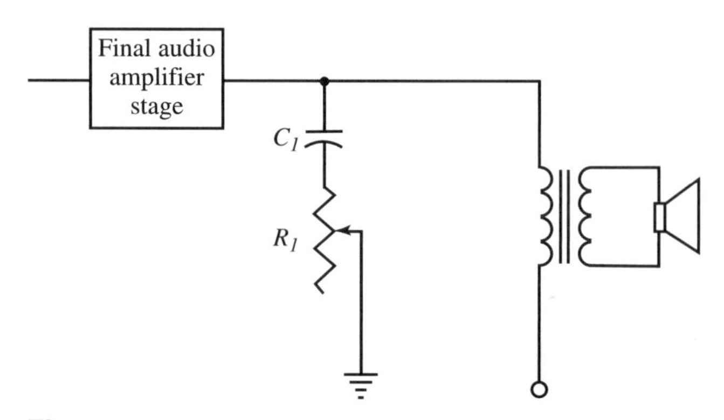

Figure 22-62. Basic tone control circuit. It removes the higher frequencies from the speaker output.

Tone Controls

Tone controls are used to adjust the amount of high or low frequencies sent out from speakers. Most radios, televisions, and audio amplifiers have these controls. One tone control circuit is shown in **Figure 22-62.**

Capacitor C_1, has a low reactance for high audio frequencies. The value of C_1 is usually 0.05μF and R_1 is 50,000 ohms.

A tone switch with three control positions can also be used. The reactance of the capacitor for each position filters or removes the high frequencies. **Figure 22-63** shows the tone switch control system with three capacitors of different values.

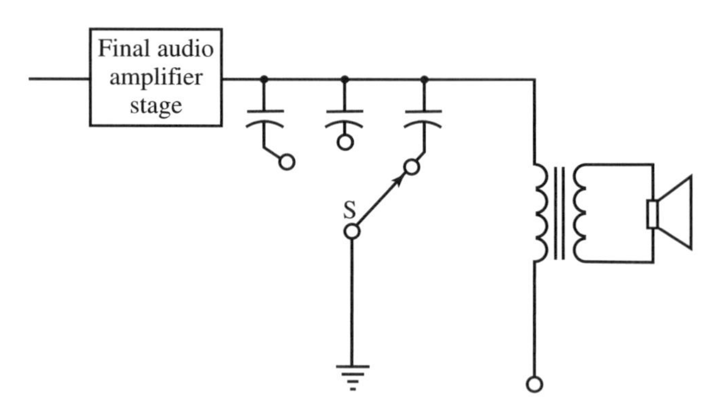

Figure 22-63. This tone control has three positions. The tone is selected by the listener.

Review Questions for Section 22.9

1. What is a transducer?
2. Make a sketch showing the operating principles of the PM speaker.
3. Low frequency speakers are called __________.
   High frequency speakers are called __________.
4. Name the basic types of headphones.

Summary

1. A basic radio receiver consists of an antenna, a ground, a tank circuit, a diode, a filter, and a speaker.
2. A radio wave is an electrostatic radiation of energy produced by an oscillator circuit.
3. Two types of transmitted waves are ground waves and sky waves.
4. A ground wave follows the surface of the earth to the radio receiver and has three parts: the surface wave, the direct wave, and the ground reflected wave.
5. Radio waves travel at the speed of light (186,000 miles per second or 300,000,000 meters per second).
6. The relationship between frequency and wavelength is:

$$\lambda = \frac{V}{f}$$

7. Basic transmitters include continuous wave, modulated continuous wave, amplitude modulation, and frequency modulation.
8. Some basic types of microphones are the carbon, dynamic, crystal, and velocity.
9. Modulation is the process of adding, or superimposing, audio waves to carrier waves.
10. Modulation percentages refer to the amount a carrier wave has been varied or modulated.
11. Amplitude modulation is a process in which an audio wave is combined or superimposed on a carrier wave and the amplitude of the carrier wave is made to vary at the audio signal rate.
12. With frequency modulation, a constant amplitude continuous wave is made to vary in frequency at the audio frequency rate.
13. The tuned radio frequency (TRF) receiver picks up a transmitted RF wave, amplifies it, detects or demodulates it, and amplifies the audio wave.
14. High frequency rectification of the RF wave is called detection.
15. The superheterodyne receiver converts the tuned RF signal to an intermediate frequency (IF) signal so that this signal can be further amplified and refined.
16. Selectivity is the ability of a receiver to select a single frequency and reject all others.
17. Sensitivity is the ability of a receiver to respond to weak incoming signals.
18. Some common intermediate frequency signals are 455 kHz for AM and 10.7 MHz for FM.
19. Transducers are devices that convert one form of energy to another. Some common transducers are microphones and speakers.

Test Your Knowledge

Please do not write in the text. Place your answers on a separate sheet of paper.

1. Make a sketch showing the direction of travel of a wave transmitted from a horizontal antenna.
2. The signal created by an oscillator is a(n) __________ wave.
3. What is the wavelength (in meters) of a signal with a 1200 kHz frequency?
4. What are the standard frequencies for the following receivers?
   a. AM receiver.
   b. FM receiver.
   c. TV receiver.
5. What is the purpose of a microphone?
6. __________ is a process in which an audio wave is superimposed on a carrier wave.
7. What is the formula for computing percent of modulation?
8. What power is required to modulate a transmitter having a dc input of 250 watts to 50 percent?
9. The amplitude of a modulating signal has a maximum deviation of 15 kHz and a frequency of 1500 Hz. What is the modulation index?
10. The process of removing the audio wave from the modulated RF wave is:
    a. demodulation.
    b. detection.
    c. Both of the above.
    d. None of the above.
11. What two characteristics determine the quality of a tuning circuit?
12. A(n) __________ __________ circuit changes the resonant frequency range of a circuit.
13. The RF amplifier:
    a. is the first stage that the received signal is fed to from the antenna.
    b. provides amplification to the incoming signal.
    c. improves the selectivity of a receiver.
    d. All of the above.
14. Sketch a block diagram for a superheterodyne receiver.
15. What is the advantage of a separate local oscillator attached to a mixer?
16. __________ __________ __________ automatically adjusts all volume levels to a constant level.

17. What is the purpose of alignment?
18. Name two types of FM detectors.
19. Explain the operation of a PM speaker.
20. What is the purpose of a tone control?

For Discussion

1. How can you determine from an antenna if the transmitted waves are horizontally or vertically polarized?
2. Explain the basic block diagram of a transmitter.
3. How does the tuning circuit select a desired frequency?
4. Discuss the operation of the following microphones: carbon, dynamic, crystal.
5. What is the primary difference between AM and FM?
6. How does an AM receiver differ from an FM receiver?
7. Discuss two basic methods of FM detection.
8. Explain the process of heterodyning signals.

Basic laws of electricity come into play when performing complex tasks. (Techtronix)

Television and Video Display Units

Objectives

After studying this chapter, you will be able to:

- Explain the steps in the transmission of a television signal.
- Discuss the scanning process.
- Identify circuits in both black-and-white and color television receivers and explain their functions.
- Identify the size and makeup of a television channel.
- Discuss a variety of television innovations including video cassette recorders, remote control, and satellite television.
- List the benefits of HDTV as compared to analog television.
- Explain the difference between multicasting and datacasting.
- Discuss the compression technique of MPEG2.
- Discuss the various flat-panel technologies.

Key Words and Terms

The following words and terms will become important pieces of your electricity and electronics vocabulary. Look for them as you read this chapter.

active-matrix
advanced television systems committee (ATSC)
aspect ratio
brightness control
charged coupled device (CCD)
datacasting
deflection yoke
digital video disc
electro-luminescence
enhanced definition television (EDTV)
feedhorn
fine tuning
focal point
frame
frame rate
geostationary orbit
high definition television (HDTV)
liquid crystal display (LCD)
MPEG2
multicasting
national television standards committee (NTSC)
passive-matrix
picture element
pixel
polarized light
progressive scanning
raster
resolution
scanning
shadow mask picture tube
synchronization (sync) pulse
thin film transistor (TFT)
tuner
ultra high frequency (UHF)
very high frequency (VHF)
video detector
video head

Television is a giant in the field of communications. Through television, entertainment, education, information, and advertising are available to millions of people. And through satellite links, television can instantaneously bring the entire world into our home or office.

This text will not address the detailed electronic circuits involved in the production, transmission, and reception of television signals. However, it is important for all persons interested in the science of electronics to have a basic knowledge of television and related information.

23.1 TELEVISION SIGNALS

Taking a picture from one location and reproducing it in your home is a combination of several processes. First, a television camera must record the images to be transferred. Next, those images must be turned into radio waves and sent out into the air, or turned into electrical signals and transmitted through cables. Finally, the signals must be received and translated back into pictures.

Television Cameras

What looks like a solid picture is really an extremely large number of dots. In a black-and-white picture, these dots are varying degrees of black and white. They are called ***picture elements.*** Look closely at a photo in a daily newspaper and you will see that these picture elements are clearly visible.

As a television camera views a studio scene, it "sees" the scene as a combination of these picture elements. The scene is focused on a photosensitive mosaic in the camera that consists of many photoelectric cells. Each cell responds to the scene by producing a voltage that is in balance with the strength of the light. These voltages are amplified and used for modulation of the AM carrier wave. This wave is transmitted to the home receiver.

A line drawing of a camera tube called an image orthicon is shown in **Figure 23-1.** The scene in front of the camera focuses on the photo cathode through a standard camera lens system. The varying degrees of light cause electrons to be emitted on the target side of the cathode. These form an electronic image of the scene. The target plate operates at a high positive potential. The electrons from the cathode are attracted to the target. The target is made of low resistance glass and has a transparency effect. The electron image appears on both sides of the target plate.

At the right in Figure 23-1 is an electron gun, which produces a stream of electrons. The speed of the stream is increased by the grids (at the top and bottom in the figure). The beam scans left to right and top to bottom and is controlled by the magnetic deflection coils around the tube. The moving electron beam strikes the target plate and the electrons return to the electron multiplier section. The strength of the electron stream returning to the multiplier is balanced with the electron image. It provides the desired signal current for electronic picture reproduction.

Scanning

Scanning is the point-to-point examination of a picture. In the camera, the electron beam scans the electron image. It responds to the point-to-point brilliance of the picture.

The scanning system used in the United States is the interlace system. It consists of 525 scanning lines. The beam starts at the top left hand corner of the picture. It scans the odd numbered lines of the scanning pattern (lines 1, 3, 5, 7, etc.) from left to right. At the completion of 262 1/2 lines, the electron beam is returned to the starting point by vertical deflection coils. The beam then scans the even numbered lines.

One scan of 262 1/2 lines represents a ***field.*** One set of the odd and even numbered fields represents a ***frame.*** A representation of the interlace scanning pattern is shown in **Figure 23-2.**

The frame frequency has been set by the Federal Communications Commission (FCC) at 30 hertz. This means your TV receives 30 complete frames per second, or 60 picture fields of alternate odd and even lines per second. To the human eye, this appears as a constant, nonflickering picture.

A device called a horizontal deflection oscillator causes the beam to move from left to right. To produce 525 lines per frame at a frame frequency of 30 Hz, the horizontal deflection oscillator must work at a frequency of 15,750 Hz (525 × 30).

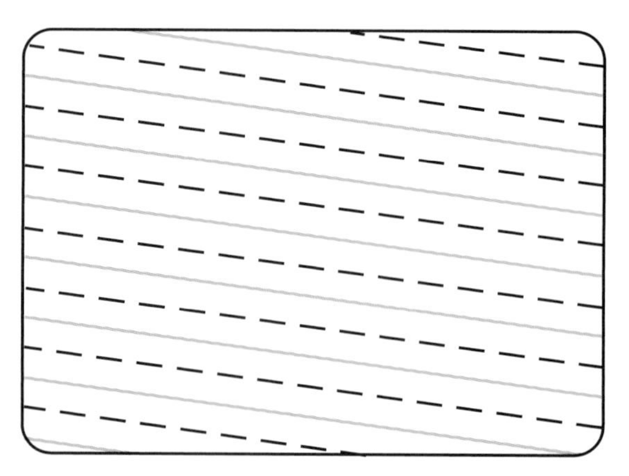

Figure 23-2. This is a portion of the interlace scanning system used in television. There are actually 525 lines.

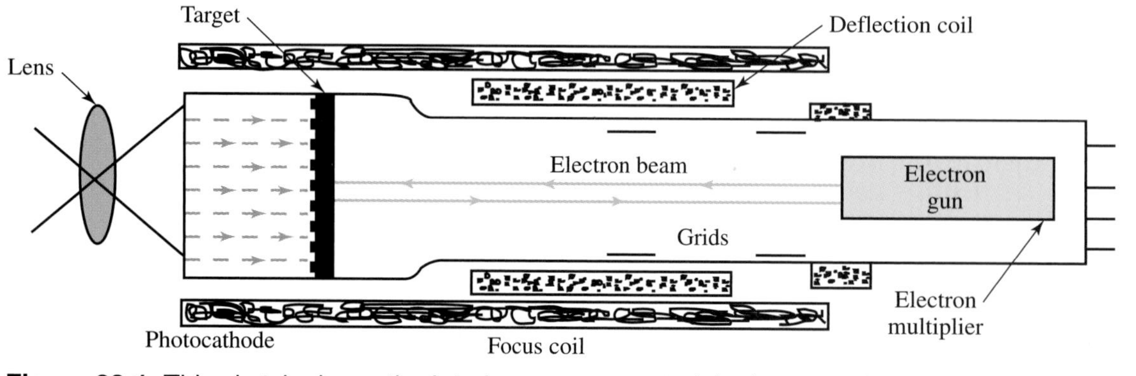

Figure 23-1. This sketch shows the interior arrangement of the image orthicon tube used in television cameras.

The vertical deflection oscillator causes the beam to move from top to bottom. The vertical deflection oscillator must have a frequency of 60 cycles per second. This is a field frequency.

Closer study of the scanning process reveals that the beam scans as it moves from left to right. After it has read one line, it quickly retraces to the left and starts reading the next line in the same way we read a book. The retrace time is very rapid, but still shows a line in the picture. Therefore, the picture must be black during this retrace time. Also, when the beam reaches the bottom of the picture, the beam must be returned to the top to scan again. The picture must also be black during vertical retrace, or trace lines would be visible.

The oscillators that make the scanning and retrace voltages for both the horizontal and vertical sweep must produce a waveform as shown in **Figure 23-3.** This is a sawtooth waveform. Notice the gradual increase in voltage during the sweep and the rapid decrease during retrace. These voltages are applied to coils that surround the picture tube. These coils are called the ***deflection yoke.*** An increase in the strength of the magnetic fields in the coils causes the electron beam to move. See **Figure 23-4.**

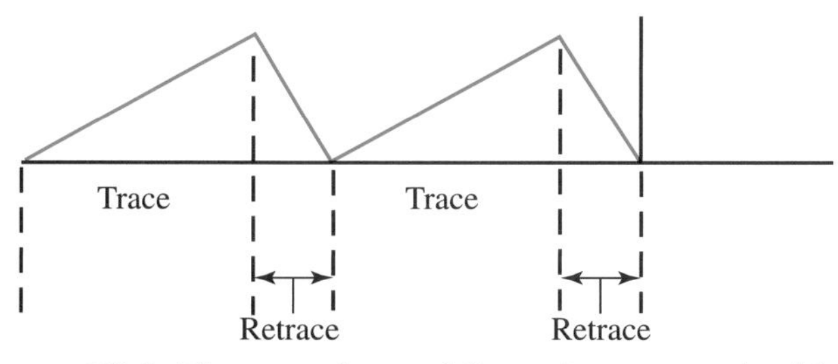

Figure 23-3. The waveform of the voltages required for scanning and retrace. It is called a sawtooth waveform.

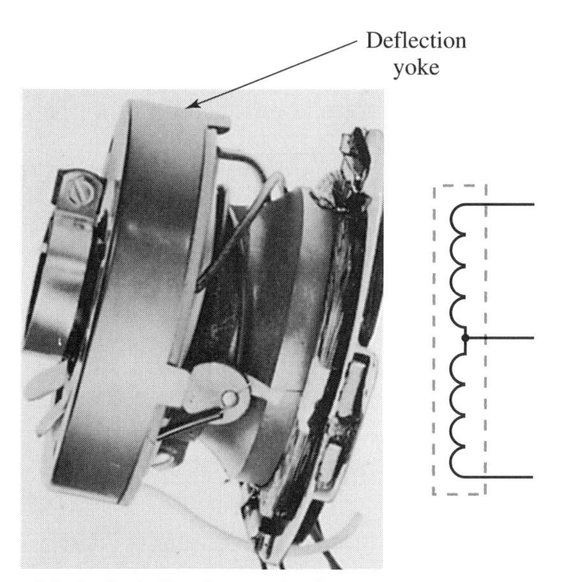

Figure 23-4. A deflection yoke fits around the neck of the television picture tube. (Triad)

Scanning at the studio and scanning on a television must be in step. For this to occur, a pulse generator triggers the horizontal and vertical oscillators at the studio. This same pulse is also sent over the air and received by the television set. This pulse, known as the ***synchronization pulse,*** or ***sync pulse,*** triggers the oscillators in the receiver and keeps them at exactly the same frequency. Older televisions have ***horizontal hold*** and ***vertical hold*** controls. These controls are used to make slight, adjustments so that the oscillators can lock in with the sync pulses. Newer televisions make the adjustments automatically.

Composite Video Signals

All video signals are formed in the same way so that a television can be used in any geographic area. These standards are set by the FCC and are used by all TV broadcasting stations.

The television signals received by TVs contain picture (video) and sound (aural) information. The video information is an AM signal. The audio information is an FM signal.

The amplitude of the modulation is divided into two parts. The first 75 percent is used to transmit video information. The remaining 25 percent is for the sync pulses, **Figure 23-5.** Also, a system of negative transmission has been adopted. This means that the higher amplitudes of video information produce darker areas in the picture. At 75 percent the picture is completely *black.* Look at Figure 23-5 again, the percentage is shown on the left.

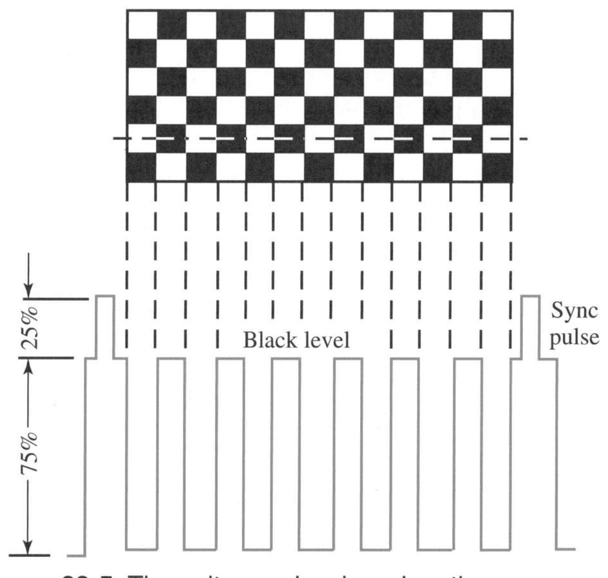

Figure 23-5. The voltages developed as the camera scans one line across a checkerboard. The sync pulses are transmitted at the end of each line.

As the beam scans each black bar, a similar action takes place. At the end of the line, the screen is driven back to the *pedestal,* or *blanking level.* During this blanking pulse, beam flyback occurs and a sync pulse is sent in the blacker-than-black, or infrablack, region (upper 25 percent) for oscillator synchronization. A second line to be scanned would be an exact copy of the first unless the picture is changed. At the bottom of the picture, a series of pulses trigger the vertical oscillator and keep it synchronized.

Figure 23-6 shows a composite (complete) video signal. The video information for one line between the blanking pulses is varying degrees of black and white.

Compare the two video signals in **Figure 23-7.** Signal A is made up mostly of bright objects. The average overall brightness of the scene is another form of the information sent to a TV receiver from the transmitter. It can be detected from the composite video signal.

Basic Cathode Ray Tube Controls

The cathode ray tube (CRT) is used to produce images in most television sets. Let's review the theory of vacuum tube operation. The control grid determines the flow of electrons through the tube. In the CRT this is also true, **Figure 23-8.** At zero bias the CRT is at maximum current; therefore, the screen is bright or white. At cutoff bias, the current is zero and the screen is black. The tube operates at a selected bias on the control grid. This bias can be controlled by the knob on the TV or a button on the TV remote called the ***brightness control.***

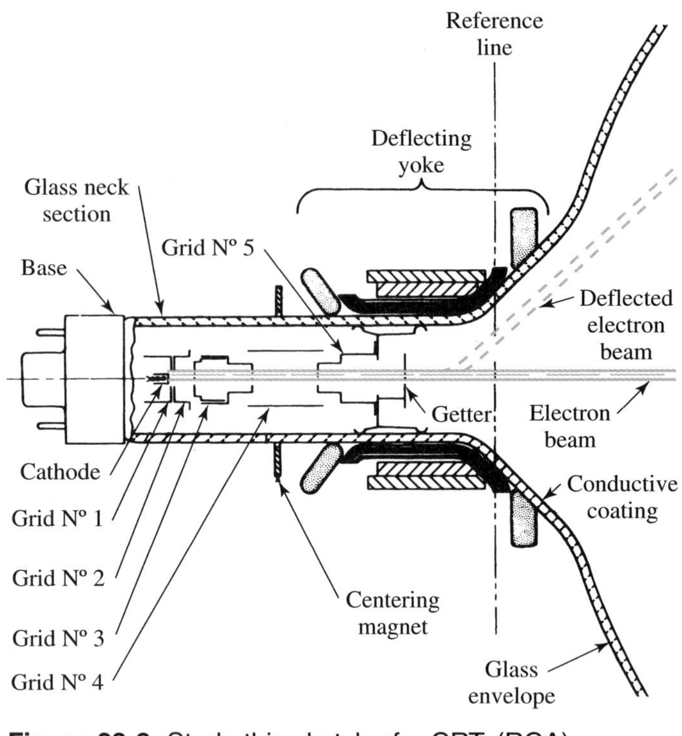

Figure 23-8. Study this sketch of a CRT. (RCA)

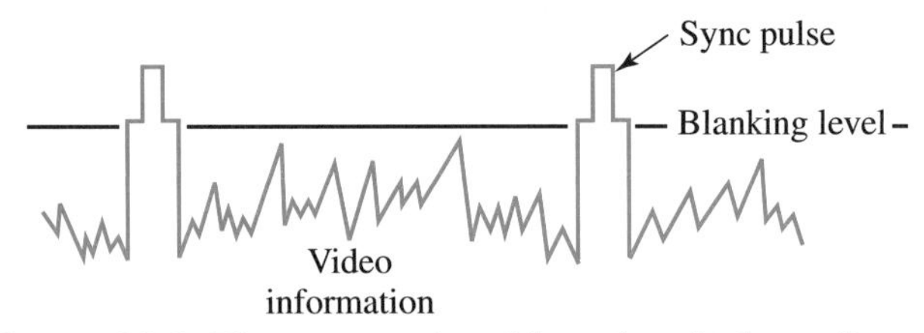

Figure 23-6. The composite video signal of one line scanned by the television camera.

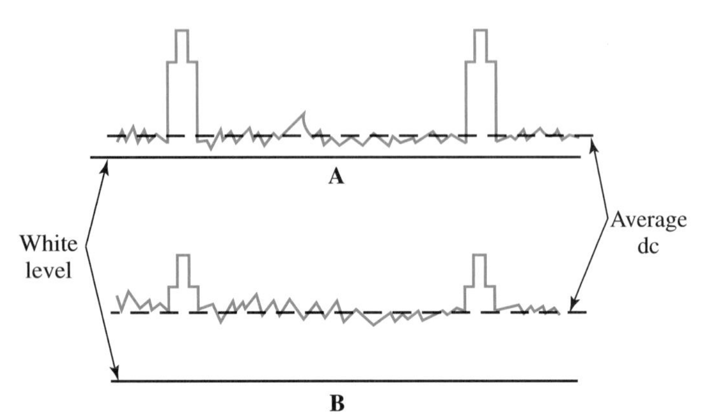

Figure 23-7. A comparison of a dark and a light picture as they appear in the video signal. A–Light signal. B–Dark signal.

When no picture is being received on the TV, the scanning electron beam can be seen in the form of lines on the TV screen. This is called the ***raster.*** Turn a TV to a vacant channel and observe this raster. Now adjust the brightness control from black to bright. The incoming, detected video signal is applied to the grid of the CRT (sometimes to the cathode, depending on polarity of signal). The video signal adds to, or subtracts from, the bias on the tube. This action results in a modulated electron stream that conforms to the picture information in the video signal. The picture is produced on the fluorescent screen.

The sharpness or focus of the electron beam can be adjusted by changing the voltages of the focusing grids. From time to time, they may require adjustments. Refer again to Figure 23-8. Find the centering magnet. Slight adjustments on this magnet will correct a picture that is off center.

Review Questions for Section 23.1

1. Briefly explain how an orthicon television camera works.

2. __________ is the point-to-point examination of a picture.
3. One scan of 262 1/2 lines represents a(n) __________.
4. What is a deflection yoke and what is its purpose?
5. The _________ triggers the oscillators in the receiver and keeps them at exactly the same frequency.
6. What is a raster?

23.2 TELEVISION RECEIVERS

The television receiver is a fairly complex electronic device. As you work through the block diagram of the receiver shown in **Figure 23-9,** take note of the similarities between your television set and the radio receivers you just studied in Chapter 22.

Black-and-White Television Receiver

Figure 23-9 shows the parts of the television receiver. This block diagram shows the links between the parts of the television. Trace the signal path through the stages. The purpose of each group of components will be apparent. The name of each block reflects its purpose in the circuit. The following text takes you stage by stage through a television receiver.

The ***RF amplifier*** serves a function similar to that in the superheterodyne radio. The incoming television signal is chosen by switching fixed inductors into the tuning circuit. These tuned circuits provide constant gain and selectivity for each television channel. In this stage, the video signal, with all its information, is amplified and fed to the mixer.

In the ***mixer*** stage, the incoming video signal is mixed with the signal from a local oscillator to produce an intermediate frequency. The commonly used IF is 45.75 megahertz. When a channel is chosen with the channel selector, the tuning circuit is changed. The frequency of the oscillator is also changed so that it is always producing the IF of the correct frequency. A ***fine tuning*** changes the frequency of the oscillator slightly in order to provide the best response.

The RF amp, mixer, and oscillator are combined in one unit. The unit is called the ***tuner*** or ***front end*** of a television. These units are usually put together in the factory. Adjustments should not be made on these units unless you have the correct instruments and thorough knowledge of procedures.

The ***PIX-IF amplifiers*** amplify the output of the mixer stage. This includes the 45.75 megahertz intermediate frequency, the video, and the aural information. To provide maximum frequency response for each stage up to 45.75 MHz, each stage must amplify a broad band of frequencies. The voltage gain of each stage is reduced. More stages of IF amplification are required. In this system, the

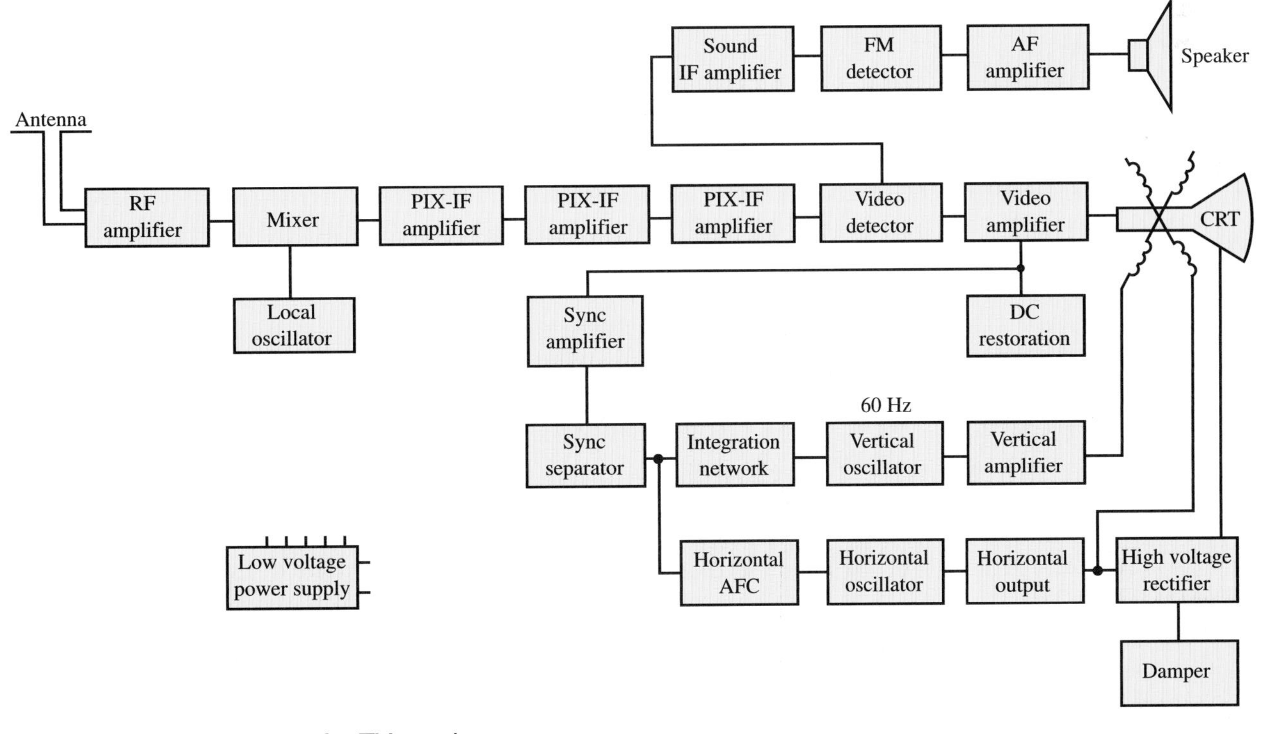

Figure 23-9. Block diagram of a TV receiver.

sound is passed through the IF amplifier with the video signal. It is called the intercarrier system.

The output from the last IF stage is fed to the ***video detector*** or ***demodulator.*** The detection process is the same as in the radio. The video signal used to amplitude modulate the transmitted carrier wave is separated and fed to the next stage.

In the ***video amplifier*** stages, the demodulated video signal is amplified and fed to the grid (sometimes cathode) of the CRT. This signal modulates or varies the strength of the electron beam and produces the picture on the screen.

The FM sound signal is amplified in the ***sound IF amplifier.*** Later in this chapter, you will learn that the FM sound of the television program is separated from the video carrier wave by 4.5 megahertz. This produces a 4.5 MHz FM signal at the output of the video detector, which is coupled to the sound IF amplifiers.

The ***FM audio detector*** detects the frequency variations in the modulated signal and converts them to an audio signal.

AF amplifiers are the same as those used in the conventional radio or audio system discussed in the previous chapter. The audio signal is amplified enough to drive the power amplifier and the speaker.

The output of the video amplifier is fed to the ***sync separator.*** This circuit removes the horizontal and vertical sync pulses that were transmitted as part of the composite video signal. These sync pulses trigger the horizontal and vertical oscillators and keep them in step with the television camera.

The ***sync amplifier*** is a voltage amplifier stage that increases the sync pulses.

In the ***horizontal AFC,*** the horizontal oscillator frequency is compared to the sync pulse frequency. If they are not the same, voltages are developed that change the horizontal oscillator to the same frequency.

The ***horizontal oscillator*** operates on a frequency of 15,750 Hz. It provides the sawtooth waveform needed for horizontal scanning.

The ***horizontal output*** stage correctly shapes the sawtooth waveform for the horizontal deflection coils. It also drives the horizontal deflection coils and provides power for the high voltage rectifier.

The output of the horizontal oscillator shocks the ***horizontal output transformer (HOT).*** The high ac voltage developed by this autotransformer is rectified by the ***high voltage rectifier*** and is filtered for the anode in the CRT. See **Figure 23-10.**

The ***damper*** stage dampens out oscillations in the deflection yoke after retrace.

The output of the sync amplifier is fed through a vertical integration network to the ***vertical oscillator.*** The integrator is an RC circuit designed with a time constant. The integrator builds up a voltage when a series or groups of closely spaced sync pulses occur.

Figure 23-10. Horizontal output transformer (HOT). (Triad)

The transmitted vertical pulse is a wide pulse. It occurs at a frequency of 60 Hz. The vertical oscillator is triggered by this pulse and, therefore, keeps on frequency.

The output of the vertical oscillator provides the sawtooth voltage to the deflection coils. The deflection coils move the beam from the top to the bottom of the screen and the flyback from bottom to top, at the end of each field.

The ***vertical output*** amplifier is used by the output of the oscillator to provide the proper currents in the deflection yoke for vertical scanning.

Special purpose circuits

The basic television circuit has been improved in many ways. Automatic controls have been designed for many functions that would be tiresome for the TV viewer to manage. Two of these improvements are dc restoration and automatic gain control.

DC restoration. The average darkness or brightness of a TV picture is transmitted as the average value of the dc component of the video signal. If the video amplifiers are RC coupled, the dc value of the signal is lost. In this case, the average value of the video signal is taken from the detector and used to set the bias on the CRT.

Automatic gain control. The AGC serves the same function as it did in the radio receiver. Its purpose is to provide a fairly constant output from the detector by varying the gain of the amplifiers in previous stages. This is done by rectifying the video signal to produce a negative voltage. This voltage is applied to the bias of the previous amplifiers to change their gain.

Color Television Receiver

Color television was developed in the late 1940s. The system currently used in the United States was pioneered by RCA Laboratories. In March, 1950, a color television demonstration was given in Washington, DC to FCC personnel, reporters, and other interested people. As a result of this demonstration, color television development was launched.

An invention that made color television possible was the ***shadow mask picture tube,*** **Figure 23-11.**

The three basic colors used in color television are red, blue, and green, **Figure 23-12.** By combining these colors, any color can be produced on the screen.

The first color picture tube produced for retail sale was the ***delta-type*** tube. It was invented in 1950, and was basically the same tube used in the color television demonstrations given by RCA Laboratories in Washington, DC.

The delta-type tube uses three electron guns placed in the neck assembly of the picture tube. The electrons are emitted by the three guns toward the screen. The screen is filled with hundreds of thousands of dots containing the colors red, blue, and green. In between the electron gun and the color producing screen, the three electron beams are focused through an aperture or shadow mask, **Figure 23-13.** This shadow mask ensures that the electron beams strike the dots properly.

A line is made by one complete scan (from left to right) of all three electron beams hitting all the dots across the screen. If all electron beams are adjusted properly, the result will be a white line. A color line is made by mixing the electron beams.

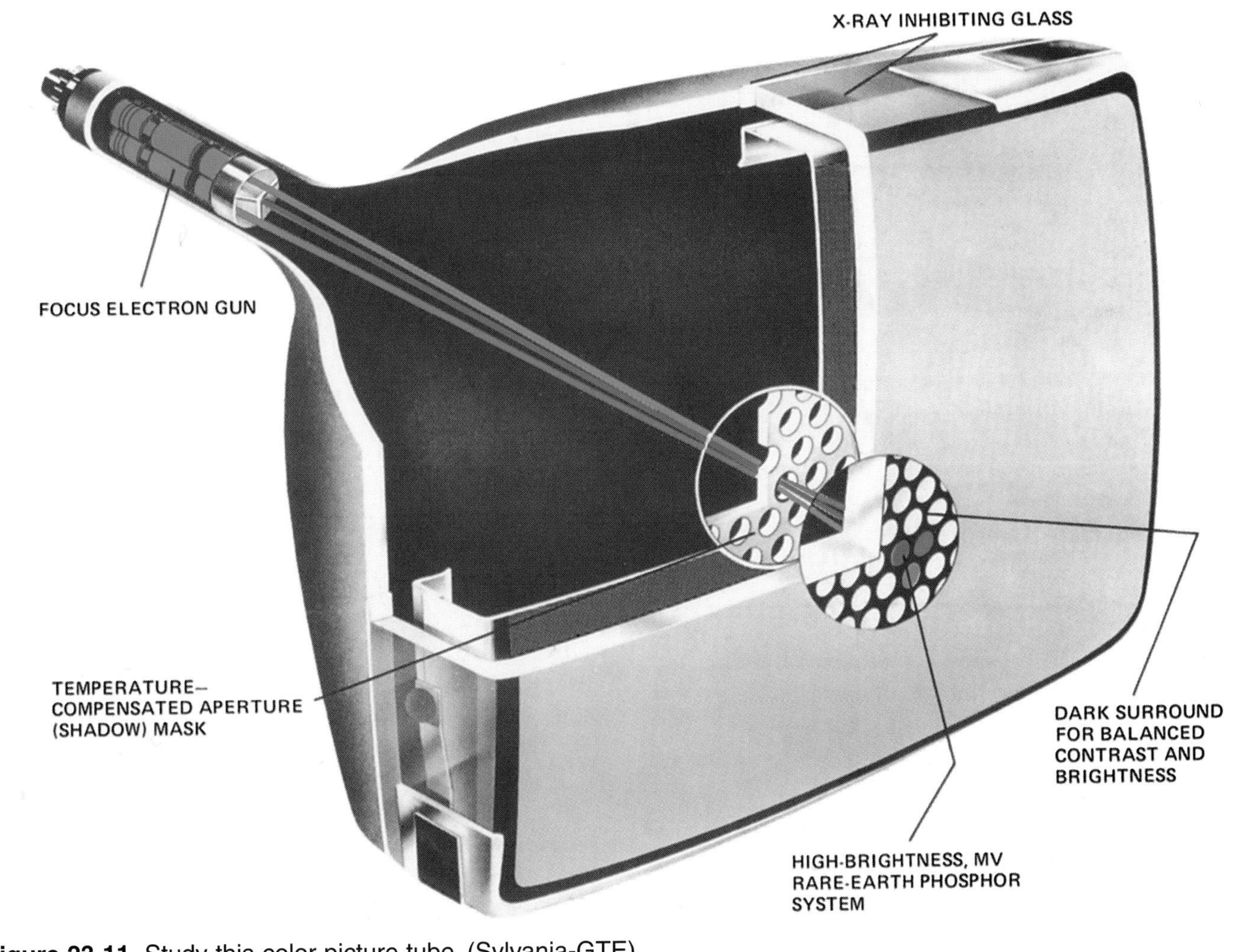

Figure 23-11. Study this color picture tube. (Sylvania-GTE)

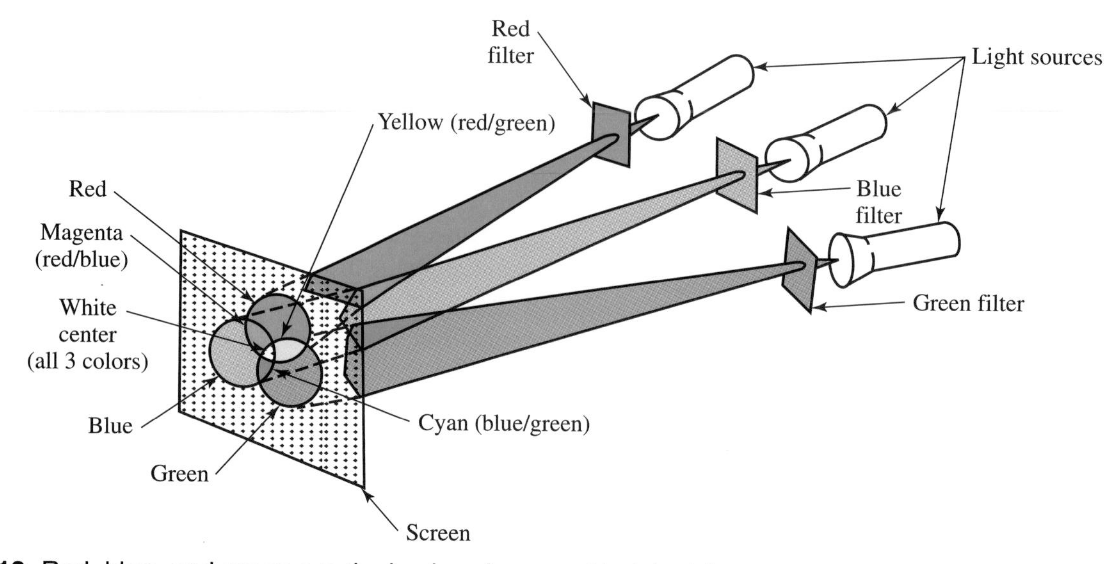

Figure 23-12. Red, blue, and green are the basic colors used in television.

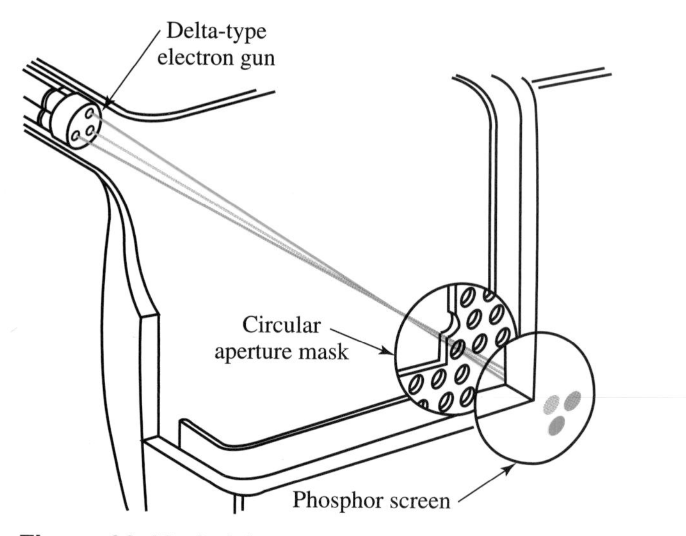

Figure 23-13. A delta-type gun, shadow mask, and tri-dot screen arrangement. (Sylvania-GTE)

In-line gun assemblies were invented after the shadow mask tube. **Figure 23-14** shows four in-line gun assemblies with different aperture grill and screen patterns.

A color television receiver is a very complex instrument. It has to be able to produce both color and black-and-white pictures. But not all televisions are color televisions, so the incoming signal must also work with black-and-white sets. **Figure 23-15** shows a block diagram of a color television receiver.

Television Channel

The FCC has assigned a portion of the radio frequency spectrum for each television channel. There are two types of television channels: ***very high frequency (VHF)*** and ***ultra high frequency (UHF)***. Each channel is 6 MHz wide. The VHF channels, 2 to 13, are listed in **Figure 23-16** along with frequency bands and carrier frequencies.

Examine channel 4 in **Figure 23-17.** The basic video carrier frequency is 67.25 MHz. Recall that when an RF carrier wave is amplitude modulated, sideband frequencies appear. These stand for the sum and difference between the carrier frequency and the modulating frequencies.

To send a very clear, sharp picture, frequencies of at least 4 MHz are needed for modulation. These frequencies combine in channel 4 for a band occupancy of 63.25 MHz (67.25 – 4 MHz) to 71.25 MHz (67.25 + 4 MHz). It is a total channel width of 8 MHz. However, the FCC allows only 6 MHz, therefore, a compromise must be made.

In commercial TV broadcasting the upper sideband is transmitted without attenuation. The lower sideband is partly removed by a vestigial-sideband filter at the transmitter. The curve in **Figure 23-17** shows the basic response traits of the TV transmitter. The sound is transmitted as a frequency-modulated signal at a center frequency 4.5 MHz above the video carrier. In channel 4 the sound is at 71.75 MHz.

The ultra high frequency (UHF) television band covers from channel 14 to 83. As in VHF, the bandwidth of each channel is 6 MHz. The same frequency bandwidths are used for the picture carrier as VHF.

The UHF channels used for commercial TV are set by the FCC. See **Figure 23-18.**

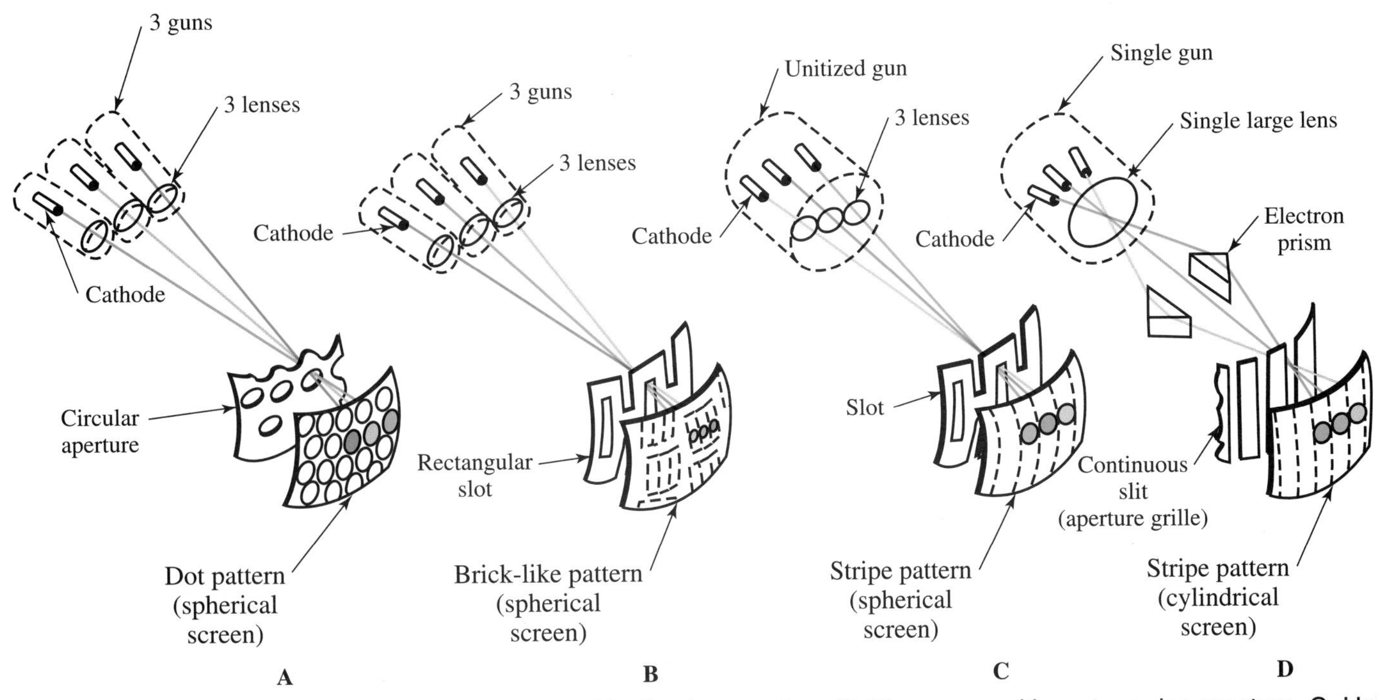

Figure 23-14. In-line color tubes. A–Three gun with circular aperture. B–Three gun with rectangular aperture. C–Unitized gun with slot aperture. D–Sony Trinitron with one gun and grille aperture.

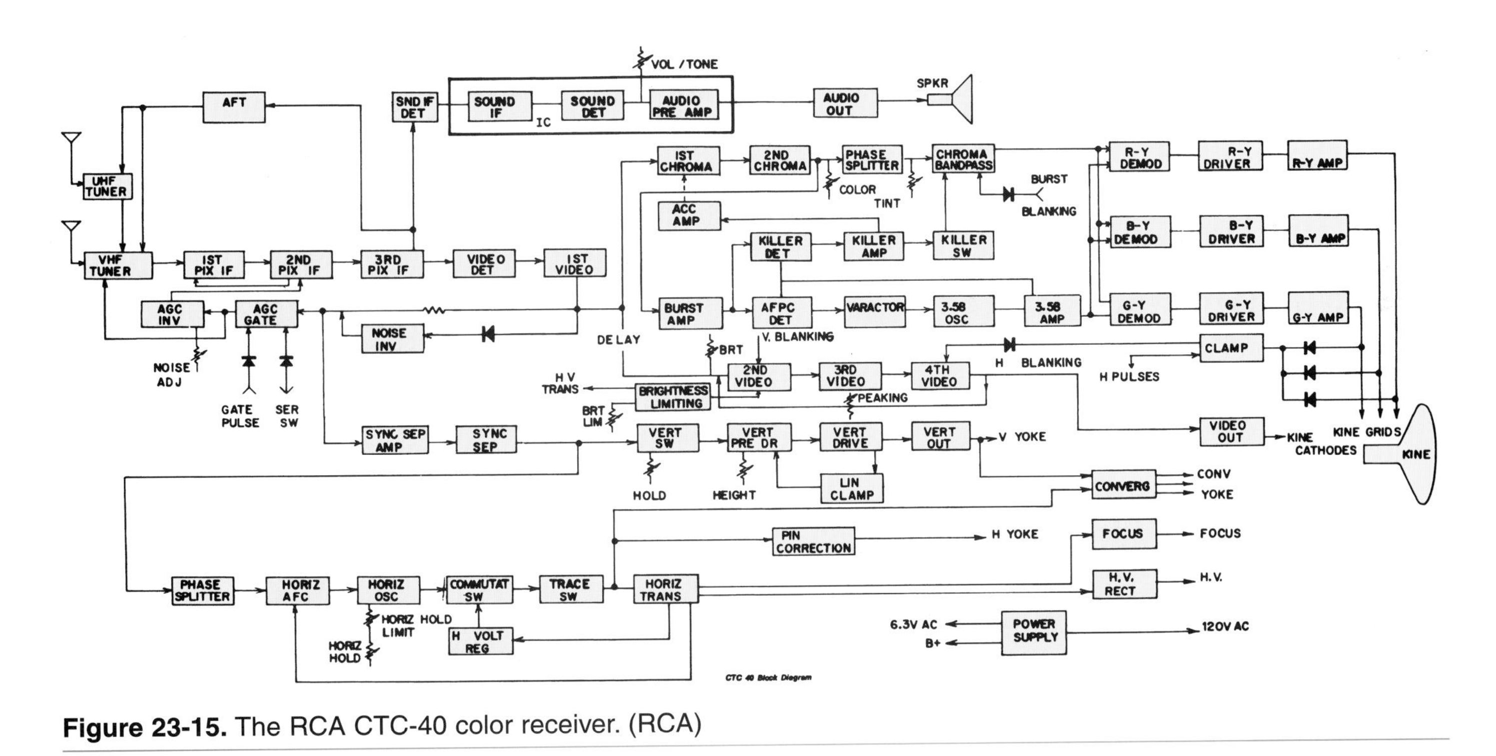

Figure 23-15. The RCA CTC-40 color receiver. (RCA)

Review Questions for Section 23.2

1. What is the purpose of an RF amplifier in a television receiver?

2. The output of the mixer stage is amplified by the:
   a. PIX-IF amplifier.
   b. video amplifier.
   c. AF amplifier.
   d. None of the above.

Channel number	Frequency band (MHz)	Video carrier frequency	Aural carrier frequency
2	54–60	55.25	59.75
3	60–66	61.25	65.75
4	66–72	67.25	71.75
5	76–82	77.25	81.75
6	82–88	83.25	87.75
7	174–180	175.25	179.75
8	180–186	181.25	185.75
9	186–192	187.25	191.75
10	192–198	193.25	197.75
11	198–204	199.25	203.75
12	204–210	205.25	209.75
13	210–216	211.25	215.75

Figure 23-16. VHF frequency assignments.

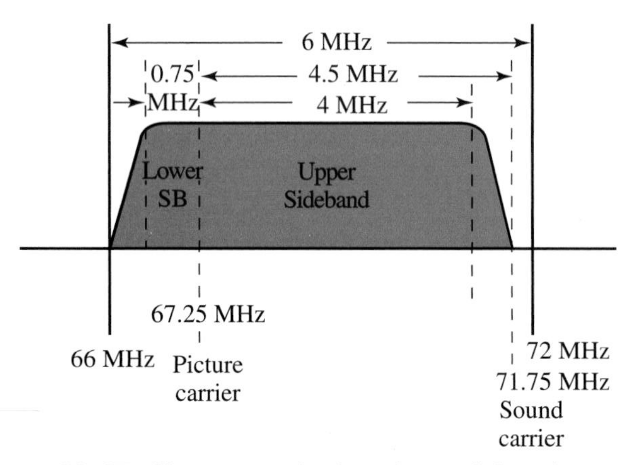

Figure 23-17. Shown are the locations of the picture carrier and sound for channel 4, VHF. The channel is 6 MHz wide.

Channel number	Frequency band (MHz)
14	470–476
15	476–482
16	482–488
17	488–494
18	494–500
19	500–506
•	
•	
•	
•	
83	884–890

Figure 23-18. Examples of some ultra high frequency (UHF) channels and where they are located in the frequency band.

3. The __________ __________ stage correctly shapes the sawtooth waveform for the horizontal deflection coils.
4. What is the purpose of a shadow mask?
5. A Sony Trinitron color tube uses a(n) __________ aperture.
6. Each TV channel bandwidth is:
   a. 6 Hz.
   b. 60 MHz.
   c. 6 MHz.
   d. None of the above.

23.3 TELEVISION INNOVATIONS

In addition to the electronics of television itself, there are a number of other electronic innovations that have been created to work with television. A few of these innovations are detailed here.

Video Cassette Recorders

Video cassette recorders (VCRs) are very popular and useful. VCRs can be used to play rented videotapes. They also can be used to record and play back both regular television broadcasts and home movies made with a video camera. Once, there were two types of video tape formats available: VHS and Beta. Now VHS is the only format available for home use.

While some cheaper VCR models are still available with two heads, most VCRs have four heads. A ***video head*** is a tiny electromagnet that reads information from the recorded tape during playback. It writes information onto the tape during recording. VCRs can have many heads. These extra heads provide better sound and picture quality.

Look at **Figure 23-19.** Recording information on a magnetic tape is simple. The tape is plastic with a thin coating of iron oxide on one side. The iron oxide is a magnetic material. The recording head is a tiny electromagnet. The voice or picture message is converted to electrical impulses. These electrical impulses are applied to the coil winding on the recording head. The fluctuations of the electrical impulses make the magnetic recording head fluctuate at the same rate as the electrical impulses. The magnetic head induces a magnetic pattern on the metal-oxide tape. The magnetic patterns uniquely match the original voice or picture patterns. This same method is also used to record digital signals on magnetic disks, such as the disks used by computers.

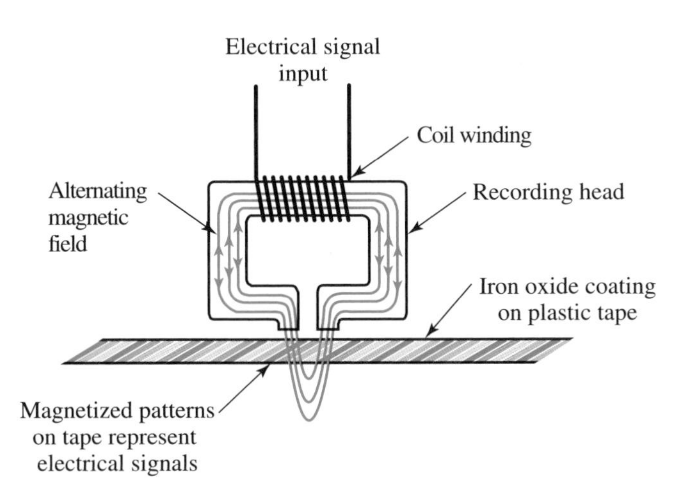

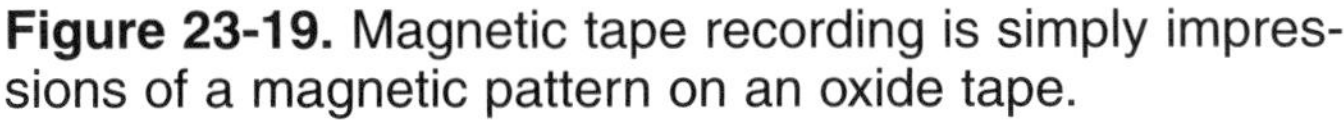
Figure 23-19. Magnetic tape recording is simply impressions of a magnetic pattern on an oxide tape.

The quality of recording is not governed by the speed of the tape alone. To get a quality video image, the magnetic impulses must be recorded much faster than a sound recording. To achieve a higher recording speed, the VCR head and tape are intersected at an angle, and the magnetic paths are recorded across the tape at an angle. This technique allows for much more information to be recorded or played back without requiring the tape to run at an exceptionally high speed. Look at **Figure 23-20** to see how the angle of recorded information is developed by the interaction of the tape head and VCR recording tape.

The video tape not only records voice and video but also speed information, end of tape location, copyright and anticopy coding.

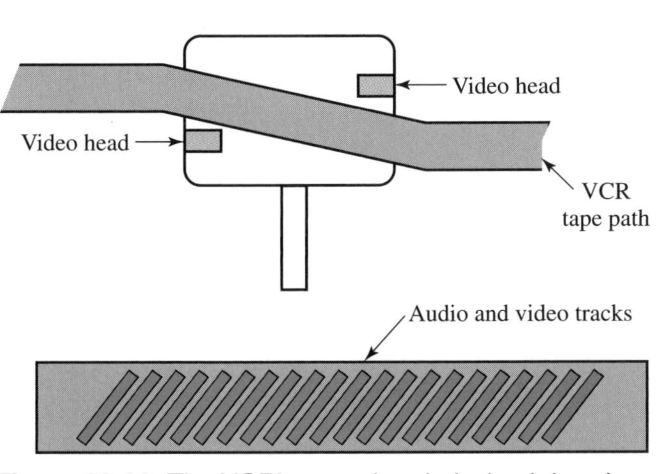

Figure 23-20. The VCR's record and playback head rotates across the tape.

Remote Control

A TV or VCR remote control is an interesting application of infrared light and digital techniques. When a button on the remote control is pushed, a digital code is sent out of the remote control to an infrared sensor on the TV. The sensor on the TV amplifies and decodes the signal. See **Figure 23-21.**

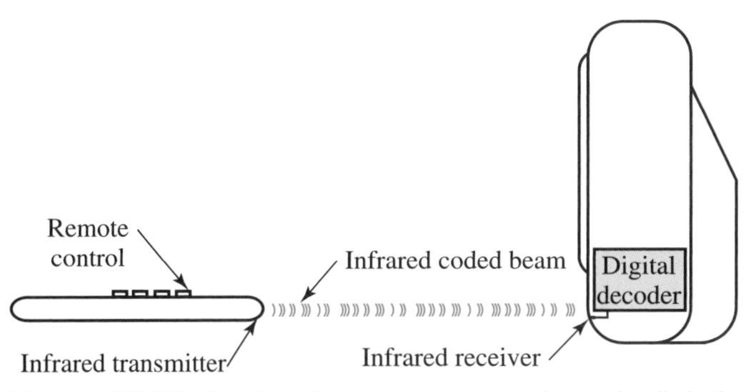

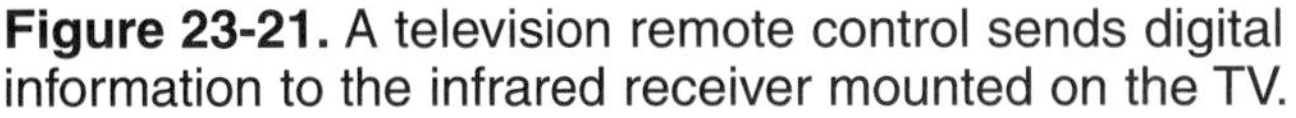
Figure 23-21. A television remote control sends digital information to the infrared receiver mounted on the TV.

Large Screen Projection TV

Most large screen projection TVs use a special electron gun assembly that projects three separate images onto a screen. Early projection TVs had screens with a significant curvature. As these televisions became more popular and more advanced, engineers were able to develop a flat screen projection. **Figure 23-22** shows how the electron gun is assembled.

The main performance problem with large screen or projection TV is the loss of clarity of the video image. Remember there are only 525 lines per frame of display.

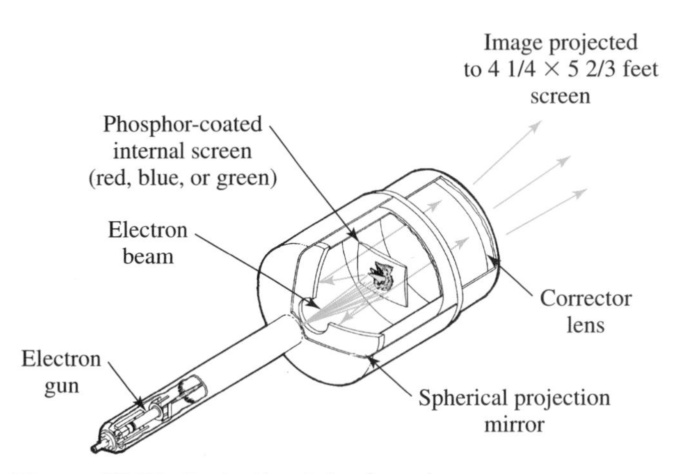

Figure 23-22. Projection tube for a large screen television.

As the picture is increased in size, the lines become a distraction. The image does not magnify; it simply gets larger. Some large projection TVs offer a slight improvement in video image by scanning the same lines twice for a total of 1050 lines across each frame. These sets do have a sharper, more appealing video image to the human eye, but no real magnification has taken place.

Satellite TV

Arthur C. Clarke first introduced the idea of launching satellites to improve communications. He did this in an article in the fall, 1945 issue of *Wireless World* magazine. He stated that if satellites could be launched high enough (35,880 kilometers or 22,300 miles) above the equator, they would be in geostationary orbit. ***Geostationary orbit*** means an object rotates with the earth.

The first communications satellite, Telestar I, was launched by the National Aeronautics and Space Administration (NASA) in 1962. It was a small, experimental satellite that only operated a few hours each day. This satellite made communication between the United States and Europe possible. In April, 1965, NASA launched the first commercial satellite, Early Bird. This satellite was owned by the International Telecommunications Satellite Organization (INTELSAT), a group created in 1964. Now there are many satellites in orbit allowing for television, telephone, radio, data, and other communications messages.

Rockets and space shuttles place satellites into space. There they are deployed and the circuits are activated, **Figure 23-23. Figure 23-24** shows an SBS communications satellite now in orbit. This satellite was designed to provide voice, video teleconferencing, data, and electronic mail services to U.S. businesses.

From its geostationary, or synchronous, orbit 22,300 miles above the equator, AUSSAT, Australia's first national communication satellite, links that entire country and Papua, New Guinea, through an advanced telecommunications system. See **Figure 23-25.** When the satellite is in orbit, the antennas point south, making the spacecraft look upside down if viewed from earth.

Refer to **Figure 23-26.** It shows the inside of a satellite. A traveling wave tube amplifier increases the strength of the communication signal for its broadcast

Figure 23-23. The launching of the Hughes communication Leasat 4 satellite. (Hughes Communications, Inc.)

Figure 23-24. A Satellite Business Systems (SBS) satellite being prepared for launch. (Hughes Communications, Inc.)

Figure 23-25. A communications satellite used for communication in Australia. (Hughes Communications, Inc.)

Figure 23-26. Looking inside a communications satellite. (Hughes Communications, Inc.)

back to earth. It is being adjusted by an engineer. The amplifier is onboard a communication satellite. This satellite is built to carry both standard traveling wave tubes and solid state power amplifiers. This type of satellite is reliable and has a long life.

The diagram of the parts of a satellite are shown in **Figure 23-27. Figure 23-28** shows satellites in orbit over North America.

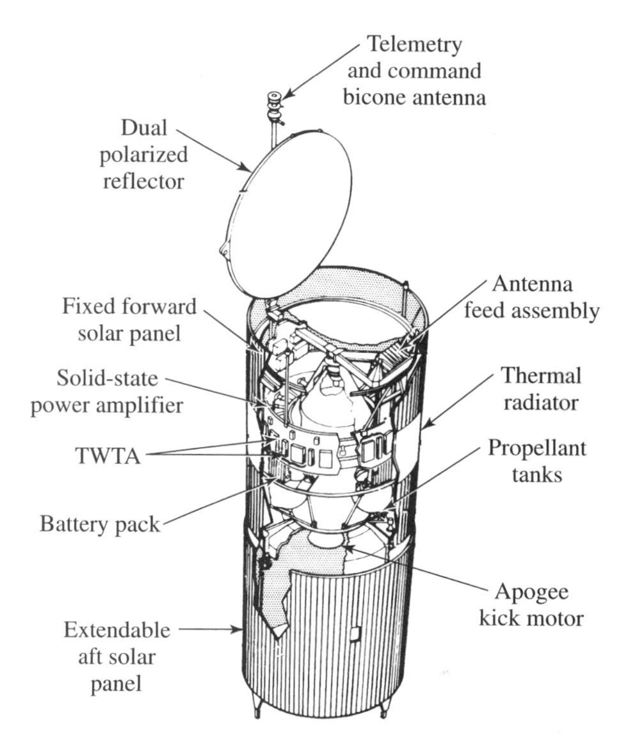

Figure 23-27. Parts of the Telstar III satellite. (Hughes Communications, Inc.)

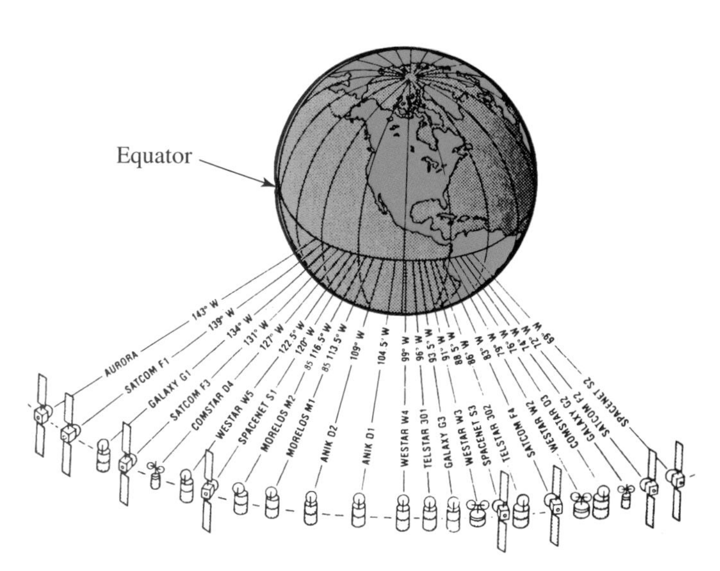

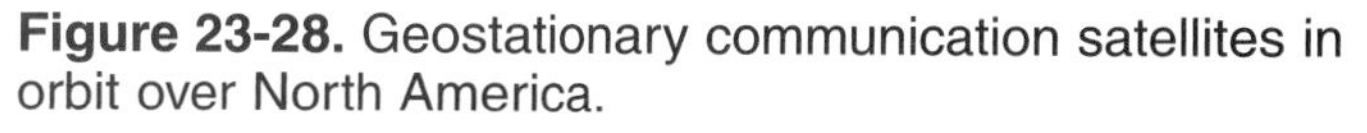

Figure 23-28. Geostationary communication satellites in orbit over North America.

Satellite transmission

Once a signal is made by a communication station on earth, it is beamed up to the satellite. The satellite picks up the signal on its receiving antenna, amplifies the signal, and then sends it back down to earth. See **Figure 23-29.** The signal sent up from the studio to the satellite is a narrow, targeted signal. The signal sent back down from the satellite is a wide signal designed to cover a large area of the earth.

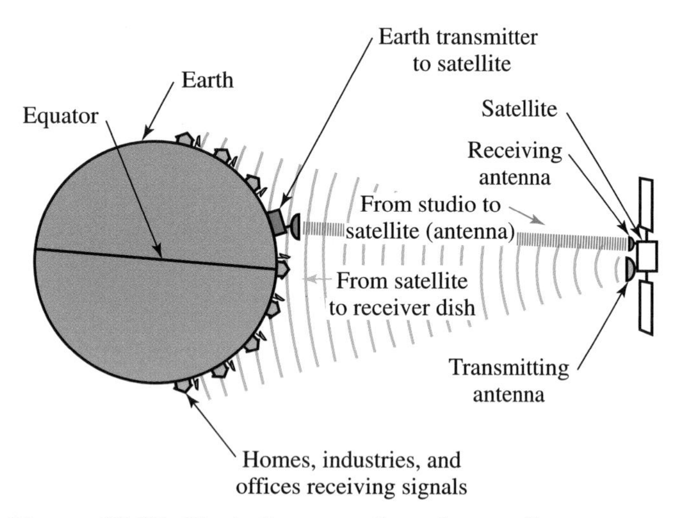

Figure 23-29. Study the operation of a satellite.

The signal transmitted by the satellite is picked up on Earth by a receiving dish or parabolic antenna. The dish focuses the received signals into a small area called the ***focal point.*** The ***feedhorn,*** which acts as a receiver, is located here, **Figure 23-30.** Located near the feedhorn is a low noise amplifier (LNA) that amplifies the received signal. The signal is then fed through a piece of electrical coaxial cable to the TV receiver. A coaxial cable has a conductor inside another conductor. The two conductors are insulated from each other.

Figure 23-31 shows four dish designs. Motors are often used to move dishes. This way, signal from more than one communication satellite can be received or the dish can be lined up with a particular satellite.

A satellite receiver is shown in **Figure 23-32.** This receiver can be programmed with infrared remote control. Complex circuitry allows the user to store satellite positions, polarity, frequencies, and tuning voltages into memory. The programmed information can then be recalled from the unit's front panel or the remote control unit. Other remote control functions include volume control with mute, direct or scan channel selection, and video fine tuning.

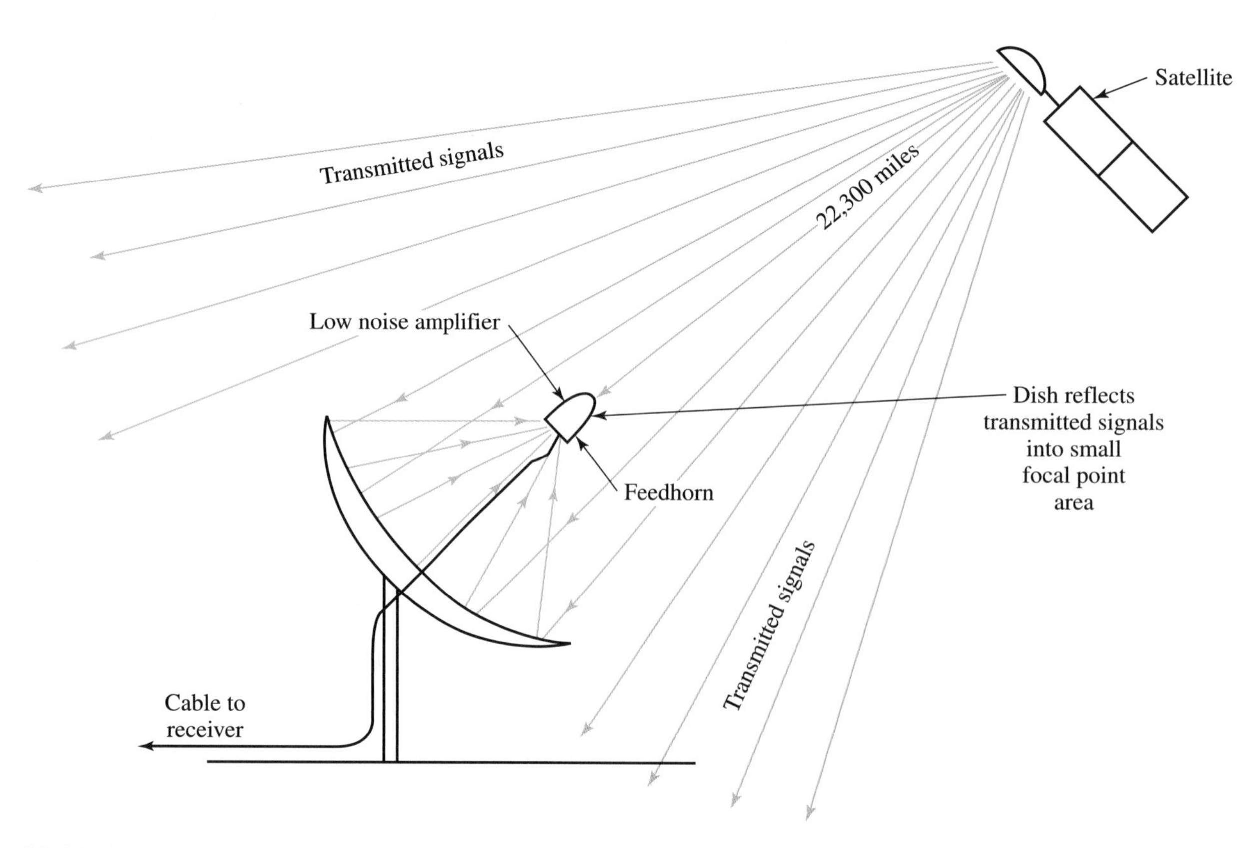

Figure 23-30. A parabolic, or dish, antenna. Study the parts.

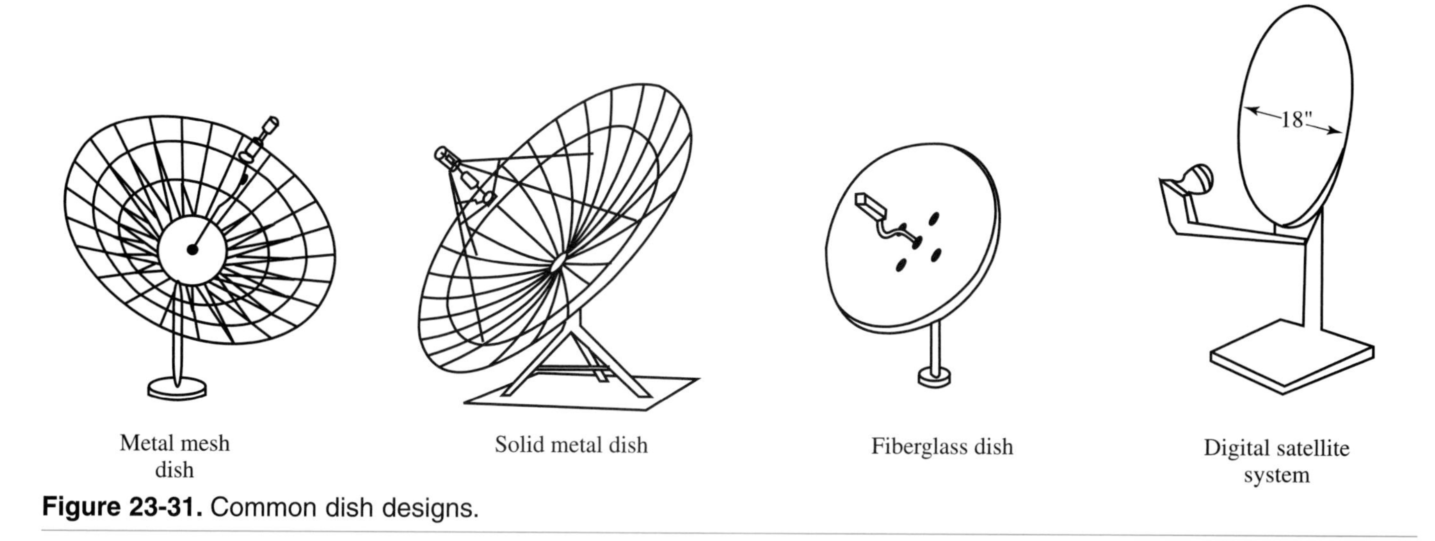

Figure 23-31. Common dish designs.

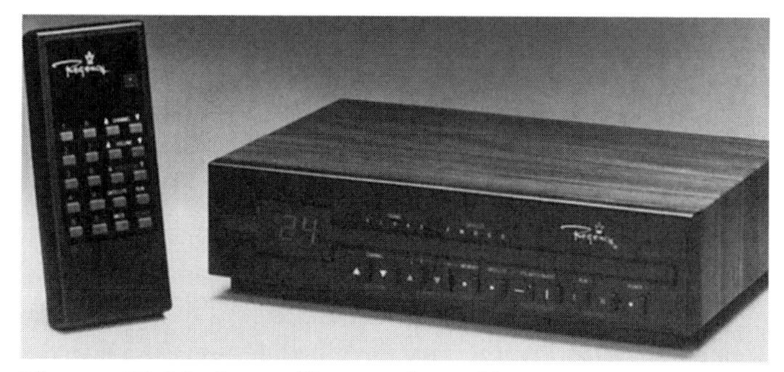

Figure 23-32. A satellite receiver. (Regency Electronics)

Review Questions for Section 23.3

1. A(n) __________ pattern is left on a recording tape.
2. A remote control sends a(n) __________ __________ to the TV.
3. How is an image projected onto a large screen TV?
4. Why is the resolution limited for a typical large screen TV?
5. Who first proposed the concept of communication satellites?
6. How does a receiving dish work?

23.4 HIGH DEFINITION TELEVISION (HDTV)

The analog television display system had remained the same for over fifty years without any major improvements to the quality of the transmitted image. This condition may have continued if not for the development of the computer monitor. The computer monitor had a greater image resolution than the analog television. By merging the analog television system with the digital system, many features that were not possible with the traditional analog system could be added. When converting to the digital system, there was greater ability to manipulate screen images. For example, since there is greater color-depth control in a digital system, images could be easily re-apportioned as the image horizontal to vertical ratio changed between digital and analog television reception.

The FCC's approval in 1996 of a digital television standard was the first step toward an improved and a higher quality picture. Although the switch to digital has progressed slowly, with more and more television stations switching to digital broadcasts and many digital television formats emerging, ***high definition television (HDTV)*** has become the dominant digital television technology.

HDTV allows for higher resolutions and a wider display screen than an analog display system. HDTV uses digital broadcasting techniques, allowing more information-rich data to be transmitted by airwaves than by analog broadcasts. Digital broadcasting can broadcast multiple channels in the same bandwidth as that used for one analog channel. Broadcasting multiple channels is referred to as ***multicasting***. Multicasting allows not only for the image to be broadcast, but also for two to four channels to be broadcast in the same single-channel bandwidth. The additional channels can be used to transmit additional images, resulting in "picture-in-picture" and information such as stock prices, weather reports, sports scores, or background information about the actors in the movie being viewed. Any information found on the Internet can be incorporated into the display screen. When additional information is transmitted along with the video image, it is referred to as ***datacasting***. You will soon likely be able to integrate a digital camera into the system so that you can see the baby sleeping in the next room while watching your favorite television show.

To be considered a complete HDTV system, three major system components are required, **Figure 23-33:**

- A digital camera to record the images at the higher resolution.
- A digital receiver (HDTV tuner) to convert received broadcasts into image and sound.
- A display unit capable of producing images at the high definition TV resolution.

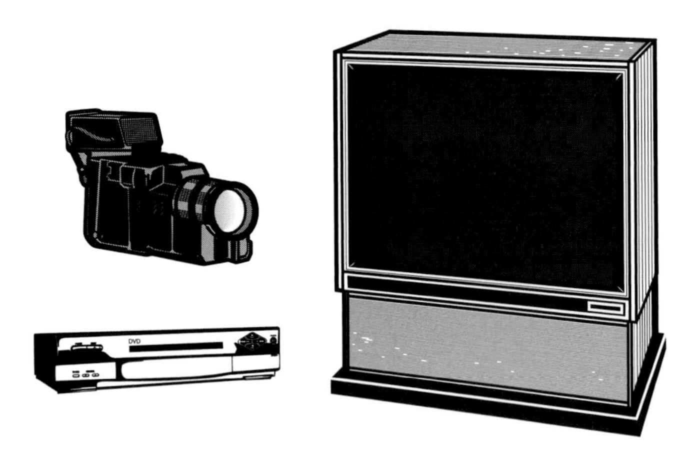

Figure 23-33. Major components required for a complete HDTV system.

If any of the three major parts are missing, the HDTV system is incomplete and will not produce an HDTV picture. There are many television variations that use HDTV terminology but do not produce the desired HDTV effect. For example, a television system may be capable of receiving a transmitted HDTV broadcast, but incapable of displaying the higher resolution HDTV image.

Digital Camera Technology

Traditional analog television uses vacuum tube imaging to capture images, while HDTV uses the ***charged coupled device (CCD)***. The CCD is an integrated circuit consisting of an array of photo sensors that convert light from a camera's focused image to electrical energy, **Figure 23-34.** The level of electrical energy is directly proportional to the level of light captured by the photo sensors. The CCD converts the individual packets of electrical charge into a series of analog signals representing the level of light amplitude at each photo sensor location. An analog-to-digital converter (ADC) converts the series of analog signals to digital signals. The digital pattern can then be sent to a block of computer memory to be stored as a still image, recorded to CD or DVD, transmitted across a computer network, or broadcast using the existing assigned television bands.

For full color images and higher resolutions, three sets of CCD sensors are used. A beam splitter inside the camera separates the incoming light into three colors: blue, red, and green, **Figure 23-35.** Each color is sent to a corresponding CCD. The three images are then overlaid, producing a picture rich in color. Since the full array

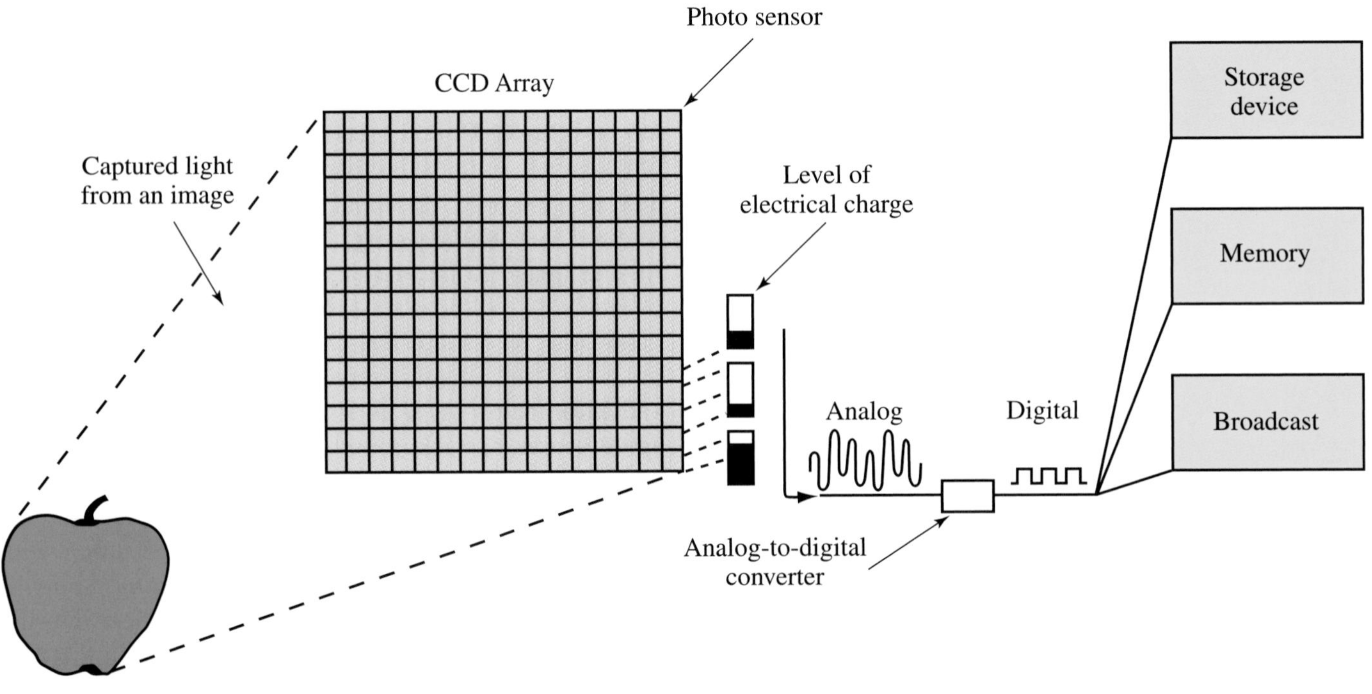

Figure 23-34. A CCD as it captures an image.

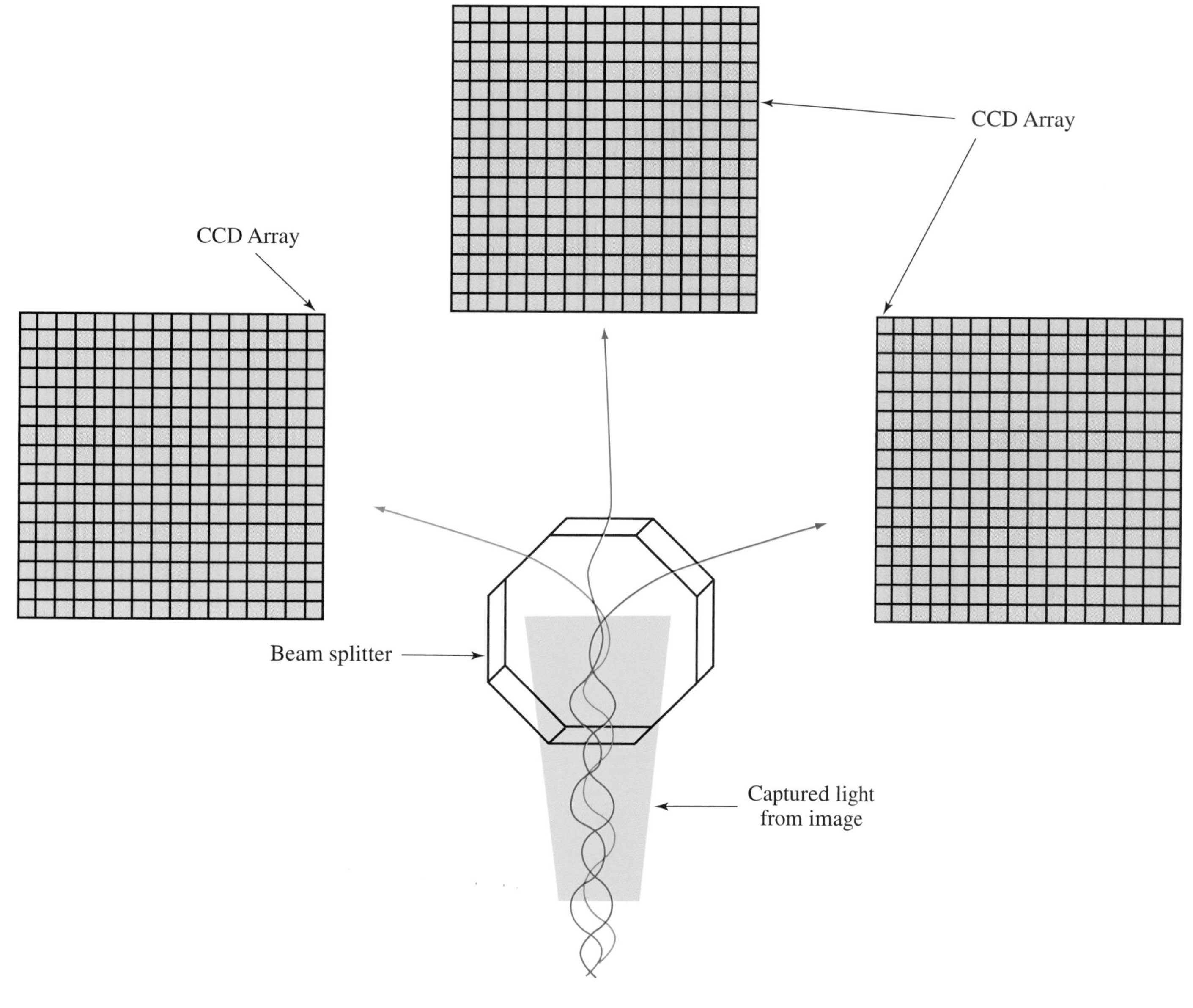

Figure 23-35. A three-CCD system.

of each CCD is used for each color, the three-CCD camera is capable of higher HDTV resolutions.

HDTV Picture Quality

The most impressive attribute of HDTV is the picture's visual quality. To compare the HDTV image to the analog television image, we must first convert the typical analog image to an equal digital resolution.

The ***National Television Standards Committee (NTSC)*** formulated the standards for analog television and video in the United States. The NTSC standard calls for 525 scan lines at a 60 Hz refresh-rate based on the interlace technique. NTSC is not compatible with most computer video systems and must be converted before it can be displayed. The ***Advanced Television Systems Committee (ATSC)*** was established in 1983. The committee spent years developing standards that were eventually adopted by the FCC for digital television broadcasting and receivers. These standards have been designed to eventually replace the NTSC standards.

There are 18 scanning formats described in the ATSC standards. Variations in the standards are derived from concerns about interlace scanning and progressive scanning, frame rates, and aspect ratio. Earlier, in the section about the analog television system, you learned about interlacing. Interlace scanning is a two-step process of transmitting the odd lines of the scan and then going back over the image, filling in the even lines to make a complete image, **Figure 23-36**. ***Progressive scanning*** is the capture and transmission of the entire image at one

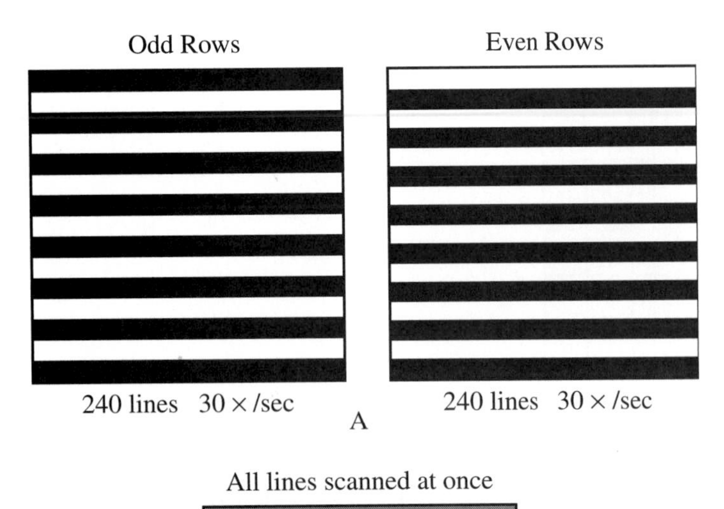

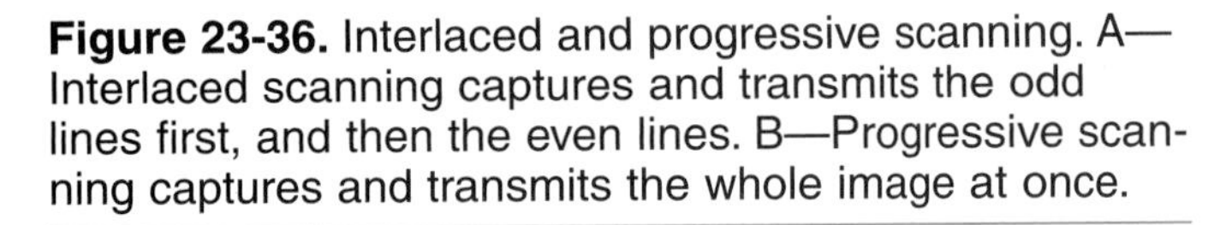

Figure 23-36. Interlaced and progressive scanning. A—Interlaced scanning captures and transmits the odd lines first, and then the even lines. B—Progressive scanning captures and transmits the whole image at once.

time. Each line is placed on the screen progressively in one sweep.

Frame rate is how often the image is updated on the screen. Currently, three frame rates (in frames per second) exist: 60, 30, and 24. ***Aspect ratio*** is the relationship of the horizontal to vertical screen presentation measurements. Aspect ratio standards can be either 4:3 or 16:9. The 16:9 is a wide screen aspect ratio similar to common movie theaters. The 4:3 aspect ratio is a standard television rectangle. The various factors of aspect ratio, frame rates, and scanning method combine to form the 18 different screen standards. **Figure 23-37** lists the 18 ATSC digital TV compression formats.

HDTV has a vertical scanning rate equal to 720p (progressive) and 1080i (interlaced) vertical lines. The actual display may have a higher vertical scan rate than the 1080i standard. This is especially important as the size of the display area increases.

To compare the quality of analog television to HDTV, you must convert scan lines to maximum number

Horizontal pixels	Vertical pixels	Aspect ratio	Picture rate
1920	**1080**	**16:9**	**60i**
1920	**1080**	**16:9**	**30p**
1920	**1080**	**16:9**	**24p**
1280	**720**	**16:9**	**60p**
1280	**720**	**16:9**	**30p**
1280	**720**	**16:9**	**24p**
704	480	16:9	60p
704	480	16:9	60i
704	480	16:9	30p
704	480	16:9	24p
704	480	4:3	60p
704	480	4:3	60i
704	480	4:3	30p
704	480	4:3	24p
640	480	4:3	60p
640	480	4:3	60i
640	480	4:3	30p
640	480	4:3	24p

Figure 23-37. ATSC digital TV compression formats. Those listed in bold are HDTV formats.

of ***pixels***. A pixel is the smallest unit of an image on a graphic display. It can be thought of as a single dot in the entire image.

Analog television approximates a screen composed of 480 × 440 pixels, producing 211,200 total pixels. HDTV approximates a screen composed of 1920 × 1080 pixels, producing 2,073,600 total pixels and by far a greater detailed image than the analog system.

The Moving Picture Experts Group developed the ***MPEG2*** image compression standard to increase the amount of video data transmitted in an HDTV system. By compressing the broadcast video data, more information could be broadcast in the same amount of bandwidth. The MPEG2 compression technique can reduce the image information by as much as 97 percent, but an average of 50 percent is typical. The compression technique is based on the fact that the majority of the video images on a television screen do not change from frame to frame. For example, a news broadcast has a persistent image, such as a background, with very little movement requiring new data. Parts of the image that are persistent do not need to be broadcast in each frame. This reduces the total amount of image information that has to be transmitted. The same technique is used for DVD, CD-RW, and still cameras.

Because more information can be transmitted using a completely digital system, sound quality has also greatly improved. HDTV incorporates the 5.1 channel surround sound system into its standard. The 5.1 system is composed of a left, right, center, left surround, right surround, and a sub-woofer signal. This is the same quality audio used in the best stereo systems available.

There are some misleading terms used when describing advanced television systems. The fast evolution of these systems and the terminology used by advertisers can often lead to disappointed consumers. ***Enhanced definition television (EDTV)*** is a system that receives digital transmissions and displays images at 480p or higher. The fact that it can receive high definition television transmissions and decode them makes it an enhanced system. However, actual HDTV displays images at 720p or 1080i. You may have an HDTV receiver connected to a display unit that cannot produce the higher display quality, thus defeating the purpose of HDTV. Some systems simply take the existing NTSC system and double the number of scanning lines, but this does not provide any new image information. This is like using a photocopier to double the size of an original image. Since no new image information has been added, picture quality cannot be enhanced.

Flat-Panel Displays

Flat-panel displays have been associated with portable computer systems for some years now. As electronic display technology evolves, display units for computers, televisions, and other forms of communication are merging. Many televisions now use flat-panel technologies instead of picture tube technologies. Flat-panel displays are lightweight, thin, and have more applications than the bulky CRT, **Figure 23-38.**

Figure 23-38. The flat-panel ViewSonic VPW 425" plasma TV is capable of both television and computer applications.

While a CRT uses a mask to isolate the individual pixels on a screen display, the flat-panel display does not. The flat-panel display controls the individual pixels electrically. Flat-panel displays typically sandwich a thin film of phosphorescent material or liquid crystal between two thin surfaces, **Figure 23-39.** One surface is covered with vertical conductors and the other with horizontal conductors, forming a grid. At each point on the grid is a pixel. Each pixel can produce a dot of light on the display unit.

Gas-plasma displays

Gas-plasma displays operate on the principle of electro-luminescence. ***Electro-luminescence*** is the display of light created when a high frequency passes through a gas to a layer of phosphor, resulting in the release of photons. The electrical energy from releasing photons is better known as producing light.

A gas-plasma display consists of millions of tiny cells sandwiched between two glass plates. See **Figure 23-40.** Each cell contains an inert gas and is coated with a phosphorous material of red, blue, or green. Transparent electrodes run horizontally behind the front panel on top of the cells, and address electrodes run vertically along the rear glass panel beneath the cells. When the address electrode and its corresponding transparent electrode are energized, the gas, in an excited, plasma state, releases an ultraviolet light. The ultraviolet light strikes the phosphorus coating inside the cell, causing the cell to release a light corresponding to its color. By varying the pulses of current, the entire light spectrum can be duplicated.

Liquid crystal display (LCD) panels

The most common flat-panel display is the ***liquid crystal display (LCD)***. The liquid crystal display (LCD) operates on the principle of polarized light passing through tiny crystals of liquid. Light is composed of many different light waves. Each light wave travels at different angles. A voltage applied to the crystal causes the crystal to warp and, in turn, determine the amount of polarized light passing through to the display screen.

The LCD is classified according to the electronic circuitry and method used to apply light to the display's surface. There are two categories of LCD: active and passive. Passive displays are more affordable than active displays because they require fewer transistors and are less complex. Active-matrix displays are costly because they use one or more transistors at every pixel.

Both active and passive displays are made up of groups of individual screen areas referred to as pixel areas. Each pixel area is made up of three color dots or pixels: red, green, and blue. The combined effect of the

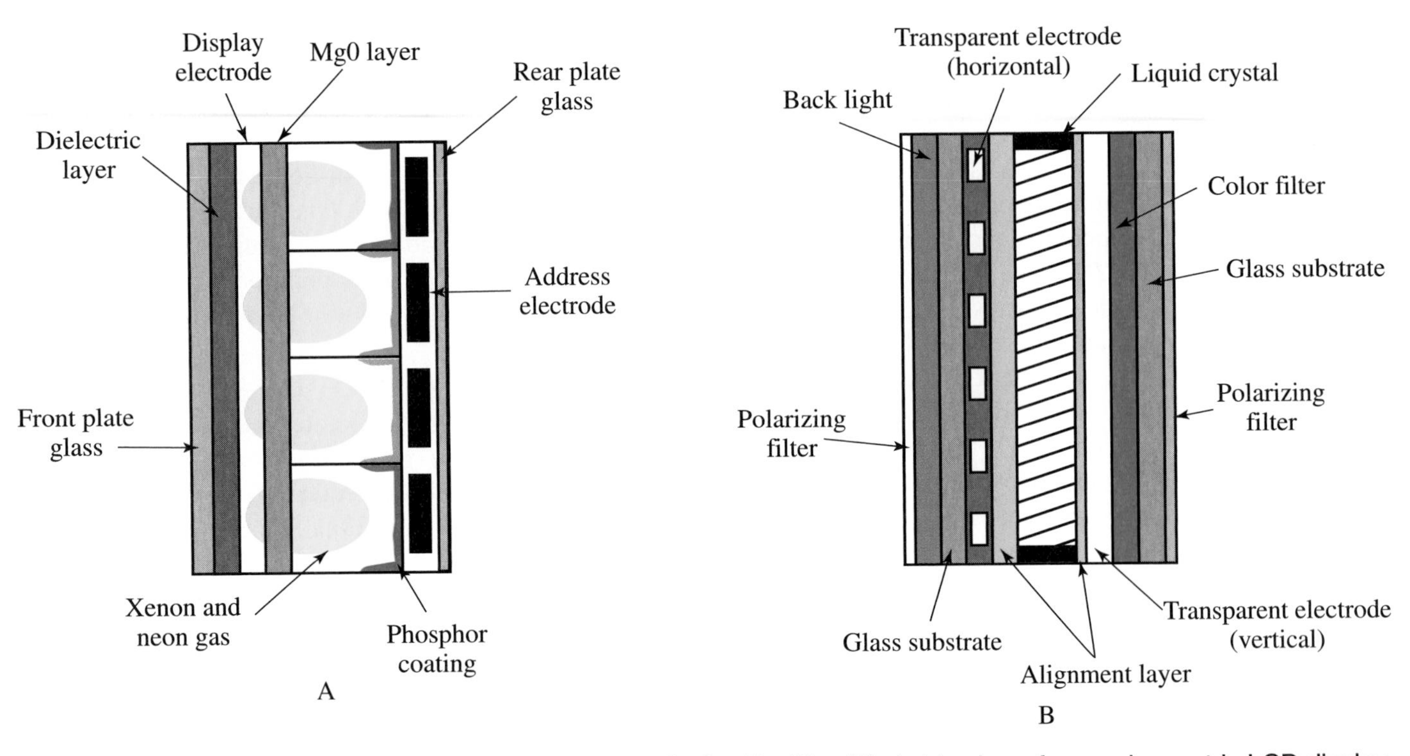

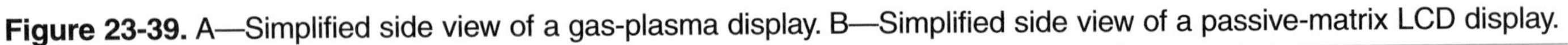
Figure 23-39. A—Simplified side view of a gas-plasma display. B—Simplified side view of a passive-matrix LCD display.

three pixels produces pixel areas representing different colors. Varying the intensity of each pixel can produce millions of colors. Combined with the surrounding areas, the pixels form an image on the display screen.

How color liquid crystal displays work

To understand how LCD technology works, follow along with **Figure 23-41.** The typical LCD panel is simple in construction. A backlight is required to generate light. The light passing through the first filter results in ***polarized light***. Polarized light consists of light waves all the same shape and of a single frequency rather than the entire spectrum of light frequencies generated by the backlight. The light passing through the first filter consists only of vertical waves.

A liquid crystal sits between the two filters. When a voltage is applied to the liquid crystal, the molecules in the crystal rotate from the vertical position. When the vertical light passes through the energized liquid crystal, it too rotates, changing into a horizontal light wave that is blocked by the second filter. The second filter allows only vertical waves to pass through. The amount of voltage applied to the crystal determines the amount of rotation from the vertical position to a horizontal position. The more voltage applied, the less light that will pass through the second filter.

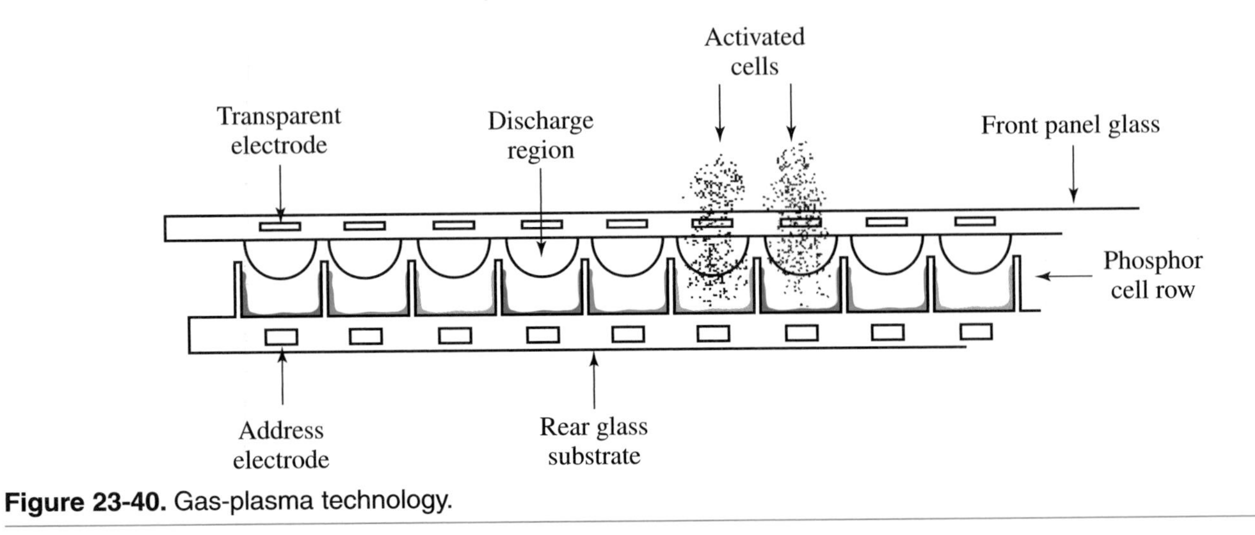

Figure 23-40. Gas-plasma technology.

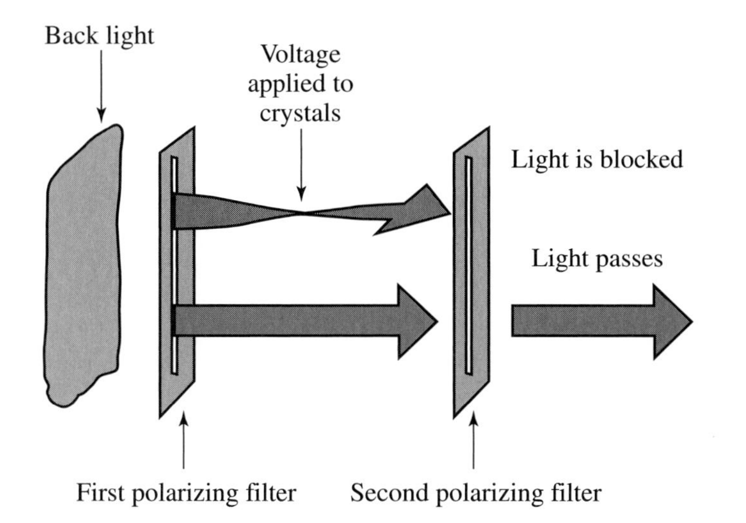

Figure 23-41. A typical LCD panel is based on the principles of polarized light waves.

Passive-matrix display

There are two types of electrical circuitry used to energize the crystal area, active and passive, **Figure 23-42.** In a ***passive-matrix*** display, a grid of semi-transparent conductors run to each crystal used as part of the individual pixel area. The grid is divided into two major circuits, columns and rows. Transistors running along the top and the side of the display unit head the columns and rows. A ground applied to a row and a charge applied to a column activates a pixel area. The voltage is applied briefly and must rely on screen persistence and a fast refresh rate. Because current must travel along the row and column until it arrives at the designated pixel, response time is slow.

Active-matrix display

In an ***active-matrix*** display, each individual pixel in the grid has its own individual transistor. The active-matrix provides a better image than the passive-matrix. The active-matrix image is brighter because each cell can have a constant supply of voltage.

The most common active-matrix display is the ***thin film transistor liquid crystal display (TFT-LCD).*** Often, this type of display is referred to simply as a TFT display. The TFT display consists of a matrix of thin film transistors spread across the entire screen. Each transistor controls a single pixel on the display. There are over one million transistors in a display, three transistors at each pixel area, one transistor for each color pixel, **Figure 23-43.** The liquid crystals in the TFT display are energized in a pattern representing the data to be displayed.

The conventional television has used the CRT to display images because the original LCD design had limitations that could not compete with larger display units. As the size of the display unit grew to over 18 inches, problems developed with the brightness of the display and in converting the analog television signal to a digital signal and to a wide-angle viewing area without image distortions. These problems were solved with the introduction of thin film transistor LCD technology.

Advantages of LCD displays over CRT displays

- LCDs can be constructed much smaller and are lighter in weight than CRT displays.
- LCDs are more economical to run because they require less power.

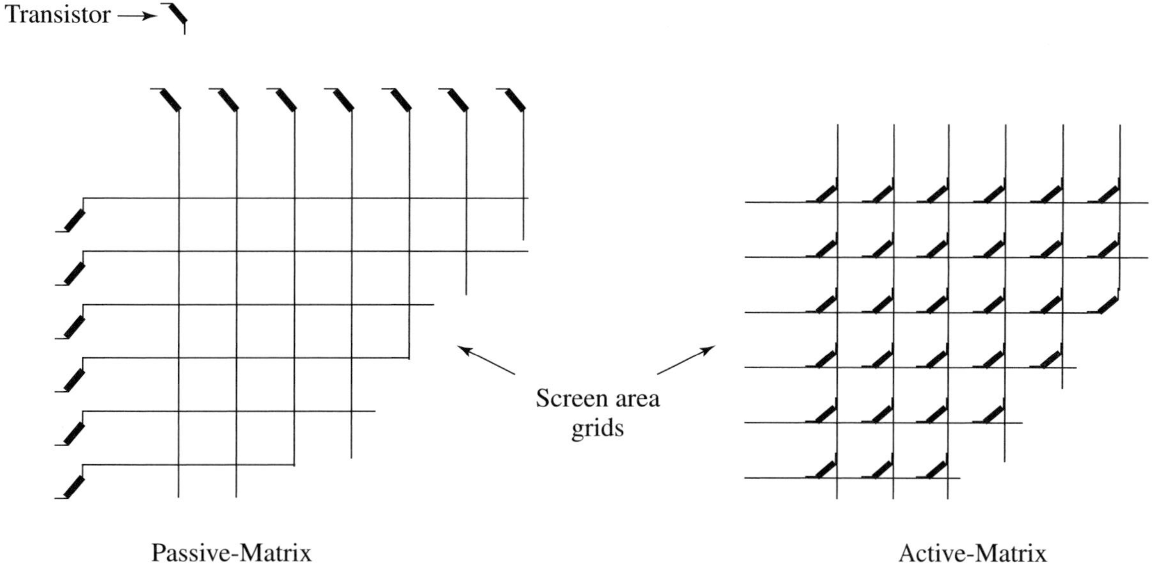

Figure 23-42. In an active-matrix display, each individual cell in the grid has its own individual transistor. The active-matrix provides a better image than does the passive.

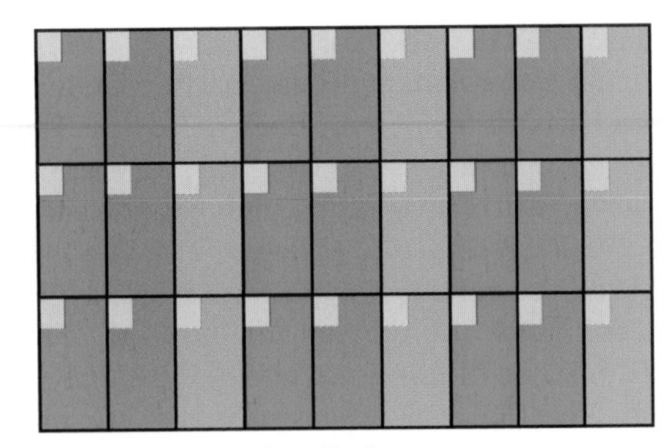

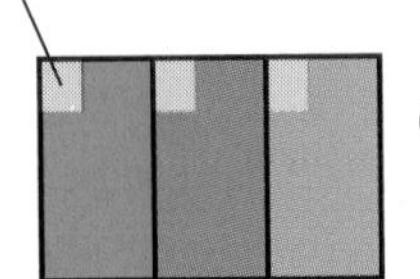

Figure 23-43. Each pixel area on the TFT display consists of three transistor-controlled color fields. The three color fields—red, green, and blue—are combined to form various shades and hues of color.

Figure 23-44. Shown are two different resolutions of the same image. The top picture is low resolution. The bottom picture is a higher resolution.

- LCDs generate less heat.
- LCDs create images that are more detailed.
- LCDs produce less electromagnetic interference (EMI).

Disadvantages of LCD displays as compared to CRT displays

- Lack of an industry-wide standard.
- Higher cost for a comparable size.
- Complexity of scaling images without distortion.

Display Resolution

Resolution is a measurement of an image's quality. The higher the resolution, the higher the quality or more detailed the image, **Figure 23-44.** The term resolution can also be used to describe the detail produced by printers, digital cameras, and any similar type of graphic equipment and is measured in dots per inch. Display resolution is the amount of detail a monitor is capable of displaying and is measured in pixels. A display unit with a maximum resolution of 1920 × 1080 has 1920 pixels along the horizontal axis and 1080 along the vertical axis.

A display unit capable of a high resolution can display at lower resolutions, but, since the display uses fewer pixels per area at lower resolutions, the image loses its sharpness.

A display unit designed to operate at one frequency and resolution is much simpler in design than a display for multiple frequencies and resolutions. Typically, the electronic system controlling a multiple resolution and frequency display uses microprocessor technology similar to that found in a computer system. In fact, the modern digital television system resembles a computer more than the earlier analog television systems.

Display manufacturers design displays to be used by multiple applications. This means that a single display can be used for analog and digital television systems, as well as computer systems. This is achieved by integrating a controller board capable of handling various inputs. The Philips SXGA triple input display controller is one such unit designed to accept input from many different sources. It allows a display to be used with television systems and PC systems as well.

Philips SXGA Triple Input Display Controller

To illustrate how one display system can be used in multiple applications, let's look at the Philips SXGA triple-input TFT display system controller, **Figure 23-45.** The triple input display system controller accepts input from analog, digital, and parallel sources. The analog input accepts the traditional UHF and VHF frequencies. The digital input accepts HDTV broadcasts over cable, as well as from personal computer systems. The parallel interface accepts input from other sources such as USB connection devices like a camera or a recorder.

The block diagram for the SXGA triple input is different from traditional television. This system requires a microprocessor and special modules to process digital information. At the opposite end there is just one output, the panel port.

You will notice horizontal and vertical sweep controls typical in a CRT imaging system are absent in this unit. Since the image is digital, there is no need for such circuitry. There are several items in the display controller that are more commonly associated with a computer system.

Standards Organizations

There are several organizations that are presently creating standards for LCD type panel displays. They are as follows: Video Electronics Standards Organization (VESA), Digital Flat Panel (DFP), and Digital Visual Interface (DVI). The variance has caused much confusion concerning video display standards, not only for screen display resolution, but also for standard connector construction. Look at the table in **Figure 23-46** for a brief summary of the three major standards groups involved in the development of a standard for their individual interest. The VESA workgroup is headed by the VESA standards organization. The DFP is led by Compaq, and the DVI is led by Intel. Each group consists of members from across the television and computer industry.

Display Connector Types

There are many different connector types developed for HDTV and various other displays. The display connectors are designed either for digital or analog transmissions, or for a combination of both digital and analog. Some manufacturers have tried to cut production costs by designing connectors to work with both television and computer systems. While making multiple-application connections is reasonable, it has caused some physical and electrical incompatibility between designs due to a lack of one general standard for all manufacturers. As long as different standards exist, compatibility issues will arise between the devices from different manufacturers. Look at the various connection designs in **Figure 23-47.**

Display systems, still evolving, have yet to designate one universal standard. This evolution is similar to when the video tape recording systems introduced the Beta and VHS taping systems. While the Beta system was technically

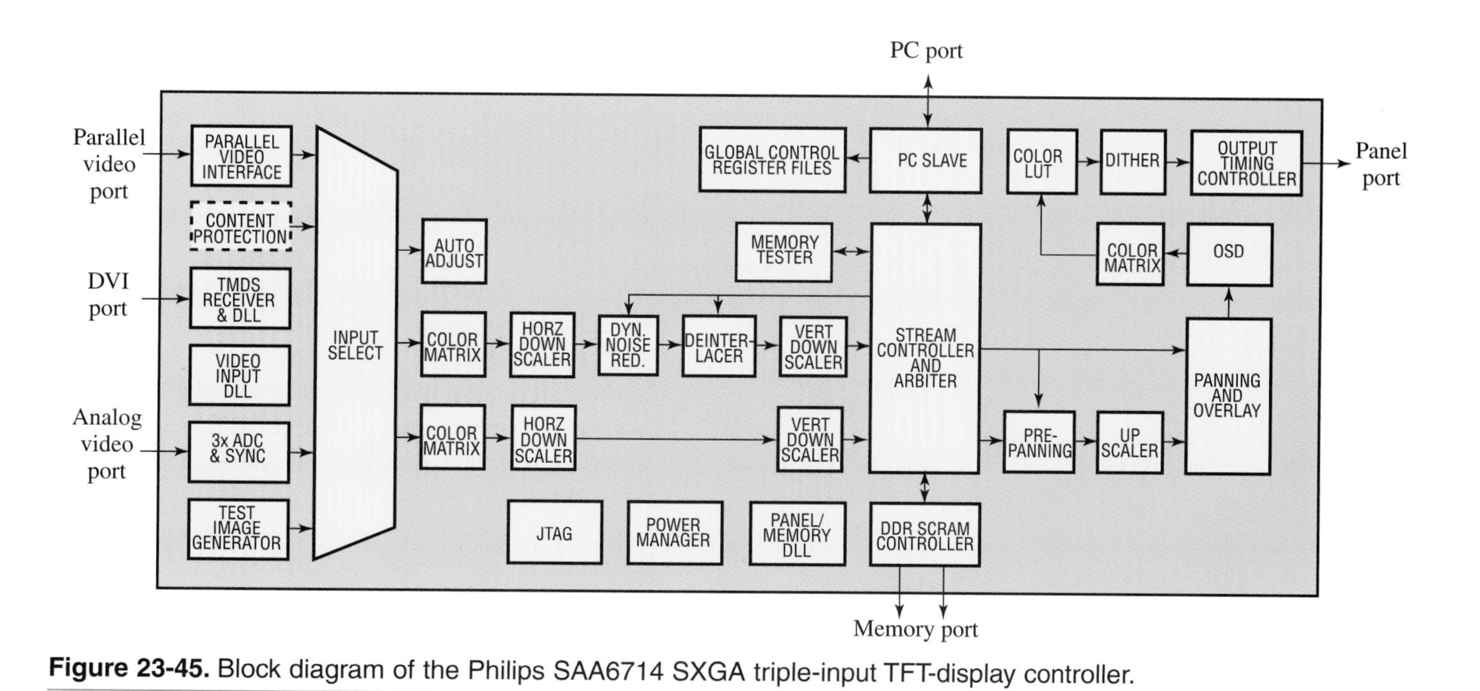

Figure 23-45. Block diagram of the Philips SAA6714 SXGA triple-input TFT-display controller.

Standards for TFT	VESA	DFP	DVI
Max resolution	SXGA 1280 × 1024	SXGA 1280 × 1024	HDTV 1920 × 1080
Connection signal	Analog, USB, IEEE 1394	Digital only	Analog and VESA

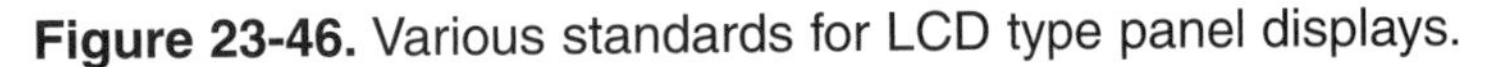

Figure 23-46. Various standards for LCD type panel displays.

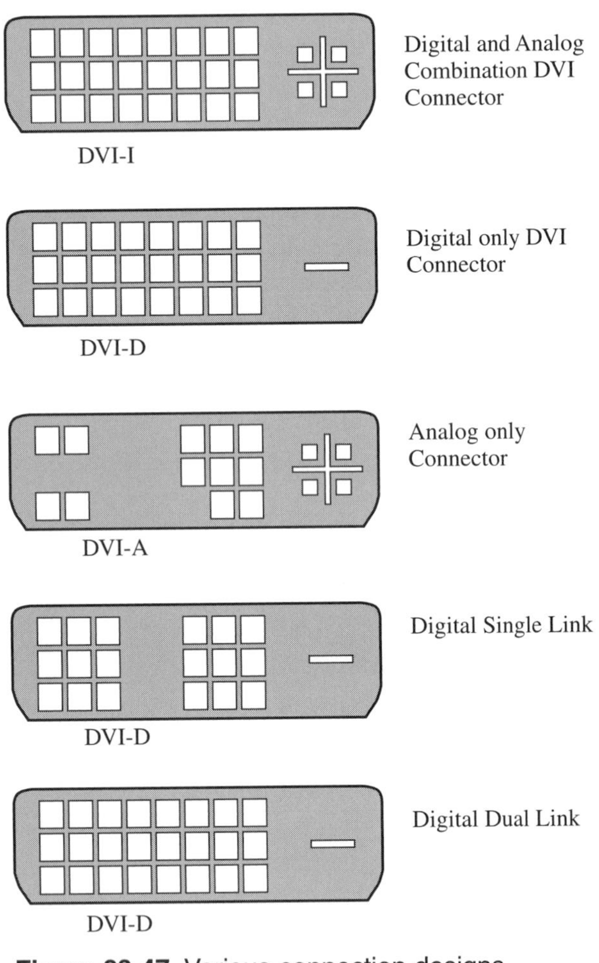

Figure 23-47. Various connection designs.

superior to the VHS system, the VHS system became the designated standard because of its convenience.

Review Questions for Section 23.4

1. What three things are needed to have a complete high definition television system?
2. What is multicasting?
3. What is datacasting?
4. What does the acronym ATSC represent?
5. What does the acronym NTSC represent?
6. What do the lowercase letters i and p represent in association with frame rates?
7. What are the two common display formats for HDTV?
8. What does the 5.1 channel surround sound system consist of?
9. How many pixels are required for each color display's pixel area?
10. What are the two classifications of LCD displays based upon the number of transistors in relation to the number of points in the image?

Summary

1. A television picture is produced by scanning an image captured from a television camera onto a cathode ray tube.
2. Scanning is the point-to-point examination of a picture.
3. The scanning system used in the United States is the interlace system. It consists of 525 scanning lines. Scanning starts at the top left hand corner of the picture. It scans from left to right on the odd numbered lines and then scans the even numbered lines. All of this takes place 30 times per second.
4. Black-and-white signals are made by a single color picture tube. Color signals are made by a three color (red, blue, and green) picture tube.
5. The bandwidth of a TV signal is 6 megahertz. The VHF band covers channels 2 through 13. The UHF band covers channels 14 through 83.
6. Video cassette recorders are used for recording and playback of magnetic tapes.
7. The image on a large screen TV is made by projecting three color images onto a screen.
8. Satellite TV is used for communication worldwide through the use of satellites that orbit the earth in a geostationary (synchronous) position.
9. Traditional analog television uses vacuum tube imaging to capture images, while HDTV uses CCD cameras to capture images and convert them to digital information.
10. A digital camera uses a charged coupled device (CCD) to capture an image and convert it into digital signals that can be stored or transmitted.
11. A pixel is a single color element of a color pixel area.
12. HDTV is associated with display formats 720p and 1080i.

13. MPEG2 is the most commonly used video compression standard for HDTV.
14. Dolby surround sound consists of 5.1 channels.
15. The most popular LCD technology for high-quality color images is the TFT-LCD.

Test Your Knowledge

Please do not write in the text. Place your answers on a separate sheet of paper.

1. The dots that make up a picture are called __________ __________.
2. Briefly explain the interlace scanning system.
3. What causes an electron beam to move from left to right?
4. The speed of an electron stream from an electron gun is increased by:
   a. grids.
   b. target plates.
   c. scanning.
   d. None of the above.
5. A composite video signal contains the:
   a. video information.
   b. sound information.
   c. Both of the above.
   d. None of the above.
6. The __________ __________ __________ is used to produce images in most televisions.
7. What are the three colors used to produce a color TV image?
8. What is the purpose of having more than two heads on a VCR?
9. _____ allows for two to four channels to be broadcast in the same single-channel bandwidth.
10. To be considered a complete HDTV system, three main components are required: a digital _____, digital _____, and a _____ _____ capable of producing images at the high definition TV resolution.
11. The _____ _____ _____ _____ developed standards for digital television broadcasting and receivers.
12. _____ is a wide screen aspect ratio similar to that found in common movie theaters.
13. The Moving Picture Experts Group developed the _____ image compression standard to increase the amount of video data transmitted in an HDTV system.
14. _____ percent of a typical transmitted picture does not change from picture frame to picture frame.
15. The most common active-matrix display is the _____ - _____.

Matching Questions

Match the following terms with their correct definitions.

a. Sync separator.
b. Sync amplifier.
c. Mixer.
d. Vertical oscillator.
e. Video amplifier.
f. Horizontal AFC.

16. The output of this device provides the sawtooth voltage to the deflection coils.
17. Removes the horizontal and vertical sync pulses transmitted as part of the composite video signal.
18. Compares the frequencies of the horizontal oscillator and sync pulse.
19. Amplifies the demodulated picture signal and feeds it to the grid of the CRT.
20. Combines the incoming video signal with a local oscillator signal.
21. A voltage amplifier stage in which the sync pulse is increased.

For Discussion

1. Discuss the function of each of the following controls found in a TV receiver. State the circuit that is regulated by each control.
   a. Horizontal hold.
   b. Brightness.
   c. Vertical hold.
   d. Fine tuning.
   e. Channel selector.
   f. Vertical linearity.
   g. Horizontal linearity.
   h. Height control.
   i. Contrast.
   j. Width control.
2. Explain the process of negative transmission used in television in the United States.
3. How does the vertical integration network separate the vertical sync pulse?
4. If dc amplifiers were used in the video section, would dc restoration be necessary?
5. Why are both UHF and VHF channels needed for television?
6. Research and discuss the various types of video display.
7. Discuss what you think television will be like in the year 2020.
8. How did development of the shadow mask picture tube promote the reality of color TV?
9. Why would a digital television image be shaper than a conventional television image?

The parabolic shape of a satellite dish focuses the received signal to the feedhorn. Before reaching the receiver, the received signal passes through a low noise amplifier.

Fiber Optics and Lasers

Objectives

After studying this chapter, you will be able to:

- State the advantages and disadvantages of fiber-optic systems.
- Explain light theory.
- Explain causes of light energy losses in fiber-optic systems.
- Explain the transmission of light as data.
- Explain how light is received and changed into data.
- Explain how lasers operate.
- List safety precautions to be taken when working with lasers.

Key Words and Terms

The following words and terms will become important pieces of your electricity and electronics vocabulary. Look for them as you read this chapter.

attenuation
buffer
cladding
coherent light
core
dispersion
extrinsic loss
fiber optics
fresnel reflection loss
incoherent light
laser
laser dioide
lasing
optical time domain reflector (OTDR)
photon
pigtail splice
PIN diode
plenum
ruby laser
scattering

In the simplest terms, fiber optics involves the transmission of analog or digital information in the form of light waves through a glass or plastic core cable. The use of fiber optics is becoming common place in long distance telephone, computer communications, medical systems, as well as the military. In this chapter, you will discover the reasons why.

24.1 FIBER OPTICS

Fiber optics is a relatively recent development in the electronics world and has met widespread acceptance. ***Fiber optics,*** as it is used in the electronics field, is the controlled transmission of light as a signal. Look at **Figure 24-1.** The AM signal is converted to light and transmitted through the fiber-optic cable. At the cable destination, the fluctuating light signal is converted back to an AM signal. You have already learned that copper wire as a conductor will accomplish the same goal. Let's look at some of the reasons why optic fibers might be a better medium.

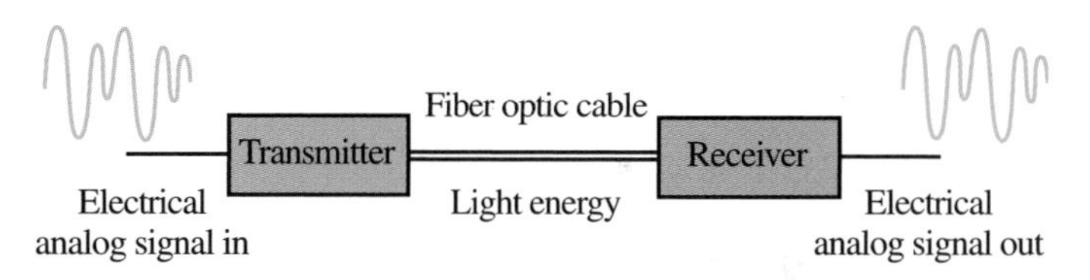

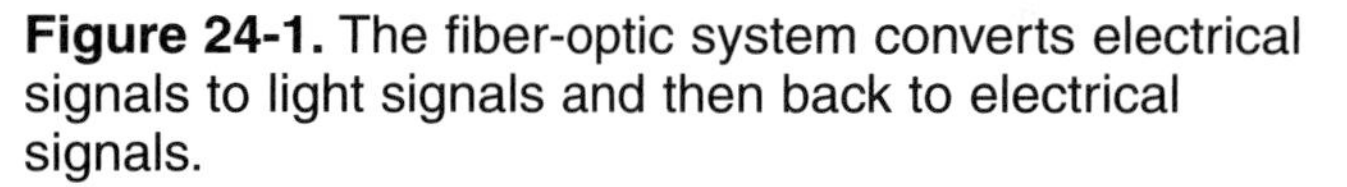

Figure 24-1. The fiber-optic system converts electrical signals to light signals and then back to electrical signals.

Fiber Optics, Why?

It may seem like extra work to convert an electronic signal to light and then convert it back again to an electronic signal. One could question why the use of copper wire, where these conversions are unnecessary, would not be more efficient.

There are quite a number of reasons a fiber-optic transmission medium might be chosen over another conductor.

- Fiber-optic cable is lightweight and very small in diameter.
- Fiber-optic cable is corrosion and water resistant.
- Fiber-optic cable provides data security.
- Fiber-optic cable is immune to electromagnetic interference.

- Fiber-optic cable provides a great deal of safety from fire and explosion.
- Fiber-optic transmission has a wide bandwidth.

These advantages are all important in different applications, so we will discuss each one further.

Weight and size

Fiber-optic cables are exceptionally lightweight when compared to traditional copper wiring systems. This is important in general wiring applications but a much greater advantage is gained when it is used in applications where weight is a critical factor. Some of these uses are in aircraft and ships. The lighter the weight of the plane or ship, the more cargo it can carry. Fiber-optic cables are approximately 1/10th the weight of a comparable copper wiring system.

Fiber-optic cable also has a smaller overall diameter than conventional cable systems. This is extremely important when running communication lines. Many more communication lines can be provided in the same conduit using a fiber-optic cable. In telephone lines, over 1000 fiber-optic lines can easily fit into the space where a system of 100 conventional cables once resided.

Corrosion and water resistance

The very nature of glass or plastic make it resistant to most corrosives. Water does not effect the light conduction capabilities of a properly installed cable system. Fiber-optic cables have been run under all of the world's oceans and are expected to last for many, many years.

Security

In today's society, information is a profitable and, in some cases, priceless commodity. It is so important in business, industry, and in military applications that it must be protected against unethical people who would seek to gain from stealing it. It is virtually impossible to tap into a fiber-optic cable without being detected on a secure communication line.

Any cut made into the cable or cladding will disturb the light signal. Cutting into the cable completely disrupts the transmission of the signal. Removal of any of the cladding results in a loss of signal strength.

At the present, fiber-optic lines are secure. However, many systems throughout history have been viewed as virtually impenetrable only to be compromised later. It should be emphasized that applied creative human intellect can develop methods to compromise even the best security system. But for now, the fiber-optic cable is far more secure than copper cable.

Immunity to electromagnetic energy

Most of us have the annoying experience of having our television or radio signals interfered with by lightning strikes or the operation of an electric motor nearby. Fiber-optic cable conducts light instead of electricity. This makes it immune to the electromagnetic interference generated by motors, radio signals, lighting, and other sources of electromagnetic energy. Conventional copper cable must be shielded to prevent electromagnetic interference. Adding shielding to a cable results in a cable that is much more expensive to produce, much larger in overall diameter, and a lot heavier.

The military has large investment in fiber optics and began research in the field many years ago. The detonation of nuclear weapons creates a powerful magnetic field. This magnetic field is strong enough to destroy communication systems based on copper wire. The destruction caused by this magnetic field can occur many miles from an actual nuclear explosion area. Since fiber-optic cable can withstand the electromagnetic field generated by even a nuclear weapon, military communications could remain intact through its use.

Safety

Fiber optics can be installed near explosive vapors and dust such as in petroleum factories or food processing plants. This use eliminates the fear of an electrical spark caused by a shorted or grounded circuit using conventional conductors.

In addition, light for illumination can be directly transmitted through fiber optics in place of conventional lamps. When using electrical lamps for illumination in hazardous industries, the lamp fixtures must be a very expensive, explosion-resistant type. Fiber-optic lamps are safer and cheaper.

Bandwidth

The bandwidth available on a fiber-optic cable has superior merit when compared to copper. Light is transmitted at a much higher frequency than typical electrical signals. More signals can be transmitted at one time with fiber optics than with wire cable systems. Conventional cable systems have limitations and losses due to inductive reactance. Copper cables lose their conduction capabilities at extremely high frequencies. Fiber-optic cable can handle high frequencies with little to no problem.

The Nature of Light

There are two main theories of light. One theory addresses light as a wave, and the other treats light as a particle. When light energy is traveling through air, water, or a vacuum, its nature is explained in terms of waves, similar to the electromagnetic waves of radio and TV. When light interacts with a solid, it is explained as a particle, and is referred to as a ***photon.*** The correct theory remains to be seen. For now, the best of both theories are used to explain the phenomena of light and its interaction with the surrounding world we live in.

Look at **Figure 24-2.** As you can see light is described in terms of wavelength. Light is usually measured in nanometers (nm). Visible light has wavelengths between 390 nm and 770 nm. Below visible light, larger wavelengths such as UHF and VHF television waves, and radar waves exist. Above visible light rays, are the much shorter wavelength X-rays, gamma rays, and cosmic rays. The actual spectrum of visible light is quite small when compared to all other forms of electromagnetic radiation.

Most fiber-optic systems use infrared wavelengths between 850 and 1500 nm. At these wavelengths, there is less loss of signal. A typical LED has a wavelength that is between 800 and 900 nm or 1250 and 1350 nm.

Cable Construction

A fiber-optic cable serves as a wave guide for light. The construction of a fiber-optic cable is simple. Look at **Figure 24-3.** The fiber-optic cable is composed of a glass or plastic core surrounded by a cladding. The glass or plastic ***core*** is the medium for the movement or transfer of the light energy. The ***cladding*** keeps the light wave contained to the core. The cladding surrounds the entire core and causes the transmitted light to remain in the core. Without the cladding on the core, the light would be lost though the sides of the core material. The cladding and core are surrounded by a buffer area. The ***buffer*** is wrapped around the core and cladding to provide physical protection.

In addition to the buffer area, water proofing materials can be inserted under the cable sheath. The cable sheath is similar to the insulated jacket on typical conductors. The outer sheath construction and materials used depend on the environment in which the fiber-optic cable will be installed. Some sheaths are designed to be oil or water resistant. Other sheaths are designed to be installed in salt water or buried directly in the earth. Some cables are be installed in a building plenum area. A ***plenum*** is an air space above a drop ceiling or under a raised floor, such as those in computer rooms. See **Figure 24-4.**

When fiber-optic cable is installed in plenum areas, the installation usually falls under the jurisdiction of the National Electrical Code. The NEC has strict regulations about the type of insulation that can be used on the outer covering of fiber-optic cables installed in plenum areas. Fumes produced if this covering were to burn in a fire can be dangerous to personnel in the building. In many

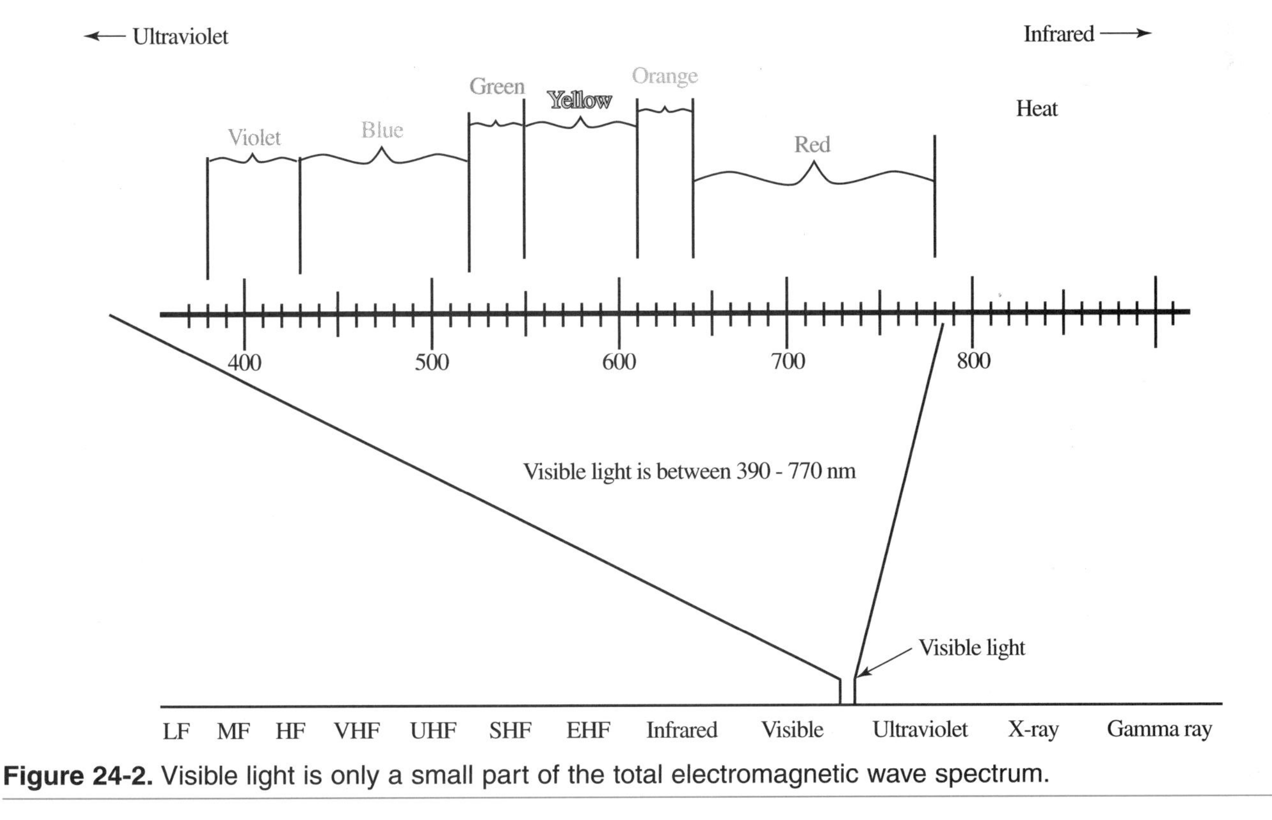

Figure 24-2. Visible light is only a small part of the total electromagnetic wave spectrum.

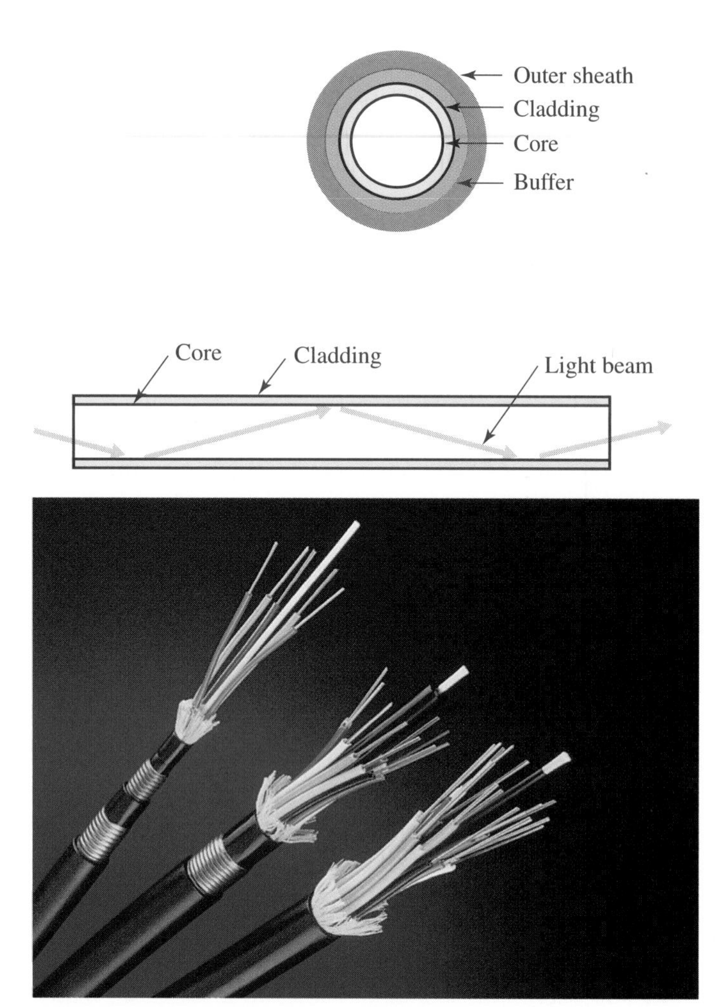

Figure 24-3. A fiber-optic cable is constructed of a glass or plastic core surrounded by a reflective cladding. (Siecor Corporation, Hickory, NC)

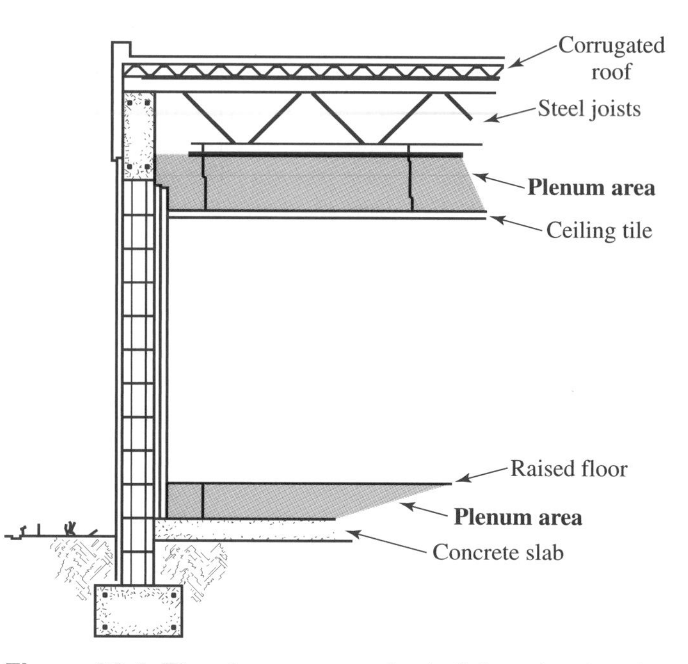

Figure 24-4. The plenum area of a building structure is located above the dropped ceiling and under a raised floor.

buildings, the plenum area also contains the air-conditioning system. This ducting would make fumes and vapors present in the plenum area harmful to personnel throughout the building.

Attenuation

Typical metallic conductors have a defined resistance value and a predictable voltage drop. They are also classified by the amount of current they can safely handle. Since fiber-optic cable handles light energy and not electrical energy, a new set of terms must be used to describe the transmission characteristics of the cable. The size of fiber-optic cable is expressed in μm (micromils).

The losses in transmission of signal power from one end of the cable to the other is called ***attenuation.*** Attenuation is expressed in decibels (dB), or decibels per kilometer (dB/km). Decibel is a relative measurement for signal strength. Look at the chart in **Figure 24-5.**

Decibel rating is patterned after the sense of hearing. This measurement is not linear. In fiber optics the measurement is similarly nonlinear, much like the scale of an analog ohmmeter. Take notice of the dB rating for 0.00 dB, 10 dB, 20 dB, and 30 dB. Compare the dB rating to the equivalent percentage. The loss is similar to powers of ten.

The three major causes of signal loss can be characterized as:

- Scattering.
- Dispersion.
- Extrinsic loss (bends, splices, and connectors).

Scattering

Scattering is the loss of signal strength due to impurities in the core material. Signal strength (light energy) losses in the line are due to impurities in the core material. No core material can be made 100% pure. There will always be a microscopic amount of impure material remaining as part of the glass or plastic core. These impurities cause the light to reflect, or scatter. The total amount of impurities along the length of the cable is cumulative. The longer the cable, the more signal loss there is. Glass cores are a better transmitter of light than plastic, but they are more expensive. To reduce attenuation, glass cores are used on very long runs of fiber-optic cable.

dB	%	dB	%
0.00	100.0	1.6	69.2
0.01	99.8	1.7	67.6
0.02	99.5	1.8	66.1
0.05	98.9	1.9	64.6
0.10	97.0	2.0	63.1
0.15	96.6	3.0	50.1
0.20	95.5	4.0	39.8
0.25	94.4	5.0	31.6
0.3	93.3	6.0	25.1
0.4	91.2	7.0	20.0
0.5	89.1	8.0	15.8
0.6	87.1	9.0	12.6
0.7	85.1	10.0	10.0
0.8	83.2	15.0	3.2
0.9	81.3	20.0	1.0
1.0	79.4	25.0	0.3
1.1	77.6	30.0	0.1
1.2	75.9	35.0	0.03
1.3	74.1	40.0	0.01
1.4	72.4	45.0	0.003
1.5	70.8	50.0	0.001

dB	mW	µW	nW
0	1.0	1000	–
-3	0.5	501	–
-6	0.25	25	–
-9	0.13	126	–
-10	0.1	100	100,000
-12	0.063	63	63,100
-15	0.032	32	31,600
-18	0.016	16	15,800
-20	0.010	10	10,000
-21		7.9	7940
-24		3.9	3980
-27		1.95	1990
-30		1.0	1000
-33		0.50	501
-36		0.25	251
-39		0.13	126
-40		0.10	100
-42		0.063	63
-45		0.032	31.6
-48		0.016	15.8
-50		0.010	10.0

Figure 24-5. Decibel conversion charts. Wattage has been calculated assuming a 1 mW signal strength.

Dispersion

Dispersion is a distortion of the optical signal caused by light waves arriving at the termination point at different times. The light beams do not travel in perfectly straight lines, but rather reflect off the core cladding. See **Figure 24-6.**

When a signal is transmitted through a long fiber-optic cable it will suffer dispersion. Dispersion is a form of signal distortion. The signal tends to flatten out as it travels the length of the cable. As light is transmitted down the core, it is reflected from side to side. On a short run of fiber-optic cable there is no real problem with dispersion, but on a long run, not all the light arrives at the end of the cable at the same time. The slight variation in the time the reflected light arrives can be critical at high frequency rates of transmission.

A typical digital pulse is transmitted into one end of a fiber-optic cable. Due to the reflection of the light along the core of the cable, the light pulses at the cable termination at different times. This factor causes the original square wave to be distorted in shape. It appears flattened and elongated.

A receiver, equipped with a digital gate, can quickly reshape the signal back to a square shape again. However, the process of regenerating an original amplitude frequency audio or video signal is not always so successful.

Extrinsic loss

Extrinsic losses are caused by physical factors outside the normal core such as bends in the optical wave guide, splices, and connectors. Fiber-optic cables have a minimum radius of bend. If this radius is exceeded, attenuation of the signal will result. Splices in fiber-optic line also need special attention. Splices, couplings, and connectors are the main reasons for transition signal losses in normal fiber-optic cable runs.

Splices and Connectors

Small fiber-optic systems can include all the cabling necessary for the system made by the supplier prior to installation. Measurements are made, and all cable splices and connections are prepared prior to installation. This type of installation ensures that the highest quality control standards were applied to the preparation of the cable system. However, these types of installations are not always possible. In the field, factors such as dirt,

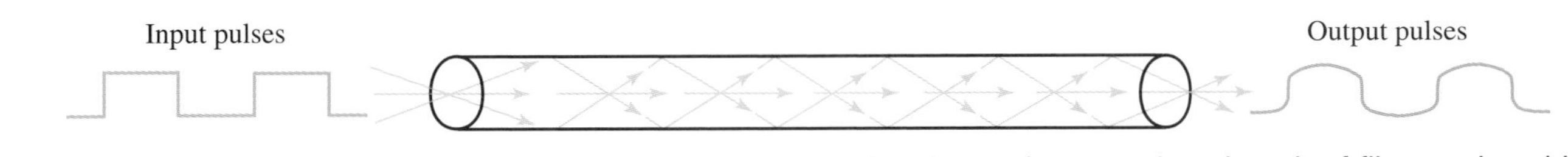

Figure 24-6. Optical dispersion is the distortion of a light signal as it travels over a long length of fiber-optic cable.

dust, and chemicals can hamper the cable splicing and termination connector installation.

Splices couplings and connections require special equipment and techniques. The cores must be in near-perfect alignment when splices are made. Misalignment or other improper splicing will result in loss of signal strength. See **Figure 24-7.**

A glass core fiber-optic cable is cleaved, not cut like a regular wire conductor. To ***cleave*** the fiber core, it must be scribed with a sharp cleaving tool edge made of diamond or ceramic material. Once the core has been scribed, pressure is applied to the scribed area until the core breaks. The scribe and breaking action produces a clean, clear surface at the end of the core. Traditional cutting tools used for wire cable would badly scar the ends of the core. **Figure 24-8** shows a high quality fiber cleaver.

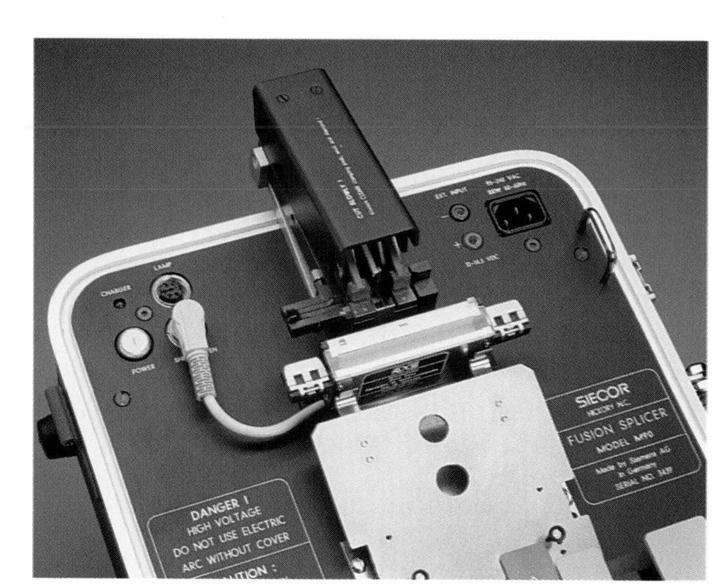

Figure 24-8. At the top of this fiber splicer is a fiber cleaver with a diamond scribe. (Siecor Corporation, Hickory, NC)

Plastic core cables need not be cleaved, but an extremely sharp cutter must be used to obtain a clean, clear cut. Marks or scratches on the ends of the core material will result in a tremendous loss of light signal due to reflection and refraction. The light will scatter in many different directions rather than transfer directly to the next cable or connector. Special equipment is utilized for making splices.

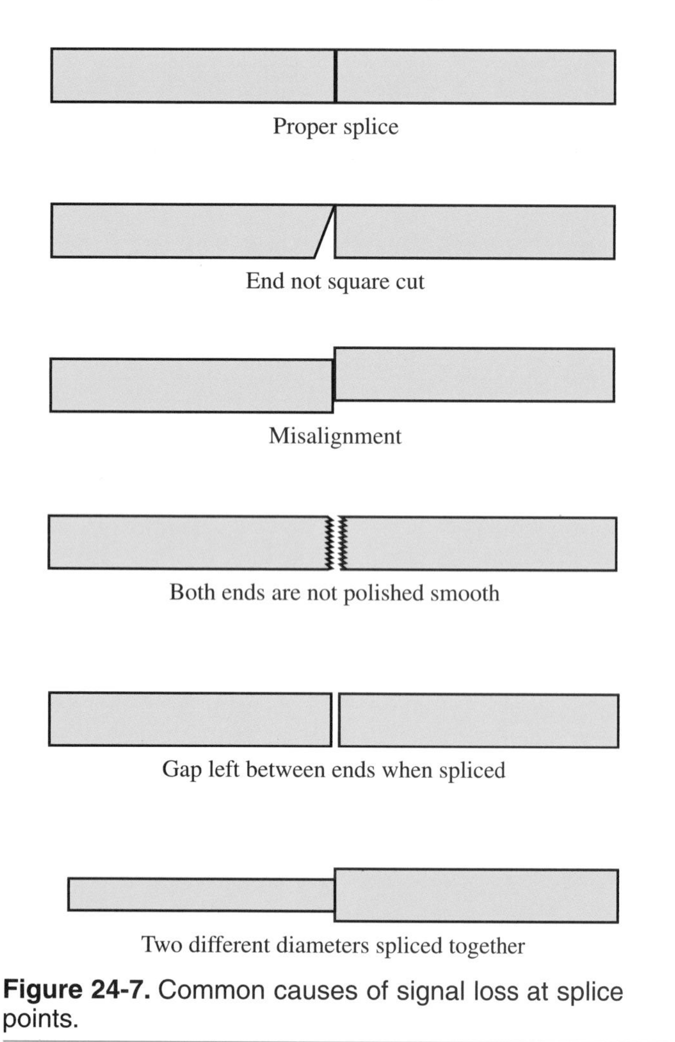

Figure 24-7. Common causes of signal loss at splice points.

The splicing equipment has an eyepiece located at the top of the tool. Since the cable diameter is very small, a microscope with at least 30× magnification is necessary to inspect the cut and splice. This tool is usually found in combination with tools that cut the fiber, polish the ends, and then fuse the ends together for a near-flawless splice called a fusion splice.

A ***fusion splice*** is the joining together of two cores using heat to fuse or melt the materials together. A fusion splicer is shown in **Figure 24-9.** This fusion splicer performs all the functions just described and more. It is equipped with video cameras and a liquid crystal display for a close, simultaneous inspection of the splice at two different angles.

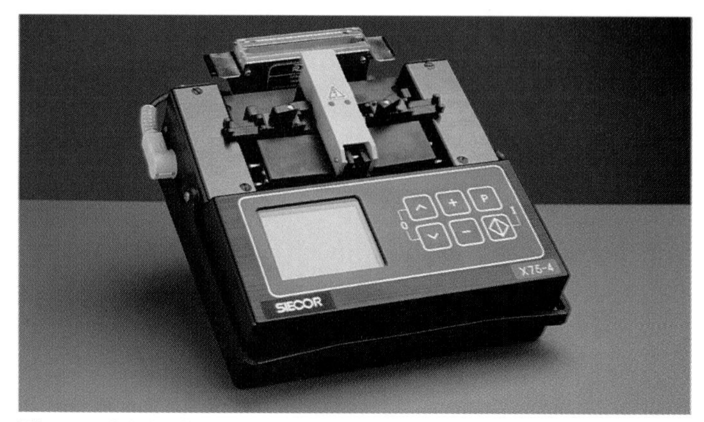

Figure 24-9. A microprocessor-controlled fusion splicer. (Siecor Corporation, Hickory, NC)

Where two different materials are joined together, there will be a loss. This loss is referred to as a ***fresnel reflection loss.*** Fresnel losses are due to differences in refraction. They commonly occur at connection points due to the differences of refraction properties of cable core material, the connector materials used for sealing the connector, or air. All glass and plastic core materials, sealing materials, and air have different refraction indices or properties.

This type of loss can be minimized by using a sealing material with a refraction index that closely matches the index of the glass or plastic fiber-optic core material. A mismatch of sealing agent to core material can cause a grossly exaggerated signal loss.

Care must also be taken to protect the cladding on the core when preparing cables for splicing. Damage to the cladding will result in signal loss.

At times, a temporary splice may be needed. This type of splice does not need to be made to such a high degree of accuracy. In this case, a quick easy convenient method is needed for testing and laboratory purposes. One such lab splice is shown in **Figure 24-10.** This lab splice can be used many times. It is used to conveniently make connection to metering devices and lab experiments. It is intended solely for use as a temporary splice.

Fiber-optic cable cores are sized in micrometers (μm). The termination connector must be sized in accordance to the diameter of the core material it is terminating. Termination of cables in the field are usually made by utilizing a pigtail splicing method. A ***pigtail splice*** is a factory-made connection on one end of a short piece of fiber-optic cable. The pigtail can be easily spliced into a fiber-optic cable inside a cable connection box. At times, it is impossible to pull cables and their individual end connectors through conduit. The pigtail splice system is a viable solution, **Figure 24-11.**

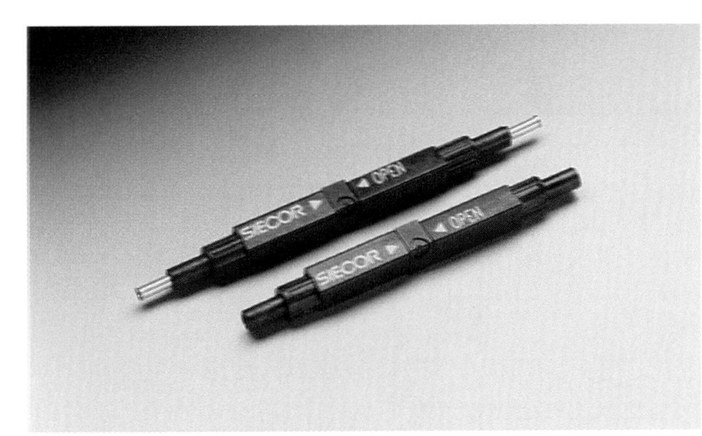

Figure 24-10. Lab splices are used as a temporary splice for lab or test purposes. (Siecor Corporation, Hickory, NC)

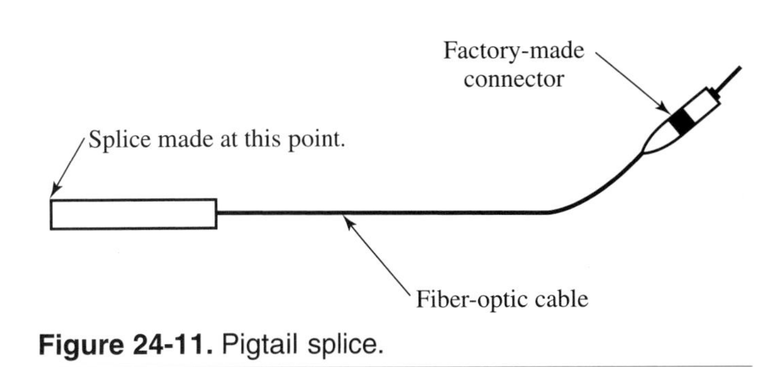

Figure 24-11. Pigtail splice.

Fiber-Optic Transmitters

Fiber-optic transmitters are usually LEDs or laser diodes. These diodes produce the beam of light that carries the signal. Many various electronic systems are used to convert the electrical signal that drives the LED or laser diode. The light signal can be either AM amplitude modulated or pulse modulated. FM modulation does not work well with fiber optics and is still being refined. At present, most signals are either AM or digital.

In Chapter 17, Introduction to Semiconductors and Power Supplies, you became familiar with LEDs, so here we will look at the construction of a laser diode, **Figure 24-12.**

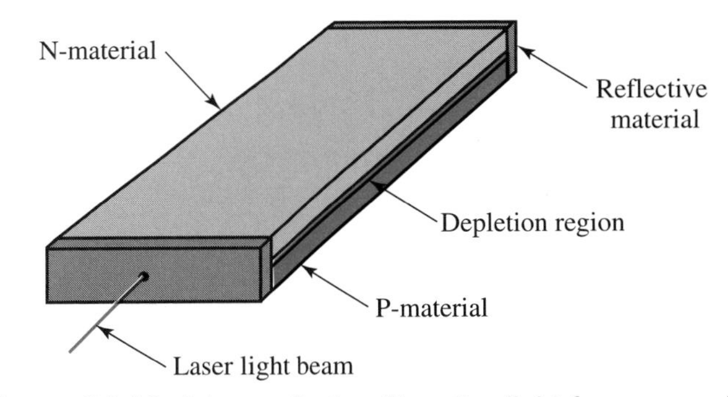

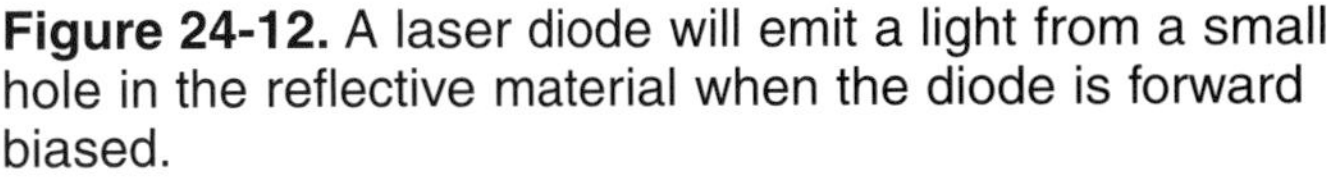

Figure 24-12. A laser diode will emit a light from a small hole in the reflective material when the diode is forward biased.

The construction of a semiconductor laser diode is simple. The ***laser diode*** is similar to a conventional diode in construction. A P-type material and a N-type material are sandwiched together. A forward biased diode emits radiation at its junction, or depletion region. One end of the diode is coated with a reflective material, allowing the light to be radiated out the other end.

Laser diodes used in fiber optics have a wavelength around 1300 nm. Laser diodes can transmit data at frequencies as high as one GHz (gigahertz, or one billion hertz). LEDs can transmit data at rates as high as 300 MHz, but most are less than 200 MHz.

Fiber-Optic Receivers

Receivers use photodiodes to convert the light energy back to electrical energy. Some of these diodes are the avalanche photodiode, the PIN diode, or a phototransistor and photodarlington, **Figure 24-13.** Phototransistor and photodarlington are not the preferred devices for high speed receivers. The reaction time of these devices is slow in comparison to the avalanche and PIN diode.

The avalanche diode reacts to light in much the same way as its name implies. The avalanche diode is similar in construction to the laser diode. When light strikes the P-type material, both a hole and an electron are caused in the depletion region. The reverse bias voltage applied to the avalanche diode is quite high in comparison to other forms of detector diodes. As a result of the high reverse bias voltage, only a few photons are needed to cause the diode to change rapidly to full conduction. The advantage of the avalanche diode is its ability to quickly go into a conductive mode. The fact that the avalanche diode requires a fairly high voltage for conduction is its only disadvantage.

The ***PIN diode*** derives its name from the way it is constructed. The PIN stands for *positive, intrinsic, negative.* These terms describe the layers of the diode. The main difference between the photodiode and the PIN diode is that the PIN diode has a wider intrinsic layer (depletion layer) than the photodiode. By widening the depletion area of the diode, the capacitance of the diode is decreased. This allows the PIN diode to react rapidly to a change in light intensity.

The circuit shown in **Figure 24-14** is a typical transmitter and receiver. The transmitter is a simple LED and LED driver. The receiver consists of a PIN diode, a transimpedance amplifier, and a post amplifier. The PIN diode converts the light wave pulses into a varying current. The transimpedance amplifier is a preamplifier that converts the varying current into a corresponding varying voltage. The postamplifier is the final amplifier stage, which receives the signal from the transimpedance amplifier.

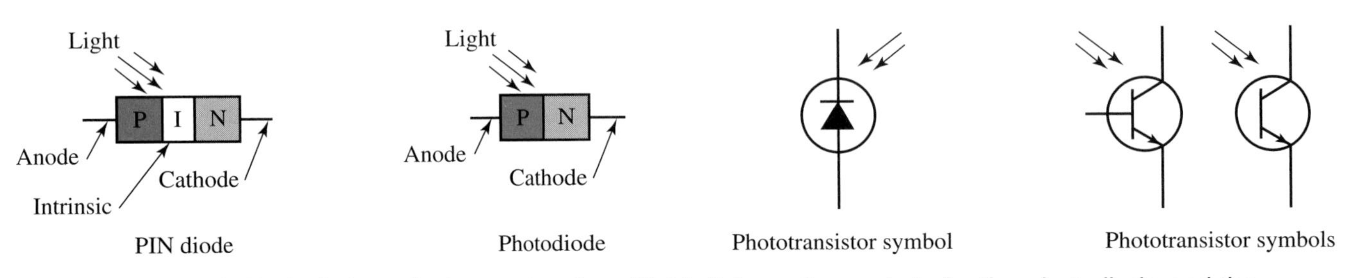

Figure 24-13. Left. Typical photodiode construction. Right. Schematic symbols for the photodiode and the phototransistor.

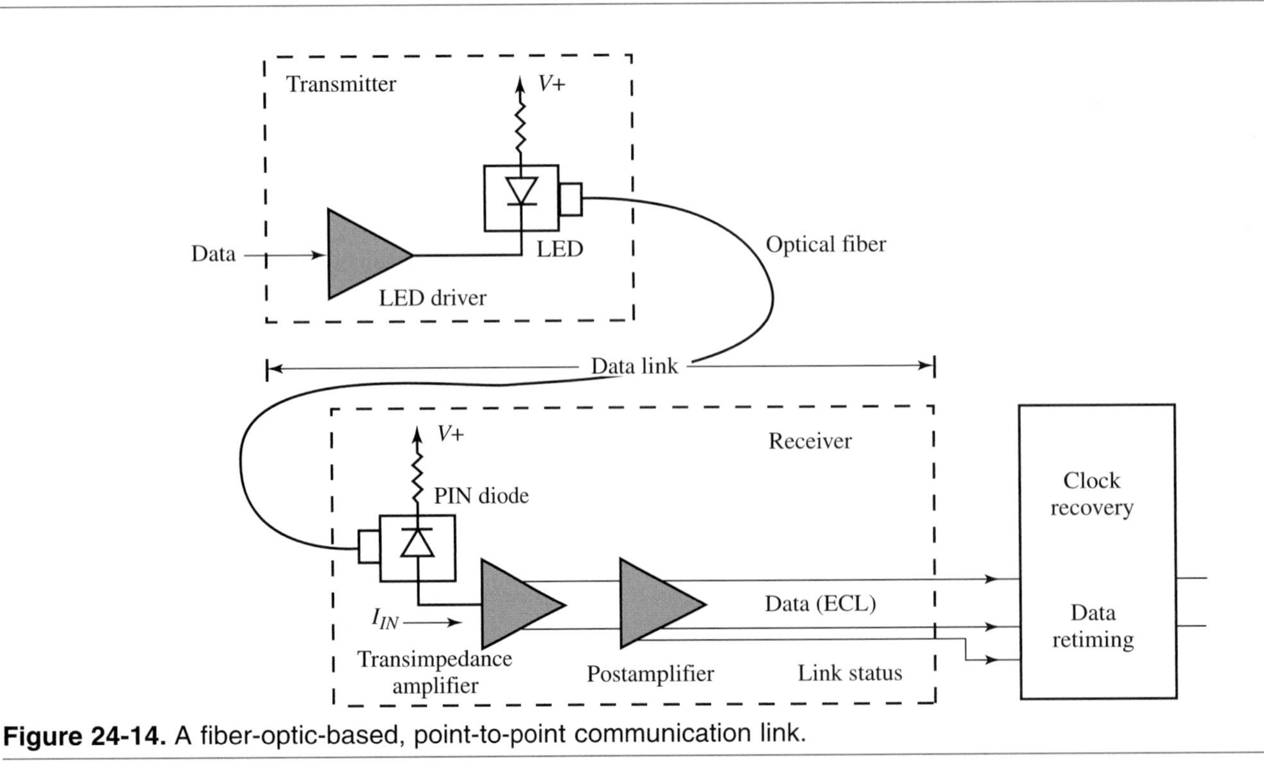

Figure 24-14. A fiber-optic-based, point-to-point communication link.

Troubleshooting Fiber-Optic Systems

Troubleshooting fiber-optic systems is as unique as the system itself. Two basic components are needed to troubleshoot the system. A light transmitter and a light receiver. The transmitter can be of LED or laser type depending on the characteristics of the fiber being tested. LEDs with a wavelength of 850 to 1300 nm are generally used as the source. For a visual inspection, an LED at a wavelength of approximately 650 nm (red) is used.

The transmitter and receiver are first calibrated for the type of cable being tested. The core diameter and core construction vary among manufacturers. A short sample of the cable, or jumper, is used as a reference. Once the receiver has been calibrated to the reference cable, testing of the system can begin. Look at **Figure 24-15.**

The receiver can be quite simple in design. It is used to measure the loss, or attenuation, in dB. An acceptable loss for a cable is approximately two to three dB per 1000 feet.

A rather unique piece of test equipment used in fiber optics is the ***optical time domain reflector (OTDR).*** The OTDR consists of a laser transmitter and a receiver. Light wave pulses are transmitted into the cable and reflected back to the OTDR receiver. A detailed analysis of the entire cable system can be achieved from one location. The receiver has a screen looking much like the screen of an oscilloscope. **Figure 24-16** is a sketch of a readout of a sample fiber-optic system.

The display of the cable system on the screen is known as a signature trace. In the illustration, an entire run of cable including splices and connectors are displayed on the screen. The sloping line is the indication of the attenuation of a section of optical fiber. The vertical spikes indicate reflective events caused by splices or connections. The vertical drops indicate the losses at the splices or connectors. The signature trace can be stored and or connected to a computer to print a hard copy of test results.

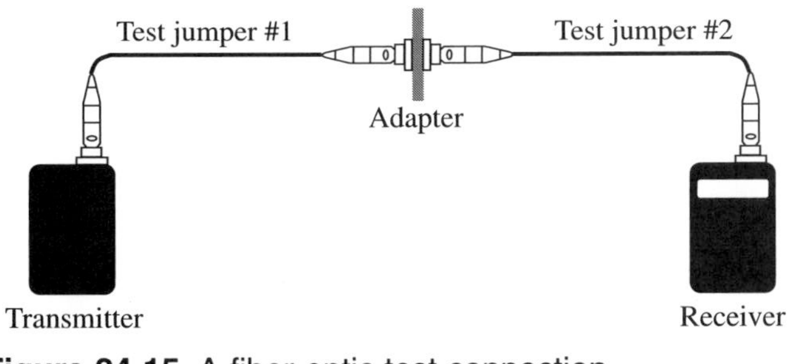

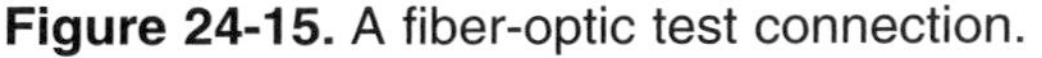

Figure 24-15. A fiber-optic test connection.

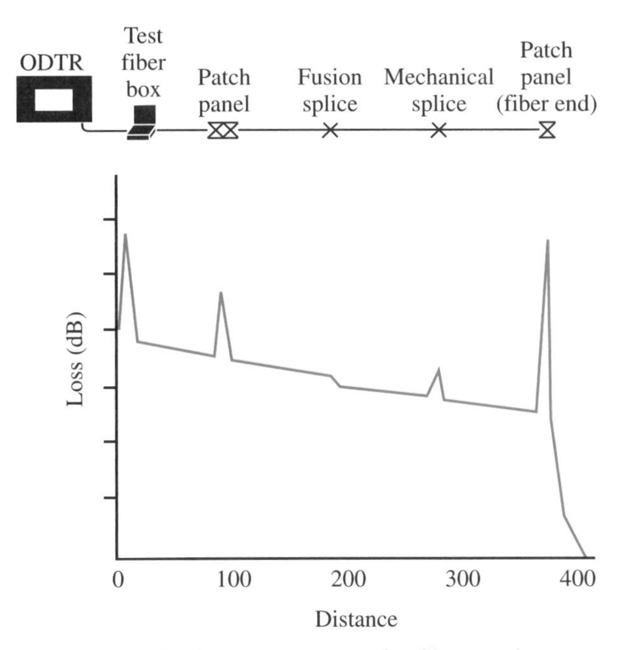

Figure 24-16. A signature trace of a fiber-optic system.

Review Questions for Section 24.1

1. List five advantages of fiber-optic cable over copper cable.
2. What are the two theories of light?
3. Define attenuation.
4. Give three examples of attenuation.
5. Where does most of the signal loss occur in a properly installed fiber-optic system?

24.2 LASERS

The word ***laser*** is an acronym for *l*ight *a*mplification by *s*timulated *e*mission of *r*adiation. Laser light is different from other forms of light because light produced by a laser is in phase. To understand what it means for light to be in phase, refer back to the chart in Figure 24-2. You can see that each type of light has a definite wavelength. Light consisting of all the same wavelength is called ***coherent light,*** and is said to be ***in phase.*** Light consisting of many different wavelengths is called ***incoherent light.*** Incoherent light is not in phase. An incandescent light consists of many different wavelengths and is incoherent light. See **Figure 24-17.**

Laser Construction

Look at **Figure 24-18.** This is a typical laser. It consists of a light source such as the strobe and a ruby

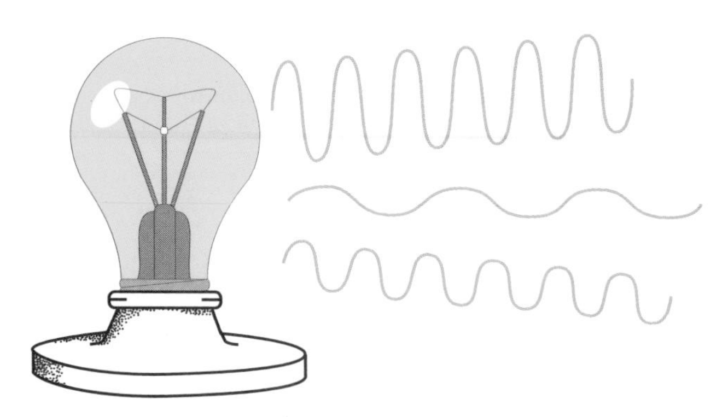

Figure 24-17. Incandescent light is incoherent light, it consists of many different wavelengths.

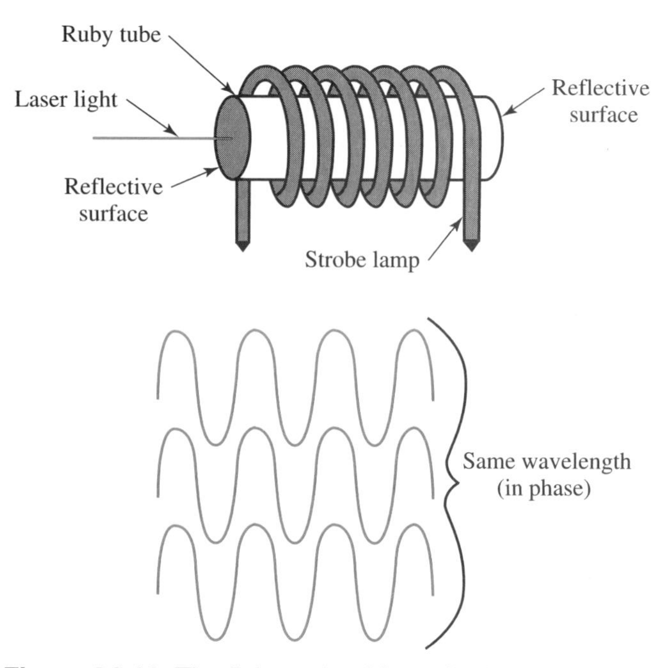

Figure 24-18. The light emitted from the laser is all the same wavelength and is in phase.

cylinder. One side of the cylinder is totally reflective. The other side is almost totally reflective, but there is a small area that permits the laser light to leave the tube. The strobe emits a light beam that is absorbed by the ruby laser tube. This is the *stimulation* portion of the laser. When the light enters the tube, it tries to leave the tube at the ends. Since the ends of the tube have reflective material, the light reflects back and forth through the tube. This back and forth reflection causes additional photons to be released. This is the *amplification* portion of the laser.

The light resonates inside the tube, which is sometimes referred to as an optical resonator cavity. The resonating light forces the light inside the tube to resonate at the same frequency, thus becoming coherent light. The light inside the tube is in phase. The light escapes through the small hole in the reflective material at one end of the tube. This escape is known as ***emission of radiation.*** When the light reaches the level to produce the stimulation of more photons, it is referred to as ***lasing.*** This is the basic operating principles behind the laser.

Types of Lasers

There are many different types of laser material. The optical resonator can be made from solids, liquids, or gases. A ***ruby laser*** is not produced with the gem that is found in mining operations. The ruby laser uses a manufactured ruby consisting of an aluminum oxide compound and chromium. As discussed in the previous section, one end of this ruby is totally reflective and the other is almost totally reflective except for a small opening in the reflective material. The opening allows the laser light to leave the resonant cavity of the tube.

The light from a ruby laser is usually pulsed rather than continuous. A continuous beam from a ruby laser would cause a dangerously high heating of the tube. Ruby lasers that operate as a constant beam use a cooling system such as liquid nitrogen. These lasers that require cooling are very powerful. They are used for applications such as welding and cutting hard materials such as ceramics, metal, or diamonds.

Gas lasers are quite common today, **Figure 24-19.** With a ***gas laser,*** the resonant cavity is filled with a gas such as helium-neon, CO_2 (carbon dioxide), or argon. Electrical terminals are placed inside the tube and a high voltage is applied to the gas. The gas ionizes and produces light. This begins the laser action. The laser action releases additional photons that result in an even more powerful light.

Different gas lasers are used for a variety of tasks. Argon lasers are used in the medical field to remove tattoos and perform eye surgery. CO_2 lasers are very powerful and are found in industrial environments. They are used for cutting and drilling metal. The CO_2 laser produces a beam with an infrared wavelength that can produce a great deal of heat. Infrared light is not normally reflected, which makes it ideal for metal work.

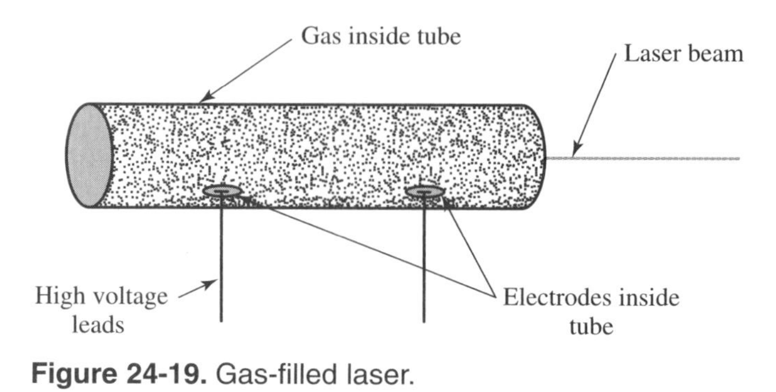

Figure 24-19. Gas-filled laser.

Night Vision Devices

The military began developing night vision techniques as far back as World War II. The original technology required an external *infrared* light source (such as a flashlight or searchlight), to enhance available light. Today's night vision devices are very sensitive and do not require this type of illumination. Instead, modern devices use electronics to amplify the light from the moon, stars, or nearby lights.

The viewing process involves three basic steps: focusing onto a sensing screen, multiplying the electrons, and focusing the electrons on a phosphor screen. The *photocathode* starts the initial change from photon energy to electron energy. Next, a *microchannel plate (MCP)* multiplies the electrons. Then the electronic focus manipulates the beam of electrons toward an eyepiece or a *charged coupled device (CCD)*. Electrons strike the phosphorous screen, releasing light energy. The light energy can then be viewed directly through the eye piece or on a digital display unit.

All night vision devices use *tube* technology to pick up the image and display it, but they vary in how they amplify the electrons. In original models, the light was converted to electrons and then the optics system focused the electrons through an anode and accelerated them so they hit a phosphor screen, creating a visible image. Accelerating the electrons in this manner caused image distortion and decreased the tube life.

First generation devices combined the three stages of night vision viewing into a single structure, the *intensifier tube*. Improved materials were used to convert light to electrons and the units operated at lower light levels. However, these devices still had the same drawbacks of distortion and decreased tube life as the original.

The second generation provided a major breakthrough with the development of the microchannel plate (MCP). The MCP is a glass disk, 18 mm in diameter. Fiber optics are used to create millions of microscopic holes (channels) through the disk. As electrons travel through these holes, they strike the sides, setting thousands of additional electrons in motion through the channel. The MCP multiplies the number of electrons thousands of times, producing a clearer image without the distortion characteristic of the original technology.

Third generation devices use sensitive materials such as gallium arsenide (GaAs) as the photocathode. This allows more efficient conversion of light to electrons at lower light levels. Also, the MCP in these devices is coated with an ion barrier film to increase tube life. Third generation devices are more sensitive, smaller, and provide better resolution than their predecessors.

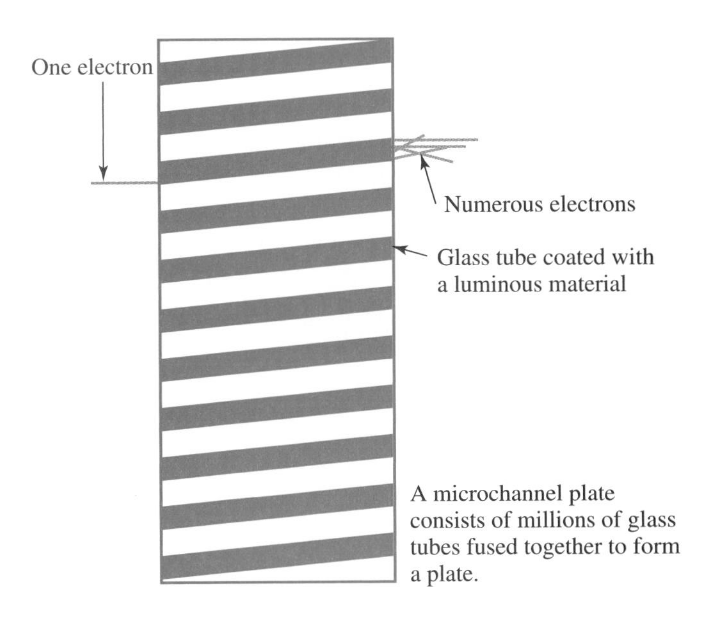

When night vision devices are worn as goggles, the electrons are displayed directly on a coated eye piece. When used as a separate unit in a plane, helicopter, or tank, a CCD is used. The CCD transmits digital data to be displayed on a monitor or saved to a computer.

Early night vision devices did not have a bright source protection, so users could be momentarily blinded by a bright flash of light. This sort of flash could overload the tube and disable the unit. Most devices now contain a feature that reduces the voltage to the photocathode when the device is exposed to bright light sources.

Although color night vision technology exists, commercially available devices are limited to monochrome images with a display phosphor color (usually green). Viewing distance depends on ambient light levels. However, if conditions warrant, the user can see an object such as an island from very far away when navigating in coastal waters. (ITT Night Vision)

The gas in the gas laser can be replaced easily by pumping a different gas into the resonant cavity. Different gases produce different colored beams. This is how many different colors are achieved in a laser light show. Argon lasers are commonly used in light shows because of their blue-green color spectrum. The helium-neon laser is the classical thin red beam laser you've seen commonly in movies and TV.

Another type of laser uses a liquid dye as its lasing medium. Many different colors of liquid dye can be pumped into the laser tube to produce different wavelengths of light.

Laser Applications

Lasers have many different applications. They are in industry, medicine, business, and entertainment.

Construction

The laser can be used for measuring distances and alignment. Laser systems are commonly used as surveying instruments. When the laser technology is combined with a computer, the distances and angles can be recorded and stored. Calculations can be made by the computer, and a diagram or layout can be printed very easily and accurately.

The laser can be used as a construction level. The beam can be rotated to illuminate a line of light on a wall to assist installation of a drop ceiling or to determine the grade when removing or smoothing land. Lasers have been used to align pipes in underground systems such as utility or storm drains. Here, the light is simply positioned in one section of pipe and a target is set inside each pipe as it is placed. The laser alignment system ensures the straightness and slope of the pipe.

Medicine

Lasers are also used in conjunction with fiber optics to perform many types of surgery. Three different fiber-optic systems can be inserted into a patient. One system provides regular low level light for vision. A second fiber is used for viewing inside the patient, like a camera. The third fiber is used to direct the laser beam to the targeted surgical area.

Lasers are also used for some types of angioplasty. *Angioplasty* is the surgical removal of plaque inside arteries. These techniques have saved countless lives and the pain of conventional, invasive surgery.

Supermarket checkout

Another commonly used laser known as a Nd:YAG (pronounced en-dee-yag). The ***Nd:YAG laser*** uses yttrium aluminum garnet, a solid, for the laser medium. You have seen these low power lasers many times at supermarket checkout counters. Beneath a piece of glass in checkout counters, or in a hand-held gun, is a laser system used to identify product codes on labels. The familiar black parallel lines on labels are universal product codes (UPC) that can be digitized and transmitted to a computer easily. See **Figure 24-20.**

In this device, a laser is used to produce a narrow beam of light. The beam of light is shined on a rotating mirrored surface composed of many flat surfaces. The laser beam is reflected off the rotating mirrored surface causing it to flash at an accelerated rate across and through the glass top.

The light strikes the bar code and is either absorbed by the black bars or reflected back down to the receiver unit by the gaps between the bars. The receiver unit is a light sensitive diode or transistor. The electrical impulse generated at the receiver is transmitted through wires to a computer where it is interpreted. The bar code data identifies the item as a number. The number identifies the item and its price. It also enters the transaction into a computer data base to keep track of store inventory.

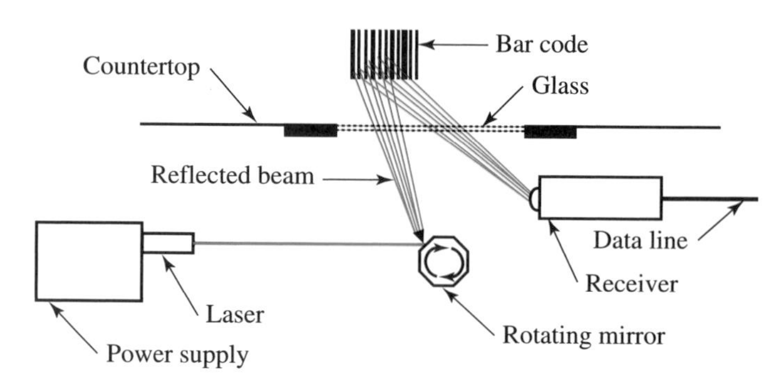

Figure 24-20. A typical checkout counter uses a laser to scan a bar code from a product label.

CD player

Another common application for a low power laser, such as a laser diode, is the compact disc player (CD). Look at **Figure 24-21.** The beam from a laser diode is directed to the CD through an optical lens. The laser beam is reflected off the CD to a mirror and then to a photoreceiver. The photoreceiver receives the beam and converts it to electrical impulses.

The CD itself contains many tiny pits smaller than a human hair. The quantity and length of the pits varies over the disc. This information represents the recorded sound pattern. The laser light hitting the photoreceiver varies in intensity depending on whether the beam is reflecting from a pit or reflecting from the flat area of the disc. Changes in intensity (moving from a pit to a flat area or vice versa) represent 1s. Lack of change over a period of time represents a 0. This light intensity data is converted to an electrical signal. These digital electrical impulses are converted to analog signals, amplified, and then sent on to a speaker to be converted to sound

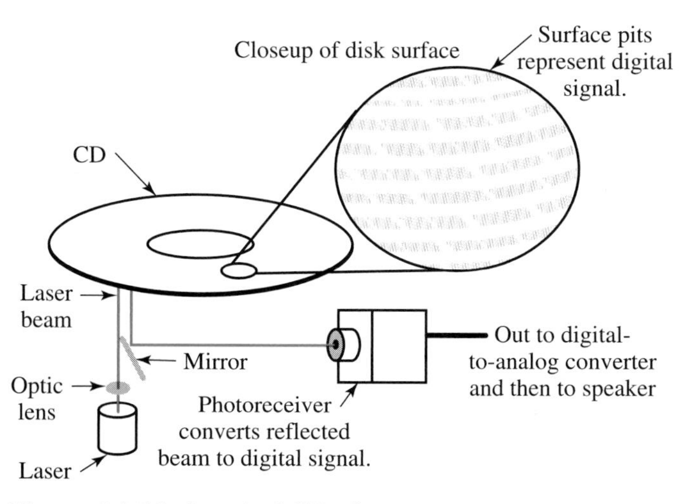

Figure 24-21. A typical CD player arrangement.

Other uses

The use of lasers is expanding every day. Experiments are constantly being conducted by private industry, the government, in education, and in the military. We have only scratched the surface in the use of lasers. In the very near future, you will witness inventions and applications that are not even dreams now.

Laser Safety

We cannot place enough emphasis on safety when using lasers. Laser light can be very deceiving. At times, laser damage can be minimal and only cause temporary irritation. Other times, a laser that may appear of very low power can do irreversible damage to the eye.

The light from the laser does not have to be in the visible spectrum to do damage. Ultraviolet light is not detectable by the human eye, but it can severely burn the cornea. Visible light is magnified and absorbed by the retina.

The amount of exposure time required for eye damage is as fast a the blink of an eye—laser fast. In addition, the laser beam does not have to shine directly into the eye to cause damage. A laser beam can reflect off many different types of surfaces and result in a direct flash into the eye. Eye protection should be worn *at all times* when in the same room as a laser.

Laser Classification

Lasers are classified by their maximum possible output during normal operation. Wattage is not the only consideration for rating power of a laser. Beam width is also of prime consideration. Laser classifications are as follows.

Class I. These beams are fairly safe to work with. These devices include the lasers in pointers, checkout counters, and disc players.

Class II. Visible laser beams only. These will cause damage if the beam strikes the eye for over 1/4 second.

Class III. These beams will cause eye damage regardless of the length of exposure time.

Class IV. Exposure to these beams will not only severely damage the eye, but skin damage can also occur. These laser beams can cause combustible materials to ignite. Extreme caution need be used.

Eye protection must be worn when working with Class II or higher lasers.

Review Questions for Section 24.2

1. Lasers can be made from __________.
   a. solids
   b. liquids
   c. gases
   d. All the above.
2. The letters in the word laser form the acronym l__________ a__________ by s__________ e__________ of r__________.
3. How is a laser similar to an oscillator circuit?
4. Explain how a CD player utilizes a laser.

Summary

1. Fiber-optic cables are lightweight and corrosion resistant.
2. Fiber-optic circuits provide security.
3. Fiber optics are not affected by magnetic fields as are conventional conductors.
4. Fiber optics do not pose fire hazards.
5. Fiber optics provide a wider bandwidth than conventional conductors.
6. Light theory is a composite of wave theory and particle theory.
7. Light encompasses more than just the visible spectrum.

8. Attenuation is signal loss in fiber-optic cable. It is measured in decibels.
9. Scattering is loss of signal due to impurities in the core material.
10. Dispersion is distortion of the signal caused by the light waves arriving at the receiver at different times.
11. Fresnel loss is caused by differences in light refraction between different materials.
12. LEDs and laser diodes are commonly used as fiber-optic transmitters. PIN diodes, avalanche diodes, phototransistors, and photodarlingtons are commonly used in fiber-optic receivers.
13. An optical time domain reflector (OTDR) is an instrument used to make a signature trace of a fiber-optic cable signal.
14. Laser is an acronym for light amplification by stimulated emission of radiation.
15. Laser light is a coherent light.
16. Lasers are used in industry, medicine, business, and entertainment.
17. Eye protection must be worn when working with Class II or higher lasers.

Test Your Knowledge

Please do not write in the text. Place your answers on a separate sheet of paper.

1. Name five advantages of fiber-optic cables over copper cables.
2. Name the defects that can cause a splice in fiber-optic cables to result in signal loss.
3. What is attenuation?
4. According to wave and particle theories, light acts like a(n) __________ when it comes into contact with solid material such as a phototransistor. It acts like a(n) __________ when transmitted through the air.
5. Name two common transmitters of light through a fiber-optic cable.
6. Explain how laser light differs from incandescent light.
7. How does a laser amplify light?
8. List three applications of laser light.
9. Name a common use of CO_2 lasers.
10. What precautions must be taken when working with lasers?

For Discussion

1. Discuss the type of person who would be interested in tapping into a communications system. What type of information might they be seeking and what would be the uses of such information?
2. Could other mediums such as a pipe filled with water be used to transmit data?
3. What is attenuation and how is it measured?
4. List the causes for signal loss in a fiber-optic cable.
5. Discuss field installations of fiber-optic cable and its challenges.
6. What causes the greatest signal loss in a typical fiber-optic installation?
7. Why are most fiber-optic cable termination connectors not made in the field?
8. How far might a laser light project?
9. Could a voice be transmitted by laser light?
10. Would it be possible to light the interior of a building using fiber-optic cable?
11. Could a laser be used to determine air quality such as smog content?

25 Introduction to the Personal Computer (PC)

Objectives

After studying this chapter, you will be able to:

- Define field replacement unit.
- Explain the function of BIOS.
- Describe the power-on self-test (POST).
- Describe how data is stored as magnetic media.
- Identify various types of computer ports.
- Identify various factors that affect processor speed.

Key Words and Terms

The following words and terms will become important pieces of your electricity and electronics vocabulary. Look for them as you read this chapter.

application
basic input output system (BIOS)
central processing unit (CPU)
chipset
digital video disk
direct memory access (DMA)
double data rate-DRAM (DDR-DRAM)
dual inline memory module (DIMM)
dual inline package (DIP)
dynamic RAM (DRAM)
field replacement unit (FRU)
file allocation table (FAT)
graphical user interface (GUI)
local area network (LAN)
microprocessor
operating system
random access memory (RAM)
read only memory (ROM)
single inline memory module (SIMM)
single inline package (SIP)
software
static RAM (SRAM)
synchronous DRAM (SDRAM)
universal asynchronous receiver/transmitter (UART)
wide area network (WAN)

The personal computer (PC) has evolved over many years. Computers of today are not only useful, but also decorative and portable. Computers have evolved from machinery that took up the space of a large room, to the size of our desktop, palm-sized, **Figure 25-1**, or smaller. Today's computers allow us to perform a variety of tasks in places we could never before imagine and affect nearly every aspect of our daily lives.

Some fields computers assist us in are education, **Figure 25-2**, the sciences, entertainment, transportation, communication, defense, manufacturing, finance, and business. Computers are used in schools and universities to educate students in various academics. Computers help scientists predict weather patterns and major geologic shifts. They help those in the entertainment industry edit and produce music and movies. They are also a major source of entertainment in nearly every home to those playing games, sending and receiving e-mail, and exploring the Internet. Computers allow for safer travel in the air, and on the highways. They are a must in today's world of high-speed communications. They play an important role in the country's defense system and have been accredited with contributing to great advances in the civilian world as a result. Computers help make manufactured products more dependable and less expensive. They help us calculate faster, and with much more ease. Computers have become a basic tool in nearly all offices and businesses today, making a more efficient workforce. Computers perform far more tasks in our world than can possibly be mentioned.

This chapter will introduce you to the basic concepts of the PC based on the IBM compatible, **Figure 25-3**, or the Intel 80XX family of processors. Although the Apple Macintosh is not directly discussed, there are many similarities between the principles of the IBM compatible and the Apple computer or Motorola processor based computers.

To cover the PC in-depth requires many hours of instruction. After completing this introductory course in electricity and electronics, you may wish to go on to study PC repair as a career or simply to gain a more in-depth knowledge of PC operations and capabilities.

Figure 25-1. A—An IBM Wearable computer. This man is using a wireless computer as he travels across the city. The 10.5-ounce system operates by voice command. (International Business Machines Corporation) B—Technicians at an aerospace manufacturer use a wearable computer to access schematic diagrams, repair manuals, and maintenance records. Wearable computers allow the job to be done quickly, keeping airplanes on schedule and safe for passengers. (Xybernaut Corporation)

Figure 25-2. This child is using a wearable computer to assist with learning. This computer is designed particularly for children with disabilities and helps kids with problems like autism and cerebral palsy to learn more effectively. (Xybernaut Corporation)

Figure 25-3. IBM IntelliStation Z Pro. (International Business Machines Corporation)

25.1 PC REPAIR

PC repair originally fell under the topic of electronics repair; however, computer repair has become its own profession. While a background in electronics can be a major asset to a computer technician, this study is not an absolute necessity. Although, many PC technicians often return for more instruction in electronics basics.

The PC is a complex assortment of digital electronics controlled by operating system software. On the surface, it may appear that a PC is quite easy to repair or upgrade. In fact, in many cases this can be true; however, most times PC repair or upgrades require the expertise of a highly trained PC technician. Computer problems may be complex and interrelated to other issues regarding hardware, software, networking, or connectivity. Do not be fooled by the simplistic approach of the repair to simply "swap out" modules. Many times the problem is not hardware at all, but is actually a software problem. The PC technician must be able to accurately diagnose the problem to complete a repair and ensure that the same problem does not reoccur. Computer problems can be broken down into three major areas: hardware, software, and users.

Common Points of Hardware Failure

Hardware does fail, but not as often as people might think. Two of the most common hardware components to fail are the hard disk drive and the modem. The hard disk drive is in constant use and is a combination of an electronic/mechanical device. See **Figures 25-4** and **25-5**. Mechanical devices wear out much faster than electronic components, but not all mechanical devices cause hard disk drive failures. Many hard disk drives are attacked by a computer virus or worm program that causes the computer to appear completely dead or to behave erratically.

Modems often fail because of lightning strikes to telephone lines. Since the modem is connected directly to the telephone system, the lightning strike often hits the modem, even when the PC is turned off. A surge protector with telephone protection should always be used. See **Figure 25-6**.

Software can also cause computer problems. Software is designed to go hand in hand with personal computers, but even if the software documentation says it is made to work with a particular operating system, compatibility issues still arise. Users actually account for the majority of computer-related problems. Most computers have many options and settings. These can be changed accidentally or intentionally. Once a user accidentally creates a problem, they typically cannot reverse their own error. In fact, many users cannot even replicate the problem to show a PC technician what they have done. Many so-called "viruses" are actually problems created by a user accidentally changing system or software settings. To properly troubleshoot a PC, the technician must learn to be patient with users, while also understanding the electronic, mechanical, software, and hardware functions of a PC.

Originally, technicians performed component-level repair on computers. Today, all the parts in a typical PC are considered ***field replacement units (FRU)***. An FRU is a module that can be quickly and easily replaced, rather than repaired. The entire computer can be broken down into FRUs. Some examples of FRUs are the hard drive, floppy drive, CD-ROM, DVD, memory chips, power supply, and motherboard. In practical application, there is no one item in a modern PC system that must be replaced in a computer repair shop; although, it is not uncommon for a PC technician to have to order an FRU or to remove the computer from the site for such replacement. These issues depend on the condition of the repair environment. Such repairs may take a computer out of service while a replacement unit is found.

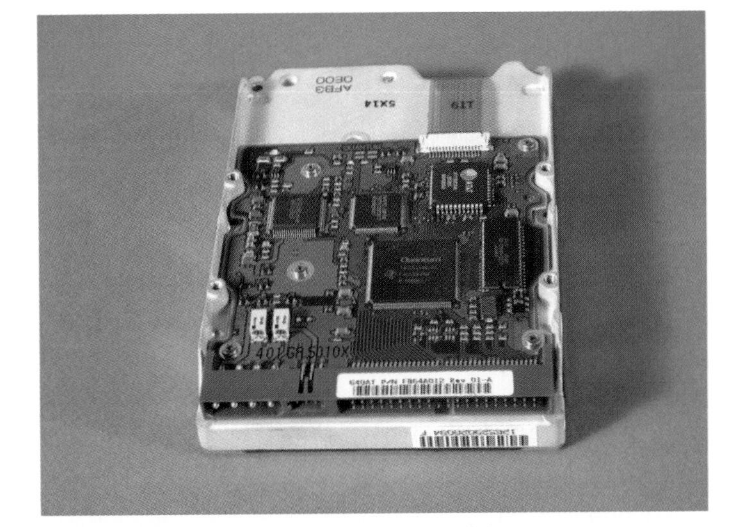

Figure 25-4. The hard drive is one of the most common computer devices to fail. This picture illustrates the hard drive's electronic components.

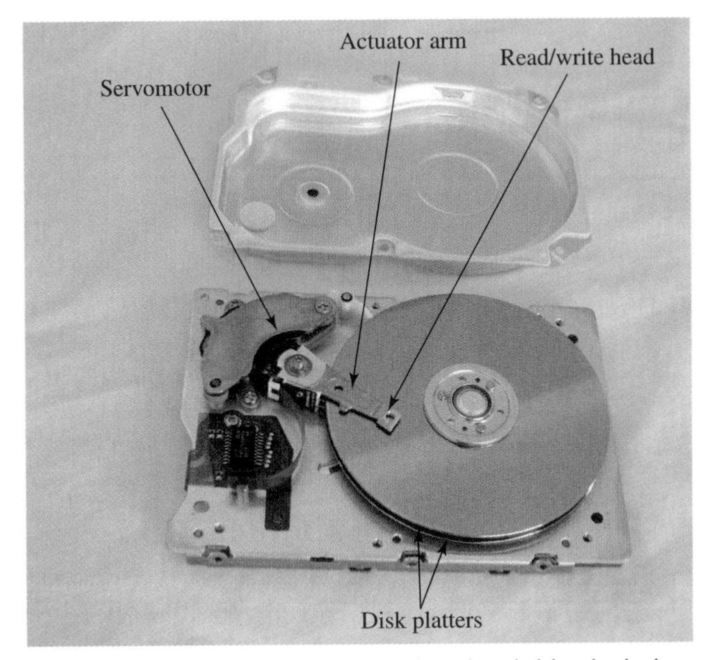

Figure 25-5. Mechanical devices in a hard drive include the servomotor and the actuator arm.

Figure 25-6. This surge protector, connected to a broadband modem, includes network protection. Should a surge come in through the modem's power line or Internet connection, the surge protector prevents damage to components and computers throughout the network.

Data Recovery

It may be interesting to note that there are data recovery businesses that specialize in recovering data from failed and damaged hard disk drives. Data recovery is not limited to software utility programs. Many times, hard disk drive failures are caused by electronic component and drive motor failures. Data recovery companies can still salvage data after the drive mechanics have failed. It takes a highly skilled electronics technician to repair/replace the electronic components associated with a hard disk drive. Once the hard disk drive controller board or drive motor has been replaced, data can usually be recovered.

Computer Technology Certifications

There are many advanced technical certifications available for computer hardware and software operation and repair. One such certificate available through the CompTIA organization is called the A+ certification. CompTIA is a not-for-profit organization designed to certify a technician in nonvender specific (generic) computer maintenance and repair. In other words, CompTIA does not certify any one particular computer brand, but offers a general examination that tests candidates to ensure they are qualified to perform common maintenance, repairs, and support on computers. Manycomputer manufacturers require CompTIA certification before allowing students or employees entry to specialized training on their products. Many companies such as Nortel Networks require their technicians to become certified, not only in PC repair, but also in basic networking. This is because there is virtually no single modern piece of equipment that does not rely on some computer technology.

More and more, electronic equipment such as radar systems, cable television, biomedical equipment, security systems, fire alarm systems, commercial airplanes, military weapons, and diesel engines rely on basic computer and network technology. Computer certification will become a common requirement for most technician jobs in the future. Any person interested in pursuing a career in electronic/computer technology should acquire a good foundation in computer hardware and software, and obtain one or more computer certifications.

The PC as a Tool

The personal computer has become an effective tool for electronics technicians as well as technicians in most other fields of study. Resources such as parts information, electronic theory simulations, electronic calculations, schematic diagrams, and schematic drawing tools have become commonplace through the use of CD-ROMs and Internet sites. Special computer adapter cards can be inserted into a PC, allowing the computer to display oscilloscope images of live circuits or to serve as a multimeter. The PC is now a vital tool for the study of electronics as well as a resource for troubleshooting and learning new technologies. As a student of electricity and electronics, you have a head start on many of the PC concepts because of the background obtained thus far. However, there is still a lot to learn.

Review Questions for Section 25.1

1. What are the two most common computer components to fail?
2. The hard disk drive is a(n) _____ device.
3. In the field of PC repair, what are the parts within a typical PC considered?

25.2 COMPUTER DESIGN, COMPONENTS, AND TECHNOLOGIES

This section covers the computer's physical design and components, as well as technology relating to data storage and file formats. By becoming familiar with the computer's components and technologies, you will have laid a foundation on which to build further knowledge, either in your studies as a PC technician or in casual reading about new computer technologies.

The overall design of a computer can be divided into four major systems: input, output, processing, and storage. See **Figure 25-7.** Nearly all computers used today still use the same basic four types of equipment designed in the mid-1940s. Computers contain equipment pieces but are designed and programmed to work as a single unit. The basic design of every computer includes:

- An ***input*** device for entering data into the system.
- An ***output*** device for retrieving data from the system.
- A ***central processing unit (CPU)*** for processing the data.
- A ***storage*** device for storing programs and data.

Input can be any type of data, such as numbers, words, and symbols. Processing is done in a preset sequence as the computer follows a program's instructions. Programs and data are stored in memory. Output can be printed material, data sent to other equipment, or communication over the phone line. The remainder of this chapter will contain an overview of the various computer components and functions. An in-depth study of each of the items covered is required to become a proficient PC technician.

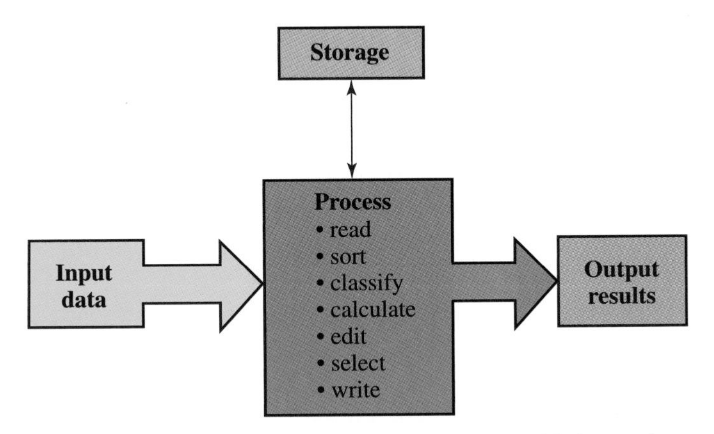

Figure 25-7. This diagram shows the flow of information through a computer.

The Case

There are three classic case styles used for PCs: desktop, tower, and mini-tower, **Figure 25-8**. The desktop, as its name suggests, generally sits on top of the user's desk. The desktop's wide surface makes an excellent platform for a monitor, and the case's minimal height, about eight inches for a full-sized desktop, conveniently places the monitor at eye level with the user. The tower, representative of its name, is a tall, narrow unit, designed to set on the floor, beside a desk, or inside of a cabinet; although, it can be placed on top of the desk if the user prefers. The mini-tower is a modified version of a standard tower. It is slightly shorter than a typical tower model.

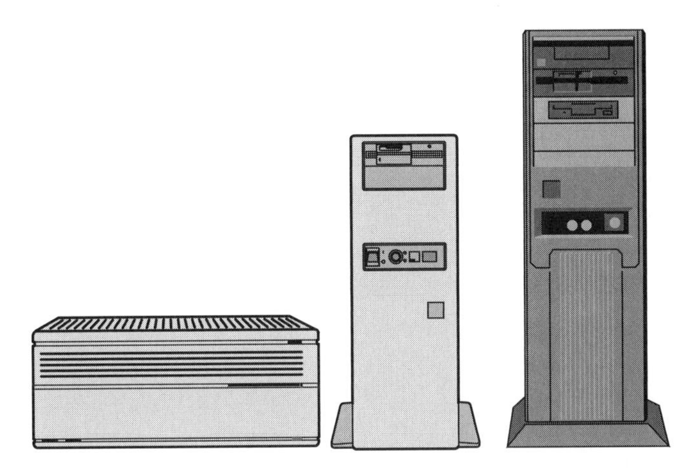

Figure 25-8. The three most common case styles are the desktop, mini-tower, and tower.

Case styles are left to the user's taste, desk design, space limitations, or sometimes the requirements of the job. However, one of the important aspects of selecting a case is the number of bays available for mounting devices such as a CD-ROM, floppy drive, DVD, and hard drive. Some case styles have a limited number of bays, which could cause expansion or diversification issues with advanced use. See **Figure 25-9**. External bays provide a location for installing devices such as floppy drives, DVD players, tape drives, and other accessories that are intended to be physically accessed by the user. Internal bays provide locations inside the computer for devices such as hard drives, which do not need to be physically accessed by the user. Unused external bays typically have a cover over each of the device bay areas. The cover can be easily removed, leaving the bay area exposed for installation of a device.

Form Factor

The case, motherboard, and power supply have what is known as a form factor. Some common form factors are XT, AT, Baby AT, ATX, and Mini ATX. The form factor defines the physical attributes of the item, such as the length, width, and height. In other words, the case, motherboard, and power supply form factors must match to be physically compatible with each other.

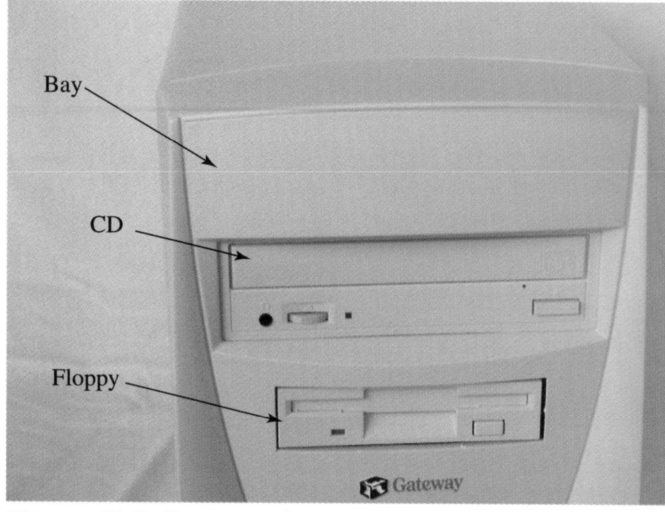

Figure 25-9. Computer bays and their use.

Input Devices

Input devices allow entry of data into the computer. The device is controlled by program instructions that send a command to an input or output (I/O) channel. Examples of these commands may be to enter a letter pressed on the keyboard or to move the cursor across the screen. Input devices are also known as peripherals. Some examples of input devices include keyboards; an optical or wireless mouse; a roller ball or touch pad; and a game controller. **Figure 25-10** lists examples of some common input devices.

Power Supply

The power supply converts standard 120-volt ac electrical energy into dc voltage levels. The typical dc voltage levels are +12, +5, +3.3, -12, -5 and zero or ground. The power supply provides power to many devices used as part of the computer, such as the hard disk drive, floppy disk drive, and CD-ROM, **Figure 25-11**. It also supplies power to the motherboard. The ATX power supply connection on the motherboard is a single unit that is "keyed." This means that the connection on the motherboard is designed so that the power supply's connector can fit into it only one way. Earlier models, such as the AT and XT, used a pair of motherboard power connectors. Although these connectors were also "keyed," technicians had to use care in connecting them to the motherboard because they could be easily switched, resulting in damage to the motherboard. A memory aid in correctly connecting two power connectors to the motherboard is "black goes back-to-back." This means the black wire on each power connector should be next to the other when plugged into the motherboard.

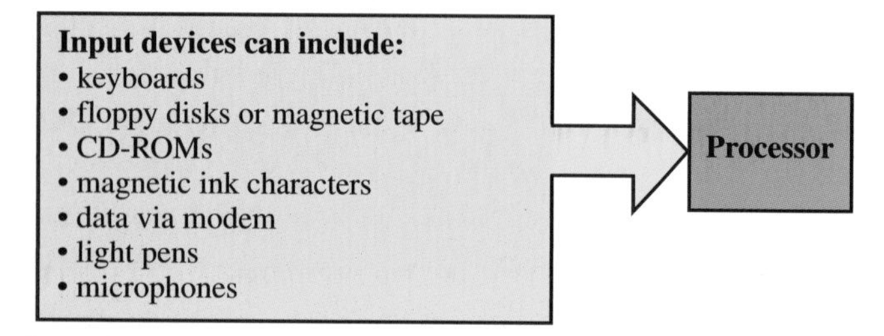

Figure 25-10. Listed are some examples of input devices.

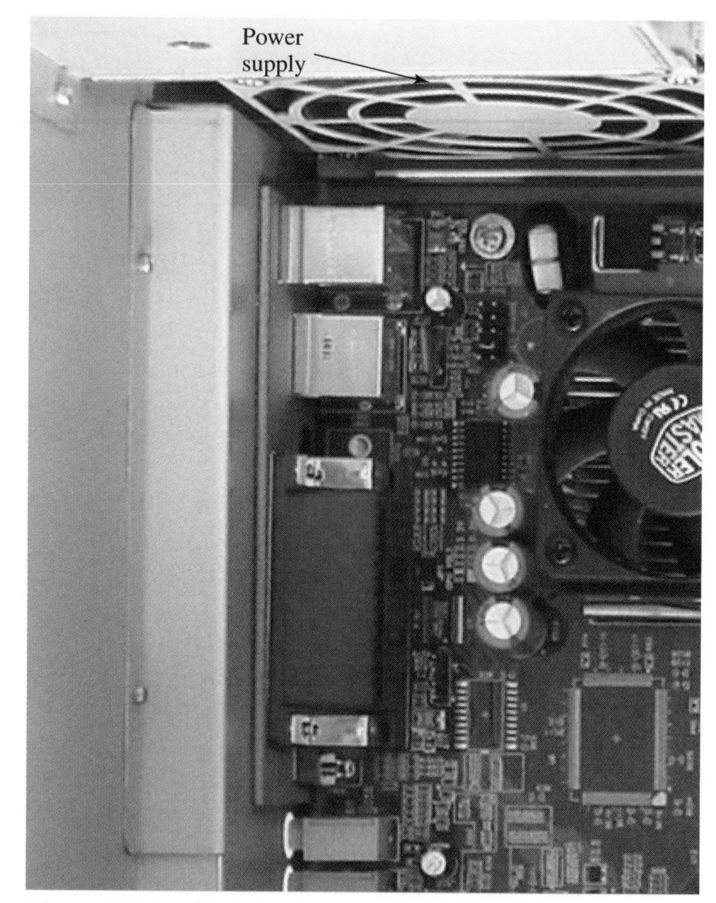

Figure 25-11. The power supply provides power to the motherboard and to other computer components.

Motherboard

The motherboard is a sophisticated electronic circuit board that provides the communication pathways between all major computer components. The motherboard houses the CPU, memory, expansion slots, chipsets, BIOS, CMOS battery, oscillator circuit, and other electronic components, **Figure 25-12.** Small circuit paths called traces, running in parallel across the motherboard, **Figure 25-13**, provide a path for data, control

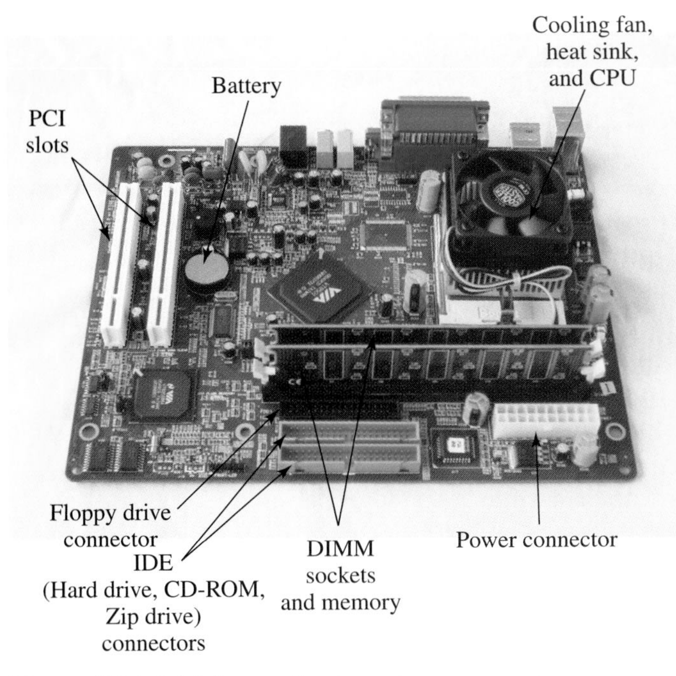

Figure 25-12. Major motherboard components.

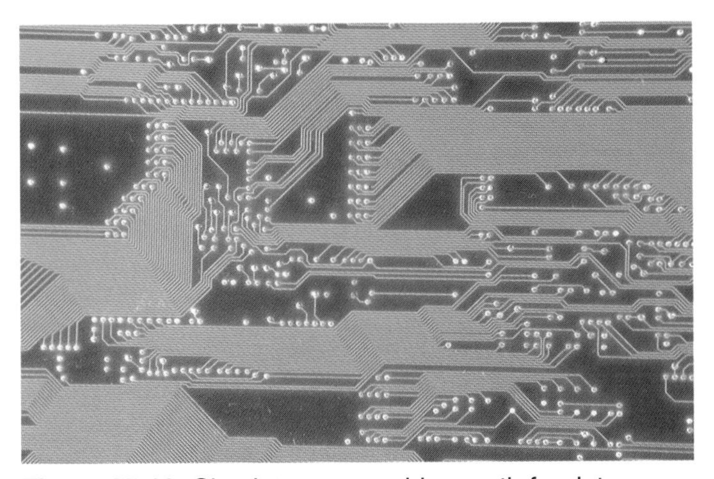

Figure 25-13. Circuit traces provide a path for data, control signals, and power.

signals, and power. The power supplied to the motherboard is limited to the low wattage needs of the motherboard's electronic components. Components such as hard disk drives and CD drives, which require larger volumes of electrical energy, connect directly to the power supply.

The motherboard also provides slots for expansion cards. The slots provide a quick and easy way to connect to the motherboard bus system without the use of solder. An expansion card can be simply inserted into the slot. The two main types of slots encountered are Industrial Standard Architecture (ISA) and Peripheral Component Interface (PCI). ISA is one of the oldest types of slots and was developed many years ago. It is not manufactured into boards any longer.

The PCI type of slot is standard today. A single Accelerated Graphics Port (AGP) may be encountered only for video card use. Video cards are designed to offer greater enhancements than standard video circuits built into a typical motherboard. Video cards often contain their own BIOS, additional RAM, and a processing unit, such as an accelerator or a coprocessor, to support high-speed graphics and to operate with minimum CPU intervention. See **Figure 25-14**.

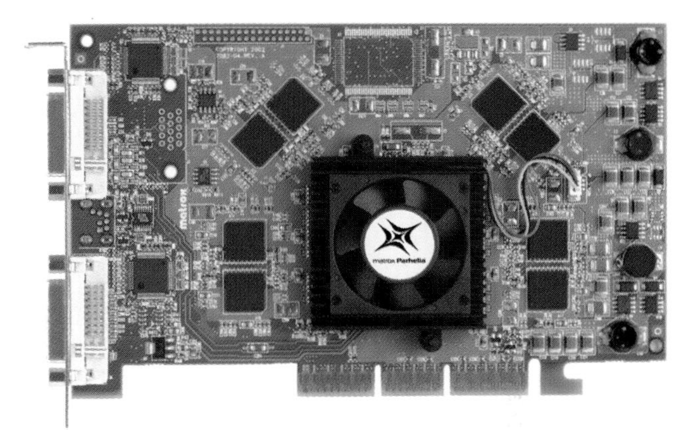

Figure 25-14. Matrox Parhelia graphics accelerator card. (Matrox Graphics, Inc.)

CPU

The ***central processing unit (CPU)***, or microprocessor, processes all software program instructions and is responsible for manipulating data. It also contains an arithmetic/logic unit, which handles all mathematical computations. The arithmetic/logic unit will be covered in detail in Chapter 26—Microcontrollers. The CPU is the most expensive integrated circuit in the computer. A typical CPU contains millions of transistors and integrated circuits that are programmed to process information. CPUs are generally compared by the working frequency, such as 1.2 GHz or 2 GHz. However, speed is only one factor in measuring CPU performance. Other factors that greatly improve the performance of a CPU are the CPU's instructional command set, on-chip cache, **Figure 25-15**, and the parallel processing of instructions. The CPU instruction set is the actual set of commands the CPU understands. These commands are used to process data through the CPU.

On-chip cache is a small amount of memory used to store data that is frequently accessed by the CPU. Because the memory is integrated into the chip and does not need to transfer data across the motherboard, the processing speed of the CPU is increased. Parallel processing is a technology that is designed to process more

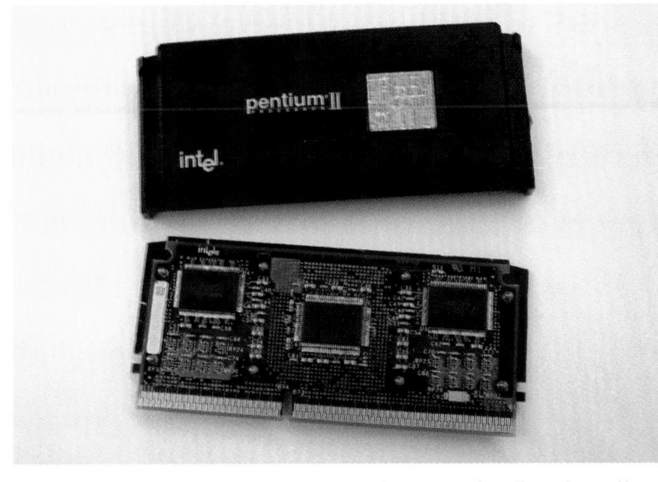

Figure 25-15. CPU and cache inside of a Pentium II cartridge.

than one set of instructions or data through a CPU at the same instant. By processing more than one piece of data, speed appears to be double the rated frequency of the unit. There are other factors that affect overall performance of a computer such as the motherboard bus speeds, the amount of RAM, the version of operating system, as well as the types of physical interfaces used to connect to other components.

Modern CPUs run at very high frequencies when compared to their early predecessors. The high frequencies produce a great deal of heat inside the CPU structure. To prevent damage to the CPU caused by the excessive heat, cooling technologies, such as fans and heat sinks, must be applied.

Memory

There are two main types of memory used in a computer, RAM and ROM. Both are classified by their ability to retain program data and instructions after power is no longer supplied to the chip.

The contents of ***read only memory (ROM)*** are retained after the power to the computer is turned off. The ROM chip located on the motherboard is called the BIOS. The BIOS contains programming that is vital to the computer's operation. It would be impossible for the computer to run without a BIOS. The BIOS will be covered in detail later in this section. Other types of ROM will be discussed in Chapter 26—Microcontrollers.

Random access memory (RAM) is a temporary storage area that loses content when electrical power is interrupted. The contents of RAM constantly change as you load and unload different programs. When you are finished with one program and start another, the new program is loaded into RAM, replacing the last program. It is also possible to load more than one program into RAM and then switch between the programs. This is called multitasking.

There are two main types of RAM used in computers, dynamic and static. ***Dynamic RAM (DRAM)*** is composed of millions of transistors and capacitors, **Figure 25-16**. The capacitors are used to maintain a charge at the base of the transistor. The charge of the capacitor determines if the transistor is open or closed across the emitter and collector. The open or closed condition represents data stored as a binary one or zero. The capacitors hold electrical charges in patterns matching the binary data patterns. While large capacitors can hold an electrical charge for a long period of time, the microscopic-sized IC capacitors cannot. Because the capacitors lose their charge quite rapidly, they must be constantly recharged. The constant recharging is referred to as refreshing, and how often it occurs is called the refresh rate. Typically, DRAM needs to be refreshed thousands of times per second.

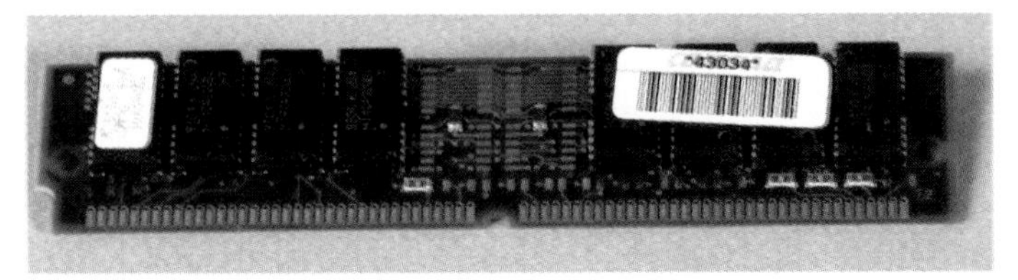

Figure 25-16. Basic DRAM found in older PCs.

Static RAM (SRAM) is composed mainly of digital flip-flops that retain their state of one or zero until the power is removed from the chip, **Figure 25-17**. SRAM is much faster and more expensive than DRAM. SRAM is typically used when small amounts of memory are incorporated into other devices to enhance their performance. For example, the CPU has a small amount of RAM incorporated into the chip. This small amount of RAM is called cache (pronounced cash) and is composed of SRAM. Synchronous DRAM ***(SDRAM)*** is an enhanced

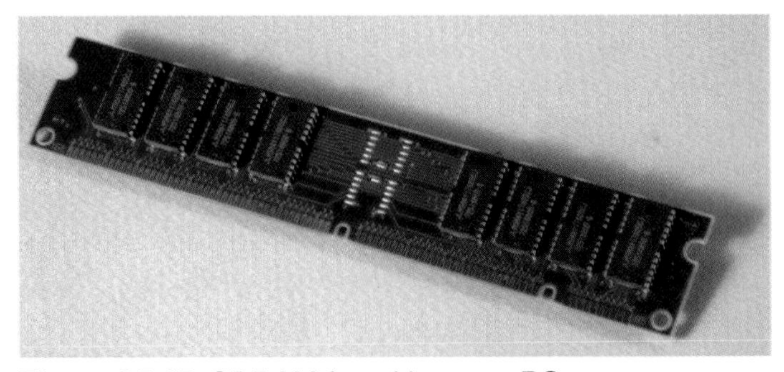

Figure 25-17. SDRAM found in many PCs.

version of DRAM found in many computers. The S represents synchronous, not static. The SDRAM runs in synchronization with the CPU, producing a higher overall memory performance than simple DRAM. Because the SDRAM data transfer is coordinated to match the CPU timing, there is no delay before transferring data to and from the main memory as with earlier versions of memory technologies.

Another form of special DRAM is ***double data rate-DRAM (DDR-DRAM)***. DDR-DRAM is designed to transfer data at twice the normal data rate transfer speed. Until DDR-DRAM, data was transferred at the same rate as one complete digital signal cycle. DDR technology transfers data on both the rising and falling edge of a digital pulse. See **Figure 25-18**.

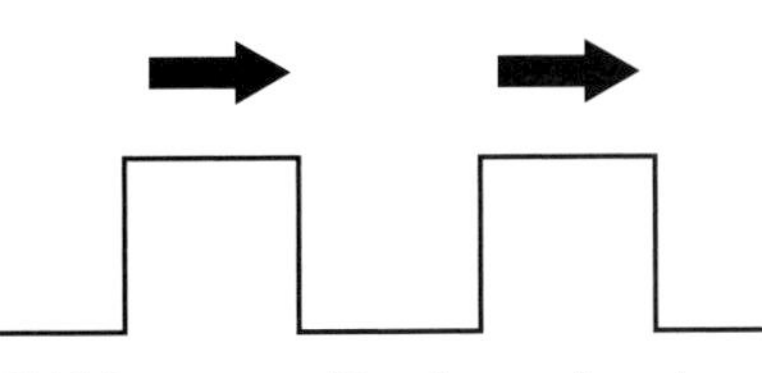

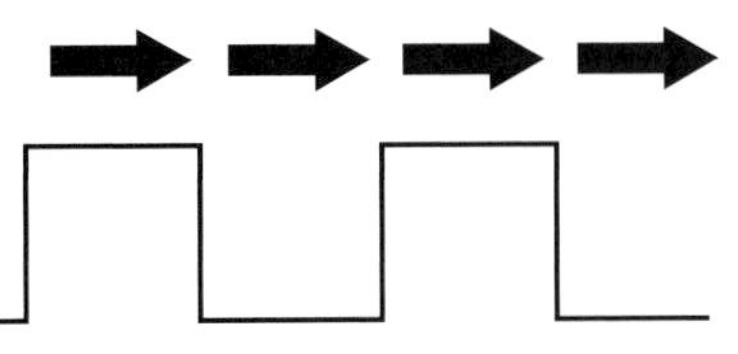

Figure 25-18. DDR-DRAM can achieve data rate transfers twice as high as conventional DRAM by transferring data on both the rising and falling edge of the digital pulse.

SIP, DIP, SIMM, and DIMM are PC memory chips that are packaged in several different styles.

SIP

A ***single inline package (SIP)*** is, as the name implies, a single row of connections running the length of the chip, **Figure 25-19**. This type of module is not used often today because of the high physical profile that it creates.

DIP

A ***dual inline package (DIP)*** is a chip that has two rows of connections, one on each side of the chip, **Figure 25-20**. This style is commonly used for cache memory or memory that needs to be mounted permanently on the board. You can see memory of this type on older model motherboards mounted in rows near the CPU.

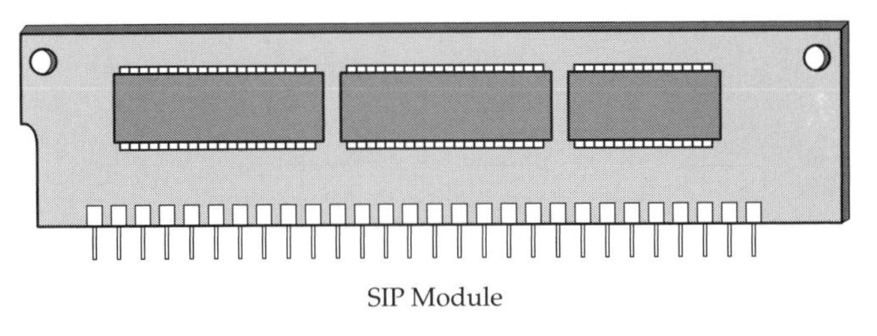

Figure 25-19. Early PC memory modules were SIPs.

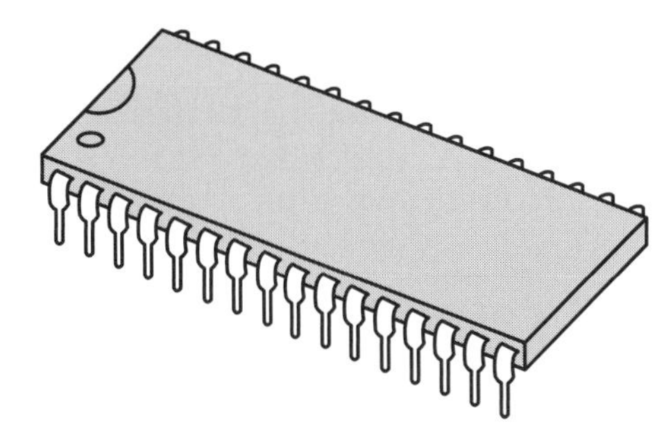

Figure 25-20. DIP modules, serving as cache memory, can still be found on some older motherboards.

SIMM

A ***single inline memory module (SIMM)*** is a row of DIP memory chips mounted on a circuit board that is inserted into memory slot sockets on the motherboard. The SIMM is designed so that, when plugged into the memory socket, each side of the edge connector is the same circuit.

DIMM

A ***dual inline memory module (DIMM)*** is constructed similarly to a SIMM, **Figure 25-21**. The major difference is that the edge connectors directly across the circuit board from each other do not connect electrically. They are not the same electrical connection as they are on a SIMM. The DIMM design allows for more electrical connections per inch than the SIMM design.

BIOS

The computer ***basic input output system (BIOS)*** contains the programming necessary to start the computer and then hand control over to the operating system. The BIOS program performs several crucial functions. See **Figure 25-22**. First, it starts the computer when power is applied to the PC. The BIOS performs a power-on self-test (POST). The POST checks to see if major components, such as memory chips, CPU, power supply, video

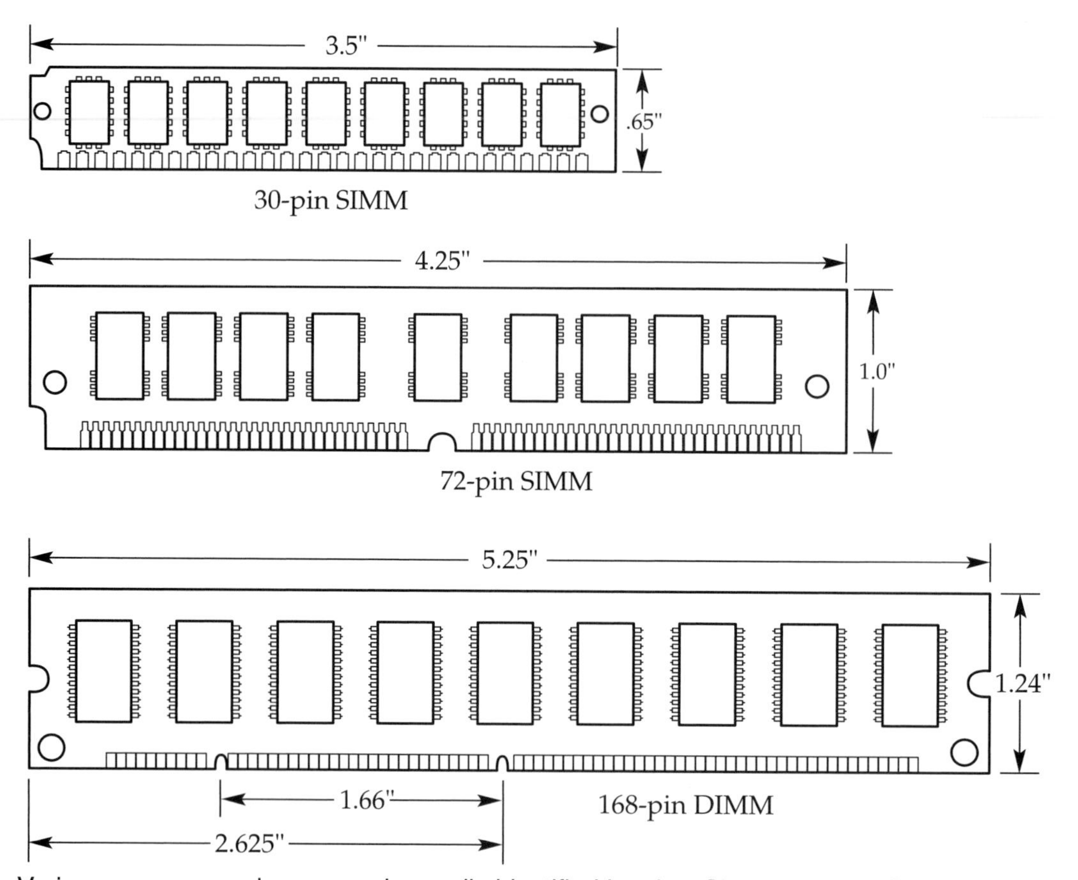

Figure 25-21. Various memory packages can be easily identified by size. Shown are two SIMMs and the more modern DIMM.

circuitry, keyboard, and hard disk drive, are in working order. It does not perform detailed checks, but rather a basic set of tests to identify that the components are installed, have power, and appear to be in working order. Later, when the POST is complete and the actual operating system is installed, a more detailed check is performed. The BIOS also automatically detects any new hardware that has been added to the system since the last time the computer was started. Detecting and installing the communications necessary for a new piece of hardware is referred to as plug and play (PnP) technology. Plug and play means that most hardware devices designed today can be automatically detected at startup, install all necessary software programs, and begin working immediately. For example, if you install a new sound card, the system should automatically detect it and install necessary software to finish setting up the card. It should be noted that plug and play does not automatically detect and configure all hardware. Therefore, hardware still often requires human intervention for setup.

The BIOS also contains the CMOS (pronounced seemoss) setup program. This program, normally accessed during bootup, allows a user to configure system settings such as date and time, drive configurations, boot order, and built-in peripherals (serial, parallel, USB, etc.). The settings are not saved in the BIOS, but are stored in CMOS. CMOS, which stands for complementary metal oxide semiconductor, is a type of memory that can hold its data even after the computer is turned off, as long as it receives power from a lithium battery. In the earlier computer systems, CMOS was located in the real time clock (RTC) chip. Today CMOS is located in the chipset.

The original BIOS chip was a true ROM, meaning that a program could only be written once to the chip. The ROM could not be erased and rewritten to. Today's more modern BIOS is actually referred to as flash memory because a new program can be flashed into, or written in, the BIOS chip.

Chipsets

A ***chipset*** is an integrated circuit on the motherboard containing control circuitry that interfaces to the computer's subsystems, such as memory, CPU, and input and output devices, **Figure 25-23**. Chipsets are designed to carry out instructions that do not require CPU intervention. The earliest computers processed all instructions and all data directly through the CPU. As the computer

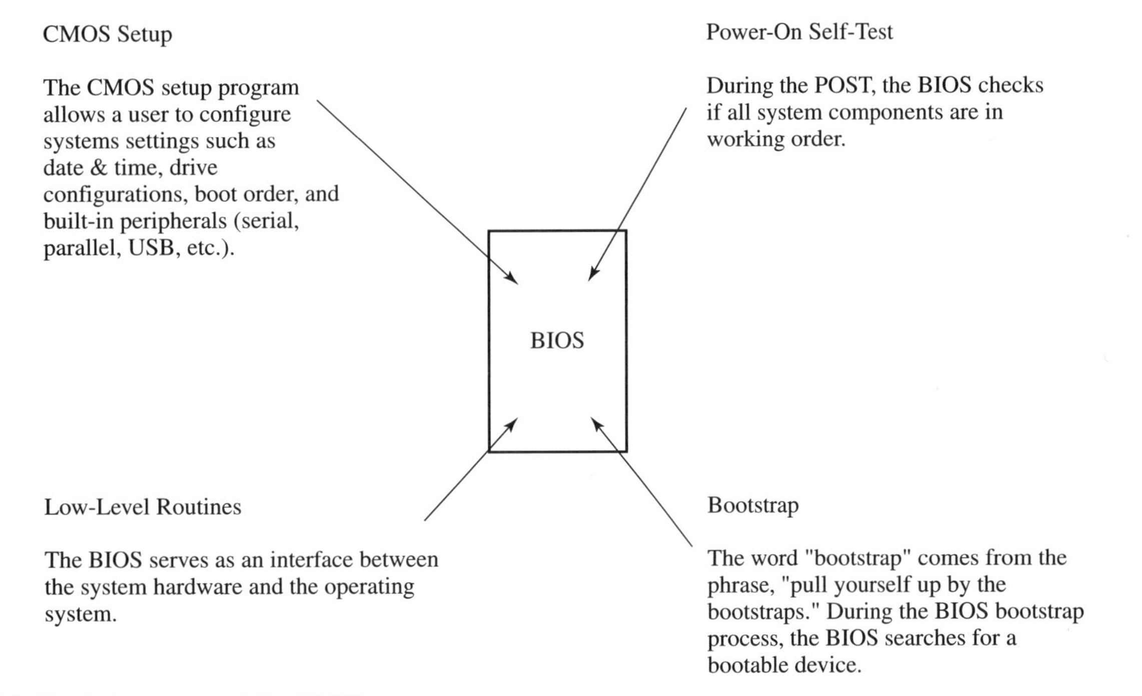

CMOS Setup
The CMOS setup program allows a user to configure systems settings such as date & time, drive configurations, boot order, and built-in peripherals (serial, parallel, USB, etc.).
Power-On Self-Test
During the POST, the BIOS checks if all system components are in working order.
BIOS
Low-Level Routines
The BIOS serves as an interface between the system hardware and the operating system.
Bootstrap
The word "bootstrap" comes from the phrase, "pull yourself up by the bootstraps." During the BIOS bootstrap process, the BIOS searches for a bootable device.

Figure 25-22. Basic functions of the BIOS.

evolved, it became apparent that many redundant processes did not require the continuous intervention of the CPU. For example, when a program loads or transfers data from a floppy disk or the hard drive, it needs to go directly to RAM, where it will be executed. Since each bit of data or program does not need to be processed by the CPU, ***direct memory access (DMA)*** was developed so that the data could go directly to the RAM from the storage media, such as a floppy disk or the hard drive. Today, chipsets inserted into key points along the motherboard bus support the direct movement of data from the floppy or hard disk drive to the RAM. See **Figure 25-24**.

A similar technology, known as bus mastering, was developed to move data between any two major components in a computer, such as a modem to the hard drive, a floppy drive to a modem, or a network card to the hard

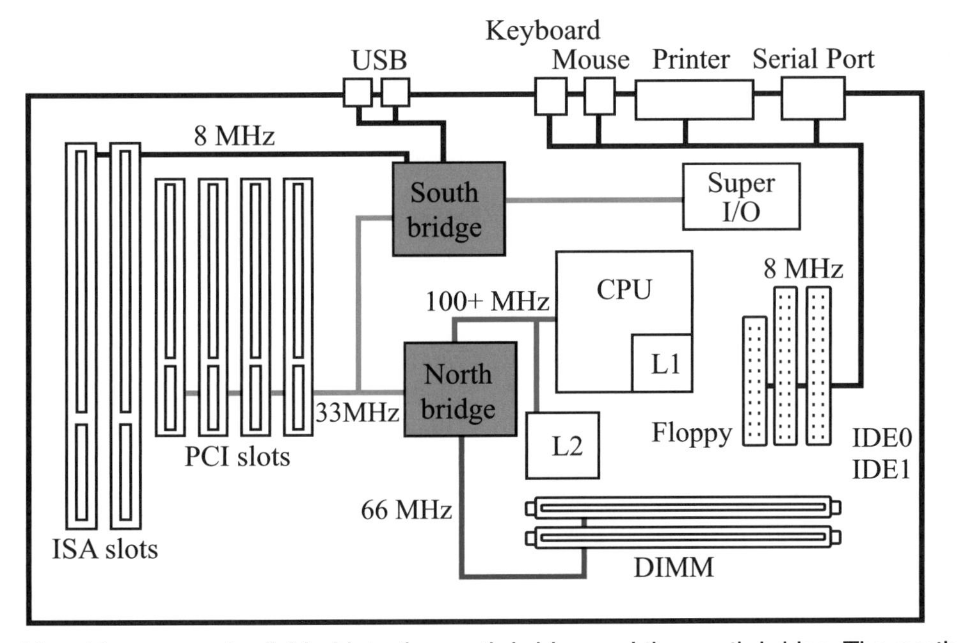

Figure 25-23. The chipset is commonly divided into the north bridge and the south bridge. The north bridge is used to transfer and control higher data speed systems, such as graphics and DVD hardware. The south bridge controls the slower devices associated with the PCI and ISA buses.

drive. Typically, bus mastering moves data from a port to a device, such as a serial port to a hard drive. Bus mastering also relies on a special set of chips built into the motherboard that permit data to move between two devices without going directly through the CPU.

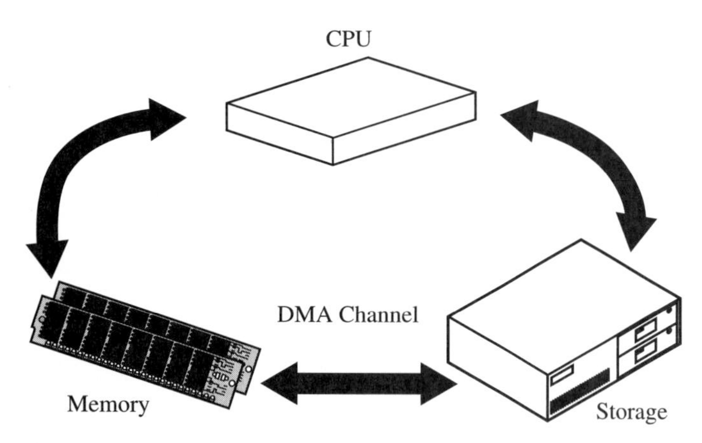

Figure 25-24. In older computers, all data and instruction had to flow through the CPU. With direct memory access (DMA), the CPU allows data to flow directly to RAM from many different computer devices. This allows the CPU to concentrate on other duties, producing a more efficient system.

Mass Storage Devices

Two main technologies used to store massive amounts of data are magnetic storage devices and optical, or laser, devices. The hard disk drive, floppy drive, ZIP drive, and tape drive use magnetic principles for storing data. CD-ROM, CD-RW, and DVD use laser light technology to store and retrieve data.

Two of the most common magnetic storage devices associated with computers are floppy disk drives and hard disk drives. Each utilizes data storage based on magnetic principles with which you are already familiar. As you know from earlier studies, electricity can produce magnetism, and magnetism can produce electricity. Using these principles, magnetic patterns representing digital patterns of ones and zeros can be stored and retrieved from special disks, **Figure 25-25**. The disks are coated with an iron oxide material that is easily magnetized.

A hard disk drive is very simple in construction. It contains one or more platters that are coated with a special iron oxide. The iron oxide retains magnetism induced by the read/write head attached to the end of the actuator arm. The magnetic patterns are in response to digital pulses sent to the read/write head from the computer motherboard and CPU. The digital pulses represent computer data. Converting the digital pulses into magnetic patterns on the disk surface is called writing data to the disk. The operation can be reversed to retrieve data from the disk. The fast moving magnetic patterns on the disk cause magnetic lines of force to cut across the read/write head, producing digital pulses representing the stored data on the disk. The process of retrieving data from the disk platter is called *reading*.

The ***file allocation table (FAT)*** is used by the operating system to keep track of files on a disk. The FAT includes information such as the name, date, and time of the creation or modification of a file; the file's location; and the number of sectors used by the file. The file allocation table also records the location of every area on the disk that is used for storage. Storage areas are referred to as sectors.

The FAT is the oldest and longest used file system for personal computers. The original FAT16 file system was developed for the early disk operating system (DOS). Although DOS has been virtually replaced by recent Windows operating systems, the FAT16 system can still be used by these systems.

The two common FAT systems are FAT16 and FAT32. FAT16 is so named because it uses a maximum of 16 bits to identify each sector on a disk. FAT32 uses a maximum of 32 bits for sector identification. The new technology file system (NTFS) was developed for the Windows NT operating system. The file system is structured similarly to FAT16 and FAT32, but it is greatly enhanced to provide better security and large disk access.

CD Drives

CD-ROM (read only), CD-R (recordable), CD-RW (read/write), and DVD technology are all very similar. All of these technologies use laser diodes to read the discs. See **Figure 25-26**. CD-R and CD-RW drives can also write to a disc and store data on the disc as a series of lands and pits. A land is in reference to the disc's surface, and pits are actual pits in the CD-ROM disc. The lands and pits symbolize the ones and zeros that represent the data. CD-R and CD-RW technologies are similar except that the land and pit areas are simulated through the use of a photosensitive material inside the disc. The photosensitive material can be made to resemble the land and pit areas used in the traditional disc system by applying higher-than-normal laser light to the organic material. When the higher-than-normal laser light strikes the organic material, the organic material will no longer reflect light. By no longer reflecting light, it represents data, the same way the pit areas do on the DVD.

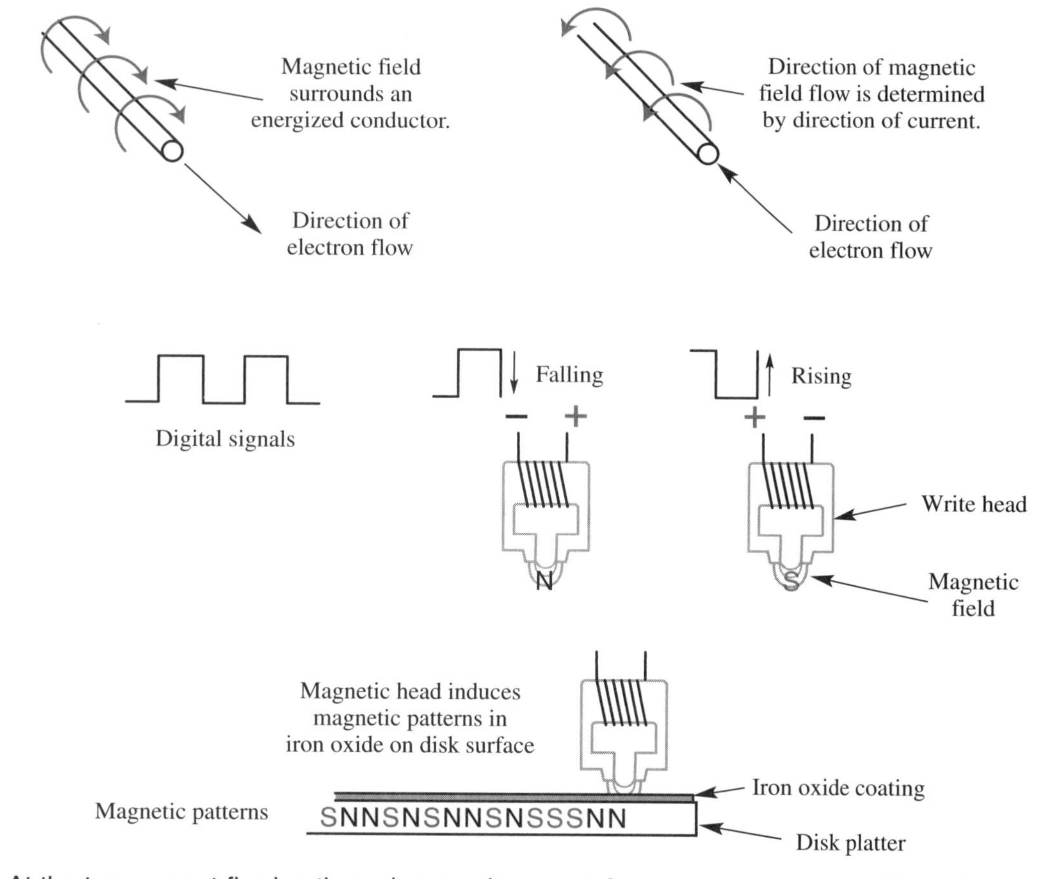

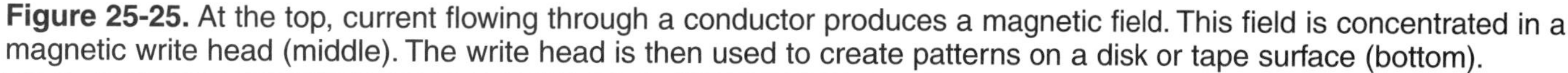

Figure 25-25. At the top, current flowing through a conductor produces a magnetic field. This field is concentrated in a magnetic write head (middle). The write head is then used to create patterns on a disk or tape surface (bottom).

The DVD players use laser optic technology to play back the data stored on the disc. Rather than using the conventional infrared laser diode as a source of light, the DVD system uses a new laser diode that produces light in the blue spectrum. This means the light has a shorter wavelength. This makes the system capable of reading a smaller pattern of pits on the disc, **Figure 25-27**. This allows more data to be stored on a disc. The double sided, dual layer disc can store up to 17 gigabytes of data, or 8.5 gigabytes on each side of the disc. See **Figure 25-28**.

The ***digital video disc*** is a high-capacity storage medium, which is used to store full-length movies, computer programs, and other information. While the typical compact disc can store 650 megabytes of information, the DVD can store up to 17 gigabytes. This is over 25 times the storage capacity on the same size disc.

In addition to the storage capacity, DVD uses the MPEG format for compressing the video data. The combination of the greater storage capacity and the MPEG compression allows for several hours of video play.

Note that the term disc is spelled differently. When talking about CD and DVD technology, the correct spelling is "disc". When talking about magnetic storage, the correct spelling is "disk."

A typical CD-ROM encodes data in a spiraling path from the center of the disc to its outer edge, **Figure 25-29**. Like magnetic storage devices, CD devices also use a special file format similar in design to the original FAT. The CD-ROM file system (CDFS) was originally introduced with Windows 95. Some other CD file formats are High Sierra Format, ISO9660, and Universal Disk Format (UDF). The Optical Storage Technology Association (OSTA), formed by a group of manufacturers, created a file system structure called UDF that could be used for CD-RW, magneto-optical (MO) disc, and DVD technologies. UDF is the successor to the ISO9660 file structure. One of the greatest improvements in UDF is that it allows for a bootable disc.

There are many compatibility issues today due to the many different CD file formats that are not downward compatible. For example, many CD-RW files cannot be accessed or read from older CD-ROM drive systems because of the way the file structure is organized and the intensity of the laser used. Additionally, there are file systems designed especially for music and video, which only add to the compatibility problems between different CD drives.

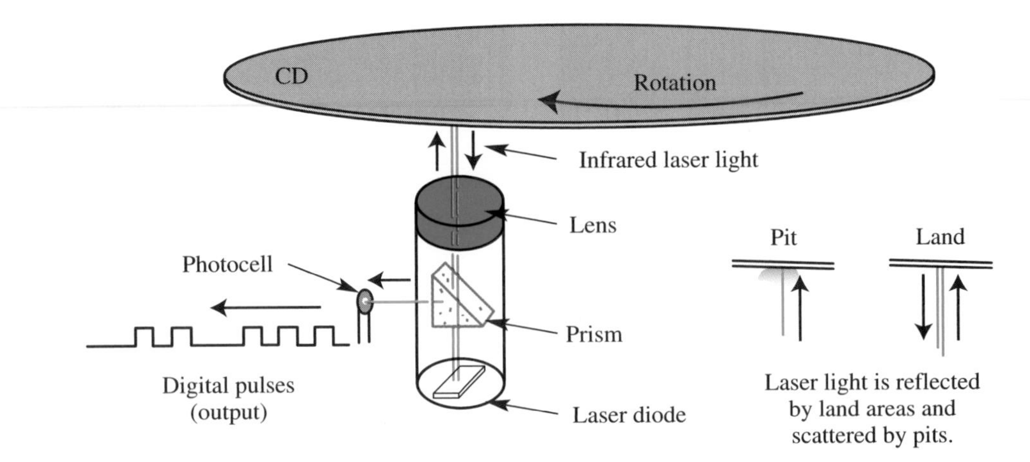

Figure 25-26. A laser diode sends a beam of light through a prism and onto the CD. If the light hits a land, the light is reflected back. Pits disperse the light. The reflected light travels back through the prism, which sends some of the light to a photocell. The photocell turns the light into digital electrical pulses.

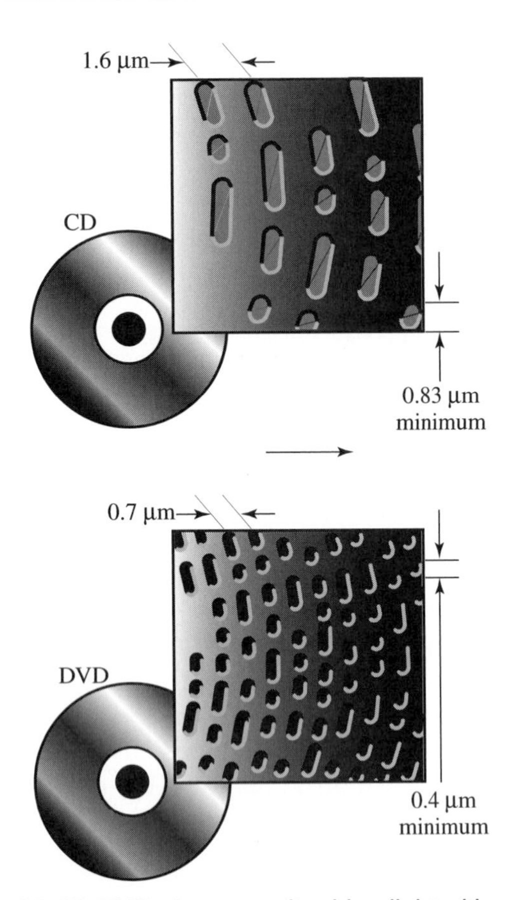

Figure 25-27. DVD players emit a blue light with a shorter wavelength than conventional CD players. This allows smaller and more densely packed pits to be imprinted on the discs.

Modem

The modem is used to support communications across telephone lines. Telephone lines were originally designed to transmit analog signals for voice, not digital, communication. Since high frequency digital signals will not pass through a typical telephone line leaving a home, digital data and commands must first be converted by the modem from digital format to analog. In fact, the term modem is a contraction of two electronic terms that describe the general function of the modem—modulate and demodulate. A modulated signal is generated and encoded with data. The modulated signal is an analog signal, not digital. When the modulated signal reaches its final destination, such as a modem on a distant computer, the modulated signal is then demodulated. That is, the encoded data in the analog signal is extracted and then converted back into digital codes. See **Figure 25-30.**

Long spans of telephone networks are, in fact, digital. To be completely accurate, the digital signal in the PC is first converted to an analog signal. The analog signal is then converted back to a digital signal before it encounters telephone-switching gear located at the telephone company hardware. The digital signal is transmitted across vast distances and then converted back to an analog signal when it is near its destination. At the destination computer, the modem converts the analog encoded signal to a digital signal. The major chip responsible for how the modem works is called a ***universal asynchronous receiver/transmitter (UART)***, **Figure 25-31**. The UART converts the parallel digital data into a series stream of digital data and then encodes the digital signal into a modulated or analog signal.

Review Questions for Section 25.2

1. A type of memory designed to permanently retain its contents is _____.
2. BIOS is a form of _____ memory.

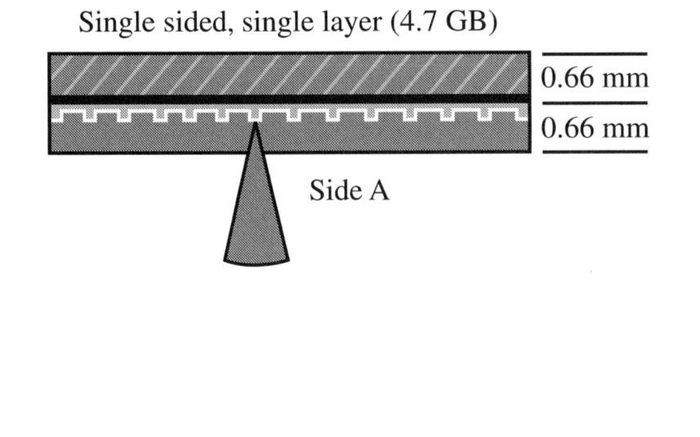

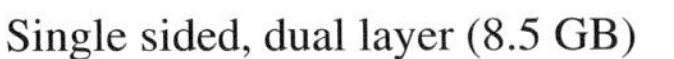

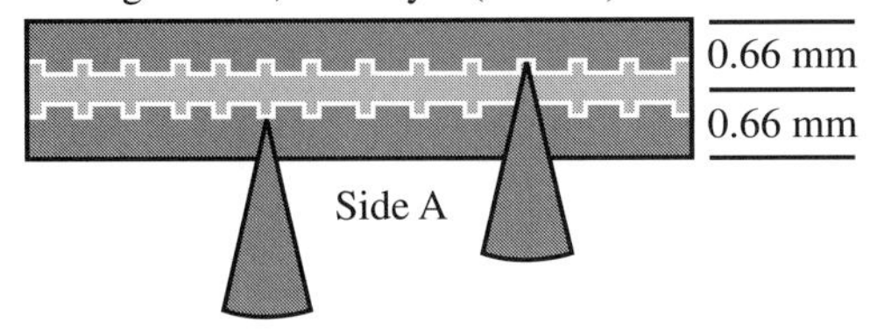

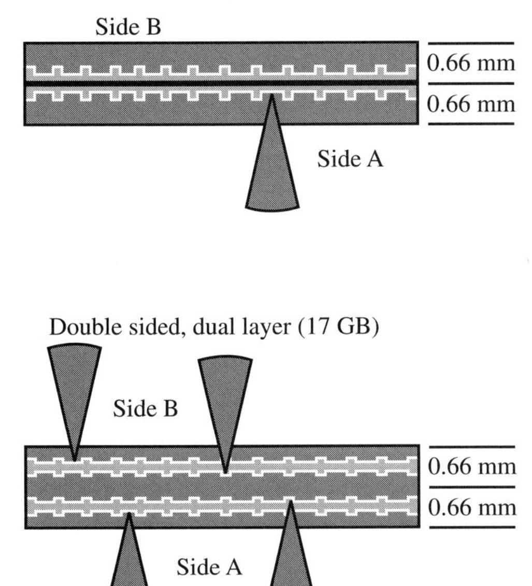

Figure 25-28. Four different disc layouts. Single sided, single layer discs can hold approximately 4.7 gigabytes of information. Double sided, double layer discs can hold about 17 gigabytes of information.

3. A type of memory that is only temporary and is used to hold loaded programs while they run is _____.
4. Name three differences between SRAM and DRAM.
5. What is SDRAM?

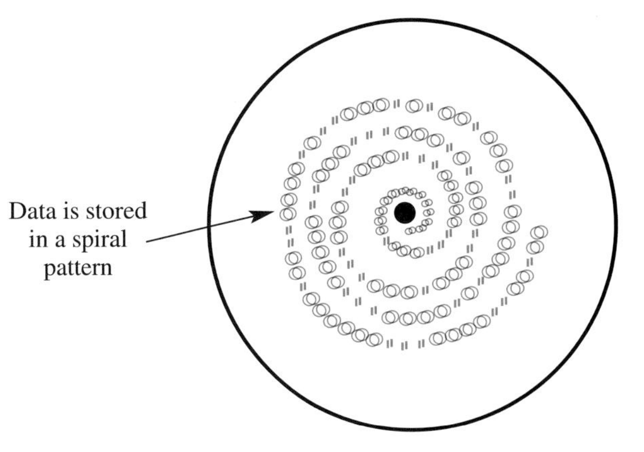

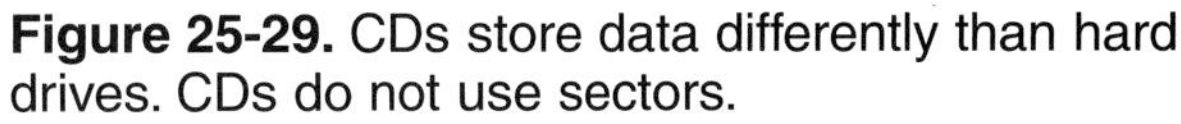
Figure 25-29. CDs store data differently than hard drives. CDs do not use sectors.

6. Automatically detecting new hardware and setting it up with the correct software is referred to as _____.
7. A(n) _____ is used to retain information about the system configuration identified by BIOS.
8. The two main technologies used to store data are _____ and _____.
9. At the end of a hard disk drive's actuator arm is the _____ head.
10. Converting digital pulses to magnetic patterns on a disk is referred to as _____ to the disk.
11. CD-ROMs and DVDs use _____ _____ to retrieve data.
12. What does the UART do?

25.3 SOFTWARE

Software is a set of instructions used to run a computer. Software can be broken down into two broad classifications: operating systems and applications. An ***operating system*** provides the utilities needed to communicate and carry out general user interactions. Typical user interactions are storing and retrieving files or data; using communication input devices, such as a mouse and keyboard; or displaying information on a monitor or printer. An ***application*** is written for tasks that are more specific. Some software applications are word processing, spreadsheets, databases, games, multi-media software, flight simulators, educational tutorials, and Internet browsers. Let's take a closer look at how an operating system and some other software programs work together to form a complete computer communication system that supports user commands.

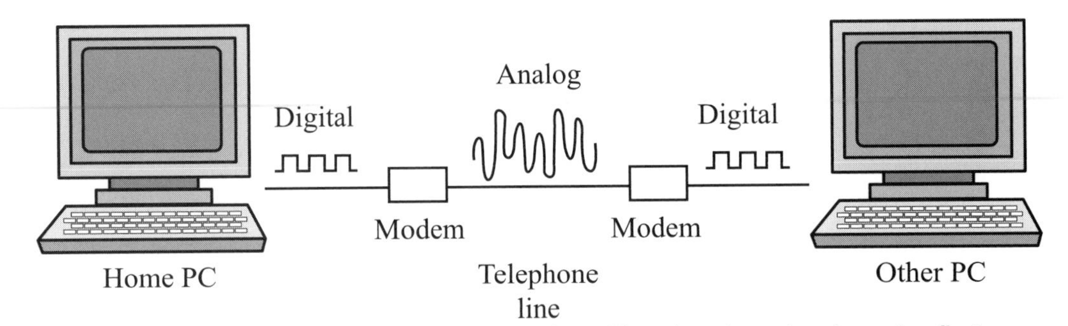

Figure 25-30. These two computers are connected by a modem. The signal coming from the first computer is digital until it passes through its modem. The modem converts the digital signal to an analog signal and passes it through the phone lines. The second computer's modem receives the analog signal and converts it back to a digital signal that the computer can read.

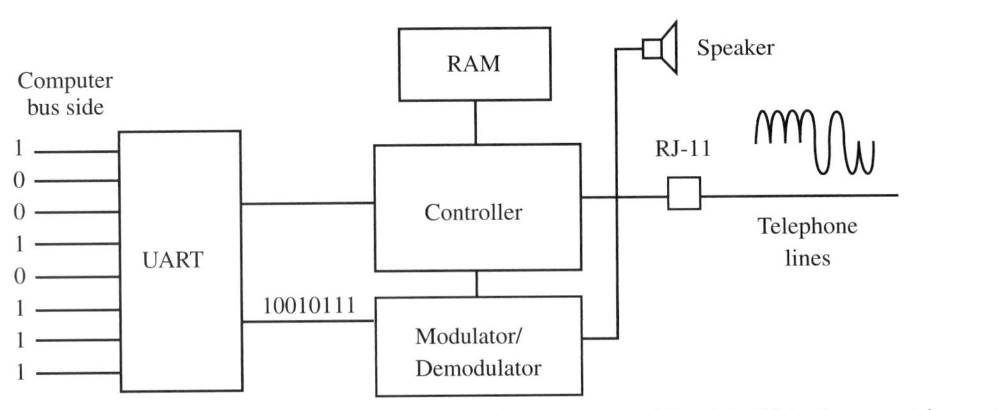

Figure 25-31. There are only a few major components in a modern modem. The UART is the most important.

BIOS Software

The BIOS is a combination of software and hardware referred to as firmware. A software program is written and then installed in the BIOS chip.

The BIOS is responsible for communication between the operating system and the computer system hardware. Originally, computers were designed without a BIOS. Software operating systems, as well as application software, were written to communicate with specific models of computers. Computer systems, as well as operating system software and application software, were proprietary. There was little compatibility. Today, because of the BIOS, operating systems, as well as software applications, are normally compatible. Compatibility issues usually (but not always) arise with limitations regarding a specific processor family, such as the Intel 80XX processor family. As long a PC is equipped with an 80XX processor or a compatible clone system, the software operating system and software application will work satisfactorily. Look at **Figure 25-32** to see the relationship of various software programs and PC hardware. Refer back to this illustration as more software systems are explained.

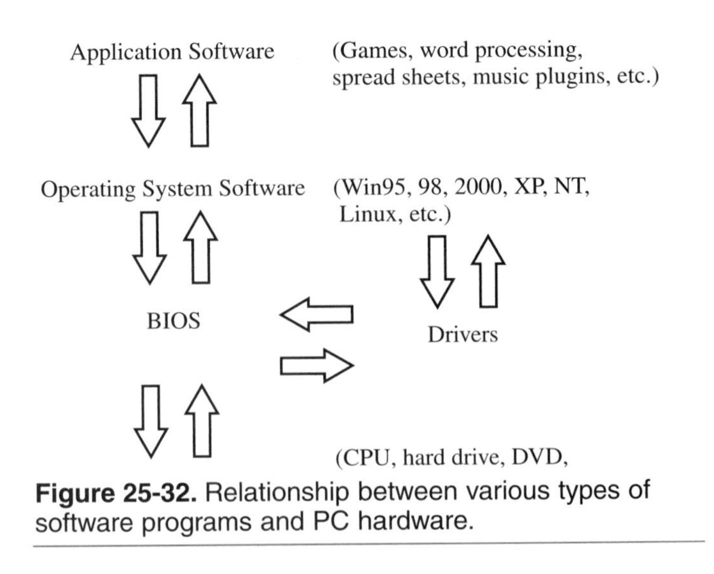

Figure 25-32. Relationship between various types of software programs and PC hardware.

Operating System (OS)

The operating system (OS) is a type of software that determines how a user interacts with the computer. It is responsible for how data is stored and retrieved. The OS also determines the type of security to be used. Today's modern operating systems use a ***graphical user interface (GUI)***, which is a drawing that appears across the screen allowing a user to simply pick and click on a desired function. See **Figure 25-33**. For example, when using the Windows family of operating systems, you can simply click on the floppy drive symbol (icon) to see the files contained on the disk. **Figure 25-34** illustrates a directory listing in a GUI interface. To open a file, you can click or double click on it. A single-click or double-click will depend on the operating system being used and how it is configured.

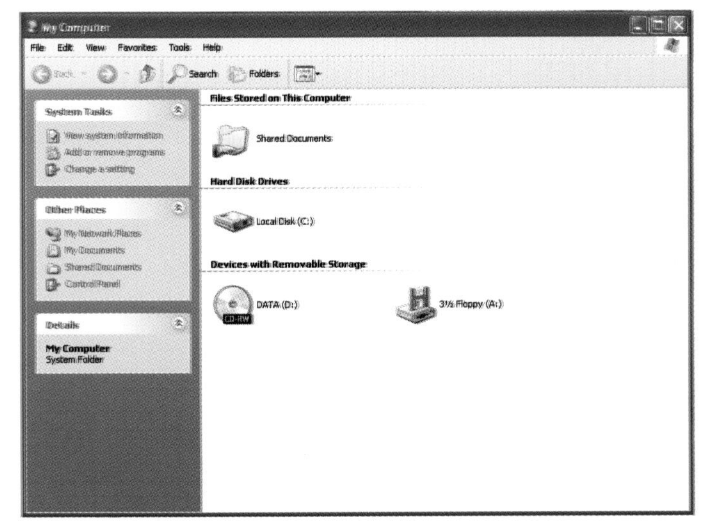

Figure 25-33. Typical GUI for Windows. While the mechanics behind the interface can be very different between the various versions of Windows, the appearance of the GUI has remained fairly constant.

Figure 25-34. File structure as seen through a GUI. In a GUI, file folder icons are usually used to represent directories.

Some of the operating systems found today are Windows 98, Windows 2000, Windows Me, Windows XP, Windows NT, and various forms of Linux such as Red Hat Linux and Caldera. These systems are much easier to use than older systems.

One of the most important characteristics of an operating system is the method it uses to store data. Stored data is typically referred to as "files." Data storage involves such issues as the maximum size hard disk drive the operating system can access, the maximum number of files the operating system can store, and the file formats (including security, encryption, and compression) the operating system can use.

Drivers

Drivers are a special classification of software required to support communication between the computer system and a specific peripheral. A driver is required to translate commands from operating system and application software, to the BIOS and to specific hardware devices. There are thousands of different peripheral devices that can be added to a computer. While many operating systems contain drivers for previously released or commonly used equipment, no operating system can possibly be designed to communicate with all the various hardware devices available. In fact, after a particular operating system is released, many new peripherals come on the market very quickly. A driver is required to support the translations to and from the peripheral. One of the classic examples of a need for a driver is when a new printer has been designed and released after a particular OS has been released. For the operating system to communicate correctly with the new printer, a driver must be installed. Most peripherals come packaged with a driver disc, **Figure 25-35**. If the correct driver is not installed, the printer will print gibberish or may not run at all. All peripherals such as mice, keyboards, monitors, printers, and scanners require a driver.

It is interesting to note that most computer crashes associated with the Windows family of operating systems are associated with an application software package that directly accesses the system hardware. This is a common occurrence in gaming software. Designers of games, as well as applications that demand high-system speed, such as some computer aided animation or drawing programs, write code that communicates directly with the machine hardware. While in most instances it will work, it does cause the computer to lock up when software memory usage clashes. A computer crash commonly occurs when two software programs try to occupy the same memory location at the same time. Microsoft started a program years ago that required software venders to submit the program code and software for testing before Microsoft

Figure 25-35. Many new peripherals come with their own driver disc.

would allow the Microsoft logo to be displayed on the software package, **Figure 25-36.** This system was designed to assure software compatibility between the Microsoft operating system and the application.

Figure 25-36. This software box is stamped with Microsoft's logo signifying that it is certified for use with Windows 2000 Professional, Windows NT Workstation 4.0, and Windows 98.

Viruses

While computer viruses have become more controllable than in the early days, there are still some very destructive programs written for the sole purpose of causing problems within a network of computers or an individual computer. The problems can range from simple nuisances, to complete destruction of data files and software. Viruses often come from diskettes that are exchanged between users and, more commonly, from e-mail sent over the Internet. See **Figure 25-37**.

To avoid viruses, care should always be taken when accepting software, data, or e-mail from a source that is not prepackaged or is unknown to you. Only download files from respectable on-line service providers. Even with precaution, there is no guarantee that a diskette or program will be virus free. Always keep an up-to-date antivirus program on your computer.

There will always be new viruses and worms that are created for destructive purposes. All smart computer users should be aware of them.

Review Questions for Section 25.3

1. What are the two broad categories of software?
2. What is firmware?
3. What purpose does a driver serve?
4. What is a common cause of system lockups?

25.4 COMPUTER DEVICES

It is important to understand that computers today are capable of supporting peripherals that used to be stand-alone devices. Computers can output data in many formats and accept data from many pieces of equipment. This data may be used separately or incorporated into other programs that the computer is currently working with. As a computer's capability expands, so will the number and type of peripherals available to enhance the computer's versatility. Many of these include a wide variety of printers, scanners, and cameras.

Dot Matrix Printer

A dot matrix printer derives its name from the pattern, or matrix, of tiny dots used to create text and images. See **Figure 25-38**.

The dot matrix printhead consists of a line of small metal rods called print pins. They look similar to small

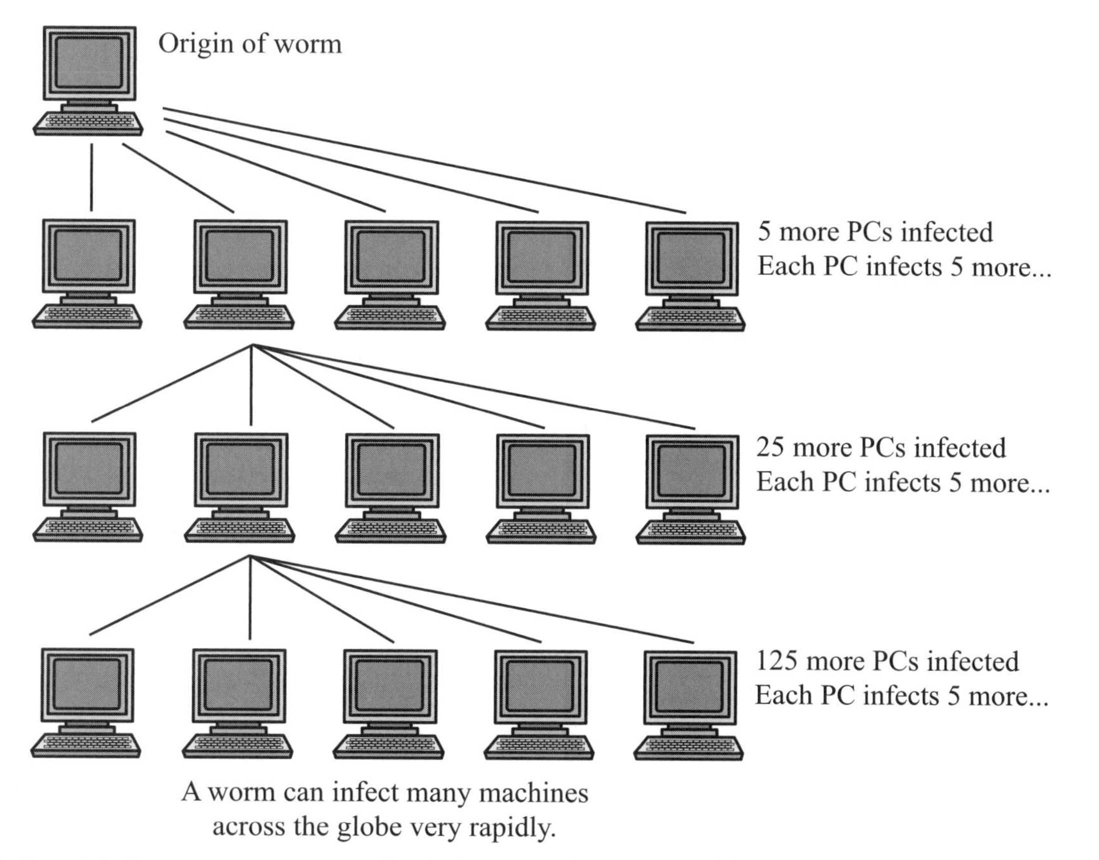

Figure 25-37. E-mail is the most common method of transmitting a worm. Using this method, the worm multiplies at an exponential rate.

nails. An electrical coil, called a solenoid, controls each print pin. The solenoid creates a magnetic field when energized. The print pin is constructed with a magnet at one end. When the solenoid is energized, the magnetic end of the pin is attracted to the coil. When the solenoid is energized, the print pin moves forward, rapidly striking the ink ribbon. Each energized solenoid causes the individual print rod to strike the ribbon, leaving a dot of ink on the paper. To form a letter, such as the letter "A," a series of electrical pulses are sent to the printhead. The pulses are coded in a sequence that forms the letter "A." The letter is printed as the entire printhead moves horizontally across the paper. The two common printhead styles are nine pin and twenty-four pin. Dot matrix printers are noisy and are limited today for use with text applications, not for graphics. When graphics with fine detail are desired, other printers such as inkjet or laser are used.

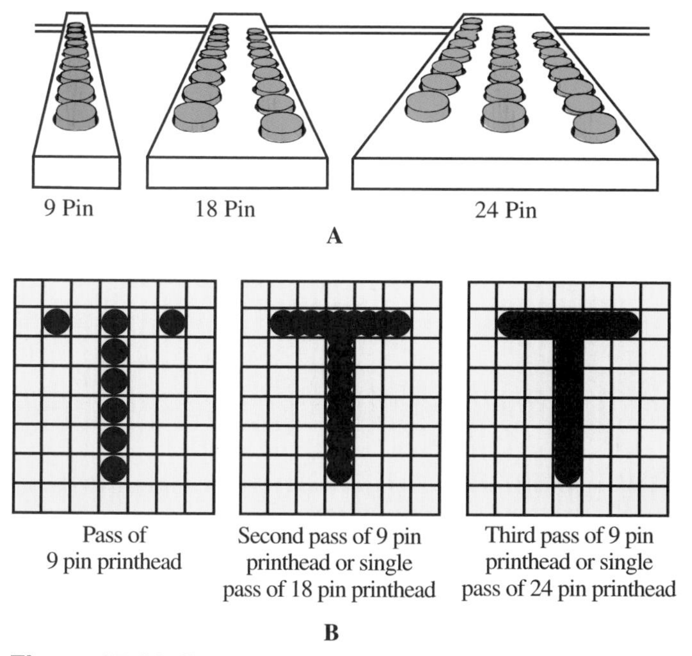

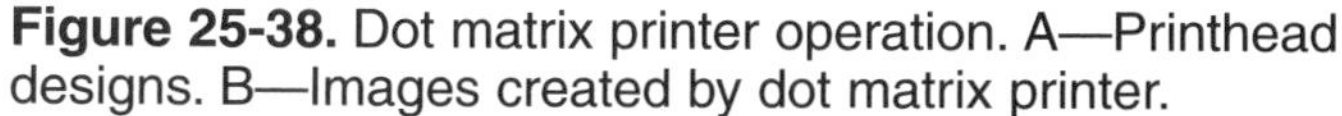
Figure 25-38. Dot matrix printer operation. A—Printhead designs. B—Images created by dot matrix printer.

Inkjet Printer

An inkjet printer uses a specially designed cartridge that moves horizontally in front of the paper spraying fine mists of ink. See **Figure 25-39**.

The two main types of inkjet printers use either a thermo resistor or a piezocell crystal to produce the force that fires the drop of ink through the nozzle. The thermo resistor type uses heat to expel the ink droplet, whereas

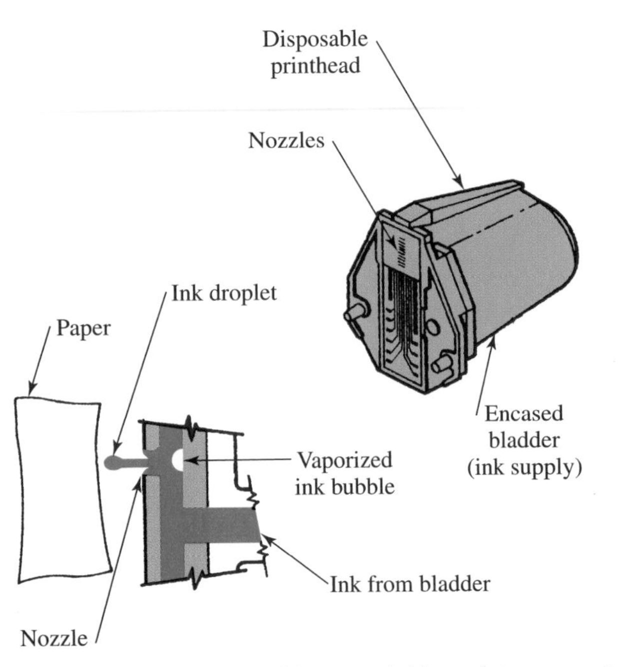

Figure 25-39. Drawing of how an inkjet printer operates.

the piezocell uses vibration. Printers that use the thermo resister technology are often referred to as bubble jets. **Figure 25-40** illustrates a single nozzle disassembled into its major pieces: the nozzle, firing chamber, and thermo resistor. The operation is simple. The ink flows into the firing chamber from the ink cartridge reservoir. When an electrical pulse from the computer enters the thermo resistor, the ink in the firing chamber is instantly heated to over 600°C. The extreme heat expands the ink in the chamber causing it to fire through the nozzle onto the paper. As the chamber cools, a vacuum is created which draws more ink into the firing chamber from the ink cartridge reservoir. An inkjet printer can fire its chamber many times a second. The single ink droplet is quite small. A typical low cost inkjet printer can produce 720 × 720 drops per square inch (dpi). A high quality inkjet printer can produce over two million dpi or 1400 × 1400 dpi.

The piezocell is a crystal that deflects or bends when energized by an electrical pulse. When a piezocell is used in place of a thermo resistor, the piezocell flexes every time an electrical pulse is sent. When the piezocell crystal flexes, the ink droplet is fired through the nozzle. When the electrical pulse stops, the crystal returns to its original shape. This creates a vacuum in the firing chamber and draws in more ink. A color inkjet printer uses four hues of ink: cyan, magenta, yellow, and black. This is a standard combination of colors often referred to simply as CMYK. The "K"—rather than "B"—is used to signify black. See **Figure 25-41**.

Laser Printer

The printing process used by a laser printer is called electrophotographic process (EP). Traditional printing is based on wet ink applied to paper using several different processes. Electrophotographic process is a combination

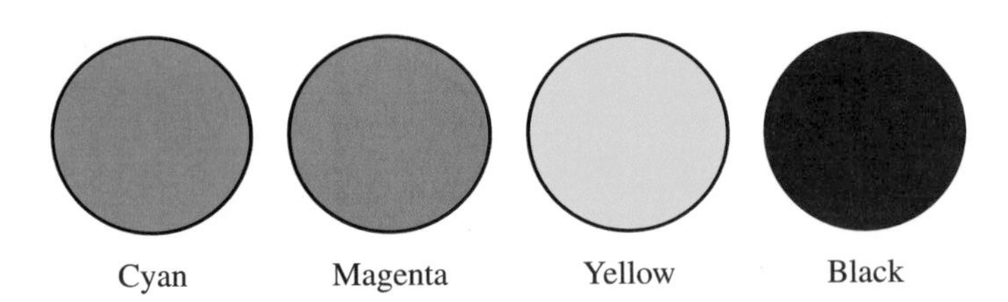
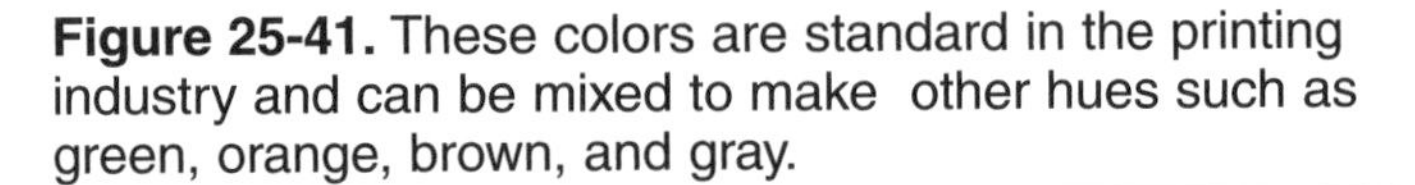

Figure 25-41. These colors are standard in the printing industry and can be mixed to make other hues such as green, orange, brown, and gray.

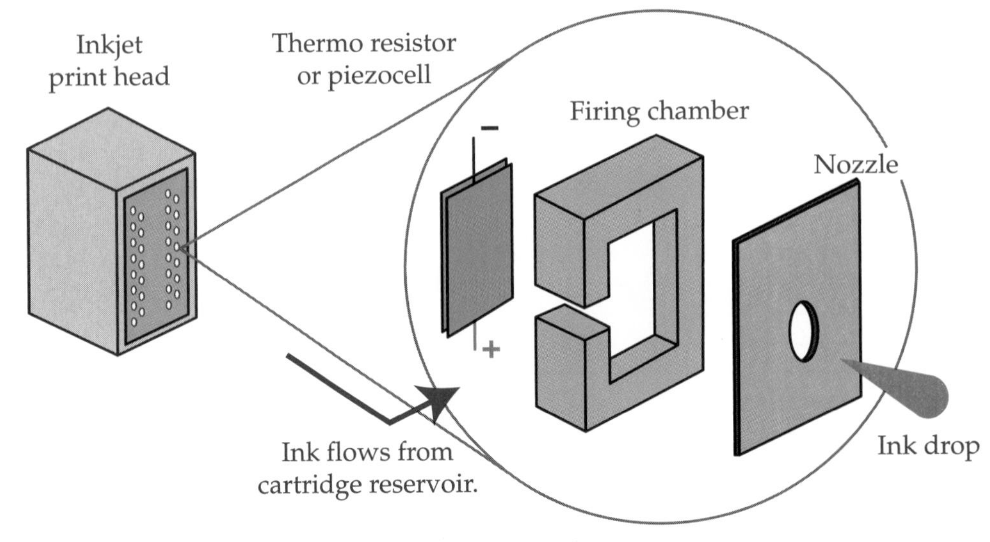

Figure 25-40. Exploded diagram of a typical color inkjet printhead.

of static electricity, light, dry chemical compound, pressure, and heat. Static electricity is used to charge the printer drum to over 600 volts. After the drum is charged, an image is written on the drum using laser light from a laser diode. A spinning, pentagon-shaped mirror is struck with the laser light to form an image on the drum. Intense light from the laser reduces the static charge on the drum wherever it strikes the drum surface. The image is formed from the various levels of static electricity charges on the drum. The static image is a representation of the image data sent to the printer from the computer. The drum rotates the static image past the toner cartridge, picking up bits of toner with the static charge. See **Figure 25-42**.

The paper leaving the paper tray picks up a charge from the transfer corona wire. This charge is stronger than that of the toner on the drum. As the paper passes under the drum, it attracts the charged toner to the paper. The paper, which now holds the toner image, passes through the fuser. The fuser produces very high heat using a quartz lamp. The toner, which consists of plastic and tiny metallic particles, absorbs the heat from the lamp. This heat melts the plastic onto the paper's surface. The paper then exits to the output tray and reveals an image similar to the one on the computer screen, and of that sent to the laser printer.

CAUTION:

The laser printing process utilizes extremely high voltage levels. Never attempt to repair or disassemble a laser printer. Repairing a laser printer requires special training and safety precautions.

Scanner

Scanners are used to copy graphics and text into the computer's memory. Once the picture is captured, it can be manipulated with software. The picture can be reduced, enlarged, rotated, darkened, lightened, sharpened, or touched up in any other way. The scanner has had a tremendous impact on desktop publishing. **Figure 25-43** shows how a scanner works.

The two main parts of a scanner are a light source and a charged coupled device (CCD). In Chapter 23—Television and Video Display Units, you learned that the

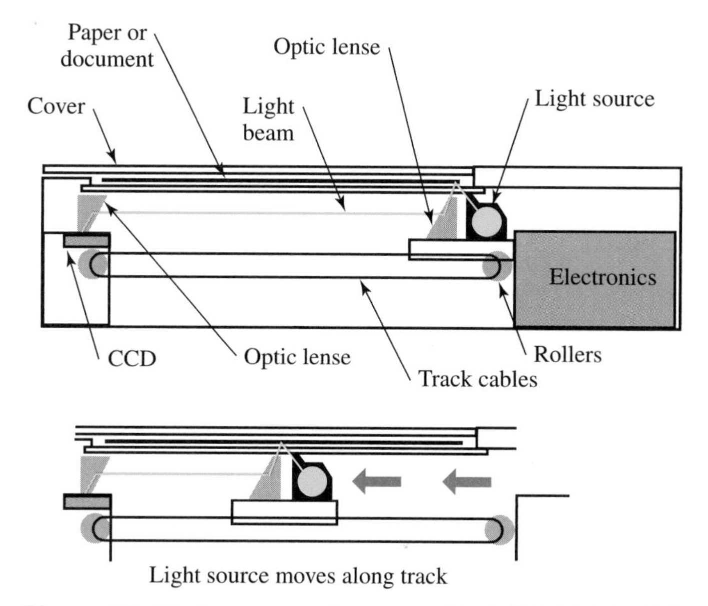

Figure 25-43. A scanner changes reflected light into digital signals.

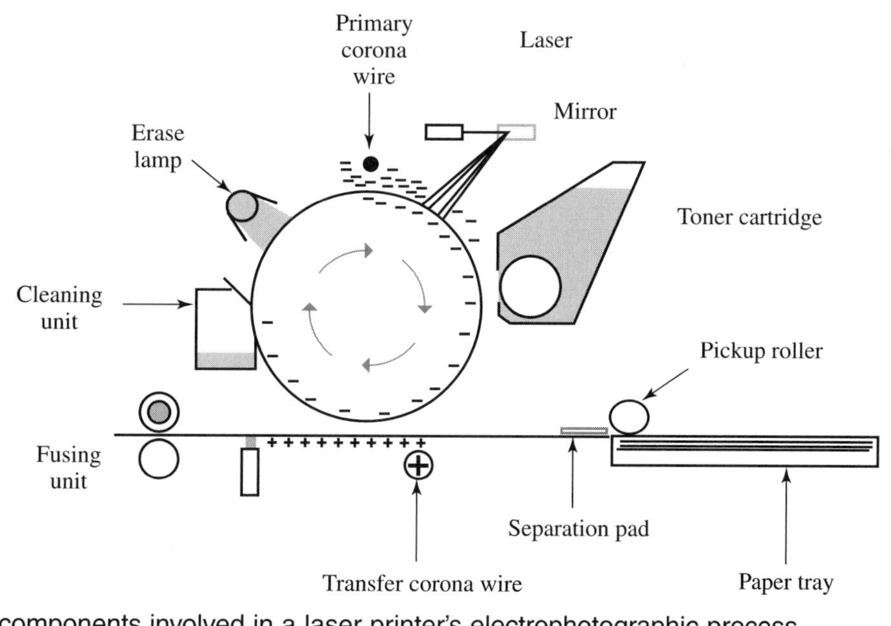

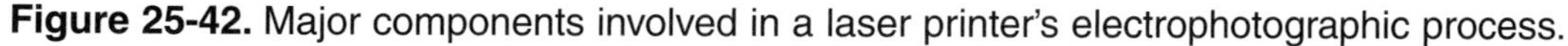

Figure 25-42. Major components involved in a laser printer's electrophotographic process.

CCD was an integrated circuit containing an array of photo sensors used to capture images. The same technology used in the digital camera is also used in the scanner.

A scanner has a stepping motor that pulls a light source along a track of wire cables. At each step along the scanner track, the CCD scans and receives the light beam. The amount of charge the CCD receives is in direct proportion to the intensity of light and dark. The CCD then transmits a series of digital signals that represent the light and dark contrast of that line, **Figure 25-44**, and transports the charges to a storage register. A storage register is like a series of flip-flops. After one line of data has been completely captured and transferred to the storage register, the data is then transferred in bytes to a section of computer memory. This process is repeated every time the scanner stops along the document being scanned. The CCD sensors in the array are very small and close together. The individual sensors can be spaced 15 mm from center to center or closer.

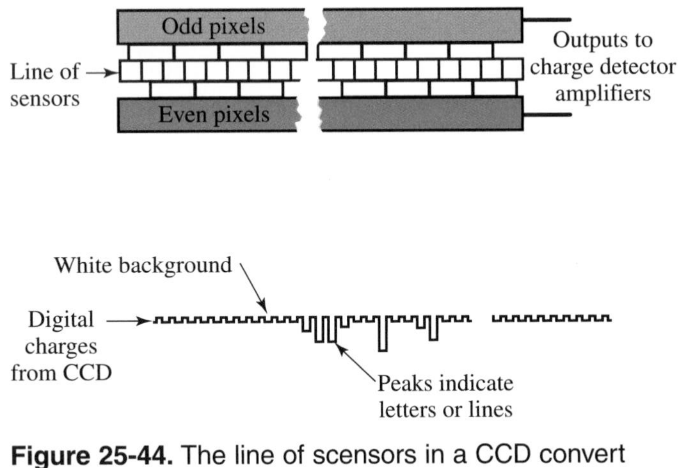

Figure 25-44. The line of scensors in a CCD convert black and white levels into a digital pattern.

As the popularity of the digital camera increases, scanners are used less for converting photographic images into digital images. Scanners are still used for converting printed text materials into digital text objects that can be imported into word processing and desktop publishing programs.

Digital Camera

A digital camera uses the same technology, CCD, as the scanner. In the digital camera, the CCD captures all the light from the image at once. The individual packets of charge are then converted into a series of digital signals. The digital signals are transferred to a block of computer memory. Once in the computer memory, the data can be manipulated by software. The popularity of the digital camera is rapidly rising. Memory sizes, storage capabilities, and resolution improvements are being made almost daily. This technology will soon be commonplace.

The web-camera, or web-cam as it is commonly known, can be incorporated into a computer system for use in video and audio transmission. Web-cams are used for a multitude of purposes such as security, monitoring multiple stations, personal use, and entertainment. Applications for this technology include use by educational institutions for distance learning and by corporations for video conferencing.

Computer CRT

You are already familiar with the operation of a CRT from Chapter 23—Television and Video Display Units. Computer CRT screens have a much higher resolution than conventional television screens. A computer CRT uses a shadow mask with openings placed at much closer intervals than those in a television. Also, computer CRT monitors produce sharp images that are rated by dot-pitch. Dot-pitch is the distance, measured in millimeters, between two dots formed on the display area or screen—the lower the dot-pitch number, the crisper the image. The typical dot-pitch range is from 0.15 mm to 0.30 mm.

Computer monitors are described in literature with acronyms such as SVGA, UVGA, and XGA. The acronyms are actually a description of the video standard or technology and not necessarily the quality of the image being produced. All computers are capable of producing standard VGA resolution. This resolution has been incorporated throughout the industry for compatibility and to aid in troubleshooting. Video systems are memory intensive. Consumers demand high-resolution images that exceed the standard VGA resolution. See **Figure 25-45** for a table of some of the various resolutions used today. Many are not industry standard, but rather proprietary.

The typical PC motherboard has all the necessary electronics incorporated to produce VGA as well as higher resolutions such as SVGA. You will notice in the chart that there are several different SVGA resolutions, because it is not a true industry standard.

Multimedia Card/Video Adapter Card

Some computers are equipped with a video adapter card, rather than having the video circuitry built into the motherboard. An adapter card is especially designed to increase video performance and to handle all the multimedia for the computer system. The card fits into any

Video Standard	Horizontal	Vertical
VGA	640	480
SVGA	800	600
	1024	768
	1280	1024
	1600	1200
SXGA	1280	1024
SXGA+	1400	1024
UXGA	1600	1200
HDTV	1920	1080
UXGA Wide	1920	1200
QXGA	2056	1536
SXGA Wide	1600	1024

Figure 25-45. VGA standards and their resolutions.

available expansion slot on the motherboard and connects directly to the computer monitor. Special software drivers must be installed to provide communication between the operating system and the monitor via the video card. Some video cards are referred to as multimedia cards. They not only increase the quality of the video display, they also support a direct connection to the CD-ROM for sound and video. They are equipped with speaker connections and a microphone connection. These cards are often very expensive. They have their own RAM, BIOS, and processing unit so the card can work as independent as possible from the CPU.

Review Questions for Section 25.4

1. Name three common types of printers.
2. What colors do the letters CMYK represent?
3. What do the letters EP represent in relation to laser printers?
4. Name two items that commonly use a CCD?
5. What is the main difference between a television CRT and a computer CRT?
6. What is dot-pitch?

23.5 LAN, WAN, AND THE INTERNET

Computers can be connected, or networked, to one another through cable or wireless technologies, enabling communications and data and resource sharing. A network can be limited to a single building, or spread across the entire globe.

LAN

A ***local area network (LAN)*** is a group of computers connected together to share files and other resources such as printers. All computers or workstations in the system are usually connected to a powerful computer known as the file server. See **Figure 25-46.** The file server provides the software and data files to which the workstations commonly need access. A network system is designed to save money and time. Databases are easily shared and files are easily accessed from a single location. Networks are common in businesses and educational settings.

In addition to the network operating system software on the server, network client software is required on each workstation to complete communications between the workstations and the file server, printers, and modems on the LAN. The network operating system software also provides security. A network administer uses special utility programs to set up user passwords and to limit access to certain resources and files.

WAN

A ***wide area network (WAN)*** is simply an expansion of a LAN. LANs are typically set up in a single building and connected by copper or fiber optic cable. A WAN network can have stations many miles apart. The WAN network depends on routers, a hardware or software device that determines the best route for the information to take, **Figure 25-47**. The router sends signals over telephone lines or via satellite to tie the computers and information together. Through the WAN, users have instant access to all programs and information contained in the network. A network administrator can also manage WAN security.

Internet

We have studied some significant electronic products that have made an impact on our society. Many of these products have greatly influenced the way we live and work. The telegraph, radio, television, telephone, satellite, and the computer are all examples of such products.

The *Internet*, or *World Wide Web (WWW),* is one such technology. Most of us use it daily. As an electronic technician you will need to use it often for product information, specifications, research, and for many other segments of your work. The Internet is quickly becoming the publication of choice for many manufacturers. These manufacturers now provide, or require for download, information that had been previously available only in

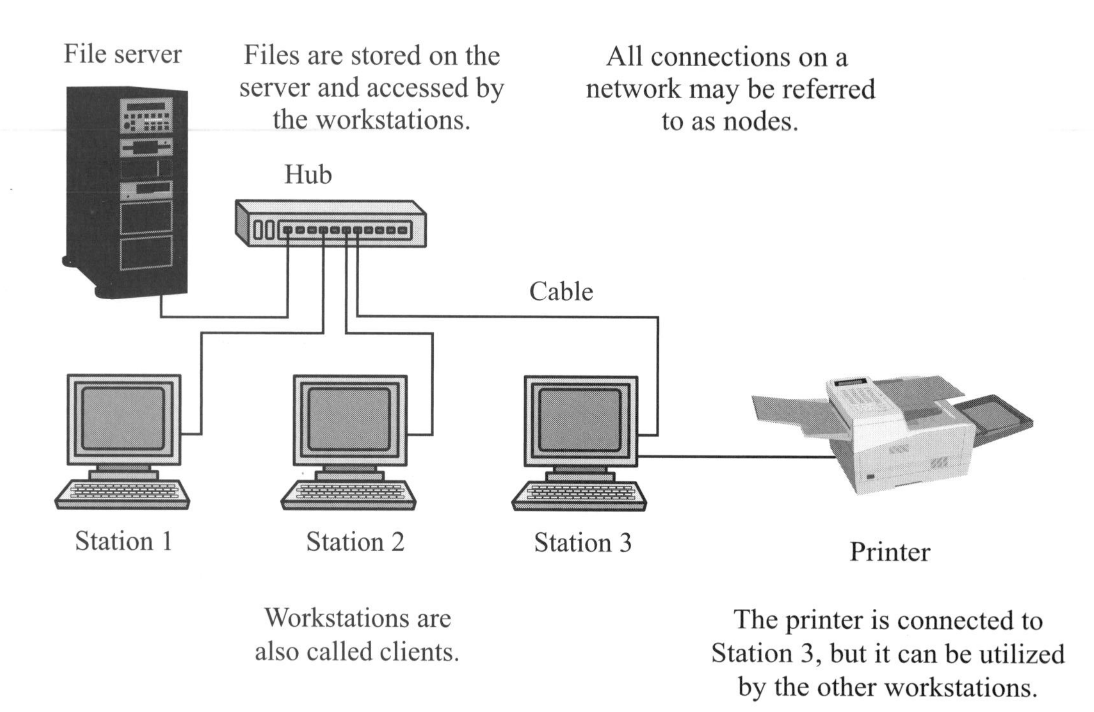

File server
Files are stored on the server and accessed by the workstations.
All connections on a network may be referred to as nodes.
Hub
Cable
Station 1
Station 2
Station 3
Printer
Workstations are also called clients.
The printer is connected to Station 3, but it can be utilized by the other workstations.

Figure 25-46. A simple client/server network.

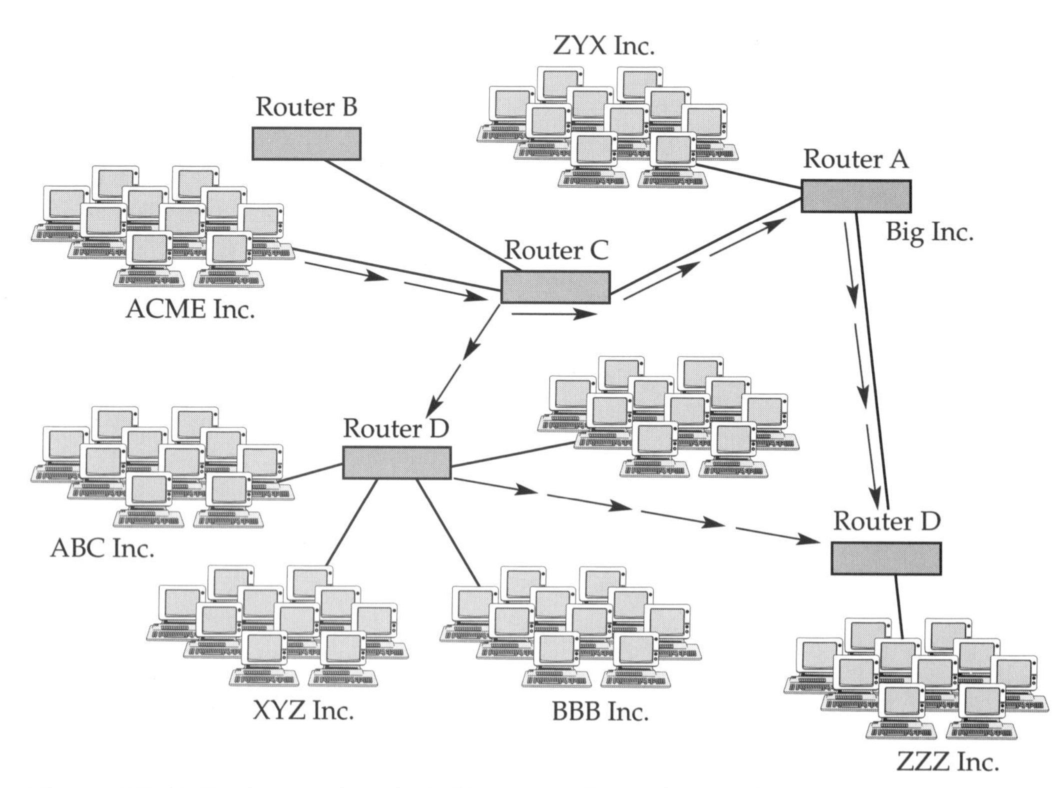

Figure 25-47. Routers route data traffic across the entire world. As you can see, there are several routes that can be chosen to send a data packet from ACME, Inc. to ZZZ, Inc. The job of the router is to select the best route. This is determined by the amount of data to be sent, the status of the other networks the router is connected to, and then the quickest route.

hard copy. The Internet allows material to be freely exchanged by willing participants. It allows access to museums, libraries, research facilities, electronic suppliers, people, and services throughout the U.S. and the world.

The name "World Wide Web" is descriptive of a web of cable and satellite technology across the globe that allows people to exchange a wide variety of information and services, **Figure 25-48**. Through the Internet, news stories are available as they occur. Weather reports, satellite images, and advertising can be accessed instantly. The Internet can be used for purely entertainment purposes too, **Figure 25-49**. Games, movies, videos, and music are all available. The resources available through the Internet are virtually unlimited. To access the Internet, a user needs only a service provider, a modem for transmission, a web browser (a program that allows the user to navigate, view, and interact with the Internet), and some basic knowledge of how the system works. Many Internet providers give free access time to allow a user to try the system.

Wireless technology has made the Internet and many of its functions completely portable. Business can be conducted from nearly anywhere in the world, and transactions, messages, and many other types of communications can be accomplished from remote locations.

Figure 25-48. Royal Caribbean uses the IBM NetVista Internet Appliance to provide in-cabin access to the Internet. (International Business Machines Corporation)

Review Questions for Section 25.5

1. What is a LAN?
2. What does a file server provide?
3. What is a WAN?

Figure 25-49. The IBM NetVista Kiosks system is used to preview thousands of CDs, DVDs, and console games on-line. (International Business Machines Corporation)

Summary

1. The overall design of a computer system is a collection of input, output, central processing, and storage devices.
2. The computer motherboard, case, and power supply must conform to the same form factor.
3. The central processing unit (CPU) processes all software programming instructions and performs data manipulation tasks.
4. Cache is a small amount of computer memory incorporated into a computer device.
5. There are two main classifications of memory based upon their ability to retain program data and instructions after power is no longer supplied to the chip: random access memory (RAM), and read only memory (ROM).
6. There are two main classifications of RAM: static RAM (SRAM), and dynamic RAM (DRAM).
7. DDR-DRAM is twice as fast as DRAM.
8. The BIOS contains the programming necessary to start the computer as well as perform basic diagnostics, and interface software with hardware.
9. Chipsets allow devices to communicate and exchange data without using the CPU.

10. The file allocation table (FAT) is used to locate files stored on magnetic storage devices such as hard disk drives and a floppy disk.
11. The new technology file system (NTFS) was an improvement over the FAT system of file storage.
12. The CD universal disc format (UDF) file system allows for bootable CD-ROMs and is intended to be a way to universally exchange data between different CD technologies.
13. A modem is used to convert digital signals into analog signals thus supporting computer communications over existing telephone lines.
14. The UART is the main component of a modem.
15. Inkjet printers use piezocell as well as thermo resistors to force ink from ink cartridges.
16. Laser printers use the electrophotographic process (EP) for developing images.
17. Computer CRT displays use the same principle as television CRTs but at much higher screen resolutions.
18. A LAN is a collection of computers connected to a file server for sharing data, and software.
19. The Internet is a system using both hard wire and wireless technology connecting together computers all over the world.

Test Your Knowledge

Do not write in this text. Place your answers on a separate sheet of paper.

1. What are the three major areas of computer failure?
2. What two PC hardware devices are prone to failure?
3. What causes the majority of computer problems and failures?
4. Electrical components that are readily changed in the field are referred to as _____.
5. Of what four major systems are computers comprised?
6. What is a form factor?
7. What voltage levels are provided by a typical ATX power supply?
8. Name six devices commonly found on a system motherboard.
9. What is the most common slot style on motherboards today?
10. What is cache?
11. What one major factor is used to compare CPU performance?
12. What other factors influence CPU performance?
13. What are two broad classifications of memory based upon the user's ability to readily change its contents?
14. What are two broad classifications of memory based upon their electronic construction?
15. Approximately how much faster is DDR-DRAM than DRAM?
16. What is the name of the motherboard component that contains the POST?
17. What is P-n-P technology?
18. What is a motherboard chipset?
19. What are the two most common magnetic storage media?
20. How does a floppy disk keep track of file locations?
21. What is the name of the file system designed as an improvement over FAT?
22. What mass storage device uses laser diode technology?
23. What computer device supports communication over a telephone line?
24. What chip is used in a modem?
25. What is a combination hardware and software component, such as a BIOS chip, referred to?
26. What is the name used for a software program that is required for communication between a hardware device and a computer system?
27. What is a common cause of computer system lockups?
28. What printer technology utilizes a piezocell or thermo resistor?
29. What is the technical name for the laser printer process?
30. What component is used to capture the light from an image and convert it into a digital signal?

For Discussion

1. What do you think computers will be like in 2050?
2. Discuss the ways in which computers affect your daily life.
3. Is technology affecting our lives in a positive way? Explain your answer.
4. Research a specific event or invention in the evolution of the computer.
5. Research five different computer operating system commands, and explain what each one does.
6. Locate another PC information site in addition to the ones listed in the textbook, and share information about it.
7. What types of data could be stored on a personal computer hard drive to warrant an expensive repair involving hard disk drive electronic repair?

Microcontrollers

Objectives

After studying this chapter, you will be able to:
- Define microcontroller.
- Explain the difference between a computer and a microcontroller.
- Explain the function of main microcontroller parts.
- Identify various types of ROM.
- Name the general types of programming languages.
- Name some common components found on a microcontroller module.

Key Words and Terms

The following words and terms will become important pieces of your electricity and electronics vocabulary. Look for them as you read this chapter.

arithmetic/logic unit (ALU)
assembler program
assembly language
editor
electrically erasable programmable read only memory (EEPROM)
erasable programmable read only memory (EPROM)
high-level language
interpreter
library
machine code
machine language
macro
masked ROM
microcontroller
microcontroller interface
object code
opcode
operand
port
program counter
programmable read only memory (PROM)
register
source code
stack
stack pointer
watchdog timer
word size

This chapter introduces you to the microcontroller. The microcontroller is a programmable integrated circuit. This means that the microcontroller can be programmed to perform a set of specific functions. The microcontroller is, therefore, the merging of programming languages and digital logic components. Complex controls that once required thousands or even millions of individual components are duplicated in a single chip. The information in this chapter builds a bridge of technical knowledge between software programs and digital ICs. You will discover how computer software can be used to control integrated circuit devices to achieve the same results as control systems that require many discrete devices. The microcontroller has dramatically changed the way control circuitry is designed.

26.1 OVERVIEW OF THE MICROCONTROLLER

A ***microcontroller*** is an integrated circuit (IC) that can be programmed to perform a set of functions to control a collection of electronic devices. Being programmable is what makes the microcontroller unique. Often thought of as a "tiny computer" on a single chip, the microcontroller is used in many applications and can be found in almost every electronic device we encounter daily. In this section, you will learn what makes the microcontroller comparable to the computer. You will also learn why the microcontroller has replaced many relays and solid-state devices in industry.

Comparison and Contrast of the Microcontroller and the Computer

You will find that many of the terms you will learn in this chapter can also be applied to computers. This is because the microcontroller and the computer share many common components:
- A microprocessor to process instructions.
- Memory locations to store data and programming information.
- Input/output facilities to move data between itself and another device.

See **Figure 26-1**. Although the microcontroller and the computer share many common components, there are many distinct differences between them. The major differences between a computer and a microcontroller are

- The limitations of the microprocessor.
- The amount of memory available for data and program manipulation.
- The type of storage available to permanently store a program.

A microprocessor inside a computer is more powerful and more versatile than that in a microcontroller. The computer's microprocessor allows the computer to run a variety of programs and applications. A single computer can be used to perform a variety of functions such as word processing, playing a DVD, playing an interactive game over the Internet, monitoring security systems, and making a CAD drawing.

A microprocessor in a microcontroller has a limited instruction set that allows it to run a specific application. In other words, the microcontroller can only understand a limited number of commands, and these commands are specific to the functions the processor is designed to handle. For example, a microcontroller may be designed specifically to monitor and control a fuel injection system on an automobile. In fact, every car manufactured today has many microcontrollers incorporated into its electrical system—antilock brakes, fuel control, air conditioning, heating, and keyless security locks. Each microcontroller on a vehicle is designed to handle a specific function.

The microcontroller, like the computer, uses RAM to store programs and data temporarily. The amount of RAM built into a microcontroller is minimal and is usually enough for its intended use. The RAM in a computer is ample enough to run a variety of memory intensive applications.

All computers and microcontrollers must have some sort of ROM that contains a permanent set of instructions so that they can communicate with other devices. The microcontroller stores in its ROM the program it needs to perform a specific job. The computer has a ROM, called the BIOS, that contains BASIC programming functions. The computer stores the rest of its programs, operating system and applications, in other storage devices such as a hard drive. A larger capacity for program storage makes the computer more versatile and powerful than a microcontroller and allows it to run many different programs.

Although computers are versatile and powerful, the microcontroller has its advantages. Microcontrollers are small and can fit inside other devices like an appliance or a vehicle. Microcontrollers cost less to produce, and they

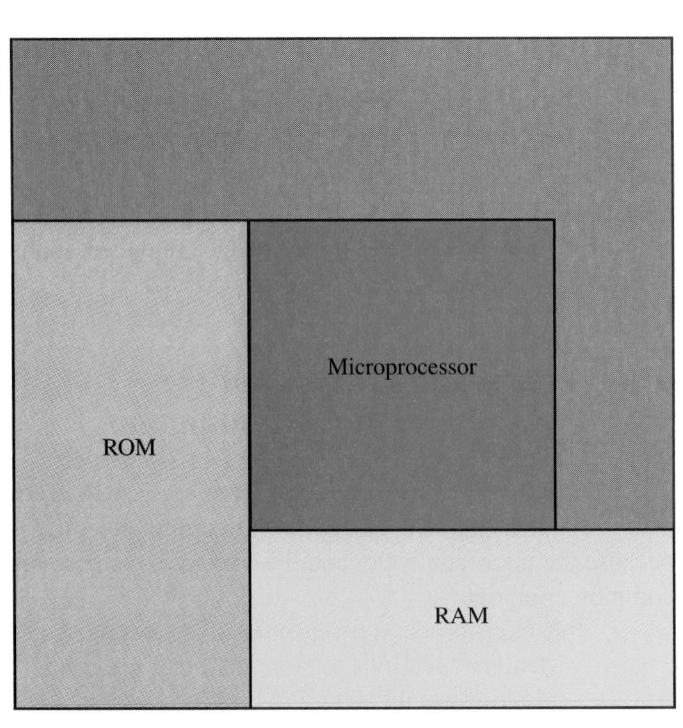

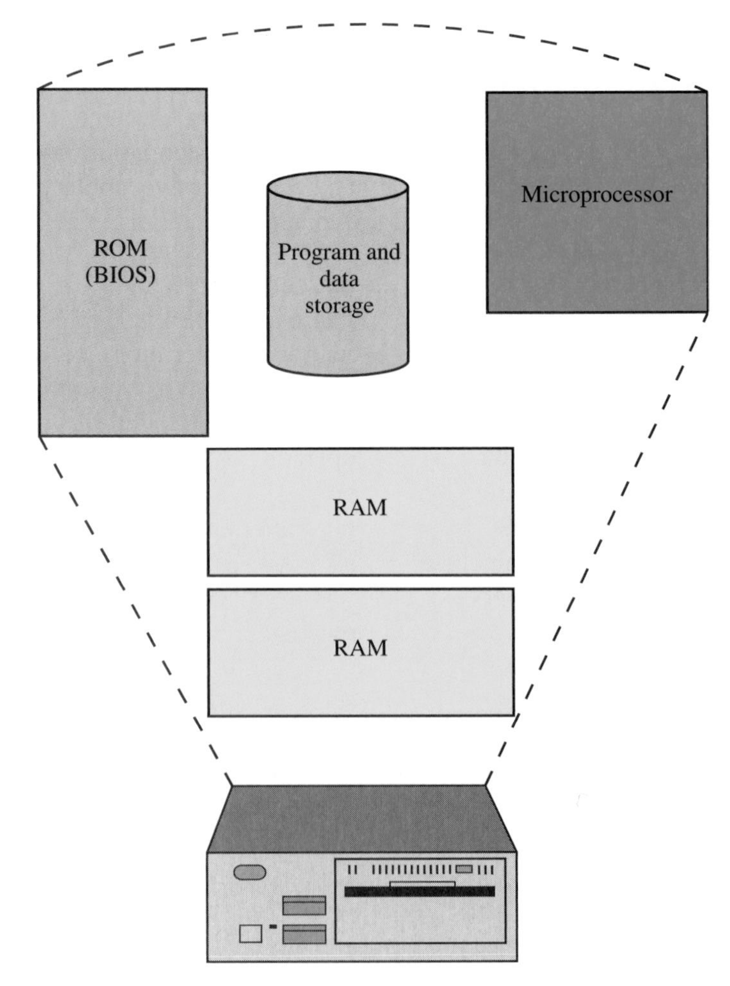

Figure 26-1. The microcontroller and the computer share similar components.

consume less power. Microcontrollers are used in almost all types of electronic equipment, from coffeepots to laser printers. They are even incorporated into computer systems. They can be found in floppy drives, CD-ROM drives, and video cards. A microcontroller is often embedded into the electronic circuit boards of these computer devices. A microcontroller is usually a single chip, but when incorporated into a large control system, it is referred to as an embedded controller, **Figure 26-2**. An embedded controller often depends on other components in the system, such as additional memory, to perform its function.

Microcontrollers are also used heavily in industry. They have taken the place of relays, solid-state devices, and other discrete components.

Figure 26-2. The microcontroller inside this refrigerator contributes to energy and cost savings by monitoring temperature and controlling the compressor. Courtesy of Motorola, Inc.

Advantages of Using Microcontrollers in Industry

There are a number of advantages to using microcontrollers in industry. Some of the major advantages of microcontrollers are that they are reusable, dependable, cost effective, and energy efficient.

Reusable

The typical microcontroller is programmable, which means it is reusable. This is especially advantageous for prototyping control circuitry. When developing a complex control system, it is not unusual for it to fail when first applied. As a matter of fact, a complex control project may need to be rewritten and/or rewired many times before it meets design expectations. The fact that the control circuit can be modified by programming rather than rewiring is very advantageous for fast project prototype development.

Dependable

Integrated circuits, such as the microcontroller, are much more dependable than relays. Before microcontrollers, control circuitry relied on many electromechanical relays and timers to control the system. Relays depend on electromagnets to move armature and contact parts, so they eventually wear out due to mechanical friction.

Relays are also susceptible to damage caused by dust, dirt, corrosion, rust, insects, and other contaminants that can interfere with the moving parts. Microcontrollers have no moving parts. This provides a much higher rate of reliability. Relays and high-power transistors can be incorporated for final applications to motors, but the actual timing and control logic does not need to rely on the mechanical action of relays.

Cost effective

Microcontrollers can be produced at lower costs than their electromechanical predecessors. Also, microcontrollers can be reprogrammed if the designed application does not work correctly or if the application for its use changes.

Energy efficient

Because the majority of the circuitry is made from integrated circuits, the energy cost of using a microcontroller is much less than if using individual components of a relay-type logic circuit. Relay logic uses numerous relays wired in series and parallel to form control circuit conditions similar in function to logic gates. A microcontroller consumes less electrical energy than conventional electromechanical devices.

Disadvantages of Using Microcontrollers in Industry

There are a few disadvantages to using microcontrollers. The two most prominent disadvantages are the need for skilled programmers and the sensitivity of the controllers to electrostatic charges.

Programming complexity

Special skills are required to program the microcontrollers. This requires a higher level of training for some personnel. In addition, there are many different programming languages to choose from. This can lead to a compatibility problem when attempting to merge two dissimilar systems into one control system.

Electrostatic sensitivity

Most microcontrollers are composed of complimentary metal-oxide semiconductor (CMOS) integrated circuitry. CMOS can be damaged easily by a static charge. Static precautions must be strictly obeyed.

Review Questions for Section 26.1

1. Name three differences between a computer and a microcontroller.
2. What types of devices are controlled by a microcontroller?
3. What is an embedded controller?
4. What are four advantages of using microcontrollers in industry?
5. Name two disadvantages of using microcontrollers in industry.

26.2 MICROCONTROLLER COMPONENTS

The microcontroller consists of thousands of digital circuits. These digital circuits are combined into areas that provide specific functions. Examine **Figure 26-3**. The simplified block diagram illustrates how the major sections inside the microcontroller work together to process the program instructions. The parts of the microcontroller are used to save data and programs, perform math and logic functions, and generate timing signals. The different areas are connected by a bus system. The bus system contains tiny parallel circuits that carry the digital pulse patterns from section to section. The ROM stores the program required for the microcontroller to function. The ROM controls how the chip components operate and how data and instructions flow through the chip.

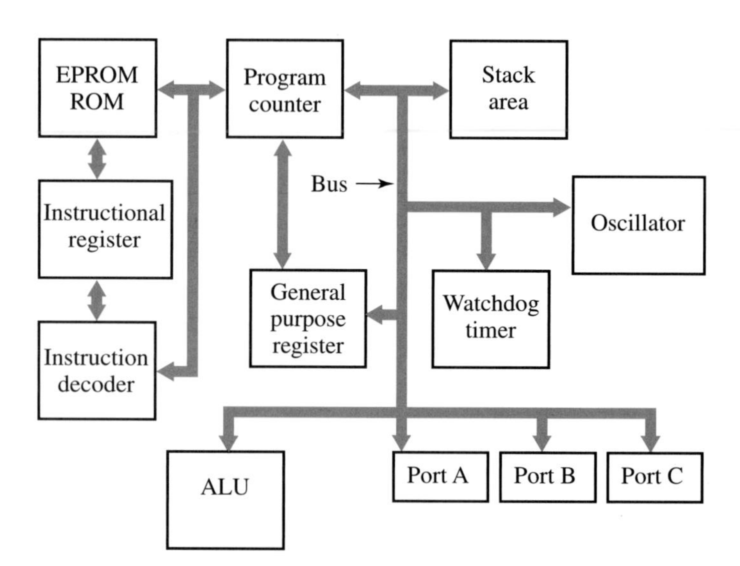

Figure 26-3. A simplified block diagram of a typical microcontroller. All the major components are connected together by a bus system. Data and instructions are moved around the chip via the bus.

Ports and Registers

There are many different types of registers and ports found inside a microcontroller chip. ***Ports*** and ***registers*** are special memory locations dedicated to a specific function such as a hardware location or a place to manipulate data. Some ports correspond to the input and output pin assignments of the chip. By placing a one or zero into a specific port address, you can change the pin assignment of the microcontroller from an input pin to an output pin. Ports can also contain information sent to the pins on the microcontroller chip from sensors or switches. The contents can then be stored in memory or remain in the port.

A register can be used to hold the contents of data being manipulated. For example, when performing a mathematical operation, the data being manipulated must be stored in two different registers and the result placed in a third register. See **Figure 26-4**.

In the illustration, you see three registers labeled as A, B, and C. The content of register A and the content of register B are added together and then stored in register C. Another typical use of a register is to store a value to be compared to another register value. For example, compare the content of register A to the content of register B. If the values are identical, an action is initiated such as turning on a valve, activating a door switch, or turning on an alarm. The contents of the two registers, A and B, could compare a temperature rise to a preset value, or the distance moved by a robotic arm. As you work with programs and memory, the use and applications for registers will become more apparent.

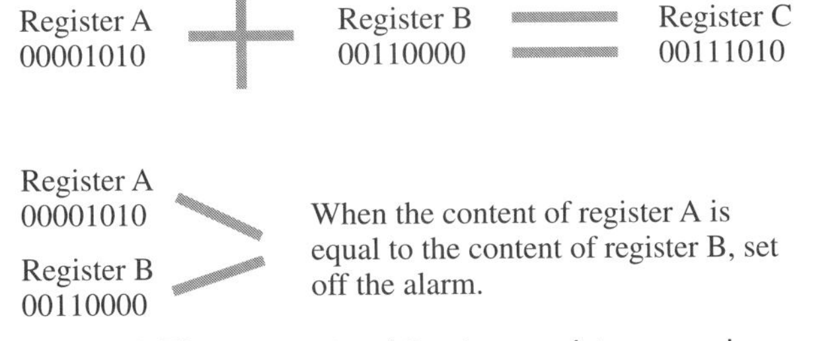

Figure 26-4. The contents of the two registers can be compared to determine if an action should begin.

Stack Pointer and Program Counter

The area in memory that is used to store data and program information is called the ***stack***. The most common use of the stack is to store the program address of the next instruction to be executed. Since the processor processes commands sequentially, one after the other until complete, it relies on the stack to temporarily store the address of the next instruction in case the main program is interrupted. A special device called the ***stack pointer*** keeps track of the last stack location used while the processor is busy manipulating data values, checking ports, or checking interrupts. Look at **Figure 26-5**.

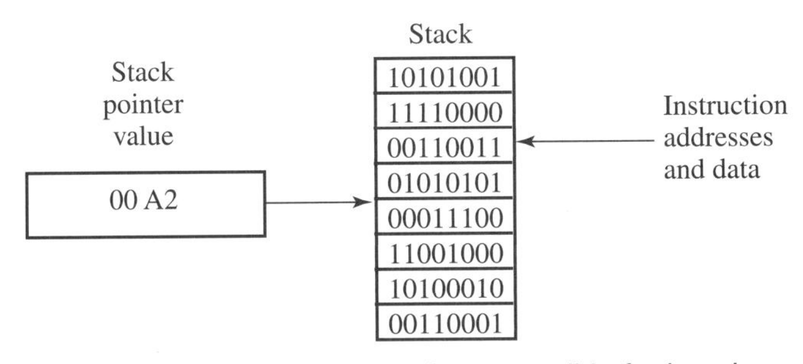

Figure 26-5. The stack pointer is responsible for keeping track of the sequence of memory locations, referred to as the stack, while the computer is manipulating data and checking interrupts.

Interrupts are, as the name implies, a way to interrupt the processor while it is processing instructions and data to perform a different task. The interrupts come in two varieties: hardware interrupts and software interrupts. An example of a hardware interrupt is when input pin number 3 receives an electrical signal from a push button, and the microprocessor stops what it is doing to respond to the signal. While it is responding to the push button signal (a hardware interrupt), the contents of the ***program counter*** are pushed onto the stack. The stack pointer bookmarks the location of the last item pushed onto the stack. When the processor finishes servicing the interrupt, the counter address is popped from the stack and returned to the program counter, and the microprocessor continues processing the instructions in order.

ALU

The ***arithmetic/logic unit (ALU)*** performs common mathematical and logical operations on data. The logical operations are similar to the digital devices you have studied earlier in this textbook. Some typical logic functions are similar to the ADD, OR, NOT, NOR, and NAND digital gates.

Oscillator

The oscillator is a complex digital device that requires a steady digital pulse for timing. All of the separate functions are controlled by one central timing system. The timing pulse provides the basis for proper sequence of all the separate sections of the microcontroller chip. The source of the steady pulse rate is the oscillator circuit.

Watchdog Timer

A specialized program often found as part of the microcontroller is called a watchdog timer. The ***watchdog timer*** is designed to prevent the microcontroller from halting or "locking up" because of a user-written program. Remember that a processor processes instructions step-by-step. It will wait until one instruction is completed before processing the next. A problem can arise if a command is issued to a processor that cannot be accomplished. For example, a program is written by a user and contains an instruction calling for checking the input value of I/O pin 36. If there is no I/O pin 36, the microcontroller will wait forever for input. It will not process the next command until it receives input on pin 36. Without the watchdog timer, the microcontroller would halt and thus fail to process further instructions.

The watchdog timer uses a routine that is based on timing. If a program has not been completed or repeated as a loop within a certain amount of time, the watchdog timer issues a reset command. A system reset sets all the register values to zero. The reset feature allows the controller to recover from the crash. It releases the program and sets the controller to start over again.

Memory

As stated earlier in this chapter, microcontrollers and computers use various types of memory. These types of memory are usually referred to by acronyms. The two broad classifications of memory are random access memory (RAM) and read only memory (ROM). These two types of memory are classified according to their function as memory units.

You have already learned about RAM in detail in Chapter 25—Introduction to the Personal Computer (PC). Random access memory is used in the microcontroller just as it is in the computer, to temporarily store programs and data. It is volatile memory, which means it loses its contents when electrical power is no longer applied to the memory chip. Power can be lost due to power failures or by simply turning off the power switch.

Read only memory is used to permanently store a program or data and retain the information even when the power is disconnected from the unit. It is described as nonvolatile memory. In Chapter 25 you learned about ROM only as it applied to the computer's BIOS. In this chapter, we will take a closer look at the many types of ROM.

Programmable Memory

ROM is classified as to how it is constructed and how the program code inside the chip can be altered. ***Masked ROM*** (usually referred to as just ROM) is a special type of memory that is permanently programmed during the manufacturing process. Masked ROM memory cannot be reprogrammed. To change the program in a system using masked ROM, you must use an entirely different masked ROM chip.

Programmable read only memory (PROM) is memory that is programmed after it is manufactured. It is manufactured with thousands of empty memory cells. A program writer "burns" a pattern into the empty cells, forming a program in the blank PROM chip. Like masked ROM, a PROM chip program is permanent for the life of the chip.

Erasable programmable read only memory (EPROM) is a special type of PROM that can be reprogrammed many times. When manufactured, a small window is left on the chip. Any program entered into the chip can be erased at a later time by shining an ultraviolet light through the window on the chip. The chip can then be reprogrammed using a programming device.

Electrically erasable programmable read only memory (EEPROM) is a type of memory chip that can be erased electrically and then reprogrammed. It is commonly used for the computer BIOS chip.

A microcontroller is usually constructed with a PROM on the chip itself containing the basic set of instructions for running the microcontroller and its internal parts. This program is permanent and not alterable. In addition to the basic ROM, another section of memory is used to store programs written by the user. It typically contains a small EEPROM memory area. If larger program storage is required, additional EEPROM chips can be connected to the microcontroller.

Review Questions for Section 26.2

1. The special function areas of a microcontroller chip are connected using an _____ system.
2. The _____ keeps track of the data held in the stack area.
3. Math and logic functions are provided by the _____.
4. The _____ circuit provides a steady pulse for timing.
5. A(n) _____ prevents the microcontroller from crashes caused by bad programming.

26.3 PROGRAMMING LANGUAGES AND TERMINOLOGY

A variety of languages are used to program controllers and computers. These languages range from the most basic level of machine language to a wide selection of high-level languages. Some languages used to program microcontrollers are C, C++, BASIC, Visual BASIC, Quick BASIC, and numerous proprietary languages such as PBASIC for the Parallax microcontroller. Software programs especially written for programming microcontrollers are usually referred to as ***editors***. Look at the illustration in **Figure 26-6**. Use this drawing as a reference while we discuss the various languages used to program microcontrollers.

Machine Language, Assembly Language, and High-Level Languages

One language that can be used to program a microcontroller is machine language. ***Machine language,*** or ***machine code,*** is a computer language constructed of ones and zeros that represent binary codes and digital voltage pulse levels. The computer understands machine code without further conversion. There is no form of code that is closer to the machine itself, hence the name machine code. Machine code is also referred to as *object code* or *executable code*. Machine code matches the type of processor with which it is communicating. For example, an Intel processor cannot use a machine code written for a Motorola (Apple) processor, or vice versa.

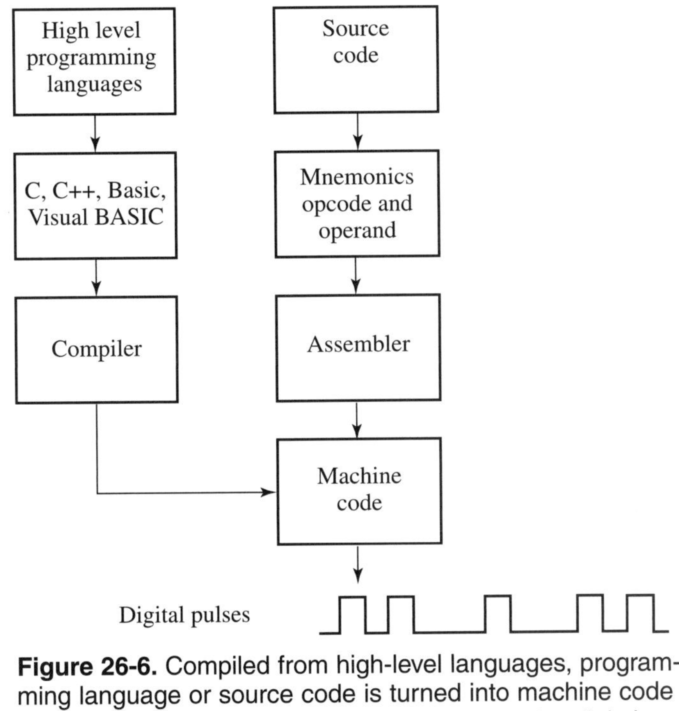

Figure 26-6. Compiled from high-level languages, programming language or source code is turned into machine code by a compiler. Machine language is closest to the digital signals found moving through the system circuitry. Binary ones and zeros can be used to illustrate the digital pulse patterns. Compilers and assemblers convert high-level programming languages and assembly language into machine language.

Machine code is difficult to learn and inefficient for writing programs. The main advantage of using machine language programs is that they run more quickly than other types of programming.

Machine language programming was the original way to program a microprocessor. Programming in this language was very tedious, and it took many hours to write even the simplest program. Soon, a faster programming language was invented called an assembly. ***Assembly language*** uses a series of words, called mnemonics, combined with hexadecimal data values to create a program. After the program is written, it is assembled. An ***assembler program*** converts the series of instructions into a machine code that can be read by the processor. Programs written in assembly language are very compact, efficient, and fast. However, simple programs still can take hours to write using assembly language.

Most programmers write programs using ***high-level languages*** such as C, C++, BASIC, Visual BASIC, and Quick BASIC because it is much easier and faster. Very large programs that would be almost impossible to write in assembly language can be written using a high-level language. The high-level language can then be assembled into machine language code by using a program called a *compiler*.

The main differences between using a high-level language and using assembly language is the amount of code stored after being converted into machine language, and the speed of its execution. Assembly code is difficult and time-consuming to program in, but the code uses less memory and executes faster than high-level code.

The assembly code is much more exacting than high-level programming languages. For example, when using an assembly, you must indicate the exact memory location in which to place the data, and each command must be in sequence followed by the proper amount of data bits. This is more work for the programmer. A high-level language is not as restrictive. The programmer must simply indicate that data needs to be stored in a file. However, when high-level language is compiled into the low-level machine language, allowances must be made for the fact that no specific memory location was specified. The program must use an additional set of commands to locate a free block of memory that does not conflict with another data block. Higher-level languages are also not as restrictive about putting commands in exact sequence for execution. Assembly code must be in exact sequence, or it will lock up (crash) the system when implemented.

Programming Terminology

One of the most difficult aspects of programming is learning all of the new terminology. Many new words have been created, and many common words, such as bit, word, and library, have been given new meanings. Knowledge of these terms will help you understand the manuals that come with microcontrollers.

The size of a normal group of bits processed at one time through a computer system is referred to as its ***word size***. The word size typically matches the bus width of the chip. Some common word sizes are 8, 12, 14, 16, 24, 32, and 64 bits in length. A word usually contains an instruction and some data to be acted on.

The ***opcode*** is the part of a word that contains the instruction to be carried out. Instructions are typically followed by data referred to as the ***operand***. This is a repeating sequential pattern used in programming. Together, the opcode and the operand combine to form the ***source code.*** Source code is simply another name for the program written by the programmer. The source code is then converted into machine code by an assembler or a compiler. When source code is assembled or compiled, it becomes what is known as ***object code*** and is saved as a file. Object code is also called *executable code* or *machine code*. Source code can also be run directly by an ***interpreter***. The interpreter runs the source code without compiling the code into machine code, **Figure 26-7**.

Many times programmers will save the programs they have written so that they can use the code again on another control system. As you work with programmable devices, you will notice how much of the code can be reused for another project. A ***library*** is a collection of code modules. For example, the robot project that is covered later in this chapter can be programmed to perform in a number of ways. It can be programmed to run while connected to a PC. The keyboard arrow keys can be used to control the robot's movement. The robot can also be programmed to follow a line drawn on a piece of cardboard without being connected to a PC. In addition, the robot can be programmed to use infrared sensors to detect objects in front of it such as a wall. When the object is detected, the robot makes a 90-degree right turn. In every case, you can reuse some of your code blocks rather than rewriting the complete program from scratch. The parts that are reusable can be saved to a code library. Libraries are sometimes referred to as *code modules*. Some software systems allow you to call the code modules out of a storage memory from the main program.

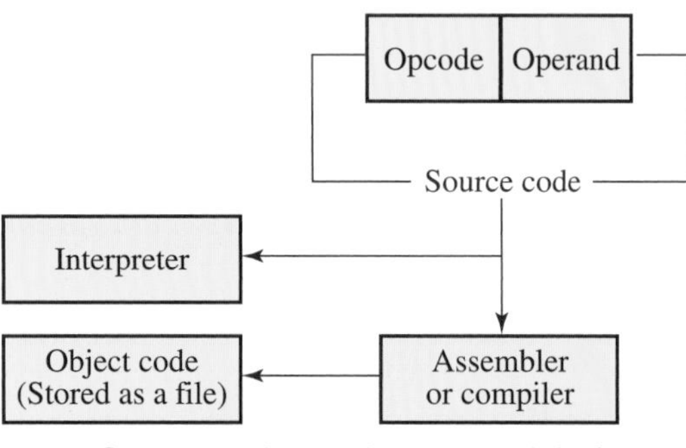

Figure 26-7. Source code can be run straight from an interpreter or from object code.

A ***macro*** is a collection of assembler instructions represented by a single word. A macro assembler converts macros into machine language. The term *macro* and *mnemonic* are often used interchangeably when talking about programming. Mnemonics, discussed earlier with assembly language, are instructions that can be converted directly into machine language. The mnemonic itself looks like a short abbreviation of the actual command being issued. For example, MOV might be used to represent the instructions to move data.

Review Questions for Section 26.3

1. The most basic language, which is composed of ones and zeros, is called _____.
2. Name three high-level languages.
3. Which portion of the source code contains the data, the opcode or the operand?

26.4 THE MICROCONTROLLER MODULE

A microcontroller can consist of a single chip, or it can be composed of several components collected together on a circuit board to function as a single unit. When a microcontroller is mounted on a circuit board with other components, it is sometimes referred to as a *module* or a *microcontroller board*.

A microcontroller module typically consists of a microcontroller, a power source, an interface for connecting to a programming device, I/O ports, and additional memory. A power source is required to power the microcontroller and any accompanying components located on the printed circuit board. An interface is needed to communicate with the controller. The ***microcontroller interface*** is usually an older serial connection or a universal serial bus (USB) connection matching the ones found on most PCs. A set of input/output (I/O) ports is used to send and receive signals from the devices the microcontroller is designed to control. The microcontroller's I/O pins can be programmed as output or input pins. When programmed as an output pin, each pin can output digital signals. When programmed as an input pin, each pin can receive digital signals. Devices commonly controlled through the I/O ports are relays, power transistors, servos, stepper motors, LEDs, and solenoids. Devices that commonly send electrical signals to the I/O ports are photocells, piezocells, thermistors, and thermocouples. LEDs are often used to indicate the presence of power or that communication is taking place, **Figure 26-8**.

Depending on the microcontroller's intended use, you might find timing oscillators. The oscillator circuit can be a separate component or incorporated into the microcontroller chip. When constructed separately from the chip, a crystal or ceramic resonator is used as well as capacitors and resistors. See **Figure 26-9**. One advantage of using a separate oscillator circuit is that it can produce a number of frequencies rather than a fixed frequency. By having the ability to produce a wide range of frequencies, the microcontroller can match the frequency required of a large control circuit in which it may be embedded.

Additional components such as digital-to-analog and analog-to-digital converters can be used to change the digital pulses into analog signals. If you need to drive a device requiring a higher output current, you can simply

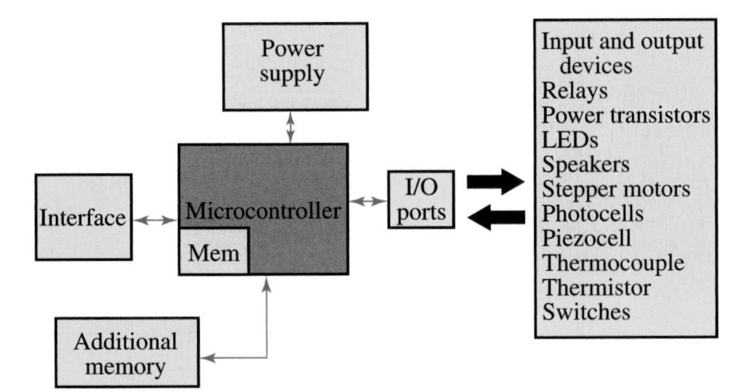

Figure 26-8. The basic components found on a microcontroller module and the typical devices to which the microcontroller interfaces.

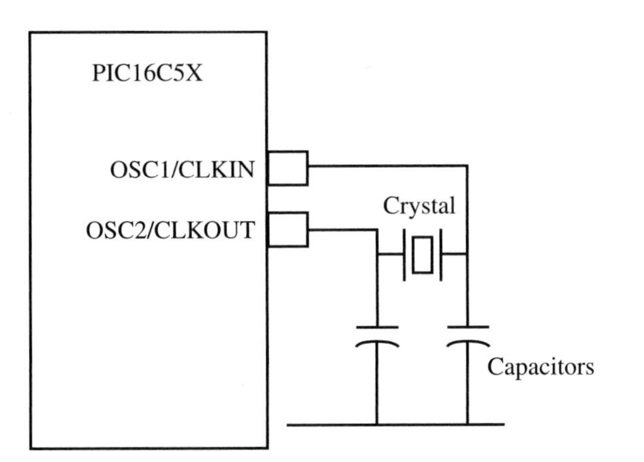

Figure 26-9. The PIC exterior oscillator circuit is a simple crystal with two capacitors to form a tank circuit. The exact values of the crystal and capacitors will vary depending on the desired frequency and the exact model of PIC being used.

incorporate a small mechanical or solid-state relay between the output pin and the load device.

A microcontroller chip usually contains a relatively small amount of programmable memory. If a microcontroller is going to be used to run a large or complicated program, or if much data is to be stored, additional RAM and ROM can be mounted next to the microcontroller. A reset button can be added to clear the processor if a program fails.

Microcontroller Relay Circuit

Many times, the device or devices to be controlled by a microcontroller far exceed the electrical power limits of the controller. A method must be used to increase the power output capabilities of the microcontroller. This can be accomplished many different ways. One common way is using mechanical or solid-state relay devices. **Figure 26-10** shows an application for controlling a device when the electrical requirements far exceed the output power of the microcontroller I/O pins.

To drive a load of higher current, a special circuit must be built and inserted between the microcontroller module and the load device. The circuit in the schematic is simple in design. The output pin of the microcontroller connects directly to an optic coupler. The optic coupler consists of an LED and a photosensitive transistor. The optic coupler is designed to prevent higher voltage levels and short circuits in the load device circuit from damaging the sensitive microcontroller chip through a backfeed situation. This could happen if a typical transistor was used.

When the LED inside the optic coupler is energized by the output of the microcontroller, the light forces the photosensitive transistor into conduction, thus completing the circuit to the relay. When the relay energizes, the relay contact closes and completes the circuit to a heavy-duty device such as an appliance, some lighting, or a welder. Note the fact that the heavy appliance load can use any type of power supply, such as 120 Vac, 240 Vac, 12 Vdc, and 48 Vdc.

The components, such as the transistor, are sized according to the relay coil requirements. The diode placed in parallel with the relay coil is used to prevent an inductive kick from damaging circuit components. As you recall from the study of inductance reactance, when a coil is energized or de-energized, the induction voltage created can be quite high. This can damage circuit components such as transistors. The diode prevents the high level by short-circuiting the induction created by the relay coil. When ac is used as the applied voltage in the load circuit, capacitors can also be incorporated into the circuit to prevent damage induced by the ac circuit. In place of a mechanical or solid-state relay, high power transistors, SCRs, or even a TRIAC can be used to increase the output capabilities of a microcontroller.

Microcontroller Interface

Most microcontrollers interface with an IBM compatible computer through the serial or parallel port. See **Figure 26-11**. This figure shows the connections for a standard 9-pin serial port PC connection and the PIC16C5X microcontroller, from Microchip Technology. The 9-pin serial port connection is also called a DB-9. The Pins 1, 2, 3, and 4 on the microcontroller are tied directly to pins 2, 3, 4, and 5 on the 9-pin serial port connection. Note that pins 6 and 7 are connected together on the serial port of the PC. This allows the PC to

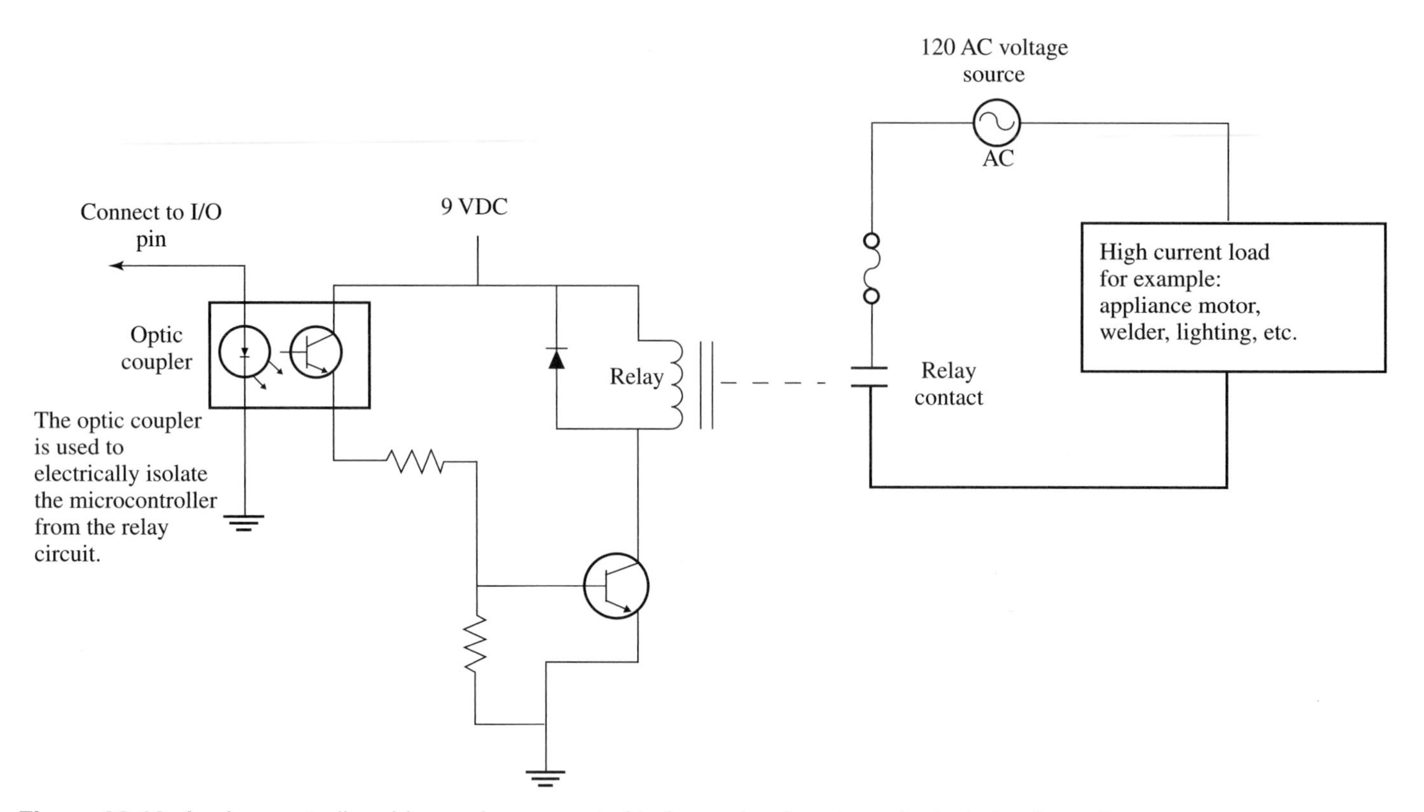

Figure 26-10. A microcontroller chip can be connected to heavy loads using a typical circuit as shown.

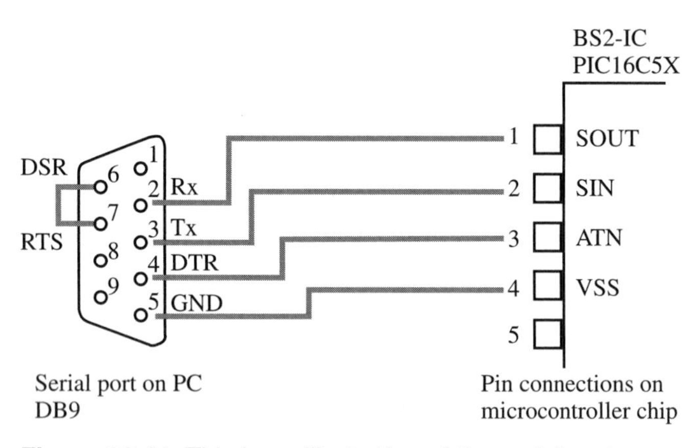

Figure 26-11. This is an illustration of the serial port connections using a BS2-IC. The connection for the BS1-IC is different. Always check the technical specifications of the microcontroller chip.

automatically detect communications on the port. Also, be aware that this illustration only refers to a connection to a PIC16C5X microcontroller. It is important to always check the technical specifications sheet of the particular microcontroller you are using. Microcontroller pins for different models have different pin functions. In addition, the parallel port, as well as the USB port, can be used for some microcontroller communications.

Review Questions for Section 26.4

1. A microcontroller that consists of several components is called an _____.
2. Name four major parts of a microcontroller module.
3. What is one advantage of using a separate oscillator circuit?
4. Which ports found on an IBM compatible PC can be used to download a program to a microcontroller?
5. Do all microcontrollers use the exact same pin location to connect to a serial port?

26.5 GETTING STARTED WITH MICROCONTROLLERS

One of the most used microcontrollers by students and hobbyists are the PICmicro® microcontrollers by Microchip Technology. There are a variety of educational kits that incorporate the PICmicro chips. Two of the most popular kits are the BASIC Stamp 1 & 2 starter kits from Parallax and the PIC BASIC Compiler Bundle from microEngineering Labs.

What makes the BASIC Stamp microcontroller unique is that it comes with a built-in interpreter. As you recall, an interpreter reads one line of code at a time and processes the command. Therefore, you do not need to compile the source code to run the program. You can program the BASIC Stamp with Parallax's own BASIC language called PBASIC. The BASIC Stamp is sold as a complete package with the BASIC Stamp controller, carrier board, Stamp Software, and cable or as separate components. The cable for the BASIC Stamp 1 connects to the parallel port of your PC, and the cable for the BASIC Stamp 2 connects to the serial port of your PC; although, the BASIC Stamp software communicates serially through both these ports as it downloads the program to the BASIC Stamp.

The PicBASIC Compiler Bundle from microEngineering Labs comes with the PicBASIC Compiler, a software package that compiles your BASIC program and downloads it to the PICmicro microcontroller. Included in the Compiler Pro Bundle is the PicBASIC Compiler Pro software that is compatible with BASIC Stamp 2 commands. Both kits come with the EPIC Plus Programmer and a 25-pin cable that connects the programmer board to the parallel port of your PC.

Review Questions for Section 26.5

1. What is one the most popular microcontrollers used by students and hobbyists?
2. The language used to program the BASIC Stamp is ______.

Summary

1. A microcontroller is an integrated circuit that can be programmed to perform a specific function.
2. When a microcontroller is part of a larger control assembly, it is often referred to as an embedded controller.
3. A microcontroller is perfect for prototyping because it can be easily reprogrammed.
4. A watchdog circuit is designed to prevent a microcontroller from locking up.
5. Machine language is the lowest-level programming language and the language closest to the hardware.
6. Most programmers use high-level languages to write programs for computers as well as microcontrollers.
7. Computers are used to program microcontrollers, and the program is transferred to the microcontroller using serial, parallel, or USB ports.

Test Your Knowledge

Please do not write in the text. Place your answers on a separate sheet of paper.

1. Explain the difference between a microcontroller and a personal computer.
2. What is the purpose of the watchdog timer?
3. What does the acronym PROM represent?
4. What does the acronym EPROM represent?
5. What does the acronym EEPROM represent?
6. The programming language constructed of ones and zeros that represent binary codes and digital voltage pulse levels is called _____ language.
7. Name three or more high-level programming languages.
8. A(n) _____ is used to convert a high-level language into machine language.
9. A word is a combination of _____ and _____.
10. Word size is typically equal to the _____ width of the chip.
11. _____ and _____ are special memory locations dedicated to a specific function such as an I/O port.
12. A microcontroller that consists of several components is called an _____.

For Discussion

1. Discuss the differences between a computer and a microcontroller and describe some applications for each.
2. Discuss the different kinds of read only memory.
3. Describe some of the functions of a microcontroller.
4. Describe some applications that you can think of for a microcontroller.
5. Discuss your experience with the exploration of technical web sites and your future plans for accomplishing this important task.

Closeup of an eraseable programmable read only memory (EPROM) module. Shining an ultraviolet light through the EPROM's tiny window can erase a program that has been written to the EPROM.

Career Opportunities in Electronics

Objectives

After studying this chapter, you will be able to:

- ❑ Assess your interest in electronics careers.
- ❑ Distinguish between a goods-producing and a service-producing business.
- ❑ Discuss a variety of electronics careers and their educational requirments.
- ❑ Define entrepreneur and entrepreneurship.
- ❑ List career information sources.
- ❑ Outline ideas for a successful job search.
- ❑ List the qualities needed to be successful in your career.

Key Words and Terms

The following words and terms will become important pieces of your electricity and electronics vocabulary. Look for them as you read this chapter.

apprenticeship	professional
aptitude	semiskilled worker
engineer	service-producing business
entrepreneur	skilled worker
goods-producing business	technician

27.1 FINDING A CAREER

How do you want to earn a living after you complete your education? What vocation will you choose to assure an interesting and rewarding future? What type of education will you need to prepare for your chosen field of study? Do you have the interest, desire, and ability to use the educational opportunities within your reach?

These are questions only you can answer. There are many people, however, willing to help you answer these questions. Parents, school guidance counselors, teachers, and friends in business and industry can supply information gained from years of experience. Do not hesitate to ask them about their careers.

Perhaps you are considering a career in electronics. The electronics industry offers many outstanding career opportunities, **Figure 27-1.** Review the following

Figure 27-1. People with electronics backgrounds may be called upon to do a variety of tasks every day. (Siemens)

guidelines. They may help you to make decisions regarding a career in this industry.

Electronics is a science. If you are interested in scientific subjects in high school, chances are you would enjoy an electronics career.

Mathematics is the tool of scientists and engineers. Do you like mathematics and do well in it? If your answer is yes, then you may be well-suited to an electronics career.

Many scientists and engineers began electronics as a hobby. If electronics is one of your hobbies, this may indicate career success in the electronics industry.

Discuss your abilities and aptitudes with your school guidance counselor. ***Aptitudes*** are your talents and natural abilities. Take an inventory test. These tests measure your interest and capacity for learning in specific areas.

Have you enjoyed this electronics textbook? The fact that you are studying electronics is a good indicator of an interest in this area. Have you taken other classes in electronics? Did you like them? Did you do well?

Give serious thought to your responses to the above questions and guidelines. Your responses may start you on the way to achieving success in your chosen area.

27.2 CAREERS IN THE ELECTRONICS INDUSTRY

Discoveries and developments in science and technology are made every day. Electronics and other fields of service play a major role in daily life.

Think of common devices that have been improved with more complex machines. For example, the communication powers of the telephone, television, and radio have been increased through the use of satellites. The safety of boats and airplanes has been improved through the use of sonar and radar devices.

Where else is the use of electronics apparent? Industry is electronically controlled. Robots and machines are constantly displacing production workers, **Figure 27-2.** In many businesses, computers can complete complex tasks in only a few minutes. Electronics has also entered the field of medicine. Using laser technology, surgeons can now perform surgery without cutting any skin.

With each new device that is developed, many highly skilled technicians must be trained to maintain and service it. The need for people with electronics training will continue to grow into the next century.

Figure 27-2. With the help of a computer programmer, this robot moves picture tubes in a manufacturing plant. (Hirata Corporation of America)

Types of Electronics Industries

In most industries, there are two types of electronics businesses: those that sell products and those that sell services. Some businesses combine the business of sales and service. In nearly every industry today, there is some sort of job set that requires the skill of an electrician or an electronics technician. The automotive industry, the manufacturing industry, the processing industry, and others have become dependent on electrically based processes and equipment. In addition, nearly each of these industries have become dependent on at least a basic level computer technician. Nearly all of the products and services we use today include some sort of computer, microcontroller, or part that requires a skilled service technician to repair or replace. The electronics field is growing constantly and quickly.

Goods-producing businesses sell books, steel, cars, etc. These products can be seen and touched. They are *tangible.* Electronic goods include such things as resistors, capacitors, televisions, compact disc players, and computers.

Service-producing businesses provide useful labor that results in a need or want being satisfied. Electronic services include installation, maintenance, repair, and updating of electronic systems (computers, telephone lines, etc.). See **Figure 27-3.**

The United States Bureau of Labor Statistics estimates that employment has been shifting from goods producing industries to service producing industries over the past few years. In addition, they predict that in 2010 nearly four out of five jobs will be in industries that provide services, **Figure 27-4.**

Figure 27-3. There are many high tech jobs in the service field. This technician is using an infrared camera to perform preventative maintenance.

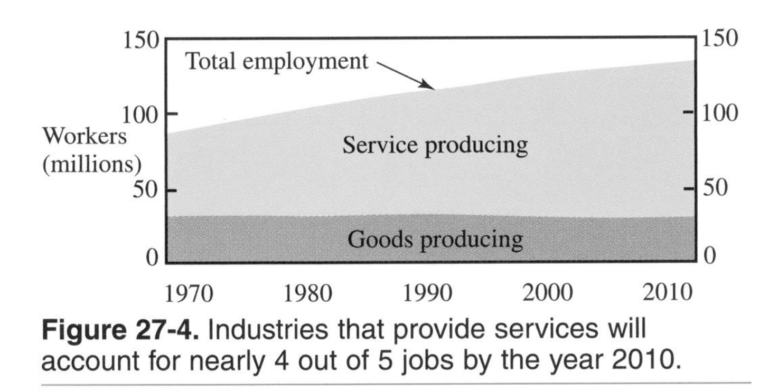

Figure 27-4. Industries that provide services will account for nearly 4 out of 5 jobs by the year 2010.

Careers

In both goods-producing and service-producing electronics industries, four general classes of workers exist: semiskilled, skilled, technical, and professional. In addition to these four types, there are also the entrepreneurs. These are people who run their own businesses. Career opportunities exist in all of these areas. The training needed to perform jobs in each of these areas varies. Likewise, employment outlooks and salaries vary.

Semiskilled workers

Semiskilled workers perform jobs that do not require a high level of training. These workers are most often found working on assembly lines. Most semiskilled jobs are limited to a certain type and number of tasks. These tasks are often simple and repetitive (repeated). In general, semiskilled work is routine. Advance study and training are required to move up from a semiskilled position. There are very few electronics jobs done by semiskilled workers.

Skilled workers

Skilled workers have thorough knowledge of, and skill in, a particular area. This knowledge and these skills are gained through advanced study, **Figure 27-5.** Advanced study is obtained through apprenticeship programs or community college programs. Some skilled workers also obtain advanced training from the military services.

Figure 27-5. Skilled workers are needed in all areas of electronics. (Intel Corp.)

An ***apprenticeship*** is a period of time spent learning a trade from an experienced, skilled worker. This training is done on the job. This training is usually combined with special classes or self-study courses. Four years is the usual length of time for an apprenticeship program.

Many community colleges offer courses and programs in the electronics field. Courses can be taken to learn more about a particular subject. Or a program can be followed to gain a particular skill.

The military services offer many specialized areas of study in the electronics field. The opportunities for learning a trade are very good in the military services.

Many jobs in the electronics field are skilled positions. Some of these jobs are maintenance and construction electricians, assemblers, and quality control inspectors. Assemblers wire and solder various parts for televisions, stereos, computers, etc. Quality control inspectors check the finished work of the assemblers.

Technicians

Technicians are specially trained workers capable of doing complex, technical jobs. Many technicians receive their training in two-year programs at community colleges, **Figure 27-6.**

Figure 27-6. Technicians may be required to do some very complex tasks. Advanced classes and on-the-job training are usually required. (Tektronix)

Technicians work with electronic equipment and assist engineers. They have the training needed to service and repair complex machines and components. Engineers rely on technicians to help them conduct research, test machines and components, and design new devices. Therefore, technicians must keep up to date on developments in the electronics industry.

Careers for technicians include broadcast technicians, robotics technicians, and computer technicians.

Professionals

Nearly all ***professional*** workers have four years of college training. Many have more advanced degrees, such as masters and doctorates. Professionals have excellent opportunities for advancement.

One of the best known professional positions in the electronics industry is the engineer, **Figure 27-7.** ***Engineers*** design and monitor the building of new equipment. Their goal is to design equipment that runs smoothly and does its job. Once this goal is met, technicians are assigned to maintain the equipment.

Figure 27-7. This research engineer is testing a device that her team designed. (Siemens)

Engineers must have a solid background in math and science. This background allows them to visualize designs before putting them down on paper.

Teaching is another professional position in the electronics industry. Teachers of electronics have the opportunity to challenge students interested in electronics. They can share their knowledge and interest in electronics with their students. The rewards of teaching are many.

Entrepreneurs

Entrepreneurs own and operate their own businesses. These small businesses make up 97% of businesses in the United States. They provide 58% of the jobs in America.

Entrepreneurships usually start with an idea for filling a hole in the marketplace where a new product or service is needed. Then a business plan is made. This plan outlines goals for the business, along with timetable for meeting those goals. This plan is vital if the business is to succeed.

In addition to a sound business plan, a successful entrepreneur possesses certain skills. The entrepreneur has knowledge of a certain industry, service, or product. This knowledge allows the owner to make smart business decisions about what is being sold. For instance, an appliance service technician needs knowledge of the appliance being serviced. If this is not the case, the business will fail.

The successful entrepreneur also has sound management skills. These skills allow the owner to manage money, time, and employees.

Entrepreneurial skills are also very important for the successful entrepreneur. These skills allow the business owner to control the business and move it in the right direction.

Entrepreneurship opportunities are vast in the electronics industry. With the growth in consumer electronics products, similar growth has occurred in servicing these products. Support of the office products industry also allows for many business opportunities. And servicing of home appliances continues to be a sound business in the electronics industry.

Consulting is yet another growing business in the electronics industry. Consultants work for clients on projects. The specific job they do depends on what work is needed. The consultant is paid by the client. When the job is completed, the consultant is free to move on to a new job and client.

Career Information Sources

An excellent reference on careers in many industries is the *Occupational Outlook Handbook.* This book is published by the United States Department of Labor and Bureau of Statistics. Most public and secondary school libraries have copies of this book. The book can also be viewed on-line at www.blg.gov/emp.

School guidance counselors are another outstanding source of career information. They can help you find information on particular careers, two- and four-year colleges that offer programs in areas that you are interested, training programs through trade schools and the military services, etc. Guidance counselors are well informed and always ready to help when asked.

Education

The educational requirements for jobs in the electronics industry vary, **Figure 27-8.** However, a minimum of a high school education is a solid foundation on which to build. Good basic math and reading skills are a must in order to succeed in the field of electronics.

Some high school graduates enter industry directly and receive specialized education in the training programs offered by large companies. Many of these workers do not stop at this point, however. They continue to study and read to keep abreast of all the changes and new technologies that develop in this industry.

Specialized training can also come from a two-year college offering an associate degree program in technology. Trade schools and the military services also offer programs in electronics technology. The newest statistics

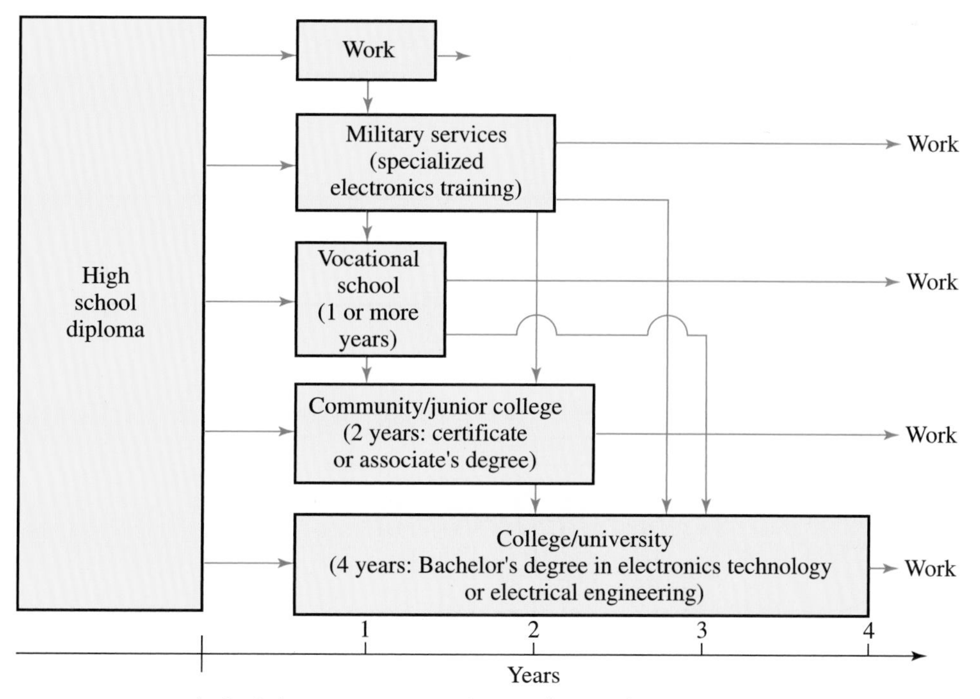

Figure 27-8. There are many roads to take on your way to an electronics career.

from the U.S. Department of Labor predict that 90% of all technical jobs in the year 2010 will require only a two-year degree or technical certificate.

Careers in science and engineering require a college degree. Advanced degrees are a good method of advancing. Many state and private universities offer engineering and electronics degrees. Talk with your school counselor to learn locations and entrance requirements. Many universities have web sites for students. These sites contain complete outlines of electrical engineering courses. You may be interested in seeing the requirements first hand.

The chart in **Figure 27-9** shows major job categories and educational requirements. Study this chart carefully.

Review Questions for Section 27.2

1. What is the difference between a goods-producing business and a service producing business?
2. Nearly all __________ workers have four years of college training.
3. After entrepreneurs pick a product to market, what is their next step?
4. Give two sources of career information.

27.3 JOB SEARCH IDEAS

Finding a job is a time-consuming and difficult task. The *Occupational Outlook Handbook* has excellent tips for conducting a job search. Start by talking to your parents, neighbors, teachers, or guidance counselors. These people may know of job openings that have not been advertised. Read the want ads in the newspaper. Look through the Yellow Pages. Companies are grouped according to industry in the Yellow Pages. You may see companies to contact. City, county, or state employment services can also provide useful job leads. Private employment agencies might provide leads also. However, they often charge a fee if a job is found.

An excellent source for job information is the Internet. Many manufacturers post job positions that can be accessed from web sites.

Once you have secured an interview, it is important to be prepared for it. Read the following tips for a successful interview.

Preparation:

Learn about the organization.
Have a specific job or jobs in mind.
Review your qualifications for the job(s).
Prepare to answer broad questions about yourself.
Review your resume.
Arrive before the scheduled time of your interview.

Personal Appearance:

Be well groomed.
Dress appropriately.
Do not chew gum or smoke.

The Interview:

Answer each question concisely.
Be prompt in giving responses.
Use good manners.
Speak properly and avoid using slang.
Convey a sense of cooperation and enthusiasm.

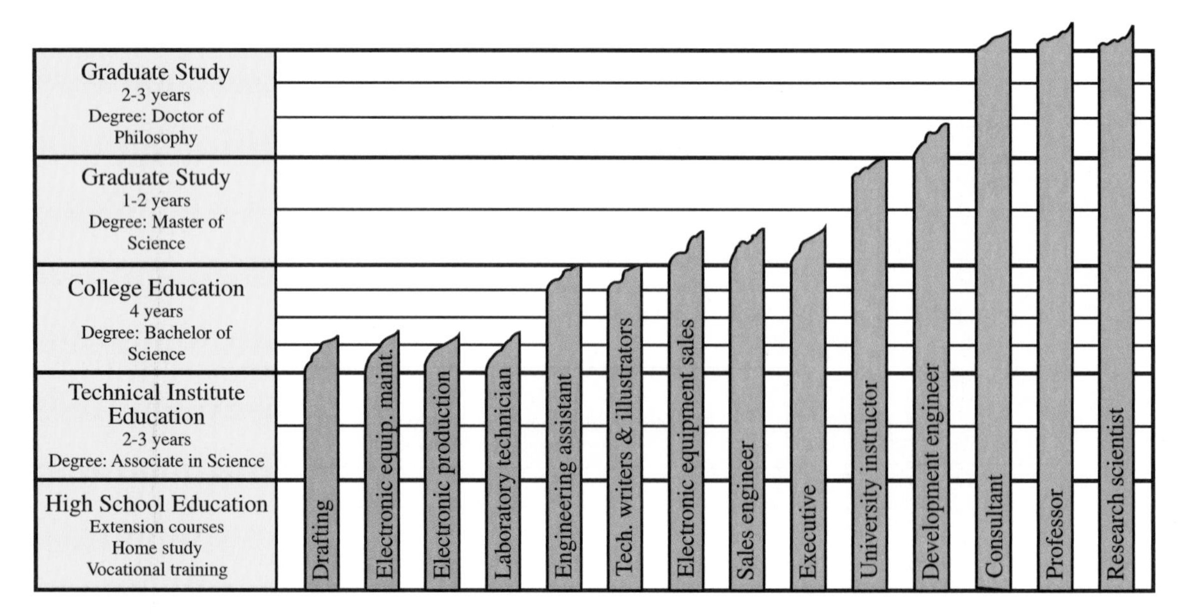

Figure 27-9. This chart shows the educational requirements for various electronics careers. (Electronics Industries Association)

Ask questions about the position and the organization.

Test (if employer gives one):

Review your basic electronic and electrical principles.
Listen carefully to instructions.
Read each question carefully.
Write legibly and clearly.
Budget your time wisely and don't dwell on one question.

Information to Bring to an Interview:

Social Security number.
Drivers license number.
Resume. See **Figure 27-10.**
Three references. Get permission from people before using their names. Try to avoid using relatives. For each reference provide a name, address, telephone number, and occupation.

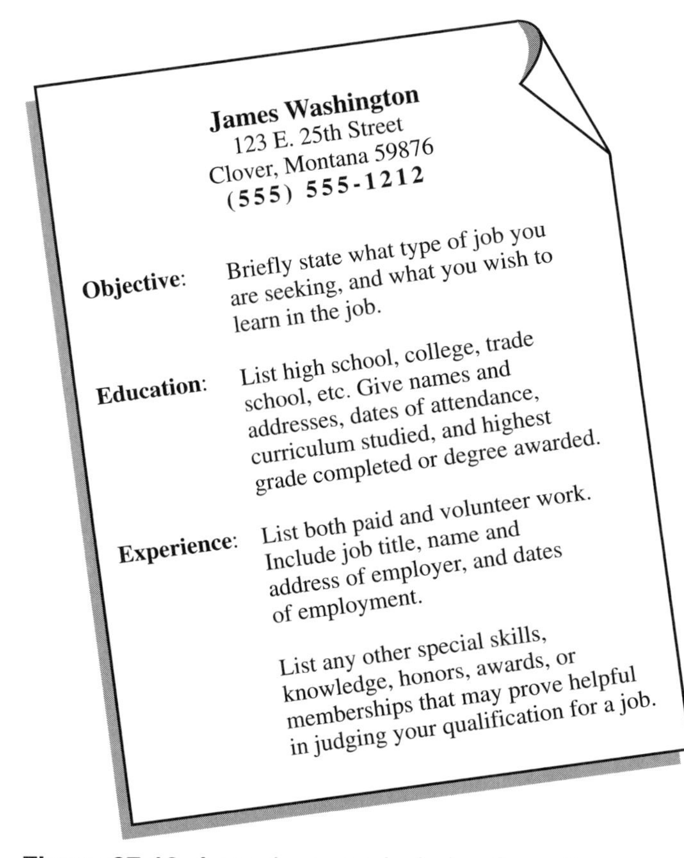

James Washington
123 E. 25th Street
Clover, Montana 59876
(555) 555-1212
Objective:
Briefly state what type of job you are seeking, and what you wish to learn in the job.
Education:
List high school, college, trade school, etc. Give names and addresses, dates of attendance, curriculum studied, and highest grade completed or degree awarded.
Experience:
List both paid and volunteer work. Include job title, name and address of employer, and dates of employment.
List any other special skills, knowledge, honors, awards, or memberships that may prove helpful in judging your qualification for a job.

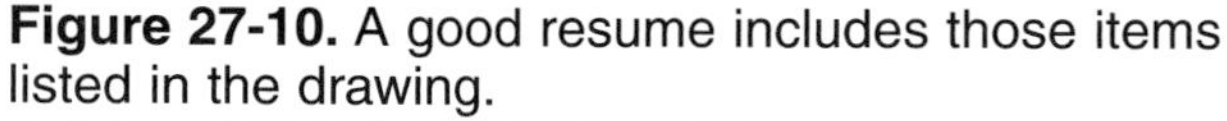
Figure 27-10. A good resume includes those items listed in the drawing.

27.4 WORK HABITS AND SKILLS

Once you have been hired into your chosen profession, it is essential that you employ positive work habits and skills. Employers will expect you to demonstrate the qualities that make up professionalism and leadership. Exhibiting these qualities will lead to a successful and rewarding career.

Professionalism

Professional behavior is required for success and advancement in the workplace. A good employee will demonstrate:

- **Cooperation**. An employee must cooperate with supervisors, other employees, and customers.
- **Dependability**. A dependable employee is timely, completes all assignments, and sets realistic goals for completing projects.
- **Good work ethic.** Good employees put an honest effort into their work. They do not spend their work time having extended casual conversations with their coworkers or visiting inappropriate Internet sites.
- **Respect.** In order to be respected, employees must show respect for others, the company, and themselves.

Leadership

Leadership is the ability to guide or direct. Employees who manage or supervise others must have strong leadership skills. However, even employees who do not manage others benefit from developing their leadership skills. Possessing leadership skills is a requirement for career growth. The following are some leadership skills:

- **Motivation**. A leader must be able to motivate others. Treating others fairly, praising others when appropriate, maintaining a positive attitude, and involving others in decisions all help to motivate people.
- **Goal accomplishment**. A leader must be able to define goals and then see that they are accomplished. Assigning tasks according to each team member's abilities and limitations helps to complete projects more efficiently.
- **Problem solving**. A leader must be able to address problems as they occur. A leader may

involve others to help solve the problems, but must see that a solution is found.

- **Recognizing limitations**. Leaders must recognize their own weaknesses, and be willing to rely on others to help in those areas.
- **Role model.** A leader must serve as a role model for other employees through both words and actions. Leaders must always behave in a manner deserving of respect.

The general skills listed under professionalism and leadership are as important as specific job skills. Developing these skills is essential to being successful in any career.

After you have put forth your best effort, you will want to make sure that you are always treated fairly in the workplace. Every employee should be familiar with the federal laws that protect them from discrimination and harassment, and ensure their equality. You can view these laws on-line at the U.S. Equal Employment Opportunity Commission web site www.eeoc.gov.

Review Questions for Section 27.4

1. List four qualities of professionalism.
2. List some attributes of a dependable employee.
3. What are some ways in which a leader can motivate others?
4. Why is it important for an employee to exhibit professionalism and leadership?

Summary

1. Goods-producing businesses produce tangible products.
2. Service-producing businesses produce useful labor that results in a need or want being filled.
3. Semiskilled workers perform job tasks that are often simple and repetitive. These workers do not require high levels of training.
4. Skilled workers have thorough knowledge of, and skill in, a particular area.
5. Apprenticeship is a period of time spent learning a trade from an experienced, skilled worker.
6. Technicians are specially trained workers capable of doing complex, technical jobs.
7. Professionals are highly trained workers. Nearly all have four-year college degrees and many have masters and doctorates.
8. Entrepreneurs are people who own and operate their own businesses.
9. Three skills required by the successful entrepreneur are: knowledge of a certain industry, service, or product, management skills, and entrepreneurial skills.
10. Professional behavior is required for success and advancement in the workplace.
11. Leadership is the ability to guide or direct.

Test Your Knowledge

Please do not write in the text. Place your answers on a separate sheet of paper.

1. List three new electronic devices that will require maintenance by specially trained technicians.
2. What are tangible products?
3. In 2010, nearly four out of five jobs will be in industries that provide __________.
4. A(n) __________ is a period of time spent learning a trade from a skilled, experienced worker.
5. What skills do successful entrepreneurs possess?
6. What qualities do successful employees possess?

For Discussion

1. What type of career would you like to have? How much training and education will this career require?
2. What traits do you think are required of successful entrepreneurs?
3. Identify the ways in which professionalism and leadership will play a role in the career you would like to have.

Authors' Acknowledgements

We would like to acknowledge and express our appreciation and thanks to those associates who so willingly supplied their assistance and knowledge in the preparation of this text.

1964, 1968, and 1975 editions: Robert Mathison, Gary Van, James Collins, Leon Pearson, and John Lienhart, electronic technicians at Chico State College; Bobbie Fikes and Moore Smalley, graduate students in electronics at San Jose State College; Charles Tyler, San Jose City College, for illustrations; Virtue B. Gerrish (wife of Mr. Gerrish) for typing the original manuscript.

1977 and 1980 editions: Neal S. Hertzog, Marketing Office Manager, Buck Engineering Co.; Gerald J. Kane, Director of Engineering, Buck Engineering Co.; Shirley Miles and Carol Keeton for typing the manuscript; El Mueller, Quasar Electronics Corp.; Carl Giegold for photography; special thanks to Carrie, Ed, Cammie, and Toy Dugger for their special support and patience.

1988 edition: David R. Milson of Dick Smith Electronics for granting permission to use certain projects from Fun Way into Electronics; sincere gratitude and love to Carrie Dugger, whose support and help were steadfast during the preparation of this edition, and for typing this manuscript.

1996 edition: Knight Electronics for the loan of equipment for the text cover.

1999 edition: Tandy Corp. for granting permission for use of a project. Baldor Electric Co., Fermi National Accelerator Laboratory, Fluke Corp., Gibson Guitar Corp., ITT Night Vision, Knight Electronics, Mechtronix Systems Inc., Motorola, Siecor Corp., Siemens Medical Systems, Inc., and the U.S. Air Force for supplying additional photographs. Fluke Corp., Knight Electronics, Lab-Volt Systems, and LG Precision for the loan of their equipment.

Thanks to the many industries whose generous supply of illustrations and product information made this text possible. Special credit is due:

Allen-Bradley, Co.; Advent; Apple Computers; AP Products; AT&T; Ballentine Laboratories, Inc.; Varley and Dexter Laboratories, Beckman Industrial Corp.; Bell Laboratories; Boonton Electronics; British Airways; Bud Radio, Inc.; Burr-Brown; Centralab, Globe Union, Inc., Cincinnati Milicron; Dale Electronics; Daystrom, Weston Instruments Div.; Delco Products, General Motors; Delco-Remy Div., General Motors; DeVilbiss Co.; Dick Smith Electronics; Dynascan Corp.; Electric Storage Battery Corp.; Electronic Industries Association; Electronic Instrument Co. (EICO); ESB Brands, Inc.; First Class Peripherals; J.A. Fleming; John Fluke Mfg. Co.; Fosdick Machine Tool Co.; General Electric Co.; Gould-A/M/I Semiconductors; Graymark Enterprises, Inc.; Gulton Industries, Inc. Hammarlund Mfg. Co., Inc.; Hayes Microcomputer Products; Heath Company; Heathkit; Hewlett Packard, Inc.; Hickok Teaching Systems, Inc.; Honeywell; Hughes Communications, Inc.; Hughes Optical and Data Systems Group; Hughes Solid State Products; IBM; International Rectifier Corp.; Iwatsu Instruments; Johnson Co.; Kepco, Inc.; Jack Kilby; Koss Corp.; Lab-Volt, Buck Engineering Co.; Lafayette Radio Electronics; Lattice Semiconductors; Lindberg Engineering Co.; Liquid Xtal Displays Inc.; Mallory Battery Co.; Marantz; Maxell Corp.; James Millen Co.; J.W. Miller Co.; Minneapolis-Honeywell Regulator Co.; MJR Co.; Montgomery Ward; Motorola Semiconductor Products, Inc.; Murdock Corp.; National Automatic Tool Co.; National Semiconductor; NEC Information Systems, Inc.; Ohmite Mfg. Co.; Omega Group Co.; Optima Enclosures; Pacific Telephone Co.; Panasonic Battery Sales Division; Perma-Power; Philco Corp.; Plastoid Corp.; Potter & Brumfield, AMF Inc.; Quasar Electronics Corp.; Radio Shack Corporation; Raytheon Company; RCA; Regency Electronics; Science-Electronics, Inc.; Shure Bros., Inc.; Simpson Electric Co.; Siemens; Society of Audio Consultants; Sony Corp. of America; Sprague Products Co.; L.S. Starrett Co.; Stromberg-Carlson; Sublogic Corp.; Superior Electric Co.; Sylvania Electric Products, Inc. (GTE); Tektronix; Texas Instruments Inc.; Triad Transformer Corp.; Tripplett Corp.; Ungar Electric Tools; Union Carbide Consumer Products Co.; Union Switch & Signal Co.; Unisys; United States Bureau of Labor Statistics; United States Office of Education; United Transformer Corp.; Vactec Inc.; Viz Mfg. Co.; Westinghouse Electric Corp.; Welch Scientific Co.; Xybernaut Corp.; Zenith Data Systems; Zenith Electronics Corp.

Material Safety Data Sheets

As an electronics/electrical technician, you will often have to classify, identify, handle, and dispose of materials and waste that are hazardous to you and to the environment. Some items you might come in contact with are batteries, lead-based solder, or circuit boards constructed with lead-based solder, toner, and cleaning fluids. Hazardous materials such as these must be handled and disposed of properly. Always read the material's Material Safety Data Sheet (MSDS) if the material has one. An MSDS provides important information about a material, such as the material's name (common, synonym, and formula), known hazards, precautions, and disposal requirements and methods. Otherwise check with the manufacturer for proper handling instructions. For disposal procedures, check with the manufacturer or the Environmental Protection Agency (EPA).

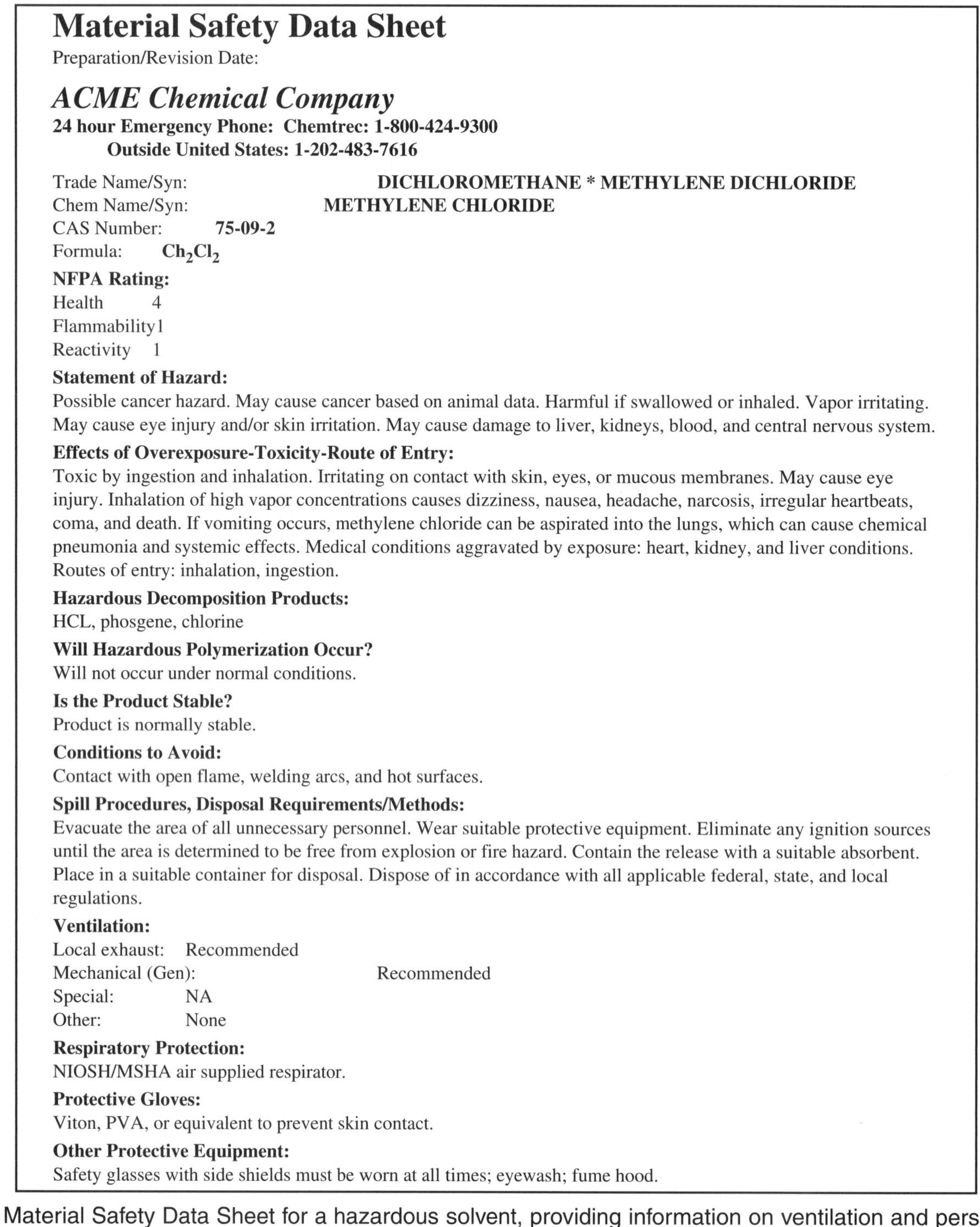

Material Safety Data Sheet

Preparation/Revision Date:

ACME Chemical Company

24 hour Emergency Phone: Chemtrec: 1-800-424-9300
Outside United States: 1-202-483-7616

Trade Name/Syn: **DICHLOROMETHANE * METHYLENE DICHLORIDE**
Chem Name/Syn: **METHYLENE CHLORIDE**
CAS Number: **75-09-2**
Formula: **Ch_2Cl_2**

NFPA Rating:
Health 4
Flammability 1
Reactivity 1

Statement of Hazard:
Possible cancer hazard. May cause cancer based on animal data. Harmful if swallowed or inhaled. Vapor irritating. May cause eye injury and/or skin irritation. May cause damage to liver, kidneys, blood, and central nervous system.

Effects of Overexposure-Toxicity-Route of Entry:
Toxic by ingestion and inhalation. Irritating on contact with skin, eyes, or mucous membranes. May cause eye injury. Inhalation of high vapor concentrations causes dizziness, nausea, headache, narcosis, irregular heartbeats, coma, and death. If vomiting occurs, methylene chloride can be aspirated into the lungs, which can cause chemical pneumonia and systemic effects. Medical conditions aggravated by exposure: heart, kidney, and liver conditions. Routes of entry: inhalation, ingestion.

Hazardous Decomposition Products:
HCL, phosgene, chlorine

Will Hazardous Polymerization Occur?
Will not occur under normal conditions.

Is the Product Stable?
Product is normally stable.

Conditions to Avoid:
Contact with open flame, welding arcs, and hot surfaces.

Spill Procedures, Disposal Requirements/Methods:
Evacuate the area of all unnecessary personnel. Wear suitable protective equipment. Eliminate any ignition sources until the area is determined to be free from explosion or fire hazard. Contain the release with a suitable absorbent. Place in a suitable container for disposal. Dispose of in accordance with all applicable federal, state, and local regulations.

Ventilation:
Local exhaust: Recommended
Mechanical (Gen): Recommended
Special: NA
Other: None

Respiratory Protection:
NIOSH/MSHA air supplied respirator.

Protective Gloves:
Viton, PVA, or equivalent to prevent skin contact.

Other Protective Equipment:
Safety glasses with side shields must be worn at all times; eyewash; fume hood.

Part of a Material Safety Data Sheet for a hazardous solvent, providing information on ventilation and personal protective equipment. Always check the MSDS before working with any chemical.

Reference Section

Appendix 1

Scientific Notation

In the study of electronics, you will work with very small quantities. As a student of mathematics, you understand that multiplication and division of numbers having many zeros and decimal points requires your concentration. Otherwise, errors will occur. The "powers of ten" or scientific notation is a useful method for performing these tasks.

$1 \times 10^0 = 1$
$1 \times 10^1 = 10$
$1 \times 10^2 = 100$
$1 \times 10^3 = 1000$
$1 \times 10^4 = 10{,}000$
$1 \times 10^5 = 100{,}000$
$1 \times 10^6 = 1{,}000{,}000$
$1 \times 10^{-1} = 0.1$
$1 \times 10^{-2} = 0.01$
$1 \times 10^{-3} = 0.001$
$1 \times 10^{-4} = 0.0001$
$1 \times 10^{-5} = 0.00001$
$1 \times 10^{-6} = 0.000001$

Follow these examples and you will learn to use the scientific notation method.

To express a number:

$47{,}000 = 47 \times 10^3$
$0.000100 = 100 \times 10^{-6}$
$0.0025 = 25 \times 10^{-4}$
$3{,}500{,}000 = 3.5 \times 10^6$

To multiply, add the exponents of ten:

$47 \times 10^3 \times 25 \times 10^{-4} = 47 \times 25 \times 10^{-1}$
$100 \times 10^{-6} \times 3.5 \times 10^6 = 100 \times 3.5$

To divide, subtract exponents:

$$\frac{3.5 \times 10^6}{25 \times 10^{-4}} = \frac{3.5 \times 10^{10}}{25} \quad [6 - (-4) = +10]$$

$$\frac{100 \times 10^{-6}}{25 \times 10^{-4}} = \frac{100 \times 10^{-2}}{25} \quad [-6 - (-4) = -2]$$

To square, multiply exponent by 2:

$(25 \times 10^2)^2 = 25^2 \times 10^4$
$(9 \times 10^{-3})^2 = 81 \times 10^{-6}$

To extract square root, divide exponent by 2:

$$\sqrt{81 \times 10^4} = 9 \times 10^2$$

$$\sqrt{225 \times 10^{-8}} = 15 \times 10^{-4}$$

See Appendix 3 for using scientific notation in conversions.

Appendix 2

Color Codes

A standard color code has been adopted to determine the values of resistors and capacitors. Figure A2-1 shows those values.

Resistors

Secure several resistors from stock. Hold the resistor so that the color bands are to the left, Figure A2-2. The band colors are (from left to right) brown, black, green, and silver.

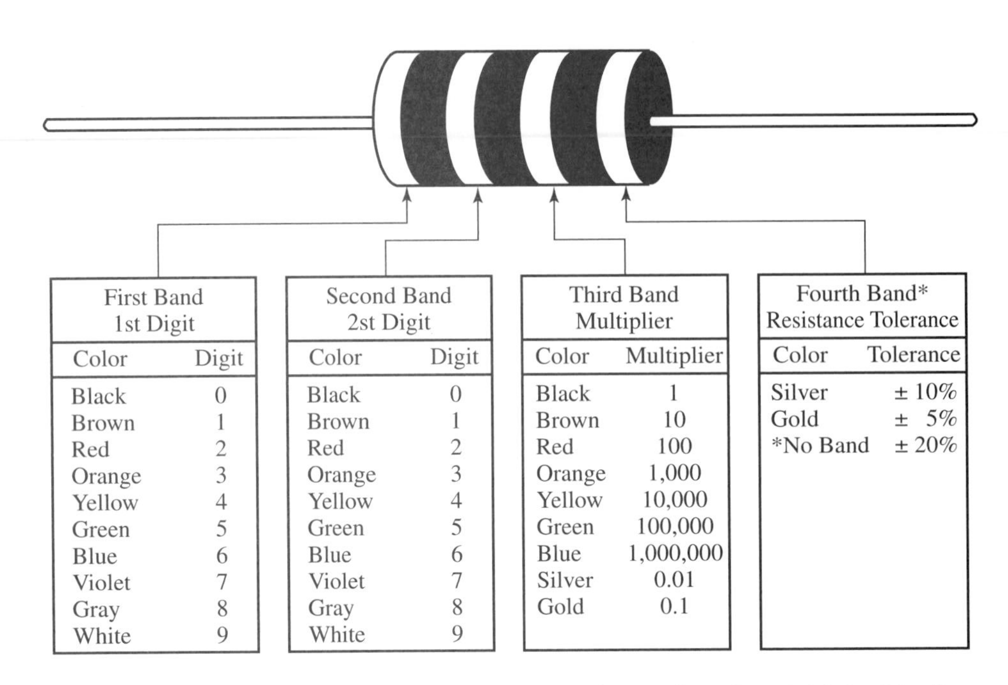

First Band 1st Digit	
Color	Digit
Black	0
Brown	1
Red	2
Orange	3
Yellow	4
Green	5
Blue	6
Violet	7
Gray	8
White	9

Second Band 2st Digit	
Color	Digit
Black	0
Brown	1
Red	2
Orange	3
Yellow	4
Green	5
Blue	6
Violet	7
Gray	8
White	9

Third Band Multiplier	
Color	Multiplier
Black	1
Brown	10
Red	100
Orange	1,000
Yellow	10,000
Green	100,000
Blue	1,000,000
Silver	0.01
Gold	0.1

Fourth Band* Resistance Tolerance	
Color	Tolerance
Silver	± 10%
Gold	± 5%
*No Band	± 20%

Figure A2-1. Standard color code for resistors and capacitors is established by the Electronic Industries Association (EIA).

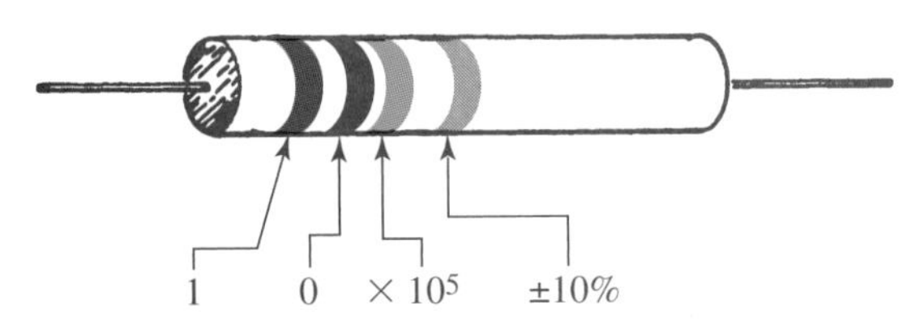

Figure A2-2. When reading color code, hold the resistor with the bands at the left. This resistor value is 1,000,000 Ω or 1 MΩ.

The brown band is 1. The black band is 0. The third band is the multiplier band. Green means multiply by 10^5. The silver band tells us that the actual resistance value is within the limits of ±10 percent of the color code value.

Review the following examples:

Yellow, Violet, Brown, Silver	470 Ω	±10%
Brown, Black, Red, Gold	1000 Ω	±5%
Orange, Orange, Red, None	3300 Ω	±20%
Green, Blue, Red, None	5600 Ω	±20%
Red, Red, Green, Silver	2,200,000 Ω	±10%

Capacitors

The practice of marking capacitors with numbers is fairly common. Some capacitors can be read directly. Others, however, are coded. There are many special marking systems. A detailed manual describing all codes should be consulted when necessary.

In Figure A2-3 we will use the color code shown in Figure A2-1. The upper left-hand corner dot is white. The dot representing the first significant figure is red. The dot representing the second significant figure is green. This makes the first two numbers 25. The multiplier is brown. Multiply by 10^1. The value of the capacitor is 250 pF. The tolerance value and class can also be found. However, their meanings will not be covered in this textbook.

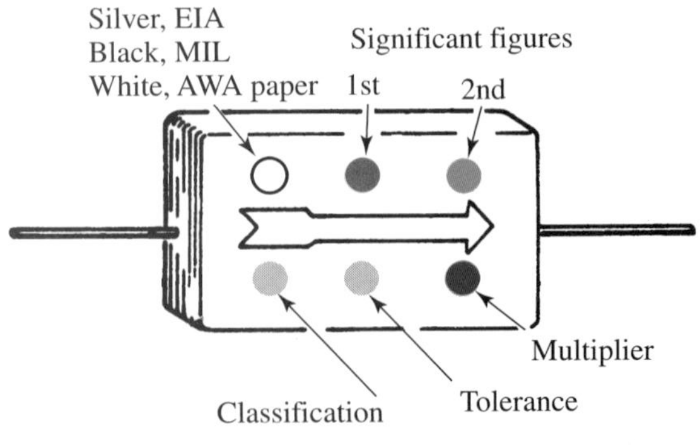

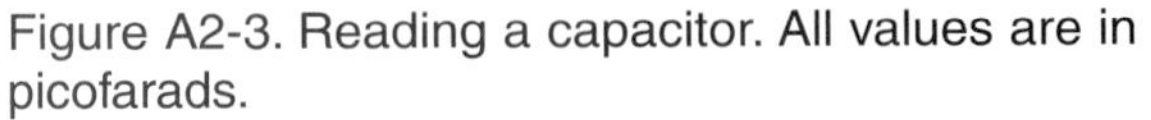

Figure A2-3. Reading a capacitor. All values are in picofarads.

Appendix 3

Conversions

When using formulas to find unknown values in a circuit, the formula may be given in basic units, while the quantities you wish to use are given in larger or smaller units. A conversion must be made. For example, Ohm's law states that:

$$I \text{ (in amperes)} = \frac{E \text{ (in volts)}}{R \text{ (in ohms)}}$$

If R were given in megohms, such as 2.2 MΩ, it would be necessary to change it to 2,200,000 Ω. If E were given in microvolts, such as 500 μ V, it would be necessary to change it to 0.0005 volts.

Scientific notation offers a simple way to make conversions. First, however, examine the prefixes shown in Figure A3-1.

Compare the meaning of these prefixes to scientific notation in Appendix 1.

10,000 ohms = $10{,}000 \times 10^{-3}$ kilohms
47 kilohms = 47×10^{3} ohms
950 kilohertz = 950×10^{3} hertz
= 950×10^{-3} megahertz
100 milliamperes = 100×10^{-3} amperes
= 100×10^{3} microamperes
0.01 microfarads = 0.01×10^{-6} farads
= 0.01×10^{6} pF
250 pF = 250×10^{-6} mF
= 250×10^{-12} F
8 mF = 8×10^{-6} F
= 8×10^{6} pF
75 milliwatts = 75×10^{-3} watts

A simple rule will also help you. It states that if the exponent of ten is negative, the decimal point is moved to the left in the answer.

$447 \times 10^{-3} = 0.447$
$250 \times 10^{-6} = 0.00025$

If the exponent is positive, the decimal point is moved to the right.

$447 \times 10^{3} = 447{,}000$
$250 \times 10^{6} = 250{,}000{,}000$

Here is a problem. What is the reactance of a 2.5 MH choke at 1000 kHz?

$X_L = 2\pi fL$
$X_L = 2\pi f \text{ (in Hz)} \times L \text{ (in henrys)}$

Two conversions must be made:

1000 kHz = 1000×10^{3} Hz
2.5 mH = 2.5×10^{-3} henrys

Therefore:

$X_L = 2 \times 3.14 \times 1000 \times 10^{3} \times 2.5 \times 10^{-3}$
$X_L = 6.28 \times 2.5 \times 10^{3}$ (1000 changed to 10^{3})
$X_L = 15.6 \times 10^{3} = 15{,}600$ ohms

Appendix 4

Trigonometry

Trigonometry is a part of the "tool kit" of the electronics technician. Using trigonometry simplifies finding the solutions of alternating current problems. Trigonometry finds many uses in designing and understanding electronic circuits.

Prefix	*Symbol*	*Multiplication Factor*
exa	E	10^{18} = 1,000,000,000,000,000,000
peta	P	10^{15} = 1,000,000,000,000,000
tera	T	10^{12} = 1,000,000,000,000
giga	G	10^{9} = 1,000,000,000
mega	M	10^{6} = 1,000,000
kilo	k	10^{3} = 1,000
hecto	h	10^{2} = 100
deca	da	10^{1} = 10
(unit)		10^{0} = 1
deci	d	10^{-1} = 0.1
centi	c	10^{-2} = 0.01
milli	m	10^{-3} = 0.001
micro	μ	10^{-6} = 0.000001
nano	n	10^{-9} = 0.000000001
pico	p	10^{-12} = 0.000000000001
femto	f	10^{-15} = 0.000000000000001
atto	a	10^{-18} = 0.000000000000000001

Figure A3-1.

Basically, trigonometry is the relationship between the angles and sides of a triangle. These relationships are called functions. They represent the numerical ratio between the two sides of the right triangle.

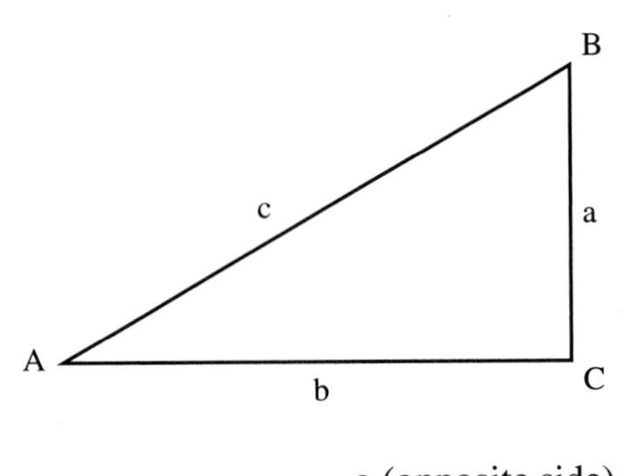

$$\text{Sine of angle A} = \frac{\text{a (opposite side)}}{\text{c (hypotenuse)}}$$

$$\text{Cosine of angle A} = \frac{\text{b (adjacent side)}}{\text{c (hypotenuse)}}$$

$$\text{Tangent of angle A} = \frac{\text{a (opposite side)}}{\text{b (opposite side)}}$$

There are other functions, but these three are widely used in solving problems in electronics. Using these equations, if two values are known, the third can be found.

For example, in a triangle, side a = 6 and b = 10. Find the tangent of angle A.

$$\tan A = \frac{a}{b} = \frac{6}{10} = 0.6$$

Look at the table of natural trigonometric functions, Figure A4-1. Find the angle whose tangent is 0.6. It is 31 degrees.

Try this problem. Angle A is 45 degrees and side a is 6. What is the value of side c? Use the sine equation.

$$\sin 45° = \frac{6}{c}$$

Look up the sine of 45°. It is 0.707. Then,

$$c = \frac{6}{0.707} = 8.5 \text{ (approx.)}$$

In electronics, the right triangle used for the previous examples can be labeled other ways. But the problems are worked out in the same manner.

For example, a circuit contains 40 Ω resistance (*R*) and 50 Ω reactance (*X*). What is the circuit impedance (*Z*) in ohms?

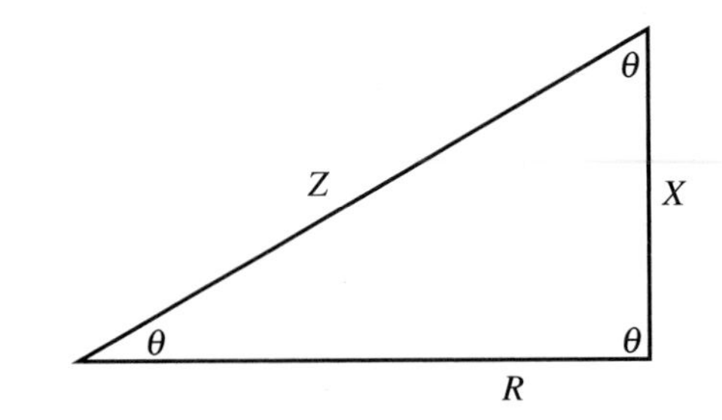

Find angle *θ*:

$$\tan \theta = \frac{X}{R} = \frac{50}{40} = 1.25$$

Look up in the table:

$$\theta = 51° \text{ (approx.)}$$

Find *Z*:

$$\sin \theta = \frac{X}{R}$$

Look up in the table:

$$0.777 = \frac{50}{Z}$$

$$Z = \frac{50}{0.777} = 64.3\ \Omega$$

These applications of trigonometry must be thoroughly understood. The best way to gain this understanding is through practice. Problems have been included in the text for this purpose.

APPENDIX 5

Standard Symbols and Abbreviations

±	plus or minus
= or ::	equals to
≡	identity
≈ or ≅	is similar to
≠	does not equal
>	is greater than
<	is less than
≥	greater than or equal to
≤	less than or equal to
∴	therefore
∠	angle
Δ	increment or decrement (change in)
⊥	perpendicular to
‖	parallel to
∞	infinity

Natural Trigonometric Functions							
Angle	Sine	Cosine	Tangent	Angle	Sine	Cosine	Tangent
1°	0.0175	0.9998	0.0175	46°	0.7193	0.6947	1.0355
2°	0.0349	0.9994	0.0349	47°	0.7314	0.6820	1.0724
3°	0.0523	0.9986	0.0524	48°	0.7431	0.6691	1.1106
4°	0.0698	0.9976	0.0699	49°	0.7547	0.6561	1.1504
5°	0.0872	0.9962	0.0875	50°	0.7660	0.6428	1.1918
6°	0.1045	0.9945	0.1051	51°	0.7771	0.6293	1.2349
7°	0.1219	0.9925	0.1228	52°	0.7880	0.6157	1.2799
8°	0.1392	0.9903	0.1405	53°	0.7986	0.6018	1.3270
9°	0.1564	0.9877	0.1584	54°	0.8090	0.5878	1.3764
10°	0.1736	0.9848	0.1763	55°	0.8192	0.5736	1.4281
11°	0.1908	0.9816	0.1944	56°	0.8290	0.5592	1.4826
12°	0.2079	0.9781	0.2126	57°	0.8387	0.5446	1.5399
13°	0.2250	0.9744	0.2309	58°	0.8480	0.5299	1.6003
14°	0.2419	0.9703	0.2493	59°	0.8572	0.5150	1.6643
15°	0.2588	0.9659	0.2679	60°	0.8660	0.5000	1.7321
16°	0.2756	0.9613	0.2867	61°	0.8746	0.4848	1.8040
17°	0.2924	0.9563	0.3057	62°	0.8829	0.4695	1.8807
18°	0.3090	0.9511	0.3249	63°	0.8910	0.4540	1.9626
19°	0.3256	0.9455	0.3443	64°	0.8988	0.4384	2.0503
20°	0.3420	0.9397	0.3640	65°	0.9063	0.4226	2.1445
21°	0.3584	0.9336	0.3839	66°	0.9135	0.4067	2.2460
22°	0.3746	0.9272	0.4040	67°	0.9205	0.3907	2.3559
23°	0.3907	0.9205	0.4245	68°	0.9272	0.3746	2.4751
24°	0.4067	0.9135	0.4452	69°	0.9336	0.3584	2.6051
25°	0.4226	0.9063	0.4663	70°	0.9397	0.3420	2.7475
26°	0.4384	0.8988	0.4877	71°	0.9455	0.3256	2.9042
27°	0.4540	0.8910	0.5095	72°	0.9511	0.3090	3.0777
28°	0.4695	0.8829	0.5317	73°	0.9563	0.2924	3.2709
29°	0.4848	0.8746	0.5543	74°	0.9613	0.2756	3.4874
30°	0.5000	0.8660	0.5774	75°	0.9659	0.2588	3.7321
31°	0.5150	0.8572	0.6009	76°	0.9703	0.2419	4.0108
32°	0.5299	0.8480	0.6249	77°	0.9744	0.2250	4.3315
33°	0.5446	0.8387	0.6494	78°	0.9781	0.2079	4.7046
34°	0.5592	0.8290	0.6745	79°	0.9816	0.1908	5.1446
35°	0.5736	0.8192	0.7002	80°	0.9848	0.1736	5.6713
36°	0.5878	0.8090	0.7265	81°	0.9877	0.1564	6.3138
37°	0.6018	0.7986	0.7536	82°	0.9903	0.1392	7.1154
38°	0.6157	0.7880	0.7813	83°	0.9925	0.1219	8.1443
39°	0.6293	0.7771	0.8098	84°	0.9945	0.1045	9.5144
40°	0.6428	0.7660	0.8391	85°	0.9962	0.0872	11.4301
41°	0.6561	0.7547	0.8693	86°	0.9976	0.0698	14.3006
42°	0.6691	0.7431	0.9004	87°	0.9986	0.0523	19.0811
43°	0.6820	0.7314	0.9325	88°	0.9994	0.0349	28.6363
44°	0.6947	0.7193	0.9657	89°	0.9998	0.0175	57.2900
45°	0.7071	0.7071	1.0000	90°	1.0000	0.0000	

Figure A4-1. The table of natural trigonometric functions.

Letter symbols

Δ	a change in
ω	angular velocity
β	beta
X_C	capacitive reactance
C	capacitor
G	conductance
I	current
f	frequency
A	gain
Z	impedance
L	inductance
X_L	inductive reactance
Φ	magnetic flux
μ	micro
M	mutual inductance
Ω	ohms
μ	permeability
θ	phase displacement
Q	quality factor
X	reactance
R	resistance
f_0	resonant frequency
t	time
E or V	voltage
λ	wavelength

Standard abbreviations

ac	alternating current
AWG	American Wire Gage
amp	ampere
AM	amplitude modulation
AF	audio frequency
AFC	automatic frequency control
AVC	automatic volume control
CRT	cathode ray tube
CW	continuous wave
cps	cycles per second
dB	decibel
dc	direct current
DPDT	double-pole, double-throw
DPST	double-pole, single-throw
emf	electromotive force
F	farad
FET	field-effect transistor
FM	frequency modulation
gnd	ground
H	henry
Hz	hertz
HF	high frequency
hp	horsepower
IF	intermediate frequency
kH	kilohertz
kΩ	kilohm
kV	kilovolt
kWh	kilowatt hour
LF	low frequency
mmf	magnetomotive force
max	maximum
MHz	megahertz
MΩ	megohm
μ A	microampere
μ F	microfarad
μ H	microhenry
μ μF	micromicrofarad
μ V	microvolt
mH	millihenry
mA	milliampere
mV	millivolt
mW	milliwatt
min	minimum
osc	oscillator
p	pico
pF	picofarad
pot	potentiometer
PF	power factor
RF	radio frequency
rpm	revolutions per minute
rms	root mean square
SPDT	single-pole, double-throw
SPST	single-pole, single-throw
UHF	ultra high frequency
VTVM	vacuum tube voltmeter
VHF	very high frequency
V	volts
W	watts

Common Schematic Symbols

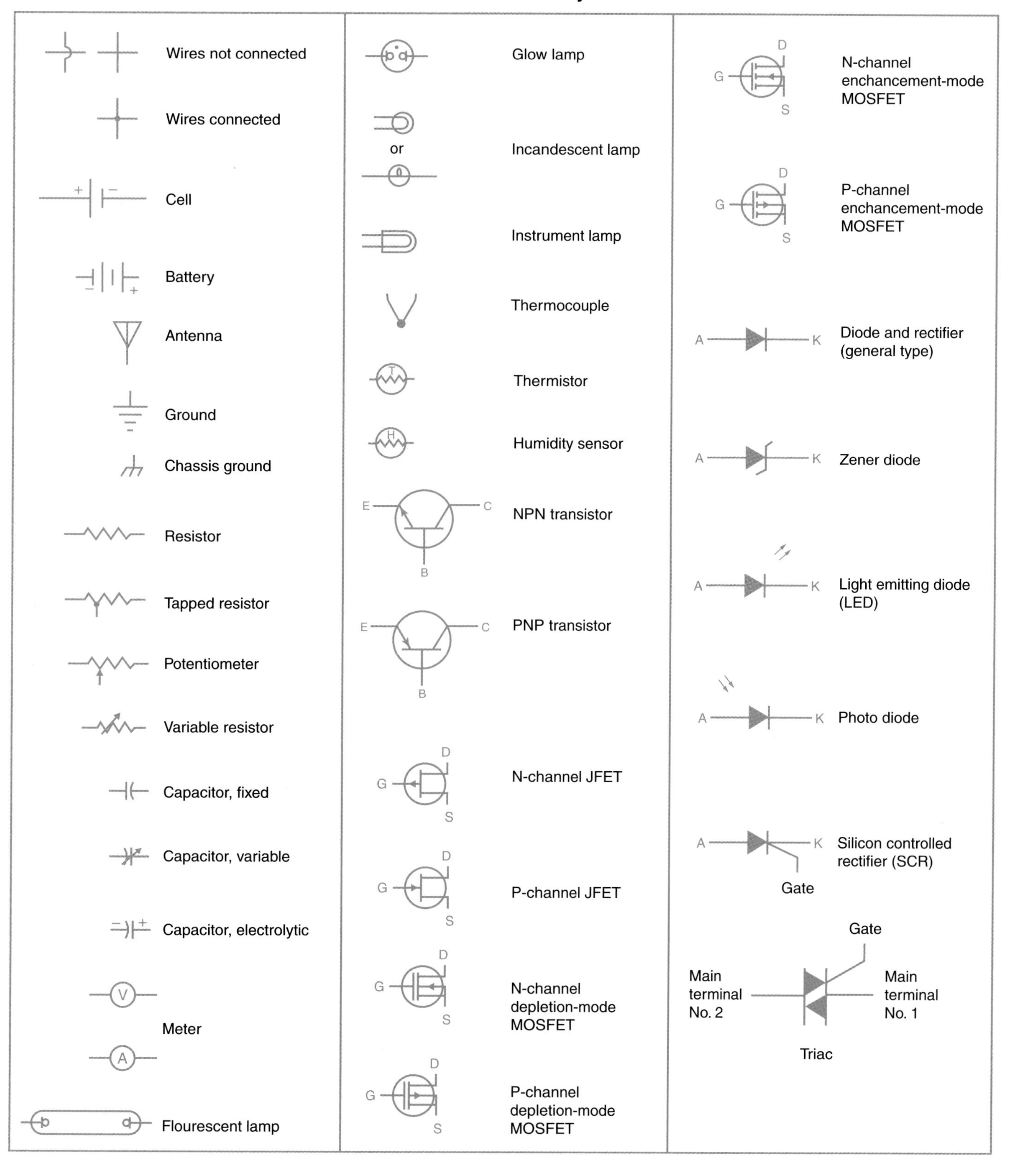

Wires not connected
Wires connected
Cell
Battery
Antenna
Ground
Chassis ground
Resistor
Tapped resistor
Potentiometer
Variable resistor
Capacitor, fixed
Capacitor, variable
Capacitor, electrolytic
V
Meter
A
Flourescent lamp
Glow lamp
or
Incandescent lamp
Instrument lamp
Thermocouple
T
Thermistor
H
Humidity sensor
E
C
B
NPN transistor
PNP transistor
G
D
S
N-channel JFET
P-channel JFET
N-channel depletion-mode MOSFET
P-channel depletion-mode MOSFET
N-channel enchancement-mode MOSFET
P-channel enchancement-mode MOSFET
A
K
Diode and rectifier (general type)
Zener diode
Light emitting diode (LED)
Photo diode
Silicon controlled rectifier (SCR)
Gate
Main terminal No. 2
Main terminal No. 1
Triac

Common Schematic Symbols

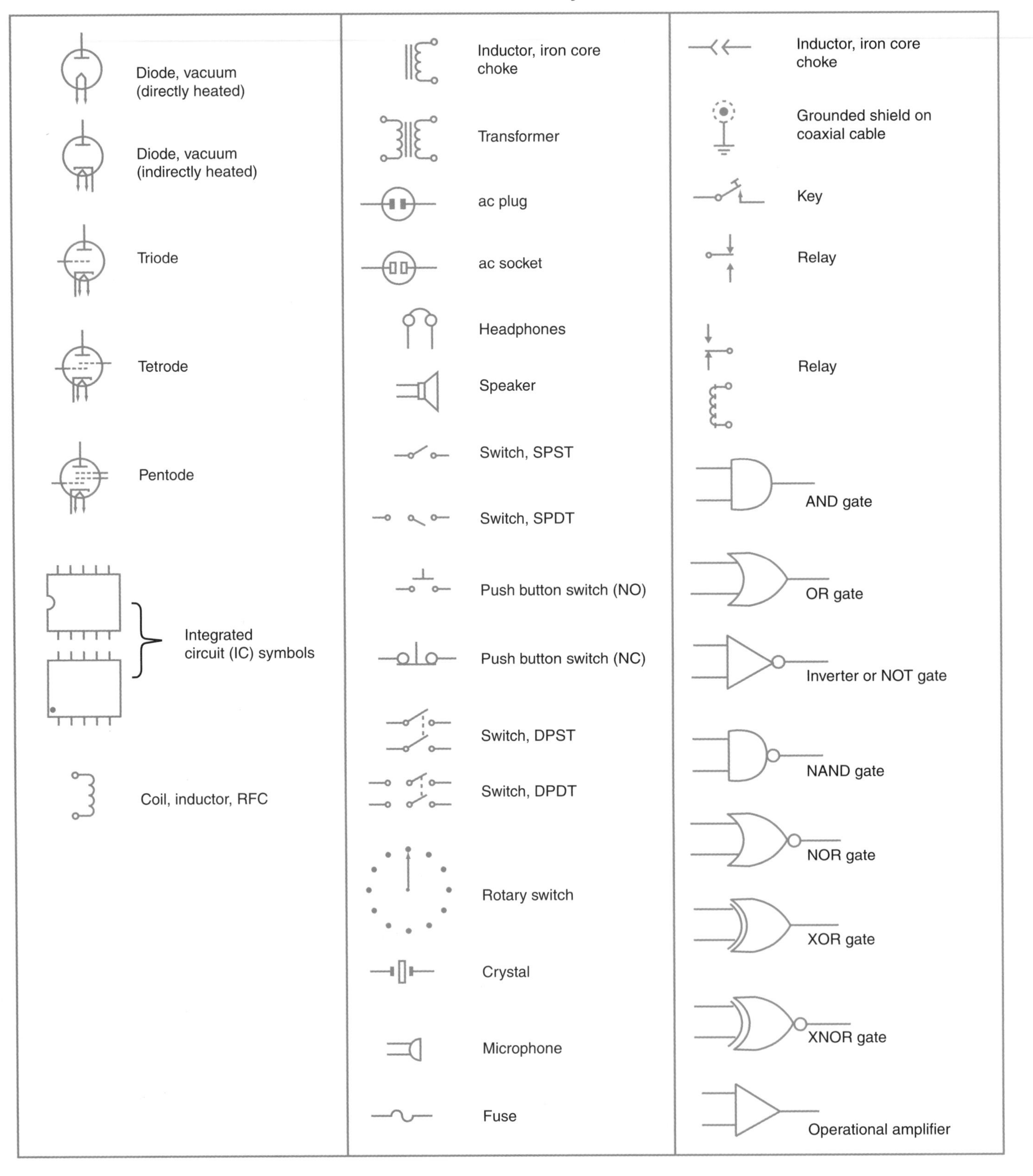
Diode, vacuum (directly heated)
Diode, vacuum (indirectly heated)
Triode
Tetrode
Pentode
Integrated circuit (IC) symbols
Coil, inductor, RFC
Inductor, iron core choke
Transformer
ac plug
ac socket
Headphones
Speaker
Switch, SPST
Switch, SPDT
Push button switch (NO)
Push button switch (NC)
Switch, DPST
Switch, DPDT
Rotary switch
Crystal
Microphone
Fuse
Inductor, iron core choke
Grounded shield on coaxial cable
Key
Relay
Relay
AND gate
OR gate
Inverter or NOT gate
NAND gate
NOR gate
XOR gate
XNOR gate
Operational amplifier

Appendix 6

Electromagnet Demonstrations Kit

A number of experiments in this text require the use of one or two electromagnets. These electromagnets can be purchased or assembled. The following instructions contain the parts needed to build a number of the projects involving electromagnets illustrated and described in preceding chapters of this text. Projects that can be built from the following parts include experiments involving: permanent magnetic fields, electromagnetic fields, electromagnets, relays, circuit breakers, chimes, buzzers, series dc motors, shunt dc motors, ac motors, reactance dimmers, induction coils, series circuits, and parallel circuits.

Items in the electromagnet demonstrations kit are listed below. They include items like sockets, leads, bolts, and nuts that are purchased along with brackets, coils, etc. Construction details for student-made parts are given in the drawings, Figure A6-1.

The electromagnets and other required equipment can also be purchased. The equipment photographed in the experiments in this text are part of an electronics training package from Lab-Volt (Electricity and Electronics Training System Models 556 & 557).

The Electromagnet Demonstrations Parts List

1—base, 1/8 ×7 1/2 ×10 pegboard
4—base feet, rubber 3/8
4—feet bolts, 3/8 ×8-32 round head with nuts
1—base support, wood (Plan, Detail M)
2—coils (Plan, Detail F)
4—coil brackets, brass 1/32" (Plan, Detail G)
2—coil cores, 1/2 ×2 1/2 mild steel (Plan, Detail E)
1—solenoid core, mild steel (Plan, Detail H)
2—magnets, 1/2 ×2 1/2 round
1—armature bracket, brass 1/32" (Plan, Detail J)
1—dc armature (Plan, Detail A)
1—ac armature (Plan, Detail B)
1—brush support (Plan, Detail L)
1—relay armature (Plan, Detail N)
1—Armature plate (Plan, Detail K)
1—circuit breaker stop (Plan, Detail P)
2—contact brackets (Plan, Detail R)
1—Induction coil (Plan, Detail D)
1—.01 mF paper capacitor, mount on subbases.
1—push button switch
1—toggle switch
3—subbase
6—jiffy leads, 8"
3—miniature sockets with bulbs, 6 V
1—doorbell transformer
1—50 Ω potentiometer with subbase for potentiometer (2 W)
1—terminal base
3—battery holders, D cell
1—chime, hard steel tube or bar
1—galvanometer, 500-0-500 μ A
12—bolts, 3/8 ×8-32 round head
12—nuts, hex. 8-32
4—stove bolts, 3/16 ×2 1/2
8—nuts, 3/16"

Construction hints:

1. The two coils, F, are interchangeable in all projects. The coil brackets are bolted to the pegboard base. The saw kerf at each end of coil fits snugly into bracket.
2. The permanent magnets are interchangeable with the iron cores when used for magnetic field demonstrations or in the dc motor.
3. Induction coil D is made to slide over coil F. Wind with more than 8 layers of No. 30 wire if higher voltages are desired. Capacitor is connected across breaker points.
4. If difficulty is experienced with the ac motor, try starting it by winding a string around the armature shaft and pulling it. Its speed must be synchronized with the 60 hertz line voltage for operation.
5. All motors are constructed by placing armatures between plate R, bolted to base, and armature bracket as the top bearing.
6. The brush support L is constructed so that it can be locked in several positions to demonstrate a change in commutation angle.

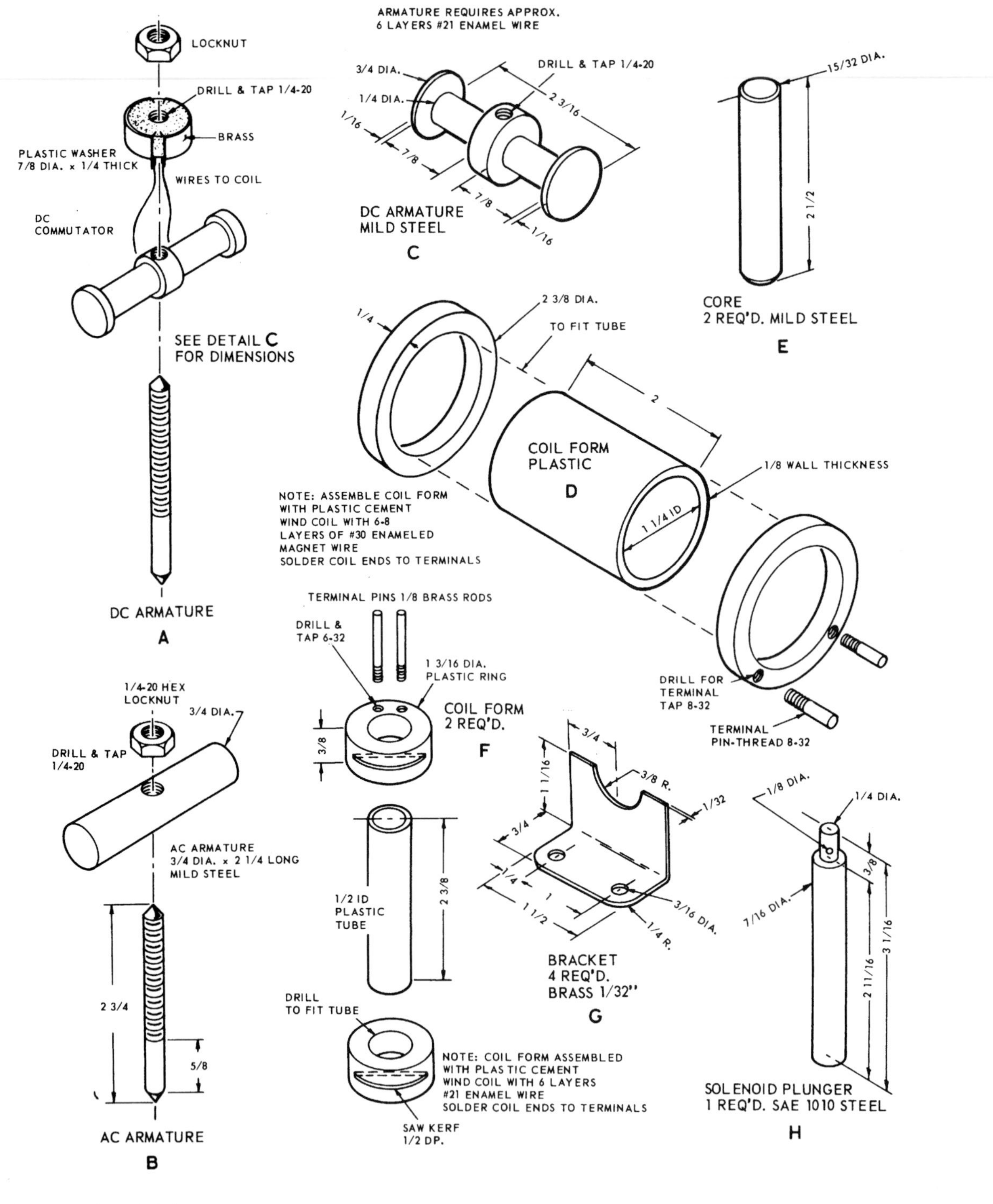

Figure A6-1. Construction details.

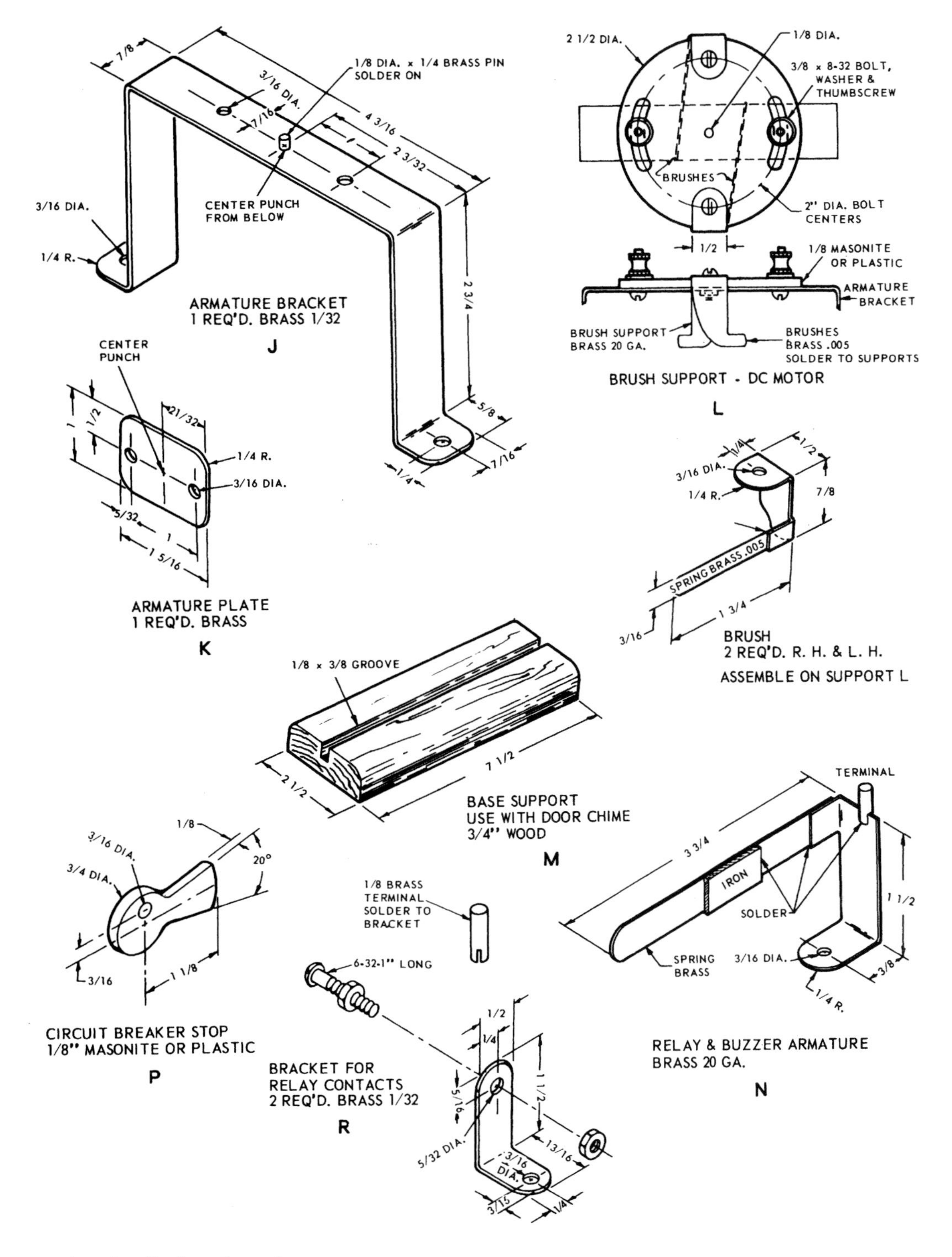

Figure A6-1. Construction details (continued).

Glossary of Technical Terms

A

abacus: An ancient calculating device with movable beads used to perform basic mathematical functions.

A battery: A battery used to supply heater voltage for electron tubes.

ac: Abbreviation for alternating current.

ac generator: A generator that uses slip rings and brushes to connect armature to an external circuit. Output is alternating current.

ac plate resistance: A variational characteristic of a vacuum tube, representing the ratio of change of plate voltage to change in plate current, while grid voltage is constant. Its symbol is RP.

acceptor circuit: A series-tuned circuit at resonance. It accepts a signal at resonant frequency.

acceptor impurity: Impurity added to semiconductor material that creates holes for current carriers.

active-matrix: A display in which each individual cell in the grid has its own individual transistor.

actuator: A mechanical device that causes a circuit to open and close.

adjustable resistor: A type of wire-wound resistor in which one side of the wire is exposed.

advanced television systems committee (ATSC): A committee, established in 1983, that developed standards for digital television broadcasting and receivers. These standards, adopted by the FCC, have been designed to eventually replace the NTSC standards.

aerial: An antenna.

AF amplifier: Used to amplify audio frequencies.

AGC: Abbreviation for automatic gain control.

air-core inductor: 1. Inductor wound on an insulated form without a metallic core. 2. A self-supporting coil without a core.

alignment: The process of transformer adjustment.

alignment tool: A special fiber or plastic screwdriver used to adjust trimmer capacitors.

alkaline battery: A primary cell composed of manganese dioxide for the positive plate, powdered zinc for the negative plate, and caustic alkali for the electrolyte.

alnico: A special alloy made of aluminum, nickel, and cobalt that is used to make small, permanent magnets.

alpha: The Greek letter α, representing a current gain of a transistor. It is equal to a change in the collector current caused by a change in the emitter current for constant collector voltage.

alpha cutoff frequency: Frequency at which the current gain drops to 0.707 of its maximum gain.

alternating current (ac): A current of electrons that moves first in one direction and then in the other.

alternator: An ac generator.

amalgam: A compound or mixture containing mercury.

amalgamation: The process of adding a small quantity of mercury to zinc during manufacturing.

ambient temperature: The temperature of the material (air, water, or soil) surrounding the conductor.

American Radio Relay League (ARRL): The association of radio amateurs.

American Wire Gauge (AWG): A numbering system used for sizing wire.

ammeter: A meter used to measure current.

ampacity rating: An indication of how much current a switch can safely handle.

ampere (A): Electron flow representing one coulomb per second past a given point in a circuit.

ampere-hour: The capacity-rating measurement of batteries. A 100 ampere-hour battery will theoretically produce 100 amperes for one hour.

ampere turn (At): The unit of measurement of magnetomotive force. It represents the product of amperes times the number of turns in the coil of the electromagnet.

amplification: Ability to control a relatively large force by a small force.

amplifier: An electronic circuit that uses a small input signal to control a larger output signal.

amplifier, power: An amplifier used to increase power output. Sometimes called a current amplifier. There are several classes of power amplifiers:

- **Class A:** An amplifier biased so that an output signal is produced for 360 degrees of the input signal cycle.
- **Class B:** An amplifier biased so that an output signal is reproduced during 180 degrees of input signal. The class B amplifier is biased close to cutoff.
- **Class C:** An amplifier biased so that less than 180 degrees of output signal is produced during 360 degrees of input signal.

amplitude: The maximum extent or range of a quality.

amplitude distortion: Deviation in amplitude caused by the nonlinear operation of an amplifier when peaks of input signals are reduced or cut off by either an excessive input signal or incorrect bias.

amplitude modulation (AM): Varying of the radio wave amplitude at an audio frequency rate.

analog devices: Electronic devices that have variable outputs controlled by variable inputs. Also called linear integrated circuits.

analog meter: A meter with a continuously variable scale.

AND gate: A logic gate that is used to determine the presence of yes signals or 1s.

angle of compensation: A correcting angle applied to a compass reading to compensate for local magnetic influence.

angle of variation: The angle between true north and magnetic north. Also called declination or the angle of declination.

angular phase: Position of a rotating vector with respect to a reference line.

angular velocity (ω): Speed of a rotating vector in radians per second.

anode: A positive terminal.

antenna: A device for radiating or receiving radio waves.

aperture mask: A thin sheet of perforated material placed directly behind the viewing screen in a three-gun color picture tube.

apparent power: Power apparently used in a circuit as a product of current times voltage.

application: A software program written for specific tasks. Some examples of software applications are word processing, spreadsheets, databases, games, multimedia software, flight simulators, drafting, educational tutorials, e-mail, and Internet browsers.

apprenticeship: A period of time spent learning a trade from an experienced, skilled worker.

aptitude: Talents and natural abilities.

arithmetic/logic unit (ALU): A part of the central processor that performs arithmetic operations such as subtraction, and logical operations such as true/false comparisons.

armature: The revolving part in a generator or motor. The vibrating or moving part of a relay or buzzer.

armature reaction: Effect on the main field of a generator by an armature acting as an electromagnet.

Armstrong oscillator: An inductive feedback oscillator.

artificial magnets: Manufactured magnets.

aspect ratio: The ratio of width to height of a television picture. Aspect ratio standards can be either 4:3 or 16:9.

assembler program: A program that converts the series of instructions written in assembly language into machine language.

assembly language: A low-level computer programming language closest to machine code that uses a series of words called mnemonics to create a program.

astable: One of two main modes of operation for the 555 timer in which the timer puts out a repetitive signal from the output pin.

A supply: Voltages supplied for heater circuits of electron tubes.

at-cut crystal: Crystal cut at approximately a 35° angle with a Z axis.

atom: The smallest particle of an element.

atomic number: The number of protons in the nucleus of a given atom.

atomic weight: The mass of the nucleus of an atom relative to carbon, which has a weight of 12.01115.

ATSC: *See advanced television systems committee.*

attenuation: A decrease in amplitude or intensity.

attenuator: Networks of resistance used to reduce voltage, power, or current to a load.

automatic gain control (AGC): A circuit employed to vary gain or amplitude in proportion to input signal strength so output remains at a constant level. Also called automatic volume control.

automatic volume control: Another name for automatic gain control.

autotransformer: A transformer with a common primary and secondary winding. Step-up or step-down action is accomplished by taps on the common winding.

AVC: Abbreviation for automatic volume control.

average value: The value of an alternating current or the voltage of a sine wave form, found by dividing the area under one alternation by the distance along the X axis between 0° and 180°.

B

back emf: *See counter electromotiveforce.*

ballast: Fine wire used to limit the current inside the phosphor tube of a fluorescent lamp.

band: A group of adjacent frequencies in a frequency spectrum.

band-pass filter: A filter circuit designed to pass currents in a continuous band of frequencies, but to reject or attenuate frequencies above or below the band.

band-reject filter: A filter circuit designed to reject currents in a continuous band of frequencies, but to pass frequencies above or below the band.

band switching: Employing a switch to change coils to allow a receiver to switch to a different frequency range of reception.

bandwidth: The band of frequencies allowed for transmitting a modulated signal.

barrier region: The potential difference across a PN junction due to the diffusion of electrons and holes across the junction.

base: A semiconductor between the emitter and the collector of a transistor.

basic input output system (BIOS): A ROM chip that contains the programming necessary to handle all input and output functions of a computer or microcontroller.

bass: Low-frequency sounds in an audio range.

battery: Several voltaic cells connected in series or in parallel and usually contained in one case.

baud rate: The approximate number of bits per second transmitted by a modem.

B battery: A group of series cells in one container producing high voltage for plate circuits of electronic devices.

beat frequency: The frequency obtained by combining two frequencies in a receiver.

beat frequency oscillator: An oscillator whose output is beat with a continuous wave to produce a beat frequency in an audio range.

bel: A unit for measuring gain. It is equivalent to a 10:1 ratio of power gain.

beta: The Greek letter β, representing the current gain of a common-emitter connected transistor. It is equal to the ratio of change in the collector current to change in the base current, while the collector voltage is constant.

bias, types:

- **cathode self:** Bias created by a voltage drop across the cathode resistor.
- **fixed:** Voltage supplied by a fixed source.
- **forward:** Connection of potential to produce current across the PN junction. The source potential is connected so it opposes potential hill and reduces it.
- **reverse:** A connection of potential so little that no current will flow across the PN junction. Source potential has same polarity as potential hill and adds to it.

biasing: Setting up the correct dc operating voltages between input leads of a transistor.

binary: A number system having a base of 2, using only the symbols 0 and 1.

BIOS: *See basic input output system.*

bipolar junction transistor (BJT): A transistor that consists of three layers of impure semiconductor crystals.

bit: One binary digit (0 or 1).

black box: A box containing an unknown and possibly complicated circuit.

blanking pulse: Pulses transmitted by a TV transmitter, which are used in a receiver to cut off the scanning beam during retrace time.

bleeder: A resistor connected across the power supply to the discharge filter capacitors.

blocking capacitor: A capacitor used to block dc.

branch circuit: The wiring from an electrical circuit panel to the last device connected in that circuit.

branch circuit protection: Installation of a fuse or breaker to protect the individual circuit feeding a motor.

breadboard: An easily constructed circuit board used for experiments or prototypes. Also called a *protoboard.*

bridge circuit: A circuit with series-parallel groups of components connected by a common bridge. The bridge is frequently a meter in measuring devices.

bridge rectifier: A full-wave rectifier circuit employing four rectifiers in a bridge configuration.

brightness: The overall intensity of illumination.

brush: A sliding contact, usually made of carbon, between the commutator and the external circuit in a dc generator.

B supply: Voltages supplied for the plate circuits of electron tubes.

buffer: A type of amplifier placed between the oscillator and the power amplifier to isolate them from the load.

bus: An electrical pathway on which current flows.

bus lines: Small copper strips that provide the voltage from the power supply to the computer components.

BX cable: Cable sheathed in metallic armor.

bypass capacitor: A fixed capacitor that bypasses unwanted ac to ground.

byte: Two binary nibbles or eight binary bits.

C

cable: Stranded conductor or a group of single conductors insulated from each other.

cable TV: A communications system that sends television signals using coaxial cables.

cache: Temporary memory storage unit.

candela: A brightness rating based upon the amount of light generated by one candle.

capacitance: 1. An inherent property of an electric circuit that opposes change in voltage. 2. A property of a circuit whereby energy may be stored in an electrostatic field.

capacitive coupling: Coupling resulting from a capacitive effect between components.

capacitive reactance (X_C): Opposition to ac as a result of capacitance.

capacitor: A device that possesses capacitance. A simple capacitor consists of two metal plates separated by an insulator.

capacitor input filter: A filter employing a capacitor as its input.

capacitor motor: A modified version of a split-phase motor, employing a capacitor in series with its starting winding, to produce phase displacement for starting.

capacity: The ability of a battery to produce current over a given length of time. The capacity of a battery is measured in ampere-hours.

carrier: 1. Usually a radio frequency continuous wave to which modulation is applied. 2. Frequency of a transmitting station. 3. In a semiconductor, electrons or holes that flow during conduction.

carrier wave: Signal for an ac sine wave created by an oscillator with a continuous wave transmitter.

cascade: Arrangement of amplifiers where the output of one stage becomes the input of the next throughout a series of stages.

cascode: Electron tubes connected so that the second tube acts as a plate load for the first. Used to obtain higher input resistance and retain low noise factor.

cathode: Negative terminal.

cathode follower: Single-stage Class A amplifier, whose output is taken from across an unbypassed cathode resistor.

cathode ray tube (CRT): Vacuum tube in which electrons emitted from the cathode are shaped into a narrow beam and accelerated to high velocity before striking the phosphor-coated viewing screen.

cat whisker: Fine wire used to contact a crystal.

C battery: Battery used to supply grid bias voltages.

CCD: *See charged coupled device.*

cemf: Abbreviation for counter electromotiveforce.

center frequency: Frequency of transmitted carrier wave in FM when no modulation is applied.

center tap: Connection made to the center of a coil.

central processing unit (CPU): The computer component that directs or controls computer operations. It is composed of an arithmetic/logic unit, a control unit, and internal memory.

ceramic capacitor: A small capacitor made of a special ceramic dielectric.

channel: A path from the source to the drain in a junction field-effect transistor.

characteristic curve: Graphic representation of the characteristics of a component, circuit, or device.

charged coupled device (CCD): An integrated circuit consisting of an array of photo sensors that convert light from an image into electrical energy. CCDs are commonly found in scanners and digital cameras.

chassis: A metallic surface upon which electronic components are mounted.

chip: A complete circuitry package contained on a small piece of silicon. The circuit often consists of transistors, diodes, resistors, capacitors, and connector wires.

chipset: A set of circuits integrated onto one chip, designed to perform a specific function or a set of functions. On the motherboard, it is an integrated circuit containing the control circuitry that interfaces to all of the computer subsystems such as memory, CPU, and input and output devices.

choke input filter: Filter employing choke as its input.

circuit: A complete pathway on which a current flows.

circuit breaker: Safety device that automatically opens a circuit if it is overloaded.

circuit designer: A person who designs a complete integrated circuit.

circular mil (cmil): Cross-sectional area of a conductor one mil in diameter.

circular-mil-foot: Unit conductor one foot long with a cross-sectional area of one circular mil.

circulating current: Inductive and capacitive currents flowing around a parallel circuit.

cladding: A layer that keeps the light wave contained to the core of a fiber-optic cable.

clocked R-S flip-flop: A flip-flop in which the output changes when there is a change in the R or S input and a pulse appears at the clock input.

coaxial line: Concentric transmission line in which the inner conductor is insulated from the tubular outer conductor.

coefficient of coupling (K): Percentage of coupling between coils, expressed as a decimal.

coherent light: Light consisting of waves that are all the same length.

collector: Third terminal of a transistor. The other two are the base and the emitter.

collector junction: One of two junctions in a transistor; it is located between the collector and the base.

Colpitts oscillator: A basic type of oscillator characterized by tapped capacitors in the tank circuit.

combination circuit: A series-parallel circuit.

common base: Transistor circuit in which the base is common to the input and output circuits.

common collector: Transistor circuit in which the collector is common to the input and output circuits.

common emitter: Transistor circuit in which the emitter is common to the input and output circuits.

commutating pole: *See interpole.*

commutation: Process of reversing current in armature coils and conducting direct circuit to an external circuit by means of commutator segments and brushes.

commutator: 1. Group of bars providing connections between armature coils and brushes. 2. Mechanical switch to maintain current in one direction in an external circuit.

compact discs: Thin, metallic disks that can store large amounts of data.

comparator: A mode of operation of an op amp in which the inverting and noninverting inputs are compared to each other.

complementary metal-oxide semiconductor (CMOS): A digital circuit arrangement that uses field-effect transistors for its logic circuits.

complimentary: A principle behind the operation of a flip-flop that results in one high output when the other is low.

compound: Combination of two or more elements.

compound generator: Uses both series and shunt windings. A source of energy that converts mechanical energy to electrical energy.

compound generators, types:

- **flat:** When no-load and full-load voltages have the same value.
- **over:** Full-load voltage is higher than no-load voltage.
- **under:** Full-load voltage is less than no-load voltage.

computer: A machine that accepts data, processes it according to a stored program of instructions, and outputs the results.

computer virus: A destructive program written for the sole purpose of causing problems within a computer system.

condenser: Older name for a capacitor.

conductance: Ability of a circuit to conduct current. The symbol is *G*. It is equal to amperes per volt and is measured in siemens.

conduction band: Outermost energy level of an atom.

conductivity, N-type: Conduction by electrons in an N-type crystal.

conductivity, P-type: Conduction by holes in a P-type crystal.

conductor: A low-resistance material through which electric current can easily flow.

conservation of energy: A law of physics that states energy can neither be created nor destroyed.

constant speed motor: A motor in which the field windings shunt across, or parallel to, the armature.

continuous wave (CW): Uninterrupted sinusoidal RF wave, with all wave peaks equal in amplitude and evenly spaced along the time axis.

contrast: Relative difference in intensity between black and white in a reproduced television picture.

control grid: 1. Grid in a vacuum tube closest to the cathode. 2. Grid to which an input signal is fed to the tube.

control unit (CU): Directs and coordinates data moving throughout the central processing unit.

conventional current flow theory: A theory that stated electricity had both positive and negative polarities and electrical current flowed from positive to negative.

converter: Electromechanical system for changing ac to dc.

coordinates: Horizontal and vertical distances used to locate a point on a graph.

copper losses: Heat losses in motors, generators, and transformers as result of resistance of wire. Sometimes called the I^2R loss.

core: Medium for the movement or transfer of the light energy in a fiber-optic cable.

coulomb: Quantity of electrons representing 6.24×10^{18} electrons.

counter electromotiveforce (cemf): Voltage induced in a conductor moving through a magnetic field that opposes source voltage.

counter emf (cemf): *See counter electromotiveforce.*

coupling: Percentage of mutual inductance between coils. Also called linkage.

covalent bond: Atoms joined together, sharing each other's electrons to form a stable molecule.

crossover frequency: Frequency in a crossover network at which an equal amount of energy is delivered to each of two speakers.

crossover network: Network designed to divide audio frequencies into bands for distribution to speakers.

cross talk: Leakage from one audio line to another that produces objectionable background noise.

crystal diode: Diode formed by small semiconductor crystal and cat whisker.

crystal lattice: Structure of material when outer electrons are joined in a covalent bond.

crystal microphone: A type of microphone that uses the piezoelectric effect of certain crystals.

crystal oven: Maintains a constant temperature of crystals used to control transmitter frequency.

C supply: Voltages, usually negative, supplied for grid bias of electron tubes.

cumulative compound motor: A motor in which the magnetic field on the series winding reinforces the magnetic field of the shunt winding.

current: Transfer of electrical energy in a conductor by means of electrons moving constantly and changing positions in a vibrating manner. Its symbol is *I.*

current ratio: The ratio between the currents in the primary and secondary windings in a transformer.

cutoff: A condition, similar to an open switch, in which a transistor is not conducting.

cutoff bias: Value of negative voltage applied to a transistor that will cut off current flow.

cycle: 1. Set of events occurring in sequence. 2. One complete reversal of an alternating current from positive to negative and back to the starting point.

D

daisy wheel: Type of printhead on a computer printer.

damper: Stage that dampens out oscillations in the deflection yoke after retrace in a television set. Used to prevent oscillations in the horizontal output transformer.

damping: Gradual decrease in the amplitude of oscillations in a tuned circuit due to the energy dissipated in resistance.

damping resistor: Broadens the frequency response of the circuit by carrying a part of the line current which cannot be canceled at resonance.

D'Arsonval movement: Movement that consists of a permanent-type magnet and a rotating coil in a magnetic field. An indicating needle is attached to the rotating coil.

data: Information given to the computer for processing.

datacasting: A digital broadcasting technique that allows for the transmittal of additional information along with the video image. Stock prices, weather reports, sports scores, background or human interests stories about the actors in the movie being viewed, or any information found on the Internet can be incorporated into the display screen.

dB: Abbreviation for decibel.

dc: Direct current.

dc amplifier: Directly coupled amplifier that amplifies without loss of dc components.

dc component: A dc value of an ac wave that has an axis other than zero.

dc generator: Generator with connections to the armature through a commutator. Output is direct current.

dc machine: A device that can be used as either a dc generator or motor.

DDR-DRAM: *See double data rate-dynamic random access memory.*

decade counter: A common digital counter that counts in base ten.

decay: Expresses gradual decrease in values of current and voltage.

decibel (dB): A unit of relative measurement. The unit is used to express the ratio of two amounts, often output power as compared to input power.

decimal system: A numbering system in which ten digits are available.

declination: Angle between true north and magnetic north.

dedicated line: Refers to a circuit that is used for only one purpose.

decoder: Part of a communications system that changes a coded message into an uncoded message.

deflection: 1. Deviation from zero of the needle in a meter. 2. Movement or bending of an electron beam.

deflection angle: Maximum angle of deflection of an electron beam in a television picture tube.

deflection yoke: Coils that surround a picture tube.

degenerative feedback: Feedback 180° out of phase with the input signal, so it subtracts from the input.

delta connection: A series connection of windings in motors and transformers.

delta-type electron gun: An electron gun arrangement where the guns are placed 120° apart in a color picture tube.

demodulation: Process of removing the modulating signal intelligence from the carrier wave in a radio receiver.

depletion layer: In a semiconductor, region in which the mobile carrier charge density is insufficient to neutralize the net fixed charge of donors and acceptors.

depletion-mode MOSFET: MOSFET in which the current through the source drain circuit is reduced by gate voltage.

depolarizer: Chemical agent, rich in oxygen, introduced into a cell to minimize polarization.

detection: *See demodulation.*

D flip-flop: Semiconductor device similar to the J-K flip-flop, except it does not require two inputs.

diaphragm: Thin disk used in an earphone for producing sound.

dielectric: Insulating material between plates and capacitor.

dielectric constant: Numerical figure representing ability of a dielectric or insulator to support electric flux. Dry air is assigned number 1.

dielectric field of force: *See electrostatic field.*

differential compound motor: A motor with two windings that oppose each other magnetically.

diffusion: Movement of carriers across semiconductor junction in the absence of external force.

digital circuits: Electronic circuits that handle digital information (on or off) using switching circuits.

digital encoder: A type of feedback system that sends pulses back to the control unit rather than varying the voltage.

digital integrated circuit: A switching-type (on or off) integrated circuit.

digital meter: An electronic meter in which the output is displayed as a number rather than indicated by a pointer.

digital multimeter (DMM): Uses modern electronic circuitry to take electrical measurements and display values, usually on a liquid crystal display screen.

digital video disc (DVD): A high-capacity storage medium used to store full length movies, computer programs, and other information.

DIMM: *See dual inline memory module.*

diode: *See semiconductor diode.*

diode detector: Detector circuit utilizing unilateral conduction characteristics of a diode.

diode-transistor logic (DTL): An arrangement of digital circuits using diodes, resistors, and transistors to produce logic gates.

DIP: *See dual inline package.*

dip switch: A microminiature switch, made in an in-line assembly (having multiple switches), used in computers for attaching peripherals, such as printers.

direct current (dc): Flow of electrons in one direction.

direct memory access: The method of moving data directly from a storage media, such as a hard drive, to the RAM without going through the CPU. On a motherboard, chipsets are inserted into key points along the bus to support the direct movement of data from the floppy or hard disk drive to the RAM.

discharge lamp: Produces light by energizing a gas such as argon, neon, helium, or a vapor of mercury or sodium.

discriminator: A type of FM detector.

dispersion: Distortion of an optical signal caused by light waves arriving at the termination point at different times.

distortion: Deviations in amplitude, phase, and frequency between input and output signals of an amplifier or system.

domain theory: Theory concerning magnetism, assuming that atomic magnets produced by the movement of electrons around the nucleus have a strong tendency to line up in groups. These groups are called domains.

donor impurity: Impurity added to semiconductor material that creates negative electron carriers.

doping: Adding impurities to semiconductor material.

double data rate-dynamic random access memory (DDR-DRAM): A memory module designed to transfer data at twice the normal data rate transfer speed. DDR technology transfers data on both the rising edge and the falling edge of a digital pulse.

double-pole: Switch with two common positions.

double-throw: Switch makes contact in either of two positions.

drain: One of three main parts of a junction field-effect transistor.

DRAM: *See dynamic random access memory.*

dressing: Shaping the end of the brush to match the surface of the commutator or slip ring.

dry cell: Nonliquid cell composed of a zinc case, carbon-positive electrode, and paste of ground carbon, manganese dioxide, and ammonium chloride as the electrolyte.

dual inline memory module (DIMM): A memory module constructed similarly to a SIMM, but that has more electrical connections per inch—168 pins compared to 72.

dual inline package (DIP): A chip that has two rows of connections, one each per side of chip. This style is commonly used for cache memory or memory that needs to be mounted permanently on board.

dynamic memory: A type of memory chip composed of transistors and capacitors that constantly need to be refreshed in order to retain its data.

dynamic plate resistance: *See ac plate resistance.*

dynamic random access memory (DRAM): A type of memory chip composed of millions of transistors and capacitors that needs to be refreshed thousands of times per second in order to retain its data.

dynamic speaker: Speaker that produces sound as result of the reaction between a fixed magnetic field and the fluctuating field of the voice coil.

dynamometer: Measuring instrument based on opposing torque developed between two sets of current-carrying coils.

dynamotor: Motor/generator combination using two windings on a single armature. Used to convert ac to dc.

E

eddy current: Induced current flowing through the core of an electromagnetic device.

eddy current loss: Heat loss resulting from eddy current flowing through resistance of a core.

Edison cell: Cell using positive electrodes of nickel oxide and negative electrodes of powdered iron. The electrolyte is a diluted solution of sodium hydroxide.

Edison effect: Effect first noticed by Thomas Edison, in which emitted electrons were attracted to the positive plate in a vacuum tube.

EEPROM: *See electrically eraseable programmable read only memory.*

EDTV: *See enhanced definition television.*

effective value: That value of alternating current of sine wave form that has the equivalent heating effect of a direct current ($0.707 \times E_{peak}$). Also called root mean square (rms) value.

efficiency: Ratio between output power and input power.

electric field of force: *See electrostatic field.*

electric motor: Converts electrical power into rotating mechanical power.

electrically erasable programmable read only memory (EEPROM): A type of PROM that does not lose its data when power is removed and can be electrically erased and reprogrammed.

electricity: The flow of electrons.

electrodynamic speaker: Dynamic speaker that uses an electromagnetic fixed field.

electrodynamometer movement: A meter with movement similar to the D'Arsonval movement, but in which the permanent magnetic field is replaced with coils from an electromagnet. The meter is used to measures watts.

electro-luminescence: The display of light created when a high frequency signal passes through a solid or liquid resulting in the release of photons.

electrolyte: Acid solution in a cell.

electrolytic capacitor: Capacitor with a positive plate of aluminum and a negative plate of dry paste or liquid. Dielectric is a thin coat of oxide on an aluminum plate.

electromagnet: Coil wound on soft iron core. When current runs through the coil, the core becomes magnetized.

electromotive force (emf): Force that causes free electrons to move in a conductor. Unit of measurement is the volt.

electron: Negatively charged particle.

electron flow theory: Electrons flow from negative to positive.

electron tube: Highly evacuated metal or glass shell that encloses several elements.

electronics: The study of the flow of electrons in active devices, such as transistors, semiconductors, diodes, or integrated circuits.

electrostatic field: Space around a charged body in which the body's influence is felt.

electrostatic precipitator: Equipment that uses static electricity to reduce air pollution.

electrostriction: Piezoelectric property of some elements in which they change in shape and size when voltage is applied; conversely, they produce voltage when subjected to pressure or stress.

element: A distinct substance, which in combination with other elements, makes up all matter in the universe.

emission: Escape of electrons from a surface.

emission, cold cathode: Phenomenon of electrons leaving the surface of a material due to high potential field.

emission of radiation: The escape of light through the small hole in the reflective material at one end of a laser tube.

emission, types:

- **A0:** Continuous wave, no modulation.
- **Al:** Continuous wave, keyed.
- **A2:** Telegraphy by keying modulating audio frequency.
- **A3:** Telephony.
- **A4:** Facsimile.
- **A5:** Television.
- **F0:** Continuous wave, no FM.
- **F1:** Telegraphy by frequency shift keying.
- **F2:** Telegraphy by keying modulating audio frequency.
- **F3:** Telephone, FM.
- **F4:** Facsimile.
- **F5:** Television.

emitter: 1. In a transistor, a semiconductor section, either P or N, which emits minority carriers. 2. The cathode in a vacuum tube from which electrons are emitted.

emitter biasing: A method for setting forward bias of the emitter junction.

emitter junction: One of two junctions in a transistor; it is between the emitter and the base.

encoder: Part of a communications system that changes the information source into coded form.

energy: That which is capable of producing work.

energy efficiency rating (EER): Efficiency rating of equipment such as heating, cooling, and cooking appliances, expressed as a percent.

engineer: Someone who designs and monitors the building of new equipment.

enhanced definition television (EDTV): A system that receives digital transmissions but displays images at 480p or higher. The fact that it can receive high definition television transmissions and decode them makes it an enhanced system.

enhancement-mode MOSFET: MOSFET in which current through the source drain circuit is increased by the gate voltage.

ENIAC: World's first large electronic computer, developed in 1946.

entrepreneur: Someone who owns and operates his or her own business.

EPROM: *See erasable programmable read only memory.*

erasable programmable read only memory (EPROM): A type of PROM that can be programmed electrically and then erased using an ultraviolet light.

equivalent resistance: Total resistance of a series, parallel, or combination circuit.

exciter: Small dc generator used to excite or energize the field windings of a large alternator.

exclusive NOR (XNOR) gate: A logic gate that provides a logic high output (1) only if all inputs are logic high or logic low.

exclusive OR gate (XOR): A logic gate that provides a high output (1) whenever any, but not all, inputs are logic high.

extrinsic losses: Losses caused by physical factors outside the normal core of fiber-optic cable, such as bends, splices, and connectors.

extrinsic semiconductor: The result of adding impurities to pure semiconductor material. A pure semiconductor is intrinsic.

F

FAT: *See file allocation table.*

farad: Unit of measurement of capacitance. A capacitor has a capacitance of one farad when a charge of one coulomb raises its potential one volt.

feedback: Transferring voltage from the output of a circuit back to its input.

feedhorn: A receiver located in the focal point of a satellite dish.

field-effect transistors (FET): A voltage device that controls current in the source-drain circuit through the amount of potential applied at the gate.

field-effect transistor-VOM (FET-VOM): A device used to measure ac and dc voltage, ac and dc current, resistance, and decibel ratings.

field magnets: Electromagnets that make magnetic field motors or generators.

field replacement unit (FRU): A module in a computer that can be quickly and easily replaced on-site, without the need to take the computer into a repair shop.

field windings: Electromagnets that can replace permanent field magnets in a generator motor to make it more powerful.

figure of merit: Link between inductive reactance and resistance in a circuit. Also called the *Q* or quality factor.

filament: 1. Heating element in vacuum tube coated with emitting material so it also acts as a cathode. 2. Element in incandescent lighting that gives off light and heat.

file allocation table (FAT): A file located on the zero sector of a drive that includes information such as the name, date, and time of the creation or modification of a file, the file's location, and the number of sectors used by the file.

file server: A powerful computer that provides software and data files to other computers that are part of the local area network.

filter: Circuit used to attenuate a specific band or bands of frequencies.

filter network: A device used for reducing the pulsations at the output of a half-wave or full-wave rectifier.

fine tuning: Slightly changes the frequency of the oscillator to provide the best response for an incoming signal.

fixed paper capacitor: A capacitor made of layers of tinfoil. The dielectric is made of waxed paper.

flat-compounded generator: A compound generator that maintains the same voltage either at no-load or full-load conditions.

flip-flop: A digital device based on the operation of combined logic gates.

floppy disk: A small, flexible disk used with computers as an external data storage device. The disk is coated with magnetic oxide and encased in a hard plastic cover.

fluorescent: Property of a phosphor that indicates radiated light will be extinguished when electron bombardment ceases.

flux density: Number of lines of flux per cross-sectional area of a magnetic circuit. Its symbol is *B*.

flywheel action: Periodic current changes in a circuit.

FM audio detector: Detects the frequency variations in the modulated signal and converts them to an audio signal.

focal point: The small point in a satellite dish that receives signals.

focus: Control used to make the wave on a CRT screen appear sharper.

frame: One set of the odd- and even-numbered fields in a scanning pattern.

frame rate: The number of times per second an image is updated on a video display screen. Currently, three frame rates exist: 60, 30, and 24 frames per second.

frequency: Number of complete cycles per second measured in hertz (Hz).

frequency bands, types:

- **EHF:** Extremely high frequencies, 30,000–300,000 MHz
- **HF:** High frequencies, 3 – 30 MHz
- **LF:** Low frequencies, 30 – 300 kHz
- **MF:** Medium frequencies, 300 – 3000 kHz
- **SHF:** Super high frequencies, 3000 – 30,000 MHz
- **UHF:** Ultrahigh frequencies, 300 – 3000 MHz
- **VHF:** Very high frequencies, 30 – 300 MHz
- **VLF:** Very low frequencies, 10 – 30 kHz

frequency departure: Instantaneous change from the center frequency in FM as result of modulation.

frequency deviation: The amount of frequency variation from each side of the center frequency.

frequency distortion: Deviation in frequency caused by the signals of some frequencies being amplified more than others, or when some frequencies are excluded.

frequency meter: Meter used to measure the frequency of an ac source.

frequency modulation (FM): A radio wave frequency is made to vary at an audio frequency rate.

frequency response: Rating of a device indicating its ability to operate over a specified range of frequencies.

fresenel reflection loss: Loss occurring

when two different fiber-optic materials, with different refraction properties, are joined.

front end: In a television, the RF amp, mixer, and oscillator combined into one unit. Also called the tuner.

FRU: *See field replacement unit.*

fuel cell: Constructed much like a battery cell with two metallic electrodes designed to allow hydrogen and oxygen gases to combine with the electrolyte of potassium hydroxide.

full-wave rectifier: Rectifier circuit that produces a dc pulse output for each half-cycle of applied alternating current.

fundamental: 1. A sine wave that has the same frequency as a complex periodic wave. 2. Component tone of the lowest pitch in a complex tone. 3. Reciprocal of a period of a wave.

fundamental frequency: The basic frequency produced by the oscillator in a continuous wave transmitter.

fuse: Safety device that opens an electric circuit if it becomes overloaded.

fusion splice: Joining two fiber-optic cores using heat to fuse or melt the materials together.

G

gain: Ratio of the output of ac voltage to the input of ac voltage.

galvanometer: Meter that indicates very small amounts of current.

ganged capacitor: Two capacitors controlled by a single shaft.

gap: The distance between the core of the electromagnet and its armature.

gas-filled tube: Tube designed to contain a specific gas in place of air, usually nitrogen, neon, argon, or mercury vapor.

gas laser: A laser in which the resonant cavity is filled with a gas such as helium-neon, CO_2 (carbon dioxide), or argon.

gate: One of three main parts of a junction field-effect transistor.

gauss: Measurement of flux density in lines per square centimeter.

generator: Rotating electric machine that provides a source of electrical energy. A generator converts mechanical energy to electric energy.

geostationary orbit: An object in orbit above, and rotating at the same speed as, the earth.

ghost: In television, a duplicate image of a reproduced picture, caused by multipath reception of reflected signals.

gigabyte: A measure of computer memory, equal to 1,000,000,000 bytes.

gilbert: Unit of measurement of magnetomotive force.

goods-producing business: Sells tangible products.

graphical user interface (GUI): A format that takes advantage of the computer's graphic capabilities to make programs easier to use.

grid: Fine wire placed between the cathode and plate of an electron tube.

grid bias: Voltage between the grid and cathode, usually negative.

grid current: Current flowing in the grid circuit of an electron tube when the grid is driven positive.

grid dip meter: A test instrument for measuring resonant frequencies, detecting harmonics, and checking the relative field strength of signals.

grid leak detector: Triode amplifier connected so it functions like a diode detector and an amplifier. Detection takes place in a grid circuit.

grid modulation: Modulation circuit where modulating signal is fed to the grid of the modulated stage.

grid voltage: Bias or C voltage applied to the grid of a vacuum tube.

ground: 1. The common return circuit in electronic equipment whose potential is zero. 2. A connection to earth by means of plates or rods.

ground wave: A radio wave that follows the surface of the earth to the radio receiver.

H

half-power points: Points at which power is half that of the maximum.

half-wave rectifier: Rectifier that permits one-half of an alternating current cycle to pass and rejects reverse current of the remaining half-cycle. Its output is pulsating dc.

hardware: The physical components that make up a computer.

harmonic frequency: Frequency that is a multiple of the fundamental frequency.

Hartley oscillator: A basic type of oscillator that has a tapped oscillator coil.

head window: The opening in the cover of a data disk through which information can be read and written.

heat sink: Mass of metal used to carry heat away from a component.

heater: Resistance heating element used to heat the cathode in a vacuum tube.

henry (H): Unit of measure of inductance. A coil has one henry of inductance if an emf of one volt is induced when current through an inductor is changing at rate of one ampere per second.

hertz (Hz): Unit of measure for frequency. One hertz equals one cycle per second.

heterodyne: Process of combining two signals of different frequencies to obtain a different frequency.

high-definition television (HDTV): Any television system capable of transmitting a picture resolution greater than 525 lines per screen.

high leg: The hot conductor in a delta-connected transformer bank that is not common to the neutral. The high leg exhibits a higher voltage to ground than the other hot conductors.

high-level language: A programming language closer than a low-level language to human language. High-level languages are assembled into machine language code using a program called a compiler.

high pass filter: Passes high-frequency current while rejecting low-frequency current.

high voltage rectifier: Rectifies the high ac voltage developed by the horizontal output transformer in a television receiver.

hole: Positive charge. A space left by a removed electron.

hole injection: Creation of holes in semiconductor material through the removal of electrons by a strong electric field around point contact.

horizontal AFC: In a television receiver, the horizontal oscillator frequency is compared to the sync pulse frequency. If they are not the same, voltages are developed that change the horizontal oscillator to the same frequency.

horizontal hold: In older televisions, the control that makes slight horizontal adjustments so the oscillators can lock in with the sync pulses.

horizontal oscillator: Provides the sawtooth waveform needed for horizontal scanning in a television receiver.

horizontal output: Correctly shapes the sawtooth waveform for the horizontal deflection coils in a television receiver. Also drives the horizontal deflection coils and provides power for the high-voltage rectifier.

horizontal output transformer (HOT): Shocked by the output of the horizontal oscillator, this autotransformer develops high ac voltage in a television receiver.

horizontal polarization: An antenna positioned horizontally so its electric field is parallel to the earth's surface.

horsepower (hp): One horsepower equals 33,000 ft.-lb. of work per minute, or 550 ft.-lb. of work per second. Also 746 watts equals 1 hp.

hot: An electrically charged connection.

hum: Form of distortion introduced in an amplifier as a result of coupling to stray electromagnetic and electrostatic fields or insufficient filtering.

hydrometer: Bulb-type instrument used to measure the specific gravity of a liquid.

hysteresis: Property of a magnetic substance that causes magnetization to lag behind the force that produced it.

hysteresis loop: Graph showing the density of a magnetic field as the magnetizing force is varied uniformly through one cycle of alternating current.

hysteresis loss: Energy loss in a substance as molecules or domains move through the cycle of magnetization. Loss is due to molecular friction.

I

image: One of two possible signals that can be mixed to result in the intermediate signal. It is not the desired signal.

impedance (Z): Total resistance to the flow of an alternating current as a result of resistance and reactance.

incandescent lamp: Light and heat are created from current flowing through a filament.

incoherent light: Light consisting of many different wavelengths.

indirectly heated: Refers to an electron tube employing a separate heater for its cathode.

induced current: Current that flows as a result of induced electromotive force.

induced emf: Voltage induced in the conductor as it moves through a magnetic field.

inductance: 1. Inherent property of an electric circuit that opposes a change in current. 2. Property of a circuit whereby energy may be stored in a magnetic field.

induction motor: An ac motor operating on the principle of a rotating magnetic field produced by out-of-phase currents. The rotor has no electrical connections but receives energy by transformer action from the field windings. Motor torque is developed by interaction of the rotor current and the rotating field.

inductive circuit: Circuit in which a noticeable emf is induced while the current is changing.

inductive reactance (X_C): Opposition to an ac current as a result of inductance.

inductor: An electronic component used to produce inductance in a circuit.

input: 1. Data put into a computer. 2. The process of putting data in a computer.

instantaneous polarity: Part of a sine wave that is represented at the output terminals of each generator at that instant of time.

instantaneous value: Any value between zero and maximum, depending upon the instant selected.

insulation: A protective coating that keeps the electron flow contained to the conductor path.

insulation resistance: Resistance to current leakage through and over the surface of the insulating material.

insulator: A material that possesses a high resistance to current flow (electricity).

integrated circuit (IC): A packaged electronic circuit containing resistors, transistors, diodes, and capacitors, with their interconnected leads. These are usually processed from a chip of silicon.

intensity: Magnetizing force per unit length of magnetic circuit.

intercalation: A compound inserted between or among existing elements.

interelectrode capacitance: Capacitance between metal elements in an electron tube.

interlace scanning: Process of scanning all odd lines and then all even lines to reproduce a television picture.

intermediate frequency (IF) amplifier: Used to increase the power output of audio frequencies.

intermediate range: In digital circuits, the area between value logic high and value logic low that acts as a buffer range. Also called invalid value range.

internal resistance: Refers to the internal resistance of a source of voltage or emf.

internet: A system of connecting computers worldwide.

interpoles: Auxiliary poles located midway between main poles of a generator to establish flux or satisfactory commutation.

interpreter: A program that reads and carries out source code one line at a time without compiling the code into machine code.

interrupted continuous wave (ICW): Continuous wave radiated by keying the transmitter into long and short pulses of energy (dashes and dots), conforming to a code, such as Morse.

interstage: Existing between stages, such as an interstage transformer between two stages of amplifiers.

intrinsic semiconductor: Semiconductor with electrical characteristics similar to a pure crystal.

invalid value range: *See intermediate range.*

inverted diode: A rectifier that produces a negative voltage. It is made by reversing the diode in the rectifier circuit.

inverter: *See NOT gate.*

ionization: The loss or gain of one or more electrons.

ionization potential: Voltage applied to a gas-filled tube at which ionization occurs.

ionized: Electrically unbalanced.

ionosphere: Atmospheric layer from 40 to 350 miles above the earth, containing a high number of positive and negative ions.

IR drop: *See voltage drop.*

iron vane meter: Meter based on the principle of repulsion between two concentric vanes placed inside a solenoid.

isolation: Electrical separation between two locations.

isolation transformer: A transformer with the specific purpose of isolating the primary circuit from the secondary circuit.

I^2R loss: Power loss resulting from resistance in the windings. Also called copper loss.

J

J-K flip-flop: A clock-driven flip-flop, similar to the R-S flip-flop except that it retains its output status when two lows are present at its inputs.

joule: Unit of energy equal to one watt-second.

junction diode: PN junction, having unidirectional current characteristics.

junction field-effect transistor (JFET): A voltage device that controls current in the source-drain circuit by the amount of potential applied to the gate.

K

key: Manually operated switch used to interrupt RF radiation of a transmitter.

key click filter: Filter in the keying circuit of a transmitter to prevent surges of current and sparking at key contacts.

keying: Process of causing a CW transmitter to radiate an RF signal when key contacts are closed.

kilo: One thousand.

kilobyte: A measure of computer memory equal to 1000 bytes.

kilowatt: One thousand watts.

kilowatt-hour (kWh): 1000 watts per hour. Common unit of measurement of electrical energy for home and industrial use.

Kirchhoff's current law: At any junction of conductors in a circuit, the algebraic sum of the currents equals zero.

Kirchhoff's voltage law: In a simple circuit, the algebraic sum of voltages around a circuit equals zero.

L

lambda: Greek letter λ. Symbol for wavelength.

laminations: Thin sheets of steel used in the cores of transformers, motors, and generators.

large scale integration (LSI): A complete computer circuit with up to 1000 components built on a single silicon chip.

laser: Acronym for Light Amplification by Stimulated Emission of Radiation. A highly intense beam of coherent light, not necessarily visible.

laser diode: A type of fiber-optic transmitter that produces a beam of light that carries the signal.

lasing: Basic operating principle behind the laser. Light must reach a certain level to produce the stimulation of more photons.

lattice crystalline: When the atoms in a covalent bond are each bonded to their own nucleus and to each other.

laws of magnetism: Like poles repel; unlike poles attract.

layout designer: Person who creates a detailed technical drawing from a schematic diagram.

LCD: *See liquid crystal display.*

lead acid cell: Secondary cell that uses lead peroxide and sponge lead for plates, and sulfuric acid and water for the electrolyte.

Leclanche cell: Scientific name for a common dry cell.

left hand rule for a coil: A method of determining the polarity of an electromagnetic field or the direction of electron flow. Grasp the coil with your left hand in such a manner that your fingers circle the coil in the direction of current flow. Your extended thumb will point to the north pole of the coil.

left hand rule for a conductor: Rule used to determine the direction of the magnetic field around a current carrying conductor. Grasp the conductor with your left hand, extending your thumb in the direction of the current. Your fingers curling around the conductor indicate the circular direction of the magnetic field.

Lenz's law: Induced emf in any circuit is always in such a direction as to oppose the effect that produces it.

library: A collection of code modules.

light emitting diode (LED): Special-function diode that emits light when connected in the forward bias direction.

limiter: A stage or circuit that limits all signals at the same maximum amplitude.

linear: Continuously variable.

linear amplifier: An amplifier whose output is in exact proportion to its input.

linear device: Electronic device or component whose current-voltage relation is a straight line.

linear integrated circuit: An amplifying (variable output) integrated circuit.

linear meter scale: A meter scale used to interpret ampere and voltage values.

linearity: Velocity of the scanning beam. It must be uniform for good linearity.

line-operated bridge circuit: When bridge rectifiers are used in circuits without transformers.

lines of force: Graphic representation of electrostatic and magnetic fields showing direction and intensity.

liquid crystal display (LCD): 1. A digital or alphanumeric display unit that can be used in visual outputs. 2. A type of display that operates on the principle of polarized light passing through tiny crystals of liquid.

lithium battery: A primary cell that has a long life.

load: Resistance connected across a circuit that determines current flow and energy used.

load line: Line drawn on characteristic family of curves of a transistor. Used for load line analysis.

load line analysis: A graphic display that plots the maximum current value of the collector against the maximum voltage across the emitter-collector in a transistor.

load resistor: Completes the basic power supply circuit.

local action: Defect in voltaic cells caused by impurities in zinc, such as carbon, iron, and lead.

local area network (LAN): A group of computers in a single building that are connected and share files.

local oscillator: Oscillator in a superheterodyne receiver, whose output is mixed with the incoming signal to produce an intermediate frequency.

locked rotor: A rotor that is not turning while power is being applied to the motor because of a physical impedance.

lodestone: Natural magnet. Called a "leading stone" by early navigators.

logic gate: An integrated circuit used mostly as a digital switching device.

low-pass filter: A circuit intended to pass low-frequency current and oppose high-frequency current.

lower sideband: Frequency below the carrier frequency as result of modulation. Equal to the difference between the carrier and the modulating frequency.

L-section filter: Filter consisting of a capacitor and an inductor connected in an inverted L configuration.

lumen: A measure of the amount of light generated by a lighting system.

lumens per watt (LPW): The amount of light produced for each watt of energy used.

M

machine code: A computer language constructed of ones and zeros representing binary codes and digital voltage pulse levels. Machine code is the language that the computer understands without further conversion.

machine language: A computer language constructed of ones and zeros that represent binary codes and digital voltage pulse levels.

macro: A collection of code modules.

magnet: Substance that has the property of attracting iron and producing an external magnetic field.

magnet poles: Point of maximum attraction on a magnet; designated as north and south poles.

magnetic amplifier: Transformer-type device employing a dc control winding. Control current produces more or less magnetic core saturation, thus varying the output voltage of the amplifier.

magnetic circuit: Complete path through which magnetic lines of force may be established under the influence of a magnetizing force.

magnetic field: Imaginary lines along which magnetic force acts. These lines emanate from the north pole and enter the south pole, forming closed loops.

magnetic flux: Entire quantity of magnetic lines surrounding a magnet. Its symbol is Φ (phi).

magnetic induction: Moving a conductor through a magnetic field to induce voltage (electricity) in a conductor.

magnetic lines of force: Magnetic lines that make up a magnetic field.

magnetic materials: Materials such as iron, steel, nickel, and cobalt, which are attracted to a magnet.

magnetic pickup: Phonocartridge that produces an electrical output from an armature in a magnetic field. The armature is mechanically connected to a reproducing stylus.

magnetic poles: Concentrations of magnetic force at the ends of a magnet.

magnetic saturation: A condition that exists in magnetic material when further increase in magnetizing force produces very little increase in flux density.

magnetic shielding: Conducting magnetic lines around an object.

magnetization: Graph produced by plotting the intensity of a magnetizing force on the X axis and relative magnetism on the Y axis.

magnetizing current: Current used in a transformer to produce a transformer core flux.

magnetohydrodynamic (MHD): A method of generating electricity by passing an ionized gas through a magnetic field.

magnetomotive force (mmf): Force that produces flux in a magnetic circuit.

mainline current: Total current in a parallel circuit.

major carrier: Conduction through a semiconductor as a result of a majority of electrons or holes.

masked ROM: Usually referred to as just ROM, it is a special type of memory that is permanently programmed during the manufacturing process and cannot be reprogrammed.

matched pair: Refers to the use of semiconductor diodes to ensure that each diode in a discriminator has equal conduction capability.

matter: Anything that occupies space or has mass.

maximum power transfer: A condition that exists when the resistance of a load equals the internal resistance of the source.

maximum value: Peak value of a sine wave in either a positive or negative direction.

maxwell: The cgs electromagnet unit of magnetic flux.

mega: Prefix meaning one million.

megabyte: A measure of computer memory equal to 1,000,000 bytes.

memory: Storage facilities of a computer.

mercury vapor rectifier: Hot cathode diode tube that uses mercury vapor instead of high vacuum.

metal oxide semiconductor field-effect transistor (MOSFET): A voltage device used extensively in digital circuits and memory circuits in computers. It has a very thin film of insulation (silicon dioxide) between the gate and channel area.

metallic rectifier: Rectifier made of copper oxide, based on the principle that electrons flow from copper to copper oxide, but not from copper oxide to copper.

mho: Old unit of measurement of conductance. Usually called a siemens.

mica capacitor: Capacitor made of metal foil plates separated by sheets of mica.

micro: Prefix meaning one millionth.

microcomputer: A computer whose central processing unit is contained on a single chip.

microcontroller: A programmable integrated circuit (or an IC) that can be programmed using a computer to perform a set of functions.

microcontroller interface: The connection point through which a microcontroller interfaces with a computer or other external device. The connection point is usually a serial or a universal serial bus (USB) port.

microcontroller relay circuits: A circuit used to increase the power output capabilities of the microcontroller when the electrical requirements of the device to be controlled far exceed the output power of the microcontroller I/O pins. This can be accomplished many different ways. One common way is using mechanical or solid-state relay devices.

microfarad (μF): One millionth of a farad.

microhenry (μH): One millionth of a henry.

microphone: Energy converter that changes sound energy into corresponding electrical energy.

microprocessor: A single integrated circuit that does the processing in a microcomputer.

microsecond (μs): One millionth of a second.

midrange: A speaker used to reproduce intermediate frequencies.

mil: One thousandth of an inch (0.001").

milli: Prefix meaning one thousandth.

milliammeter: Meter that measures current values in milliamperes.

millihenry (mH): One thousandth of a henry.

millisecond (ms): One thousandth of a second.

minor carrier: Conduction through a semiconductor opposite to a major carrier. For example, if electrons are a major carrier, then holes are a minor carrier.

minus: Negative terminal or junction of a circuit. Its symbol is –.

mismatch: Incorrect matching of a load to a source.

modem: A circuit that changes computer data in such a way that it can be transmitted and received over telephone lines.

modulated continuous wave (MCW): Carrier wave amplitude modulated by a tone signal of constant frequency.

modulation: Process by which the amplitude or frequency of sine wave voltage is made to vary according to variations of another voltage or current.

modulation index: The relationship between the maximum carrier deviation and the maximum modulating frequency.

molecule: Smallest division of matter. If further subdivision is made, matter will lose its identity.

motor: Device that converts electrical energy into mechanical energy.

motor reaction: Opposing force to rotation developed in a generator and created by a load current.

motor, types:

- **compound:** Uses both series and parallel field coils.
- **series:** Field coils are connected in series with armature circuit.
- **shunt:** Field coils are connected in parallel with armature circuit.
- **MPEG2:** An image compression standard, developed by the Moving Picture Experts Group, used to increase the amount of video data that can be transmitted in an HDTV system.

mu (μ): Greek letter representing one millionth. Also the symbol for permeability.

multicasting: A digital broadcasting technique that allows not only for multiple channels to be broadcast, but also for two to four channels to be broadcast in the same single-channel bandwidth.

multimeter: Combination volt, ampere, and ohm meter.

multiplier: Resistance connected in series with meter movement to increase its voltage range.

multiplier resistor: In a voltmeter, a switch placed in series with the meter movement coil to measure higher voltages.

multiunit tube, types:

- **heptode:** Seven elements with five grids.
- **hexode:** Six elements with four grids.
- **octode:** Eight elements with six grids.
- **twin diode:** Two diodes in one envelope.
- **twin diode tetrode:** Diode and tetrode in one envelope.
- **twin diode triode:** Diode and triode in one envelope.
- **twin pentode:** Two pentodes in one envelope.

multivibrators, types:

- **astable:** A free-running multivibrator.
- **bistable:** Has two stable states.
- **free-running:** Frequency of oscillation depends on the value of circuit components. Oscillation is continuous.
- **monostable:** One trigger pulse is required to complete one cycle of operation.
- **one-shot:** Same as monostable.

mutual inductance: Two coils located so the magnetic flux of one coil can link with turns of the other coil. The change in flux of one coil will cause an emf in the other.

N

NAND gate: A negative AND logic gate.

National Electrical Code (NEC): A collection of electrical standards that must be followed to ensure safety to personnel and prevent the possibility of an electrical fire.

national television systems committee (NTSC): A committee that formulated the standards for analog television and video in the United States. The NTSC standard calls for 525 scan lines at a 60 Hz refresh-rate based on the interlace technique. NTSC is not compatible with most computer video systems and must be converted before it can be displayed.

natural magnet: Magnets found in a natural state in the form of a mineral called magnetite.

narrow band FM: When maximum deviation of a carrier wave is limited so the FM wave occupies the same space as an AM wave carrying the same message.

Nd:YAG laser: Uses yttrium aluminum garnet for the laser medium.

negative ion: Atom that has gained electrons and is negatively charged.

network: Two or more components connected either in series or in parallel.

neutral: Refers to a grounded connection.

neutron: An electrically neutral particle.

nibble: One-half of a byte, or four bits.

nickel cadmium cell: Alkaline cell with paste electrolyte that is hermetically sealed.

noise: Any undesired interference to a signal.

noise spike: Noise and interference in radio reception due to amplitude variations.

no-load voltage: Terminal voltage of a battery or supply when no current is flowing in an external circuit.

nonlinear device: Electronic device or component whose current-voltage relation is not a straight line.

nonlinear scale: A scale, such as the ohms scale, whose markings are not evenly spaced.

nonvolatile: Describes the contents of read-only memory. Even if the power supply is turned off or the computer is unplugged, memory is retained.

NOR gate: A negative OR logic gate.

normally closed (NC): Describes a relay closed in the de-energized position.

normally open (NO): Describes a relay open in the de-energized position.

north pole: The pole that points toward the north when a magnet is freely suspended.

NOT gate: An inverter that changes the polarity of an incoming signal in the output.

NPN: A type of bipolar transistor; a P-type crystal placed between two N-type crystals.

NTSC: *See national television systems committee.*

nucleus: Core of an atom.

O

object code: Compiled source code.

oersted: Unit of magnetic intensity equal to one gilbert per centimeter.

offset null: A calibration feature of the op amp that prevents the output from generating a signal when there is no intentional input.

ohm: Unit of measurement of resistance. Its symbol is Ω.

ohmmeter: Meter used to measure resistance in ohms.

Ohm's law: Mathematical relationship between current, voltage, and resistance, discovered by George Simon Ohm.

ohms per volt: Unit of measurement of sensitivity of a meter.

opcode: The part of a word that contains the instruction to be carried out.

open circuit: Circuit broken or load removed. Load resistance equals infinity.

operand: The data following the opcode in a word. The opcode and the operand make up the source code.

operating system (OS): A program that serves as the interface between user or application and the BIOS. The operating system provides the utilities needed to communicate and carry out general user interactions. Typical user interactions are the storing and retrieving files or data, using communication input devices such as a mouse and keyboard, and displaying information on a monitor or printer.

operational amplifier (op amp): A type of linear integrated circuit used as basic amplifier circuit.

optical time domain reflector (OTDR): A special piece of fiber-optic test equipment that can trace and record attenuation of a fiber-optic cable.

OR gate: A logic gate that will provide an output signal if there is a signal on either of its inputs.

OS: *See operating system.*

oscillator: An electron tube generator of alternating current voltages.

oscillators, types:

- **Armstrong:** Uses a tickler coil for feedback.
- **Colpitts:** Uses a split tank capacitor as a feedback circuit.
- **crystal-controlled:** Controlled by piezoelectric effect.
- **Hartley:** Uses the inductive coupling of a tapped tank coil for feedback.
- **power:** The collector load of each transistor is the primary of the transformer.
- **RC:** Depends upon the charge and discharge of a capacitor in series with resistance.

oscilloscope: Test instrument that uses a cathode ray tube to observe a signal.

out of phase: When the current wave leads or lags the voltage wave.

output: Processed results of a computer.

output device: Equipment that records or displays the processed results of a computer.

overcompounded: Describes a generator that increases the output voltage at full-load conditions.

overmodulation: Condition when a modulating wave exceeds the amplitude of a continuous carrier wave, resulting in distortion.

P

parallel circuit: Contains two or more paths for electrons supplied by a common voltage source.

parallel resonance: Parallel circuit of an inductor and capacitor at a frequency when inductive and capacitive reactances are equal. Current in capacitive branch is 180° out of phase with the inductive current, and their vector sum is zero.

parasitic oscillation: Oscillations in the circuit resulting from circuit components or conditions occurring at frequencies other than desired.

passive-matrix: A type of liquid crystal display that has a grid of semi-transparent conductors running to each crystal used as part of the individual pixel.

peak: Maximum value of a sine wave.

peak inverse voltage (PIV): Rating indicating the maximum value of voltage that can be applied in the reverse direction across a diode.

peak reverse voltage: Same as peak inverse voltage.

peak to peak: Measured value of a sine wave from the peak in a positive direction to the peak in a negative direction.

peak value: Maximum value of an alternating current or voltage.

pentavalent: 1. Semiconductor impurity having five valence electrons. 2. Donor impurities.

pentode: Electron tube with five elements including a cathode, plate, control grid, screen grid, and suppressor grid.

percent of modulation: Maximum deviation from normal carrier value as a result of modulation.

percent of speed regulation: A ratio of the speed under no-load conditions to the speed under full-load expressed as a percentage of the full-load speed.

percent of voltage regulation: The output voltage of a power supply under load compared to the output voltage with no load.

period: Time for one complete cycle.

peripheral: A device that works as part of a computer but is joined to the outside of the process unit.

permanent magnet: Bars of steel and other substances that have been permanently magnetized.

permeability: Relative ability of a substance to conduct magnetic lines of force as compared with air. Its symbol is μ.

phase: Relationship between two vectors with respect to angular displacement.

phase angle (θ): The angle between vector Z and vector R in an ac circuit.

phase displacement: Equal to the angle θ between the two polar vectors when a current wave and a voltage wave are out of phase.

phase distortion: A deviation in phase resulting from the shift of phase of some signal frequencies.

phase inverter: Device or circuit that changes the phase of a signal 180°.

phase splitter: Amplifier that produces two waves that have exactly opposite polarities from a single input waveform.

photoelectric emission: Escape of electrons as a result of light striking the surface of certain materials.

photoengraving: Placing a photomask over the N-type layer in an integrated circuit.

photographically reduced: Process used to reduce a detailed drawing of an integrated circuit many times.

photomask: Working masks made from the photographically reduced technical drawing of an integrated circuit.

photon: Quantum of radiant energy.

photoresist: A light-sensitive emulsion.

photoresistive cell: A light-sensitive resistor.

photosensitive: Describes the characteristic of a material that emits electrons from its surface when energized by light.

phototube: Vacuum tube employing photosensitive material as its emitter or cathode.

photovoltaic cell: A crystalline silicon cell that directly converts sunlight into electricity.

pi-section filter: Filter consisting of two capacitors and an inductor connected in a pi (π) configuration.

picture element: Small areas or dots of varying intensity from black to white that contain the visual image of a scene.

piezoelectric effect: Property in which certain crystalline substances change shape when an emf is impressed upon a crystal. The action is also reversible.

pigtail splice: A factory-made connection on one end of a short piece of fiber-optic cable.

PIN diode: Three layers of the diode: positive, intrinsic, and negative. It has a wider intrinsic layer (depletion layer) than the photodiode.

pitch: Property of musical tone determined by frequency.

PIX-IF amplifier: Amplifies the output of the mixer stage.

pixel: The smallest unit used to designate an area on a graphic display.

plate: Anode of a vacuum tube. Element in a tube that attracts electrons.

plenum: Space above a drop ceiling or under a raised floor in a building. It is used as part of the air-conditioning duct system and to run cables.

plus: Positive terminal or junction of a circuit. Its symbol is +.

PM speaker: Speaker using a permanent magnet as its field.

PN junction: A piece of N-type and a piece of P-type semiconductor material joined.

PNP: A type of bipolar transistor. Has an N-type crystal placed between two P-type crystals.

point contact diode: Diode consisting of a point and a semiconductor crystal.

polarity: Property of a device or circuit having poles, such as north and south, or positive and negative.

polarization: 1. Defect in a cell caused by hydrogen bubbles surrounding a positive electrode and effectively insulating it from chemical reaction. 2. Producing magnetic poles or polarity.

polarized light: A filtered light consisting of light waves that are the same shape and of a single frequency rather than the entire spectrum of light.

polyphase: Consisting of currents having two or more phases.

port: A special memory location dedicated to a specific function such as a hardware location or a place to manipulate data. Some ports correspond to the input and output pin assignments of the chip.

positive feedback: Energy fed back to the tuned circuit. Also called regenerative feedback.

positive ion: Atom that has lost electrons and is positively charged.

potential: Electrical pressure or voltage.

potential barrier or **potential hill:** Sometimes referred to as the barrier region. It is the space between two types of joined semiconductor material.

power (*P*): Rate of doing work. In dc circuits, $P = I \times E$.

power detector: Detector designed to handle signal voltages having amplitudes greater than one volt.

power factor (PF): Relationship between the true power and apparent power of a circuit.

power gain (A_P): Product of voltage gain (A_V) multiplied by current gain (A_I) to tell how much a signal has been amplified.

power supply: Electronic circuit designed to provide various ac and dc voltages for equipment operation. Circuit may include transformers, rectifiers, filters, and regulators.

preamplifier: Sensitive, low-level amplifier with sufficient output to drive a standard amplifier.

pre-emphasis: Process of increasing the strength of signals or higher frequencies in FM at a transmitter to produce a greater frequency swing.

prefix: A term added to the beginning of a base word to denote an increase or decrease in value.

primary: The input winding in a transformer.

primary cell: Cell that cannot be recharged.

primary winding: Coil of a transformer that receives energy from an ac source.

printed circuit board (PCB): Made from very thin layers of conductive material (such as copper) adhered to a plastic backing. The circuit is etched to leave active pathways for the circuitry.

printhead: Pins that produce images on a dot matrix printer.

professional: A worker who has four years of college training or more advanced degrees.

program: A set of instructions used to run a computer.

program counter: A register or memory location in the microprocessor that contains the address of the next instruction to be executed.

programmable integrated circuit: An integrated circuit (IC) that can be programmed using a computer to perform a set of functions. It can consist of a single chip, or it can be composed of a few chips and discrete components collected together to function as a single unit.

programmable read only memory (PROM): A type of memory chip that is programmed after it is manufactured. Like the masked ROM, once a program is burned into the chip, it cannot be erased or reprogrammed.

progressive scanning: A method of constructing an image on a display by capturing and transmitting the entire image at one time. With this method, each line is placed on the screen progressively in one sweep, rather than in two sweeps as in interlaced scanning.

proton: Positively charged particle in the nucleus of an atom.

pulsating direct current: An intermittent flow of dc.

pulse: Sudden rise and fall of a voltage or current.

pure direct current: A constant flow of electrons.

pyrometer: An indicating device that includes a meter and a thermocouple.

Pythagorean theorem: The square of the length of the hypotenuse of a right triangle equals the sum of the squares of the lengths of the other two sides.

Q

***Q*:** Letter representation for quantity of electricity (coulomb).

***Q*, figure of merit:** The link between inductive reactance and resistance in the circuit. Also called the quality factor.

quanta: Definite amount of energy required to move an electron to a higher energy level.

quiescent: At rest; inactive.

quiescent point (Q-point): Optimum performance point for a transistor, typically one-half of the supply voltage.

R

radio frequency choke (RFC): Coil that has high impedance to RF currents.

radio spectrum: Division of electromagnetic spectrum used for radio.

radio wave: Electromagnetic radiation produced from current alternating through an antenna.

random access memory (RAM): Active memory that can read, write, and temporarily store data, when the computer is operating.

raster: Area of light produced on the screen of a television picture tube by an electron beam. Contains no picture information.

ratio detector: Type of FM detector.

RC circuit: A circuit that contains resistance and capacitance.

RCL circuit: A circuit that contains resistance, capacitance, and inductance.

RCL networks: AC circuits that have resistors, capacitors, and inductors placed in the circuit to pass, reject, or control current.

reactance (*X*): Opposition to alternating current as result of inductance or capacitance.

reactive power: Power used by a reactive component of a circuit.

reactor: Another name for inductor.

read only memory (ROM): Memory for instructions and fixed data that are stored by the manufacturer. ROM cannot be changed.

reciprocal: The reciprocal of a number is one divided by the number.

rectangular oil-filled capacitor: Hermetically sealed in a metal can and used in power supplies of radio transmitters and other electronic equipment.

rectification: Changing an alternating current to a direct current.

rectifier: Component or device used to convert ac into a pulsating dc.

regenerative feedback: Feedback in phase with the input signal so it adds to the input.

register: A memory location in the CPU used to temporarily store data that is being manipulated.

reject circuit: Parallel tuned circuit at resonance. Rejects signals at resonant frequency.

rel: Unit of measurement of reluctance.

relative conductance: Percent comparison of the conductance of a material to the conductance of silver, which is considered 100%.

relative resistance: Numerical comparison of the resistance of a material to the resistance of silver, which is assigned a value of 1.0.

relaxation oscillator: Nonsinusoidal oscillator whose frequency depends upon the time required to charge or discharge the capacitor through the resistor.

relay: Magnetic switch.

reluctance: Resistance to the flow of magnetic lines of force.

repulsion-start motor: Motor that develops starting torque by interaction of rotor currents and single-phase stator field.

reset: A device used to protect a circuit from overload and short circuit conditions. Also called a circuit breaker.

residual magnetism: Magnetism remaining in a material after the magnetizing force is removed.

resistance: Quality of an electric circuit that opposes the flow of current through it.

resistor: Component used to create desirable voltage drops and limit current values in electronic circuitry.

resistor-transistor logic (RTL): An arrangement of digital circuits using resistors and transistors to perform a logic function. This is no longer in use.

resolution: A measurement, in pixels, of the quality of an image. The higher the resolution, the higher the quality or more detailed the image.

resonant frequency (f_o): Frequency at which a tuned circuit oscillates. *See tuned circuit.*

retentivity: Ability of a material to retain magnetism after the magnetizing force is removed.

retrace: Returning the scanning beam to the starting point after one line is scanned.

reverse current cutout: Relay that permits current to flow in only one direction.

RF amplifier: Used to amplify radio frequencies.

ripple: Movement above and below the average voltage.

RL circuit: A circuit that contains resistance and inductance.

root mean square (rms) value: *See effective value.*

rotor: Rotating part of an ac generator.

Rowland's law: States the number of lines of magnetic flux is in direct proportion to the magnetomotive force and inversely proportional to the reluctance of the circuit.

ruby laser: Laser that uses a manufactured ruby consisting of an aluminum oxide compound and chromium.

run protection: Protects a motor from excessive heat damage or fire while in a normal mode of operation.

running windings: Turns of heavy wire placed in slots around the inside of a stator.

S

satellite TV: Television in which the signals are sent via satellites.

saturation point: Maximum current through the collector in a transistor.

sawtooth generator: Oscillator producing a sawtooth waveform.

sawtooth wave: Wave shaped like the teeth of a saw.

scan: Sweeping an electron beam across each element of a picture in successive order to reproduce the total picture in television.

scanner: A device used to copy graphics and text into a computer's memory.

scattering: The loss of signal strength due to impurities in the core material.

schematic: Diagram of an electronic circuit showing electrical connections and identifying various components.

screen grid: Second grid in an electron tube between the control grid and plate that reduces interelectrode capacitance.

SDRAM: *See synchronous dynamic random access memory.*

second harmonic distortion: Distortion of a wave by addition of its second harmonic.

secondary: The output winding in a simple transformer.

secondary cell: Cell that can be recharged by reversing chemical action with an electric current.

secondary emission: Emission of electrons as a result of electrons striking the plate of an electron tube.

secondary winding: Coil that receives energy from the primary winding by mutual induction and delivers energy to the load.

seed: A solid silicon particle.

selectivity: Relative ability of a receiver to select the desired signal while rejecting all others.

self bias: Provides a more stable operating point than fixed biasing and requires only one power source.

self-excited generator: A generator that uses its own leftover magnetism in place of a separate power source.

self-inductance: Emf is self-induced when it is induced in a current-carrying conductor.

selsyn unit: A transformer application designed to relay information through motor units. Also called a synchro.

semiconductor: Has resistivity in a range between conductors and insulators.

semiconductor diode: Result of fusion between a small N-type crystal and a P-type crystal.

semiconductor, N-type: Uses electrons as the majority carrier.

semiconductor, P-type: Uses holes as the majority carrier.

semiskilled worker: Performs jobs that do not require a high level of training.

sensitivity: Ability of a circuit to respond to small signal voltages.

sensitivity of meter: Indication of the loading effect of a meter. Resistance of moving coil and multiplier divided by voltage for full scale deflection. Sensitivity equals one divided by current required for full scale deflection.

separately excited generator: A generator whose field windings are excited by a separate dc source.

series circuit: Contains only one possible path for electrons.

series generator: A source of electrical energy whose field windings are connected in series with the armature and load.

series resonance: Series circuit of an inductor, capacitor, and resistor at a frequency when inductive and capacitive reactances are equal and canceling. Circuit appears as pure resistance and has minimum impedance.

service factor: The ability of a motor to withstand or avoid damage from an overload condition.

servo motor: Any motor that is modified to give feedback concerning the motor's speed, direction of rotation, and number of revolutions.

shaded-pole motor: Motor in which each field pole is split to accommodate a short-circuit copper strap called a shading coil. This coil produces a sweeping movement of field across the pole face for starting.

shading coil: A single turn of heavy wire placed in a slot that is cut in the face of the poles of a shaded-pole motor.

shading the pole: Placing a winding wire in the slot of the starter of a shaded-pole motor.

shadow mask: *See aperture mask.*

shield: Partition or enclosure around components in a circuit to minimize the effects of stray magnetic and radio-frequency fields.

short circuit: Direct connection across a source that provides a zero resistance path for a current.

shunt: 1. To connect across or parallel with a circuit or component. 2. Parallel resistor to conduct excess current around a meter moving coil. Shunts are used to increase the range of a meter.

shunt generator: Generator whose field windings are connected across the armature in shunt with a load.

side carrier frequencies: Frequencies that are equal to the sum and difference between the carrier wave frequency and the modulating wave frequency.

sideband frequency: Combination sum wave and difference wave.

sidebands: Frequencies above and below the carrier frequency as a result of modulation. Lower frequencies are equal to the difference between the carrier and modulating frequencies. Upper frequencies are equal to the carrier plus modulating frequencies.

siemens: Unit of measure for conductance.

signal generator: An electronic oscillator that generates various signals for testing.

significant sideband: Has an amplitude of 1% or more of the unmodulated carrier in FM transmission.

silicon: Special quadvalent material found

in many semiconductor materials that conducts the flow of electricity from one pathway to another.

silicon controlled rectifier (SCR): A three-junction device (anode, gate, and cathode) that is usually open until a signal on the gate switches it on.

silver oxide battery: A compact primary cell used mainly for watches.

SIMM: *See single inline memory module.*

sine wave: A curve on a graph depicting the flow of alternating current through a conductor within a given period of time.

single-battery bias: A common method for biasing transistors.

single-phase motor: Motor that operates on single-phase alternating current.

single-phasing condition: Occurs when the power supply to a three-phase motor loses one of its three lines due to an open or blown fuse.

single-pole: Refers to a switch that provides one path for electron flow. The switch can be turned on or off.

single-throw: Refers to a switch that controls only one circuit.

single inline memory module (SIMM): A memory module that has a row of DIP memory chips mounted on its circuit board. The SIMM is inserted into a memory slot socket on the motherboard. Each side of the edge connector on its circuit board is the same circuit.

single inline package (SIP): A memory module that has a single row of connections running the length of the chip.

single sideband transmission: FM transmission in which the carrier and one sideband are suppressed.

sinusoidal: Wave varying in proportion to the sine of an angle.

SIP: *See single inline package.*

skilled workers: People who have thorough knowledge of, and skill in, a particular area.

sky wave: Waves radiating toward the sky from a radio antenna.

sliding tap: An instrument that moves across the exposed surface of a portion of the wire in an adjustable resistor, allowing the resistance value to be varied.

slip: The difference in the rotor speed and the magnetic field of the stator in ac motors, usually expressed as a percent.

slip rings: Metal rings connected to rotating armature windings in a generator. Brushes sliding on these rings provide connections for an external circuit.

socket: Device for holding a lamp or electron tube.

soft tube: Gaseous tube.

software: Set of instructions used to run a computer.

solenoid: Coil of wire carrying an electric current that possesses characteristics of a magnet.

solid state: Electronic devices such as diodes, transistors, ICs, and other solid substances, as opposed to vacuum tubes or electromechanical relays.

sound IF amplifier: Where an FM sound signal is amplified in a television receiver.

source: One of three main parts of a junction field-effect transistor.

source code: The combination of an opcode and an operand. Source code is the programming code used to make the operating system.

source of supply: Generator, battery, or other device that is attached to the input of a circuit and produces electromotive force.

south pole: The pole that points toward the south when a magnet is freely suspended.

space charge: Cloud of electrons around the cathode of an electron tube.

space wave: Combination of direct wave and ground reflected wave.

speaker: Device to convert electrical energy into sound energy.

specific gravity: Weight of a liquid relative to water, which has an assigned value of 1.0.

splatter: Distortion and interference.

split ring: A device that reverses electrical connections and is used on dc generators. Also called a commutator.

split-phase motor: Single-phase induction motor that develops starting torque by phase displacement between field windings.

splitting phases: Dividing a single-phase current into a polyphase current.

square law detector: Detector whose output voltage is proportional to the square of the effective input voltage.

squirrel cage rotor: Made of bars placed in slots of the rotor core and joined at the ends. Used in induction motors.

SRAM: *See static random access memory.*

stack: A sequence of instructions and data.

stack pointer: A register or memory location that keeps track of the last stack location used while the processor is busy manipulating data values, checking ports, or checking interrupts.

standing wave: Wave in which the ratio of instantaneous value at one point to that at another point does not vary with time. Waves appear on transmission line as a result of reflections from termination of line.

standing wave ratio: 1. Ratio of effective voltage at the loop of a standing wave to effective voltage at the node. Also called effective current. 2. Ratio of characteristic impedance to load impedance.

starting windings: Many turns of relatively fine wire in an induction motor.

static charge: Negative or positive charge on a body.

static electricity: Electricity at rest as opposed to electric current.

static memory: An integrated circuit composed mainly of flip-flop type circuitry that does not require the constant refreshing like dynamic memory.

static random access memory (SRAM): A type of memory chip composed mainly of digital flip-flops, which will retain their state of one or zero until the power is removed from the chip.

stator: Stationary coils of an ac generator.

steady state: Fixed, nonvarying condition.

step-down transformers: Used to lower a voltage.

stepper motor: A motor designed to rotate in small increments, using a series of digital pulses to control rotation.

step-up transformers: Used to raise a voltage.

storage: A device for storing data on a computer.

storage battery: Common name for a lead acid battery used in automotive equipment.

stratosphere: Layer of atmosphere above the troposphere in which temperature is constant and clouds do not form.

stylus: Phonograph needle or jewel that follows grooves in a record.

subharmonic: Frequency below harmonic, usually a fractional part of the fundamental frequency.

sulfation: Undesirable condition of a lead acid battery caused by leaving it in a discharged condition. Sulfates forming on plates make the battery partially inactive.

superconductors: Conductive materials that are exposed to extremely low temperatures causing their resistance value to approach zero.

superheterodyne: Radio receiver in which the incoming signal is converted to a fixed intermediate frequency before detecting an audio signal component.

superimposed: Refers to an audio wave that is combined with a carrier wave to make an understandable message.

supersonic: Frequencies above audio frequency range.

suppressor grid: A third grid placed in the tube between the screen grid and plate to overcome the drawbacks of secondary emission.

surface alloy transistor: Silicon junction transistor, in which aluminum electrodes are deposited in shallow pits etched on both sides of a thin silicon crystal, forming P regions.

sweep circuit: Periodic varying voltage applied to deflection circuits of a cathode ray tube to move an electron beam at a linear rate.

switch: Device for directing or controlling current in a circuit.

sync amplifier: A voltage amplifier stage that increases the sync pulses in a television receiver.

sync separator: A circuit that removes the

horizontal and vertical sync pulses transmitted as part of the composite video signal in a television receiver.

synchro: Electromechanical device used to transmit the angular position of a shaft from one position to another without mechanical linkage.

synchronization (sync) pulse: Used to trigger an oscillator or circuit.

synchronous: Having the same period or frequency.

synchronous dynamic random access memory (SDRAM): An enhanced version of DRAM found in many computers that runs in synchronization with the CPU. This type of memory produces a higher overall memory performance than simple DRAM.

synchronous motor: Type of ac motor that uses a separate dc source of power for its field. It runs at synchronous speed under varying load conditions.

synchronous speed: The speed at which the rotor speed equals the speed of the rotating magnetic field.

synchronous vibrator: Vibrator with additional contact points to switch the output circuit so the current is maintained in one direction through a load.

system: An assembly of parts linked into an organized whole.

T

tank circuit: Parallel resonant circuit.

tantalum capacitor: Uses tantalum, instead of aluminum, for the electrode.

tap: Connection made to a coil at a point other than its terminals.

technicians: Specially trained workers capable of doing complex, technical jobs.

television: Method of transmitting and receiving a visual scene by radio broadcasting.

television channel: Allocation, in a frequency spectrum, of 6 MHz assigned to each television station for transmission of picture and sound information.

tetrode: Electron tube with four elements including cathode, plate, control grid, and screen grid.

TFT: *See thin film transistor.*

thermal runaway: In a transistor, a regenerative increase in collector current and junction temperature.

thermionic emission: Process in which heat produces energy for release of electrons from the surface of the emitter.

thermistor: Semiconductor device that changes resistivity with a change in temperature.

thermocouple: A device in which two dissimilar metals in contact with each other are heated, a potential difference develops between the metals.

thermocouple meter: Meter based on the principle that if two dissimilar metals are welded together and the junction is heated, a dc voltage will develop across open ends. Used for measuring radio frequency currents.

thermo-overload: An overload protection device for motors that consists of a simple ratchet wheel held in place by a metal alloy such as solder. When the overload condition generates sufficient current flow to melt the solder, the wheel is free to rotate, causing the circuit to open.

thermopile: A large number of thermocouples joined in series.

theta (θ): 1. Angle of rotation of a vector representing selected instants at which a sine wave is plotted. 2. Angular displacement between two vectors.

thimble: A type of printhead on a computer printer.

thin film transistor (TFT): A type of display that has a matrix of transistors spread across the entire screen with each thin film transistor controlling a single point or part of a pixel on the display.

thoriated tungsten: Tungsten emitter coated with a thin layer of thorium.

three-phase alternating current: Combination of three alternating currents having their voltages displaced by 120° or 1/3 cycle.

three-phase generator: The most common generator used in production of electrical power. It consists of a rotating magnetic field inside three sets of windings.

threshold of sound: Minimum frequency at which a sound can be heard.

thyrathron: Gas-filled tube in which a grid is used to control firing potential.

thyristors: Fast-triggering semiconductor devices used as switches.

tickler: Coil used to feed energy from the output to the input circuit.

time constant (RC): Time period required for the voltage of a capacitor in an RC circuit to increase to 63.2% of maximum value or decrease to 36.7% of maximum value.

time-delay off: After a relay coil is de-energized, there is a time delay before the relay contacts change their position.

time-delay on: After a relay is energized, there is a pause before the contacts close.

tolerance: A reflection of the precision of a resistor's value.

tone control: Adjustable filter network to emphasize either high or low frequencies in the output of an audio amplifier.

torque: Turning power of a motor.

transducer: A device that converts the electrical energy of audio frequencies into sound energy (or vice versa).

transfer characteristic: Relation between the input and output characteristics of a device.

transformer: Device that transfers energy from one circuit to another by electromagnetic induction.

transformers, types:

- **isolation:** Has one-to-one turns ratio.
- **step-down:** Has turns ratio greater than one. Output voltage is less than input voltage.
- **step-up:** Has turns ratio of less than one. Output voltage is greater than input voltage.

transient response: Response to momentary signal or force.

transistor: Semiconductor device capable of a wide range of operating characteristics.

transistor-transistor logic (TTL): An arrangement of digital circuits using transistors to perform logic functions.

transmission line: Wire(s) used to conduct or guide electrical energy.

transmitter: Device for converting intelligence into electrical impulses for transmission through lines or through space from a radiating antenna.

TRF: Abbreviation for tuned radio frequency.

triac: A full-wave silicon switch.

triode: Three-element vacuum tube, consisting of a cathode, grid, and plate.

trivalent: 1. Semiconductor impurity having three valence electrons. 2. Acceptor impurity.

troposphere: Lower part of the atmosphere where clouds form and temperature decreases with altitude.

true power: Actual power absorbed in a circuit.

truth table: A binary table that explains the operation of digital logic circuits.

tubular electrolytic capacitor: A small capacitor that has a metal case enclosed in an insulating tube.

tuned amplifier: Employs tuned circuits for input or output coupling.

tuned circuit: Circuit containing capacitance, inductance, and resistance in series or in parallel. When energized at a specific frequency known as its resonant frequency, an interchange of energy occurs between the coil and capacitor.

tuner: The RF amp, mixer, and oscillator combined in one unit.

turns ratio: The number of turns of the primary winding of a transformer compared to the number of turns of the secondary winding.

tweeter: A type of speaker used to accent the higher end frequencies of audio sound.

U

UART: *See universal asynchronous reciever/transmitter.*

ultra high frequency (UHF): Television frequencies that cover channels 14 to 83.

undercompounded: Refers to a generator that has a decreased voltage at full-load current.

unijunction transistor (UJT): A three-terminal transistor that has an emitter and two bases.

unity coupling: Two coils positioned so all

lines of magnetic flux of one cell cut across all turns of the second coil.

universal asynchronous receiver/transmitter (UART): An integrated circuit that converts outgoing parallel, digital data into a series stream of digital data and then encodes the digital signal into a modulated or analog signal. The UART is also responsible for converting an incoming analog signal to a parallel digital signal.

universal motor: 1. Series ac motor that also operates on dc. 2. Fractional horsepower ac/dc motor.

universal time constant chart: Graph with curves representing the growth and decay of voltages and currents in RC and RL circuits.

upper sideband: Frequency above the carrier frequency as result of modulation. Equal to the difference between the carrier and the modulating frequency.

V

vacuum tube diode: Uses the cathode and plate as electrodes.

valance electrons: Make up the outermost ring of electrons surrounding the atom.

valid logic high: Operating voltage required for a digital circuit to be in the 1, or on, position. Voltage range is usually 2 to 5 volts.

valid logic low: Operating voltage required for a digital circuit to be in the 0, or off, position. Voltage range is usually 0 to 1.5 volts.

valve: British name for vacuum tube.

variable capacitance diode: Used to tune radio circuits.

variable capacitor: Capacitor with a capacitance that can be adjusted by turning a shaft.

vector: Straight line drawn to scale, showing the direction and magnitude of a force.

vector diagram: Diagram showing the direction and magnitude of several forces, such as voltage and current, resistance, reactance, and impedance.

velocity factor: Speed of propagation of a signal along a transmission line, as compared to the speed of light.

vertical hold: In older televisions, the control that makes slight vertical adjustments so the oscillators can lock in with the sync pulses.

vertical oscillator: Receives the output of the sync amplifier through a vertical integration network in a television receiver.

vertical output: Refers to the amplifier used by the output of the oscillator to provide the proper currents in the deflection yoke for vertical scanning in a television receiver.

vertical polarization: Antenna positioned vertically so its electric field is perpendicular to the earth's surface.

very high frequency (VHF): Television frequencies that cover channels 2 to 13.

very large scale integration (VLSI): Integrated circuits with over 1000 components.

vestigial sideband filter: Removes a large portion of one sideband to improve fidelity in AM transmission.

vibrator: Magnetically operated interrupter, similar to a buzzer, that changes steady state dc to pulsating ac.

video amplifier: Used to amplify video frequencies.

video detector: Receives the output from the last intermediate frequency stage in a television receiver.

video head: A tiny electromagnet in a VCR that reads information from the recorded tape during playback.

video signal: Electrical signal from a studio camera used to modulate a TV transmitter.

voice coil: Small coil attached to a speaker cone, to which a signal is applied. Reaction between the field of the voice coil and the fixed magnetic field causes mechanical movement of the cone.

volatile: Describes the contents of random access memory (RAM), which can read, write, and temporarily store data while the computer is operating. Since the memory is only temporary, data can be lost if a power supply is turned off or interrupted.

volt (V): Unit of measurement of electromotive force or potential difference. Its symbol is *E* in electricity and *V* in semiconductor circuits.

volt-ampere (VA): Unit measure of apparent power.

volt-ampere-reactive (VAR): Unit measure of reactive power.

volt-ohm-milliammeter (VOM): A voltmeter, ohmmeter, and ammeter combined into one instrument.

voltage: The force or difference in potential that causes electrons to flow.

voltage divider: Tapped resistor or series resistors across a source voltage to multiply voltages.

voltage doubler: Rectifier circuit that produces double the input voltage.

voltage drop: Voltage measured across a resistor. Voltage drop is equal to the product of the current times the resistance in ohms.

voltage gain (A_V): Calculated by dividing the change in the output voltage by the change in the input voltage.

voltage multiplier: Rectifier circuits that produce output voltage at a multiple greater than the input voltage, usually doubling, tripling, or quadrupling.

voltage rating: Maximum voltage for which a switch is designed.

voltage ratio: The ratio between the voltages of the primary and secondary windings.

voltage regulator (VR) tube: Gas-filled, cold cathode tube that maintains a constant voltage drop over its operating range, independent of current.

voltaic cell: Cell produced by suspending two dissimilar elements in an acid solution. Potential difference is developed by chemical action.

voltmeter: Meter used to measure voltage.

VTVM: Abbreviation for vacuum tube voltmeter.

VU: A value numerically equal to the number of decibels above or below the reference volume level. Zero VU represents a power level of one milliwatt dissipated in a 600-ohm load or voltage of 0.7746 volts.

W

watchdog timer: A specialized program often found as part of the microcontroller designed to prevent the microcontroller from locking up or crashing because of a user-written program.

watt (W): Unit of measurement of power.

wattage rating: A resistor's ability to safely dissipate heat.

watt-hour: Unit of energy measurement, equal to one watt per hour.

watt-hour meter: Meter that shows an instantaneous rate of power consumption of a device or circuit.

wattless power: Power not consumed in an ac circuit due to reactance.

wattmeter: Meter used to measure power in watts.

wavelength: Length of one complete cycle of a waveform.

wave meter: Measures the frequency of a wave.

wave trap: Type of band-reject filter.

weak-signal detector: Unit that detects signal voltages with amplitudes less than one volt.

wheatstone bridge: Bridge circuit used for precision measurement of resistors.

wide area network (WAN): Connects computers across many miles.

Windows™: A typical graphical user interface.

woofers: A type of speaker used to accent lower frequency audio sound.

word size: Some common word sizes are 8, 12, 14, 16, 24, 32, and 64 bits in length. A word, which usually contains an instruction and some data to be acted upon, typically matches the bus width of the chip.

work: When a force moves through a distance, work is done. Work = force × distance.

working voltage (WV): Maximum voltage that can be steadily applied to a capacitor without arc-over.

world wide web (WWW): A system that connects computers worldwide.

wye (or star) connection: A parallel winding connection used in motors and transformers.

X

x-axis: 1. Horizontal axis of a graph. 2. Optical axis of a crystal. 3. Axis through corners of a hexagonal crystal.

x-cut crystal: Crystal cut perpendicular to the x-axis.

xerography: A copy process using powder toner, heat, and electrostatic principles to create an image.

XNOR gate: *See exclusive NOR gate.*

XOR gate: *See exclusive OR gate.*

Y

y-axis: 1. Vertical axis of a graph. 2. Axis drawn perpendicular to the faces of a hexagonal crystal.

y-cut crystal: Crystal cut perpendicular to the y-axis.

yagi antenna: Dipole with two or more director elements.

yoke: Coils placed around the neck of a television picture tube for magnetic deflection of the beam.

Z

z-axis: Optical axis of a crystal.

zener breakdown point: Point at which the zener diode is able to maintain a fairly constant voltage.

zener diode: Silicon diode that makes use of the breakdown properties of a PN junction.

zero reference level: Power level selected as a reference for computing the gain of an amplifier or system.

Index

D